XXI Fluid Mechanics Conference 2014

Journal of Physics: Conference Series Volume 530

Krakow, Poland
15 – 18 June 2014

Editors:

J. S. Szmyd
E. Fornalik-Wajs
M. Jaszczur

ISBN: 978-1-63439-311-9
ISSN: 1742-6588

Printed from e-media with permission by:

Curran Associates, Inc.
57 Morehouse Lane
Red Hook, NY 12571

Some format issues inherent in the e-media version may also appear in this print version.

Printed by Curran Associates, Inc. (2014)

Institute of Physics
Dirac House, Temple Back
Bristol BS1 6BE UK

Phone: 44 1 17 929 7481
Fax: 44 1 17 920 0979

techtracking@iop.org

Additional copies of this publication are available from:

Curran Associates, Inc.
57 Morehouse Lane
Red Hook, NY 12571 USA
Phone: 845-758-0400
Fax: 845-758-2634
Email: curran@proceedings.com
Web: www.proceedings.com

TABLE OF CONTENTS

KEYNOTE LECTURES

AERODYNAMIC AND ATMOSPHERIC SCIENCE

BIO-FLUIDS

COMBUSTION AND REACTING FLOWS

COMPUTATIONAL FLUID DYNAMICS

EXPERIMENTAL FLUIDS MECHANICS

FLOW MACHINERY

GENERAL FLUID DYNAMICS

HEAT AND FLUID FLOW

MEASUREMENT TECHNIQUES

MICRO- AND NANO- FLOW

FLOW MODELLING

NON-NEWTONIAN FLUIDS

TURBULENCE

PROFESSOR J W ELSNER'S COMPETITION

XXI FMC

XXI Fluid Mechanics Conference

AGH–UST, Krakow, Poland, 15 – 18 June 2014

Proceedings

Editors:
J.S. Szmyd, E. Fornalik-Wajs, M. Jaszczur

Preface

This Conference Volume contains the papers presented at the 21st Fluid Mechanics Conference (XXI FMC) held at AGH - University of Science and Technology in Krakow, Poland, 15 - 18 June 2014, and accepted for Proceedings published in the Journal of Physics: Conference Series.

The Fluid Mechanics Conferences have been taking place every two years since 1974, a total of forty years. The 21st Fluid Mechanics Conference (XXI FMC) is being organized under the auspices of the Polish Academy of Sciences, Committee of Mechanics. The goal of this conference is to provide a forum for the exposure and exchange of ideas, methods and results in fluid mechanics.

Conference topics include, but are not limited to Aerodynamics, Atmospheric Science, Bio-Fluids, Combustion and Reacting Flows, Computational Fluid Dynamics, Experimental Fluid Mechanics, Flow Machinery, General Fluid Dynamics, Hydromechanics, Heat and Fluid Flow, Measurement Techniques, Micro- and Nano-Flow, Multi-Phase Flow, Non-Newtonian Fluids, Rotating and Stratified Flows, Turbulence.

Within the general subjects of this conference, the Professor Janusz W. Elsner Competition for the best fluid mechanics paper presented during the Conference is organized. Authors holding a M.Sc. or a Ph.D. degree and who are not older than 35 years of age may enter the Competition. Authors with a Ph.D. degree must present individual papers; authors with a M.Sc. degree may present papers with their supervisor as coauthor, including original results of experimental, numerical or analytic research.

Six state-of-the-art keynote papers were delivered by world leading experts.

All contributed papers were peer reviewed. Recommendations were received from the International Scientific Committee, reviewers and the advisory board. Accordingly, of the 163 eligible extended abstracts submitted, after a review process by the International Scientific Committee, 137 papers were selected for presentation at the 21st Fluid Mechanics Conference, 68 papers were accepted for Proceedings published in the Journal of Physics: Conference Series.

The total number of submitted and accepted papers for this year's conference represents a significant increase over previous Fluid Mechanics Conferences, and has expanded its initial national character and borders which speaks to the great vitality of fluid mechanics.

We hope that these proceedings will be used not only as a document of the event but also to assess achievements and new paths to be taken in fluid mechanics research.

Finally, we would like to congratulate the winners of the 2014 Professor Janusz W. Elsner Competition Ruri Hidema from Japan, Fernando Tejero from Spain and Łukasz Łaniewski-Wołłk from Poland.

Janusz S. Szmyd
Elżbieta Fornalik-Wajs
Marek Jaszczur

Acknowledgments

We would like to express grateful appreciation to our colleagues from the Polish Academy of Sciences, Committee of Mechanics, as well as to the International Scientific Committee i.e. Members and the Advisory Board. Their advice and efforts have helped us to overcome the problems normally associated with organising international meetings.

Special thanks goes to the reviewers for their work in encouraging the submission of papers and the subsequent review of all papers. Their contribution cannot be overestimated.

The 21st Fluid Mechanics Conference was organised by AGH University of Science and Technology, the Polish Academy of Sciences the Committee of Mechanics and the AGH-UST Foundation. Proceedings was published in the Journal of Physics: Conference Series. The demanding work involved could not have been done without the contribution of so many individuals from all institutions as well as numerous external co-workers. Without their extremely valuable help such a meeting would have been impossible. Thank you all so much!

Janusz S. Szmyd
Elżbieta Fornalik-Wajs
Marek Jaszczur

Honorary Committee

Rector
Professor Tadeusz Słomka
AGH - University of Science and Technology

Organizing Committee

Prof. Janusz S. Szmyd - Conference Chairman
(AGH - University of Science and Technology)

Dr. Elżbieta Fornalik-Wajs - Conference Secretary
(AGH - University of Science and Technology)

Anna Inglot – Conference Manager
(Foundation for the AGH - University of Science and Technology)

Members

Dr. Grzegorz Brus (AGH - University of Science and Technology)
Dr. Adam Holda (AGH - University of Science and Technology)
Dr. Marek Jaszczur (AGH - University of Science and Technology)
Dr. Robert Kaczmarczyk (AGH - University of Science and Technology)
Dr. Remigiusz Nowak (AGH - University of Science and Technology)
Dr. Łukasz Pyrda (AGH - University of Science and Technology)
Dr. Andrzej Ziemba (AGH - University of Science and Technology)
Elżbieta Koszałka (AGH - University of Science and Technology)
Anna Kraszewska (AGH - University of Science and Technology)
Łukasz Pleskacz (AGH - University of Science and Technology)
Aleksandra Roszko (AGH - University of Science and Technology)
Anna Ściążko (AGH - University of Science and Technology)

International Scientific Committee

Chairman:
J. S. Szmyd (AGH - University of Science and Technology, Conference Chairman)
Co-Chairman:
T. A. Kowalewski (Polish Academy of Sciences, Committee of Fluid Mechanics - Chairman)

Members:
A. Bar-Cohen (USA)

K. Bajer (Poland)

N. Djilali (Canada)

P. Doerffer (Poland)

S. Drobniak (Poland)

T. Gatski (USA)

A. Gelfgat (Israel)

B. J. Geurts (The Netherland)

K. Hanjalic (The Netherland)

J. M. Hyun (South Korea)

Y. Kawaguchi (Japan)

J. S. Lee (South Korea)

P. di Marco (Italy)

J. Patterson (Australia)

M. Plesniak (USA)

A. Pollard (Canada)

M. Reeks (UK)

J. Rokicki (Poland)

A. van Steenhoven (The Netherland)

A. Solan (Israel)

K. Suga (Japan)

A. Styczek (Poland)

H. Yoshida (Japan)

Advisory Board

T. Chmielniak (Polish Academy of Sciences- Corresponding Member)

Z. Kolenda (Polish Academy of Arts and Sciences – Corresponding Member)

J. Mikielewicz (Polish Academy of Sciences – Ordinary Member)

J. Siemek (Polish Academy of Arts and Sciences – Corresponding Member, Polish Academy of Sciences - Corresponding Member)

Reviewers

D. Asendrych
J. Badur
K. Bajer
K. Banaś
A. Bartosik
A. Bogusławski
P. Di Marco
N. Djilali
P. Doerffer
S. Drobniak
M. Ekiel-Jeżewska
W. Elsner
T. Gatski
A. Gelfgat
B. Geurts
K. Hanjalic
Y. Kawaguchi
T.A. Kowalewski
A. Kucaba-Piętal
H. Kudela
M. Morzyński
R. Mosdorf
A. Olajossy
C. Oleśkowicz-Popiel
J. Patterson
A. Pollard
R. Puzyrewski
R. Rohatyński
J. Rokicki
A. Solan
K. Suga
J.S. Szmyd
A. Van Steenhoven

Peer review statement

All papers published in this volume of *Journal of Physics: Conference Series* have been peer reviewed through processes administered by the proceedings Editors. Reviews were conducted by expert referees to the professional and scientific standards expected of a proceedings journal published by IOP Publishing.

Journal of Physics: Conference Series **530** (2014) 012001 doi:10.1088/1742-6596/530/1/012001

Horace Lamb & Osborne Reynolds: Remarkable Mancunians ... and their Interactions

B. E. Launder
School of MACE, The University of Manchester, Manchester, UK.

brian.launder@manchester.ac.uk

Abstract The paper provides glimpses into the professional lives of arguably, the two outstanding fluid mechanicists of their time who were simultaneously professors at Owens College, Manchester. Their interactions with each other were sometimes amicable but, equally, sometimes testy and their views on their common professional subject differed radically. Reynolds was appointed to the Chair of Engineering in 1868 at the age of 25 against strong competition while Horace Lamb, graduating a decade after Reynolds, was appointed as the inaugural Professor of Applied Mathematics at the University of Adelaide where he stayed for nine years before being appointed to a chair at Owens College in 1885. Among their various interactions the most significant arose from Reynolds' famous "Reynolds averaging" paper. That was sent for review by Lamb who was critical of the paper but finally recommended that a revised version be published since Reynolds had essentially invented the subject. Reynolds, in his turn, criticised Lamb's patronizing reference to engineers' approach to fluid mechanics in a draft revision of his book *Hydrodynamics*. Nevertheless, on Reynolds' death in 1912, it was Lamb who attended his funeral on behalf of the University and the Royal Society and who later wrote a moving, much cited obituary of him.

1. Introduction

In the mid-19th Century, Manchester was at the centre of Britain's industrial revolution. However, while the city grew rapidly, creating employment for a wide range of people, its only further-education establishment was the Manchester Mechanics Institute founded in 1824[1] which at that time was concerned with pre-university technical instruction. There was talk among the city's political and business community, however, that the city needed a broader, higher-level educational establishment; all that was missing was the necessary financial support. This vital input was provided in 1846 by a wealthy textile merchant, John Owens, a bachelor who, on his death, bequeathed almost £100,000 – roughly equivalent to 100M€ in today's money – to trustees for the creation of a college for young men within "the parliamentary borough of Manchester or within two miles of any part of its limits", Clapp [1].

Thus, in 1851, Owens College, Manchester was opened. The college was not at the time accorded full university status but was, rather, approved as a "provincial examination centre for degree candidates of the University of London". Its premises, figure 1, were the former home of Richard

[1] Over the next 140 years the Institute underwent several re-profilings and re-namings before emerging as UMIST (the University of Manchester Institute of Science & Technology) in 1966 and finally amalgamating with the Victoria University of Manchester to form the current University of Manchester in 2004.

Cobden, the MP for Stockport (a town about 10km south of Manchester) who having achieved considerable national fame and popularity by forcing the repeal of the unpopular Corn Laws, decided to move his family to London to be closer to the centre of the nation's political power. The College embarked on the recruitment of professors across the usual range of degree-level subjects of the period. This did not include the discipline of Engineering, however. While Kings College, London had had the distinction of appointing the first full-time professor of engineering in Great Britain & Ireland in 1840, it was Irish and Scottish universities that, over the ensuing decade, had taken the initiative in adding chairs in engineering at universities in Dublin, Belfast, Cork, Galway and Glasgow (Allen [2]). There had been no further appointments in England when, in the mid-1860s a group of prominent Manchester industrialists began lobbying for the creation of a chair. Indeed, under the leadership of William Fairbairn, FRS, a senior industrialist, they raised the additional funds to support this aspiration, [2]. Thus, in early 1868 the College was able to advertise for a professor in engineering.

Figure 1. Early photograph of Owens College

2. Professor Osborne Reynolds: the first decade in post

A detailed account of Osborne Reynolds' early life and how he was appointed, less than two years after his graduation in Mathematics from Cambridge, to the Chair of Civil & Mechanical Engineering at Owens College has been provided by Launder & Jackson [3]. The following paragraph, however, perhaps provides the essential background.

Osborne Reynolds was the second of three children, born in Belfast when his father was headmaster of a local school there. Shortly after his birth, however, the family returned to England but his mother died following the birth of Osborne's younger brother. Thus, Osborne's educational and personal development was largely due to his father who never re-married. His father had an unusual background: after graduating in Mathematics from Queens' College, Cambridge he had trained to become a priest (like *his* father and grandfather), a calling he followed intermittently through his life. But he also had a good deal of interest in things practical of a mechanical nature, taking out several patents relating to farming [2], since the family had extensive agricultural estates in Suffolk. Thus, Osborne, following the personal tuition provided by his father, moved on at the age of 19 to spend rather more than a year as a trainee with a firm of boat-builders and general mechanical engineers owned by a Mr Edward Hayes of Stoney Stratford. This provided him with a general grounding in engineering manufacture and management. He then went on to read Mathematics at his father's college in Cambridge and, despite having to put much time into the study of Greek (which had not formed part of his curriculum with his father), he graduated in 1867 with high honours and immediately took up appointment with civil engineering consultants, Lawson & Mansergh in London (Allen [2]). However, once the Chair position at Owens' College came to his notice, he knew that he had to apply. On January 18th, 1868 he wrote in response to the advertisement:

> "Gentlemen, I beg to offer myself as a candidate for the Professorship in Engineering at Owen's [sic] College. I am in my twenty-sixth year. From my earliest recollection I have had an irresistible liking for mechanics; and the studies to which I have especially devoted my time are mechanics and the physical laws on which mechanics as a science are based." (University of Manchester, Osborne Reynolds Archive).

Some details surrounding the choice of the successful candidate are given in [3]. Briefly, of the 27 applicants just two were interviewed and of those, Osborne Reynolds emerged as the winner. Figure 2 shows fragments from the official document of appointment giving (upper) the terms of Reynolds' appointment and (lower) his signature and that of the chairman of the trustees of Owens College (together with those of witnesses) confirming acceptance of the terms.

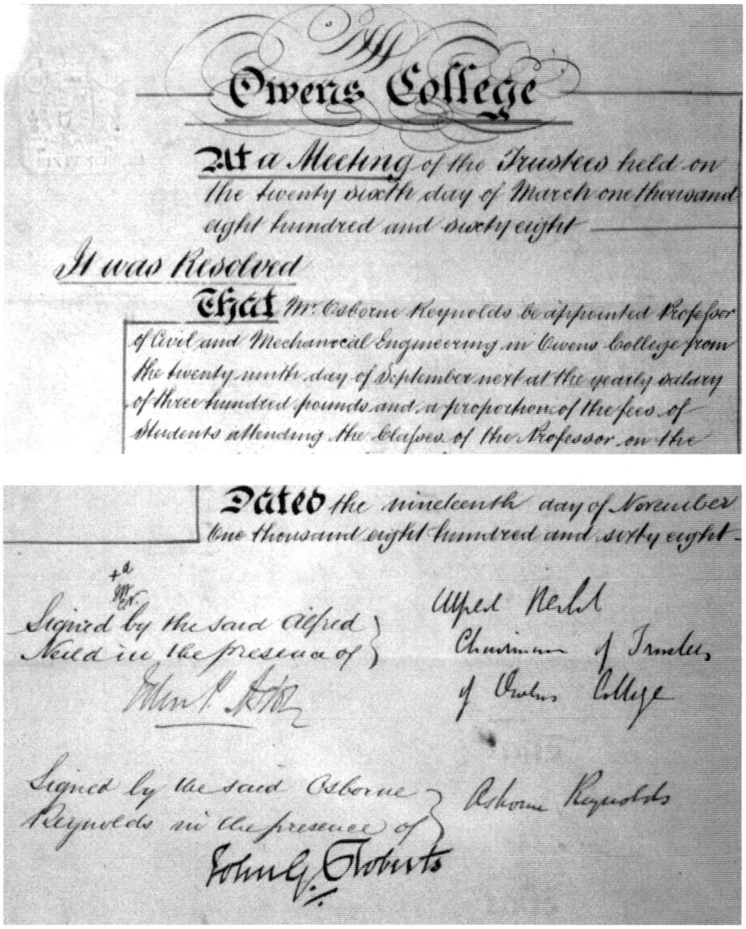

Owens College

At a Meeting of the Trustees held on the twenty sixth day of March one thousand eight hundred and sixty eight

It was Resolved

That Mr Osborne Reynolds be appointed Professor of Civil and Mechanical Engineering in Owens College from the twenty ninth day of September next at the yearly salary of three hundred pounds and a proportion of the fees of Students attending the Classes of the Professor on the

Dated the nineteenth day of November one thousand eight hundred and sixty eight

Signed by the said Alfred Neild in the presence of

Alfred Neild
Chairman of Trustees of Owens College

Signed by the said Osborne Reynolds in the presence of

Osborne Reynolds

John G. Roberts

Figure 2. Two fragments from Owens College's official appointment document. Upper: Opening statement; Lower: Confirming signatures by Osborne Reynolds and Alfred Neild, Chairman of the Trustees. (© University of Manchester, Osborne Reynolds Archive)

A photograph of Reynolds around the time of his appointment to the Chair appears in figure 3. The year 1868 would have been a very special one for him for besides his election to the Chair at Owens College he married Charlotte, a daughter of a doctor from Leeds. However, that joy was short-lived for the following year Charlotte died from a peritoneal infection twelve days after the birth of the couple's son. This clearly posed a challenging start for Reynolds but he may have been comforted by the fact that his father had had to cope with a similar misfortune, indeed with bringing up three young children alone rather than one.

Figure 3. Photograph of Osborne Reynolds from around the time of his appointment to the Chair of Civil & Mechanical Engineering at Owens College ©University of Manchester.

J. J. Thomson, winner of the Nobel Prize in Physics and subsequently President of the Royal Society, was one of the first students to enrol at the College, at the age of 14. His memoirs (Thomson, [4]) noted that, to accommodate the new Department of Engineering, the stables at the rear of the building were converted into a lecture room and the hayloft above into a drawing office. There were, however, no laboratories so Reynolds had to rely on using other science laboratories in the College or performing simple experiments at home. Thus his early papers were mainly concerned with explaining natural phenomena: for example, the tails of comets, the solar corona and the aurora; the electrical properties of clouds and the phenomenon of thunderstorms; the destruction of sound by fog; the calming of seas by raindrops and oil films; and the formation of hailstones and snowflakes. Horace Lamb in his obituary notice of Reynolds, [5], records that there were shades of disappointment noted at the academic nature of these studies by practising engineers who had worked so hard to have the Chair of Engineering created. Following the College's move in 1873 to the new campus to the south of Manchester (the site of the present University of Manchester), however, Reynolds did begin to address problems of a more pressing engineering importance beginning with two experimental papers on heat transfer, one considering the effect of the presence of air on the condensation rate of steam on a cooled surface (Reynolds [6]) and the other on the extent and action of the heating surface of steam boilers (Reynolds [7]). At about the same time Reynolds also initiated work on the 'racing' of ships' screw propellers and on the multi-staging of turbines and centrifugal pumps (the latter leading in 1875 to a detailed patent specification [2]).

In the second half of the 1870s Reynolds' research in fluid mechanics shifted towards more general phenomena such as the dispersion of surface waves in deep water and the motion of vortices, the latter made visible by dye traces in water. A fuller account of these and other studies from the period can be found in Allen [2] and Jackson [8]. In recognition of the exceptional contributions that he had made within eight years of his appointment to the Chair of Engineering, in 1877 Reynolds was elected to the Fellowship of the Royal Society.

Throughout this busy and productive period of his career he remained a devoted and involved father. However, in the summer of 1879, his son went down with a severe sickness that led to Reynolds renting accommodation for them both in St. Leonards-on-Sea, at the time a quiet community on the south coast of England. Despite his love and attention, however, his son died on September 27th. That event may be said to mark the end of the first phase of Reynolds career at Owens College.

3. Professor Horace Lamb: his full-circle route to Owens' College

Unlike Reynolds, Horace Lamb, was a 'town boy' growing up in Stockport, a community just far enough from Manchester to be considered an independent township. Born November 29th 1849, he was the son of a cotton mill foreman who had earned a reputation for devising improved operating practices. However, his father died while Horace was still a boy and when, sometime later, his mother decided to re-marry, Horace was thenceforth brought up by his mother's sister. Under her strict but loving regime Horace thrived, attending Stockport Grammar School and becoming the school captain. His specialties were mathematics and the classics. Indeed, he took the Cambridge examinations in classics and was awarded an entrance scholarship at Queens' College which, on reflection, he declined having concluded that mathematics and its application was his primary love. However, the level of mathematics he had achieved was adjudged inadequate for him to be admitted to Cambridge; thus, he attended Owens College for a year under the tutorship of Thomas Barker, the professor of pure mathematics. Horace evidently prospered under Barker's guidance for he was awarded a minor scholarship to Trinity College whence he graduated in 1872, being placed second among the two hundred or more mathematics graduates from Cambridge that year.

His outstanding degree performance led to his being immediately elected a Fellow of Trinity and appointed as a tutor in the college. Moreover, he had developed a passionate interest in the analysis of fluid mechanics and this interest he conveyed by way of what may well have been the first integrated lecture course on the mechanics of fluids delivered to final year undergraduates in 1874. Of his lectures one of those students was later to write "His lectures were a revelation … Lamb, in his own inimitable manner, revealed the mysteries of [vortex rings] and made the properties of a liquid in rotational motion clear to us" (Glazebrook, [9]). One might thus have supposed that the enthusiasm with which his lectures were received would have marked the start of a long and distinguished career at Cambridge – but it didn't! Horace had remained on cordial terms with the headmaster of Stockport Grammar School and through that link met and became romantically entwined with the head's sister-in-law. The Cambridge colleges at that time required that junior staff should be unmarried; so Lamb had to find an alternative employer.

On learning of Lamb's predicament, a friend and former teacher at Stockport Grammar School who had subsequently emigrated to South Australia alerted him to the fact that the newly established University of Adelaide was creating a Professorship of Pure & Applied Mathematics. Lamb thereupon sought out further details about the University and the position from the Agent General for South Australia (based in London) and, having thus satisfied himself, submitted an application. The twenty-odd applications were progressively whittled down via postal communications by two distinguished British mathematicians and finally Lamb was selected for the position. Thus, Horace Lamb and his wife, Elizabeth (née Foot), set out for Australia in January 1876, arriving in Adelaide in March, just in time for the official opening of the University.

Lamb threw himself with enthusiasm into the work of the new university. Besides his teaching load in mathematics, he voluntarily taught a course on experimental physics while in the evenings, one day per week, he ran classes of a popular nature that the public at large were invited to attend: basic courses on such topics as electricity and sound, figure 4. Moreover, it was not simply in the lecture room that Lamb contributed to the new university's development. As one of just four professors on the staff he was deeply involved in the academic, administrative and developmental activities of the University, serving as Dean from early 1878.

He also found time to pursue research. The first papers from his antipodean period were published in London-based journals in 1877 with a further one or two articles appearing in most years thereafter. Moreover, his lectures on fluid mechanics delivered orally in Cambridge in 1874 were now formally shaped and expanded (to include more recent contributions) into a 258-page textbook, Lamb [10]. The quality and number of his publications was such that Lamb was recommended in November 1883 by seven Fellows, including Lord Rayleigh and Professor Cayley, for election to The Royal Society, a proposal that was accepted the following June.

However, despite his achievements within the university, his numerous publications and the enthusiasm of the students for his teaching, all was not well for Lamb. Adelaide was so remote: he desperately missed the interaction with other academics on research matters that had been an every-

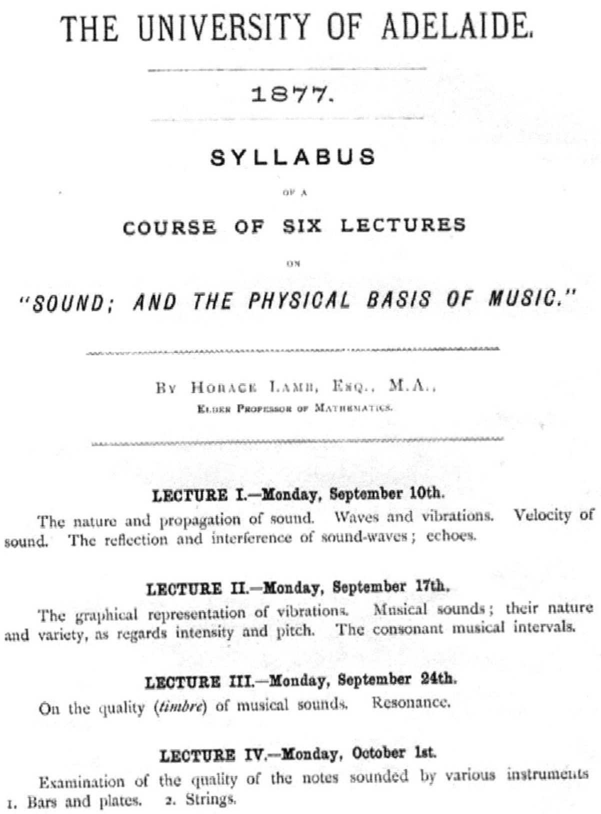

THE UNIVERSITY OF ADELAIDE.

1877.

SYLLABUS

OF A

COURSE OF SIX LECTURES

ON

"SOUND; AND THE PHYSICAL BASIS OF MUSIC."

BY HORACE LAMB, ESQ., M.A.,

ELDER PROFESSOR OF MATHEMATICS.

LECTURE I.—Monday, September 10th.

The nature and propagation of sound. Waves and vibrations. Velocity of sound. The reflection and interference of sound-waves; echoes.

LECTURE II.—Monday, September 17th.

The graphical representation of vibrations. Musical sounds; their nature and variety, as regards intensity and pitch. The consonant musical intervals.

LECTURE III.—Monday, September 24th.

On the quality (*timbre*) of musical sounds. Resonance.

LECTURE IV.—Monday, October 1st.

Examination of the quality of the notes sounded by various instruments 1. Bars and plates. 2. Strings.

Figure 4. First page of leaflet announcing a short popular course of lectures on 'Sound; and the physical basis of music' by Horace Lamb. (© University of Adelaide, Horace Lamb Archive).

day part of his life during his Cambridge years. So, he determined to press the University to allow him to take a paid sabbatical leave for a year to enable him to return to England and immerse himself in the currents of research and ideas that were then at the forefront of applied mathematics. The Registrar brought Lamb's proposal to Council members who were unsympathetic, [11]. There was very much the feeling that Lamb would simply use the period of leave to seek out a new position in Europe. The exchanges went on for several months and, while these were in mid-stream, Lamb wrote a letter to his former tutor at Cambridge, H. M. Taylor from which relevant portions appear below (University of Manchester Archive, Horace Lamb chair application):

"Adelaide, January 31st 1885

My dear Taylor,

I am going to ask you a very embarrassing question and so will acknowledge at once that I do not hope for any very satisfactory answer. However, I shall be grateful for anything you can say on the subject.

I would like to know, if possible, what chances I should have of getting employment in England. My position here is not uncomfortable, I get £1000 a year (which is, however, far from equivalent to the same sum at home), I have not too much work, although my time is a good deal cut up. On the other hand, most of the work is very elementary; and the sense of isolation is at times most painful. … I have hardly a soul to speak to who has any understanding of or sympathy with my pursuits. I confess too to feeling that I am fit for something better than the work I have here; but this kind of sentiment … is probably no more than vanity and conceit. I have been on the whole very happy and comfortable here, but after all, I think, 10 years is a long enough time in Australia, if one has any chance of getting out."

That letter was to have profound consequences! The aforementioned Thomas Barker, having become seriously wealthy from astute stock-market dealings, decided to resign his chair of pure mathematics to follow a passionate lifelong hobby on the collection and study of mosses. Owens College thus promptly advertised the vacancy, figure 5.

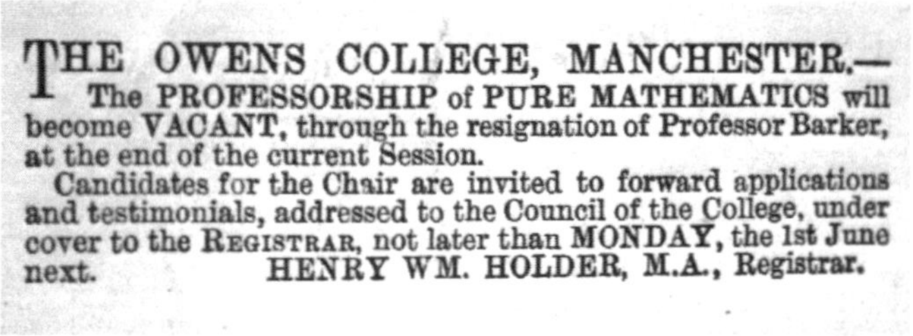

THE OWENS COLLEGE, MANCHESTER.—
The PROFESSORSHIP of PURE MATHEMATICS will become VACANT, through the resignation of Professor Barker, at the end of the current Session.
Candidates for the Chair are invited to forward applications and testimonials, addressed to the Council of the College, under cover to the REGISTRAR, not later than MONDAY, the 1st June next.
HENRY WM. HOLDER, M.A., Registrar.

Figure 5. Advertisement for Chair of Pure Mathematics at Owens College

On reading this, Taylor cabled Lamb to ask whether he would be interested in the position and, on receiving an affirmative response, promptly set about making an application on Lamb's behalf and soliciting letters of reference from the Cambridge mathematics community. The committee assigned to assess applications for the chair, among whose members was Osborne Reynolds, concluded that Lamb was indeed the best candidate. The College thus invited Lamb to provide letters of reference from senior figures at Adelaide University and attend for formal interview in Manchester. At that stage Adelaide must have realized they had probably lost him: he was granted sabbatical leave and provided warm letters of support by four senior members of Council. Thus Lamb and his family (by then comprising his wife and six children) came to Manchester where, following interview, his appointment was confirmed, Horace assuming the chair on September 29th, 1885. Even before taking up his chair responsibilities, Lamb presented himself to the Royal Society to be admitted as a Fellow. Figure 6 shows both the official photograph of this event with the reverse of the photo containing Lamb's signature. His style 'Prof'r', with the postnomial letters 'FRS', suggest his contentment with his new estate!

4. Reynolds and Lamb at Owens College

Thus, by 1885 Owens College had on its staff two of the leading Fluid Mechanicists in the country, both under 45 and at the height of their creativity. It is true that Reynolds, unlike Lamb, had moved only slowly to a position where Fluid Mechanics had become the dominant theme of his life. Like the trio of distinguished mathematical physicists, Sir George Stokes, Lord Kelvin and Lord Rayleigh who made substantial contributions to fluid mechanics among other major themes, Reynolds in his first decade at Owens College had been peripatetic in the subjects on which his fertile mind threw light. Following the death of his son, however, he evidently sought a new beginning.

Journal of Physics: Conference Series **530** (2014) 012001 doi:10.1088/1742-6596/530/1/012001

Figure 6. Horace Lamb on his admission to the Royal Society with his signature on the reverse.(© The Royal Society)

In December 1881 Reynolds had re-married; his wife was Annie Charlotte Wilkinson, just 22 years old who, we may suppose, took over the tasks of household management as Reynolds plunged more deeply into detailed observations and measurement of fluid motion. In 1883 his famous paper determining the conditions under which turbulent flow in a pipe decayed to laminar was published, Reynolds [12]. A few years later, an equally famous paper on film-lubrication appeared, Reynolds [13]. About, the latter, 32 years after its publication, Lord Rayleigh felt able to remark “it includes most of which is now known on the subject” [2].

As noted above, Reynolds had served on the selection committee leading to Lamb’s appointment to the chair of pure mathematics. Soon after Lamb’s arrival they were asked to work together in a way that, unexpectedly (for Reynolds at least), was also to have a significant impact on Lamb’s role within the College. In 1888 the senior professor of physics, Balfour Stewart, FRS, had died in post at a time when the College was conscious of the need to constrain its budget. The Principal thus appointed a committee comprising Reynolds and Lamb plus the junior professor of physics (an astronomer) and Arthur Schuster, FRS (the professor of applied mathematics) to consider whether the College could manage to avoid making a replacement chair appointment for Balfour Stewart without damage to its reputation. Lamb, who had been Dean at the University of Adelaide and was evidently a shrewd strategist, quickly took charge of the committee’s directions of thought. Its recommendation, presented to and approved by Senate, was that Schuster should move to take over Balfour Stewart’s chair as the Langworthy Professor of Physics and that the departments of pure and applied mathematics should merge under Lamb’s headship as the Beyer Professor of Pure and Applied Mathematics. Thus, at a stroke, Lamb’s responsibilities within the College were enlarged to embrace fully his interests and expertise as well as to increase his salary (since a significant proportion of his emolument arose from the fees paid by students attending his classes).

As the above example perhaps illustrates, Lamb was well-organized and decisive in his thinking, composed in his demeanour and lucidly precise in his writings. Reynolds, while sharing with Lamb a capacity for intense hard work, was in other respects almost the opposite of his colleague. While he had an engaging charm he would, for the most trivial of reasons or even for none at all, divert from the subject of his discourse leaving his listeners floundering. And he was forgetful! Turning again to the memoirs of J. J. Thomson, [4], we find as an example:

> "The Professor I had most to do with was Osborne Reynolds. He never did anything or expressed himself like anybody else. The result was that it was very difficult to take notes at his lectures so that we had to trust mainly to Rankine's text books. Occasionally in the higher classes he would forget all about having to lecture and after ten minutes or so we sent the janitor to tell him that the class was waiting. He would come rushing into the room pulling on his gown as he came through the door, take a volume of Rankine from the table, open it apparently at random, see some formula or other and say that it was wrong. He then went up to the blackboard to prove this. He wrote on the board with his back to us talking to himself, and every now and then rubbed it all out and said that it was wrong. He would then start afresh on a new line and so on. Generally, towards the end of the lecture he would finish one which he did not rub out and say that this proved Rankine was right after all."

Nevertheless, it is for his contributions to research rather than his communication foibles that Reynolds is remembered. His celebrated 1883 paper noted above, [12], in which he showed that (what we now know as) the Reynolds number determined the transition between laminar and turbulent flow in a smooth straight pipe, had received wide acclaim. Nevertheless, Lord Rayleigh (one of the referees of that manuscript), while recommending publication and singing its praises in his address as President of the British Association, had ended his referee's report with the remark: "In several passages the Author refers to theoretical investigation whose nature is not sufficiently investigated". Just how heavily that observation weighed on Reynolds' mind is of course unknown. It is likely that he didn't need any such reminder; but for him to start to explain the phenomenon of transition was such a major undertaking that his ideas needed time for successive refinement. Then finally, on May 24th, 1894, he presented an oral version of his protracted considerations at a meeting of The Royal Society and later that year had a number of copies of the manuscript type-set and printed, several of which he submitted to the *Philosophical Transactions of the Royal Society* to be considered for publication.

In fact, by then Lord Rayleigh had become editor of the journal and he immediately chose Sir George Stokes (who had reviewed the 1883 paper and had acquired a reputation for being a pioneer in the use of that new fangled device, the typewriter) to referee this latest manuscript. However, this time Stokes' response was not in the least helpful; his typed letter, shown in figure 7, is a classic example of a sitting-on-the-fence assessment. Rayleigh pressed him to be more decisive but in the mean time went to the only other fluid mechanicist in Britain with both the insight into the topic in question and the incisiveness to make plain his conclusions: Horace Lamb! Curious as it may seem for him to be asked to assess his own colleague's work, Lamb did not demur. On November 21st, 1894 he sent his longhand assessment which began with the brisk summarizing statement:

> "I think the paper should be published in the *Transactions* as containing the views of its author on a subject which he has to a great extent created, although much of it is obscure and there are some fundamental points which are not clearly established."
> (Royal Society Archive)

However, there followed, three pages of detailed and rather severe criticisms.

Journal of Physics: Conference Series **530** (2014) 012001 doi:10.1088/1742-6596/530/1/012001

Figure 7. Sir George Stokes' initial review of Reynolds' 1895 paper

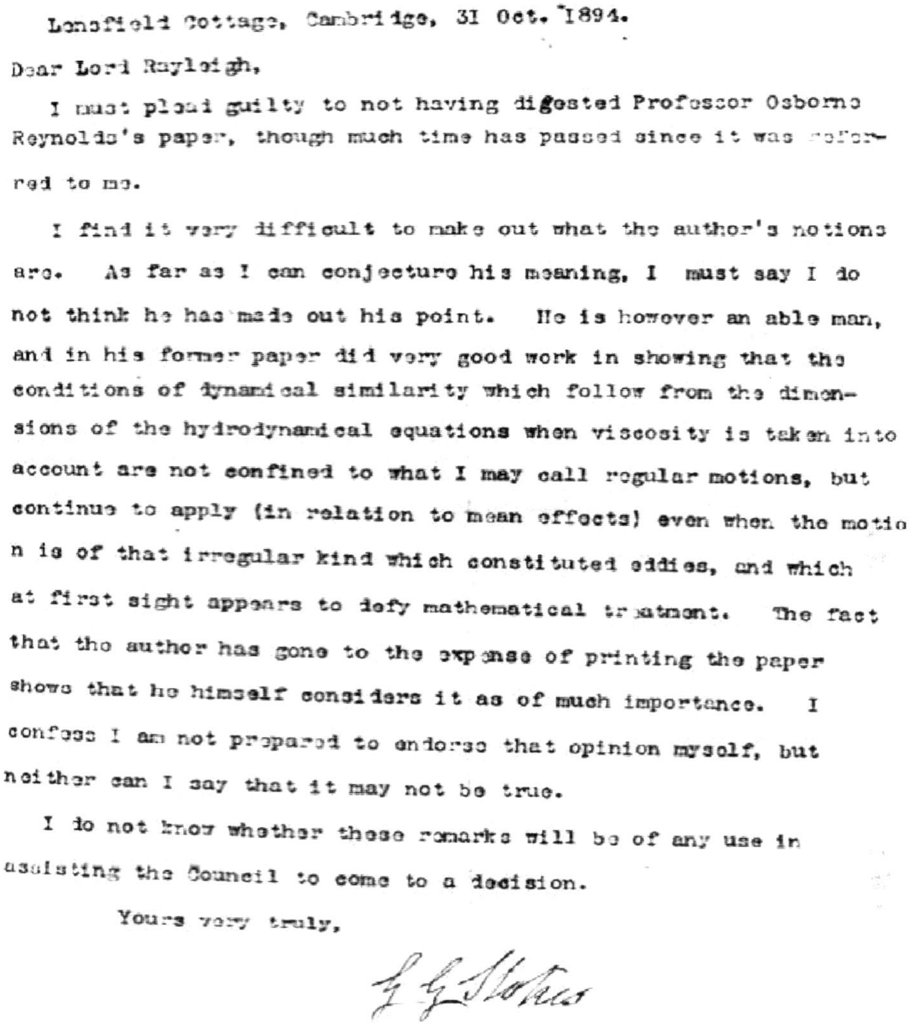

Lensfield Cottage, Cambridge, 31 Oct. 1894.

Dear Lord Rayleigh,

I must plead guilty to not having digested Professor Osborne Reynolds's paper, though much time has passed since it was referred to me.

I find it very difficult to make out what the author's notions are. As far as I can conjecture his meaning, I must say I do not think he has made out his point. He is however an able man, and in his former paper did very good work in showing that the conditions of dynamical similarity which follow from the dimensions of the hydrodynamical equations when viscosity is taken into account are not confined to what I may call regular motions, but continue to apply (in relation to mean effects) even when the motion is of that irregular kind which constituted eddies, and which at first sight appears to defy mathematical treatment. The fact that the author has gone to the expense of printing the paper shows that he himself considers it as of much importance. I confess I am not prepared to endorse that opinion myself, but neither can I say that it may not be true.

I do not know whether these remarks will be of any use in assisting the Council to come to a decision.

Yours very truly,

G G Stokes

Figure 7. Sir George Stokes' initial review of Reynolds' 1895 paper

Eventually, Stokes was coaxed (presumably by Rayleigh) into writing a report. Thereafter, Stokes and Lamb were persuaded (again presumably by Rayleigh) – to provide a joint report setting out their collective concerns about the paper. Stokes' contribution appears to have been in typing the covering letter while the report itself is in Lamb's handwriting. The first page of this joint report (which largely repeats Lamb's original assessment) appears in figure 8; but in case Lamb's longhand is not easily deciphered, a transcription follows.

“Professor Reynolds’s Paper

The referees have found great difficulty in following the arguments of this paper; partly in consequence of the fact that such terms as “mean-mean motion” and “relative mean motion” are used without any precise definition. There is a well known distinction between molecular and molar motion; but it is not clear in the case of molar motion how any distinction is to be drawn between what is “mean” and what is “relative”.

The Introduction might be greatly shortened as a good deal of it can only be understood after reading the rest of the paper. The purport of §5(a) is not evident. The author’s view does not appear to be different from that generally held but it is insisted upon as something new.”

The faint handwritten instruction “copy this” in Rayleigh’s handwriting is presumably directed at office staff to transcribe the review prior to forwarding it to Reynolds.

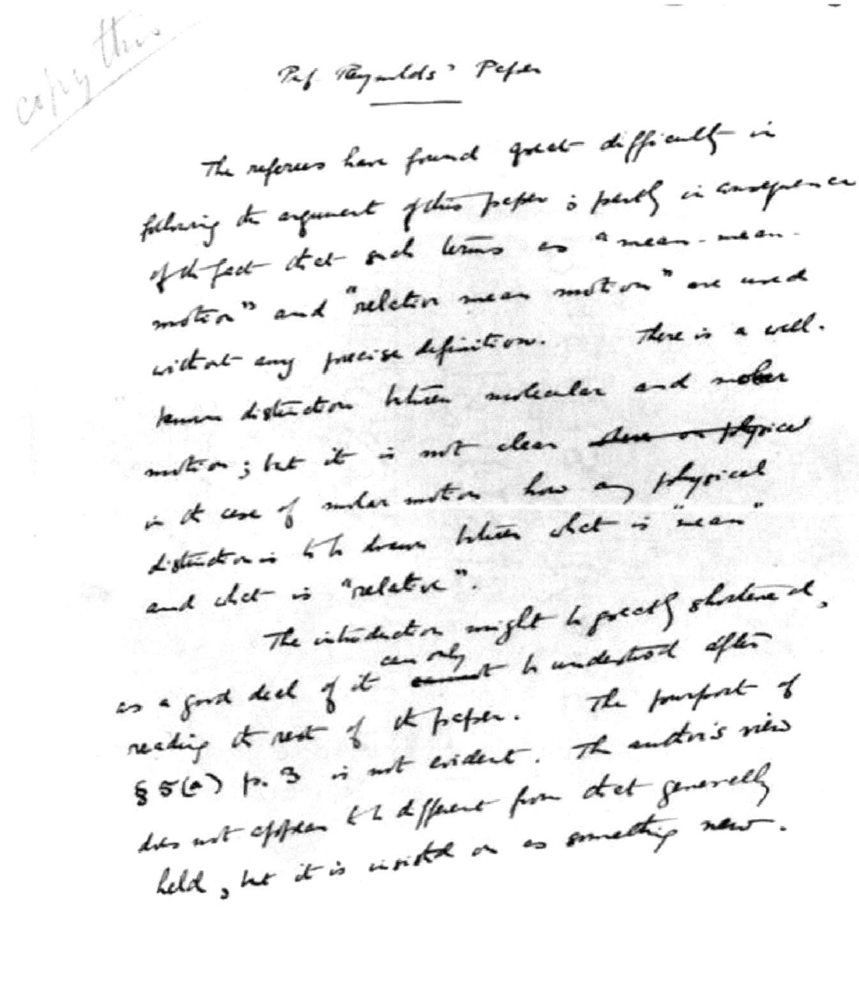
copy this

Prof Reynolds' Paper

The referees have found great difficulty in following the argument of this paper; partly in consequence of the fact that such terms as "mean-mean motion" and "relative mean motion" are used without any precise definition. There is a well known distinction between molecular and molar motion; but it is not clear in the case of molar motion how any physical distinction is to be drawn between what is "mean" and what is "relative".

The introduction might be greatly shortened, as a good deal of it can only be understood after reading the rest of the paper. The purport of § 5(a) p. 3 is not evident. The author's view does not appear to be different from that generally held, but it is insisted on as something new.

Figure 8 Joint report (in Lamb’s handwriting) on Reynolds’ 1895 paper by Professors Lamb and Stokes (Royal Society Archive)

On receiving the referees’ report from Lord Rayleigh, Reynolds evidently reflected on the criticisms and made what he judged to be appropriate adjustments to his text. Then, on February 19th,

1895, he wrote from his home a two-page reply and sent it along with a revised version of the manuscript to the editor. The opening of the letter began as follows:

> "Dear Lord Rayleigh,
>
> From the copy of the remarks on my paper on the criterion which you sent me, it is clear that the referees have found great difficulty in understanding the drift of the main argument, namely that which relates to the geometrical separation of the components $\underline{u}$, $\underline{v}$, $\underline{w}$ into mean components $\overline{\underline{u}}, \overline{\underline{v}}, \overline{\underline{w}}$ and relative components $\underline{u}', \underline{v}', \underline{w}'$ and as to the conditions of distribution of $\overline{u}, \overline{v}, \overline{w}$ under which such separation is possible.
>
> I am very glad to know of these difficulties and of the opportunity it afforded me of improving the paper in this particular."

The principal alteration to the manuscript appears to have been in the Introduction. While Lamb had proposed that that section should be greatly shortened, Reynolds added a passage of nearly four printed pages that appears within square parentheses ending with the date 'Feb. 18, 1895', i.e. the day before he wrote the above letter to Lord Rayleigh. In the absence of any archival material to the contrary, it appears that the manuscript thus revised was accepted for publication.

Readers having some familiarity with Reynolds' paper will recognize that the referees' report had been entirely concerned with preliminary aspects of the analysis and had not at all ventured into the startling physical implications that Reynolds' mathematical examination had revealed. Today, however, researchers whose interests lie in high-Reynolds-number phenomena are likely see the 1895 *Phil Trans* paper as the pinnacle of Reynolds' contributions to Fluid Mechanics. With the perspective of more than 100 years, it still seems an extraordinarily innovative paper. Not only did he introduce for the first time the concept of decomposing the motion into mean and fluctuating parts, he did so within the framework of mass-weighted averaging (70 years before the paper that is generally cited as the first advocacy of that strategy, Favre [15]). Only later in the paper did he limit attention to uniform-density flows (since his earlier experiments had employed water as the working fluid) and thus arrived at the variables we nowadays regard as 'Reynolds averaging'. He then went on to obtain the turbulent kinetic energy equation, identifying the physical significance of each of the terms. The conceptual innovation entailed in this step can be appreciated from the fact that the corresponding (albeit simpler) equation for the mean-square temperature fluctuations was only published nearly 60 years after (Corrsin [16])! It should also be noted that, in Reynolds' time, tensor subscript notation had not been devised so an analysis that today might take only a few lines would at that time stretch over pages. Finally, the rationale for all these 'preliminaries' was to determine why and when the transition from laminar to turbulent flow was at risk of occurring. He showed that for flow between parallel planes (a less taxing configuration than flow in a circular pipe), as the Reynolds number was progressively raised, the spatially averaged rate of turbulence energy generation by mean shear interacting with (what we now call) the Reynolds stresses increased more rapidly than the rate of viscous dissipation. He proposed that the critical Reynolds number was the value at which these two agencies became equal. Because of the assumptions made about the velocity distribution, this calculated critical value was far too low. But that 'failure' was immaterial to the importance of the paper; indeed, the paper served as the basis for computational research in engineering-oriented fluid mechanics for the subsequent century!

Horace Lamb was, at around the same time, also engaged in a work that was to be *his* principal and lasting legacy to students and teachers of fluid mechanics. His textbook *A Treatise on the Mathematical Theory of the Motion of Fluids*, originally published in 1879 while he was still in Adelaide, was in need of expansion to include the many advances in the mathematical analysis of fluid motion over the ensuing years (including many that he himself had made). During the course of shaping the new material he contacted Reynolds on a matter concerning the latter's earlier 1883 paper. This drew from Reynolds a somewhat tetchy response. In a letter to Lamb of April 5th, 1895 (Allen[2]) he wrote:

> "You have not noticed that just above the critical velocity the resistance *dp/dz* varies nearly as the cube of the velocity until *dp/dz* is about double what it is at the critical velocity. Of course these dimensional facts … are the definite clues to the physics and mechanics of the problem and the gist of the later research".

Reynolds then went on to make a further (though in his mind perhaps the principal) complaint:

> "Nor is it polite or true to speak of the 'empirical formula adopted by Engineers' since it is engineers who have done the scientific investigations which alone have given us accurate data".

Then, possibly recognizing that his remarks were at least ungracious, given that the purpose of Lamb's enquiry was simply to present what Reynolds had found as accurately as possible in the new edition of his book, he added the final remark, Allen [2]:

> "I am obliged for the trouble [you] have taken to bring my work in[to your new edition]. I fear my criticism will bother you … but you will take what notice of it you like."

Lamb, ever the calmer of aggrieved sensibilities, duly replaced the above offending phrase by 'practical formula adopted by writers on hydraulics' [3]. When, later that year, the new edition was published, [17], its length had more than doubled (from 279 to 604 pages) while its title had been shrunk to just a single word, *Hydrodynamics*, which when coupled with the surname of its author, still has resonance for fluid mechanicists well over a century after its publication.

5. The Final Years

Publication of the second of Reynolds' major works on turbulent flow did not mark the end of his scientific outpourings. In 1887 he gave the Bakerian Lecture to the Royal Society (Reynolds & Moorby [18]) reporting his measurements of the mechanical equivalent of heat. Of this huge experimental programme, in which he obtained the equivalence within 0.2% of modern determinations, Lamb [5] later wrote "This whole investigation is a model of scientific method and may claim to rank among the classical determinations of physical constants."

Osborne Reynolds' final years in Manchester were marked by his intense efforts to provide a mechanical theory of matter which culminated in his work *The Sub-Mechanics of the Universe* being published as Volume 3 of his collected works (Reynolds [19]). However, as his colleague was later to observe, in what must be seen as a kind understatement, "unfortunately illness had begun gravely to impair his powers of expression and the memoir as it stands is affected with omissions and discontinuities which make it unusually difficult to follow" (Lamb [5]). By 1905 Reynolds had begun to suffer so seriously from this degenerative neurological illness – possibly what today would be diagnosed as Alzheimer's disease – that he retired from what had by then become the Victoria University of Manchester. Figure 9 shows a portion of the fine retirement portrait of him painted by the distinguished portraitist, John Collier. In 1908 he moved with his wife and daughter to occupy the vicarage at St Decuman's, a village community on the hill just above Watchet, a small coastal port some 300km south of Manchester. There on the 21st day of February, 1912 he died from influenza. His funeral in St Decuman's church was attended by Horace Lamb and Reynolds' remains are buried in the churchyard beneath a beautifully engraved *art nouveau* cross.

Figure 9. From the 1904 retirement portrait of Osborne Reynolds by John Collier © The University of Manchester

By then the time was approaching for Lamb himself to retire from his professorial responsibilities. As what may have been seen as a prelude to a normal-age retirement, in 1913 Ernest Rutherford, FRS had presented the University with a portrait of Lamb painted by Lamb's son Henry, a photograph of which appears in figure 10. However, the outbreak of World War I in 1914 caused Lamb to postpone any thought of retirement until the end of hostilities. He had by then, among numerous other publications, produced a further edition of *Hydrodynamics* in 1906 and a new textbook on sound, Lamb [20]. Moreover, yet another edition of *Hydrodynamics* would appear in 1916. When, in 1920, Lamb's actual impending resignation was announced it was met with considerably more than the usual expressions of appreciation and regrets. A retirement banquet was organized in his honour to express the University's thanks for his thirty-five years of outstanding service and the University conferred on him the honorary degree of doctor of science. Even then Lamb did not retire in the accustomed sense for he then removed to Cambridge where, as the honorary Rayleigh Lecturer, he relentlessly pursued his scholarly and public-service activities for more than a dozen years. During this time he completed two further editions of *Hydrodynamics* (the final one appearing in 1932, the year after he was knighted) and served on the national Aeronautical Research Committee from 1921-27. The latter role led, in turn, to his appointment as editor of a major project to provide, through the ARC, an up-to-date account all aspects of aerodynamics with a bearing on flight. Alas, he died some years before *Modern Developments in Fluid Mechanics* was finally published in 1938, his role being taken over by Sydney Goldstein who had succeeded him at Manchester as the Beyer Professor of Applied Mathematics. This two-volume publication [21] is, however, dedicated to Horace Lamb's memory.

Figure 10 Portrait of Horace Lamb by his son, Henry Taylor Lamb, R.A.
© The University of Manchester

Acknowledgements
I am indebted to Nicholas Higham, Professor of Applied Mathematics, University of Manchester, for his photograph of figure 10 and to Jeff Hurst for skilfully optically removing from my photograph a plethora of librarians' markings that had been applied over the years to the reverse of figure 6.

References

[1] Clapp B W 1965 *John Owens: Manchester Merchant* Manchester University Press.
[2] Allen J 1970 The life and work of Osborne Reynolds In *Osborne Reynolds and Engineering Science Today* (Ed. D M McDowell & J D Jackson) 1-82 Manchester University Press.
[3] Launder B E and Jackson J D 2011 Osborne Reynolds: a turbulent life. In *A Voyage Through Turbulence* (Ed. P A Davidson et al.) 1-39 Cambridge University Press.
[4] Thomson J J 1936 *Recollections and Reflections* G. Bell & Sons, London.
[5] Lamb H 1913 Osborne Reynolds, 1842-1912 *Obituary Notices Proc. Roy. Soc.*, **88**, xv-xxi.
[6] Reynolds O 1873 On the condensation of a mixture of air and steam upon cold surfaces, *Proc. Roy. Soc.*, **21**, 274-281.
[7] Reynolds O 1874 On the extent and action of heating surfaces of steam boilers, *Proc. Manchester Lit. & Phil.* **14**, Session 1884-5.
[8] Jackson J D 1995 Osborne Reynolds: scientist, engineer and pioneer *Proc. Roy. Soc.* **451A**, 41-86.
[9] Glazebrook R T 1935 Sir Horace Lamb, *Obituary Notices of Fellows of the Royal Society* **1** 374-392 (with additional material by A. E. H. Love).

Journal of Physics: Conference Series **530** (2014) 012001 doi:10.1088/1742-6596/530/1/012001

[10] Lamb H 1879 *A Treatise on the Mathematical Theory of the Motion of Fluids*, Cambridge University Press.
[11] Launder B E 2012 Horace Lamb and the circumstances of his appointment at Owens College *Notes Rec. R. Soc.* **67** 139-158.
[12] Reynolds O 1883 An experimental investigation of the circumstances which determine whether the motion of water shall be direct or sinuous and the law of resistance in parallel channels *Phil.Trans. Roy. Soc.* **174** 935-982.
[13] Reynolds O 1886 On the theory of lubrication and its application to Mr Beauchamp's tower *Phil. Trans. Roy. Soc.* **187** 157-234.
[14] Reynolds, O 1895 On the dynamical theory of incompressible viscous fluids and the determination of the criterion Phil. Trans. Roy. Soc. 186A 123-164.
[15] Favre A 1965 Équations des gaz turbulents compressibles *J. de Mécanique* **4** 361-390.
[16] Corrsin S C 1952 Heat transfer in isotropic turbulence *J. Appl. Phys.* **23** 113 118.
[17] Lamb H 1895 *Hydrodynamics* (2nd edition of Ref [10]) Cambridge University Press.
[18] Reynolds O and Moorby W H 1897 On the mechanical equivalent of heat *Phil. Trans. Roy. Soc.* **190A** 301-422.
[19] Reynolds O 1903 *Papers on Mechanical and Physical Subjects – the Sub-Mechanics of the Universe, Collected Works Volume III* Cambridge University Press.
[20] Lamb H 1910 *The Dynamical Theory of Sound* Edward Arnold, London.
[21] Goldstein S (Editor) 1938 *Modern Developments in Fluid Mechanics Vol. I and II* Oxford University Press.

Advances In The Analysis And Prediction Of Turbulent Viscoelastic Flows

T B Gatski[1]
Center for Coastal Physical Oceanography
and Ocean, Earth and Atmospheric Sciences
Old Dominion University, Norfolk, Virginia 23529 USA
L Thais and G Mompean
Université de Lille Nord de France, USTL, F59000 Lille, and
Laboratoire de Mécanique de Lille, CNRS, UMR 8107,
F59655 Villeneuve d'Ascq, France

E-mail: gatski@ccpo.odu.edu

Abstract. It has been well-known for over six decades that the addition of minute amounts of long polymer chains to organic solvents, or water, can lead to significant turbulent drag reduction. This discovery has had many practical applications such as in pipeline fluid transport, oil well operations, vehicle design and submersible vehicle projectiles, and more recently arteriosclerosis treatment. However, it has only been the last twenty-five years that the full utilization of direct numerical simulation of such turbulent viscoelastic flows has been achieved. The unique characteristics of viscoelastic fluid flow are dictated by the nonlinear differential relationship between the flow strain rate field and the extra-stress induced by the additive polymer. A primary motivation for the analysis of these turbulent fluid flows is the understanding of the effect on the dynamic transfer of energy in the turbulent flow due to the presence of the extra-stress field induced by the presence of the viscoelastic polymer chain. Such analyses now utilize direct numerical simulation data of fully developed channel flow for the FENE-P (**F**inite **E**xtendable **N**onlinear **E**lastic – **P**eterlin) fluid model. Such multi-scale dynamics suggests an analysis of the transfer of energy between the various component motions that include the turbulent kinetic energy, and the mean polymeric and elastic potential energies. It is shown that the primary effect of the interaction between the turbulent and polymeric fields is to transfer energy from the turbulence to the polymer.

1. Introduction

Since the pioneering experiments by Toms [1], it is known that the addition of minute amounts of long polymer chains to organic solvents, or water, can lead to significant turbulent drag reduction. This discovery has had many practical applications such as in pipeline fluid transport, oil well operation, vehicles design and submersible vehicle projectiles, and more recently arteriosclerosis treatment. The numerical simulation of this phenomenon is more recent. The availability of high-performance computers has allowed direct numerical simulations (DNS) of polymer drag reduction flows from the mid-nineties. Sureshkumar *et al* [2] were the first to perform direct numerical simulations of drag

[1] To whom any correspondence should be addressed.

reduction with a viscoelastic fluid. They used the FENE-P model (Finitely Extensible Non-linear Elastic in the Peterlin approximation), appropriate for dilute polymer solutions. Afterwards, Dimitropoulos *et al* [3] performed a detailed analysis of the budget of kinetic energy and streamwise enstrophy for the FENE-P solution at low Reynolds number, and showed that the inhibition of vortex stretching was a possible mechanism responsible for drag reduction.

More recently, other DNS were carried out with the FENE-P fluid. Dubief *et al* [4] reproduced the low and high drag reduction regimes. At low drag reduction (DR < 40%), the log law region is simply moved away from the wall. At high drag reduction, the log region is again moved away from the wall but its slope is increased with respect to the Newtonian flow. Dimitropoulos *et al* [5] considered the development of a zero-pressure-gradient flat plate turbulent boundary layer, later extending this study to inhomogeneous spatial distribution of polymer concentration [3]. It should be mentioned that while DNS has helped to provide more insight into the physics of polymer turbulence interaction, the numerical workload required by viscoelastic DNS make them only amenable to relatively modest Reynolds numbers on large scale mainframe computers. From the engineering point of view, it is advantageous to envisage viscoelastic turbulent closure models for such flows. In this spirit, there have been recent attempts towards RANS (Reynolds Averaged Navier-Stokes) modeling. Such models have included zero-equation [6], one-equation [7], and two-equation $k-\varepsilon$ [8] models.

From a phenomenological point of view, there exists a longstanding controversy between the two major theories which have been proposed to explain polymer drag reduction, and which still prevails in spite of the large amount of numerical data now available. Some authors [9,10,11] support Lumley's [12] viscous theory, which explains the phenomenon by an increase of the effective viscosity in the buffer layer. Others [13,14] gravitate around the elastic theory proposed by Tabor and De Gennes [15], which states that Kolmogorov's energy cascade is interrupted due to the elastic behavior of the polymer. We are convinced that studying viscoelastic turbulence may more generally lead to a better understanding of the physics of wall turbulence, in particular its self-sustaining mechanisms.

The presentation here will focus on the development of the FENE-P constitutive model followed by a discussion of the energetic exchange, based on DNS study at a friction Reynolds number of 1000, between the turbulent flow and polymeric fluid as represented by the FENE-P model.

2. Non-Newtonian and Viscoelastic Fluids

The theory of non-Newtonian and viscoelastic fluids flourished in the second half of the last century with the developments of (molten and dilute) polymers and the growth of materials science and engineering that generated many new products and applications. From a fluid and flow dynamic perspective, it is useful to provide a brief description of the hierarchy of such non-Newtonian and viscoelastic models. Modern constitutive equations have evolved by taking into account more and more of the microscopic features of polymeric fluids. The interplay between micro- and macro-scale is of importance and leads to a multi-scale approach towards the goal of incorporating more physics into the models that range from linear constitutive equations through to more sophisticated relationships. The theory of constitutive equations for non-Newtonian and viscoelastic fluids has been described in numerous books, and the reader is referred to the following references [16-23]. Of particular relevance is the recent text by Deville and Gatski [23] which serves as a guide for the model development remarks.

2.1. *Mechanical models*

At the introductory limit of non-Newtonian constitutive equations are the generalized Newtonian fluids. The constitutive equation for such fluids is generated by allowing the dynamic viscosity to depend on the shear rate $\dot{\gamma}$ such that $\mu = \mu(\dot{\gamma})$. A very common model is the power law

$$\mu(\dot{\gamma}) = K\dot{\gamma}^{m-1} \tag{2.1}$$

with K the consistency factor and m the index of the power law ($m=1$, $K=\mu$ for the Newtonian case). For many polymers, blood or food liquids, m is between 0.3 and 0.6. Fluids with $m<1$ are pseudo-plastic fluids; those with $m>1$ are dilating fluids. The viscosity μ decreases with respect to increasing shear rate $\dot{\gamma}$ for a pseudo-plastic fluid so that the viscosity is shear thinning. In the opposite case, where μ increases with increasing $\dot{\gamma}$, the viscosity is shear thickening.
A slightly more complex constitutive equation describes so-called yield fluids. These fluids exhibit a yield stress. This means that the shear stress has to be higher than a threshold yield stress for the fluid to flow. Below this threshold, the fluid behaves like a solid, and above it may behave as Newtonian or power-law like. An example of such a fluid that occurs in our everyday lives is toothpaste. These yield fluids produce plug flows in channels or pipes since near the symmetry axis the shear stress vanishes and therefore does not reach the threshold value.

A characterizing trait of these generalized Newtonian models was the relationship between a function of the shear rate and shear stress. The next level of complexity involves not only the stress field but also a stress rate. The original concepts for the design of linear viscoelastic models were framed with the aid of mechanical parts involving springs and dashpots. A simple example illustrates the development of such a model. Consider a spring with constant stiffness k for the elastic part and a dashpot of viscosity μ . If these two mechanical elements are connected in series, the resultant stress is described by the Maxwell model. In a one-dimensional extensional flow, for example, the stress-strain relation for the spring is $\Sigma = k\epsilon_1$ and for the dashpot $\Sigma = \mu\dot{\epsilon}_2$. The strains ϵ_1 and ϵ_2 are respective measures of an extensional motion. The same stress (force), Σ , applied to the model generates a displacement Σ / k in the spring and a velocity Σ / μ in the dashpot. The total velocity is then

$$\dot{\epsilon} = \dot{\epsilon}_1 + \dot{\epsilon}_2 = \frac{\dot{\Sigma}}{k} + \frac{\Sigma}{\mu} \quad \text{or} \quad \Sigma + \lambda\dot{\Sigma} = \mu\dot{\epsilon} \tag{2.2}$$

where $\lambda = \mu / k$ is a characteristic time of the model and $\dot{\epsilon}$ is the strain rate. Note that if the stress strain relation is rewritten as

$$\frac{d}{dt}(\mathbf{\Sigma} e^{t/\lambda}) = \left(\frac{\mu}{\lambda}\dot{\epsilon}\right)e^{t/\lambda} \tag{2.3}$$

and if the constraint that $\mathbf{\Sigma}$ remains finite when $t \to \infty$ is imposed, the integration in time yields [9]

$$\mathbf{\Sigma}(t) = \int_{\tau=-\infty}^{t} \frac{\mu}{\lambda}\exp\left[-\frac{(t-\tau)}{\lambda}\right]\dot{\epsilon}(\tau)\,d\tau \tag{2.4}$$

with the quantity in the integrand,

$$G(t-\tau) = \frac{\mu}{\lambda}\exp\left[-\frac{(t-\tau)}{\lambda}\right] \tag{2.5}$$

called the relaxation modulus. This expression shows that in the Maxwell model the stress at the present time depends on the history of the deformation through the strain rate. In addition, the fluid possesses an intrinsic property of fading memory that gives more weight to the recent past than to the distant past, with the decay rate driven by the material relaxation time. The Newtonian case is recovered with $G = \mu\delta(t-\tau)$ and a linear elastic solid for $G = G_0$.

These relatively simple ideas can be generalized and comprise the group of constitutive equations of the rate type. Stress rate models like the Oldroyd-B fluid, used for dilute solutions are far from the industrial fluids of polymer type that offer shear-thinning viscosity, but are nevertheless satisfactory in many problems and form a fundamental building block of non-Newtonian modeling. It is assumed that the polymers are diluted in a Newtonian solvent so that the Cauchy stress tensor can be split into two parts

$$\boldsymbol{\Sigma} = -p\boldsymbol{I} + 2\mu_0 \boldsymbol{S} + \boldsymbol{\Xi}_{\mathrm{v}} \tag{2.6}$$

where the first part is the Newtonian solvent which consists of an isotropic part and a deviatoric part characterized by the viscosity μ_0 and rate of strain tensor $\boldsymbol{S}$, and the second part, the viscoelastic stress $\boldsymbol{\Xi}_{\mathrm{v}}$. Now consider a similar generalization of the simple 1D Maxwell model given in (2.2) by replacing $\boldsymbol{\Sigma}$ with the stress $\boldsymbol{\Xi}_{\mathrm{v}}$ and the strain rate $\boldsymbol{S}$. This yields a transport equation for $\boldsymbol{\Xi}_{\mathrm{v}}$ given by

$$\boldsymbol{\Xi}_{\mathrm{v}} + \lambda \frac{D\boldsymbol{\Xi}_{\mathrm{v}}}{Dt} = 2\mu_1 \boldsymbol{S} \tag{2.7}$$

In order to ensure the objectivity of (2.7), an objective stress rate needs to be introduced. The Oldroyd-B fluid constitutive equation is obtained by replacing the material time derivative of the extra-stress in (2.7) with the Oldroyd upper-convected derivative,

$$\boldsymbol{\Xi}_{\mathrm{v}} + \lambda \overset{\nabla}{\boldsymbol{\Xi}}_{\mathrm{v}} = 2\mu_1 \boldsymbol{S} \tag{2.8}$$

or

$$\boldsymbol{\Xi}_{\mathrm{v}} + \lambda\left(\frac{D\boldsymbol{\Xi}_{\mathrm{v}}}{Dt} - \nabla\boldsymbol{v}\boldsymbol{\Xi}_{\mathrm{v}} - \boldsymbol{\Xi}_{\mathrm{v}}(\nabla\boldsymbol{v})^{\mathrm{T}}\right) = 2\mu_1 \boldsymbol{S} \tag{2.9}$$

where $\nabla\boldsymbol{v}$ is the solvent flow velocity gradient.

2.2. *Dumbbell models*

It has long been recognized that such mechanical models, formulated by assembling springs and dashpots, and simply using tensor generalizations to obtain corresponding multi-dimensional forms was inadequate. The latter day approach that developed was to base the model development on the nanoscale. However, it does not make sense to work at this level, especially since the intent is to design a model that is computationally tractable and efficient, and that takes some microstructure features into account. Fortunately, the complexity of the micro-macro description is reduced by a statistical mechanics argument. It is not intended here to be all inclusive in both detail and scope on the development of such models since there are a wide variety of viscoelastic models that have been developed (see [23] for a more inclusive list). The viscoelastic fluid model development will be limited polymeric fluids and to the FENE-P model for the viscoelastic/polymeric stress ($\boldsymbol{\Xi}_{\mathrm{v}} = \boldsymbol{\Xi}_p$). This is the most popular model currently used in numerical simulations of viscoelastic turbulent flows.

The reader is referred to the numerous texts available on polymeric fluid dynamics for a more complete description of such model development (for example [16-23])

In general, a polymer chain may be composed of $N+1$ beads connected by N massless springs, each of length b as sketched in figure 1. The beads are assumed not to deform and may rotate with respect to each other with an end-to-end vector $\boldsymbol{r}$ connecting bead 1 to bead $N+1$ at the ends of the chain.

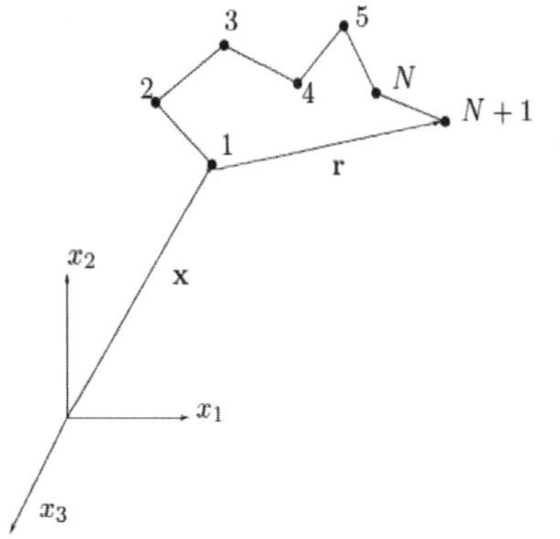

Figure 1. Polymer as chain of beads (from [23])

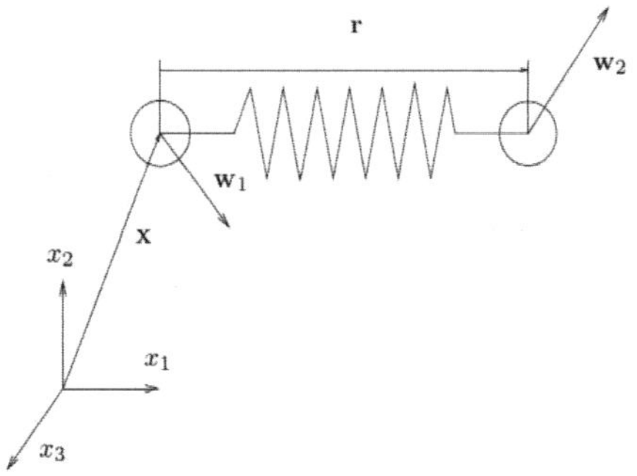

Figure 2. Elastic dumbbell (from [23])

The required interplay between the micro and macro behavior of polymeric chains can be represented by a Langevin equation for the vector connector $\boldsymbol{r}$ containing the (macro) drag and friction forces and the (micro) Brownian forces,

$$\zeta\left[\frac{\mathrm{d}\boldsymbol{r}}{\mathrm{d}t}-(\nabla \boldsymbol{v})\boldsymbol{r}\right]-\mathbf{f}_s=\mathbf{B} \tag{2.10}$$

where inertial forces have been assumed small, the local solvent velocity field is homogeneous with $\nabla \boldsymbol{v}$ the local velocity gradient, $\boldsymbol{\zeta}$ $(=6\pi\mu_s a)$ is the friction coefficient associated with the Stokes drag on the beads (each of readius a), and the Brownian force is $\mathbf{B}=-n_p k_B\boldsymbol{\theta}\ \partial\ln\psi/\partial\mathbf{r}$ with $\boldsymbol{\theta}$ the absolute temperature. The probability density function $\psi(\boldsymbol{r},t)$ is associated with the movement of the dumbbell and satisfies the Fokker-Planck equation,

$$\frac{\partial\psi}{\partial t}+(\nabla \boldsymbol{v})\boldsymbol{r}\cdot\frac{\partial\psi}{\partial \boldsymbol{r}}-2\frac{k_B\boldsymbol{\theta}}{\zeta}\frac{\partial^2\psi}{\partial \boldsymbol{r}^2}-\frac{2}{\zeta}\frac{\partial}{\partial \boldsymbol{r}}\cdot(\psi\mathbf{f}_s)=0 \tag{2.11}$$

The Cauchy stress of the fluid is assumed to be an ensemble average over a volume $\mathcal{V}$ that can be decomposed into the solvent volume $\mathcal{V}_s$, and the polymeric volume $\mathcal{V}_p$. The stress in the solvent part is the classical Newtonian stress involving the pressure, the rate of deformation tensor $\boldsymbol{S}$ and the viscosity μ_s. These volume averages can also be represented by integration over the configuration space of the polymer using the *pdf* ψ .

The polymeric contribution to the stress is due to the drag force of the beads exerted on the solvent. As (2.10) shows, the drag force is balanced by the Brownian force and the nonlinear spring force so that the polymeric stress is then given by

$$\Xi_p = \int (\boldsymbol{r} \otimes \mathbf{B}) \psi(\boldsymbol{r},t)\,\mathrm{d}\boldsymbol{r} + \int (\boldsymbol{r} \otimes \mathbf{f}_s) \psi(\boldsymbol{r},t)\,\mathrm{d}\boldsymbol{r} \tag{2.12}$$

where $\otimes$ denotes the tensor product of two vectors. Since the *pdf* ψ vanishes on the polymer surface, the Brownian force contribution is isotropic and the resulting polymeric stress is

$$\begin{aligned} \Xi_p &= -n_p k_B \boldsymbol{\theta} \boldsymbol{I} + n_p H f(\Xi) \int (\boldsymbol{r} \otimes \boldsymbol{r}) \psi(\boldsymbol{r},t)\,\mathrm{d}\boldsymbol{r} \\ &= -n_p k_B \boldsymbol{\theta} \boldsymbol{I} + n_p H f(\Xi) \langle \boldsymbol{r} \otimes \boldsymbol{r} \rangle = -n_p k_B \boldsymbol{\theta} \boldsymbol{I} + n_p H f(\Xi) \boldsymbol{c} \end{aligned} \tag{2.13}$$

where for the FENE-P model $f(\Xi)$ is the Peterlin function accounting for the nonlinear spring behaviour, $H\left(=3k_B\boldsymbol{\theta}/Nb^2\right)$ is an entropic spring constant, n_p is the number density of polymer chains per unit volume, and $\boldsymbol{c}\left(=<\boldsymbol{r}\otimes\boldsymbol{r}>\right)$ is the conformation tensor. Equation (2.13) is known as Kramers equation.

From (2.10) there remains one other contribution to the polymeric stress, that is, the drag force contribution. It is easier to evaluate this contribution to the stress by the polymeric chain by representing the complex chain by an elastic dumbbell model as shown in figure 2. The model corresponds to two spherical beads connected by either a nonlinear (entropic) spring. The single spring of the elastic dumbbell represents the net effect of the N springs of the polymer chain and the two beads represent the net effect of the $N+1$ beads of the chain with each bead moving with velocity $\boldsymbol{w}_{1,2}$. The contribution from the drag force to the polymer stress can be represented by

$$\begin{aligned} \Xi_p &= n_p \zeta \int \left[\boldsymbol{r} \otimes \left(\frac{\mathrm{d}\boldsymbol{r}}{\mathrm{d}t} - (\nabla \boldsymbol{v}) \boldsymbol{r} \right) \right] \psi(\boldsymbol{r},t)\,\mathrm{d}\boldsymbol{r} \\ &= \frac{n_p \zeta}{4} \left[\frac{\mathrm{d}}{\mathrm{d}t} \langle \boldsymbol{r} \otimes \boldsymbol{r} \rangle - (\nabla \boldsymbol{v}) \cdot \langle \boldsymbol{r} \otimes \boldsymbol{r} \rangle - \langle \boldsymbol{r} \otimes \boldsymbol{r} \rangle \cdot (\nabla \boldsymbol{v})^{\mathrm{T}} \right] \\ &= \frac{n_p \zeta}{4} \frac{\delta \boldsymbol{c}}{\delta t} \end{aligned} \tag{2.14}$$

where $\delta/\delta t$ is the Oldroyd (upper convective) derivative. Its appearance is necessary to ensure that the polymeric stress tensor is objective. Equation (2.14) is known as the Giesekus equation. From (2.10), the contributions to the polymeric stress from the drag forces should balance those from the spring and Brownian forces. Thus, the polymeric stress contributions from (2.13) and (2.14) can be equated yielding an evolution equation for the conformation tensor

$$\lambda \frac{\delta \boldsymbol{c}}{\delta t} = -\frac{k_B \boldsymbol{\theta}}{H} \boldsymbol{I} + f(\Xi) \boldsymbol{c} \tag{2.15}$$

where $\lambda = \zeta/(4H)$ is a relaxation time associated with the polymer chain. Equation (2.15) coupled with (2.13) provides the necessary mathematical formulation for the polymer chain motion and the polymeric stress link with the equations of fluid motion.

The Peterlin function, $f(\Xi) = L^2/\left(L^2 - \{\boldsymbol{c}\}\right)$ where L is the maximum extension length of the polymer chain, is an important component in the interaction between the turbulent and polymeric energies. (For the cases of interest in most DNS studies $L \sim O\left(10-10^2\right)$.) Before completing this section on the development of the FENE-P model, both the conformation evolution equation and polymeric stress equation can be non-dimensionalized with $k_B\boldsymbol{\theta}/H$ for the conformation tensor $\boldsymbol{c}$ and a rescaling of the conformation tensor to ensure that at equilibrium $\left(\boldsymbol{\delta c}/\boldsymbol{\delta t}=0\right)$ so that $\boldsymbol{c}=\boldsymbol{I}$ (see [24] for a thorough discussion of this rescaling). The final non-dimensional set of equations for the conformation tensor and polymeric stress tensor are given by

$$\lambda \frac{\delta \boldsymbol{c}}{\delta t} = -\boldsymbol{I} + f\left(\{\boldsymbol{c}\}\right)\boldsymbol{c} \tag{2.16}$$

and

$$\boldsymbol{\Xi}_p = \frac{\mu_{p0}}{\lambda}\left[f\left(\{\boldsymbol{c}\}\right)\boldsymbol{c} - \boldsymbol{I}\right] \tag{2.17}$$

where μ_{p0} is a polymeric viscosity such that $n_p k_B \boldsymbol{\theta} \approx \mu_{p0}/\lambda$, and the Peterlin function is now redefined as

$$f\left(\{\boldsymbol{c}\}\right) = \frac{L^2-3}{L^2-\{\boldsymbol{c}\}} \tag{2.18}$$

A consequence of this rescaling is the fact that at equilibrium $\boldsymbol{c}=\boldsymbol{I}$ so that $f\left(\{\boldsymbol{c}\}\right)=1$ and $\boldsymbol{\Xi}_p=0$. The vast majority of studies done in connection with DNS of polymeric viscoelastic flows have utilized the FENE-P model. Such models are representative of nonlinear dumbbells, which preclude the infinite extendability of the linear dumbbell models represented by the Oldroyd-B fluids.

3. Numerical Simulations: Focus and Formulation

As with the case of Newtonian fluids, a primary motivation for using numerical simulations in the study of flow physics is the absence, or near-absence, of dynamic modeling. In the case of viscoelastic fluids, as the previous section showed, model development for the Cauchy stress introduced by the polymeric chains required careful consideration of the phenomenological behavior of a polymer chain. When such polymers are embedded in a turbulent flow, the multi-scale dynamics is quite complex. Until recently, and even today, the state-of-the-art is such that many aspects of the dynamic interchange between the turbulent flow stress fields and viscoelastic (polymeric) fluid stress fields is not completely understood.

The large majority of DNS studies of polymeric flows are performed in a fully developed channel flow. The obvious advantage is the simplification of boundary conditions and computational resources in both the streamwise and spanwise flow directions. This additionally enhances any spectral analysis of the flow dynamics since two directions can be Fourier analyzed. Fortunately such flows display a broad range of drag reduction (DR) features so there is no compromise from a dynamics standpoint.

Since such studies focus on the inner layer dynamics of the channel flow, wall scaling is usually used and is based on zero-shear rate variables with the length and time scaled by ν_{tot}/u_τ and ν_{tot}/u_τ^2, where $\nu_{tot} = \nu_s + \nu_{p0}$ is the total zero-shear kinematic viscosity, and u_τ the friction velocity. In the following, standard notations for the channel flow geometry will be used, i.e. the channel streamwise direction is $x_1^+ = x^+$, the wall-normal direction is $x_2^+ = y^+$, and the spanwise direction $x_3^+ = z^+$, with the velocity field in the respective directions $(U^+, V^+, W^+) = (U_1^+, U_2^+, U_3^+)$.

Finally, in this section the notation will change from using matrix tensors and vectors to index notation, which will facilitate the presentation of material. Using this scaling, the dimensionless conservation equations are

$$\frac{\partial U_j^+}{\partial x_j^+} = 0 \tag{2.19}$$

$$\frac{DU_i^+}{Dt^+} = \frac{\partial U_i^+}{\partial t^+} + U_j^+ \frac{\partial U_i^+}{\partial x_j^+} = -\frac{\partial P^+}{\partial x_i^+} + \frac{\partial \Sigma_{ij}^+}{\partial x_j^+}, \tag{2.20}$$

with $\partial P^+ / \partial x^+$ the pressure gradient (this includes the constant non-zero contribution driving the flow as well as a time-dependent contribution), and the total stress tensor Σ_{ij}^+ being composed of (Newtonian) solvent and (polymeric) viscoelastic contributions

$$\Sigma_{ij}^+ = 2\beta_0 S_{ij}^+ + \Xi_{ij}^{p+} \tag{2.21}$$

Here, $S_{ij}^+ = (\partial U_i^+ / \partial x_j^+ + \partial U_j^+ / \partial x_i^+)/2$ is the strain rate tensor, β_0 the ratio of the Newtonian viscosity ν_s to the total zero-shear viscosity ν_{tot}. For a FENE-P dumbbell model, the polymeric stress is given by (cf. (2.17))

$$\Xi_{ij}^{p+} = \alpha_0 \left[f(\{\boldsymbol{c}\}) c_{ij} - \delta_{ij} \right] \tag{2.22}$$

where $\alpha_0 = (1 - \beta_0)/We_{\tau 0}$, and with $We_{\tau 0} = \lambda u_\tau^2 / \nu_{tot}$ the friction Weissenberg number representing the ratio of the elastic relaxation time λ relative to the viscous timescale. The polymeric stress for a FENE-P fluid includes the Peterlin function $f(\{\boldsymbol{c}\})$ that was defined in (2.18), ensuring shear-thinning behavior and a finite elongational viscosity for a finite value of the extension rate, with L the fully-stretched polymer length and $\{\boldsymbol{c}\}$ designating the trace of the conformation tensor $\boldsymbol{c}$. The equation system is closed with an evolution equation for $\boldsymbol{c}$ (cf. (2.16))

$$\frac{Dc_{ij}}{Dt^+} = \left(c_{ik} S_{kj}^+ + S_{ik}^+ c_{kj} \right) - \left(c_{ik} W_{kj}^+ - W_{ik}^+ c_{kj} \right) - \frac{f(\{\boldsymbol{c}\}) c_{ij} - \delta_{ij}}{We_{\tau 0}}, \tag{2.23}$$

where $W_{ij}^+ = (\partial U_i^+ / \partial x_j^+ - \partial U_j^+ / \partial x_i^+)/2$ is the rotation rate tensor.

For the results to be shown here, the simulation parameters are set as $\beta_0 = 0.9$ and $Re_{\tau 0} = 1000$, where $Re_{\tau 0} = u_\tau h / \nu_{tot}$ (with h the channel half-width) is the zero-shear friction Reynolds number setting the turbulence level. Two drag reduction cases are studied: $L = 30$, $We_{\tau 0} = 50$, medium percentage DR of 30%, and $L = 100$, $We_{\tau 0} = 115$, high percentage DR of 58%. The simulation details can be found in [11,25,26].

The Peterlin function can be related to the potential energy associated with the extension of the polymer molecule. The nonlinear spring force of a finitely extendable nonlinear elastic (FENE) polymer model was given in equation (2.13) and leads to a relation for the elastic potential energy given by

$$\mathcal{E}_p^+ - \mathcal{E}_{p0}^+ = \alpha_{0L} \ln[f(\{c\})] \tag{2.24}$$

where $\alpha_{0L} = \alpha_0(L^2 - 3)/2$, and $\mathcal{E}_{p0}^+$ is a constant reference energy at equilibrium (recall $\Xi^{p+} = 0$ at equilibrium). This yields for the Peterlin function the relation,

$$f(\{c\}) = e^{(\mathcal{E}_p^+ - \mathcal{E}_{p0}^+)/\alpha_{0L}} \tag{2.25}$$

and the polymeric stress

$$\Xi_{ij}^{p+} = \alpha_0 \left[e^{(\mathcal{E}_p^+ - \mathcal{E}_{p0}^+)/\alpha_{0L}} c_{ij} - \delta_{ij} \right] \tag{2.26}$$

The polymeric energy $\Xi_{ii}^{p+}/2$ is then directly linked to the exponential of the potential energy of the extended polymer.

4. Energetic Exchange

Although the primary focus in this section will be on the energetic exchange between the turbulent flow and the polymeric chain, it is useful to see the net effect on the mean flow at different DR levels. Figure 3 shows the mean velocity across the channel for both the 30% and 58% DR cases. At $Re_{\tau 0} = 1000$ a well-defined log-law region starts to appear (cf. [4,27]). When compared to the theoretical maximum drag reduction (MDR) limit [28], it is apparent that as Reynolds number and DR levels increase both the wall layer intercept and log-law slope change. It is this variation in the sublayer/buffer layers that has motivated much of the simulation studies to date.

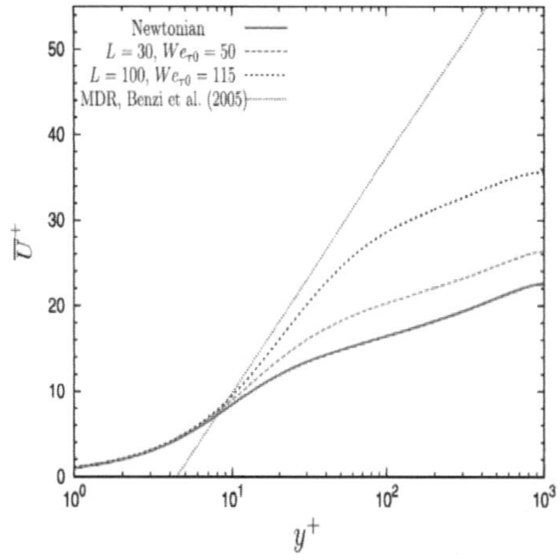

Figure 3. Mean velocity profiles for moderate and high drag reduction case (from [11])

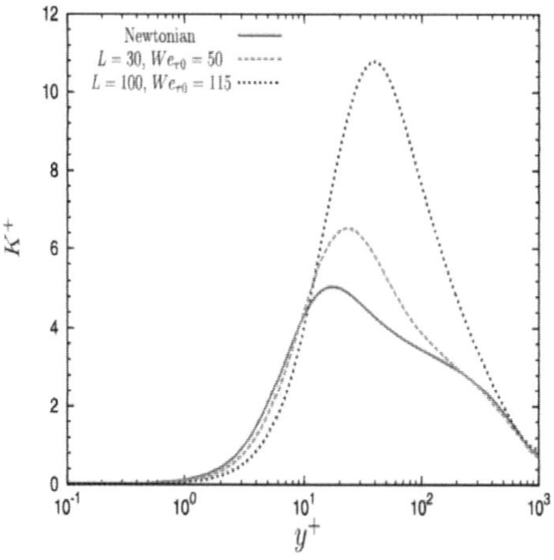

Figure 4. Turbulent kinetic energy for moderate and high drag reduction case (from [11])

As figure 4 shows, the turbulent kinetic energy K^+ peak level increases for viscoelastic flow relative to the Newtonian case. The figure further confirms the thickening of the sublayer since the peak of turbulent kinetic energy moves away from $y^+ \simeq 17$ for the Newtonian fluid to $y^+ \simeq 40$ for the high DR case. Explaining the increase in the turbulent kinetic energy level for viscoelastic flows is not that trivial. It is necessary to examine the dynamic balance associated with the turbulent kinetic equation given by

$$\frac{DK^+}{Dt^+} = P_m^+ - \underbrace{\beta_0 \overline{\left(\frac{\partial u_i'^+}{\partial x_k^+}\right)^2}}_{\varepsilon^+} + \underbrace{\beta_0 \frac{\partial^2 K^+}{\partial x_k^{+2}}}_{D_v^+} + \underbrace{\frac{\partial\left(\overline{p'^+ u_i'^+ \delta_{ik}}\right)}{\partial x_k^+}}_{D_p^+} + \underbrace{\frac{\partial\left(\overline{u_i'^+ u_i'^+ u_k'^+}/2\right)}{\partial x_k^+}}_{D_t^+} \\ - \underbrace{\overline{\Xi_{ik}'^{p+} s_{ki}'^+}}_{P_{pt}^+} + \underbrace{\frac{\partial}{\partial x_k^+}\left(\overline{\Xi_{ik}'^{p+} u_i'^+}\right)}_{D_{pt}^+} \tag{3.1}$$

where $P_m^+ = -\tau_{ik}^+ \bar{S}_{ki}^+$ is the mean flow production. From the turbulent kinetic profiles for the Newtonian and viscoelastic cases shown in figure 4, the most significant change in the energetic balance should occur in the region $10 \leq y^+ \leq 60$ where the peak turbulent kinetic energy level is clearly set. As figure 5 shows, within this region the dynamics is dominated by the combination of diffusive transport and dissipation, namely $D_t^+ + D_v^+ - \varepsilon^+$, with D_t^+ and D_v^+ both negative (although not shown growing in magnitude as viscoelasticity increases). Note that D_t^+ has the same magnitude as dissipation ε^+ at the location of peak production by the mean flow, $y^+ \simeq 25$. Since D_t^+ and D_v^+ spatially redistribute and diffuse energy within the channel rather than act as ultimate energy sinks, this suggests that the relative influence of ε^+ as compared to production diminishes with increasing viscoelasticity. As such the peak turbulent energy level in the polymeric cases can reach a higher value than in the Newtonian case.

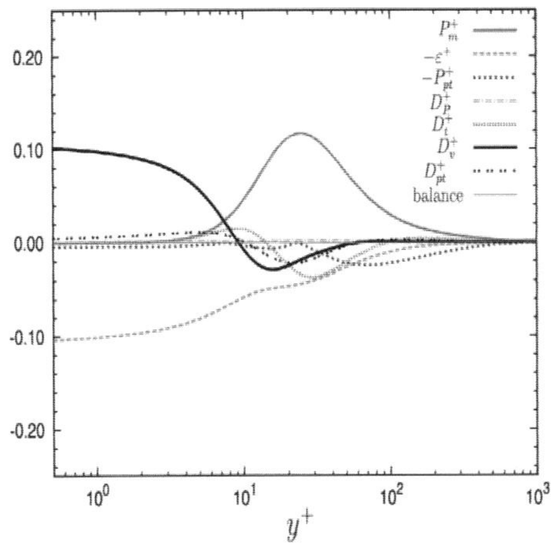

Figure 5. Turbulent kinetic energy budget for high ag reduction cases(from [11])

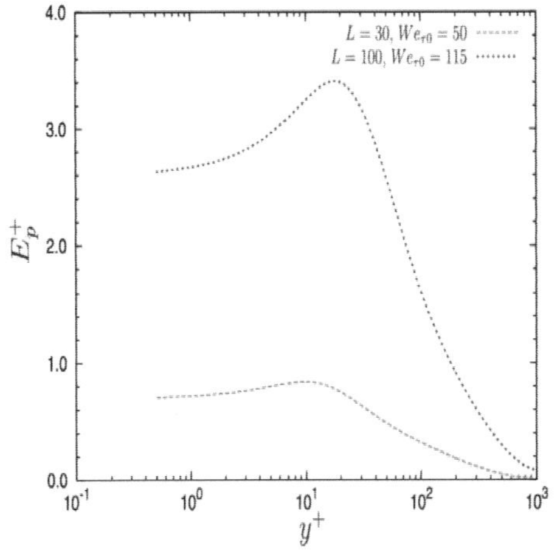

Figure 6. Elastic potential energy for moderate and high drag reduction case (from[11])

Figure 6 shows the mean elastic potential energy E_p^+ $(=\overline{\mathcal{E}}_p^+ - \mathcal{E}_{p0}^+)$ for both the medium and high DR cases. In each case a maximum exists and is located at $y^+ \approx 12$ for medium DR and at $y^+ \approx 20$ for the high DR flow. Recall that the elastic potential energy is a direct consequence of the nonlinear spring force contribution to the polymer chain force balance. As such the peak in potential energy distribution for each DR case should correspond to the location of maximum polymer chain extension $(< L)$ occurring within the channel. In addition, note that the relative peak amplitude between the two cases is a factor of 4. The fourfold increase in potential energy can be attributed to both the nonlinear behavior of the FENE-P spring (the Peterlin function) (see (2.24)), and the significant difference in the coefficient α_{0L} between the medium ($\alpha_{0L} = 0.9$) and high ($\alpha_{0L} = 4.5$) DR cases.

Since the polymeric stress (energy) is composed of the nonlinear spring force contribution and the (isotropic) Brownian motion contribution, the cross-channel distribution of the polymeric kinetic energy should be qualitatively similar to the potential energy distribution. This is shown in figure 7 for the mean polymeric energy, $K_p^+ = \overline{\Xi}_{ii}^{p+}/2$. The polymeric energies for both DR cases are larger than the corresponding potential energy. The increase is due to the exponential relationship between polymeric stress and elastic potential energy given in (2.26) as well as the difference between α_0 for the medium ($\alpha_0 = 2.0\times10^{-3}$) and high ($\alpha_0 = 0.9\times10^{-3}$) DR cases. It is interesting to observe that the turbulent kinetic energy shown in figure 4 reaches a peak value of ≈ 6 at medium DR, and a peak value of ≈ 10 at high DR. This implies that for high DR the polymeric energy and the turbulent kinetic energy are of same order of magnitude; whereas, for medium DR the turbulent kinetic energy clearly dominates the polymeric energy. This means that as DR increases the polymer takes a proportionally larger amount of energy from the turbulent kinetic energy.

The polymeric energy equation is given by

$$
\begin{aligned}
\frac{DK_p^+}{Dt^+} = P_{pm}^+ + P_{pt}^+ - \underbrace{\frac{f(\{\overline{c}\})K_p^+}{We_{\tau0}}}_{\varepsilon_p^+} \\
\underbrace{+ f(\{\overline{c}\})\left[\overline{\left(\frac{\Xi'^{p+}_{ik}}{f'(\{\overline{c}\})}\right)}\overline{S}_{ki}^+ + \overline{\Xi}_{ik}^{p+}\overline{\left(\frac{s'^+_{ki}}{f'(\{\overline{c}\})}\right)} + \overline{\left(\frac{\Xi'^+_{ik}s'^+_{ki}}{f'(\{\overline{c}\})}\right)}\right]}_{P_{pf}^+} \\
\underbrace{- f(\{\overline{c}\})\frac{\partial}{\partial x_j}\left[\frac{\overline{u'^+_j K'^+_p}}{f(\{\overline{c}\})} + \overline{\left(\frac{u'^+_j K'^+_p}{f'(\{\overline{c}\})}\right)} + \left(K_p^+ + \frac{3\alpha_0}{2}\right)\overline{\left(\frac{u'^+_j}{f'(\{\overline{c}\})}\right)}\right]}_{D_{pf}^+}
\end{aligned}
\tag{3.2}
$$

where $P_{pm}^+ = \Xi_{ik}^+\overline{S}_{ki}^+$ is the production from mean shear. The corresponding cross-channel balance of each term is shown in figure 8. The dominant balance for the polymeric energy in the inner layer ($y^+ \lesssim 15$) is between the production due to mean shear, P_{pm}^+, and the dissipation rate, ε_p^+. The remaining sinks of energy in the overall polymeric energy balance in the inner and buffer layers are from the terms that are directly dependent on the mean and fluctuating components of the Peterlin function, P_{pf}^+ and D_{pf}^+, where the latter contribution clearly dominates. In the outer layer ($y^+ \gtrsim 50$), the balance is between the turbulent exchange term P_{pt}^+ and the mean dissipation.

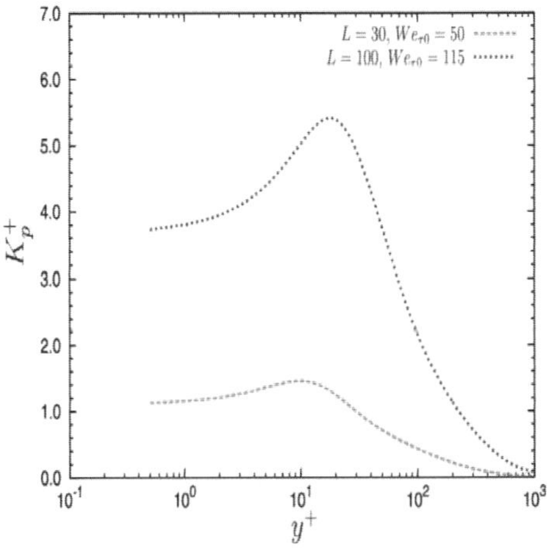

Figure 7. Polymeric kinetic energy profiles for high drag reduction case (from [11])

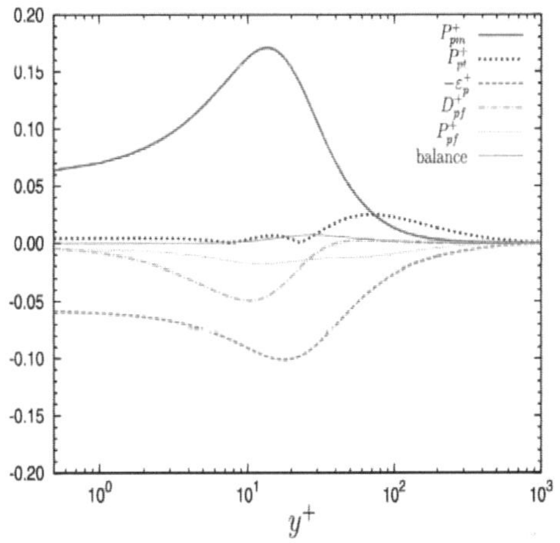

Figure 8. Polymeric energy budget for high drag reduction case (from [11])

5. On the Drag Reduction Mechanism

As the energetic budgets show, the polymer-turbulence interaction is governed by the turbulent exchange term P_{pt}^+. As figure 8 shows, this term is predominantly positive so that the primary effect of the interaction between the turbulent and polymeric fields is to transfer energy from the turbulence to the polymer. For comparison, figure 9 shows the polymeric energy budget for the low drag reduction case. Note that the magnitude of the exchange term does not change between the low and high drag reduction cases. This one-way transfer, with an amplitude independent of the drag reduction regime, is in contradiction with the purely elastic coupling which is implicit within the elastic theory of the polymer drag reduction phenomenon by Tabor and De Gennes [15]. The elastic scenario indeed implies a two-way coupling where the exchange term would change sign, and would vary in magnitude with different elasticity levels.

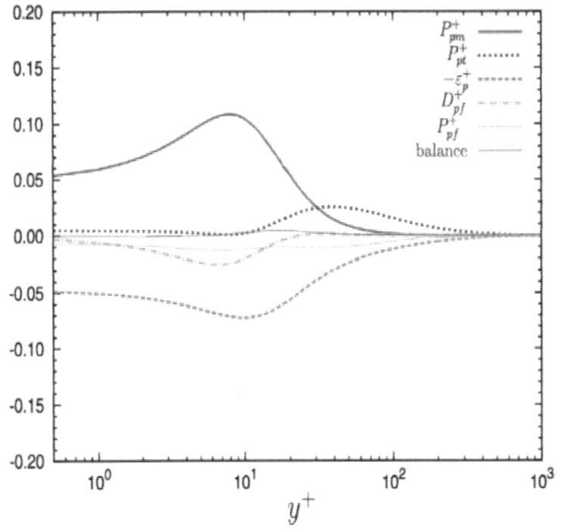

Figure 9. Polymeric energy budget for high drag reduction case (from [11])

6. Summary and Perspectives

Although in the limited space available it was not possible to pursue many of the aspects of viscoelastic fluid flows in detail, it is hoped that the material presented has provided an overview of the development of a commonly used and relatively accurate polymeric fluid stress constitutive equation – the FENE-P model. With this background in model development it is easier to see how the energetic exchange between the turbulent flow and viscoelastic fluid can occur. The solvent flow field provides a force to the polymer chains that both induces a stress which acts on the flow and a gain to the entropic springs which elongates the chains and effectively increases the polymeric viscosity. This nonlinear behaviour results in a complex energetic exchange that depends on the distance from the bounding solid surface of the channel. Additional insight could also be obtained from a spectral analysis of the energy cascade and which could provide further information relative, not only, to spatial location but also on relevant spectral scales in the cascade.

Acknowledgments
This research has granted access to the HPC resources of [CCRT /CINES /IDRIS] under the allocation i20142b2277 made by GENCI (Grand Equipement National de Calcul Intensif). The data used herein was produced on the IBM Blue Gene/P computer *Babel* at the IDRIS/CNRS computing center, Orsay, France. Cross-channel statistics from the DNS database can be publicly accessed on-line at http://lml.univ-lille1.fr/channeldata/

7. References

[1] Toms B A 1949 Some observations on the flow of linear polymer solutions through straight tubes at large Reynolds numbers Proc. of the 1st International Congress of Rheology vol 2 pp 135-141
[2] Sureshkumar R, Beris A N and Handler R A 1197 Direct numerical simulation of the turbulent channel flow of a polymer solution *Phys. Fluids* **9** 743-755
[3] Dimitropoulos C D, Dubief Y, Shakfeh E S G and Moin P 2006 Direct numerical simulation of polymer-induced drag reduction in turbulent boundary layer flow of inhomogeneous polymer solutions *J. Fluid Mech.* **566** 153-162
[4] Dubief Y, White C M, Terrapon V E, Shakfeh E S G, Moin P and S. K. Lele S K 2004 On the coherent drag-reducing and turbulence-enhancing behaviour of polymers in wall flows *J. Fluid Mech.* **514** 271-280
[5] Dimitropoulos C D, Dubief Y, Shakfeh E S G, Moin P and Lele S K 2005 Direct numerical simulation of polymer-induced drag reduction in turbulent boundary layer flow *Phys. Fluids* **17** 0105
[6] Li C-F, Gupta V K, Sureshkumar R and Khomami B 2006 Turbulent channel flow of dilute polymeric solutions: drag reduction scaling and an eddy viscosity model *J. Non-Newton. Fluid* **139** 177-189
[7] Pinho F T, Sadanandan B and Sureshkumar R 2008 One Equation Model for Turbulent Channel Flow with Second Order Viscoelastic Corrections *Flow Turbul. Combust.* **81** 337-367
[8] Resende P R, Kim K, Younis B A, Sureshkumar, R and Pinho F T 2011 A FENE-P $k-\varepsilon$ turbulence model for low and intermediate regimes of polymer-induced drag reduction *J. Non-Newton. Fluid* **166** 639-660
[9] Benzi R 2010 A short review on drag reduction by polymers in wall bounded turbulence *Physica D* **239** 1338–1345
[10] Procaccia I, L'vov V S and Benzi R 2008 Colloquium: Theory of drag reduction by polymers in wall-bounded turbulence *Rev. Mod. Phys.* **80** 225–247
[11] Thais L, Gatski T B and Mompean G 2013 Analysis of polymer drag reduction mechanisms from energy budgets *Int. J. Heat Fluid Flw.* **43** 52-61
[12] Lumley J L 1969 Drag reduction by additives *Annu. Rev. Fluid Mech.* **1** 367–384
[13] Min T, Yoo J Y, Choi H and Joseph D D 2003 Drag reduction by polymer additives in a turbulent channel flow *J. Fluid Mech.* **486** 213–238
[14] Dallas V, Vassilicos J C and Hewitt G F 2010 Strong polymer-turbulence interactions in viscoelastic turbulent channel flow *Phys. Rev. E* **82** 066303
[15] Tabor M and de Gennes P G 1986 A cascade theory of drag reduction *Europhys. Lett* **7** 519–522
[16] Bird R B, Armstrong R C and Hassager O 1987 *Dynamics of Polymeric Liquids. Vol. 1: Fluid Mechanics* (New York: Wiley)
[17] Bird R B, Armstrong R C and Hassager O 1987 *Dynamics of Polymeric Liquids. Vol. 2: Kinetic Theory* (N York: Wiley)

[18] Larson R 1988 *Consitutive Equations for Polymer Melts and Solutions* (Boston, Butterworth))
[19] Joseph D D 1990 *Fluid Dynamics of Viscoelastic Liquids* (Berlin, Springer)
[20] Huilgol R R and Phan-Thien N 1997 *Fluid Mechanics of Viscoelasticity* (Amsterdam, Elsevier)
[21] Truesdell C and Rajagopal K R 2000 *An Introduction to the Mechanics of Fluids* (Boston, Birkhauser)
[22] Phan-Thien N 2002 *Understanding Viscoelasticity Basics of Rheology* (Berlin, Springer)
[23] Deville M O and Gatski T B 2012 *Mathematical Modeling for Complex Fluids and Flows* (Heidelberg, Springer)
[24] Beris A N and Edwards B J 1994 *Thermodynamics of Flowing Systems with Internal Microstructure* (NewYork, Oxford)
[25] Thais L, Tejada-Martinez A E, Gatski T B and Mompean G 2010 Temporal large eddy simulations of turbulent viscoelastic drag reduction flows *Phys. Fluids* **22** 013103
[26] Thais L, Gatski T B and Mompean G 2011 A massively parallel hybrid scheme for direct numerical simulation of turbulent viscoelastic channel flow *Comput. Fluids* **43** 134-142
[27] Housiadas K D and Beris A N 2006 Extensional behavior influence on viscoelastic turbulent channel flow. *J. Non-Newton. Fluid* **140** 41-56
[28] Benzi R, De Angelis E, L'vov V S and Procaccia I 2005 Identification and calculation of the universal asymptote for drag reduction by polymers in wall-bounded turbulence *Phys. Rev. Lett.* **95** 194502

Transport, mixing and agglomeration of particles in turbulent flows

Michael W Reeks
School of Mechanical and Systems Engineering, Newcastle University, Newcastle upon Tyne, NE1 7RU, UK
E-mail: mike.reeks@newcastle.ac.uk

Abstract. This paper describes methods and approaches that have been used to simulate and model the transport, mixing and agglomeration of small particles in a flowing turbulent gas. The transported particles because of their inertia are assumed not to follow the motion of the large scales of the turbulence and or the motion of the small dissipating scales of the turbulence. We show how both these behaviours can be represented by a PDF approach analogous to that used in Classical Kinetic Theory. For large scale dispersion the focus is on transport in simple generic flows like statistically stationary homogeneous and isotropic turbulence and simple shear flows. Special consideration is given to the transport and deposition of particles in turbulent boundary layers. For small scale transport the focus is on how the the small scales of turbulence together with the particle inertial response enhances collision processes like particle agglomeration. In this case the importance of segregation and the formation of caustics, singularities and random uncorrelated motion is highlighted and discussed.
Keywords: inertial particles, turbulence, pdf approach, segregation, dispersion

1. Introduction

This paper is about ways of simulating and modelling the transport of small particles in a turbulent flow. More particularly it's about modelling and simulating the way the turbulence mixes and disperses these particles, the way it deposits and resuspends them from surfaces exposed to the flow and how it brings them closer together and enhances their collision processes as in particle agglomeration. Of particular concern will be the way these small particles respond to both the large and small scales of turbulence and how in turn this *inertial* response influences these various transport processes. This response is characterized by the particle Stokes number St which measures the particle response time to the typical time scale of the motion of the large or the small scales of the turbulence. We focus our attention on the motion of solid particles in a turbulent gas because the forces on the particles are simply expressed in terms of quasi-steady lift or drag dependent on the relative velocity between particle and local carrier flow. In the simple cases we wish to consider, the forces acting on the particles will simply be that derived from Stokes drag. Such particle suspensions are referred to here and elsewhere [45] as *turbulent aerosols* so that the mass loading of the aerosol particles is sufficiently low to ignore any way two-way coupling between the continuous gas[1] and particle dispersed phases.

[1] We will refer to this as the *carrier flow* or sometimes as simply the *fluid*

Journal of Physics: Conference Series **530** (2014) 012003 doi:10.1088/1742-6596/530/1/012003

Turbulent aerosols are a feature of many industrial and environmental processes: from cloud droplets, mists and fogs to mixing and combustion processes in coal fire burners, and the release and dispersal of radioactive particles in a nuclear severe accident. They influence the weather and are key factors in controlling climate; they impact upon our health and longevity. So understanding how they behave, how they can be controlled is an important area of study. In all these processes the turbulence, both the large and small scales play a crucial role in determining their behaviour. How for instance the aerosol droplets or particles are formed by nucleation, grow or evaporate, how they mix and chemically react, and the way particles interact with surfaces exposed to the flow. Such processes are controlled by either the large scales or the small scales and in some cases by both scales acting together. An area of much investigation is the mechanism for warm-rain initiation and in particular the way droplet interaction with the small scales of turbulence in clouds leads to a droplet size distribution, much broader than one would expect from a closed adiabatic parcel model of droplet formation and growth. It is the small scales of the turbulence for instance that are responsible for the segregation or un-mixing of a turbulent aerosol where not only the scale is important but also the structure of the turbulence and the mixture of straining and vorticity within those scales.

In this short review of *particles in turbulence* we consider first those processes that are controlled by the large scales of the turbulence (typical of the size of the flow containment). The large scale motion influences and controls what is commonly referred to as one-particle dispersion and in this process the influence of the local mean shear for which the influence of the local mean shear is inextricably linked to the turbulence. It is what we mean by the transport of particles in complex flows where particles are transported both by the mean shear and the turbulence produced by the shear, both being spatially non uniform. However a consideration of the transport of particles in simple generic flows like statistically stationary homogeneous and isotropic turbulence and simple shear flows provide important insights into the way particles are transported in turbulent shear flows in general. A crucial consideration is the transport of particles in a turbulent boundary layer since the turbulence has an important rate limiting influence in transporting particles to a surface and special consideration will be given to this in this review.

The influence of the small scales of turbulence is referred to as two-particle dispersion since they influence the relative motion between two particles. It is at the small scales where the phenomenon of demixing of particle suspensions occur. This process has been the subject of intensive research over the last ten years. Partly because the small scales have a more universal structure far removed from the influence and complexity of local shear and partly because they are more amenable to analysis and to the construction of simple models. Simulation and analysis have revealed some important important features such as the formation of caustics and the occurrence of singularities in the particle concentration field and the presence of random uncorrelated motion (RUM).

An important driving force in all these studies has been the need to provide a basic framework to handle the statistical nature of these processes and to use this to provide a two-fluid model of a particle-fluid mixture which treats the particle phase as a continuum and to develop a set of mass, momentum and energy equations for the dispersed phase analogous to those of the carrier phase. This is an essential element of turbulence modelling and an importnt element of the development and application of CFD. Over the last 20 years this has been the goal of the PDF approach, an approach analogous to statistical mechanics in that the quest is for a Master Equation analogous to the Maxwell-Boltzmann equation representing in this case the transport of particles in phase space - more particularly an equation that is appropriate for both one particle and two-particle transport. We begin our review of this subject by recalling the basic features of this approach, primarily because it is an approach that can in principle embody and provide models for the various processes as well in its own right provide insights into some of the long term nature of these process that in some cases are difficult to resolve in simulation and in experiment.

2. PDF development

The probability density function (PDF) approach is a rational approach to modelling dispersed particle flows in the same way that classcal kinetic theory is a rational approach for modelling gas flows. That is, there exists in both cases an underlying equation (a Master Equation) which, in a strictly formal way, can be used to derive both the continuum equations and constitutive relations of a gas or a dispersed phase of particles. In kinetic theory, the master equation is the well-known Maxwell–Boltzmann equation, whilst in the modelling of dispersed particle flows, it is known as the kinetic equation. There are currently two forms of the kinetic equation. In the first form, the pdf, as in Kinetic Theory, refers to the probability density that a particle has a certain velocity and position at a given time. This approach is referred to as the kinetic model (KM) approach. It was originally developed by Buyevich [4, 5, 6] and further developed since by a number of workers, most notably Reeks [34, 27, 28, 29], Hyland [20, 21], Swailes [41] Derevich [11], Zaichik [46] and Pozorski & Minier (1998) [26].

In contrast the second form of the PDF equation, first proposed by Simonin, Deutsch and Minier [36], refers to a more general PDF which includes the velocity of the carrier flow local to the particle as a phase space variable as well as the particle position and velocity. It is a development of the PDF approach used by Pope [17] based on a generalised Langevin model (GLM) for the equation of motion of the carrier flow encountered by a particle. It is referred to here as the GLM approach.

2.1. PDF equation for inert particles

To illustrate the differences and similarities between the KM and GLM approaches, let us consider the simplest case of the transport of inert non-reacting solid particles in a turbulent gas flow. To simplify the situation still further the drag acting on the particle is linearized with respect to the relative velocity , i.e.

$$\mathbf{F}_A \cong \underline{\overline{\eta}} \cdot (\mathbf{u} - \mathbf{v}) \tag{1}$$

where $\mathbf{v}$ is the velocity of an individual particle at position $\mathbf{x}$ at time t where the local flow velocity is $\mathbf{u}(\mathbf{x}, t)$ and $\underline{\overline{\eta}}$ is the tensor of net friction coefficients and given by

$$\underline{\overline{\eta}} = \frac{1}{2}\rho_g A C_D(\overline{Re_p})|\overline{\mathbf{u}} - \overline{\mathbf{v}}| \tag{2}$$

where $\overline{\mathbf{v}}(\mathbf{x}, t)$ is the net particle mean velocity and $\overline{Re_p}$ is the value of the particle Reynolds number based on the net relative velocity between particle and local carrier flow. The equations of motion for the particle are

$$\frac{d\mathbf{x}}{dt} = \mathbf{v} \tag{3}$$

$$\frac{d\mathbf{v}}{dt} = \underline{\beta} \cdot (\mathbf{u} - \mathbf{v}) \tag{4}$$

where $\underline{\beta}$ is the inverse particle response tensor and given by $m^{-1}\underline{\overline{\eta}}$ for a particle of mass m. In the case of Stokes drag, the elements of $\underline{\beta}$ are constants of the motion and β_{ij}^{-1} are the corresponding particle response times to changes in flow. In addition to the particle equations of motion, the equation of motion of the carrier flow velocity $\mathbf{u}$ along a particle trajectory is included, namely

$$\frac{du_i}{dt} = F_i(\mathbf{v}, \mathbf{u}, \mathbf{x}, t) \tag{5}$$

In the KM approach, one considers the continuum equations derived from an equation for the phase space probability density $\langle W\rangle(\mathbf{v},\mathbf{x},t)$ in which $\mathbf{u}(\mathbf{x},t)$ is a random function of $\mathbf{x},t$ and $\mathbf{v}$ and $\mathbf{x}$ are independent random variables. In the GLM approach, the continuum equations are derived from a conservation equation for the phase space density $\langle P\rangle(\mathbf{v},\mathbf{u},\mathbf{x},t)$ where $\mathbf{v},\mathbf{u},\mathbf{x}$ form a set of independent variables. The ensemble average $\langle..\rangle$ means an ensemble of the instantaneous phase space densities W and P over all realizations of the flow velocity field. It is convenient to resolve $u_i(\mathbf{x},t)$ and $F_i(\mathbf{v},\mathbf{u},\mathbf{x},t)$ into mean and fluctuating parts,

$$u_i = \langle u_i\rangle + u_i' \qquad F_i = \langle F_i\rangle + F_i' \tag{6}$$

so that the mean component is responsible for convection in phase space and the fluctuating part (due to the turbulence) is responsible for dispersion. Then the transport equations for mean values of W and P, namely $\langle W\rangle$ and $\langle P\rangle$, are

$$\frac{\partial}{\partial t} + \frac{\partial}{\partial x_i}\mathrm{v}_i + \frac{\partial}{\partial \mathrm{v}_i}\beta_{ij}\left(\langle u_j\rangle - \mathrm{v}_j\right)\langle W\rangle = -\frac{\partial}{\partial \mathrm{v}_i}\beta_{ij}\langle u_j' W\rangle \tag{7}$$

$$\frac{\partial}{\partial t} + \frac{\partial}{\partial x_i}\mathrm{v}_i + \frac{\partial}{\partial \mathrm{v}_i}\beta_{ij}\left(u_j - \mathrm{v}_j\right) + \frac{\partial}{\partial u_i}\langle F_i\rangle\langle P\rangle = -\frac{\partial}{\partial u_i}\langle F_i' P\rangle \tag{8}$$

To solve these equations closure relations for $\langle u_i' W\rangle$ and $\langle F_i' P\rangle$ are required. We now consider the various forms that have been derived for these terms assuming that $\beta_{ij} = \beta\delta_{ij}$.

2.1.1. Closure approximation for a non-reactive gas-particle flows Kinetic model

Based on the either the LHDI approximation (Reeks [29]) or the Furutsu Novikov formula (Swailes & Darbyshire [41]), the closure approximation for the net flux $\langle u_i'' W\rangle$ for particles with velocity $\mathbf{v}$ and position $\mathbf{x}$ at time t, is given by

$$\langle u_i' W\rangle = -\left(\frac{\partial}{\partial v_j}\mu_{ji} + \frac{\partial}{\partial x_j}\lambda_{ji}\right)\langle W\rangle + \kappa_i\langle W\rangle \tag{9}$$

where explicitly $\mu_{ij}(\mathbf{v},\mathbf{x},t)$ and $\lambda_{ij}(\mathbf{v},\mathbf{x},t)$ phase space diffusion coefficients that refer to diffusion in velocity and configuration space respectively and κ_i represents the components of a convective (body) force that depends on inhomogeneities in flow filed(i.e it is zero for homogeneous turbulence). Both μ_{ij}, λ_{ij} and κ_i depends on the flow velocity correlation measured along particle trajectories that arrive at $\mathbf{v},\mathbf{x}$ at time t (See [29, 41]. Their values depend upon the local mean shearing of the flow.

Generalized Langevin Model GLM

Simonin, Deutsch and Minier (SDM) [36] derive an equation of motion for the fluid velocity along a particle trajectory by starting from the Langevin equation which Pope [17] has used as the analogue of the Navier Stokes equation for fluid point motion. Thus along a fluid point trajectory

$$\frac{du_i}{dt} = \alpha_{ij}(\mathbf{x})\left(\langle u_j\rangle - u_j\right) + f_i(\mathbf{x}) + f_i'(t) \tag{10}$$

where $f_i(\mathbf{x})$ is the net viscous and pressure force per unit mass of fluid and $f_i'(t)$ is a white noise function of time. Both SDM and Pope consider the equation of motion in differential form because the white noise is assumed non differentiable. For convenience, it is assumed that the white noise, like all turbulence related functions is differentiable. The equation of motion has white noise properties simply because it has a time scale much shorter than the time scale over which $\mathbf{u}(t)$ varies along a fluid point trajectory $O(\alpha^{-1})$. For future reference it is noted that

$$f_i(\mathbf{x}) = \frac{D_f\langle u_i\rangle}{Dt} + \frac{\partial\left\langle u_j' u_i'\right\rangle}{\partial x_j} \tag{11}$$

where

$$\frac{D_f}{Dt} = \frac{\partial}{\partial t} + \langle u_j\rangle\frac{\partial}{\partial x_j} \tag{12}$$

SDM use this relationship to generate an equation of motion for the fluid velocity along a particle trajectory. So if $\frac{d_p}{dt}$ is the time derivative of the fluid velocity along a particle trajectory and similarly if $\frac{d_f}{dt}$ is along a fluid point trajectory, then

$$\begin{aligned} \frac{d_p u_i}{dt} &= \left(\frac{\partial}{\partial t} + \mathrm{v}_j\frac{\partial}{\partial x_j}\right) u_i(\mathbf{x},t) \\ &= \end{aligned} \tag{13}$$

SDM consider only the contribution from the gradient of the mean fluid velocity in this equation of motion for the fluid velocity along a particle trajectory. That is, they consider the equation

$$\frac{d_p u_i}{dt} = (\mathrm{v}_j - u_j)\frac{\partial\langle u_i(\mathbf{x},t)\rangle}{\partial x_j} + \alpha_{ij}(\langle u_j\rangle - u_j) + f_i(\mathbf{x}) + f_i'(t) \tag{14}$$

In effect this is equivalent to assuming that the contribution of the fluctuating fluid velocity gradient is absorbed into the white noise function $f_i'(t)$. Using the white noise function, the equation for $\langle P\rangle$ can be closed exactly; namely[2]

$$\begin{aligned} \left\langle F_i'(\mathbf{x},t)P(\mathbf{v},\mathbf{x},\mathbf{u},t)\right\rangle &= \left\langle f_i'(t)P(\mathbf{v},\mathbf{u},\mathbf{x},t)\right\rangle \\ &= -\int_0^\infty \left\langle f_i'(0)f_j'(s)\right\rangle ds\frac{\partial\langle P\rangle}{\partial u_j} \end{aligned} \tag{15}$$

Then from Eq. (14), the equation for $\langle P\rangle$ used by SDM is:

$$\begin{aligned} &\frac{\partial\langle P\rangle}{\partial t} + \frac{\partial}{\partial x_i}\mathrm{v}_i\langle P\rangle + \frac{\partial}{\partial \mathrm{v}_i}\beta_{ij}(u_j - \mathrm{v}_j)\langle P\rangle \\ &+\frac{\partial}{\partial u_i}\left[\alpha_{ij}(\langle u_j\rangle - u_j) + f_i(\mathbf{x}) + (\mathrm{v}_j - u_j)\frac{\partial\langle u_i\rangle}{\partial x_j}\right]\langle P\rangle \\ &= \int_0^\infty \left\langle f_i'(0)f_j'(s)\right\rangle ds\frac{\partial^2\langle P\rangle}{\partial u_i u_j} \end{aligned} \tag{16}$$

3. Dispersion in simple generic flows

The PDF equation can be used to obtain the continuum equations for the dispersed particle phase treated as a fluid. In the case of the GLM model an extra equation is provided for the mean carrier flow and covariance of the flow velocity seen by the particles which adds an extra layer of complication. Both approaches are equivalent to one another when the statistics of the fluid motion *seen* by the particle are Gaussian. We refer elsewhere to the explicit form for the continuum equation [32]. What is interesting is to say something about the form of the equation in simple generic flows because it gives some insights into the way particle are transported in more complex non-uniform shear flows. In particular there have been numerous experimental measurements and simulation on these basic particle-flows and it is useful to compare PDF predictions with these results.

[2] Note the closure is also exact if $\mathbf{f}''(t)$is Gaussian non-white but will include gradients of $\langle P\rangle$ in $\mathbf{x}$ and $\mathbf{v}$ as well.

3.1. Dispersion in homogeneous isotropic turbulence

Much of the early work on particle dispersion in turbulent flows was directed towards evaluating or measuring the particle diffusion coefficient in statistical stationary homogeneous isotropic turbulence or grid generated decaying turbulence which could be considered as quasi-homogeneous and isotropic. We refer specifically to the seminal experimental measurements of Wells and Stock [44] who measured the particle dispersion in a turbulence flow field in which the particles were subjected to a mean drift due to an electric field. The experiment replicated the features that one would expect if particles were falling under gravity. The beauty of the experiment was that the settling velocity could be varied for any given particle size and in fact could be reduced to zero by applying an electric force in the opposite direction to gravity. In so doing they were able to separate out settling effects from inertial effects on the particle diffusion coefficient which is something that could not be done with gravitational settling alone (inertial and settling would always be inextricably linked). The important result from the PDF approach indicates that the diffusion is bound up with the form of the momentum equation in the same was as for Brownian motion (one recalls the way Einstein was able to deduce the Brownian diffusion coefficient using a similar approach [13]). Suppose you consider a system approaching equilibrium (but not at equilibrium). In that case we might legitimately ignore the inertial acceleration term in the momentum equation so that we have a particle flux described as a gradient diffusion equation. Because the underlying flow is homogenous and terms that involved gradients of of the flow itself can be ignore. The diffusion coefficient is given in the long term $t \to \infty$by the expression

$$\epsilon = \tau_p(\langle v^2 \rangle + \lambda) \tag{17}$$

where $\tau_p = \beta^{-1}$ is the particle response time for Stokes drag and both particle mean square velocity $\langle v^2 \rangle$ and diffusion coefficient (obtained from the carrier flow velocity measured along a particle trajectory) are the long term equilibrium values. When the expression for λ and $\langle v^2 \rangle$ are substituted in this equation which both involve $\langle u'(0)u'(t) \rangle$ the carrier flow velocity auto correlation along a particle trajectory, you end up with the following remarkable result, namely

$$\epsilon = \int_0^\infty \langle u'(0)u'(t) \rangle dt \tag{18}$$

That is the diffusion coefficient does not depend explicitly upon its inertia, whatever the particle inertia, the form of the diffusion coefficient is the same. What difference there is depends on whether the carrier flow timescales along a particle trajectory are bigger or smaller than those along a fluid element trajectory. In fact particles with more inertia appear to have a diffusion coefficient that is greater than that for a fluid element or passive tracer. This is shown in Figure 1based on the results obtained by Squires & Eaton [39] for particle diffusion in DNS isotropic homogeneous turbulence. The vertical axis is the ratio of the particle diffusion coefficient compared to that of the fluid or passive scalar. The calculations were done for a range of particle Stokes numbers St from 0.06 to 0.35 where Stokes number is the ratio of the particle response time τ_p to integral timescal of the flow T_f. You can see that for all the Stokes numbers the long time values of the diffusion coefficient are greater than that of the carrier flow, the greatest value being for $St = .33$ but with little difference between this value and the value for $St = 1.09$. The formula in Eq,(18) is also appropriate for particles settling under gravity or in an electric field as with the measurements of Wells and Stock [44]. The timescales of fluid motion along a particle trajectory depends upon the time it takes for a particle to move from one eddy to another in the flow field. The faster the particle moves the shorter the timescale of the fluid motion it encounters, assuming that the eddy lifetime is much longer than the transit time of the particle. If v_g is the settling velocity and l_e the spatial length scale then $l_e/v_g \ll \tau_e$

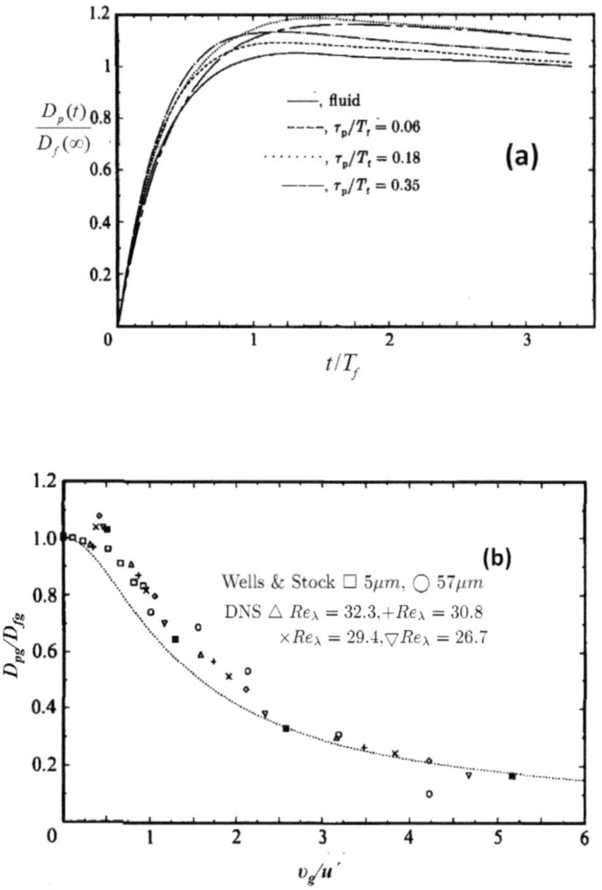

Figure 1. Ratio of particle diffusion coeffficient to long term fluid element diffusion coefficient in homogeneous isotropic turbulence: DNS and experimental measurements, the influence of inertia τ_p/T_f and settling v_g/u'. Both plots are from Squires & Eaton[39]

where τ_e is the eddy decay time. Since $\tau_e \sim l_e/u'$ this implies $u'/v_g \ll 1$ in which case

$$\epsilon = \int_0^\infty R_E(0, v_g t)dt \tag{19}$$

where $R_E(t, x)$ is the carrier flow spatial velocity correlation (in a frame moving with the settling velocity $\mathbf{v}_g$ with a separation x measured in the direction of gravity . ϵ in this case refers to the diffusion coefficient for diffusion in the direction of gravity.

This behaviour is reflected in Figure 1(b) where the long-time particle diffusion coefficient in the directiion of gravity is shown as a function of the ratio v_g/u' again taken from Squires & Eatons' DNS calculations [39] and the measurements of Wells and Stock [44]. One further prediction of the PDF approach borne out by experiments and simulation is that the diffusion flux j_D is given by a Boussinesq approximation in the long term by

$$\mathbf{j}_D = \langle \rho \boldsymbol{v} \rangle = -\boldsymbol{\epsilon} \cdot \frac{\partial \langle \rho \rangle}{\partial \boldsymbol{x}} \tag{20}$$

where $\langle\rho\rangle$ is average concentration at x the distance measured along any spatial direction, a result that reflects a Gaussian profile for the mean concentration for an instantaneous point source of particles.

Dispersion in a simple shear

An important generic flow is particles transported in a simple shear flow. In single phase flow it is used it to define what we mean by viscosity and the relation between shear stress and rate of straining of the flow. Using a simple model for the Reynolds shear stresses based on Pope's GLM model [17], exact solutions have been obtained for the particle kinetic stresses (velocity covariances) and the particle diffusion coefficients associated with an instantaneous point source located at the centre of the shear [32]. The results for the particle kinetic stresses are shown in Figures 2(a) as a function of the particle response time suitably normalized on the timescale of the turbulence which is homogenous and stationary (effectively the particle Stokes number St). x_1 is the streamwise direction and x_2 cross-streamwise direction. What is plotted is the difference between the streamwise and cross-streamwise components for the normal stresses normalized on $S^2\left\langle u'^2\right\rangle$ where S is the strain rate normalized on the time scale of the turbulence. Whats is noticeable is the difference is positive and it increases with Stokes number. As the particles cross the shear they extract turbulent energy from the mean flow – it's the work done by shear stresses in the streamwise direction which appears as an increase of turbulent kinetic energy in the streamwise direction. The second graph in Figures 2(b) is even more revealing. A calculation is made of the diffusion coefficients ϵ_{ij} which relate the diffusion current j_i to the gradients of the concentration gradient i.e.

$$j_i = \langle u_i\rangle\langle\rho\rangle - \epsilon_{ij}\frac{\partial}{\partial x_j}\langle\rho\rangle \tag{21}$$

Here we see the particle current composed of a convective term proportional to the local mean carrier flow and a long term a diffusive current. What is most interesting is the particle diffusion coefficient in the streamwise direction compared to that in the cross-streamwise direction. We note that for all Stokes numbers it is $-ve$, which includes that for a passive scalar! This does not imply however that contrary to the Second Law that a blob of particles will always contract rather than expand? These diffusion coefficients make a small contribution to the way a blob will diffuse. Particles will diffuse in the streamwise direction because as they move upwards or downwards they will experience larger positive or negative velocities according to how far away they are away from the origin. And that displacement is randomly positive or negative because it is determined by dispersion in the cross streamwise direction. So it is this process that makes the blob stretch in the streamwise direction. The $-ve$ diffusion coefficients reduce this process but never reverse it. Figure 3 shows a picture of the concentration contours and the mean particle velocities at a time $\sim$ integral times after they were released from the centre of the shear where the mean velocities are almost radial and later where they have rotated to align with he velocities of the shear during which time the contours have expanded.

4. Transport and deposition in turbulent boundary layers

There has been intensive research over the last 20 years on the near wall behaviour of particles suspended in a turbulent flow (see e.g. recent review by Li and Ahmadi [23]), in particular how particles interact with near wall structures and how this transports and deposits them at the wall. In this regard DNS has been particularly useful in simulating the behaviour of the fluid motion. Much of our understanding comes from particle tracking in these flows. Of particular note is the work of Soldati and his co workers [37] in illuminating the transport mechanisms and in developing models for the particle deposition which has been the major preoccupation from a practical point of view. The problems here have been in formulating an appropriate transport equation to account for the influence of the severe changes in the turbulence intensity

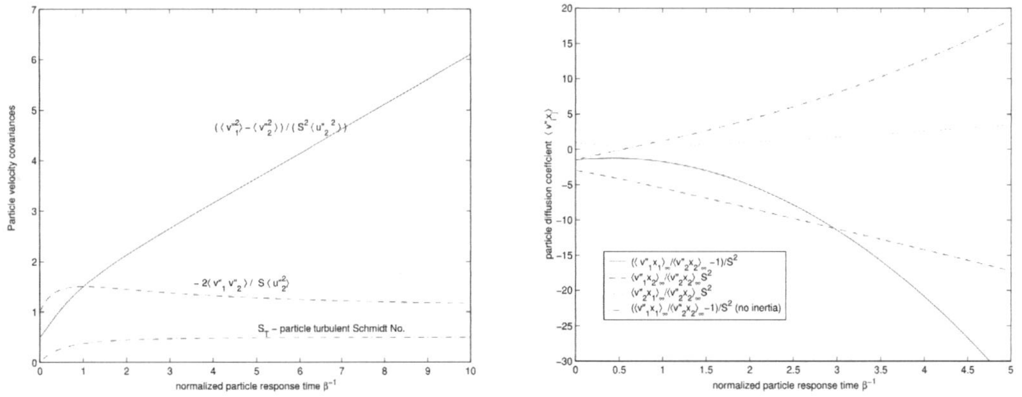

a) Particle covariance (kinetic stresses) b) Long time particle dispersion coefficients ϵ_{ij}

Figure 2. Dispersion of particles in a simple shear based on Pope's GLM versus Stokes number β^{-1} $S$is the strain rate normalised on turbulence integral time scale versus Stokes number β^{-1} . In the graphs $\epsilon_{ij} = \langle v_i'' x_j \rangle$

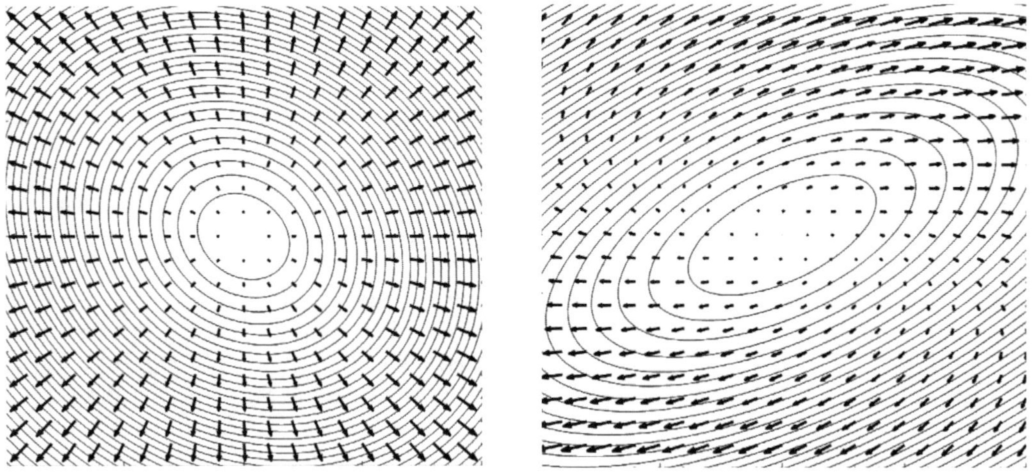

a) short term dispersion b) Long term dispersion

Figure 3. Concentration contours and particle mean velocity St = 1,normalised strain rate S =1, at (a) t = 1(b) t = 6 where t is in units of the integral timescale. Concentration contours represent a constant fraction f of the concentration at the centre of the shear, $f = 1{=}0.02n$, $n = 1, 2, ...k$

in the near wall region which in general makes the particle transport entirely non local, i.e. far from local equilibrium. In this regard the PDF approach has been particularly successful. Whilst the two-fluid equations are inappropriate near the wall, the solution of the PDF equation itself still provides a valid description. Furthermore the boundary conditions arising from the particle wall interactions are only expressible in terms of boundary conditions for a PDF equation since they involves changes in the velocity distribution at the wall. Traditional-fluid modelling is inappropriate because the boundary conditions (bc's) imposed are artificial. We note that advection diffusion models for deposition are implicit in the particle momentum equation derived from the PDF equations where the inertial term is ignored [30].

$$j = \langle \rho \rangle \boldsymbol{v}_d - \boldsymbol{\epsilon} \cdot \frac{\partial \langle \rho \rangle}{\partial \mathbf{x}} \tag{22}$$

where

$$v_d = -\tau_p \frac{\partial}{\partial \mathbf{x}} \cdot \langle \boldsymbol{vv} \rangle + \boldsymbol{\kappa} - \frac{\partial}{\partial x} \cdot \boldsymbol{\lambda} \tag{23}$$

The drift velocity v_d is composed of essentially of two terms that are of different origin. The first term arises from the gradients of the kinetic stress and reflects a balance between a drag force and the gradients of a stress, whilst the remaining terms are derived from body forces that arise directly from spatial inhomogeneous in the flow. The first term has often been referred to as turbophoresis [27]. For precise details on the application of the PDF approach to near wall behavior and the influence of natural boundary we refer to the works of Devenish et al. [12], Reeks and Swailes [33] and Darbyshire and Swailes [10].

4.1. Particle deposition in turbulent boundary layers

Deposition of particles in turbulent pipe flow has a huge literature associated with it (see [23]) - there have been many attempts at predicting the deposition as a function of particle size or response time. The problem as we have seen is the rapid decay of turbulence near the wall and the way the particles respond to that decay and the lack of local equilibrium with the flow. The particles may start out in local equilibrium far away from the wall but their inability to follow the steep changes in the fluid velocity means that at any point the particles have some memory of their previous history - their velocity has more to do with where they came from than what is happening to the fluid motion locally. The continuum equations break down. Not only that, we have to deal with the problem of boundary conditions (bc's) if we use continuum equations because perfect absorption at the wall means that the only boundary conditions is that there are no particles at the wall with velocities away from the wall. There is no means of fitting that behaviour into a traditional no slip at the wall boundary condition. PDF equations are ideally suited for this type of problem – they take account of the naturals bc's and secondly they take account of the influence of particle inertia and the steep turbulence gradients. They uniquely handle both effects together which traditional models are incapable of doing. It is useful to show some features of the near wall behaviour before considering what the deposition looks like. The results are taken from the recent analysis of vanDijk and Swailes [42] based on solution of the GLM PDF model using a discontinuous Galerkin method of solution. The Figure4(a) shows the concentration ratio as a function of distance $y+$ from the wall (denoted by x_2 in the Figure 4) for a range of values of the particle inertia or particle response time in wall units τ^+ equivalent to the Stokes number. You can see a significant peaking of the concentration near the wall as τ^+ reduces from very large values where the motion is basically ballistic (no response at all to the near wall turbulence). Figure4(b) shows the ratio of the rms of the particle wall normal velocities, v_2 to local fluid rms of the fluid u_2 as a function of distance from the wall, indicating that this ratio $\gg 1$ as you approach the wall because of the particles' inertia and the much higher velocities they have attained further away from the wall where the gradients of the turbulence are much less.

Figure 4(c) for the particle pdf $p(v_2, x_2)$shows that particles enter the domain at $x_2 = 100$ for $v_2 < 0$, and subsequently move towards the wall via, mainly, a diffusive mechanism; the pdf remains close to the local equilibrium state over a large part of the spatial domain. Closer to the wall, at about $x_2 = 20$, the fluid rms velocity decreases rapidly. The effect of this is that the particles are no longer driven by diffusion, but start moving in free flight. Most of the heavy particles ($\tau^+ = 300$, see Figure 4) have enough momentum to reach the wall, even those that have a relatively low speed. Only very slowly moving particles get trapped near the wall, and slowly drift towards adhesion at$x_2 = X_0$. This results in a small build-up near the

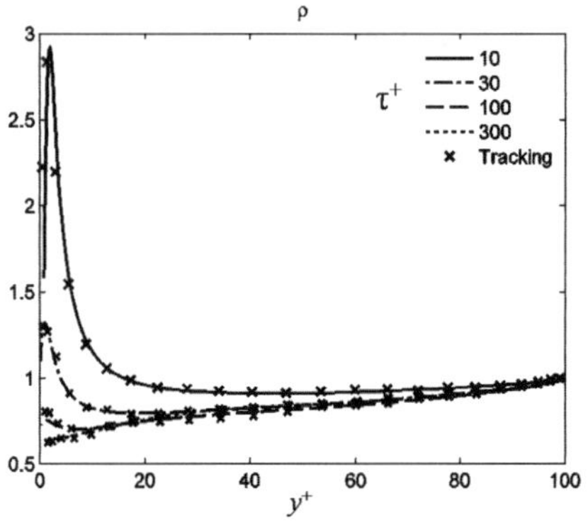

a) near wall particle concentration ρ

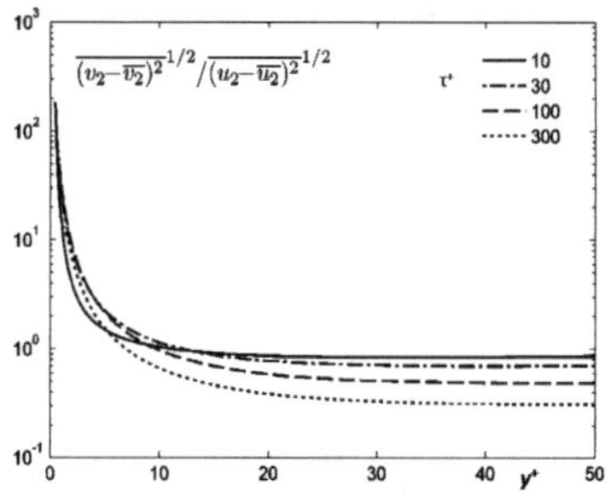

b) near wall particle normal rms normal velocities

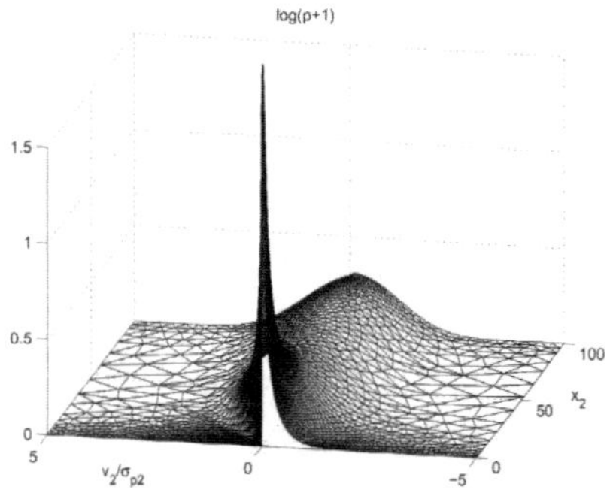

c) near wall pdf $p(v_2, x_2)$ for $\tau^+ = 10$

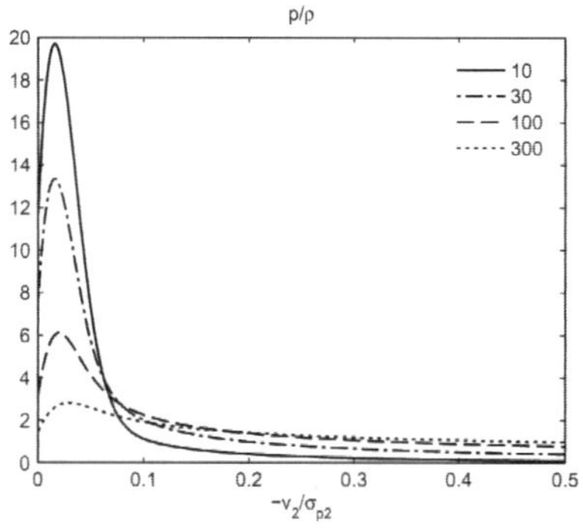

d) particle velocity distribution for $\tau^+ = 10$ at *'wall'* (one particle radius from wall).

Figure 4. Particle near wall behaviour as a function of distance from wall y^+ and particle response time τ^+ (both in wall units), van Dijk and Swailes [42] $\sigma_{p2}$is the local particle equilibrium rms velocity in the wall normal direction.

wall of particles with low velocity. For lighter particles ($\tau^+ = 10$, Figure 4(c)) only the fast particles reach the wall in free flight while most will be trapped. Hence the very sharp peak in the pdf aroud zero particle velocity. This feature that would not be picked up in the particle tracking because of the requirement of a very high resolution. This results in a strong build-up in concentration which is also clearly visible in Figure 4(a), where the particle concentration, ρ,in the boundary layer is shown. In fact recent work suggests that there is threshold velocity of τ^+below which particles are trapped in the region of almost stagnant fluid near the wall and never get deposited (see the recent alalayis of Sikovsky [35]). Similar features have been observed in particle pair collisions in homogeneous isotropic turbulence [16]. The normalised particle velocity distribution, is presented in Figure 4(d), illustrates this process as well: the distribution for τ^+=300 has a significant tail, while for $\tau^+ = 10$ the distribution is strongly localised near $v_2/\sigma_{p2} = 0$.

Finally Figure 5 shows the predictions of the particle depositions velocity as a function of particle response time τ^+obtained by Zaichik[49] compared to a range of experimental and DNS results. In particular Zaichik's predictions were obtained by solving a closed set of moment equations derived from the PDF kinetic equation. The closure is based on a quasi-normal assumption for the fourth-order moments of particle velocity and local-equilibrium closure for

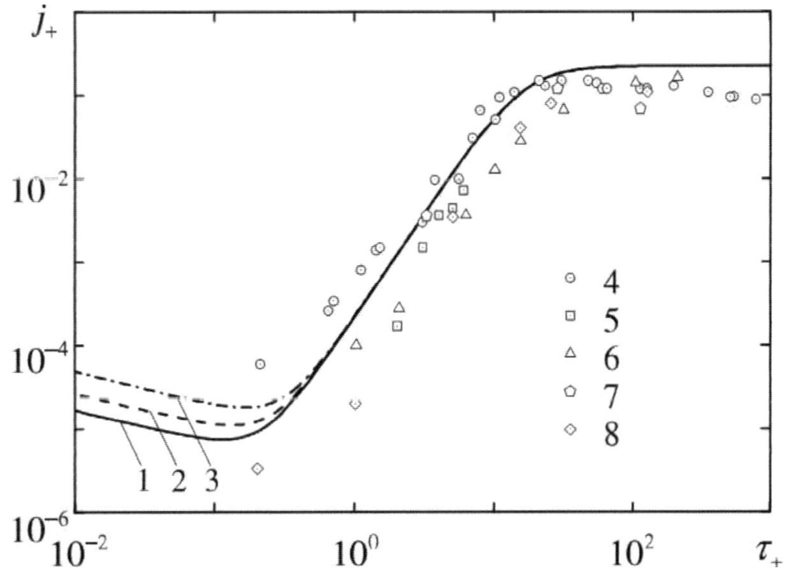

Figure 5. Particle deposition in turbulent pipe flow Zaichik [49]

3-rd order moments (neglecting their advection). The method of solution has thus much in common with the RANS approach in CFD involving Reynolds stress transport equations. Given the experimental error is very large the PDF solutions predicts the huge changes in deposition velocity from 10^{-1} to 10^{-5} for a change in τ^+ from 10 to 10^{-1}which reflects the role of particle inertia on the deposition.

5. Segregation and Agglomeration

It is by now well known that turbulence, contrary to traditionally held views, can demix a suspension of particles (see e.g. [14]). The process of segregation depends upon the ratio of the particle response time to the timescale of the turbulent structures in the flow (i.e the Stokes number, St). Early experiments and simulations (e.g. [9]) have shown that the demixing reaches a maximum when the particle response time is approximately equal to the timescale of the turbulent structure (i.e the particle Stokes number $St \sim 1$), the suspended particles being observed to segregate into regions of high strain rate in between the regions of high vorticity. This feature is best illustrated graphically in Figure 6 which shows the segregation patterns arising from a point source of particles diffusing in a 2-D random array of counter rotating vortices. The Figure shows the segregation pattern at a given value of time (in units of the integral time scale) after the source has been introduced into the flow for 3 values of the particle Stokes number $St = 0.05, 0.5, 5$. It is clear that maximum segregation occurs for the case of $St = 1$. What this Figure doesn't show however is the fact that the segregation continues in time, the patterns becoming more filamental. Whatever their Stokes number all particles will segregate, it is just that the greater the Stokes number the longer the process takes to achieve a given level of segregation. This is contrary to what many have thought that the segregation reaches an equilibrium state. Particles will segregate until they touch. In fact what happens is that the segregation forms a network of filamental *caustics* similar to the patterns of light obtained at the bottom of swimming pools as shown in Figure 6(d) where within these filamental networks it is observed that the light beams cross one another. Similarly in the case of particle segregation in a turbulent flow within the filamental caustics particle trajectories are observed to cross also. The formation of caustics in particle demixing processes was first

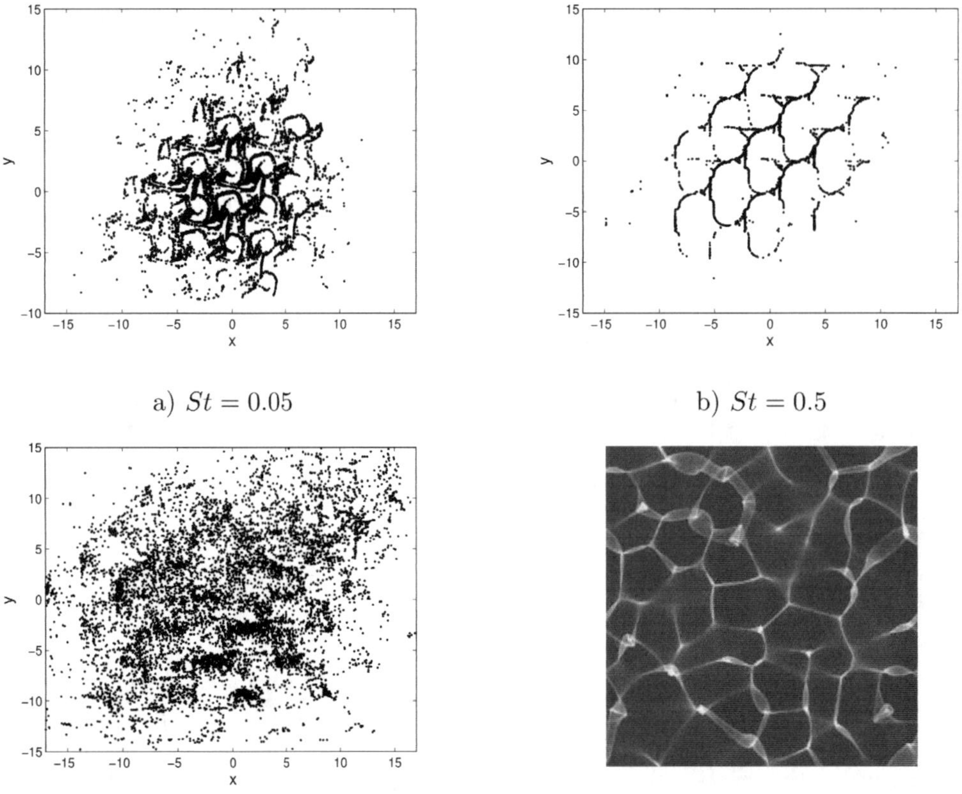

a) $St = 0.05$

b) $St = 0.5$

c) $St = 5$

(d) 'swimming pool' caustics

Figure 6. Segregations as a function of particle Stokes number St (a)-(c) based on positions of 10^4 particles after time $t = 20$ in a non-isotropic random straining flow, (d) caustic patterns observed at the bottom of a swimming pool.

recognised by Wikinson and Mehlig [45] and this crossing of trajectories within a caustic is intimately related to the occurrence of random uncorrelated motion (RUM) in flow fields that are spatially random but smoothly varying. Fevrier, Simonin & Squires [15] have observed that the spatial particle velocity field resulting from the motion of suspended particles in a direct numerical simulation (DNS) of homogeneous isotropic turbulence consists of two components: a smoothly (continuous) velocity field that accounts for all particle-particle and fluid-particle two point spatial correlations (they referred to this component as the mesoscopic Eulerian particle velocity field (MEPVF)); and a spatially uncorrelated component which we have referred to as RUM (the component of random uncorrelated motion) whose contribution to the particle kinetic energy increases as the particle inertia increases. Fevrier, Simonin & Squires attribute this feature to the ability of the particles with inertia to retain the memory of their interaction with very distant, and statistically independent eddies in the flow field.

Segregation and RUM are related to the occurrence of inter-particle collisions as follows from the seminal work of Sundaram and Collins [40]and Wang and Wexler [43] in which they demonstrate that i) segregation enhances the particle concentration of certain regions of the flow, ii) RUM, i.e. the decorrelation of velocity between particles, causes two nearby particles to collide and possibly to agglomerate. Segregation is well-known to manifest itself especially for $St \sim 1$, whereas the effect of RUM is almost invisible for small particles and becomes increasingly important for larger St. Since the interplay between these two effects determines the collision rate in a turbulent flow, it is essential to quantify segregation and RUM as accurately as possible as a function of the Stokes number and some typical flow properties in order to correctly predict the rate of inter particle collisions.

In recent years, the process of segregation of inertial particles has been studied from different viewpoints when the Stokes number is relatively small. On the one hand, [7] demonstrated a strong correlation between the positions of small inertial particles and the locations of zero-acceleration points in the carrier flow. On the other hand, [1] carried out a theoretical analysis based on the assumption that the velocity of inertial particles can be directly related to the carrier flow velocity. By doing so, they were able to show that the segregation of particles continues indefinitely in the course of time, and they showed that the concentration of inertial particles in a turbulent flow is highly intermittent, so the particles are distributed far from uniformly over space. A similar approach was chosen by Chun et al. [8] who demonstrated that the time-converged solution of the radial distribution function is of the form $g(r) \sim r^{\beta}$, where the negative number β is proportional to St^2. In addition, they confirmed this by showing results from a DNS of statistically stationary homogeneous isotropic turbulence. Most importantly the PDF/ kinetic approach has been used by Zaichik ([47] [48]) to obtain a similar result.

The understanding of dilute suspensions of inertial particles has been vastly extended by interpreting the motion of particles in terms of dynamical systems theory. The first approach in this direction was given by [38], and was later specifically applied to the motion of inertial particles in turbulent flows by [2] and [45]. [45] derived an analytical expression for the Lyapunov exponents associated with the motion of inertial particles in physical space. The derivation was based on the assumption that the typical correlation time of the carrier flow was very small, i.e. the Kubo number $Ku \ll 1$. Unfortunately, this assumption is not exactly valid in real turbulence where $Ku = \mathcal{O}(1)$, as [45] acknowledge themselves. [2] showed that if the particle clustering is fractal, the exponent in the radial distribution function is equal to $\beta = n_d - D_{corr}$, where D_{corr} is the correlation dimension introduced by [19], and n_d is the number of dimensions of the problem ($n_d = 2$ in a two-dimensional flow and $n_d = 3$ in a three-dimensional flow). [2] expressed the clustering of particle in terms of its fractal dimension in phase space and showed how this was related to the Lyapunov exponents of the $2n_d$-dimensional dynamical system. [3] obtained a correlation dimension D_{corr} by calculating the Lyapunov exponents in a Direct Numerical Simulation of turbulence for a wide range of Stokes numbers, and found that $n_d - D_{corr}$ scales with St^2, in agreement with the aforementioned results by [8].

We will describe now how a particular method known as the Full Lagrangian Method (FLM) has been used very effective;y to investigate the statistics of particle segregation in both DNS and in synthetic turbulence flows like those produced in kinematic simulation (KS). This method, originally introduced by [25] but later used by [31] and [18], consists of calculating the size of an infinitesimally small volume occupied by a group of particles, along the trajectory of one single particle. This immediately yields the concentration of particles along the trajectory, since the inverse of the volume occupied by a fixed number of particles corresponds to the particle concentration by definition. We describe how the results from the FLM are converted into statistics of the particle number density, thus providing a wealth of information on the segregation process. In particular we describe here how the FLM can be used to quantify non-uniformities in the spatial distribution of particles, the singularities in the particle concentration field and the presence of RUM.

5.1. The Full Lagrangian Approach

We consider small spherical particles acted on by Stokes drag for which the equation of motion of each particle is given by

$$\frac{d\mathbf{x}}{dt} = \mathbf{v} \;\; ; \;\; \frac{d\mathbf{v}}{dt} = \frac{1}{St}(\mathbf{u} - \mathbf{v}) \tag{24}$$

where $\mathbf{x}$ and $\mathbf{v}$ denote the position and the velocity of the particle, respectively, $\mathbf{u} = \mathbf{u}(\mathbf{x}, t)$ is the velocity of the carrier flow at the position of the particle and St is the particles Stokes number. FLM calculates the deformation of an elemental volume of particles as it moves through the flow. The deformation of such a volume is characterized by the temporal evolution of the unit deformation tensor $\mathbf{J}$, whose components J_{ij} are defined by:

$$J_{ij} \equiv \frac{\partial x_i(\mathbf{x}_0, t)}{\partial x_{0,j}}, \tag{25}$$

where $\mathbf{x}_0$ is the position of the particle at some initial time say $t = 0$. Differentiating Eq. 25 with respect to time gives:

$$\frac{\mathrm{d}}{\mathrm{d}t} J_{ij} = \frac{\partial v_i(\mathbf{x}_0, t)}{\partial x_{0,j}}. \tag{26}$$

$$\frac{\mathrm{d}}{\mathrm{d}t} J_{ij} = \frac{\partial v_i(\mathbf{x}_0, t)}{\partial x_{0,j}}. \tag{27}$$

The second derivative with respect to time is:

$$\frac{\mathrm{d}^2}{\mathrm{d}t^2} J_{ij} = \frac{\partial}{\partial x_{0,j}} \left(\frac{\mathrm{d}v_i(\mathbf{x}_0, t)}{\mathrm{d}t} \right) = \frac{1}{\tau_p} \left(\frac{\partial x_k}{\partial x_{0,j}} \right) \frac{\partial}{\partial x_k} u_i(\mathbf{x}, t) - \frac{1}{\tau_p} \frac{\partial v_i(\mathbf{x}_0, t)}{\partial x_{0,j}}. \tag{28}$$

Inserting Eq. 25 and Eq. 27 into Eq. 28 results in the equations of motion of each component J_{ij}:

$$\frac{\mathrm{d}J_{ij}}{\mathrm{d}t} = \dot{J}_{ij}, \qquad \frac{\mathrm{d}}{\mathrm{d}t}\dot{J}_{ij} = \frac{1}{\tau_p} \left(J_{kj} \frac{\partial u_i}{\partial x_k} - \dot{J}_{ij} \right), \tag{29}$$

We choose as initial conditions $J_{ij}(0) = \delta_{ij}$ and $\dot{J}_{ij}(0) = \partial u_i(\mathbf{x}_0, 0)/\partial x_j$. The volume expansion $J(t) = |\det J_{ij}|$. If the initial distribution of particles is uniform over a certain domain, the deformation in the course of time is inversely proportional to the particle number density $n(t)$ measured along the trajectory of one reference particle is related to J by

$$J = n^{-1}(t) \tag{30}$$

Thus a spatially averaged moments of the particle number density, n, can be calculated directly from the deformation $\mathbf{J}$ along sufficiently many particle trajectories.

5.2. The Statistics of the compression

The Full Lagrangian Method can thus be used to determine the compressibility of the particle phase, $\mathbb{C}$ which we define as $J^{-1}dJ/dt = dlnJ/dt$. We can relate this to the divergence of the particle velocity field $\mathbf{v}_p(\mathbf{x}, t)$ providing that the velocity field is single valued and continuous, namely

$$\mathbb{C} = d\,lnJ/dt = \nabla \cdot \mathbf{v}_p(\mathbf{x},\mathrm{t})$$

We present some results based on the FLM for a 'kinematic' flow field composed of 200 random Fourier modes (referred to as kinematic simulation (KS)) [22][24]. In addition we also present the statistics of the particle number density and the contribution that RUM makes to the kinetic energy of the transported particles. Fig. 7 shows the results of $\lim_{t\to\infty} t^{-1}\langle \ln |J| \rangle$ for

Journal of Physics: Conference Series **530** (2014) 012003 doi:10.1088/1742-6596/530/1/012003

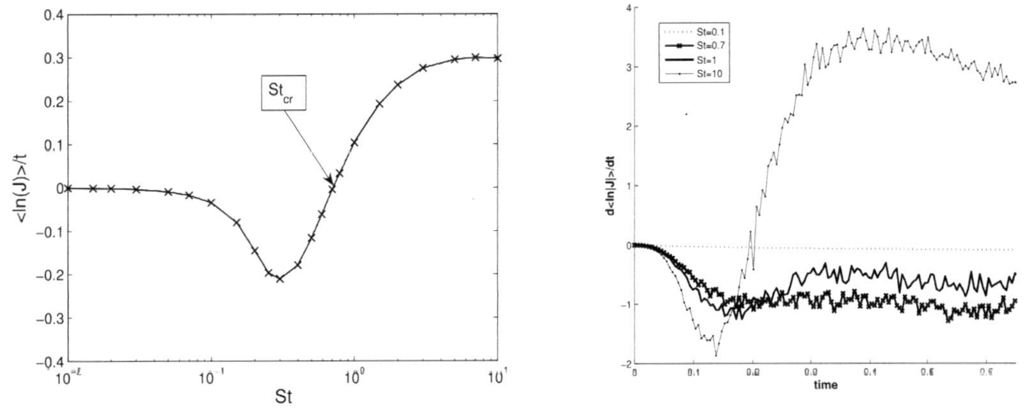

a) KS b) DNS

Figure 7. Long term net compressibility as a function of time in flow field composed of random Fourier modes (KS) and DNS of homogeneous isotropic turbulence

a wide range of values of Stokes numbers. It is noted that this value is equal to $\lim_{t\to\infty}\langle\mathbb{C}\rangle$, i.e. the time converged compressibility of the particle velocity field. We note that there is a critical value of the Stokes number St_{cr} beyond which the net compressibility is positive rather than negative.; in the present case, $St_{cr} \simeq 0.7$. If the particle Stokes number St is lower than St_{cr}, then the particles are continuously compressed into smaller volumes in the course of time and the process of segregation continues indefinitely. If, on the other hand, the Stokes number is larger than St_{cr}, the particle volumes expands or alternatively if the particles are confined they become fully mixed. This doesn't mean that there is no segregation beyond the critical Stokes number. It is clear that an average quantity only provides us with a picture of the net effect between compression/expansion and what effect prevails in thecourse of time: a compressibility – on average – does not necessarily mean that the segregation is zero everywhere. The same applies to negative compressibility. Indeed, the sign of the compressibility can be related to the topological distributionof the particle concentration field: as particles cluster, compression zones appear (in the clusters) at the same time as dilation zones (in the depleted zones).

5.3. Statistics of the particle number density

Now we investigate the statistics of the particle number density in the course of time. The moments of the particle number density $\overline{n^\alpha}$ have been determined for both the KS and DNS flow fields. The results are shown in Figure. 8, for (a) $St = 0.05$ (b) $St = 0.5$ in the case of the KS flow field and in (c) for $St = 0.4$ for a DNS flow field. In all cases the value of $\overline{n^0}$, which corresponds to $\langle|J|\rangle$, remains equal to unity for all time, as expected. The other moments of the particle number density are markedly higher than 1 and are associated with the non-uniformity in the spatial distribution of particles.

There is a qualitative distinction between the cases of small Stokes numbers such as $St = 0.05$ in Fig. 8 a), and large Stokes numbers such as $St = 0.5$ in Fig.8 b). If the Stokes number is large, it may happen that $|J| = 0$ for a particle due to the crossing of trajectories. These intermittent events, which cause $n \to \infty$, dominate the statistics of higher-order moments of the PDF at certain moments in time, as is reflected by the spikes in the curve for $St = 0.5$ in Fig. 8b). Hence, the spatial distribution of particles in a random turbulence-like flow may be highly intermittent. This is also the case for the DNS moments in Figure 8(c)

However, for sufficiently small Stokes numbers where RUM is not important (such as $St = 0.05$ in Fig.8 a)), we observe that $\overline{n^\alpha}$ depends exponentially on time and:

$$\overline{n^\alpha} \propto \exp(\gamma t), \tag{31}$$

where γ is a function of α and St As can be seen in a), the higher-order moments grow faster than the lower-order moments. This demonstrates unambiguously that the segregation process continues indefinitely in this case where $St = 0.05$.

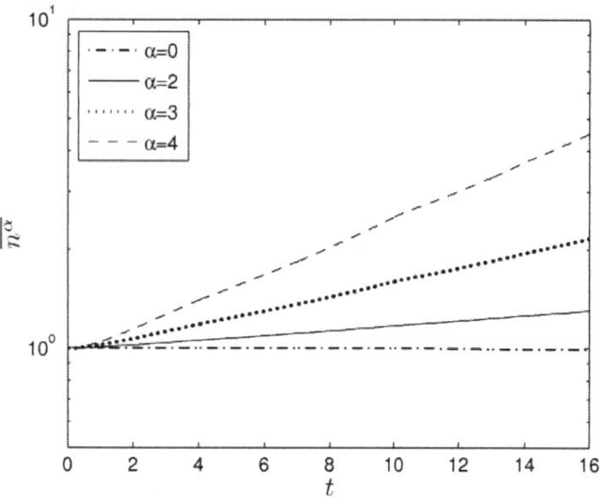

(a) KS $St = 0.05$

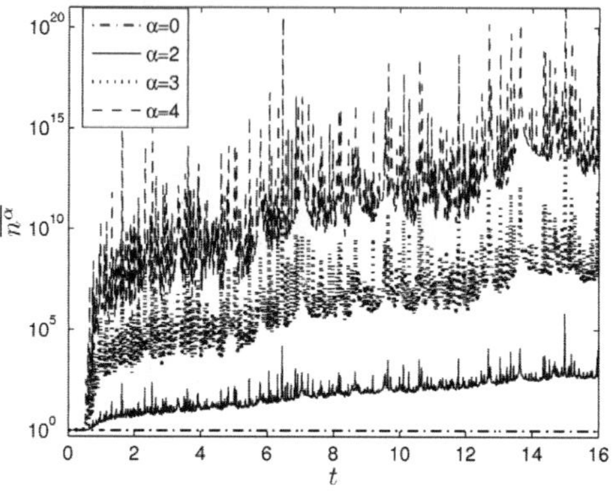

(b) KS $St = 0.5$

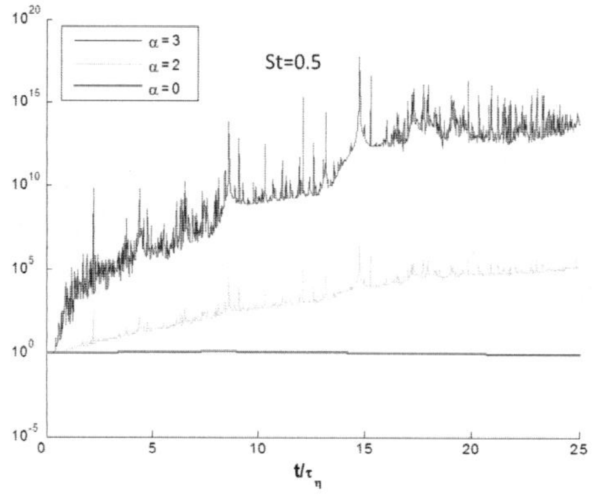

c) DNS $St = 0.4$

Figure 8. Spatially average of the α-th moment of the particle number density, $\overline{n^\alpha}$, for KS flow field a) St = 0.05; b) St = 0.5 (c) DNS $St = 0.4$

Random uncorrelated motion (RUM)

We present here results for the contribution RUM makes to the kinetic energy of the particles transported in a KS generated carrier flow field. The values are obtained by measuring the longitudinal velocity correlation between two particles separated by a distance r. Extrapolation of the correlation to $r = 0$, gives an intercept for which the RUM contribution is 1 - this quantity. The RUM contribution to the kinetic energy is shown in Figure 9(a) as a function of the particle Stokes number St. We note that for $St \to 0$, the RUM contribution $\to 0$ whilst for $St \to \infty$, it $\to$ unity i.e totally ballistic motion.

Figure 9(b) show the distribution of the fluctuating value of the compression $\mathbb{C}'$ normalised on its rms value obtained by Meneguz and Reeks [24]. Each value of $\mathbb{C}'$ has been separated

Journal of Physics: Conference Series **530** (2014) 012003 doi:10.1088/1742-6596/530/1/012003

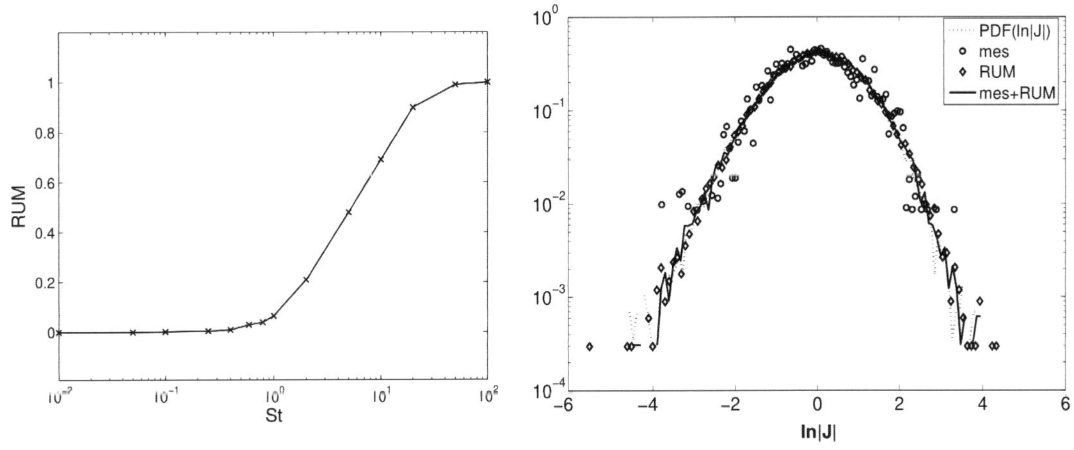

a) RUM contribution to particle kinetic energy

b) $P(\mathbb{C})$, $\mathbb{C} = lnJ$

Figure 9. RUM contribution to (a) particular kinetic energy as a function of Stokes number St and to (b) distribution P of compression $\mathbb{C}$ Both cases refer to the flow field composed of random Fourier modes.

out into its RUM components and its smoothly varying mesoscopic component. We note that negative values of $\mathbb{C}'$ in the extreme range, are almost entirely made up of the RUM uncorrelated component which is consistent with the formation of caustics of high regions of segregation and concentration.

The results presented here show that particle inertia can have major implications for the collision rates between particles. Two effects enhance the collision rate between particles: i) preferential concentration of particles in relatively few regions of the flow, ii) RUM, i.e. the decorrelation of velocities between neighboring particles so that particles are more likely to collide with one another. From Fig.7 , we know that preferential concentration manifests itself especially for $0.1 < St < 1$. The effect of RUM is most visible for $St > 0.5$ and increasingly important for larger St. Collision rates are therefore expected to be highest in the Stokes number regime $St > 0.5$. Certainly more research is needed to confirm this statement, but the results presented in the seminal paper by [40] do point in that direction, since they found a maximum collision rate for a Stokes number St (based on the smallest scale of the flow, the Kolmogorov time scale) of $2 < St < 5$, with a collision rate vanishing for $St \downarrow 0$, and a collision rate decreasing only slowly with St if $St > 5$.

6. Summary and Conclusions

This paper describes methods and approaches that have been used to simulate and model the transport, mixing and agglomeration of small particles in a flowing turbulent gas (referred to as *turbulent aerosols*). The transported particles because of their inertia are assumed not to follow the motion of the large scales of the turbulence (referred to as one point particle dispersion) and or the motion of the small dissipating scales of the turbulence (referred to as two point particle or relative dispersion). The major studies of one point particle transport have been associated with developments in two-fluid modelling where the particles are treated as a fluid in the same way as the carrier flow. The purpose has been to find suitable forms for the continuum equations of the dispersed particle phase together with the constitutive relations. The objective has been to model complex dispersed flows using much the same approach as in CFD for single phase flow. Of great value in this approach has been the application of the PDF approach involving a Master Equation analogous the Maxwell-Boltzmann equation of Classical Kinetic Theory. In this paper we looked at two PDF approaches, a kinetic approach which is concerned with the probability that a particle has a certain velocity and position and an alternative approach which considers an extra random variable, namely the carrier flow velocity measured by the particle along its trajectory. It was referred to as the GLM approach because the equation of motion for this flow velocity is based on a Generalized Langevin model (GLM) used by Pope [17]. We have described the success of these approaches in modelling simple generic flows and transport of particles in a turbulent boundary layer.

The work reported on two particle dispersion has been about methods of measuring and analyzing the statistics of particle segregation. In particular we have described the Full Lagrangian Method (FLM) which measures the deformation of an elemental volume of particles as it moves through the fluid along a particle trajectory. The statistics of the segregation process has been described in terms of the statistics of the compressibility which is shown to have a mean component and a fluctuating which is close to Gaussian except in the wings where the existence of singularities contributes to extremely large values of negative compression and the moments of the particle concentration. We have shown that these features are all compatible with the formation of caustics.

[1] E. Balkovsky, G. Falkovich, and A. Fouxon. Intermittent distribution of inertial particles in turbulent flows. *Physical Review Letters*, 86:2790–2793, 2001.

[2] J. Bec. Multifractal concentrations of inertial particles in smooth random flow. *J. Fluid Mech.*, 528:255–277, 2005.

[3] J. Bec, L. Biferale, M. Cencini, A. Lanotte, S. Musacchio, and F. Toschi. Heavy particle concentration in turbulence at dissipative and inertial scales. *Physical Review Letters*, 98:084502, 2007.

[4] Yuri Buyevich. Statistical hydromechanics of disperse systems. part 1. physical background and general equationss. *J.Fluid Mech.*, 49:489–507, 1971.

[5] Yuri Buyevich. Statistical hydromechanics of disperse systems. part 2. solution of the kinetic equation for suspended particles. *J.Fluid Mech.*, 52:345–355, 1972.

[6] Yuri Buyevich. Statistical hydromechanics of disperse systems. part 3. pseudo-turbulent structure of homogeneous suspensions. *J.Fluid Mech.*, 56:313–336, 1972.

[7] L. Chen, S. Goto, and J. C. Vassilicos. Turbulent clustering of stagnation points and inertial particles. *J. Fluid Mech.*, 553:143–154, 2006.

[8] J. Chun, D. L. Koch, S. L. Rani, A. Ahluwalia, and L. R. Collins. Clustering of aerosol particles in isotropic turbulence. *J. Fluid Mech.*, 536:219–251, 2005.

[9] C. T. Crowe, J. N. Chung, and T. R. Troutt. Chapter 18. In M. C. Roco, editor, *Particulate Two-Phase Flow*, volume 626, pages 1–1. Heinemann, Oxford, 1993.

[10] K. F. F Darbyshire and D. C. Swailes. A pdf model for particle dispersion with stochastic particle-surface interactions, fed-236, gas-solid flows. In *FED-236, Gas-Solid Flows, ASME 51-56*, pages 51–56. 1996.

[11] I. Derevich and L Zaichik. Precipitation of particles from a turbulent flow. *Izv. Akad.Nauk SSR, Mekh. Zhid. i Gaza*, pages 96–104, 1988.

[12] B. J. Devenish, D. C. Swailes, and Y. A Sergeev. A pdf model for dispersed particles with inelastic particle-wall collisions. *Phys Fluids*, 11(7):1858–1868, 1999.

[13] A. Einstein. On the theory of brownian motion. *Ann, d. Physik*, IV:549, 1905.

[14] J. R. Fessler, J. D. Kulick, and J. K. Eaton. Preferential concentration of heavy particles in a turbulent channel flow. *Physics of Fluids*, 6(11):3742–3749, 1994.

[15] P. Février, O. Simonin, and K. D. Squires. Partitioning of particle velocities in gas-solid turbulent flows into a continuous field and a spatially uncorrelated random distribution; theoretical formalism and numerical study. *J. Fluid Mech.*, 553:1–46, 2005.

[16] K. Gustavsson, E. Meneguz, M. Reeks, and B. Mehlig. Inertial-particle dynamics in turbulent flows:caustics, concentration fluctuations, and random uncorrelated motion. *New Journal of Physics*, 14:115017, 2012.

[17] D. C. Haworth and S. B Pope. A generalized langevin model for turbulent flow. *Phys Fluids*, 29:387–405, 1986.

[18] D. P. Healy and J. B. Young. Full lagrangian methods for calculating particle concentration fields in dilute gas-particle flows. *Proc. Roy. Soc. London A: Mathematical, Physical and Engineering Sciences*, 461(2059):2197–2225, 2005.

[19] H. G. E. Hentschel and I. Procaccia. The infinite number of generalized dimensions of fractals and strange attractors. *Physica D*, 8:435–444, 1983.

[20] K. E. Hyland, M. W. Reeks, and S. McKee. Derivations of a p.d.f. kinetic equation for the transport of particles in turbulent flow. *J. Phys: Math Gen.*, 32:6169–6190, 1999.

[21] K. E. Hyland, M. W. Reeks, and S. McKee. Exact analytic solutions to turbulent particle flow equations. *Phys. Fluids*, 11:1240–1261, 1999.

[22] R. H. A. IJzermans, E. Meneguz, and M. W Reeks. Segregation of particles in incompressible random flows: singularities, intermittency and random uncorrelated motion. *J. Fluid Mech.*, 653:99–136, 2010.

[23] A. Li and G. Ahmadi. Dispersion and deposition of spherical particles from point sources in turbulent channel flow. *Aerosol Sci. and Tech.*, 14:209–226, 1992.

[24] E Meneguz and M. W. Reeks. Staistical properties of particle segregation in homogeneous isotropic turbulence. *J. Fluid Mech.*, 686:338–351, 2011.

[25] A. N. Osiptsov. Lagrangian modelling of dust admixture in gas flows. *Astrophysics and Space Science*, 274:377–386, 2000.

[26] J. Pozorski and J.-P Minier. Probability density function modelling of dispersed two-phase turbulent flows. *Phys. Rev. E*, 59(1):855–863, 1998.

[27] M. W. Reeks. The transport of discrete particles in inhomogeneous turbulence. *J. Aerosol Science*, 14(6):729 –739, 1983.

[28] M. W. Reeks. On a kinetic equation for the transport of particles in turbulent flows. *Physics of Fluids*, 15:446–456, 1991.

[29] M. W. Reeks. On the continuum equations for dispersed particles in non uniform flows. *Physics of Fluids*, 446:1290–1303, 1992.

[30] M. W. Reeks. On the continuum equations for dispersed particles in nonuniform flows. *Physics of Fluids*, 5(3):750–761, 1993.

[31] M. W. Reeks. Simulation of particle diffusion, segregation, and intermittency in turbulent flows. In S. Balachandar and A. Prosperetti, editors, *Proc. of IUTAM Symposium on Computational Modelling of Disperse Multiphase Flow*, pages 21–30. Elsevier, 2004.

[32] M. W. Reeks. On probability density function equations for particle dispersion in a uniform shear flow. *J. Fluid Mech*, 522:263–302, 2005.

[33] M. W Reeks and D. C Swailes. The near wall behavior of particles in a simple turbulent flow with gravitational settling and partially absorbing wall. *J. Fluid Mech. Res.*, 22(2):31–39, 1997.

[34] M. Wi Reeks. Eulerian direct interaction applied to the statistical motion of particle. *J.Fluid Mech.*, 83:529–546, 1980.

[35] D. P. Sikovsky. Singularity of inertial particle concentration in the viscous sublayer of wall-bounded turbulent flows. In *International Conf on Turbulence Heat and Mass Transfer 2012, Palermo, Italy*, pages 51–56. 2013.

[36] O Simonin, E. Deutsch, and J.-P. Minier. Eulerian prediction of the fluid / particle correlated motion in turbulent two-phase flow. *App. Sci. Res.*, 51:275–283, 1993.

[37] A. Soldati. Physics and modelling of turbulent particle deposition and entrainment: Review of a systematic study. *Int. J. Multiphase Flow*, 35(9):827–839, 2009.

[38] J. C. Sommerer and E. Ott. Particles floating on a moving fluid: a dynamically comprehensible physical fractal. *Science*, 259:335–339, 1993.

[39] K. D. Squires and J. K. Eaton. Measurements of particle dispersion obtained from direct numerical simulations of isotropic turbulence. *Journal of Fluid Mechanics*, 226:1–35, 1991.

[40] S. Sundaram and L. R. Collins. Collision statistics in an isotropic particle-laden turbulent suspension. part 1. direct numerical simulations. *J. Fluid Mech.*, 335:75–109, 1997.

[41] D. C. Swailes and K. F. F. Darbyshire. A generalised fokker-planck equation for particle transport in random media physica. *Physica A*, 242:38–48, 1997.

[42] P. van Dijk and D. C. Swailes. Hermite-dg methods for pdf equations modelling particle transport and deposition in turbulent boundary layers. *J. Comp. Phys.*, 231(14):4094–4920, 2012.

[43] L. P. Wang, A. S. Wexler, and Y. Zhou. On the collision rate of small particles in isotropic turbulence, ii. finite inertia case. *Physics of Fluids*, 10(10):1206–1216, 1998.

[44] M. R Wells and D. E. Stock. The effects of crossing trajectories on the dispersion of particles in a turbulent flow. *J. Fluid Mech.*, 136:31–62, 1983.

[45] M. Wilkinson, B. Mehlig, S. Östlund, and K. P. Duncan. Unmixing in random flows. *Physics of Fluids*, 19:113303, 2007.

[46] L. Zaichik. Simulation of particle diffusion, segregation, and intermittency in turbulent flows. In *Eigth Intl Symp. on Turbulent Shear Flows, Technical University of Munich*, pages 10–2–1 – 10–2–6. 1991.

[47] L. I. Zaichik and V. M. Alipchenkov. Pair dispersion and preferential concentration of particles in isotropic turbulence. *Phys. Fluids*, 15:1776–1787, 2003.

[48] L. I. Zaichik and V. M. Alipchenkov. Refinement of the probability density function model for preferential

concentration of aerosol particles in isotropic turbulence. *Phys. Fluids*, 19:113308, 2007.

[49] L. I. Zaichik, N. I. Drobyshevsky, A. S. Filippov, R. V. Mukin, and V. F. Strizhov. A diffusion-inertia model for predicting dispersion and deposition of low-inertia particles in turbulent flows. *Int.J. Heat I& Mass Transfer*, 53:154–162, 2010.

Turbulence over/inside porous surfaces and challenges to its modelling

K Suga and Y Kuwata
Department of Mechanical Engineering, Osaka Prefecture University
1-1 Gakuen-cho, Naka-ku, Sakai, Osaka 599-8531, Japan

E-mail: suga@me.osakafu-u.ac.jp

Abstract. To understand turbulence over porous media, a series of PIV measurements were carried out in porous-wall channel flows. The porous walls were made of three types of foamed ceramics which had the same porosity but different permeability. For turbulence inside porous media, LES studies of fully developed flows in several model porous media were performed. Referring to these databases, a multi-scale $k-\varepsilon$ four equation eddy viscosity model for turbulence around/inside porous media was developed. The proposed model was validated with satisfactory accuracy.

1. Introduction

Turbulent flows over/inside highly porous materials are often seen in various engineering devices due to the advantages for their performance of heat and mass transfer by their large contact area [1,2]. It is readily understood that to provide gas and fuel through the porous media for chemical reaction, mass transfer enhancement over the porous wall is crucially important. Thus, understanding and predicting flows and mass transfer around/inside porous media are very important in industry. However, turbulent flow physics in such flows is not fully understood though many studies have been historically performed experimentally and numerically (e.g. Refs. [3-5]). Moreover, numerical schemes (e.g. Refs. [6,7]) for those flows are still immature compared to for single phase turbulence [8].

Therefore, the present authors' group have been performing PIV experiments [9-11] and high resolution LES studies [12,13] to understand turbulence over/inside highly porous media. Based on those studies and the knowledge in the literature, the present authors have proposed new engineering turbulence models [14,15]. In this paper, the major outcomes of those studies are summarized for serving a part of the state of the art in the studies on turbulence over/inside porous media.

2. PIV measurements of turbulence over porous walls

2.1. Experimental procedures

Turbulent flows in a channel whose bottom wall was made of a porous medium were measured by the PIV. Two flow geometries: porous wall channels [9,10] and rib-mounted porous wall channels [11], were considered. Figure 1 shows those flow geometries. The experimental setup of the flow facility is illustrated in figure 2. Tap water was used as the working fluid. The flow was fully developed in a driver channel before the test section channel. Both the channels consisted of solid smooth acrylic top and side walls and a porous bottom wall. The porous media applied were three types of foamed ceramics shown

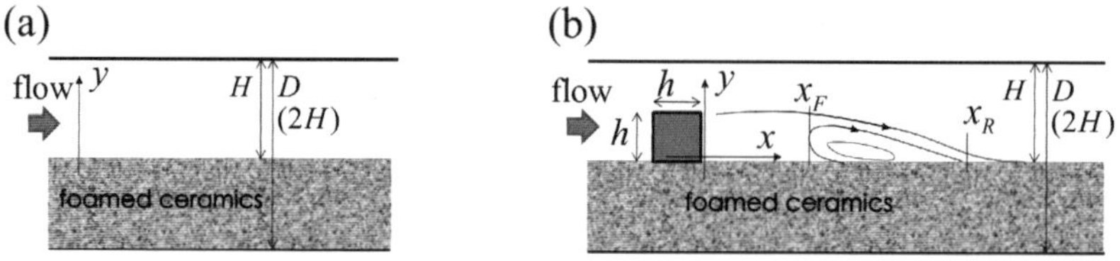

Figure 1. Flow geometries; (a) porous wall channel, (b) rib-mounted porous wall channel.

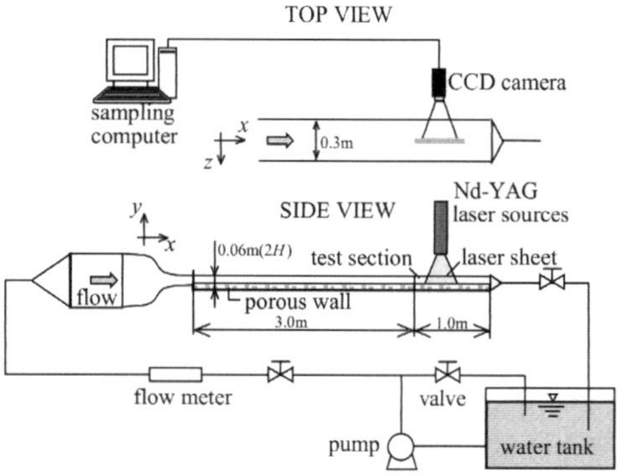

Figure 2. Experimental setup.

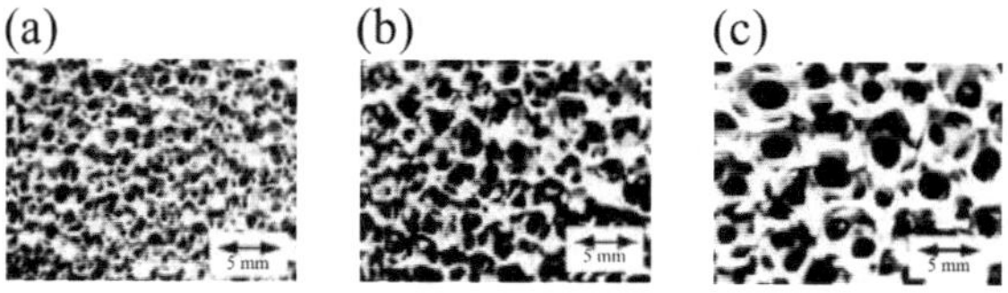

Figure 3. Surface photographs of the foamed ceramics; (a) #20, (b) #13, (c) #06.

in figure 3. Their properties are listed in table 1. The thickness of the porous wall was 0.03 m which was set to be the same as the channel height H for the clear fluid region (figure 1(a)) whose width was 0.3 m. In cases of rib-mounted channels, at 0.2 m from the beginning of the test section, a square sectioned rib whose height is h=0.015 m was mounted as shown in figure 1(b). It had been confirmed that the flow was nominally two-dimensional near the symmetry plane. The measured range of the bulk Reynolds number Re_b ($\rho U_b H / \mu$) was 1000-10000 where the fluid viscosity μ and density ρ were obtained from their relations to the water temperature. The bulk mean velocity U_b was obtained by integrating the measured sectional mean velocity profile. For the rib-mounted channel, it was performed at x=-5h where the flow was confirmed to be unaffected by the rib.

The two-component PIV system consisted of a double-pulse Nd-YAG laser with 120 mJ/pulse at a frequency of 532 nm, a CCD camera of 30 fps, a 60 mm camera lens of f/2.8 and a computer for data processing. The laser beam was formed into a sheet of approximately 1.0 mm thickness. The sheet illuminated the symmetry plane of the channel where the instantaneous images were recorded by the CCD camera. A single recorded frame covered a zone of 30×30 mm^2 with 2048×2048 pixels. For the tracer particles, acrylic colloid particles were used. The mean diameter, specific gravity and refractive index of the particles are respectively 3.1 μm, 1.19 and 1.50. The seeding density was adjusted to obtain 10-15 particle-image pairs in each interrogation window whose size was set to 64×16 or 32×32 pixels for the channel or the rib-mounted channels, respectively. Thus the measurement sampling volume was 0.8(x) ×0.2(y) ×1.0(z) or 0.4(x) ×0.4(y) ×1.0(z) mm^3. The image sampling rate was 4 Hz. The average particle displacement was set to be about 25 % length (6-8 pixels) of the interrogation window. To obtain the statistical data, in each zone, 3200-4800 image pairs were processed. The recorded data were

Table 1. Characteristics of the foamed ceramics.

Porous med. No.	Porosity φ	Mean pore diameter D_p [mm]	Permeability K [mm^2]	Forchheimer coef. c_F
#20	0.82	1.7	0.020	0.17
#13	0.81	2.8	0.033	0.10
#06	0.80	3.8	0.087	0.095

processed by commercial software (Dynamics Studio 2.0, DANTEC DYNAMICS) with the fast Fourier transform cross-correlation technique. The removed error vectors were less than 3% of the total data processed. The average number of pixels for a particle image captured by the CCD camera was counted to be about 3.1 pixels. This indicated that the particle images were well resolved in the experiments and the estimated error of the instantaneous velocity magnitude was less than 4% of the maximum velocity.

2.2. Channel flow experimental results [9,10]

Figure 4(a) shows examples of the mean velocity distributions in the porous wall channel. The position at $y = 0$ corresponds to the surface of the porous medium whilst the $y = H$ represents the top wall surface. Due to the Reynolds shear stress distribution shown later in figure 6(c), the velocity distributions look asymmetric and deviate from the parabolic profile of the laminar case (Re_b=900) as Re_b increases. This clearly indicates that the flow tends to be turbulent even at Re_b=1300. Although it is not shown here, this tendency was enhanced in the higher permeability cases. Obviously, the flow became turbulent at a lower Re_b compared with a solid wall channel flow. This suggests that porous surface enhances the onset of turbulence. At $y = 0$, finite velocities exist and their magnitude increases as Re_b increases. This tendency of the slip velocity U_w is shown in figure 4(b). The characteristic profiles of the slip velocities look fairly generalized against the permeability Reynolds number Re_K ($=u^p_\tau K^{1/2}/\nu$). This implies that Re_K can be a good measure for permeable turbulent boundary layers. Indeed, in the range of $0< Re_K <3$ a sharp increase of U_w can be seen irrespective of the porous media. It suggests that there is transition to full porous wall turbulence in that range. The friction velocity here u^p_τ was calculated from the shear stress on the porous wall obtained by extrapolating the measured Reynolds shear stress distribution to the porous wall. For turbulent flows above rough beds or submerged canopies, the log law form:

$$U^+ = \frac{1}{\kappa}\ln\left(\frac{y+d}{h_r}\right)^+ , \qquad (1)$$

has been often applied, where d is the zero-plane displacement and h_r is the equivalent roughness height. In this log law, d and h_r are empirical parameters. Note that as shown in figure 4(c), the parameter d is the depth of the zero-plane from the surface. From the semi-log velocity distribution as shown in figure 4(d), κ, d^{p+} and h_r^{p+} were obtained. Note that values with the superscripts '$p+$' and '$t+$' denote normalized values by u^p_τ and u^t_τ which are the friction velocities on the porous and top solid walls, respectively. It was found that they had close relationships with Re_K as shown in figure 5(a)-(c).

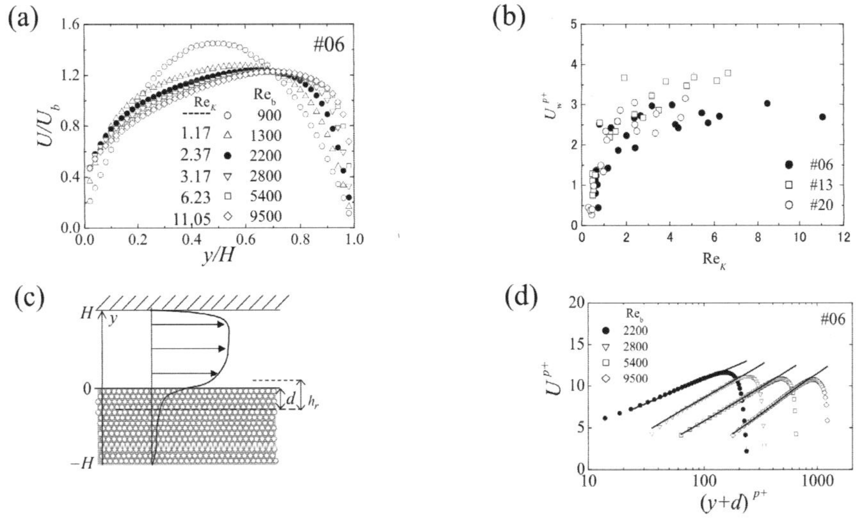

Figure 4. Time mean velocities; (a) mean velocity profiles of case #06, (b) slip velocities, (c) zero-plane displacement d and equivalent roughness h_r, (d) mean velocities of case #06 in semi-log scale.

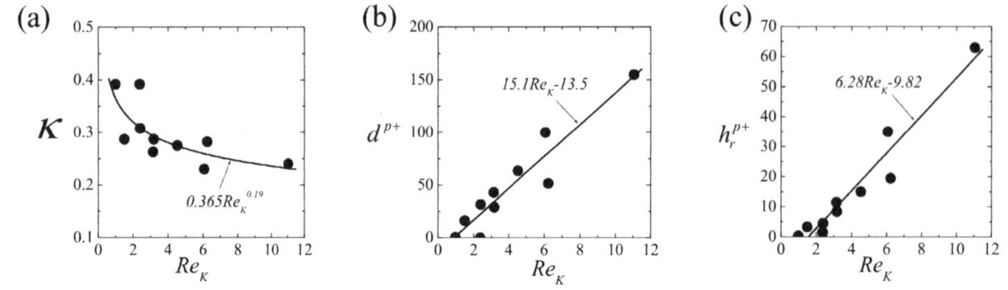

Figure 5. Distribution against the permeability Reynolds number; (a) von Kármán constant, (b) zero-plane displacement, (c) equivalent roughness height.

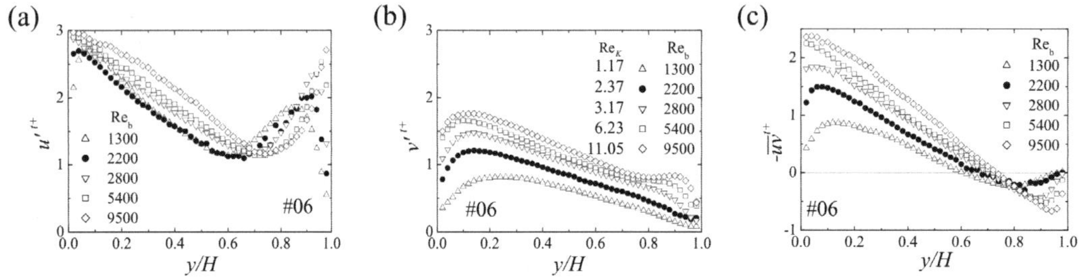

Figure 6. Turbulent quantities of case #06; (a) streamwise turbulent intensity, (b) wall normal turbulent intensity, (c) Reynolds shear stress.

To emphasise and to see the effects of the wall permeability further, u^t_τ was applied as the reference velocity for normalisation. The distribution profiles of the turbulent intensities u'^{t+} and v'^{t+} of case #06 are shown in figure 6. Although near the porous walls the peak of u'^{t+} changes a little with Re_b, v'^{t+} increases drastically as Re_b increases. It suggests that the effects of the porous wall appear in the wall normal component rather than in the streamwise component. Since wall blocking effects are weakened more due to the higher wall permeability, the wall normal fluctuation is kept at a certain level even near the wall. In other words, the wall normal fluctuation is enhanced near the porous wall due to the permeability. This tendency becomes stronger as the Reynolds number and thus Re_K increases. Figure 6(c) shows the Reynolds shear stress $-\overline{uv}^{t+}$ distribution. Due to the large contribution of v'^{t+}, $-\overline{uv}^{t+}$ monotonically increases with Re_b (and Re_K) near the porous wall. Consequently, it suggests that the higher wall permeability weakens the wall blocking effects more and the turbulent wall normal fluctuation is kept at a higher level (or enhanced) leading to the intensified turbulent shear stress near the porous wall.

Figure 7 shows the distribution of the joint probability density function of the fluctuating velocities $p(u',v')$ in the buffer regions at the position of y^+=15. Figure 7(a) corresponds to the distribution near the solid top wall whereas figure 7(b)-(d) corresponds to those near the porous walls at different conditions of Re_K =1.1-11.1. It is obvious that the distribution shape (called distribution "oval" hereafter) changes as Re_K increases. (Note that at the solid wall, Re_K is essentially zero since the wall does not have permeability.) As illustrated in figure 8(a), when Re_K becomes larger, the distribution oval becomes rounder and the angle between its major axis and the u'-coordinate becomes larger up to a certain level. Indeed, figure 7(c) looks very similar to figure 7(d) though their permeability Reynolds numbers are different: Re_K =4.5 and 11.1, respectively. Since the rounder distribution oval indicates that the turbulence is less anisotropic, figure 7 indicates that the turbulence field tends to be less anisotropic up to a certain level as Re_K increases. In order to quantitatively discuss on the shape of the distribution oval, figure 8(b) indicates the ratio of the mean deviations from the minor and major axes of the distribution oval of fluctuating velocities. Figure 8(c) indicates the difference between the major axis angles of the

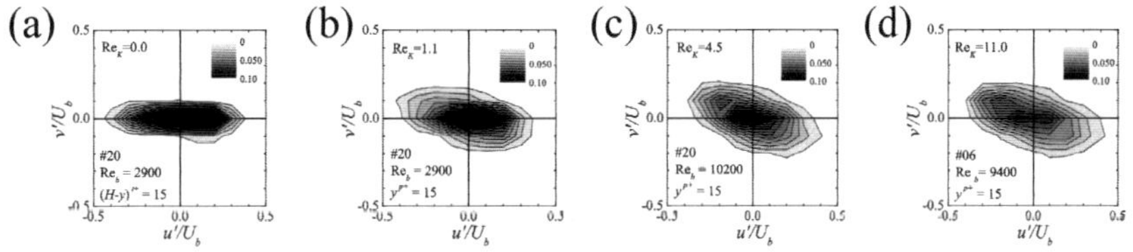

Figure 7. Joint probability density function $p(u',v')$ of fluctuating velocity at y^+=15; (a) near the solid wall side of case #20 at Re_b=2900 (Re_K =0.0), (b) near the porous wall side of case #20 at Re_b =2900 (Re_K =1.1),(c) near the porous wall side of case #20 at Re_b =10200 (Re_K =4.5),(d) near the porous wall side of case #06 at Re_b =9400 (Re_K =11.1).

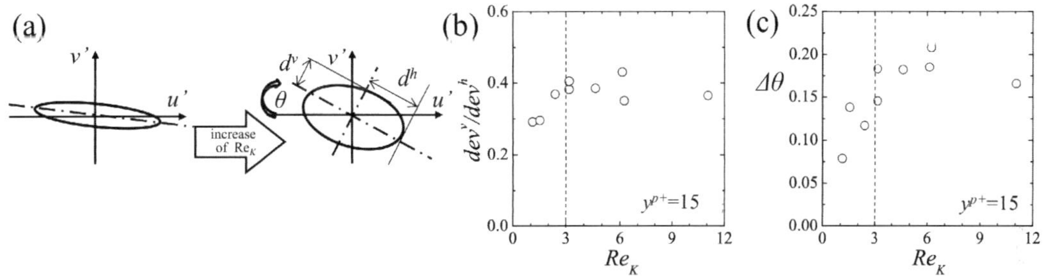

Figure 8. Measure of the distribution-oval shape; (a) distribution oval of fluctuating velocities, (b) ratio of the deviations from the minor and major axes of the distribution oval of fluctuating velocities, (c) difference between the angles of major axes.

distribution ovals near the porous and solid walls: $\Delta\theta = \theta^p - \theta^t$. The centre point, the major and minor axes of the oval were determined by averaging the distributed point positions and their angles from the u'- and v'-coordinates, respectively. Here, the mean deviations were defined as $dev^h = [\Sigma(d^h)^2 / N]^{1/2}$ and $dev^v = [\Sigma(d^v)^2 / N]^{1/2}$, where N, d^h and d^v were, respectively, the sampling number, the deviations from the minor and major axes of the distribution oval as shown in figure 8(a). When the ratio increases toward unity, the turbulence becomes less anisotropic in the u'-v' plane. As shown in figure 7(b), the ratio dev^v/dev^h of the distribution oval shows less anisotropic characteristics as Re_K increases. In the region of $Re_K \leq 3$, dev^v/dev^h tends to be smaller as Re_K reduces whilst it, however, tends to be saturated in the higher Re_K region. The distribution of $\Delta\theta$ shown in figure 8(c) also indicates the same tendency. These suggest that the turbulence structure changes depending on Re_K and thus the influence of the wall permeability on the vortex structure near permeable walls is significant.

2.3. Rib-mounted channel flow experimental results [11]

The above measurements suggested that due to the wall permeability, turbulent vortex motions over porous walls were not damped so much as near solid walls resulting in strong near-wall shear stress production. It is then understandable that if the flow includes separation near a porous wall, the formation of a separation bubble is strongly affected by the wall permeability. Thus, discussions on the turbulence separation over porous media were carried out by the PIV measurements.

Figure 9 compares streamlines obtained from the mean velocities in the solid rib-mounted channel with different wall materials: low permeability #20 and high permeability # 06 porous media, respectively. In case #20, with the increase of Re_b, it is recognisable that the reverse flow region in the clear channel tends to vanish: figure 9(a) and (b). This is considered to result from the fact that a higher proportion of the flow upstream the rib goes inside the bottom wall layer and bleeds out behind the rib as the Reynolds number increases. Also, as shown in figure 9(c), in the high permeable porous bottom wall case: #06, due to the increase of the wall permeability, the recirculation bubble behind the rib in the clear channel region looks almost extinguished. Indeed, as shown in table 1, #06 is about 4.5 times more permeable than #20. It is considered that due to the increase of the wall permeability, more amount of flow upstream the rib goes inside the bottom porous layer and bleeds out behind the rib to supply the fluid to the entrainment of a developing shear layer.

When the rib has permeability, since the flow also goes through the rib, the flow pattern and turbulence quantities change. Figure 10 compares streamlines of low and high permeability cases #20 and #06 at Re_b=3500. Due to the flow going through the rib, as shown in figure 10(a), clearly the recirculation zone behind the rib shifts downstream compared with the solid rib flow of figure 9. Also, at the high permeability case of figure 10(b), there is no recirculation zone observed and only the wake flow behind the rib can be seen.

In summary, the measurement results indicated that a turbulent recirculation or wake flow region was formed behind the rib in all the flow cases at $Re_b = 10^3\text{-}10^4$. The turbulence levels of those flows were much higher in the region around and behind the rib compared with those in the corresponding fully developed porous channel flows. However, in the solid rib-mounted channel flows, since a part of entraining fluid was supplied through the permeable bottom wall from the region upstream the rib, the recirculation behind the rib in the clear channel became smaller and eventually vanished as the wall permeability increased. The separation bubble in the solid rib flow was smaller and its reattachment length became shorter as the increase of the wall permeability and/or Re_b. Because of the reduction of the magnitude and the size of the reverse flow region in the clear channel, turbulence became weaker than those of the rib-mounted solid-wall channel flow though the turbulence levels around and behind the rib were much higher than those of fully developed porous channel flows. Due to the entrainment flow to the shear layer through the porous wall, the near wall turbulent intensities, in particular, the wall normal component tended to be large as the wall permeability increased. In the porous-rib-mounted channel flows (the rib material was the same as that of the porous wall), because of the fluid passing through the rib, the recirculation was more significantly weakened and shifted downstream. The

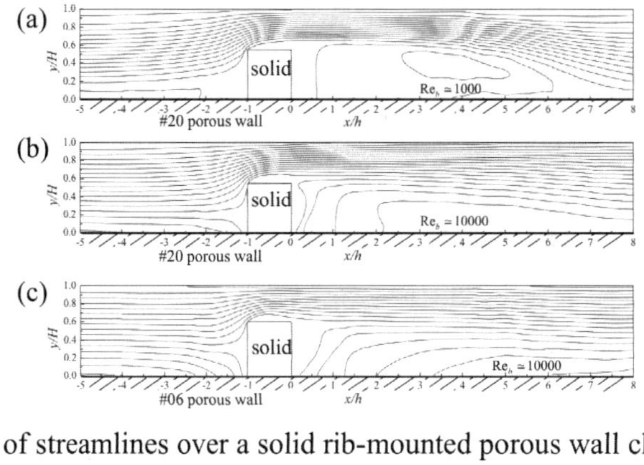

Figure 9. Comparison of streamlines over a solid rib-mounted porous wall channel; (a) case #20 at Re_b=1000, (b) case #20 at Re_b=10000, (c) case #06 at Re_b=10000.

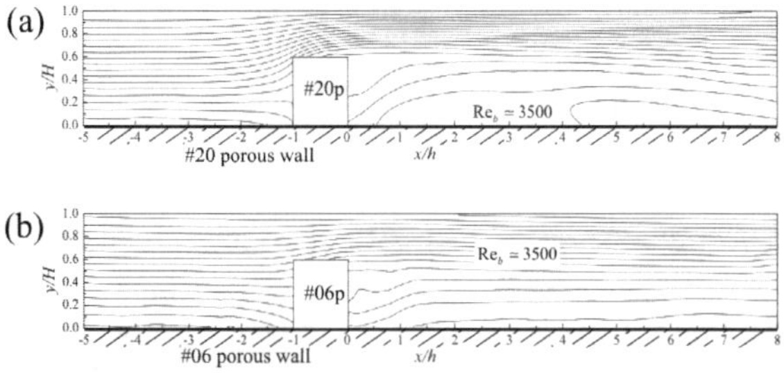

Figure 10. Comparison of streamlines over a porous rib-mounted porous wall channel; (a) case #20 at Re_b=3500, (b) case #06 at Re_b=3500.

upstream edge of the recirculation shifted further downstream as Re_b increased in the low permeability case whereas no recirculation region was observed in the clear channel at the higher wall permeability regardless of Re_b. Because of the flow through the porous rib, turbulent intensities and the Reynolds shear stress became smaller than those of the solid-rib flows.

3. Double averaged turbulence equations

So far, turbulent flows over porous media are discussed. However, to treat statistical values of turbulent flows inside porous media, so called the *double averaging* needs applying to the transport equations [16]. The double averaging consists of the conventional Reynolds and volume averaging operators. The volume averaged values are called the *superficial* and *intrinsic* (fluid phase) averaged values and defined as

$$\langle \phi \rangle = \frac{1}{\Delta V} \int_{\Delta V_f} \phi dv, \quad \langle \phi \rangle^f = \frac{1}{\Delta V_f} \int_{\Delta V_f} \phi dv, \tag{2}$$

where ΔV and ΔV_f are the REV (representative elementary volume) and the volume of the fluid phase contained within ΔV, respectively. Between them, the relation: $\langle \phi \rangle = \varphi \langle \phi \rangle^f$ exists with the porosity of the porous medium: $\varphi = \Delta V_f / \Delta V$. The volume averaging of derivatives is expressed as

$$\left\langle \frac{\partial \phi}{\partial x_k} \right\rangle = \frac{\partial \langle \phi \rangle}{\partial x_k} + \frac{1}{\Delta V} \int_A n_k ds, \tag{3}$$

where A, n_k are the superficial area of the solid phase and its unit normal vector pointing outward from the fluid to the solid phase, respectively. The dispersion is $\tilde{\phi} = \phi - \langle \phi \rangle^f$ and the interchangeable rules [17] between the Reynolds and volume averaged values are $\overline{\langle \phi \rangle^f} = \langle \overline{\phi} \rangle^f$, $\langle \phi' \rangle^f = \langle \phi \rangle^{f\prime}$, $\tilde{\phi}' = (\tilde{\phi})'$ and $\overline{\tilde{\phi}} = \tilde{\overline{\phi}}$. Then the double (volume and Reynolds) averaged momentum equation is

$$\frac{\partial \langle \overline{u}_i \rangle^f}{\partial t} + \langle \overline{u}_j \rangle^f \frac{\partial \langle \overline{u}_i \rangle^f}{\partial x_j} = -\frac{1}{\rho} \frac{\partial \langle \overline{p} \rangle^f}{\partial x_i} + \nu \frac{\partial^2 \langle \overline{u}_i \rangle^f}{\partial x_j^2} - \overline{f_i} + \overline{g_i^{\varphi}} - \frac{\partial \left(\langle \tilde{\overline{u}}_i \tilde{\overline{u}}_j \rangle^f + \langle \overline{u_i' u_j'} \rangle^f \right)}{\partial x_j}, \tag{4}$$

where f_i and $\overline{g_i^{\varphi}}$ are the drag force and hetero porous terms, respectively. The terms $\mathscr{T}_{ij} = \langle \tilde{\overline{u}}_i \tilde{\overline{u}}_j \rangle^f$ and $R_{ij}^A = \langle \overline{u_i' u_j'} \rangle^f$ are the dispersive covariance and the volume averaged (total) Reynolds stress. The latter can be further decomposed as

$$R_{ij}^A = \underbrace{\overline{\langle u_i' \rangle^f \langle u_j' \rangle^f}}_{R_{ij}} + \underbrace{\overline{\langle \tilde{u}_i' \tilde{u}_j' \rangle^f}}_{r_{ij}}, \tag{5}$$

where R_{ij} and r_{ij} are namely the macro-scale and micro-scale Reynolds stresses, respectively. The transport equations of those second moments are

$$\frac{D\mathscr{T}_{ij}}{Dt} = \mathscr{D}_{ij} + \Psi_{ij} + \mathscr{P}_{ij} - P_{ij}^d + P_{ij}^t + \mathscr{G}_{ij}^{\varphi} + \overline{f}_i \langle \hat{\overline{u}}_j \rangle^f + \overline{f}_j \langle \hat{\overline{u}}_i \rangle^f + \mathscr{E}_{ij}, \tag{6}$$

$$\frac{DR_{ij}}{Dt} = D_{ij} + \phi_{ij}^M + P_{ij} - P_{ij}^t + C_{ij}^t + G_{ij}^{\varphi} - F_{ij} - \varepsilon_{ij}^M, \quad \frac{Dr_{ij}}{Dt} = D_{ij}^m + \phi_{ij}^m + P_{ij}^m + P_{ij}^d - C_{ij}^t + G_{ij}^{\varphi m} + F_{ij} - \varepsilon_{ij}^m, \tag{7}$$

where $\langle \hat{\overline{u}}_i \rangle^f$ is the relative velocity to the solid phase, $\mathscr{D}_{ij}, \Psi_{ij}, \mathscr{P}_{ij}, \mathscr{G}_{ij}^{\varphi}$ and $\mathscr{E}_{ij}$ are the diffusion, pressure-dispersive-velocity-strain, mean shear production, hetero-porous and dissipation terms of $\mathscr{T}_{ij}$, respectively. The terms $D_{ij}, \phi_{ij}^M, P_{ij}, G_{ij}^{\varphi}$ and ε_{ij}^M are the diffusion, pressure-strain, mean shear production, hetero-porous and dissipation terms of R_{ij} while $D_{ij}^m, \phi_{ij}^m, P_{ij}^m, G_{ij}^{\varphi m}$ and ε_{ij}^m are those of r_{ij}, respectively.

Journal of Physics: Conference Series **530** (2014) 012004 doi:10.1088/1742-6596/530/1/012004

The mutual communication terms: $P_{ij}^d, P_{ij}^t\ C_{ij}^t$ and F_{ij}, are the mean dispersive shear production, turbulent shear production, macro-micro turbulence cascade and turbulent drag force terms, respectively.

4. LES of turbulence inside porous media

Due to the difficulty to access the flow fields inside porous media experimentally, numerical simulation is a better choice to investigate turbulence inside porous media. Therefore, high resolution LES studies were carried out for turbulence inside several model porous media.

4.1. Numerical methods

The microscopic LES computations [12,13] were carried out by the WALE (wall-adaptive local eddy-viscosity) model [18] using an in-house code of the D3Q27 (3 dimensional 27 discrete velocity) MRT (multiple relaxation time) LBM (lattice Boltzmann method) [19]. The lattice Boltzmann equation can be obtained by discretizing the velocity space of the Boltzmann equation into a finite number of discrete velocities ξ_α $\{\alpha = 0,\cdots,Q-1\}$. The MRT LBM transforms the distribution function **f** in the velocity space to the moment space by the transformation matrix $\mathbf{M}$. Since the moments of the distribution function correspond directly to flow quantities, the moment representation allows us to perform the relaxation processes with different relaxation-times according to different time-scales of various physical processes. The evolution equation is written as

$$\left|\mathbf{f}(\mathbf{x}+\xi_\alpha \delta t, t+\delta t)\right\rangle - \left|\mathbf{f}(\mathbf{x},t)\right\rangle = -\mathbf{M}^{-1}\hat{\mathbf{S}}\left[\left|\mathbf{m}(\mathbf{x},t)\right\rangle - \left|\mathbf{m}^{eq}(\mathbf{x},t)\right\rangle\right] + \mathbf{M}^{-1}\left(\mathbf{I}-\frac{\hat{\mathbf{S}}}{2}\right)\mathbf{M}\left|\mathbf{F}\right\rangle \delta t, \tag{8}$$

where the notations such as $|\mathbf{f}\rangle$ is $|\mathbf{f}\rangle = (f_0, f_1, \cdots, f_{Q-1})^T$, $\mathbf{I}$ is the identity matrix and δt is the time step. The term $\mathbf{F}$ is an external body force. The matrix $\mathbf{M}$ is a $Q\times Q$ matrix which linearly transforms the distribution function to the moment: $|\mathbf{m}\rangle = \mathbf{M}\cdot|\mathbf{f}\rangle$ and its collision matrix $\hat{\mathbf{S}}$ is diagonal:

$$\hat{\mathbf{S}} \equiv diag(0,0,0,0,s_4,s_5,s_5,s_7,s_7,s_7,s_{10},s_{10},s_{10},s_{13},s_{13},s_{13},s_{16},s_{17},s_{18},s_{18},s_{20},s_{20},s_{20},s_{23},s_{23},s_{23},s_{26}), \tag{9}$$

for the D3Q27 discrete velocity model. The relaxation parameters used were

$$s_4 = 1.54,\ s_7 = s_5,\ s_{10} = 1.5,\ s_{13} = 1.83,\ s_{16} = 1.4,\ s_{17} = 1.61,\ s_{18} = s_{20} = 1.98,\ s_{23} = s_{26} = 1.74, \tag{10}$$

which were from Ref.[20]. The relaxation parameter s_5 is related to the fluid viscosity, and thus for the LES, it is $\nu + \nu_{SGS} = c_s^2(1/s_5 - 1/2)\delta t$ where $c_s = 1/\sqrt{3}$ is the sound speed in the D3Q27 LBM.

In the WALE model [18], the eddy viscosity is described as

$$\nu_{SGS} = (C_w\Delta)^2 \frac{(S_{ij}^d S_{ij}^d)^{3/2}}{\left(S_{ij}S_{ij}\right)^{5/2} + (S_{ij}^d S_{ij}^d)^{5/4}}, \tag{11}$$

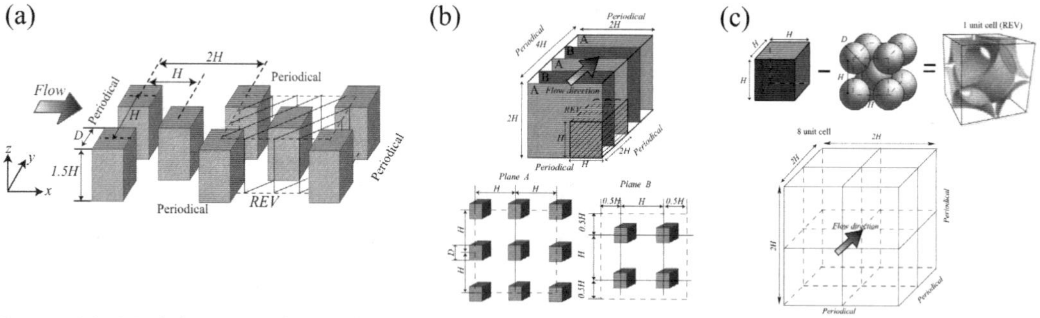

Figure 11. Model geometries and computational domains of model porous media; (a) square rod arrays, (b) staggered cube arrays, (c) BCC foam.

where $S_{ij} = (\partial \bar{u}_i / \partial x_j + \partial \bar{u}_j / \partial x_i)/2$, $S_{ij}^d = \{(\partial \bar{u}_i / \partial x_j)^2 + (\partial \bar{u}_j / \partial x_i)^2\}/2 - (\partial \bar{u}_k / \partial x_k)^2 \delta_{ij}/2$ and the value $\bar{u}_i$ is the filtered velocity. The eddy viscosity coefficient applied was $C_w = 0.1$.

The computational domain considered were $4H(x)\times H(y)\times 1.5H(z)$, $4H\times 2H\times 2H$ and $2H\times 2H\times 2H$, respectively in square rod arrays, staggered cube arrays and BCC (body centered cube) foam as shown in figure 11. After performing grid sensitivity tests, uniform computational grids of $481\times 121\times 180$, $321\times 161\times 161$ and $256\times 256\times 256$ were used respectively for those geometries. Periodic boundary conditions were applied for each domain surfaces though a pressure difference was imposed in the streamwise direction. To describe the wall surfaces, the interpolated bounce-back boundary method was applied for the distribution functions. The computed flows were in the range of 1000-3500 of the pore Reynolds number.

4.2. Budget term analysis by the LES results [11,12]

By analyzing the simulation data, the behavior of the budget terms against the porosity was investigated. In fully developed flows inside homogeneous porous media, equations (6) and (7) reduce to

$$\frac{D\mathcal{T}_{ij}}{Dt} = \Psi_{ij} - P_{ij}^d + P_{ij}^t + \bar{f}_i \langle \hat{\bar{u}}_j \rangle^f + \bar{f}_j \langle \hat{\bar{u}}_i \rangle^f + \mathscr{E}_{ij} = 0\,, \quad (12)$$

$$\frac{DR_{ij}}{Dt} = \phi_{ij}^M - P_{ij}^t + C_{ij}^t - F_{ij} - \varepsilon_{ij}^M = 0\,,\quad \frac{Dr_{ij}}{Dt} = \phi_{ij}^m + P_{ij}^d - C_{ij}^t + F_{ij} - \varepsilon_{ij}^m = 0\,. \quad (13)$$

Since the macro-scale Reynolds stress is virtually zero in the fully developed porous medium flows, the budget terms of the micro-scale Reynolds stress were focused on. (Note that in the square rod array flows, it can be detected that a very small level of the cross-beamwise macro-scale Reynolds stress component is produced by the resonance of vortex shedding [12].) Also, it was found that the turbulent shear production P_{ij}^t became negligibly small in all the cases, and hence figure 12 only shows the behaviors of the dispersive shear production P_{ij}^d. In the square rod arrays case, P_{33}^d is zero in figure 12(a) because of zero spanwise gradient of the dispersive velocity. At $\varphi = 0.5$, P_{11}^d and P_{22}^d are nearly

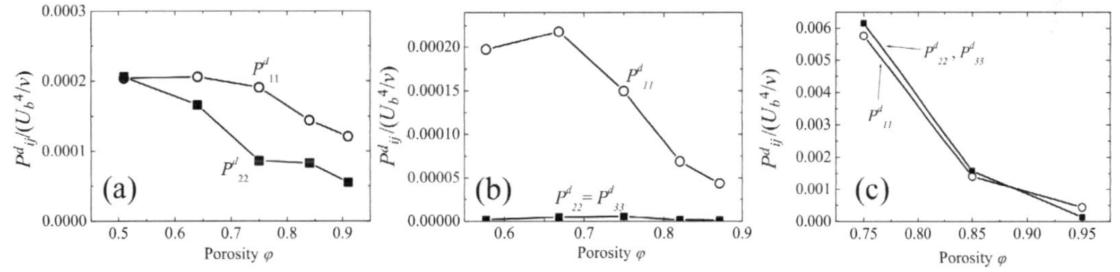

Figure 12. Dispersive shear production terms in the porous media; (a) square rod arrays, (b) staggered cube arrays, (c) BCC foam.

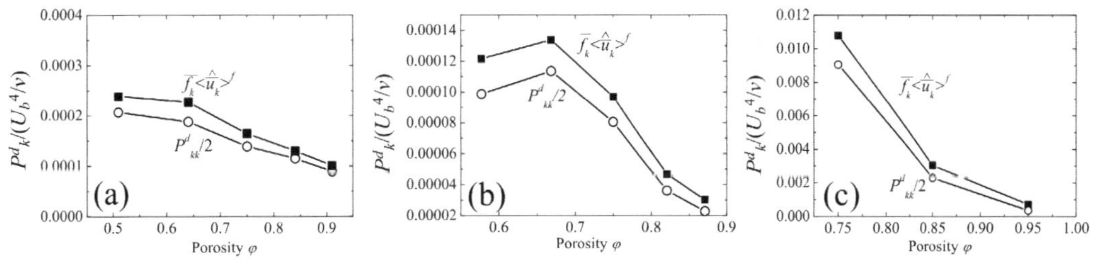

Figure 13. Production and drag terms in the porous media; (a) square rod arrays, (b) staggered cube arrays, (c) BCC foam.

the same and as the porosity increases, they tend to decrease while at $\varphi > 0.75$ the decreasing rate of P_{22}^d becomes lower. In the cube array flows of figure 12(b), P_{11}^d is always dominant while the other components are marginally small. In the BCC foam flows, P_{ij}^d is nearly isotropic as shown in figure 12(c). It is thus understood that the behavior of P_{ij}^d, which is an important term to be modeled, is so diverse depending on the structure of the porous media.

In the fully developed porous media, the following relations come out from the transport equations of the dispersive and micro-scale turbulent energy which are derived from equations (12) and (13):

$$\frac{D\mathscr{T}_{kk}/2}{Dt} \approx -P_{kk}^d/2 + \bar{f}_k \langle \hat{\bar{u}}_k \rangle^f - \mathscr{E}_{kk}/2 = 0, \quad \frac{Dr_{kk}/2}{Dt} \approx P_{kk}^d/2 - \varepsilon_{kk}^m/2 = 0, \tag{14}$$

since the macro-scale turbulence and P_{kk}^t are negligible. The drag force term $\bar{f}_k \langle \hat{\bar{u}}_k \rangle^f$ in equation (14) balances with the sum of the dispersive shear production P_{kk}^d and the dissipation $\mathscr{E}_{kk}$ terms. The behaviours of the drag force and the dispersive shear production terms are indicated in figure 13. Interestingly, the profiles of the dispersive production and the drag force terms look synchronized in each case. This fact indicates that the majority of the dispersive kinetic energy, which is proportional to the drag force, transfers to the micro-scale turbulence energy via the dispersive shear production term P_{kk}^d. Consequently, with the relation: $P_{kk}^d \approx \varepsilon_{kk}^m$, the transferred energy dissipates in the micro-scale turbulence. For the modeling issue, it suggests that the dispersive shear production term can be scaled by the drag force term $\bar{f}_k \langle \hat{\bar{u}}_k \rangle^f$ which is usually modeled using the Darcy-Forchheimer model.

5. Turbulence modelling

The present authors have been developing turbulence models considering the analysis of the LES data [14,15]. For validation of the developed models, comparison with the PIV data has been carried out. In the present paper, the development of a multi-scale four equation *k*-ε eddy viscosity (4 eqn. *k*-ε) model [15] is shown below.

5.1. Multi-scale four equation k-ε eddy viscosity model [15]

In the model, the eddy viscosity models are applied to $R^A{}_{ij}$ and r_{ij} as

$$R_{ij}^A = \frac{2}{3}k^A\delta_{ij} - \nu_t^A S_{ij},\ \nu_t^A = c_\mu f_\mu^A \frac{(k^A)^2}{\tilde{\varepsilon}^A},\ r_{ij} = \frac{2}{3}k_m\delta_{ij} - \nu_t^m S_{ij},\ \nu_t^m = c_\mu f_\mu^m \frac{k_m^2}{\tilde{\varepsilon}_m}, \tag{15}$$

where S_{ij} is the double averaged strain tensor: $S_{ij} = \partial\langle\bar{u}_i\rangle^f/\partial x_j + \partial\langle\bar{u}_j\rangle^f/\partial x_i - (2/3)\delta_{ij}\partial\langle\bar{u}_k\rangle^f/\partial x_k$.

The (total) turbulent energy $k^A(=R^A{}_{kk}/2)$, the isotropic part of its dissipation rate $\tilde{\varepsilon}^A$, the micro-scale turbulent energy $k_m(=r_{kk}/2)$ and the isotropic part of its dissipation rate $\tilde{\varepsilon}_m$ are obtained respectively by solving their transport equations. Note that the isotropic dissipation rate is defined as $\tilde{\varepsilon} = \varepsilon - 2\nu(\partial\sqrt{k}/\partial x_j)^2$. The coefficient c_μ and the damping functions $f_\mu{}^A$ and $f_\mu{}^m$ are from the Launder-Sharma *k*-ε model [21]: c_μ=0.09, $f_\mu^A = \exp[-3.4/(1+R_t^A/50)^2]$, $f_\mu^m = \exp[-3.4/(1+R_t^m/50)^2]$, $R_t^A = (k_m^A)^2/(\nu\tilde{\varepsilon}^A)$ and $R_t^m = k_m^2/(\nu\tilde{\varepsilon}_m)$.

The drag force term is modelled following the Darcy-Forchheimer term as

$$\bar{f}_i = \varphi\nu K^{ij}\langle\hat{\bar{u}}_j\rangle^f + \varphi^2 C_{ik}^F\sqrt{K^{kj}}\langle\hat{\bar{u}}_j\rangle^f\sqrt{(\langle\hat{\bar{u}}_l\rangle^f)^2}, \tag{16}$$

where $\hat{u}_i$ is the relative velocity to the solid phase, K^{ij} and $C^F{}_{ij}$ are the inverse matrix of the permeability tensor K_{ij} and the Forchheimer tensor, respectively. Since $\overline{f}_i$ balances with the pressure gradient of the momentum equation (4) in fully developed homogeneous porous medium flows, K^{ij} and $C^F{}_{ij}$ can be obtained by fitting with the pressure gradients in several different flow rate conditions for each porous medium.

5.1.1. Modelling the total turbulence. The modelled transport equations of k^A and $\tilde{\varepsilon}^A$ are

$$\frac{Dk^A}{Dt}=\frac{\partial}{\partial x_k}\left[\left(\nu+\frac{\nu_t^A}{\sigma_k}\right)\frac{\partial k^A}{\partial x_k}\right]-R_{ij}^A\frac{\partial\langle\overline{u}_i\rangle^f}{\partial x_j}+P_{k_m}^d-P_k^t+G_{k^A}^{\varphi}-\varepsilon^A, \tag{17}$$

$$\frac{D\tilde{\varepsilon}^A}{Dt}=\frac{\partial}{\partial x_k}\left[\left(\nu+\frac{\nu_t^A}{\sigma_\varepsilon}\right)\frac{\partial\tilde{\varepsilon}^A}{\partial x_k}\right]+F_\varepsilon+2\nu\nu_t^A\left(\frac{\partial^2\langle\overline{u}_i\rangle^f}{\partial x_j\partial x_k}\right)^2+\frac{c_{\varepsilon1}}{\tau^A}\left(-R_{ij}^A\frac{\partial\langle\overline{u}_i\rangle^f}{\partial x_j}+G_{k^A}^{\varphi}\right)+\frac{c_{\varepsilon2}f_{\varepsilon2}}{\tau^A}\left[\frac{\tau^A}{\tau_s}\left(P_{k_m}^d-P_k^t\right)-\tilde{\varepsilon}^A\right], \tag{18}$$

where $\sigma_k=0.5$, $\sigma_\varepsilon=0.65$, $c_{\varepsilon1}=1.44$, $c_{\varepsilon2}=1.92$ and $f_{\varepsilon2}=1-0.3\exp(-(R_t^A)^2)$. The time scales are $\tau^A=k^A/\tilde{\varepsilon}^A$ and $\tau_s=\ell_s/k_m^{1/2}$. The porous structural length scale is modelled as $\ell_s=(6.8-5\varphi)\sqrt{K_{kk}^*}$ where $K_{ij}^*=(K_{ik}L_{kj}+K_{jk}L_{ki})/2$, $L_{ij}=\ell_i\ell_j$, $\ell_i=\langle\hat{\overline{u}}_i\rangle^f/(\langle\hat{\overline{u}}_l\rangle^f\langle\hat{\overline{u}}_l\rangle^f)^{1/2}$. The hetero porous term $G_{k^A}^{\varphi}$ is modelled as

$$G_{k^A}^{\varphi}=\frac{\nu}{\varphi}\left(2\frac{\partial k^A}{\partial x_k}\frac{\partial\varphi}{\partial x_k}+k^A\frac{\partial^2\varphi}{\partial x_k^2}\right)+\frac{\nu_t^A}{\sigma_{k^A}}\frac{\partial k^A}{\partial x_k}\frac{\partial\varphi}{\varphi\partial x_k}. \tag{19}$$

For including the effects of the permeability, the term F_ε is introduced using the drag term F_k for the micro-scale turbulent kinetic energy as $F_\varepsilon=F_kc_{\varepsilon3}\left(\tau_m^{-1}-\tau_M^{-1}\right)$ where the time scales are $\tau_m=k_m/\tilde{\varepsilon}_m$ and $\tau_M=k_M/\tilde{\varepsilon}_M$ with $k_M=k^A-k_m$ and $\tilde{\varepsilon}_M=\tilde{\varepsilon}^A-\tilde{\varepsilon}_m$. The model coefficient $c_{\varepsilon3}$ is $c_{\varepsilon3}=6.5[1-\exp\{-(R_t^A/25)^2\}]\times[1-\exp\{-(R_{K^*}/9)^2\}]$ with the effective permeability Reynolds number $R_{K^*}=\sqrt{K_{kk}^*}\sqrt{k_M}/\nu$. With the assumption that the mean shear is assumed to be negligible in the REV, $P_{k_m}^d-P_k^t=\left(P_{jj}^d-P_{jj}^t\right)/2$ is estimated as $P_{k_m}^d-P_k^t\simeq\overline{f}_j\langle\hat{\overline{u}}_j\rangle^f-\mathscr{E}$. The model of the dissipation $\mathscr{E}$ of the dispersive kinetic energy $\mathscr{K}$ is shown later.

5.1.2. Modelling the micro-scale turbulence. The modelled transport equations of k_m and $\tilde{\varepsilon}_m$ are

$$\frac{Dk_m}{Dt}=\frac{\partial}{\partial x_k}\left[\left(\nu+\frac{\nu_t^m}{\sigma_k}\right)\frac{\partial k_m}{\partial x_k}\right]-r_{ij}\frac{\partial\langle\overline{u}_i\rangle^f}{\partial x_j}+P_{k_m}^d-C_k^t+F_k+G_{k_m}^{\varphi}-\varepsilon_m, \tag{20}$$

$$\frac{D\tilde{\varepsilon}_m}{Dt}=\frac{\partial}{\partial x_k}\left[\left(\nu+\frac{\nu_t^m}{\sigma_\varepsilon}\right)\frac{\partial\tilde{\varepsilon}_m}{\partial x_k}\right]+2\nu\nu_t^m\left(\frac{\partial^2\langle\overline{u}_i\rangle^f}{\partial x_j\partial x_k}\right)^2+\frac{c_{\varepsilon1}}{\tau_m}\left(-r_{ij}\frac{\partial\langle\overline{u}_i\rangle^f}{\partial x_j}+G_{k_m}^{\varphi}+\frac{c_{\varepsilon1}'\tau_mF_k}{\tau_M}\right)+\frac{c_{\varepsilon2}f_{\varepsilon2}^m}{\tau_m}\left[\frac{\tau_m}{\tau_s}\left(P_{k_m}^d-C_k^t\right)-\tilde{\varepsilon}_m\right], \tag{21}$$

where $c_{\varepsilon1}'=0.694$ and $f_{\varepsilon2}^m=1-0.3\exp(-(R_t^m)^2)$. The macro-micro turbulence cascade term and the mean dispersive production term are jointly modelled as

$$P_{k_m}^d-C_k^t=\overline{f}_j\langle\hat{\overline{u}}_j\rangle^f-\mathscr{E}-[1-\exp(-R_t^M/100)]\left\{0.2(\mathscr{E}+\varepsilon_m)-0.2(F_k+\overline{f}_j\langle\hat{\overline{u}}_j\rangle^f)-(\mathscr{T}_{ik}+r_{ik})\frac{\partial\langle\overline{u}_i\rangle^f}{\partial x_k}\right\}, \tag{22}$$

where $R_t^M=k_M^2/(\nu\tilde{\varepsilon}_M)$. The models for the drag force term F_k and the hetero porous term $G_{k_m}^{\varphi}$ are

$$F_k = \varphi \nu K^{ij} R_{ij} + \varphi^2 C_{ij}^F \sqrt{K^{jk}} R_{ki} \sqrt{\left(\langle \overline{u}_l \rangle^f\right)^2}, \quad G_{k_m}^{\varphi} = \frac{\nu}{\varphi} \frac{\partial (k^A + k_m)}{\partial x_k} \frac{\partial \varphi}{\partial x_k} + \frac{\nu}{\varphi}\left(2k_m - k^A\right) \frac{\partial^2 \varphi}{\partial x_k^2} + \frac{\nu_t^m}{\sigma_k} \frac{\partial k_m}{\partial x_k} \frac{\partial \varphi}{\varphi \partial x_k}. \quad (23)$$

5.1.3. Modelling the dispersive covariance. The dispersive covariance $\mathcal{T}_{ij}$ is the volume averaged second moment of the Reynolds averaged velocity dispersions and appears in the double averaged momentum equation irrespective of laminar and turbulent flows. However, this second moment has been often ignored and attributed to be included in the modelled Darcy-Forchheimer term, or it has been treated together with the turbulent kinetic energy. The present authors modelled the dispersive covariance through phenomenological considerations on the LES results [12,13] as follows.

In the local equilibrium condition in the REV of homogeneous porous media, the mean shear may be assumed to be ignored. Indeed, such a condition can be usually found deep inside a porous medium. Thus, the transport equations of k^A and the dispersive kinetic energy $\mathcal{K} = \mathcal{T}_{kk} / 2$ reduce to

$$\frac{Dk^A}{Dt} = P_{k_m}^d - P_k^t - \varepsilon^A = 0, \quad \frac{D\mathcal{K}}{Dt} = \overline{f}_j \langle \hat{\overline{u}}_j \rangle^f - (P_{k_m}^d - P_k^t) - \mathcal{E} = 0. \quad (24)$$

In such a flow condition, the LES results confirmed that the micro-scale turbulence was often dominant compared to the macro-scale turbulence. Hence, $\varepsilon^A \simeq \varepsilon_m$ and from equation (24), $\mathcal{E} + \varepsilon_m \simeq \overline{f}_j \langle \hat{\overline{u}}_j \rangle^f$. This combined dissipation of the dispersion energy is scaled by the porous structural length scale ℓ_s as $\mathcal{E} + \varepsilon_m \simeq \mathcal{K}^{1.5} / \ell_s$ and then the dispersive kinetic energy is obtained. As for the dissipation term $\mathcal{E}$, since it does not include velocity fluctuations and is considered to be a balancing term for the viscous drag force. Hence,

$$\mathcal{K} = \left(\ell_s \overline{f}_j \langle \hat{\overline{u}}_j \rangle^f\right)^{2/3}, \quad \mathcal{E} = \varphi \nu K^{ij} \langle \hat{\overline{u}}_i \rangle^f \langle \hat{\overline{u}}_j \rangle^f. \quad (25)$$

Since in those flow conditions, the transport equation (12) of the dispersive covariance indicates that the main production of the dispersive covariance is by the drag term $\mathcal{F}_{ij} = \overline{f}_i \langle \hat{\overline{u}}_j \rangle^f + \overline{f}_j \langle \hat{\overline{u}}_i \rangle^f$ while $(P_{ij}^d - P_{ij}^t)$ transfers a part of energy to turbulent Reynolds stress while the dissipation term $\mathcal{E}_{ij}$ does not include turbulence effects. The dispersive covariance is thus phenomenologically modelled as the sum of two parts which include the effects of the drag force and structural contributions, respectively. They are thus modelled as

$$\mathcal{T}_{ij} = \mathcal{T}_{ij,1} + \mathcal{T}_{ij,2}, \quad \mathcal{T}_{ij,1} / \mathcal{K} = 2\beta \mathcal{F}_{ij} / \mathcal{F}_{kk}, \quad \mathcal{T}_{ij,2} / \mathcal{K} = 2(1-\beta) C_{ij}^{Fs} / C_{kk}^{Fs}. \quad (26)$$

The modified structural part of the Forchheimer tensor is $C_{ij}^{Fs} = C_{ij}^{Fl} - C_{ij}^{Fl*}$ and $C_{ij}^{Fl*} = (C_{ik}^{Fl} L_{jk} + C_{jk}^{Fl} L_{ik}) / 2$ where C_{ij}^{Fl} is the molecular part of the Forchheimer tensor obtained only in laminar flows. The bridge function is modelled as $\beta = [1 + 2\exp(-6C_{kk}^{Fl*})] / 3$.

5.1.4. For interface regions. In the interface regions between porous media and clear fluids, the porosity distribution is modelled as $\varphi = \varphi_\infty + (1-\varphi_\infty)\exp(-N_\varphi y' / D_p)$, where φ_∞ is the porosity of porous media, and y' is the normal distance from the porous surface. The coefficient N_φ is arranged so that the porosity varies within the distance of the mean pore diameter D_p. Hence, N_φ =4 is applied. In the clear flow region, the permeability tensor and the Forchheimer tensor become infinite and zero, respectively. To impose those conditions, the permeability tensor and the Forchheimer tensor are respectively modified as $K_{ij} = K_{ij}^\infty / f_k$, $C_{ij}^F = f_{C_F} C_{ij}^{F\infty}$ where K_{ij}^∞, $C_{ij}^{F\infty}$ are the permeability and Forchheimer tensors of the homogeneous porous media. The model functions are designed to vary between 0 and 1 within the distance of the mean pore diameter D_p:

$$f_k = 1 - \exp[-0.08(y' / \sqrt{K^*_{jj}})^2], \quad f_{C_F} = 1 - \exp[-1.41(y' / \sqrt{K^*_{jj}})^{1/2}]. \tag{27}$$

5.2. Validation results

5.2.1. Homogeneous porous media. Figure 14 shows comparison between the LES and the 4 eqn. k-ε model prediction results of the volume averaged turbulent energy and its dissipation rate in cases SQR, CUB and BCC for various porosity conditions. It is clearly seen that the results of the 4 eqn. k-ε model well accord with the microscopic LES in all the test flow conditions. Although the model of Ref.[7] (NK08) predicts general tendency reasonably, it overpredicts turbulent energy at lower porosity conditions in cases SQR and BCC. Figure 15 compares the dispersive covariance. Note that in the test flow field geometries, the dispersive covariance tensor is diagonal. Hence, off-diagonal components do not exist. It is clear that the magnitudes of the dispersive covariance are comparable to those of the turbulent energy. The dispersive covariance is thus never negligible in any cases whilst it has not been carefully treated in many other model proposals. As can be seen, the form applied in the 4 eqn. k-ε model for the dispersive covariance performs satisfactorily, though the agreement is not perfect. It is clear that the dispersive covariance does not become isotropic even in the isotropic porous media of cases CUB and BCC.

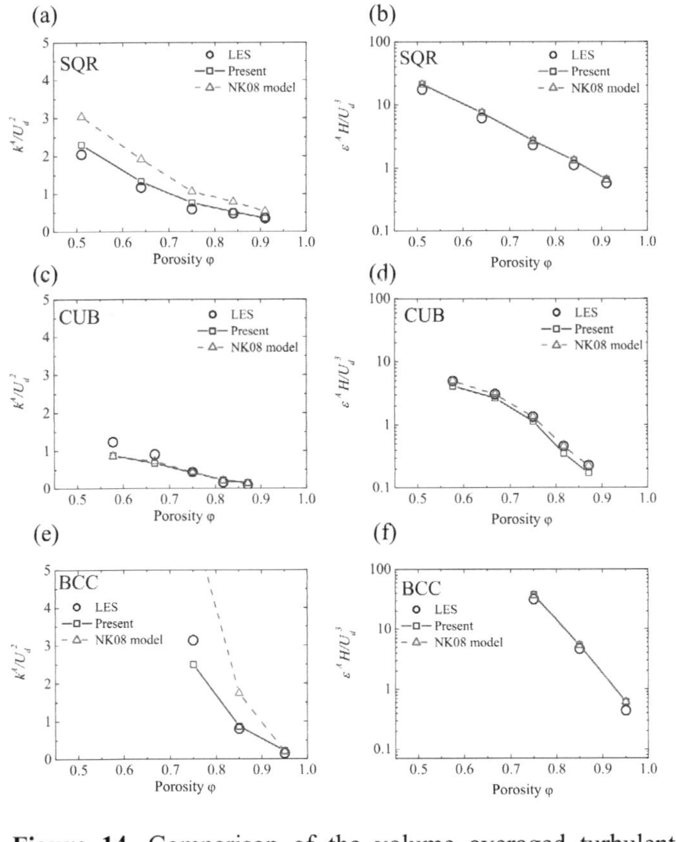

Figure 14. Comparison of the volume averaged turbulent energy and its dissipation rate between the macroscopic and LES computations; (a,b) square rod arrays (SQR) at Re_h=3500, (c,d) cube arrays (CUB) at Re_h=3000, (e,f) BCC foam (BCC) at Re_D=1000.

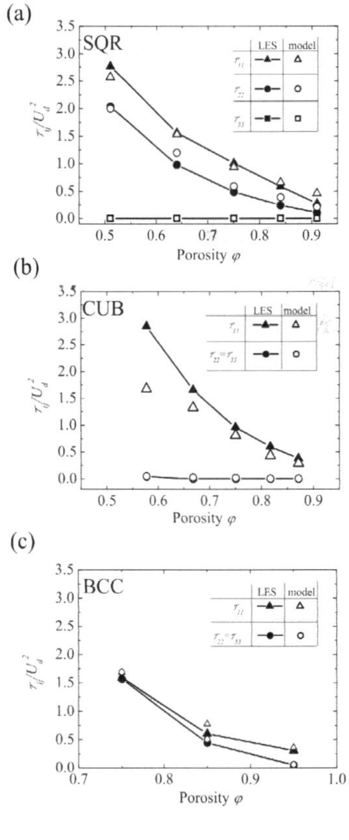

Figure 15. Comparison of the dispersive covariance between the model predictions and the LES; (a) case SQR, (b) case CUB, (c) case BCC.

5.2.2. Porous channel flows. Figures 16 and 17 compare the mean velocity and turbulent quantity profiles of the 4 eqn. *k-ε* model with those of the porous channel experiments. In those flows over porous media, flow characteristics are characterised by the permeability Reynolds number Re_K. The experiments showed that porous wall flows had two distinctive flow regimes at $Re_K < 3$ and $Re_K >> 3$. As shown in figure 16, the overall agreement in the mean velocities of the present results and the data is reasonable. In cases #20, #13 and #06, the permeability Reynolds numbers are respectively $Re_K = 1.1$, 3.1 and 11 which are from the transitional regime to the full porous wall turbulence regime. As Re_K increases, the mean velocity and the turbulent energy profiles tend to penetrate into the porous media and a larger slip velocity can be seen. The agreement in turbulence quantities shown in figure 17 is also acceptable though the peak values of the turbulent energy near the surfaces are not well predicted. Since the Launder-Sharma model, which is the mother model of the 4 eqn. *k-ε* model, also has similar performance, it is considered that its defect carries over to the proposed model.

5.2.3. Rib-mounted porous channel flows. Figure 18 (a)-(c) compares the streamwise mean velocity, turbulent energy and Reynolds shear stress profiles of the porous-rib-mounted porous channel flow. The predicted profiles by the 4 eqn. *k-ε* model well agree with those of the experiments. It is considered that the flow rate going through the rib increases as the increase of the permeability. This tendency was well captured by the 4 eqn. *k-ε* model since the agreement in the velocity distributions just behind the rib is satisfactory although different wall permeability cases are not shown here. The experiments showed that the turbulent energy and the Reynolds shear stress in the downstream region became smaller in the higher permeability case. The 4 eqn. *k-ε* model predicts this tendency well.

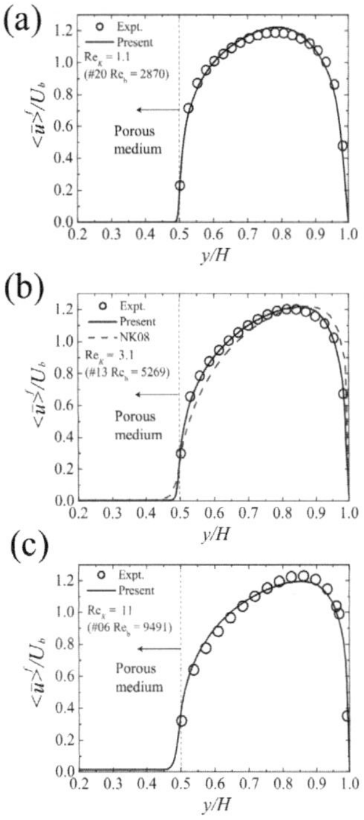

Figure 16. Comparison of streamwise mean velocity profiles between the prediction and the experiments; (a) case #20, (b) case #13, (c) case #06.

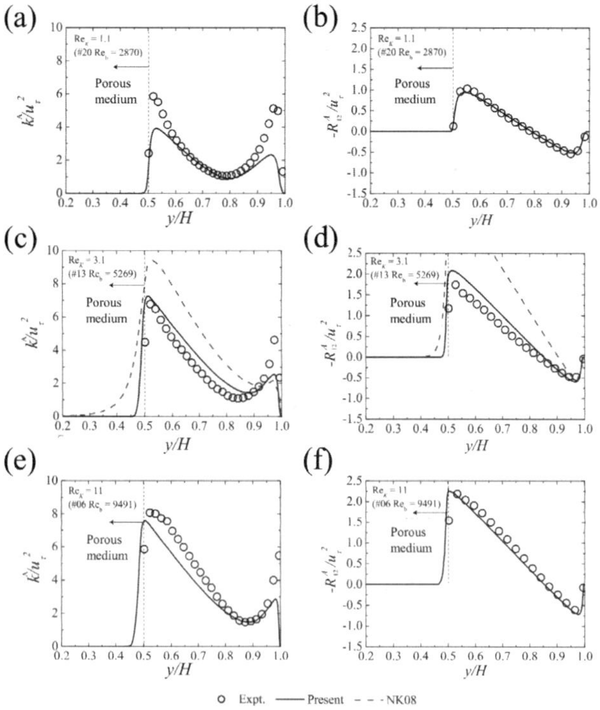

Figure 17. Comparison of turbulent energy and Reynolds shear stress between the prediction and the experiments; (a,b) case #20, (c,d) case #13, (e,f) case #06.

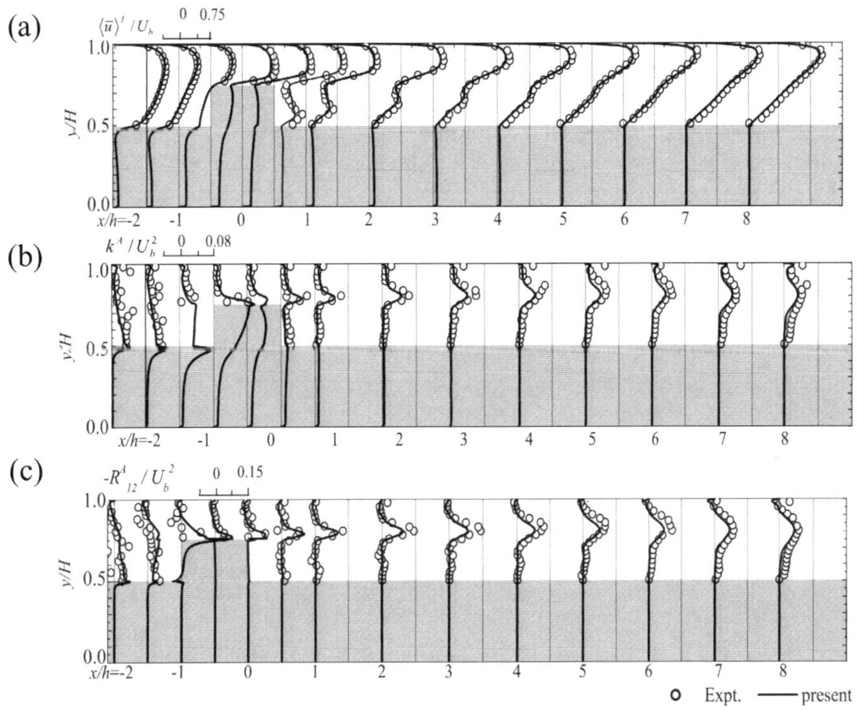

Figure 18. Comparison of the mean velocity and turbulence profiles with the rib-mounted porous channel experiments in case #06 at Re_b=10600; (a) streamwise mean velocity, (b) turbulent energy, (c) Reynolds shear stress.

6. Conclusions

By the PIV experiments, the following points were confirmed. Above a permeable wall, transition to turbulence occurs at progressively lower Reynolds numbers as the wall permeability increases. The magnitude of the slip velocity on the porous wall depends on Re_K. It drastically increases in a narrow range of Re_K <3 corresponding to laminar to turbulent transition. The flows are not in the condition comparable to the full roughness regime due to the effects of the wall permeability superimposed on the roughness effects. The wall normal component of the velocity fluctuation near a porous wall tends to be higher as the increase of the wall permeability and the Reynolds number though the streamwise component does not have such an obvious tendency. Since blocking effects of a porous wall on vortex motions are weakened more as Re_K increases, the wall normal fluctuation is kept at a higher level. This contributes to a higher level of the turbulent wall shear stress at a porous wall.

The measurement results of the rib-mounted porous channel flows confirmed the following points. In the solid rib-mounted porous channel flows, since a part of entraining fluid is supplied through the permeable bottom wall from the region upstream the rib, the recirculation behind the rib in the clear channel becomes smaller and eventually vanishes as the wall permeability increases. Because of the reduction of the magnitude and the size of the reverse flow region in the clear channel, turbulence becomes smaller. In the porous-rib-mounted porous channel flows, because of the fluid passing through the rib, the recirculation is more significantly weakened and shifts downstream. Because of the flow through the porous rib, turbulence becomes smaller than those of the solid-rib flows.

From the LES for porous medium flows, the following points were obtained. The profiles of the dispersive production and the drag force terms of the dispersive kinetic energy are synchronized. This indicates that the majority of the dispersive kinetic energy, which is proportional to the drag force, transfers to the micro-scale turbulence energy via the dispersive shear production term. Consequently, the transferred energy dissipates in the micro-scale turbulence. For the modelling issue, it suggests that the dispersive shear production term may be scaled by the drag force term.

To predict turbulence around and inside porous media, three kinds of second moments: the dispersive covariance, the volume averaged (total) Reynolds stress and the micro-scale Reynolds stress were individually modelled in the four equation k-ε model. In the proposed model, two sets of the modelled k and ε transport equations are solved for the eddy viscosity models of the total and the micro-scale Reynolds stresses. For the dispersive covariance, an algebraic form is applied after the discussions using the LES results. The validation in fully developed flows in homogeneous porous media, porous channel flows and porous rib mounted channel flows confirms that the proposed method is promising. Indeed, the model predicts the turbulent and dispersive kinetic energy profiles well in three kinds of porous media in a wide range of porosity. The results of the porous channel flows show that the prediction accuracy is satisfactory. The overall agreement between the prediction and the experiments is also satisfactory in the porous-rib-mounted porous channel flows.

Acknowledgments

K. S. thanks his former research students and colleague: Y. Matsumura, S. Tominaga, M. Mori and Dr. M. Kaneda, for their invaluable contributions to the studies summarized in this paper. A part of this study was financially supported by the research Grant (No. 24360073) of the JSPS Japan.

References

[1] Nield D A and Bejan A 2006 *Convection in Porous Media* 3rd ed (Berlin: Springer)
[2] de Lemos M J S *Turbulence in Porous Media* 2006 (Oxford: Elsevier)
[3] Zagni A F E and Smith K V H 1976 *J. Hydraul. Div* **102** 207
[4] Zippe H J and Graf W H 1983 *J. Hydraul. Res.* **21** 51
[5] Breugem W P, Boersma B J and Uittenbogaard R E 2006 *J. Fluid Mech.* **562** 35
[6] Pedras M H J and de Lemos M J S 2001 *Int. J. Heat Mass Transfer* **44** 1081
[7] Nakayama A and Kuwahara F 2008 *J. Fluids Eng.* **130** 101205
[8] Hanjalić K and Launder B E 2011 *Moldelling Turbulence in Engineering and the Environment: Second Moment Routes to Closure* (Cambridge: Cambridge Univ. Press)
[9] Suga K, Matsumura Y, Ashitaka Y, Tominaga S and Kaneda M 2010 *Int. J. Heat Fluid Flow* **31** 974
[10] Suga K, Mori and Kaneda M 2011 *Int. J. Heat Fluid Flow* **32** 586
[11] Suga K, Tominaga S, Mori M and Kaneda M 2013 *Flow Turb. Combust.* **91** 19
[12] Kuwata Y and Suga K 2013 *Trans. Japan Soc. Mech. Eng.(B)* **79** 1752
[13] Suga K and Kuwata Y 2013 *Proc. 1st Int. Conf. on Mechanics (Hsinchu, Taiwan, 8-9 Nov. 2013)*
[14] Kuwata Y and Suga K 2013 *Int. J. Heat Fluid Flow* **43** 35
[15] Kuwata Y, Suga K and Sakurai Y 2014 *Int. J. Heat Fluid Flow* doi:10.1016/j.ijheatfluidflow.2014.02.007
[16] Whitaker S 1996 *Transp. Porous Med.* **25** 27
[17] Pedras M H J and de Lemos M J S 2000 *Int. Commun. Heat Mass Transfer* **27** 211
[18] Nicoud F and Ducros F 1999 *Flow Turb. Combust.* **62** 183
[19] d'Humiéres D, Ginzburg I, Krafczyk M, Lallemand P and Luo L-S 2002 *Phil. Trans. R. Soc. A* **360** 437
[20] Takashima K, Kaneda M and Suga K 2012 *Proc. Int. Coput. Mech. Symp., JSME-CMD, (Kobe, Japan, 9-11 Oct. 2012)* paper# MS6-3-3(88)
[21] Launder B E and Sharma B I 1974 *Lett. Heat Mass Transfer* **1** 131

Air and blood fluid dynamics: at the interface between engineering and medicine

A. Pollard, F. Secretain and B. Milne*

Department of Mechanical and Materials Engineering, Queens University at Kingston,
*Dept. Anesthesiology, Kingston General Hospital and Queen's University at Kingston,
Ontario K7L-3N6, Canada

E-mail: pollard@me.queensu.ca

Abstract. The flows in the human upper airway and human heart during open heart surgery are considered. Beginning with idealized models of the human upper airway, current methods to extract realistic airway geometries using a novel implementation of optical coherent tomography modality are introduced. Complementary direct numerical simulations are considered that will assist in pre-surgery planning for obstructive sleep apnea. Cardiac air bubbles often arise during open heart surgery. These bubbles are potential emboli that can cause neurological impairment and even death. An experimental programme is outlined that uses acoustic sound to instil bubble surface oscillations that result in bubble breakup. A novel algorithm is introduced that enables a surgical team to obtain real-time *in-vivo* bubble data to aid cardiac de-airation procedures.

1. Introduction

Human bodies have multiple systems that rely on fluid transport; here, the pulmonary and arterial systems are considered with specific attention on the upper airway and the heart. In the case of the former, the fluid dynamics of air from the entrance of the oral cavity to at least the first bronchus is important because portions of it act as a vehicle for the delivery of nutrients and air as well as to enable speech. In the case of arterial systems, the heart is the major organ and during open-heart surgery air bubbles maybe introduced that have the potential to embolise. These emboli have been shown to be strongly correlated with neurological dysfunction and even death. In this paper, the fluid dynamics associated with these two physiologically critical systems are explored using theory, experiment and computations.

Note that given the breadth of the subject matter considered here, many details are not included and a reader is directed to the relevant literature.

The paper is laid out in the following manner. The human airway is first considered. Here, the research conducted in our laboratory is traced from introducing an idealized human upper airway and how we investigated the flow both computationally using lattice Boltzmann methods and RANS and experimentally using flow visualization, hot wire anemometry and particle image velocimetry. Current efforts are introduced to explore realistic upper airway geometries and a new endoscopic-type tool called long distance optical coherency tomography (OCT) that is under development by a group at the University of California, Irvine to help characterize the geometry of specific human upper airways. The images obtained can then be translated into a stereo-lithographic file (.stl), which, when imported into a grid-generator enables fluid modelling and direct numerical simulations to be performed. Importantly, OCT includes the ability to image paediatric patients who may have breathing difficulties due to pathological issues (Down's syndrome *etc.*). The second part of the paper considers the bubbles that are often encountered during cardiac surgery and developments in our laboratory, with reference to clinical practice, are introduced to both characterize the number and sizes of these bubbles in real time as well as to introduce a novel bubble destruction approach. This destruction depends on initiating air/liquid interfacial instabilities so as to cause larger bubbles to disintegrate into a cloud of smaller bubbles with the desire to make such bubbles less than 40 μm in diameter so they can pass harmlessly through the arterial system.

2. Fluid Dynamics in both idealised and real forms of the upper human airway

The human airway, defined here to include only the oral cavity, the non-nasal pharynx, the larynx and trachea, is a complex geometry. The rather tortuous pathway leading from the mouth to the lower trachea, however, provides lower resistance than breathing through the nasal orifice and thus greater volumes of oxygen/carbon dioxide can be exchanged during the inhalation-exhalation breathing cycle. Reynolds numbers within the human airway can vary $650 \leq Re \leq 13{,}000$ depending on inhalation flow rate (typically between 10 and 120 litres per minute) and location in the airway, see [21].

The fluid mechanics in the nasal passages is complicated by the turbinates, which maximise exposure of these tissues to the incoming air, thereby enabling the air to be raised in temperature, humidified and cleansed prior to exiting the nasal-pharynx, which then passes over the epiglottis and through the larynx before entering the trachea and eventually the lungs. The Reynolds number in the nasal passages is of order 1000, Doorly *et al.* [8]; while this suggests inertial dominance, there are regions in the turbinates where the flow velocities are very much lower so that diffusion controlled mass transfer takes place to the olfactory glands. In this work, the flow in the nasal passages is referred to, but is otherwise ignored; interested readers may consult that literature (see for example, Doorly *et al.* [8, 9]; Shi *et al.* [40]; Xi and Longest [47].

Physiologically, the upper respiratory tract is important because it serves as the sole route for nutrient delivery (oxygen is carried to the lungs, food is diverted through "swallowing", which engages the epiglottis, from the pharynx to the stomach via the oesophagus) as well as speech. The lower airway helps maintain homeostasis (through oxygen/carbon-dioxide exchange, acid-base balance, *etc.*), while guarding against infection (moist mucous membranes and nasal hairs trap large particles, the mucociliary escalator removes small particles, and the pharyngeal lymphoid ring guards against pathogens). Finally, it is an important site for delivery of medications (oral, sub-lingual, intra-nasal, and inhaled).

In the past, the majority of efforts have been directed to understand the basic fluid mechanics in the airway to eventually improve drug delivery to the lungs. A new, emerging driver is even more challenging: improve breathing patterns in clinically abnormal airways. This is an interdisciplinary research objective to integrate novel experimental modalities with fluid dynamic simulations to effect rapid resolution of these breathing difficulties by guiding a surgical team to maximize the benefits while minimizing failure, see for example, [22] and [46]. However, before these new directions are considered, the basic fluid dynamics of the human upper airway will be surveyed.

The human airway, as with all physiology, evolves with time. The sizes and flexibility of the airway passages change with age and disease. For example, Heenan *et al.* [17] obtained MRI scans of two subjects and due to their physiological differences, refer to them as big nose-small mouth and small nose-big mouth due to their presentation. In addition, delivery of aerosols to the lungs from an inhaler often results in poor drug delivery either by the user from improper use, particulate impaction in the throat or in the case of pathological/physiological condition, asymmetric delivery to the lungs [25].

Historically, and because of the huge variability in human airway physiology, many simplifications have been introduced when the human airway fluid dynamics are explored, which include simple curved tubes to current work on an idealised model geometry. The idealised geometry continues to be used to help establish some of the basic fluid dynamic mechanisms that should be expected in realistic geometries. Once understood, then perhaps the variations encountered from replacing idealised with real geometries will assist in our broader understanding of the effects of, for example, tissue compliance, mucous, oxygen/carbon dioxide exchange etc. as will be explored shortly.

2.1. Idealised Geometry

Stapleton *et al.* ([13]) were the first to introduce an idealised geometry that encompassed all the basic attributes of a human airway. Heenan *et al.* [17] and others (see for example [24], [5]) continue to use this idealised model, or slight variants of it to explore the rich flow features. In my group, we have considered both computational and experimental investigations using a slight variant of the University of Alberta model (Stapleton *et al.*) and a brief summary of major findings are now provided.

Johnstone *et al.* [21] performed initial flow visualization and hot wire anemometry (and endoscopic PIV, [20]) studies using steady flow over inhalation rates $10 \leq m \leq 120$ litres per minute (lpm), which are encountered physiologically. Their flow visualisation revealed complex secondary flows both in the sagittal and cross sagittal (that is transverse and coronal) planes. The hot wire data, taken only at the sagittal plane, indicated significant changes to the turbulence intensities and Reynolds stresses as flow traversed the large oral cavity and through the naso-pharynx due to the curvature and constriction of the airway. Once the epiglottis is encountered, these quantities reflected the emergence of a shear layer that gives rise to a laryngeal jet. Within these sections, from the oral cavity to beyond the larynx, flow recirculation regions that are oriented parallel to the main flow in the airway, and identified from the flow visualisation, produced data that were deemed questionable because of the directional insensitivity of the hot wires. Ball *et al.* performed RANS [2] and direct numerical simulation [3] using the same idealized geometry as Johnstone *et al.*, but only for inhalation flow rates of 10 litres per minute. RANS predictions were tolerably good compared with the hot wire data, while the DNS, which used the lattice Boltzmann method DNS provided better agreement with those same data. Subsequently, Pollard *et al.* [34] provided an overview of the flow in the idealized airway, and drew attention to the need for careful correlation between experiments and simulation due to the inherent resolution and uncertainty issues from both simulations, modelling and experiment.

Interestingly, what emerged from Ball *et. al.* [3] was the complex flow around the epiglottis. The flow around the epiglottis is preceded by its passage upstream, which had negotiated curvature and flow acceleration in the naso-pharynx region. The flow is then intercepted by the sharp, flap-like Bourda-like mouth-piece of the epiglottis that then creates a shear layer, from which the flow then evolves into what had been previously observed as a laryngeal jet. However, what had not been previously observed was the resulting complex axial helical motion of the flow distal to the epiglottis that then created an asymmetric flow that enters the first bronchus. This means that it is incorrect to assume fully developed pipe-flow both in the mean and time dependent flow structure if computational studies of flow to the lungs is contemplated.

Shinneeb *et al.* [42], [41] performed the first particle image velocimetry using a clear silicon version of the idealized model. Their data were taken in both the sagittal and trans-sagittal planes, and focussed particularly on the epiglottal region, see figure 2. They used a water-glycerin mixture with the flow rate adjusted to give the equivalent of 10 litres per minute in the airway. They determined that the epiglottis produces significant vortex shedding, see figure 3 and determined "*The flow contains a large number of counter-rotating vortical structures that are distributed throughout the pharynx/larynx region at all times particularly whose axes are*

Journal of Physics: Conference Series **530** (2014) 012005 doi:10.1088/1742-6596/530/1/012005

normal to the sagittal (x-y) plane. These structures are very energetic and subject to deformation by other structures that present in that region; and sic, that the pharynx/larynx region is characterised by bursting events (e.g. ejection, sweep, and interaction events) particularly in the epiglottal region. These events appear to be responsible for deforming and/or tearing apart the vortical structures. In addition, the limited space of the flow field in the epiglottal region as well as the relatively large number and size of the structures that present there appear to enhance the occurrence of the merging process."

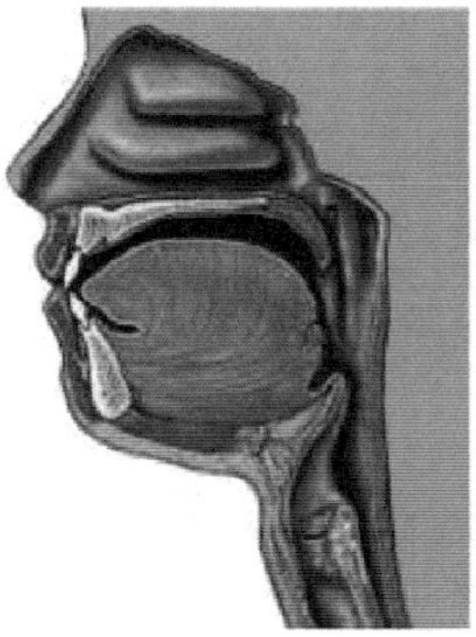

Figure 1. Sagittal view of human airway geometry, including nasal passages

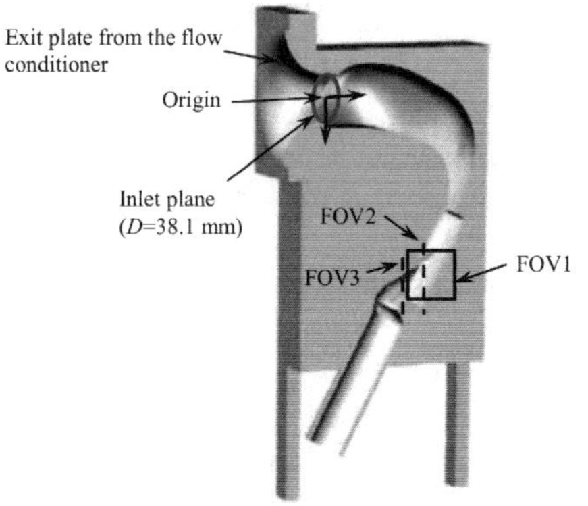

Figure 2. Saggital view of idealized human airway geometry used for PIV measurements, without nasal passages. Planes of interest in epiglottal region fields of view (FOV) [42]

The most recent work on the idealized geometry is due to Kleinstreuer *et al.* ([24]) and has provided a review of flow in the idealized airway, and complemented existing data using results from large eddy simulation.

2.2. Real geometries

The actual geometry of the human airway is highly variable because of patient specific physiology, age-dependent and pathological conditions. As referred to previously, Heenan *et al.* (2004) [17] considered two realistic geometries, derived from magnetic resonance imaging (MRI), computerized tomography (CT) to explore deposition of aerosol in the airway. These two geometries were referred to as large mouth (small nose) and small mouth (large nose) are examples of the high variability that is encountered, see figures 4 and 5. In fact, these were from healthy subjects; in the case of diseased airways, these geometries may be even more complex. More recently, de Rochefort *et al.* [7] noted in their study of the lower human airway "Finally, this combined numerical and experimental approach of flow assessment in realistic *in vivo*-based human airway geometries confirmed the strong dependence of airway flow patterns on local and global geometrical factors". While this statement confirms the expected observation that flow feature variations should be expected between subjects, the authors

Figure 4. Small mouth geometry obtained by MRI [17].

Figure 5. Large mouth geometry obtained by MRI [17].

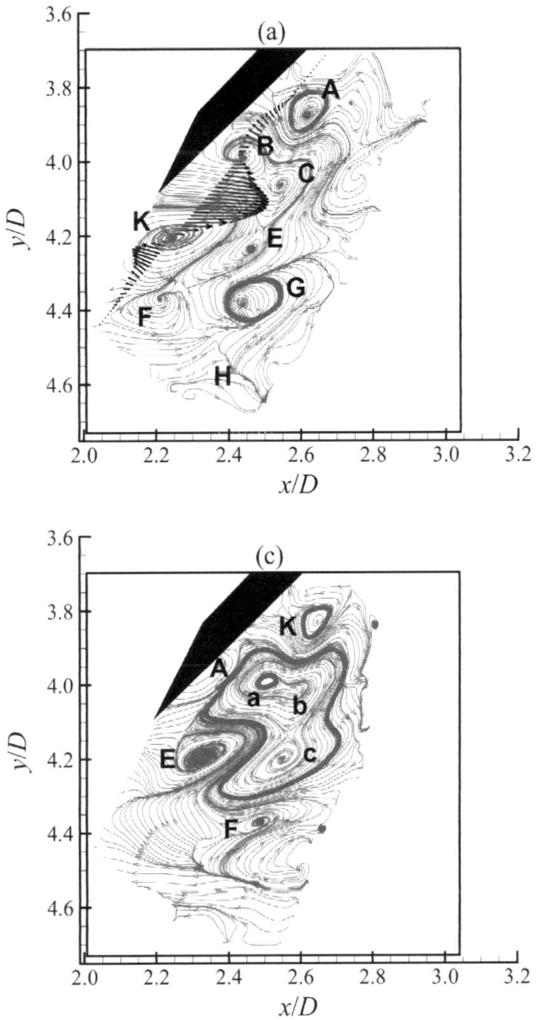

Figure 3. Four examples of POD-reconstructed fluctuation velocity fields (a), (b), (c), and (d) show vortical structures identified on the x-y plane along z/D=0. The turbulent kinetic energy recovered in these fields is 62.1% using 3 modes. Slice is from FOV1 of figure 2 [42]

seem to not appreciate the diminishing local
effects of the time dependent inlet conditions, distal to the larynx and upstream from the first bronchus, with descent from the first to the fourth bronchial branch.

2.2.1. Computational and experimental investigations of flow in real geometries While Ball *et al.* [2], [3] performed both RANS and direct numerical simulation (through the use of the lattice Boltzmann method) inside the idealized geometry, recent work has shifted to real geometries. Clearly, the complex geometries and relatively low flow rates and local Reynolds numbers suggest the use of DNS, many others continue to apply RANS as well as large eddy simulation, see for example Wootton *et al.* [46] and Mihaescu *et al.* [31]. The most recent application of DNS to real geometries has been the work of Nicolaou and Zaki [32] and Wong and Elghobashi [45]. In the former, the realistic geometries used by [17] and Grgc *et al.* [18], see figures 4 and 5, were incorporated into a structured finite-volume code, while in the latter, the lattice Boltzmann method was applied to two subjects one healthy and the other with an obstructed airway. In the geometries considered, a main conclusion was that the flow features in these geometries are more complicated than those encountered in idealized geometries while many of the main features of the flow remain unchanged.

A novel experimental method that has emerged to help interrogate flows in the airway is an adaptation of magnetic resonance imaging (MRI) to magnetic resonance velocimetry or MRV, see [12], [11]. MRI uses the frequency of excited hydrogen nuclei when they are exposed to intense magnetic fields. This technique should be able to provide the same type of data obtained using holographic PIV. However, it does not require either optical access, seeding or post processing. While the method has been considered for use as a flow investigative tool, it is limited to the use of non-magnetic models and can only, at this time, provide time average velocities. While hyperpolarised ^3He has been used *in-vitro* for lower airway velocity characterization, [7], it remains to be seen if this approach can be incorporated into the arsenal of modalities for rapidly assisting a surgical team. However, as will be noted below, there maybe more optimal and convenient modalities available.

2.2.2. Pre-surgery planning The human airway is also subject to disease and obstruction, which leads to breathing and nutrient delivery difficulties. Common issues include tonsillitis, adenoiditis (hypertrophy of the pharyngeal tonsils) and other types of inflammation. Sleep apnea is also readily encountered in the general population (about 5%), with obstructive sleep apnea (OSA) being an additional unpleasant complication (*e.g.* snoring). As noted above, CT and MRI maybe used to obtain images of the airway. However, these modalities have drawbacks when applied to OSA, especially for paediatric subjects. Moreover, these cannot identify the pathogenesis of the obstruction.

A new tool has emerged to help decrease the difficulty in acquiring requisite images. This is called optical coherence tomography (OCT), and long range OCT in particular, Jing *et al.* [22]. It uses near infrared light to obtain high resolution (of order 10 microns) images of tissue. A recent evolutionary step in this technology has enabled the frequency of image capture to be increased from between 1 to 5 frames per second to about 25. It uses an flexible endoscope (as opposed to the fixed, single fibre endoscope used by Heenan *et al.* [21]). Laser light is fed through an optical fibre to a mirror with a 45° surface, which rotates at 25 Hz. The diameter of the unit is less than $1mm$. The LROCT is positioned *in vivo* through the nasal passage. Once in position, it is slowly withdrawn all the while scanning the inner surface of the subject with an axial resolution of about 10 microns. These scans are then spliced together to give a high resolution geometry. Examples for both pre- and post operative airways are provided in figures 6 and 7.

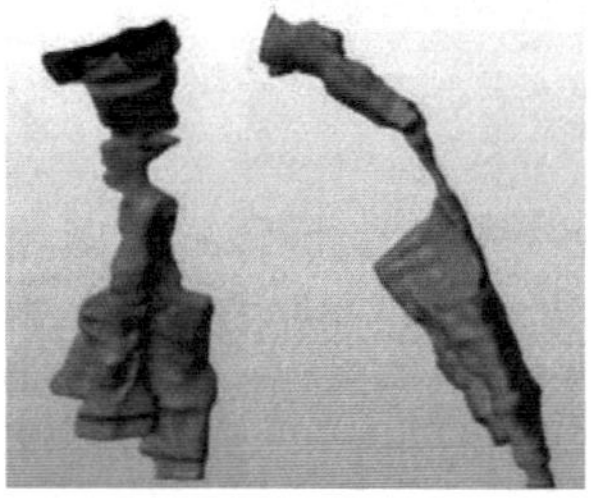

Figure 6. Reconstructed airway using optical coherence tomography, pre-operative. Note the constricted naso-pharynx region. [27]

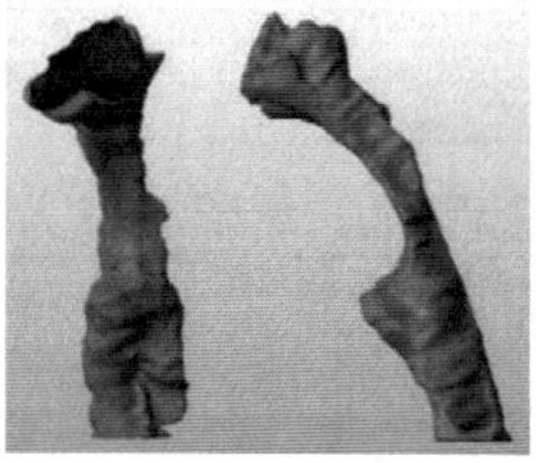

Figure 7. Reconstructed airway using optical coherence tomography, post-operative. Note the increased cross-sectional area in the naso-pharynx region compared to figure 6. [27]

A major issue with OSA is to find a quantifiable method to identify the obstruction that is the cause of the apnea while also ensuring that removal of the obstruction will provide the expected diminution in breathing difficulty. With this in mind, Wang and Elghobashi [45] working with Jing *et al.* [22] mentioned above, took the LROCT images and imported them into a lattice Boltzmann code to perform DNS. LBM was first applied to the idealised human airway by Ball *et al.* [3] with up to 144 million control elements/volumes wherein it was demonstrated to provide superior predictions to a variety of RANS approaches. The large number of CV's was required to resolve the Kolmogorov scale regions where the local Re was high. Jing *et al.*, using upwards of 210 million control volumes/elements, considered the flow entering the airway through the nasal cavity, rather than the oral cavity, for both a healthy 8 year old and a 7 year old with adenotonsillar hypertrophy (enlarged tonsils and adenoids) obstructed upper airway. They compared the time dependent velocity (see figures 10 and 11) and pressure signals at a variety of locations in the airway during both inhalation and exhalation and these computational simulations of the full breathing cycle are, to the authors' knowledge, the first of its kind in realistic airways. They could uniquely identify the obstructed region by calculating the $\partial p/\partial z$ and $\partial^2 p/\partial z^2$, where p is the mean pressure and z is the direction of the main flow. They concluded that where $\partial^2 p/\partial z^2$ became positive and was maximum correlated very well with the known location of the obstruction.

The challenge now is to move this capability into the hands of a otolaryngology and simulation team to enable reasonable turnaround between patients entering a clinic, being assessed by LROCT, incorporating the geometry into a DNS code and in an interactive environment, recompute the flow after the "team" has virtualized the removal of the obstruction to ensure the pressure drops meet physiological requirements and post-operative assessment/validation.

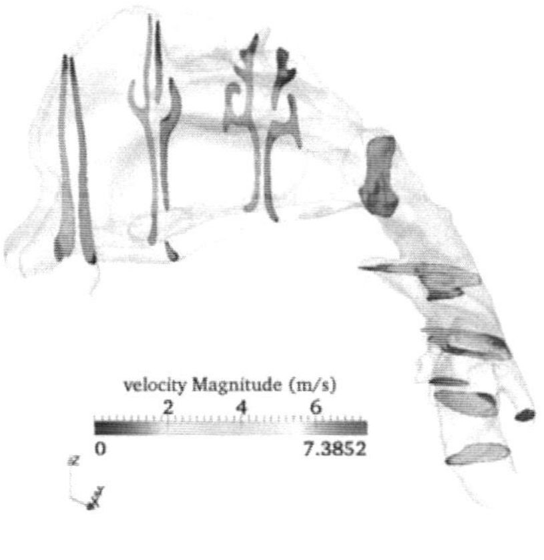

Figure 8. Instantaneous velocity contours at various cross section in *normal* airway [45]

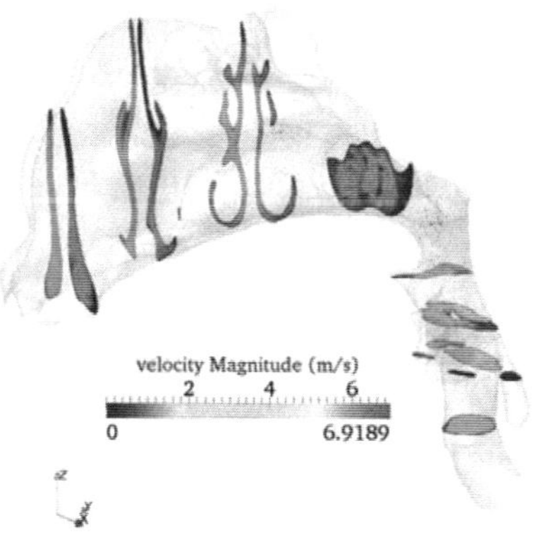

Figure 9. Instantaneous velocity contours at various cross section in *obstructed* airway [45]

2.3. Summary and prospectus for human airway research

The human upper airway presents a rich research environment for fluid dynamists providing they appreciate the constraints of the medical environment they enter. The link between clinical and computational fluid dynamic simulations using DNS continues to strengthen, but there remains many exciting opportunities and challenges, including fluid-structure interaction, incorporation of mucous onto a flexible and pliable tissue and introduction of heat and mass transfer to and from these surfaces.

Journal of Physics: Conference Series **530** (2014) 012005 doi:10.1088/1742-6596/530/1/012005

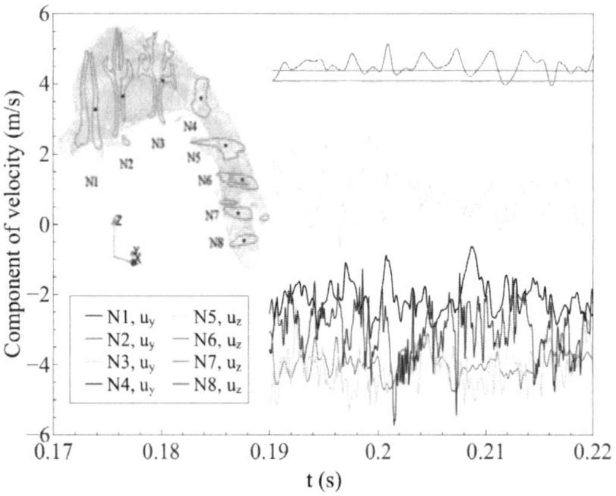

Figure 10. Time history of velocity components at various cross-sectional points in a *normal* airway [45]

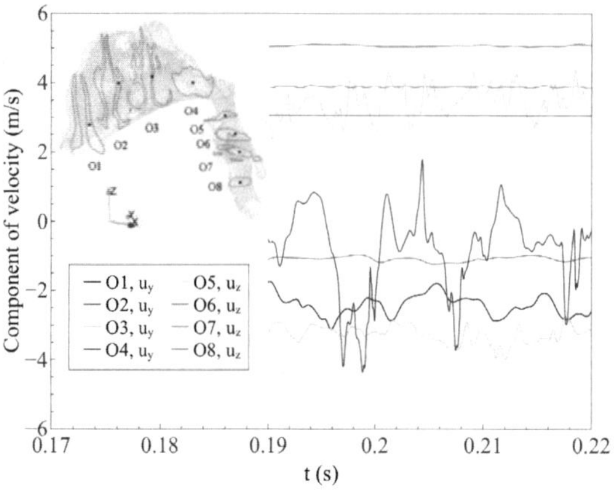

Figure 11. Time history of velocity components at various cross-sectional points in an *obstructed* airway. [45]

3. Cardiac emboli

Attention is now turned to the heart and some of the fluid dynamic issues of the introduction of air (and *i.e.* nitrogen in particular) into the cardiac cavities during open heart surgery.

Skipping the surgical details, during for example heart valve replacement, the open chest and cardiac cavity is often bathed in carbon dioxide, which is readily absorbed by blood, but more importantly assists to ensure that nitrogen absorption is minimized. Even so, during and after "de-airation" when gases in the heart cavity are expelled through a variety of non-standard methods, gaseous bubbles remain. When viewed through an ultrasound probe (trans-esophageal echocardiography (TEE)) that is positioned in the esophagus and scans the cardiac cavity to detect inter-facial surfaces, the potential gaseous emboli may appear as a "snow-storm". These bubbles vary in diameter. As blood corpuscles must be $\leq 40\ \mu m$ to negotiate the capillaries that link the arterial to venous systems, any potential emboli that is greater than 40 μm in diameter are candidates to lodge in the body with a variety of consequences, most of which place the patient at some level of risk. Complications such as neurological impairment from bubbles being transported through the carotid arteries to lodge perhaps in the brain [1, 19, 37], myocardial dysfunction [10], and even death [19] may result.

Gases trapped within the heart are assisted to escape by a variety of non-standardized techniques, which include venting, altering the patient's position to permit the air to escape or to be directed into the lower extremities of the body etc. Also, external pressure could be also applied to the carotid arteries to prevent cerebral embolism [23]. Interestingly, embolic signals have been detected in patients with a wide variety of potential embolic sources, including symptomatic and asymptomatic carotid stenosis [30, 39], atrial fibrillation [6, 43] and prosthetic cardiac heart valves [35]. They also have been detected during and after many surgical procedures, including carotid endarterectomy [15, 44] and cardiopulmonary bypass [14].

While there have been methods and appliances devised to monitor embolic activity in blood, many of these methods are used post-operatively [4]. Some systems even use a multifrequency trans-cranial Doppler (TCD) to detect the differences between solid and gaseous emboli [38]. With Doppler micro-emboli signals obtained using TCD it is not possible to distinguish two or more emboli per frame. This is not a disadvantage for most of the applications for which TCD monitoring is used; however, for the period during heart surgery prior to weaning from the cardiopulmonary bypass machine where there are numerous air emboli per frame it would be extremely difficult for Doppler ultrasonography to accurately distinguish between all emboli. Also, gas emboli post cardiac surgery are not just micro-emboli, which are considered to be $\leq 40\ \mu m$ in diameter, with gaseous potential emboli being up to $2\ mm$ or greater in diameter. All TCD systems are set up to acquire mid-cranial embolic signals but are unsuitable if emboli have diameters $\geq 40\ \mu m$ and therefore TCD is not suitable for immediate post-CPB separation. Therefore, detection of heart cavity resident potential emboli, where both large quantities and size distributions of air bubbles are encountered, will require a new method.

3.1. Emboli Bubbles and Surface Interference

As noted, gaseous potential emboli, or bubbles in blood that are $\geq 40\ \mu m$ are problematic. One possible method to ameliorate the negative effects of these larger bubbles is to break them apart. It is to this subject that attention is now turned.

3.1.1. Dimensional analysis The literature on bubbles is extensive, and will not be reviewed here; rather, the relevant material will be introduced as needed. As potential gaseous emboli tend to move through the human body using the cardiac systole - diastole wave form, it is simpler, in the first instance, to assume that the blood to be stationary and permit bubbles to rise against the gravity vector. The bubble shape is then determined from the well-known plot of the Morton number (M) as a function of the Reynolds (Re) and Eotvos (Eo) numbers (see [16])

$$Eo = \frac{g\Delta\rho D_e^2}{\sigma} \qquad M = \frac{g\mu_l^4\Delta\rho}{\rho_l^2\sigma^3} \qquad Re = \frac{\rho_l D_e U}{\mu_l} \tag{1}$$

where the meaning of the symbols should be self-evident. In the problem considered here, the log of the Morton number is between -8 and -11, but the Eotvos number is ≤ 0.3 so that spherical bubbles are expected although excursions into a wobbling mode maybe encountered.

3.1.2. Bubble surface oscillations - Lamb and Rayleigh-Plesset Lamb [26] considered a spherical drop that was subjected to small oscillations about its spherical form with only surface tension effects considered. He introduced a spherical harmonic approximation to describe the small oscillations about the mean radius of the drop:

$$r = R_o + \sum_{n,m}^{\infty} a_{nm} Y_n^m \tag{2}$$

where $r = r(\theta, \phi, t)$ is the bubble radius as a function of sphere angular oordinates, R_o is the initial bubble radius and time, $a_{nm} = a_{nm}(t)$ are the amplitudes of the spherical harmonic component of order n and degree m, and $Y_n^m = Y_n^m(\theta, \phi)$ is the normalized spherical harmonic function. The shapes so obtained are illustrated in figure 12. In these images, the colours red and blue indicate positive and negative values of the function Y_n^m, for $R_o = 0$ and $a_{nm} = 1$ from equation 2. In figure 13, the images are for an initial bubble radius $R_o = 0.6$ and $a_{nm} = 0.4$.

Lamb used the Young-Laplace equation that describes the capillary pressure difference across the interface between two fluids at rest:

$$\Delta p = \sigma(\frac{1}{R_1} + \frac{1}{R_2}) \tag{3}$$

where Δp and σ are the pressure differences across a bubble surface and the surface tension, respectively. R_1 and R_2 are the maximal and minimal radii of curvatures of the surface,

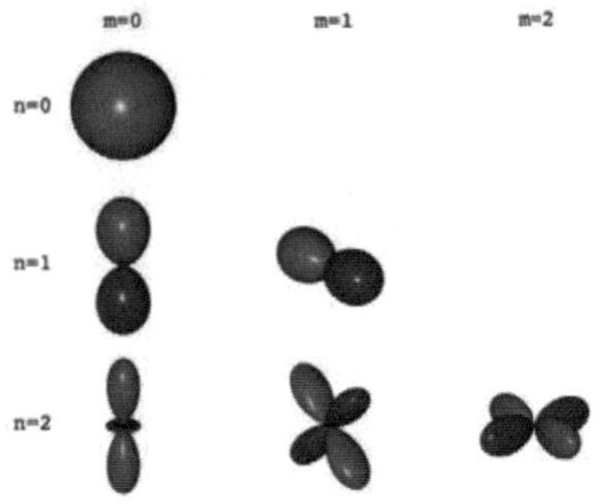

Figure 12. Spherical harmonic solution to Eqn. 2 for $R_o = 0$ and $a_{nm} = 1$. Red indicates positive and blue negative values of

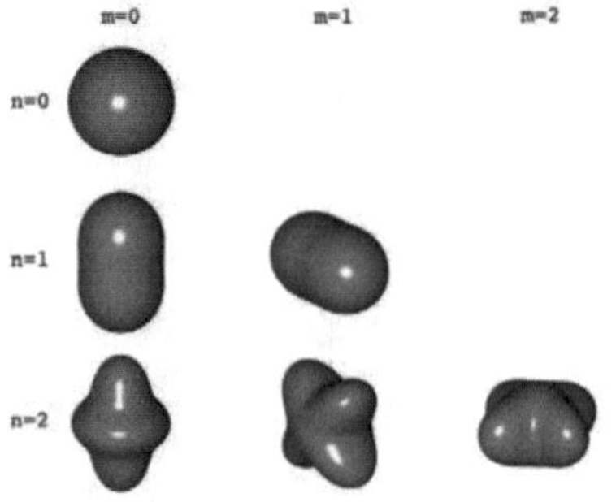

Figure 13. Example of bubble shape as a solution to Eqn. 2 for $R_o = 0.6$ and $a_{nm} = 0.4$

respectively (also known as the principal radii of curvature). Lamb employed velocity potentials and solid geometry and the spherical harmonic approximation with equation 3 to obtain the frequencies at which a bubble or drop will oscillate, this frequency being called the modal frequency. The modal frequencies are the natural frequencies of the bubble when given a particular mode of vibration (analogous to the natural frequency for a spherical bubble):

$$\omega_n = \frac{1}{2\pi R_o}\sqrt{n(n+1)(n-1)(n+2)\frac{\sigma}{((n+1)\rho_g + n\rho_l)R_o}} \tag{4}$$

where ω_n is the modal surface perturbation frequency of order n, R_o is the initial bubble radius and ρ_l and ρ_g are the densities of the liquid and gas, respectively. From equation 4 the modal frequency is only dependent on the order n and initial radius, R_o. Therefore, given an applied frequency the bubble will prefer to oscillate with a mode in accordance with equation 4. However, to oscillate the bubble with maximum amplitude with minimal power the natural frequency of the bubble must be used. Other frequencies other than the natural will tend to damp out and cause no stimulated growth of the amplitude. Attention is now directed towards the dynamics of the problem to determine the natural frequency of a bubble.

To introduce the dynamics of bubble motion, Rayleigh [36] described the pressure inside a collapsing spherical bubble and Plesset [33] extended this work to account for the stability of the interface and the following is now referred to as the Rayleigh-Plesset equation:

$$\ddot{R}R + \frac{3}{2}\dot{R}^2 + \frac{4\nu_l}{R}\dot{R} + \frac{2\sigma}{R\rho_l} - \left(\frac{p_o}{\rho_l} + \frac{2\sigma}{R\rho_l}\right)\left(\frac{R_o}{R}\right)^{3\kappa} = -\frac{p_\infty}{\rho_l} \tag{5}$$

where $\ddot{R}$ and $\dot{R}$ are the acceleration and velocity of the bubble radial surface of radius R and κ is the polytropic index that accounts for isothermal or adiabatic compression; the other symbols should be obvious as to their meaning. This equation reduces to Rayleigh's equation when $p_\infty = P_o$ (a constant external pressure), $p_g = 0$ (empty cavity) and the surface tension and viscosity are neglected. If, however, the external pressure is not constant, but fluctuating in a prescribed manner, then the bubble radius will oscillate. The natural frequency of the 0th mode (or of spherical motion) can be derived from the Rayleigh-Plesset as:

$$\omega_o = \frac{1}{2\pi R_o}\sqrt{\frac{3\kappa p_o}{\rho_l} + \frac{6\kappa\sigma}{\rho_l R_o} - \frac{2\sigma}{\rho_l R_o}} \tag{6}$$

Figure 14 presents the natural frequency and the modal frequencies as a function of bubble radius. Ideally, the shape that a bubble will take is dependent upon its original radius and the applied frequency, although the bubble will only respond to the natural frequency which leads to the discussion of the preferred mode.

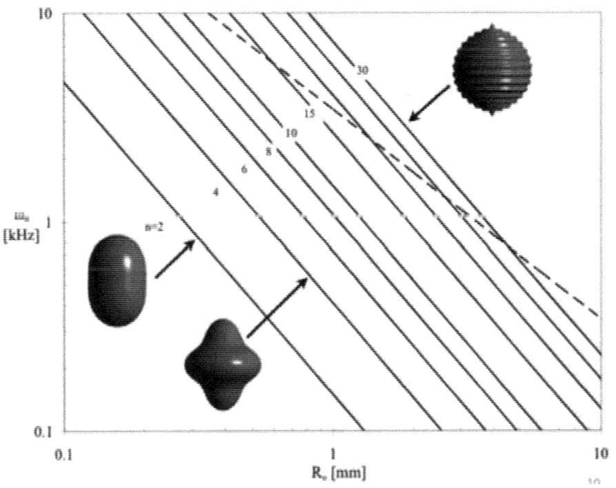

Figure 14. Surface perturbation modal frequency (ω) of the asymmetric modes of bubble oscillations of degree n (solid lines) compared with the natural or breathing mode frequency (dashed line)

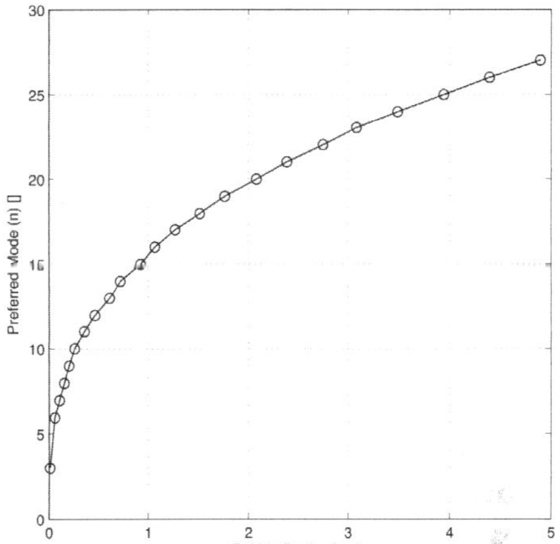

Figure 15. Preferred modes of a bubble at its natural frequency

The intercept of the natural and modal frequencies in figure 14 leads to the preferred modes of a bubble (figure 15). As a bubble is subject to it's natural frequency the bubble will not only oscillate in the spherical direction but also with a modal shape as described by the preferred mode, see figure 13.

Note, however, that equation (5) is accompanied with the *proviso*, stated by Plesset [33] "While the stability question is thereby answered, the results cannot be applied to the determination of the rate of development of interface distortions of significant amplitude." It is postulated here that with increased interfacial amplitude, however caused, will create sufficient "violent" perturbations that surface tension forces *should* enable pinch-off between the mother bubble and its daughters thereby causing a diminution in the size of the mother. Obviously, the change in diameter of the mother bubble will alter its oscillation natural frequency and it is hoped the daughter bubble diameter will be small enough to pass through the human arterial system without consequence.

From the preceding, it is evident that any approach that uses a single frequency will be of limited utility, which therefore requires the development of a methodology that is based on applying an adaptable spectrum of disturbances on bubbles to effect continuous mother-daughter evolution. It is both the current limitation in theory and the requirement to expose bubbles to a multifrequency environment that has caused us to explore this phenomenon using experiments.

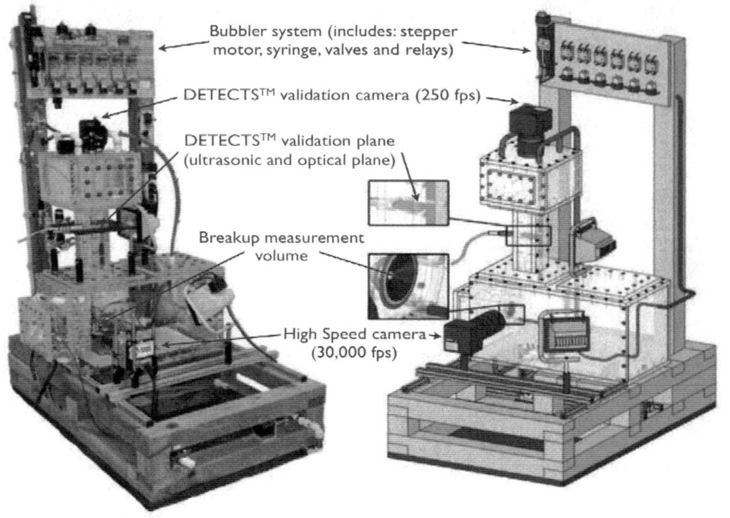

Figure 16. Photograph (L) and schematic (R) of experimental rig

The experimental rig is displayed in figure 16. This rig permits two types of experiments to be performed: (a) automatically generate single bubbles into still water permitting them to rise against the gravity vector, activate a speaker with a variety of different sound wave frequency types and obtain high speed video of each bubble evolution and (b) automatically generate bubbles and then provide cross correlation of these bubble to validate ultrasound transducers against visual records (see next section). To date, 6 Tbytes of data have been acquired, which transforms into about 700 events per day that each produces about 100 Gbytes. The frequency types include single continuous, pulsed and ramped frequencies.

The images obtained were analysed using in-house developed image processing software. Simply put, the images captured by the high speed camera were analysed. Analysis included the number and size distribution of the bubble fragments and the geometrical centre of the initial bubble. The geometric centre of the initial bubble was used to check the data to determine if the initial bubble travelled outside of the test section, in which case, those data were rejected.

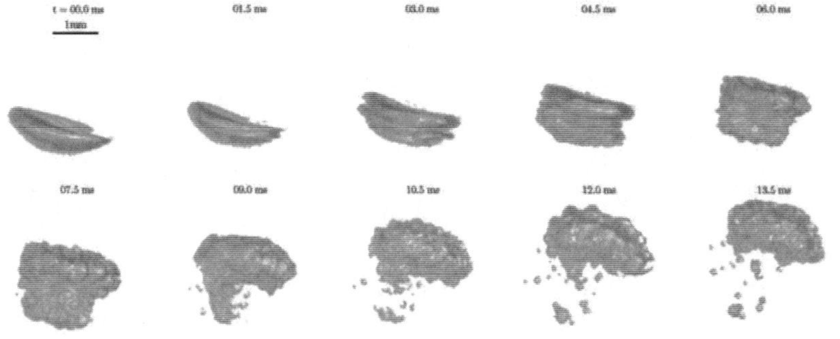

Figure 17. Sequence of bubbles: The single bubble was radiated with a frequency sweep from 1000 to 1500 Hz over 0.5 sec at 0.6 psi. For the image sequence shown, the frequency ranged from 1400 to 1427.0 Hz. over 13.5 ms

An example of the images obtained are provided in figure 17. The single bubble was radiated with a frequency sweep from 1000 to 1500 Hz over 0.5 sec. at a pressure of 0.6 *psi*. The pressure was determined using a hydrophone located in the centre of the speaker at the same distance from it as the bubble release point. The images from a frequency ramp between 1400 and 1427 Hz capture the modes of the breakup of one bubble in 1.5ms intervals. The mother bubble is seen to undergo simple harmonic oscillation for $1.5 \leq t \leq 3$ before the lower lobe undergoes an azimuthal instability that in the next image ((t=4.5 ms) can be clearly discerned as a roughening of the lower surface of the bubble. With increased time, the whole bubble evolves into a multi-modal instability. By t=9 ms surface tension forces pinch off multiple daughter bubbles of varying size, but which are very much smaller that the mother bubble.

3.2. Operating room implementation of DETECTS

We have developed a novel algorithm to detect heart resident potential emboli; it is called Detection of Emboli using Transesophageal Echocardiography for Counting, Total volume, and Size estimation or DETECTS. This software maybe used in the operating theatre during cardiac de-airation to provide real-time bubble information (as the name implies). The experimental rig, see figure 16, is equipped to validate the ultrasound images by correlating the US to high speed video images. Our work is to the best of our knowledge the first validation of information that is presented to a cardiac team and it is important to note that typical TEE or foot-print US transducers have relatively poor resolution. A three-dimensional reconstruction of the raw data was preformed to consolidate the measurement slices of the optical and ultrasound, figure 18. Bubbles from the optical data were then matched to the ultrasound bubble to minimize the error between the radius, centre of geometry and time. An example of the congruence between the optical and sound images is provided in figure 19.

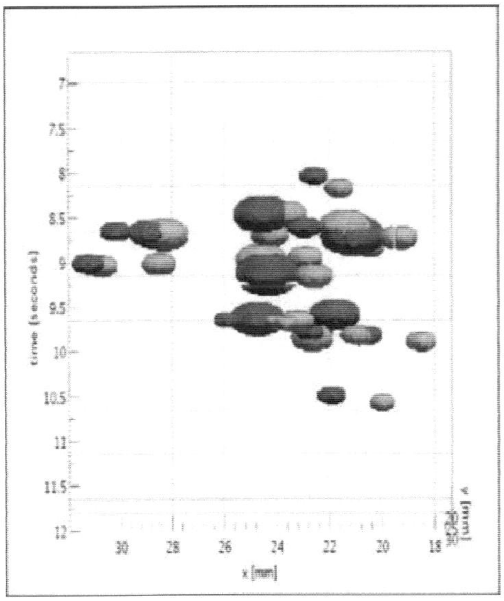

Figure 18. Comparison of optical using DETECTS (violet) and ultrasound (light blue) extracted data.

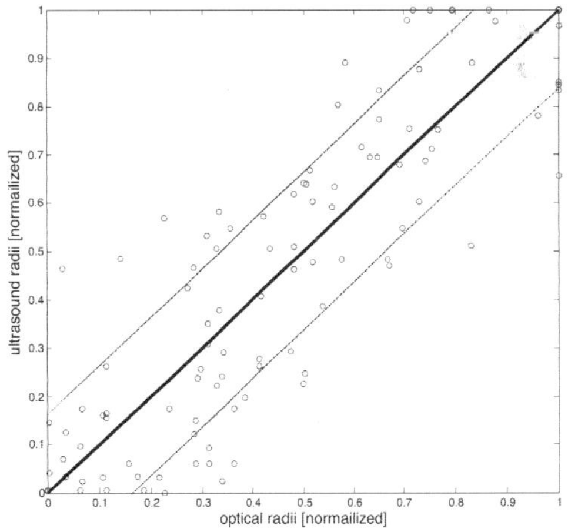

Figure 19. DETECTS extracted bubble radii *vs.* radii obtain from optical images. Lines indicate 1:1 correlation and ± 95% confidence levels.

3.3. Summary and prospectus for cardiac potential emboli research

There are many routes available to the fluid dynamics researcher who specialize in bubble dynamics. Clearly, there are interesting theoretical solutions to be considered that must be complemented by computational simulations that include fluid-structure interaction, where of course the "structure" here is the bubble interface. And, there is the need to introduce blood properties, rather than the "simple" approximation of air in water considered here. In the context of cardiac surgery, implementation of bubble destruction technology into the cardio-pulmonary system, either *in-vitro* (via the perfusion portion of the cardiac team) or preferably *in-vivo* directly, has many thermo-fluid challenges. A major obstacle for *in-vivo* application is the thermal impact on blood and tissue of what ever method that emerges as an optimal modality for bubble break-up. Additionally, a significant challenge faced by engineers is to deal with the resistance of the medical community to the introduction of non-standard procedures into their regimes.

4. Acknowledgments

The human airway research has been funded by the Natural Science and Engineering Research Council of Canada (NSERC), while that associated with the OCT [27] and DNS of [45] was supported by NIH under HL105215. AP participated in the NIH work under subcontract from a grant awarded to Dr. B. Wong, University of California, Irvine. The cardiac emboli work has been supported by NSERC through Discovery and equipment grants to AP and graduate scholarships to FS as well as through the Garfield Kelly cardio-vascular research and development fund administered by Kingston General Hospital and Queen's University awarded to BM and AP.

[1] Abu-Omar Y, Balacumaraswami L, Pigott D W, Mattews P M, and Taggart D P, 2004, Solid and gaseous cerebral emboli microembolization during off-pump, on-pump, and open cardiac surgery procedures, *The Journal of Thoracic and Cardiovascular Surgery*, **127**, 1759-1765.
[2] Ball CG, Uddin M, and Pollard A, 2006a High resolution turbulence modelling of airflow in an idealised human extra-thoracic airway, *Computers and Fluids*, **37(8)**, 943-964.
[3] Ball CG, Uddin M, and Pollard A, 2008, Mean Flow Structures Inside the Human Upper Airway, *Flow, Turbulence and Combustion*, **81**, 155-188.
[4] Brucher R and Russell D, 2002, Automatic Online Embolus Detection And Artifact Rejection With The First Multifrequency Transcranial Doppler, *Stroke*, **33**, 1969-1974.
[5] Chen J, and Gutmark E, 2014, Numerical investigation of airflow in an idealized human extra-thoracic airway: a comparison study *Biomech Model Mechanobiol*, **13**, 205-214.
[6] Cullinane M, Wainwright R, Brown A, Monaghan M and Markus H S, 1998, Asymptomatic embolization in subjects with atrial fibrillation not taking anticoagulants: a prospective study, *Stroke*, **29**, 1810-1815.
[7] de Rochefort L,Vial L, Fodil R, Matre X, Louis B, Isabey D, Caillibotte G, Thiriet M, Bittoun J, Durand E and Sbirlea-Apiou G, 2007, In vitro validation of computational fluid dynamic simulation in human proximal airways with hyperpolarized ^3He magnetic resonance phase-contrast velocimetry, *J. Appl. Physiol.*, **102**, 2012-2023.
[8] Doorly, D. J., Taylor, D. J., Franke, P., Schroter, R., 2008a, Experimental investigation of nasal airflow. Proceedings of the Institution of Mechanical Engineers, *Journal of Engineering in Medicine : Part H*, **222**(4), 439–453.
[9] Doorly, D.J., Taylor, D.J., Gambaruto, A.M., Schroter, R.C., Tolley, N., 2008b, Nasal architecture: form and flow, *Trans. R. Soc. A Math. Phys. Eng. Sci.*, **366**, 3225–3246.
[10] Eguchi S, and Bosher L H, 1962, Myocardial dysfunction resulting from coronary air embolism, *Surgery*, **51**, 103-111.
[11] Elkins CJ and Alley MT, 2007, Magnetic resonance velocimetry: applications of magnetic resonance imaging in the measurement of fluid motion, *Expts Fluids*, **43**, 823-858.
[12] Elkins CJ, Markl M, Pele N and Eaton JK, 2003, 4D magnetic resonance velocimetry for mean velocity measurements in complex turbulent flows, *Expts Fluids*, **34(4)**, 494-503.
[13] Finlay W, Stapleton K, and Yokota J, 1996, On the use of computational fluid dynamics for simulating flow and particle deposition in the human respiratory tract, *J. Aerosol Med.*, **9(3)**, 329-341.
[14] Gallagher E G and Pearson D T, 1973, Ultrasonic identification of sources of gaseous microemboli during open heart surgery, *Thorax*, **28**, 295-305.
[15] Gaunt M E, 1998, Transcranial Doppler: Preventing Stroke During Carotid Endarterectomy, **Annals of the Royal College of Surgeons of England**, **80**, 377-387.
[16] Grace J R, Wairegi T and Nguyen T H, 1976, Shapes and Velocities of single drops and bubbles moving freely through immiscible liquids, *Chemical Engineering Research and Design*, **54**, 167-173.

[17] Heenan A, Grgic B, Pollard A, Finlay W and Burnell P, 2004, An investigation of the relationship between the flow field and regional deposition in realistic extra-thoracic airways, *J. Aerosol Sci.*, **35(8)**, 1013-1023.

[18] Grgic B, Finlay W, Burnell P and Heenan A, 2004, *In vitro* intersubject and intrasubject deposition measurements in realistic mouththroat geometries, *J Aerosol Sci.*, **35**, 1025 1040.

[19] Grosset D G, Georgiadis D, Kelman A W, Cowburn P, Striling S, Lees K R, Faichney A, Mallinson A, Quin R, Bone I, Pettigrew L, Brodie E, MacKay T and Wheatley D J, 1996, Detection of Microemboli by Transcranial Doppler Ultrasound, *Texas Heart Institute Journal*, **23**, 289-292.

[20] Heenan AF, Matida E, Pollard A and Finlay WH, 2003, Experimental measurements and computational modelling of the flow field in an idealized extra-thoracic airway, *Expt. Fluids*, **35**, 70-84.

[21] Johnstone A, Uddin M, Pollard A, Heenan A, and Finlay W, 2004, The flow inside an idealised form of the human extra-thoracic airway, *Expt. Fluids*, **37(5)**, 673-689.

[22] Jing J, Zang J, Loy AC, Wong BJ, and Chen Z, 2012, High-speed upper airway imaging using full-range optical coherence tomography *J. Biomed Opt***17(11)**,110507

[23] Kalpokas M V, Nixon I K, Kluger R, Beilby D S, and Silbert B S, 2003, Carbon dioxide field flooding versus mechanical de-airing during open-heart surgery: a prospective controlled trial, *Perfusion*, **18**, 291-294.

[24] Kleinstreuer, C. and Zhang, Z., 2010, Airflow and Particle Transport in the Human Respiratory System, *Annu. Rev. Fluid Mech.*, **42**, 301–334.

[25] Labiris NR and Dolovich MB, 2003, Pulmonary drug delivery. Part I: Physiological factors affecting therapeutic effectiveness of aerosolized medications, *Br J Clin Pharmacol. doi: 10.1046/j.1365-2125.2003.01892.x* **56(6)**, 588599

[26] Lamb H, 1879, *Hydrodynamics*,Cambridge University Press.

[27] Lazarow F *et al.* 2014, Pilot Study: Intraoperative Long Range Optical Coherence Tomography as a Novel Method of Imaging the Pediatric Upper Airway Before and After Adenotonsillectomy, *to be submitted to International Journal of Pediatric Otorhinolaryngology*

[28] Leighton TG, 1994, *The Acoustic Bubble*, Academic Press.

[29] Leighton TG, Walton AJ and Pickworth MJW, 1990, *Primary Bjerknes forces*, *Eur. J. Phys.*, **11**, 47–50.

[30] Markus H S, Thomson N D and Brown M M, 1995, Asymptomatic cerebral emboli signals in symptomatic and asymptomatic carotid artery disease, *Brain*, **118**, 1005-1011.

[31] Mihaescu M, Murugappan S, Kalra M, Khosla S and Gutmark E, 2008, Large Eddy Simulation and Reynolds-Averaged NavierStokes modeling of flow in a realistic pharyngeal airway model: An investigation of obstructive sleep apnea *J. Biomechanics*, **41(10** 2279-2288.

[32] Nicolaou L, and Zaki TA, 2013, Direct numerical simulations of flow in realistic mouththroat geometries, *J. Aerosol Sci.*, **57**, 71-87.

[33] Plesset MS, 1954, On the Stability of Fluid Flows with Spherical Symmetry, *L. Applied Physics*, **25(96)**, 96-98, doi: 10.1063/1.1721529

[34] Pollard A, Uddin M, Shinneeb A-M, and Ball CG, 2012, Recent advances and key challenges in investigations of the flow inside human oro-pharyngeal-laryngeal airway, *Intl. J. Computational Fluid Dynamics* **26(6-8)**, 363-381.

[35] Ram J J, Davis D A, Lolley D M, Berger M P and Spencer M, 1993, Detection of microemboli in patients with artificial heart valves using transcranial doppler: preliminary observations, *Journal of Heart Valve Disease*, **2**, 37-4.

[36] Rayleigh L, 1917, On the Pressure Development in a Liquid During the Collapse for Spherical Cavity, *Philosophical Magazine*, *34*, 94-98.

[37] Ringelstein E B, Droste D W, Babikian V L, Evans D H, Grosset D G, Kaps M, Markus H S, Russell D and Siebler M, 1998, Consensus on Microembolus Detection by TCD, **Stroke**, **29**, 725-729.

[38] Russell D and Brucher R, 2002, Online Automatic Discrimination Between Solid And Gaseous Cerebral Microemboli With The First Multifrequency Transcranial Doppler, *Stroke*, bf 33, 1975-1980.

[39] Seibler M, Kleinschmidt A, Sitzer M, Steinmetz H and Freund H J, 1994, Cerebral microembolism in symptomatic and asymptomatic high-grade internal carotid artery stenosis, *Neurology*, **44**, 615-618.

[40] Shi, H., Kleinstreuer, C. and Zhang, Z., 2006, Laminar airflow and nanoparticle or vapor deposition in a human nasal cavity model, *J. Biomech. Eng. Trans. ASME*, **128**, 697–706.

[41] Shinneeb A-M and Pollard A, 2012, Investigation of the flow physics in the human pharynx/larynx region, *Experiments in Fluids*, **53(4)**, 989-1003

[42] Shinneeb A-M and Pollard A, 2012, Identification of vortical structures inside the human pharynx/larynx region from POD-reconstructed velocity fields, *Experiments in Fluids*, **53(2)**, 353-367

[43] Silwka U, Job F P, Wissuwa D, Diehl R R, Flachskampf F A, Hanrath P and Noth J, 1995, Occurrence of transcranial doppler high-intensity transient signals in patients with potential cardiac sources of embolism: a prospective study, *Stroke*, **26**, 2067-2070.

[44] Spencer M P, Thoma, G I, Nichollos S C and Sauvage L R, 1990, Detection of middle cerebral artery emboli

during cartid endarterectomy using transcranial doppler ultrasonography, *Stroke*, bf 21, 415-423.

[45] Wang Y and Elghobashi S, 2014, On locating the obstruction in the upper airway via numerical simulation, *Respiratory Physiology and Neurobiology*, **193**, 1-10.

[46] Wootton DM, Luo H, Persak SC, Sin S, McDonough JM, Isasi CR, and Arens R, 2014, Computational fluid dynamics endpoints to characterize obstructive sleep apnea syndrome in children, *J. Appl. Physiol*, **116**, 104–112.

[47] Xi, J.X. and Longest, P.W., 2008, Numerical predictions of submicrometer aerosol deposition in the nasal cavity using a novel drift flux approach, *Int. J. Heat Mass Transf.*, **51**, 5562–5577.

Journal of Physics: Conference Series **530** (2014) 012006 doi:10.1088/1742-6596/530/1/012006

Modeling and simulation of multi-physics multi-scale transport phenomenain bio-medical applications

Saša Kenjereš

Transport Phenomena Section, Department of Chemical Engineering, Faculty of Applied Sciences, Delft University of Technology and J. M. Burgerscentre of Fluid Dynamics, Julianalaan 136, 2628 BL Delft, The Netherlands

E-mail: s.kenjeres@tudelft.nl

Abstract. We present a short overview of some of our most recent work that combines the mathematical modeling, advanced computer simulations and state-of-the-art experimental techniques of physical transport phenomena in various bio-medical applications. In the first example, we tackle predictions of complex blood flow patterns in the patient-specific vascular system (carotid artery bifurcation) and transfer of the so-called "bad" cholesterol (low-density lipoprotein, LDL) within the multi-layered artery wall. This two-way coupling between the blood flow and corresponding mass transfer of LDL within the artery wall is essential for predictions of regions where atherosclerosis can develop. It is demonstrated that a recently developed mathematical model, which takes into account the complex multi-layer arterial-wall structure, produced LDL profiles within the artery wall in good agreement with in-vivo experiments in rabbits, and it can be used for predictions of locations where the initial stage of development of atherosclerosis may take place. The second example includes a combination of pulsating blood flow and medical drug delivery and deposition controlled by external magnetic field gradients in the patient specific carotid artery bifurcation. The results of numerical simulations are compared with own PIV (Particle Image Velocimetry) and MRI (Magnetic Resonance Imaging) in the PDMS (silicon-based organic polymer) phantom. A very good agreement between simulations and experiments is obtained for different stages of the pulsating cycle. Application of the magnetic drug targeting resulted in an increase of up to ten fold in the efficiency of local deposition of the medical drug at desired locations. Finally, the LES (Large Eddy Simulation) of the aerosol distribution within the human respiratory system that includes up to eight bronchial generations is performed. A very good agreement between simulations and MRV (Magnetic Resonance Velocimetry) measurements is obtained. Magnetic steering of aerosols towards the left or right part of lungs proved to be possible, which can open new strategies for medical treatment of respiratory diseases.

1. Introduction

Physical transport phenomena (integrated studies of mass, momentum and energy) within biological systems play a crucial role in the functioning of cells, tissues and organs. Some changes in biological transport processes can trigger a complex chain of bio-chemical reactions leading to any of a number of diseases. Typical examples include atherosclerosis, asthma, allergies, cancer, etc. The proper understanding and advanced mathematical modeling of the basic mechanisms behind transport phenomena in living systems can make it possible to predict locations where the initial stage of various diseases will occur. Then, as a next step, various optimizations of the medical drug delivery aimed precisely towards these regions can be performed. This can lead to a significant increase of the local

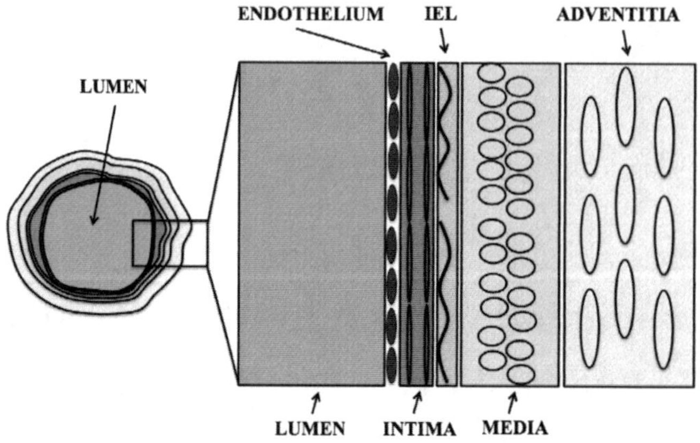

Figure 1. Schematic representation of the multi-layered structure of an artery wall (not to scale): the artery interior (lumen), the arterial wall (endothelium, intima, internal elastic layer, media and adventitia). The typical thickness of these layers is: endothelium (2 μm), intima (10 μm), IEL (2 μm), media (200 μm) and adventitia (200 μm).

medical drug concentration in the proximity of diseased areas and consequently, to reduction of side effects. In addition to these 'direct' medical drug delivery applications, numerous applications can be found also in bio-medical engineering dealing with development, design and optimization of various medical assisting devices, such as artificial heart (ventricular assist device), artificial lungs, artificial kidneys (dialysis) machines etc.

There are many challenges in dealing with transport phenomena in living systems. The typical length scales ($\mathcal{L}$) of the major transport phenomena in humans cover an immense range. Starting from the whole human body with $\mathcal{L}\approx\mathcal{O}(10^0)$ m, organs $\mathcal{L}\approx\mathcal{O}(10^{-1})$ m, cells $\mathcal{L}\approx\mathcal{O}(10^{-6})$ m, and finally, proteins $\mathcal{L}\approx\mathcal{O}(10^{-8})$ m. It is obvious that no single transport mechanism can function equally efficiently over these length scales. Diffusive transport can be very efficient for the small length scales, but is replaced by convection for transport of molecules over larger distances. For example diffusion is most important for small molecules such as oxygen (with estimated Peclet number of Pe$\approx\mathcal{O}(10^{-2})$), whereas convection is the dominant mechanism for transport of larger entities such as cells (with estimated Pe$\approx$ $\mathcal{O}(10^3)$), Truskey *et al.* [1]. The range of typical time scales ($\mathcal{T}$) is also immense: from $\mathcal{T}\approx\mathcal{O}(1)$s, for typical cardiac cycle (heartbeat), up to $\mathcal{T}\approx\mathcal{O}(10^8)$s (ten years) for development of atherosclerosis.

In addition to an immense range of length- and time-scales, the transport phenomena in bio-medical applications involve complex multi-physics interactions that include mass and heat transfer, bio-chemical reactions, phase changes, fluid dynamics, complex fluids (blood rheology), fluid-structure interactions, electromagnetism etc. To deal with such complex problems, a multi-disciplinary approach is required, which combines biology, medicine, physics, applied mathematics, bio-medical, mechanical and chemical engineering, imaging and computer sciences. In the present paper, we will address some of our most recent achievements in mathematical modeling and combined numerical and experimental studies dealing with combined mass- and momentum-transfer of blood flow in anatomically realistic human arteries and of the air flow and aerosol distribution in the human respiratory system.

2. Modeling and simulation of the initial stage of atherosclerosis

Atherosclerosis (one of the most common cardiovascular diseases) is a process of the accumulation of fatty plaques within the artery wall. It primarily takes place at curvatures and bifurcations of large and medium size arteries, e.g. the coronary or carotid arteries. The rupture of the soft fatty plaque and its release into the blood stream leads to heart attack or stroke. Based on some recent reports of the World

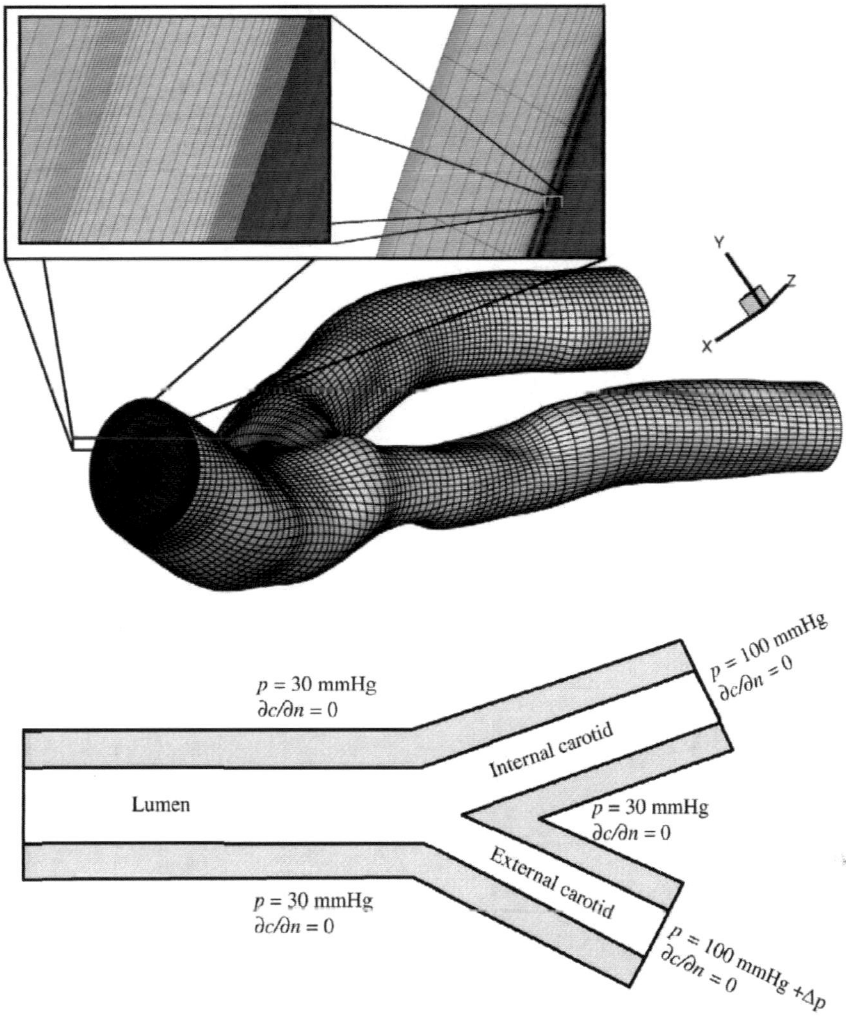

Figure 2. Top- typical numerical mesh for the patient-specific carotid artery, covering the artery interior and the multi-layered artery wall (indicated in Fig. 1). Bottom- typical boundary conditions used for the carotid artery simulations. The steady flow conditions are considered with inlet an value of Re=105, [6].

Health Organization[1], it is estimated that 17.3 million people died from the cardiovascular diseases in 2008, and that by the year 2030, more than 23 million people will die of it annually. At present, it is accepted that atherosclerosis is a chronic inflammatory disease triggered by damage of the endothelium layer (the very first layer of the arterial wall) and by an increased accumulation of the "bad" cholesterol (i.e. low-density-lipoprotein, LDL) within the interior of the artery wall. It is obvious that an early prediction of the possible locations where atherosclerosis can develop can have huge impact in treatment of this disease, Tarbell [2]. Here, we will give a short overview of a recently developed mathematical model that combines solutions of the blood flow and LDL accumulation within the multi-layered arterial wall (shown in Fig. 1) of an anatomically realistic carotid artery bifurcation (shown in Fig. 2). It should be noted that there are differences in the characteristic length-scales from a few cm (length of the artery) to a few μm (the typical thickness of endothelium layer). Equations for blood flow and LDL transfer

[1] http://www.who.int

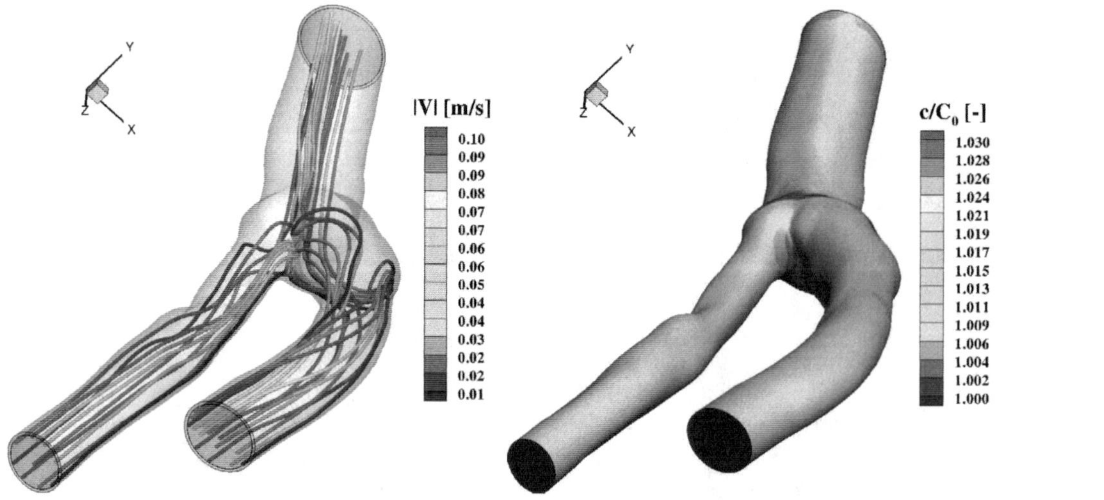

Figure 3. Top- stream-traces colored by the blood flow velocity ($|v|$); Bottom- contours of the LDL concentration at the lumen/endothelium interface; both are for the flow conditions defined in Fig. 2, [6].

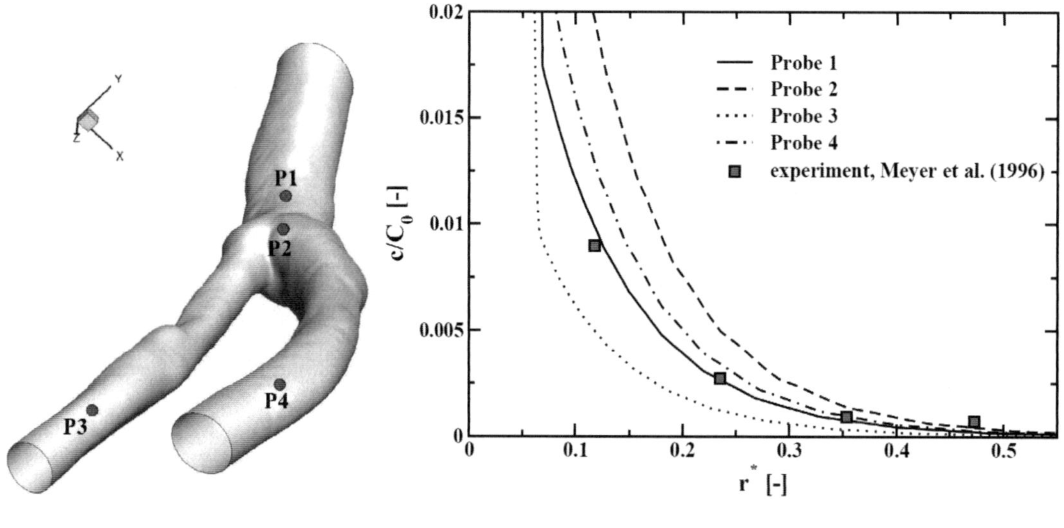

Figure 4. The radial profiles of the LDL concentration within the interior of the artery wall (media) for some characteristic locations (P1-P2-P3-P4), [6].

within the lumen (inner part of artery that is occupied with flowing blood) are:

$$\nabla \cdot \vec{v} = 0 \tag{1}$$

$$\rho \frac{\partial \vec{v}}{\partial t} + \rho\left(\vec{v} \cdot \nabla\right)\vec{v} = -\nabla p + \nabla\left(\mu \nabla \vec{v}\right) \tag{2}$$

where the LDL concentration is modeled with the standard convective-diffusive mass transport as

$$\frac{\partial c}{\partial t} + \vec{v} \cdot \nabla c = \mathcal{D} \nabla^2 c. \tag{3}$$

Here ρ is the density, μ the dynamical viscosity and $\mathcal{D}$ the diffusion coefficient. Transport of the blood plasma and of the LDL concentration within the multi-layered arterial wall is modeled by combining the porous-layer and biological membranes approach and can be written as ([3], [4], [5]):

$$\nabla \cdot \langle \vec{v} \rangle = 0 \tag{4}$$

$$\frac{\rho}{\epsilon}\frac{\partial \langle \vec{v} \rangle}{\partial t} + \frac{\rho}{\epsilon}\left(\langle \vec{v} \rangle \cdot \nabla\right)\langle \vec{v} \rangle = -\frac{\nabla \langle p \rangle}{\epsilon} + \nabla \cdot \frac{\mu}{\epsilon}\nabla \langle \vec{v} \rangle - \frac{\mu}{K}\langle \vec{v} \rangle + \sigma_d R T \nabla \langle c \rangle \tag{5}$$

where "$\langle\rangle$" are volume-averaged variables, the ϵ is the porosity (defined as a fraction of the void volume over the total volume; ϵ=0 for the totally blocked flow), K is the permeability, σ_d is the Staverman osmotic reflection term and R is the universal gas constant and T is the absolute temperature (in K). The third and fourth terms on the right side of the extended momentum equation (Eq. 5) are the flow resistance imposed by porous medium (Darcy's law) and the osmotic pressure, respectively. The osmotic pressure term makes a two-way coupling between the blood plasma momentum and the LDL concentration. The volume-averaged equation for the LDL concentration can be written as:

$$\frac{\partial \langle c \rangle}{\partial t} + (1 - \sigma_f)\langle \vec{v} \rangle \cdot \nabla \langle c \rangle = \mathcal{D}_e \nabla^2 \langle c \rangle + k_r \langle c \rangle \tag{6}$$

where $\mathcal{D}_e$ is the effective diffusivity, the σ_f is the Staverman filtration coefficient, and k_r is the effective volumetric reaction rate coefficient. To have continuous mass transport across the different interfaces, the following boundary condition is used

$$\left[(1 - \sigma_f)\, v_n c - \mathcal{D}_e \frac{\partial c}{\partial n}\right]\Big|_{+} = \left[(1 - \sigma_f)\, v_n c - \mathcal{D}_e \frac{\partial c}{\partial n}\right]\Big|_{-} \tag{7}$$

where v_n is the normal velocity to the interface. A detailed overview of the model parameters, which include density, diffusivity, viscosity, permeability, porosity and reflection coefficients for each individual layer, collected from various sources in the literature, are shown in Yang and Vafai [3].

The multi layered arterial wall layers are generated around the patient specific carotid artery geometry received from the Erasmus Medical Center in Rotterdam (the computed tomography (CT) scans), Fig. 2-top. The important geometrical features of the circumferentially averaged diameters of the common carotid artery (CCA) (the main branch before bifurcation), and of the internal (ICA) and of the external carotid artery (ICA) (branches after the bifurcation), are 6.5, 4 and 3.2 mm, respectively. The total length of the segment of the carotid artery bifurcation is about 12 cm. The numerical mesh is refined in the radial direction in the proximity of the interfaces between layers. In total, about $2{\times}10^6$ control volumes were sufficient to get grid-independent results, with a second-order linear upwind scheme used for the convective terms in the momentum and concentration equations. The boundary conditions are imposed to ensure the medically recorded mass-flow ratios through the ICA and ECA of 0.6:0.4, Groen *et al.* [7]. The characteristic transmural pressure gradient (a pressure difference between the interior and exterior of the artery) of 70 mmHg is specified to mimic a normal blood pressure. At the inlet, a fully developed laminar velocity profile is imposed to give Re=105. The inlet value of the LDL concentration of c_0=2.5 mol/m^3 (a normal value for a healthy person) is specified. We are interested in the long-time accumulation of LDL within the artery wall, so the system of equations given by Eqs.(1)-(6) is solved in a steady mode. The blood flow obtained in the lumen is shown in Fig. 3-left. It can be seen that the arterial bifurcation generates complex helical flow patterns, which is rapidly decaying farther downstream. The contours of the LDL concentration at the arterial wall (lumen/endothelium interface) are shown in Fig. 3-right. The strong variation of the LDL concentration levels is observed in the proximity of bifurcation, confirming the importance of the underlying blood flow patterns on the mass transfer within the arterial wall, as shown in Kenjereš and de Loor [6]. Note that an increase of about 3% is obtained, confirming proper capturing of the concentration polarization. The standard convective-diffusive type of mass transfer is not able to predict such a mechanism. To capture polarization effects,

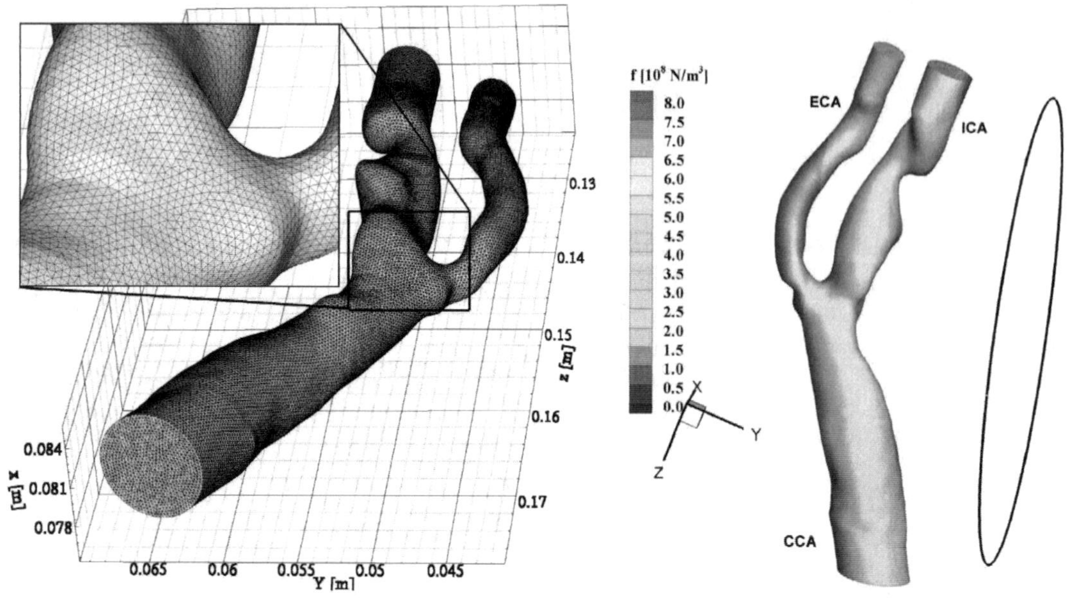

Figure 5. The numerical mesh used for the pulsating blood flow simulation in the patient-specific carotid artery bifurcation (-left) and the spatial distribution of the imposed magnetization force field to steer magnetic drug carriers to the right part of the bifurcation (-right).

it is crucial to implement biological membrane effects, as shown in Eq.(5). The locally increased values of the LDL concentration indicate locations where atherosclerosis may take place. Finally, the radial profiles of the LDL concentration at characteristic locations shown in Fig. 4-left are compared with experimental results. These experiments were performed in the rabbit artery subjected to identical values of the transmural pressure of 70 mmHG, Meyer *et al.* [8]. A good agreement between numerical and experimental results is obtained, confirming correct underlying physics of the two-way coupling between the blood flow and the "bad" cholesterol concentration under realistic biological conditions.

3. Pulsating blood flow and magnetic particle deposition in the carotid artery with stenosis

The following example addresses the pulsating blood flow within a patient-specific diseased carotid artery, as shown in Fig. 5. This can be seen as a follow-up investigation of atherosclerosis development, but now, after the disease has become established. Here, we would like to explore the possibility of delivering a medical drug primarily to the stenotic regions (regions where narrowing or constrictions of the arterial inner surface, caused by atherosclerosis, are present). In addition to the patient-specific geometry, the characteristic pulsating flow rates at the inlet of the central-carotid-artery (Q_{CCA}=1.1×10^{-5} m^3/s, Holdsworth *et al.* [9]) and at the outlets (internal/external) of the carotid artery are specified as in previous case, i.e. Q_{ICA}:Q_{ECA}=0.6:0.4, Groen *et al.* [7]. Due to the pulsating nature of the incoming velocity, flow exhibits a range of Reynolds numbers (0≤Re≤550), and it is important to include non-Newtonian behavior of the blood viscosity. Here, we adopted a generalized power law model of Ballyk *et al.* [10], which takes into account the shear-thinning of the whole blood caused by the red blood cell aggregation and dispersion at low shear rates that increases blood viscosity:

$$\mu(\dot{\gamma}) = \lambda |\dot{\gamma}|^{n-1} \quad (8)$$

$$\lambda(\dot{\gamma}) = \mu_\infty + \Delta\mu \, exp\left[-\left(1 + \frac{|\dot{\gamma}|}{a}\right) e^{-\frac{b}{|\dot{\gamma}|}}\right] \quad (9)$$

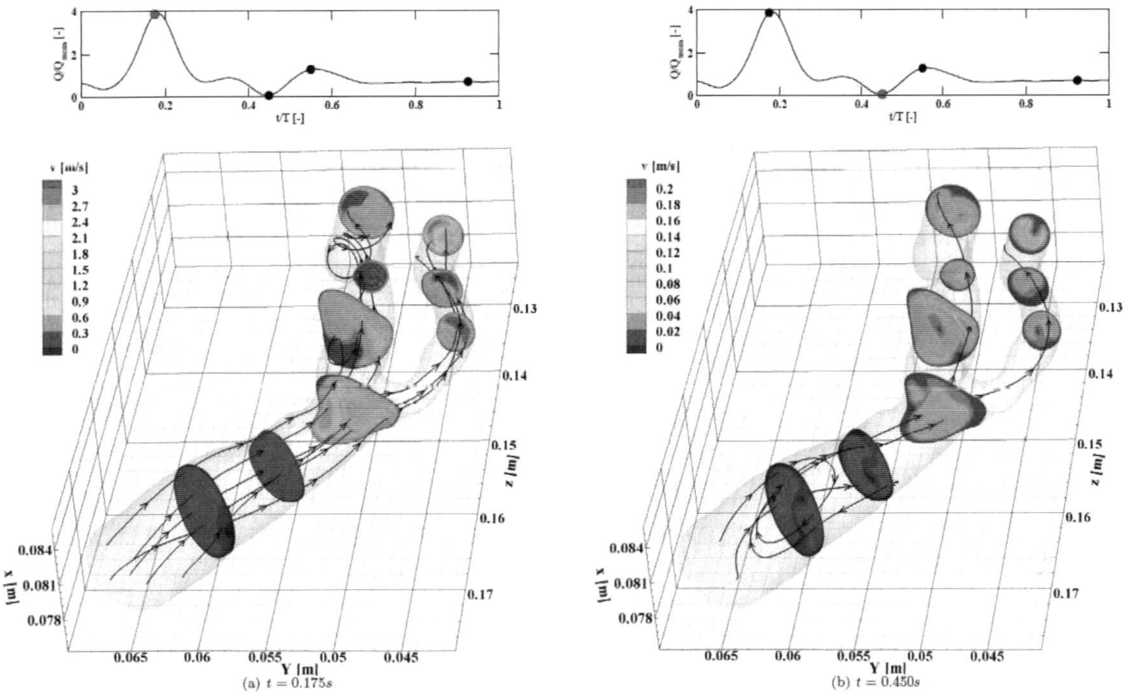

Figure 6. The calculated flow patterns at two characteristic instantaneous positions during the cardiac cycle: systole (the heart contraction, local maxima peak) (-left) and diastole (the heart relaxation, local minima peak) (-right). The carotid artery flow rate as a function of time is specified as given in Holdsworth *et al.* [9].

$$n(\dot{\gamma}) = n_\infty - \Delta n \, exp\left[-\left(1+\frac{|\dot{\gamma}|}{c}\right)e^{-\frac{d}{|\dot{\gamma}|}}\right] \tag{10}$$

$$\dot{\gamma} = \sqrt{S_{ij}S_{ji}}, \;\; S_{ij} = \frac{1}{2}\left(\frac{\partial v_i}{\partial x_j} + \frac{\partial v_j}{\partial x_i}\right) \tag{11}$$

where μ_∞=0.0035 Pa·s, $\Delta\mu$=0.025 Pa·s, n_∞=1, Δn=0.45, a=50 s^{-1}, b=3 s^{-1}, c=50 s^{-1}, d=4 s^{-1}, Ballyk *et al.* [10]. The numerical mesh is generated from the patient-specific surface geometry and contains approximately 1.2×10^6 tetrahedra, Fig. 5-left. The second-order upwind discretization scheme is used for the convective terms of the momentum equation. The second-order fully implicit time-integration scheme with a time step of Δt=0.005s is used for the transient terms. The characteristic non-dimensional parameters are Reynolds number ($16\leq$Re$=V_0D/\nu\leq512$) and Womersley number ($\alpha=D/2\,(2\pi f/\nu)^{1/2}$=5.23). To get an impression of the significant changes in the blood flow patterns during the cardiac cycle, stream-traces and contours of the velocity magnitude at characteristic cross-sections are shown in Fig. 6. It can be seen that the stenosis in the internal-carotid-artery generates the persistent recirculation zone (present during the entire pulsating cycle), Fig. 6-left. In contrast to that, during the diastolic (the lowest flow rate), a significant flow reversal occurs in the proximity of the inlet plane of the central-carotid-artery, Fig. 6-right. To check the accuracy of numerical simulations we performed the PIV and MRI experimental studies. These measurements are carried out in the PDMS-built carotid-artery phantoms based on the rapid-prototyping technique, which makes it possible to construct models from the MRI or CT scans of the vascular systems of real patients. A summary of the working fluid properties (water/glycerol mixture) and characteristic flow parameters are given in Table 1.

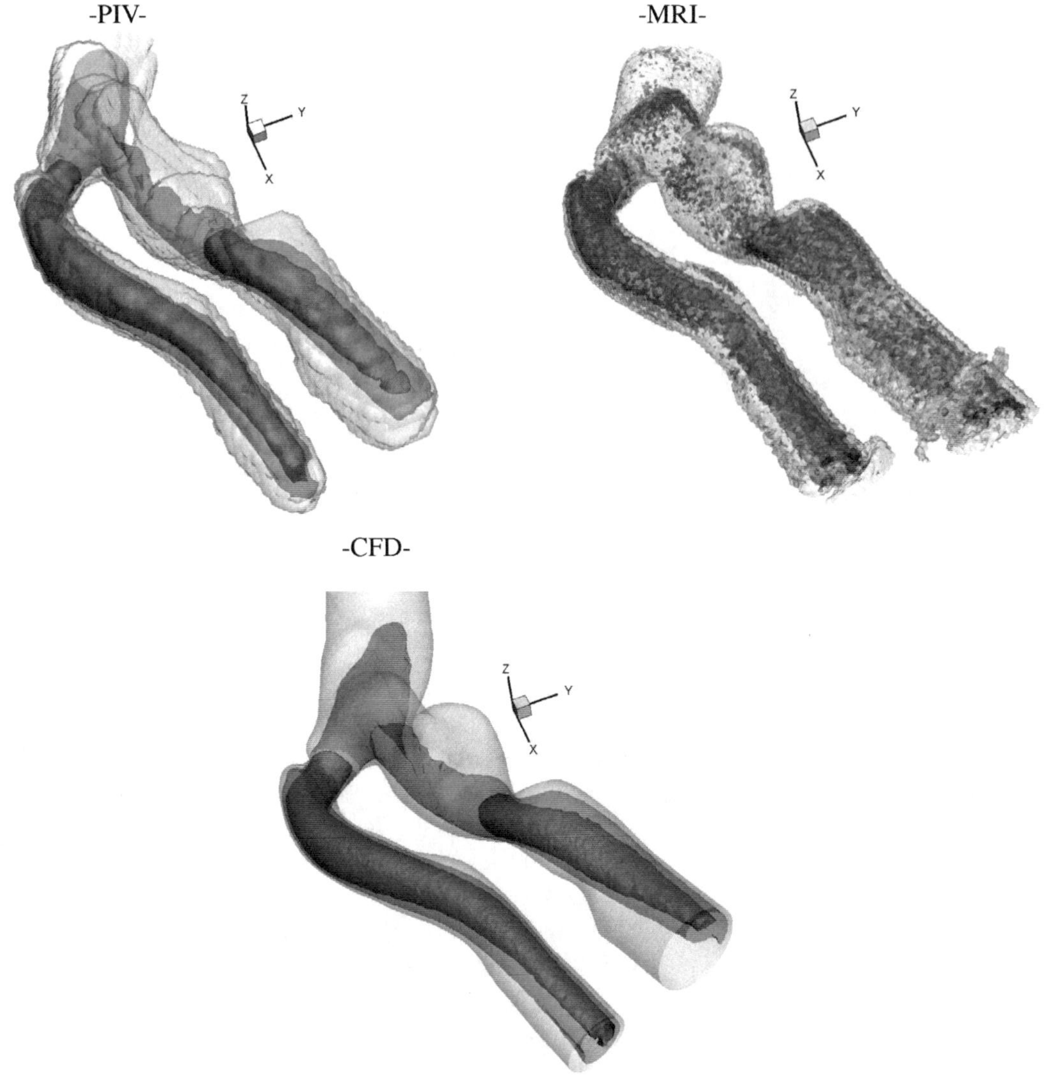

Figure 7. Isosurfaces of the non-dimensional velocity magnitude ($|v|/v_0$) for the steady flow conditions at Re=284 and outlet flow ratios of ICA:ECA=0.5:0.5; $|v|/v_0 = 0$ (-grey), 2 (-blue) and 3.5 (-red). Note that for PIV, only two velocity components are used (v_x and v_y) (from a total of 21 cross-sections in the z-direction).

The MRI measurements were performed in Leiden University Medical Centre, Leiden, The Netherlands with the Philips Achieva 7 T whole body MRI scanner. Isosurfaces of the measured and simulated velocity magnitude are shown in Fig. 7. Note that for the PIV measurements, only two available velocity components (v_x and v_y) are used, measured in 21 characteristic planes (with 1 mm increments) along the z-coordinate direction. It can be seen that overall, very good agreement between the PIV and CFD is obtained, especially in regions where the flow acceleration takes place (indicated by red iso-surfaces), whereas MRI was able to provide qualitatively reasonable results. Note that only standard clinical set-up working conditions of the MRI scanner have been used without addition of phase-contrast agents. The typical MRI scanning time was about 45 min. Typical resolutions are 0.47:0.47 mm^2, 0.35:0.35:0.7 mm^3 and 0.075:0.075:0.075 mm^3 for PIV, MRI and CFD, respectively. More detailed comparisons are shown

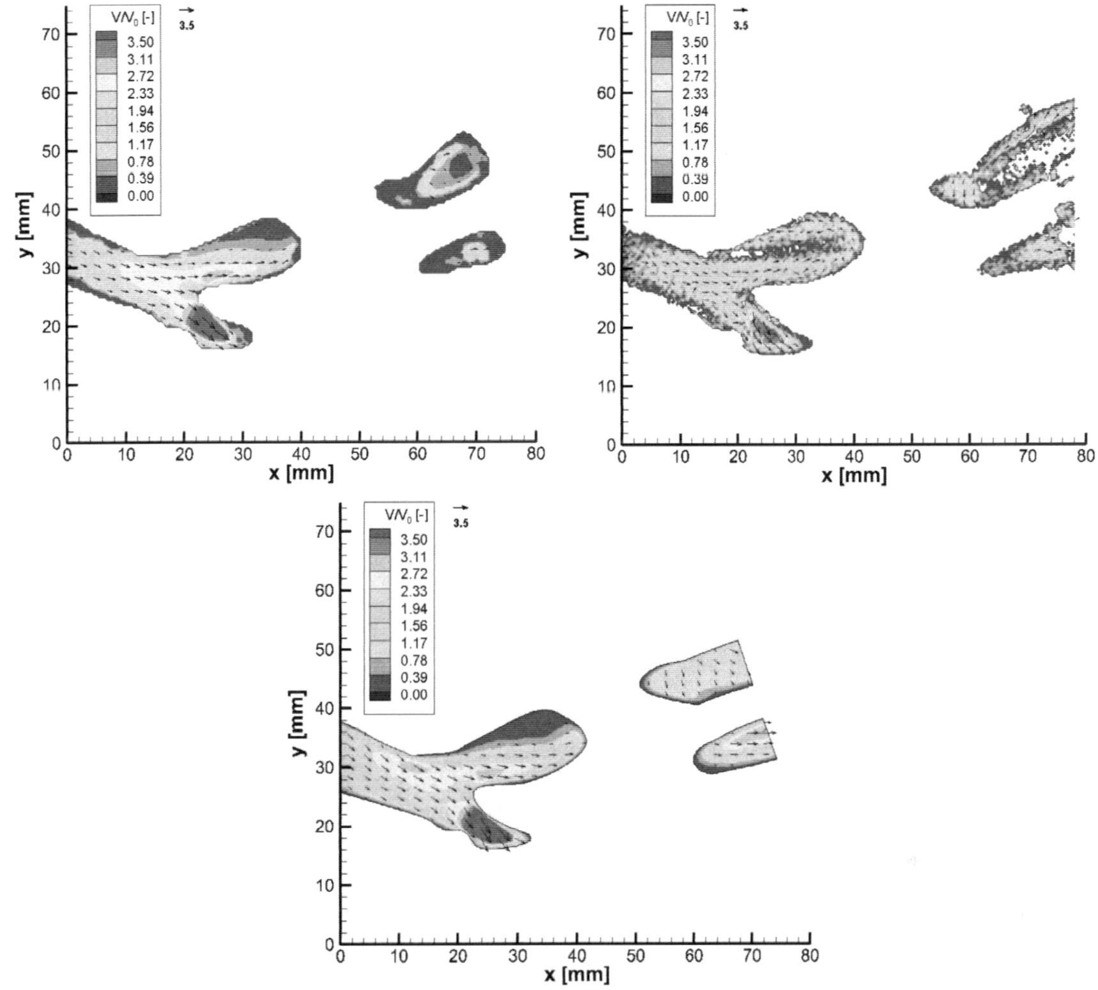

Figure 8. Comparison between the PIV, MRI and CFD results in the characteristic horizontal cross-section for the patient-specific carotid artery shown in Figs. 5 and 7.

in Figs. 8 and 9, where contours and profiles of the non-dimensional velocity magnitude are shown. Exact locations at which profiles and contours are extracted are indicated in Fig. 9-top, where the blue mask indicates plane and A1-A2 lines indicate profiles. It can be seen that PIV and CFD show very good agreement and that MRI is able to provide qualitatively overall trends in the velocity profiles (especially at the A2 location, just after carotid bifurcation), Fig. 9-bottom/right.

After obtaining reliable flow data, we investigate possibilities to impose a local delivery of the medical drug with magnetic cores. This is the essence of the Magnetic Drug Targeting (MDT) technique. The MDT showed very promising results in experimental studies on animals and in pre-clinical studies on human patients, e.g. Lübbe *et al.* [12],[13], Arnold and Parak [14], Torchilin [15]. In our previous numerical studies, we investigated various combinations of the MDT concept in simplified (Kenjereš [16], Haverkort *et al.* [17], Stuart *et al.* [18]) and complex (Haverkort *et al.* [19], Kenjereš and Righolt [20]) geometries. For all cases, a significant improvement in local capturing of the magnetic particles is demonstrated, proving the potential of MDT. The MDT approach requires introduction of the

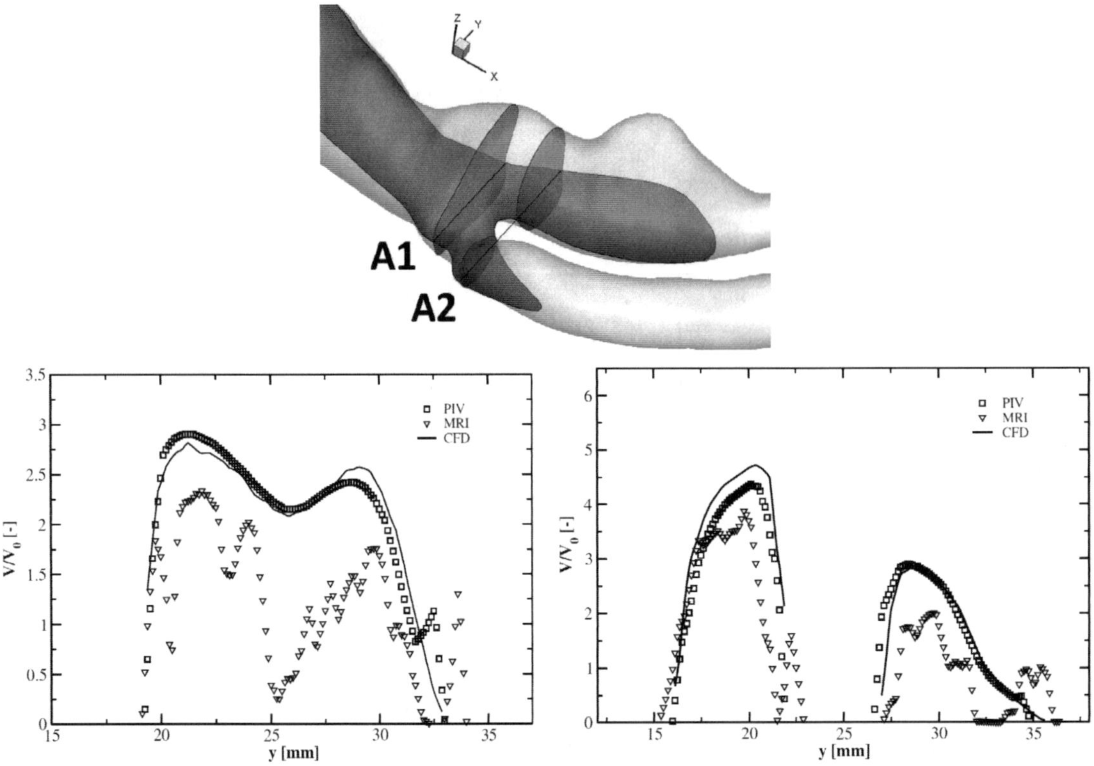

Figure 9. Comparison of the measured (PIV and MRI) and numerical (CFD) velocity profiles at characteristic locations before (A1)(-left) and after (A2)(-right) carotid bifurcation for identical flow conditions as shown in Figs. 7 and 8.

Table 1. Fluid properties and settings for PIV measurements

Parameter	Value	Unit
Glycerol concentration	64	% by wt.
Dynamic viscosity	1.5×10^{-2}	Pa s
Density	1165	kg/m^3
Temperature of fluid	25	^{0}C
Refractive index	1.413	-
Flow rate	2 - 63	ml/s
Outflow ratio ICA/ECA	50/50, 60/40	%
Inlet Reynolds number	16 - 512	-
Womersley number	5.23	-
PIV particle size	8 - 12	μm

spatial distribution of the imposed magnetic field (as shown in Fig. 5-right) and additional equations for Lagrangian particle tracking. The latter is achieved by considering a balance of all active forces acting on particles:

$$m_p \frac{\partial \vec{v}_p}{\partial t} = \vec{f}_d + \vec{f}_v + \vec{f}_m \tag{12}$$

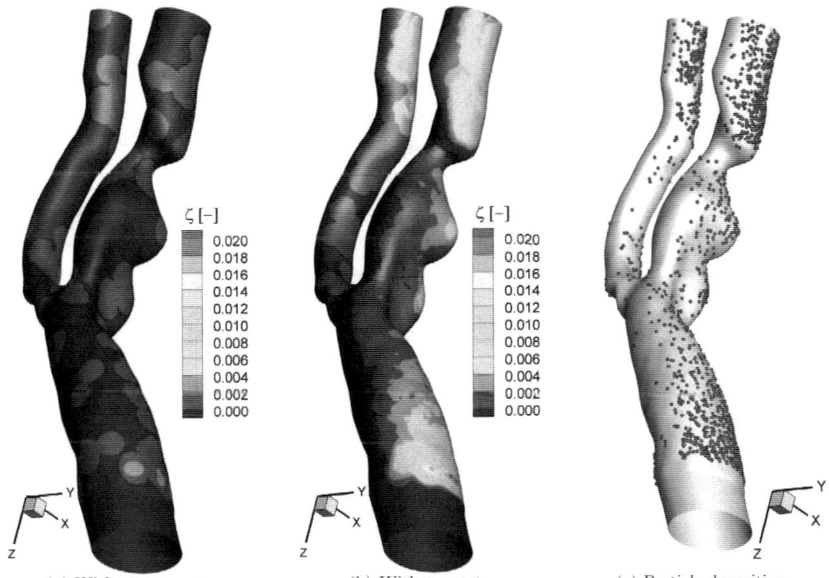

(a) Without magnet (b) With magnet (c) Particle deposition

Figure 10. Contours of the local deposition efficiency (ζ) for particles with magnetic core (d_p=d_m=2μm): no magnetic field (-left), with magnetic field (-middle). The particles deposition along the artery wall, with (red) and without (blue) an imposed magnetic field ($|b_0|_{max}$=2 T) (-right).

where $\vec{v}_p$ is the particle velocity, and $\vec{f}_d$, $\vec{f}_v$ and $\vec{f}_m$ are drag, virtual mass and magnetization force, respectively. The drag force is calculated as:

$$\vec{f}_d = \frac{1}{2}\rho A_\perp C_D(\mathrm{Re}_p)\,|\vec{v}-\vec{u}_p|\,(\vec{v}-\vec{u}_p) \tag{13}$$

$$\mathrm{Re}_p = \frac{\rho_p D_p\,|\vec{v}-\vec{u}_p|}{\mu} \tag{14}$$

where ρ_p, D_p and $A_\perp$ are the particle density, diameter and cross-section area, respectively. C_D is a drag coefficient dependent on Re_p, i.e. C_D=24/Re_p for $\mathrm{Re}_p < 0.1$, and $C_D = 3.69 + 22.73/\mathrm{Re}_p + 0.0903/\mathrm{Re}_p^2$, for $0.1 \leq \mathrm{Re}_p \leq 1$, Morsi and Alexander [11]. The virtual, or added, mass force is calculated as:

$$\vec{f}_v = \frac{1}{2}m_p\frac{\rho}{\rho_p}\frac{d}{dt}(\vec{u}-\vec{v}_p) \tag{15}$$

and finally, the magnetic force is:

$$\vec{f}_m = V_m\mu_0\vec{M}\cdot\nabla\vec{H}, \tag{16}$$

with V_m the magnetic core volume, μ_0 the permeability of free space, $\vec{M}$ the magnetization force per unit area and $\vec{H}$ the auxiliary magnetic field. For fully saturated super-paramagnetic particles, $\vec{M}$=$M_{sat}\vec{H}/|H|$, where M_{sat} is the saturation magnetization. The auxiliary magnetic field can be obtained from Biot-Savart's law for different combinations of the electric current carriers (single/multiple straight wire, single/multiple loop-wire, etc.). Since a very diluted concentration of the injected medical drug is imposed, one-way coupling between particles dynamics and blood flow is assumed.

The particles are uniformly injected at the inlet section of the CCA. The iron-core particles with saturation magnetization of M_{sat}=10^6 A/m and with density of ρ_p=6450 kg/m^3 are considered. A range of particle diameters is investigated, i.e. $0.25 \leq d_p \leq 4$ μm, where the largest value is determined by the

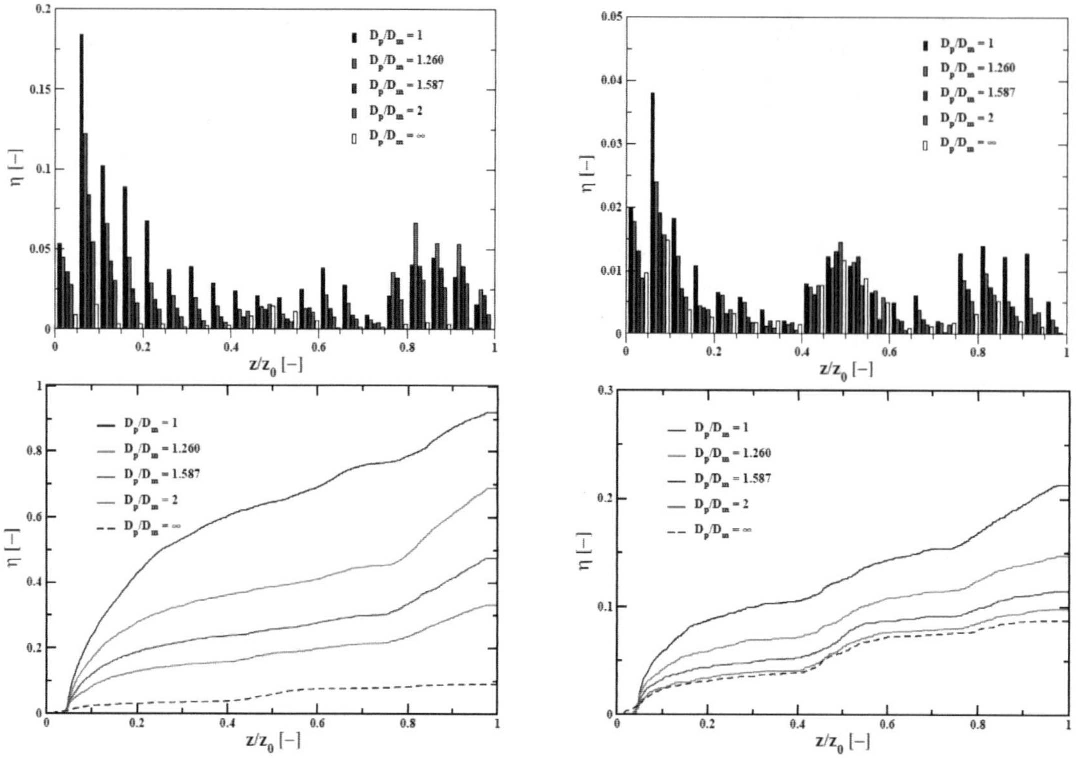

Figure 11. The local (-top) and cumulative (-bottom) capture efficiency for the CA bifurcation and particles of d_p=4 μm (-left) and 1 μm (-right) (external diameter) and different magnetic core diameters (d_m) for an imposed magnetic field of $|b_0|_{max}$=2 T (shown in Fig. 5-right).

characteristic diameter of the micro-capillary vascular system. This gives the following range of the non-dimensional Stokes numbers: $2\times10^{-7}\leq \mathrm{St}=\rho_p d_p^2 v_0/18\mu D \leq 5.5\times10^{-5}$. The magnetic-particle number is introduced, which represents a ratio between the magnetization and drag forces ($\mathrm{Mn}_p=f_m/f_d$)). The 5th order Runge-Kutta time-integration scheme is used to solve Eq.12. In addition to the total integral particle deposition, we also introduce a local particle deposition efficiency, which is calculated as:

$$\zeta(\vec{x}) = \frac{n_{wall}\left(|\vec{x}_{p,i} - \vec{x}| < \epsilon\right)}{n_{in}} \tag{17}$$

and represents the amount of particles deposited along the wall within a pre-specified neighborhood where ϵ is the characteristic sample distance (usually 1 mm). Here, n_{in} is the total number of injected particles, n_{wall} is the number of particles deposited along the wall, and $\vec{x}_{p,i}$ is the deposition location of the i-th particle. Two main conclusions can be drawn from Fig. 10. First, that it is possible to redirect the magnetic-core particles towards the internal carotid artery (where the stenotic region is present) and second, that local deposition efficiency of the particles with magnetic core can be significantly enhanced in comparison with particles without magnetic core. An overview of the local and cumulative deposition efficiencies for different classes of the particles is shown in Fig. 11. For the d_p=4μm class of particles, the cumulative deposition efficiency shows an increase from 0.1 (no magnetic-core) to 0.7 (d_p/d_m=1.26), Fig. 11-bottom/left. For smaller particles (d_p=1μm), the magnetic-capturing is less efficient but still significant, e.g. 0.09 (no magnetic core) versus 0.21 (with magnetic core), Fig. 11-bottom/right. From the standpoint of medical applications, it is particularly important to know the local deposition efficiency. Specifically, an estimation of the quantity of the injected medical drug delivered

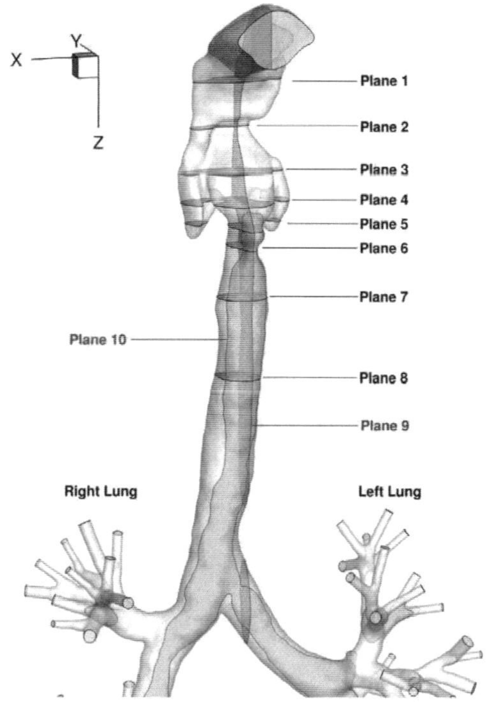

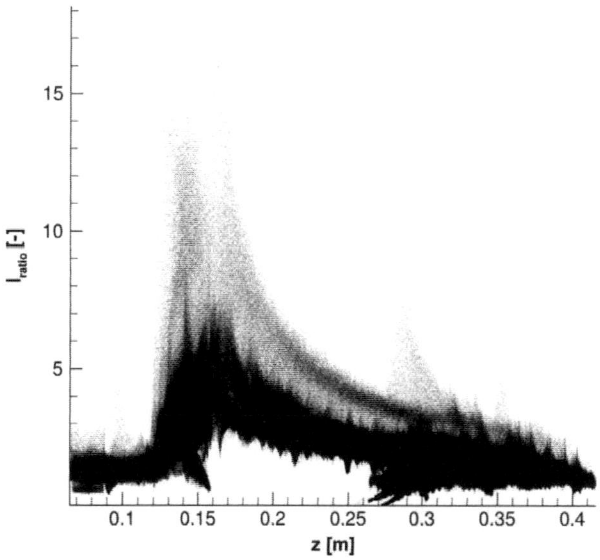

Figure 12. Left- geometry of the upper-airways with up to 8 generation of the bronchioles bifurcations in the human respiratory system; Right- the scatter-plot of the ratio between typical Kolmogorov length scales ($\eta_k = (\nu^3/\varepsilon)^{1/4}$) and characteristic control-volume based length-scales ($\Delta V^{1/3}$) for the numerical tetrahedral mesh that was generated.

to the pre-defined region(s) is essential. For the carotid-artery bifurcation, the local deposition efficiency in the proximity of the stenosis ($z/z_0 \approx 0.8$) shows a more than eight-fold and three-fold increase for particles with magnetic core ratio of $d_p/d_m = 1.26$, with $d_p = 4\mu$m and $d_p = 1\mu$m, respectively.

It can be concluded that multi-phase simulations of the pulsating blood flow with injected magnetic-core particles subjected to an external magnetic field gradient, can be powerful tool for the patient-specific optimization of the local medical-drug delivery for advanced stages of cardiovascular diseases.

4. Magnetic control of aerosol distribution in the human respiratory system

A dramatic increase of the respiratory diseases among children a result of exposure to increased environmental air pollution is reported in Chauhan and Johnston [21] and Ayres *et al.* [22]. Especially alarming is the evidence that exposure to outdoor air pollution causes lung cancer [2]. In that report, particulate matter (complex mixture of small particles and liquid droplets) was classified as carcinogenic to humans. Of particular concern are particles of 10 μm or smaller, because of their ability to enter the lungs. Numerical simulations of the upper airways in humans can be a powerful tool for prediction of the distribution of different classes of inhaled particles. In addition, computer studies can improve different strategies to deliver medical drug aerosols to specific locations within the human respiratory system to fight respiratory diseases, Kleinstreuer *et al.* [23], Grotberg [24]. Dames *et al.* [25] reported on a novel experimental technique of targeted delivery of magnetic aerosol droplets to intact mice lungs. In these studies, it is demonstrated that by imposing sufficiently strong magnetic field gradients close to the main

[2] World Health Organization, International Agency for Research on Cancer, Press release No.221, the 17th of October, 2013

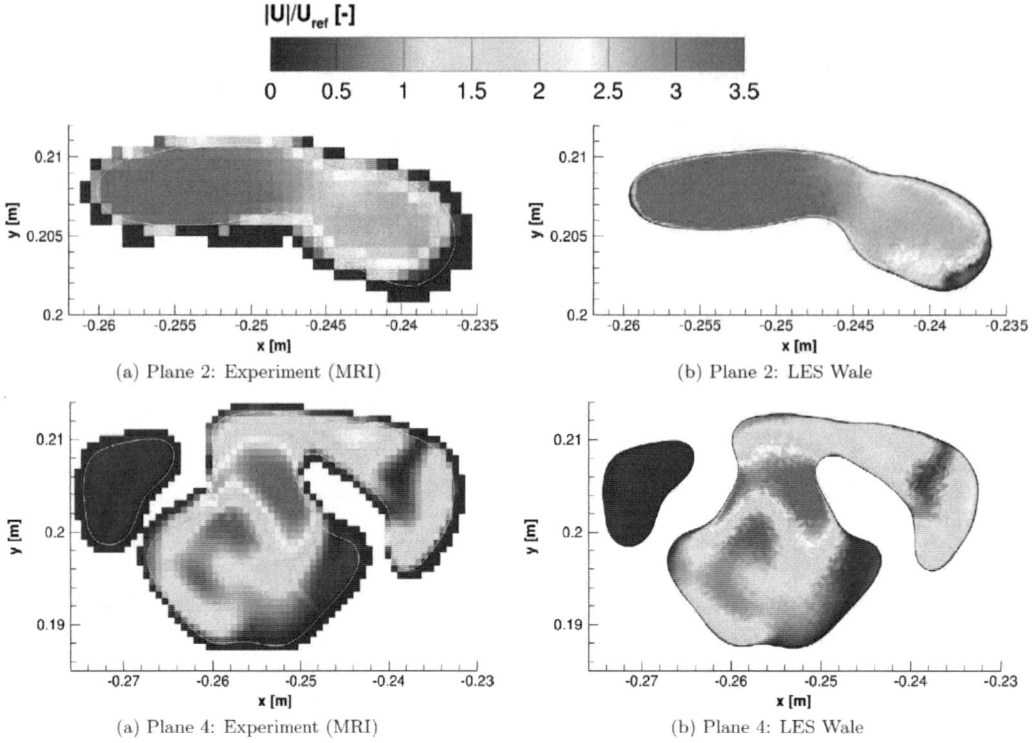

Figure 13. Comparison between the water based MRV experiments (performed by the Stanford University group, Coletti *et al.* [28]) and LES-WALE simulations for contours of the velocity magnitude in two characteristic planes shown in Fig. 12-left: Plane2 (z=0.1 m) (-top) and Plane4 (z=0.146 m) (-bottom).

trachea bifurcation, steering of magnetic droplets towards the right lung can be achieved, resulting in an overall four-fold increase in the drug deposition compared to the neutral situation. In conclusion, Dames *et al.* [25] proposed further developments and scaling-up of the presented approach to human lungs.

The next example in our study is drug-aerosol delivery in the human respiratory system, shown in Fig. 12-left. In contrast to the complex blood-rheology that should account for the non-Newtonian effects for simulations of blood flow in cardiovascular system, here we are dealing with simple air as a working fluid. The modeling challenge here lies in the significantly higher values of Reynolds number (up to 4500, for an inflow rate of 60 L/min), which in combination with the complex geometry of the human airway system, requires capturing the local turbulence generation. To achieve this, we applied two approaches: two-equation based eddy-viscosity RANS, and LES. The first one is employed because of its simplicity and relatively low computational costs. Moreover, the RANS results are used to make the preliminary estimates of the dissipation rate of the turbulent kinetic energy. This information is then used to properly redesign and redefine numerical mesh, which can be used for the reasonable resolved LES. The SST-Transient model is applied because of its suitability for flows where some transient effects can play a significant role, Menter *et al.* [26]. The steady inflow conditions of 60 L/min that correspond to the top range of lung capacity during extensive exercise is considered. The scatter-plot of the characteristic ratio between typical Kolmogorov length-scales and the characteristic control-volume based length-scale ($\Delta V^{1/3}$) is shown in Fig. 12-right. This is done for a numerical mesh containing approximately 15×10^6 control volumes. It can be seen that except for the region around z=0.15 m, this ratio is less than 5, indicating a good mesh for LES. Here, we applied LES with the wall-adapting-local-eddy-viscosity

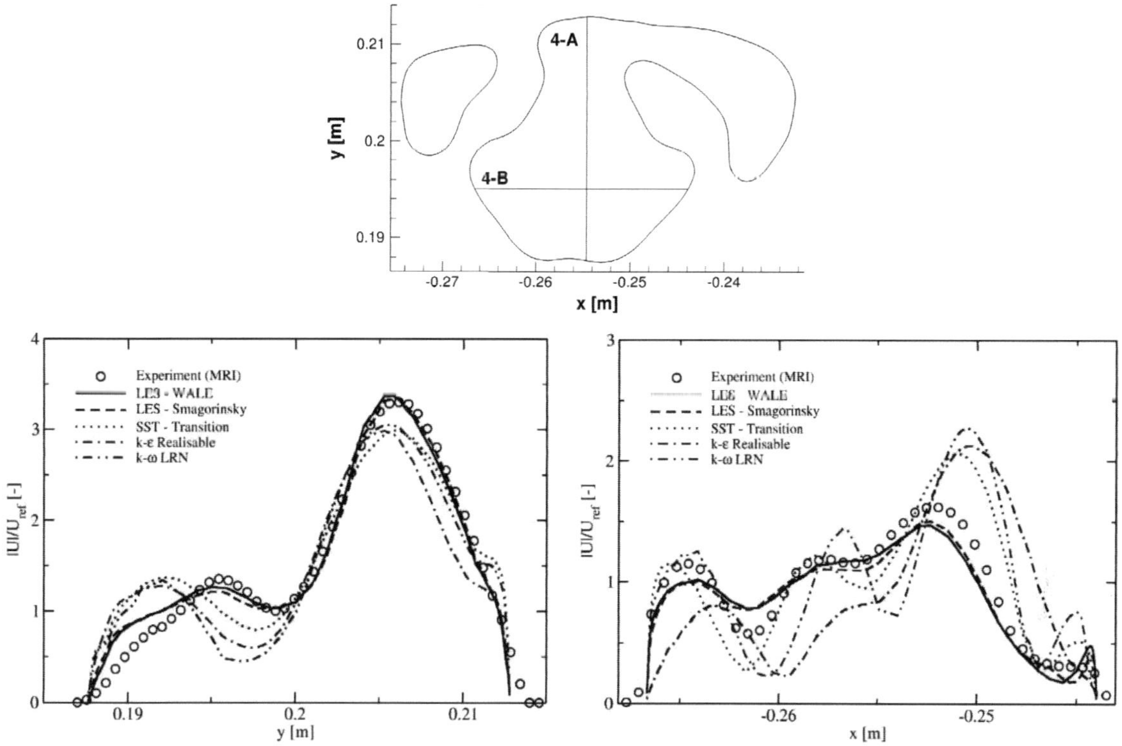

(a) Line 4 - A (b) Line 4 - B

Figure 14. Comparison of the characteristic profiles (along the 4A and 4B lines, in plane 4 shown in Fig. 12-left, z=0.146 m) of the velocity magnitude obtained in the MRV experiment (performed by the Stanford University group), various RANS and LES.

(WALE) subgrid-closure, Nicoud and Ducros [27]:

$$\nu_{SGS} = L_s^2 \frac{\left(s_{ij}^d s_{ij}^d\right)^{3/2}}{\left(\overline{s}_{ij}\overline{s}_{ij}\right)^{5/2} + \left(s_{ij}^d s_{ij}^d\right)^{5/4}},$$

$$\overline{s}_{ij} = \frac{1}{2}\left(\frac{\partial v_i}{\partial x_j} + \frac{\partial v_j}{\partial x_i}\right),\; s_{ij}^d = \frac{1}{2}\left[\left(\frac{\partial v_i}{\partial x_j}\right)^2 + \left(\frac{\partial v_j}{\partial x_i}\right)^2\right] - \frac{1}{3}\left(\frac{\partial v_k}{\partial x_k}\right)^2 \delta_{ij} \quad (18)$$

and where L_s=min($\kappa d, C_w \Delta$), with C_w=0.35 and d as the distance to the closest wall. The time integration is performed with a fully implicit second-order scheme with characteristic time step of Δt=10^{-4} s. Comparison of the velocity magnitudes of the water-based magnetic resonance velocimetry (MRV) [3] experiment and long-term averaged LES-WALE simulations in a few characteristic cross-sections is shown in Fig. 13. It can be seen that very good agreement is obtained for both locations. In addition, more detailed comparisons of the velocity magnitude profiles are shown in Fig. 14. It can be concluded that LES simulation provided the best agreement with MRV profiles for both locations. The contours of the turbulence kinetic energy in the central vertical plane are shown in Fig. 15. It can be seen that turbulence is generated just after the oropharynx region (the narrowing bend) and that it decays quite

[3] The MRV experiments were performed by the Stanford University group using a 3T General Electric whole body scanner with a spatial resolution of 0.7 mm, Coletti *et al.* [28].

Journal of Physics: Conference Series **530** (2014) 012006 doi:10.1088/1742-6596/530/1/012006

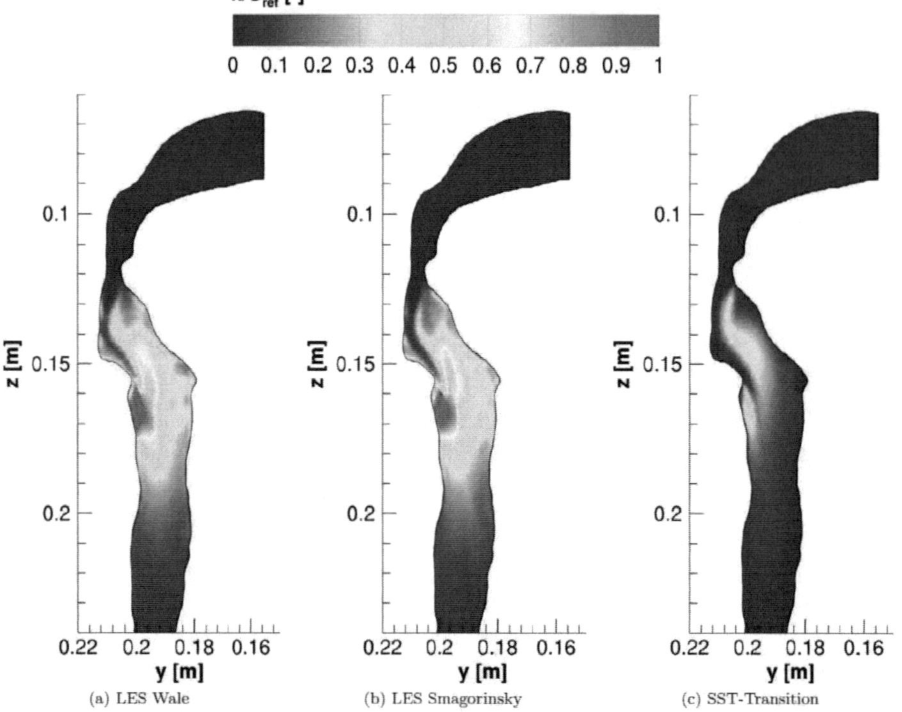

Figure 15. Contours of the turbulent kinetic energy in the central vertical plane (Plane 9, x=0.255 m, shown in Fig. 12-left) obtained with LES with two SGS models (Smagorinsky and WALE, [27]) and RANS SST-Transient model, [26].

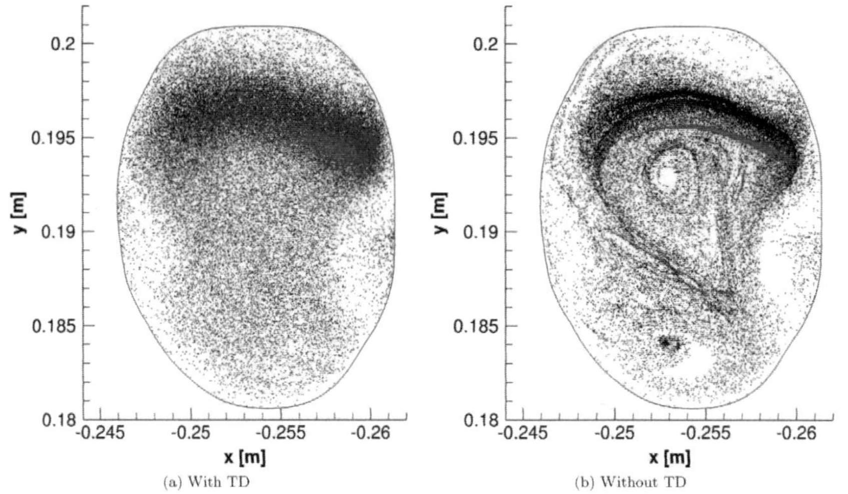

Figure 16. Particle distributions at the plane A (z=0.17 m) shown in Fig. 12-left, with (-left) and without (-right) turbulent dispersion term: St=4.8×10^{-5} (black), 4.3×10^{-2} (blue) and 1.2×10^{-1} (red), which correspond to d_p=0.1, 3 and 5 μm, respectively.

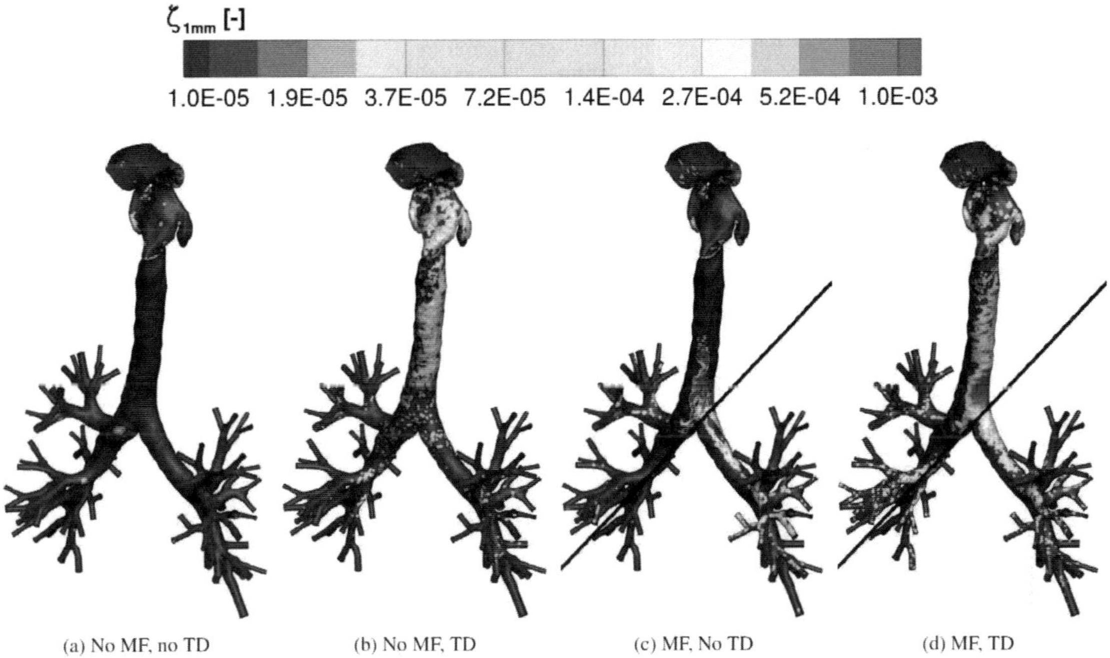

Figure 17. Contours of the local deposition efficiency in the upper-airways human respiratory system for the magnetic-core particles (St=4.3×10^{-2}, Mn_p=1.7×10^{-1}) with and without turbulent dispersion and imposed magnetic field gradient.

rapidly after that. It is also noted that LES levels of the turbulence kinetic energy are significantly higher than that of RANS which can have consequences when the drug-aerosol distribution within the human respiratory system needs to be addressed. This is illustrated in Fig. 16, where distributions of injected particles in the characteristic cross-section (Plane A), with and without turbulent dispersion terms, are compared. It can be seen that turbulent dispersion produces more homogeneous distributions.

Contours of the local deposition efficiency (ζ_{1mm}) for the St=4.3×10^{-2} class of particles are shown in Fig. 17. Situations with particles without and with magnetic cores are considered for an imposed magnetic field gradient (the current-carrying wire), as well as whether the influence of the turbulent dispersion is included or not. It can be concluded that a significantly higher deposition efficiency is obtained for the magnetically affected particles. Also, the steering of the magnetic particles towards the left side of lungs can be effectively achieved. An overview of the total deposition efficiencies for different values of St for simplified and real human upper-airways geometries is shown in Fig. 18-top. It can be seen that the present simulations agree very well with similar experimental Cheng *et al.* ([29]) and numerical simulations Kleinstreuer *et al.* ([23]) in literature. The effects of the imposed magnetic field on capturing of the magnetic-particles are shown in Fig. 18-bottom. The most significant enhancement is observed in $5\times10^{-4}\leq St\leq5\times10^{-2}$ range. For example, for St=0.01, the five- and ten-fold increases are obtained for two configurations of the imposed magnetic field.

In conclusion, it is demonstrated that LES proved to provide velocity predictions in excellent agreement with MRV measurements of the upper airways of the human respiratory system. Also, it is shown that the magnetic-drag-targeting concept can be successfully applied for this geometry as well.

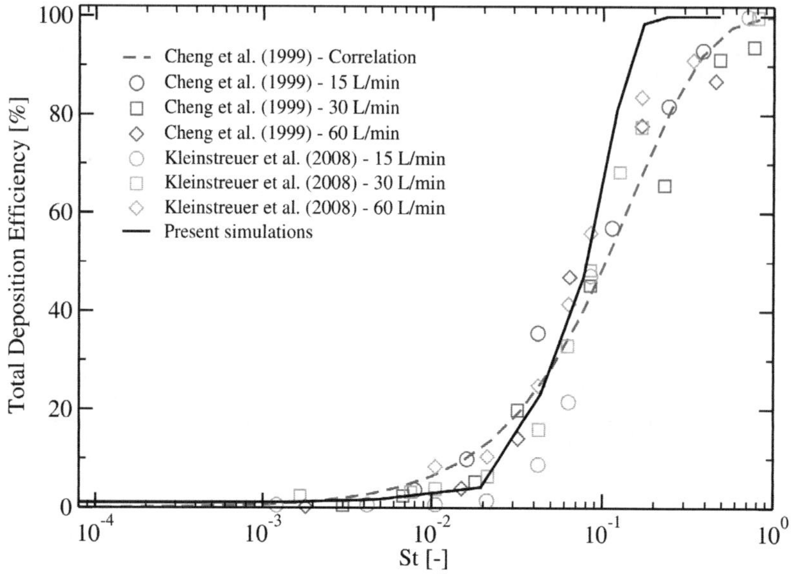

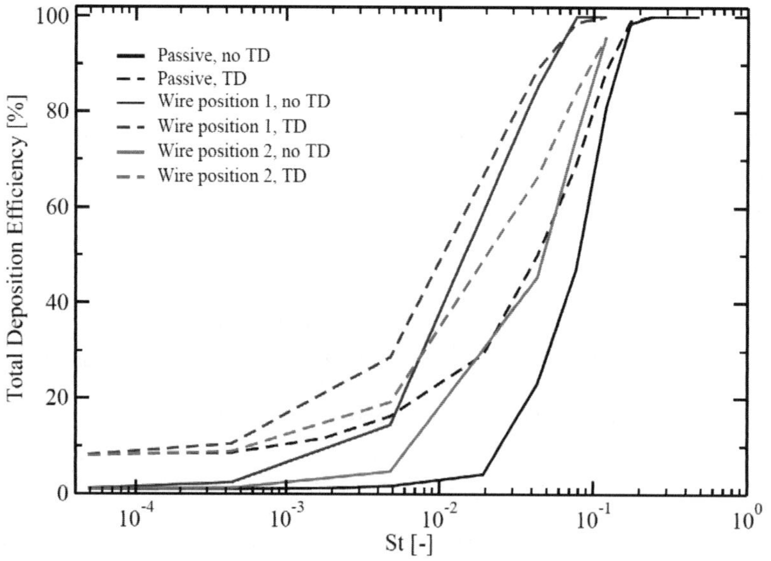

Figure 18. The total deposition efficiency for different classes of particles (top-no magnetic core; bottom- magnetic core) within the upper airways of the human respiratory system shown in Fig. 12. Comparison with some similar studies in literature performed on simplified geometries of the upper airways, Cheng *et al.* [29]. Two characteristic locations of the current-carrying wire are considered: position 1, x=-0.252 m, z=0.269 m, distance y=0.01 m; position 2, distance y=0.1 m.

5. Conclusions and outlook

We demonstrated that it is possible to perform advanced numerical simulations of the blood or air flow in patient-specific configurations. The first example addressed blood-flow simulation in the healthy and diseased patient-specific carotid artery bifurcations. A recently developed mathematical model of the LDL and plasma distribution through the multi-layer structure of the arterial wall was successfully coupled with the blood-flow predictions within the carotid artery bifurcation. By comparing results with experimental results of LDL distributions within the arterial wall of a rabbit artery under identical transmural pressure conditions, it is concluded that the present version of the model can be used to predict locations within the vascular system where atherosclerosis can develop. In the second example, we studied pulsating blood flow within diseased (advanced atherosclerosis stage) carotid artery, and tested concept of magnetic drug targeting for local delivery of the medical drug in the proximity of the stenotic (narrowing) regions. To provide an extensive dataset for validation of the numerical simulations, we performed a series of experimental PIV and MRI measurements. A good agreement between numerical simulations and experiments is obtained in predicting both temporal and spatial variations of the velocity of blood during the cardiac cycle. It is shown that application of the MDT technique can lead to significant enhancement of the local deposition efficiency of the medical drug in the proximity of diseased regions (up to ten-fold increase in comparison with the neutral case).

Finally, distributions of drug aerosols inside the upper airways of the complex human respiratory system (from the mouth to the eight bronchial generation) are performed. Special attention is devoted to proper capturing of the locally generated turbulence, which has significant impact on aerosols distribution in comparison with a laminar regime. The results of the LES with the wall-adapting local eddy-viscosity model (WALE) were in excellent agreement with MRV experiments performed in the scaled water-model. It was demonstrated that the concept of the magnetic steering can also be applied for the real scale human-respiratory system.

Future applications of combined experimental (stereoscopic PIV, PC-MRI and ultrasound measurements) and numerical studies in dealing with complex advanced physical transport phenomena in simplified generic and complex patient-specific geometries can lead to improvements of localized medical drug delivery, which can significantly increase drug concentrations at the targeted regions. Only integrated applications of the *in vivo* (living systems), *in vitro* (laboratory condition) and *in silico* (computer simulation) studies can bring insight to key aspects of the underlying biology and physics of localized medical drug delivery in the human circulatory and respiratory systems.

Acknowledgments

In the first place, I would like to acknowledge direct contributions from the 'MDT' team of students from the former Department of Multi-Scale Physics and current Transport Phenomena Section, at the Delft University of Technology that made this exciting research possible, and which results are shown in this paper: Alexander de Loor (modeling and simulations of atherosclerosis), Bernhard Righolt (CFD of the MDT in carotid artery bifurcation), Gidi van de Belt and Jasper van der Starre (for experimental studies in CA bifurcations), Jimmy Tjin (for simulations of aerosol distributions in human respiratory system). Dr. Frank Gijsen, Biomechanic Laboratory of the Erasmus Medical Centre in Rotterdam, for providing geometries of the patient-specific carotid artery bifurcations and for inspiring discussions. Dr. Thijs van Osch, Radiology Department of Leiden University Medical Centre, for his support in performing MRI measurements of the carotid bifurcations. Dr. Fillipo Coletti, Department of Mechanical Engineering, Stanford University, USA, for kindly providing unpublished MRV measurements of the flow in human respiratory system. I gratefully acknowledge assistance in several forms from a consortium of life-sciences and medical-technology partners in South Holland (Delft, Rotterdam, Leiden), within Medical Delta (http://www.medicaldelta.nl) framework.

References

[1] G A Truskey, F Yuan, D F Katz 2009 Transport Phenomena in Biological Systems, 2nd Edition, Pearson Prentice Hall Bioengineering
[2] J M Tarbell 2003 *Annual Review of Biomedical Engineering*, **5**, 79–118.
[3] N Yang, K Vafai 2006 *Int. J. Heat and Mass Transfer* **49**, 850–867.
[4] L Ai, K Vafai 2006 *Int. J. Heat and Mass Transfer* **49**, 1568–1591.
[5] U Olgac, D Poulikakos, S C Saur, H Alkadhi, V Kurtcuoglu 2009 *American Journal of Physiology - Heart and Circulatory Physiology* **296**, H1969–H1982.
[6] S Kenjereš, A de Loor 2014 *Journal of the Royal Society Interface* **11**(91), Art.No.20130941, 1–13.

[7] H C Groen, L Simons, Q J A van den Bouwhuijsen, E M H Bosboom, F H J Gijsen, A G van der Giessen, F N van de Vosse, A Hofman, A F W van der Steen, J C M Witteman, A van der Lugt, J J Wentzel 2010 *Journal of Biomechanics* **43** (12), 2332–2338.
[8] G Meyer, A Merval, A Tedgui 1996 *Circ. Res.* **79**, 532–540.
[9] D W Holdsworth, C J D. Norley, R Frayne, D A Steinman, and B K Rutt 1999 *Physiol. Meas.* **20**, 219–240.
[10] P D Ballyk, D A Steinman, C R Ethier 1994 *Biorheology* **31** (5), 565-586.
[11] S A Morsi, A J Alexander 1972 *J. Fluid Mech.* **55** (2), 193-208.
[12] A S Lübbe, C Bergemann, H Riess, F Schriever, P Reichardt, *et al.* 1996a *Cancer Research* **56** (20), 4686-4693.
[13] A S Lübbe, C Bergemann, W Huhnt, T Fricke, H Riess, J W Brock, D Huhn 1996b *Cancer Research* **56** (20), 4694-4701.
[14] W Arnold, F G Parak, 2006 *Eur. Biophys. J.* **35**, 446-450.
[15] V P Torchilin 2006 Nanoparticulates as Drug Carriers. Imperial College Press, London, U.K.
[16] S Kenjereš 2008 *Int. J. Heat and Fluid Flow* **29**(3), 752–764.
[17] J W Haverkort, S Kenjereš, C R Kleijn 2009 *Phys. Rev. E* **80**(1), Art.No.016302, 1–12.
[18] D C C Stuart, C R Kleijn, S Kenjereš 2011 *Computer and Fluids* **40**(1), 188–194.
[19] J W Haverkort, S Kenjereš, C R Kleijn 2009 *Annals of Biomedical Engineering* **37**(12), 2436–2448.
[20] S Kenjereš, B W Righolt 2012 *Int. J. Heat and Fluid Flow* **35**(3), 68–75.
[21] A J Chauhan, S L Johnston 2003 *British Medical Bulletin* **68**, 1–18.
[22] J. Ayres, R. Maynard, R. Richards 2006 Air Pollution and Health, Air Pollution Reviews - Vol.3, Imperial College Press, London, U.K.
[23] C Kleinstreuer, Z Zhang, and J F Donohue 2008 *Annual Review of Biomedical Engineering* **10**, 195-220.
[24] J B Grotberg 2011 *Physics of Fluids*, **23**, 021301, 1-15.
[25] P Dames, B Gleich, A Flemmer, K Hajek, N Seidl, F Wiekhorst, D Eberbeck, I Bittmann, C Bergemann, T Weyh, L Trahms, J Rosenecker and C Rudolph 2007 *Nature Nanotechnology* **2**(8), 495-499.
[26] F R Menter, R Langtry and S Völker 2006 *Flow, Turbulence and Combustion* **77**(1-4), 277–303.
[27] F Nicoud, F Ducros 1999 *Flow, Turbulence and Combustion* **62**, 183–200.
[28] F Colleti *et al.* 2014 private communication
[29] Y S Cheng, Y Zhou, B T Chen 1999 *Aerosol Sci. Technol.* **31**, 286–300.

Journal of Physics: Conference Series **530** (2014) 012007 doi:10.1088/1742-6596/530/1/012007

Investigation of the aerodynamics of an innovative vertical-axis wind turbine

K Kludzinska[1], **K Tesch**[1,3] **and P Doerffer**[2]
[1]Gdansk University of Technology, ul. G.Narutowicza 11/12, 80-233, Gdansk, Poland
[2]Institute of Fluid Flow Machinery, ul. Fiszera 14, 80-231, Gdansk, Poland

E-mail: krzyte@pg.gda.pl

Abstract. This paper presents a preliminary three dimensional analysis of the transient aerodynamic phenomena occurring in the innovative modification of classic Savonius wind turbine. An attempt to explain the increased efficiency of the innovative design in comparison with the traditional solution is undertaken. Several vorticity measures such as enstrophy, absolute helicity and the integral of the velocity gradient tensor second invariant are proposed in order to evaluate and compare designs. Discussed criteria are related to the vortex structures and energy dissipation. These structures are generated by the rotor and may affect the efficiency. There are also different vorticity measure taking advantage of eigenvalues of the velocity gradient tensor.

1. Introduction

The object of the analysis is the modernisation of classic design of the turbine (i.e. the Savonius wind turbine), see figure 1. This innovative design [1] is equipped with a stator which experimentally shows increased efficiency in comparison with standard design. This is true for wind tunnel experiments at least. Both turbines have the simplest design of all devices converting wind into other energy forms, which provides an opportunity for decrease of its price. The original Savonius wind turbine has numerous advantages such as low noise, simplicity of design, applicability for a wide range of wind velocities. The biggest problem of the classic design is its relatively low efficiency. The innovative design, discussed here, is devoid of this disadvantage.

Wind turbine research, such as Savonius turbine and its modifications, are in line with the general strategy of development of the Polish power industry. This strategy tends to quantitative increase the use of renewable energy sources. One of the limits, as far as the wind energy is concerned, it is high price of wind turbines available. The obtained results may increase the knowledge of the flow round a whole family of different types of rotors whose principle of operation is based primarily on the use of wind thrust accompanied with a small share of lift forces. The knowledge gained during the simulations will enable more effective designs characterised by increased efficiency and operational reliability.

The flow inside the rotor is complicated and that is why in past investigations were limited only to laboratory tests [2, 3, 4]. Another experimental method was flow visualisation [5, 6], which allows capturing flow patterns characteristics for selected instantaneous positions of rotor blades with respect

[3] To whom any correspondence should be addressed.

to the wind direction. Recently, due to the rapid development of computer hardware and software, attempts were made to analyse the structure of the flow through the Savonius rotor numerically [7, 8]. This also includes the vortex method [9, 10]. Generally, all data presented on the numerical aspect are mostly two-dimensional. Also, papers on attempts to modify the shape of the Savonius rotor are available [3, 11]. The best shapes of blade tips for a given rotor geometry and physical conditions are investigated in [12, 13, 14].

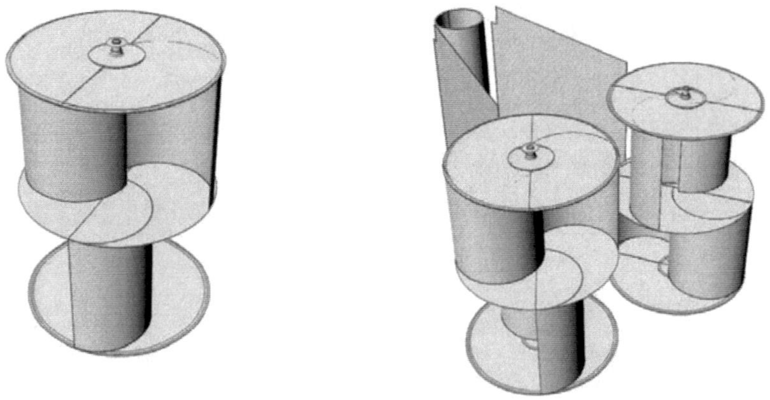

Figure 1. Classic (left) and innovative (right) designs.

2. Methodology

All the numerical calculations were performed using the commercial CFD code CFX. The turbulent flow of air was treated as an incompressible medium. The turbulence was modelled by means of the standard two-equation turbulence model $k-\varepsilon$. The reason for this choice may be explained by a need of comparison with other calculations found elsewhere. The $k-\varepsilon$ model has now become a standard for this kind of calculations.

The average form of mass conservation equation has the form

$$\nabla \cdot \langle \vec{U} \rangle = 0. \tag{1}$$

The Reynolds equation is

$$\rho \frac{\mathrm{d}\langle \vec{U} \rangle}{\mathrm{d}t} = \rho \vec{g} - \nabla p_{\mathrm{e}} + \nabla \cdot \left(2\mu_{\mathrm{e}} \langle \mathbf{D} \rangle\right) \tag{2}$$

where the effective pressure $p_{\mathrm{e}} = \langle p \rangle + 2\rho 3^{-1} k$ and the effective viscosity is composed of eddy and molecular components $\mu_{\mathrm{e}} = \mu_{\mathrm{t}} + \mu$. Two additional transport equations are those for modelled kinetic energy of velocity fluctuation k which arises from Reynolds stress transport equation

$$\rho \frac{\mathrm{d}k}{\mathrm{d}t} = 2\mu_{\mathrm{t}} \langle \mathbf{D} \rangle^2 + \nabla \cdot \left(\mu_{\mathrm{t}} \sigma_{\mathrm{k}}^{-1} \nabla k\right) - \rho \varepsilon \tag{3}$$

and dissipation of kinetic energy of fluctuation ε. This is analogous to k transport

$$\rho \frac{\mathrm{d}\varepsilon}{\mathrm{d}t} = C_{\varepsilon 1} \varepsilon k^{-1} 2\mu_{\mathrm{t}} \langle \mathbf{D} \rangle^2 + \nabla \cdot \left(\mu_{\mathrm{t}} \sigma_{\varepsilon}^{-1} \nabla \varepsilon\right) - \rho C_{\varepsilon 2} \varepsilon^2 k^{-1}. \tag{4}$$

The eddy viscosity is defined as $\mu_t = C_\mu \rho k^2 \varepsilon$. The five constants in the above equations should be deduced from experiment for a specific geometry. This 'standard' set is given by $\sigma_k = 1$, $\sigma_\varepsilon = 1$, $C_\mu = 0.09$, $C_{\varepsilon 1} = 1.44$, $C_{\varepsilon 2} = 1.92$.

The flow domain was divided into two parts: the rotating rotor and the steady wind tunnel and ambient. Both parts were merged by means of the domain interface of the so called 'transient rotor-stator' type. The time step of the transient calculations corresponded to two degree of revolution and the angular velocity of the rotating domain (rotor) corresponded to three turns per second.

The boundary conditions selected here are:

- Inlet. The mass flow rate was specified here. The specified mass flow rate corresponds to the average velocity $6\,\mathrm{m\,s^{-1}}$. The turbulence intensity defined as $\tau_t = \langle \vec{U} \rangle^{-1} \left(2k3^{-1}\right)^{1/2}$ equals 5% and the viscosity ratio $\mu_t \mu^{-1} = 10$. This represents a medium turbulent intensity.
- Opening. The so called 'far field' condition was chosen equal with prescribed constant atmospheric pressure.
- Symmetry. This means than the velocity normal component equals zero $\hat{n} \cdot \langle \vec{U} \rangle = \vec{0}$ and all the scalar values φ must fulfil $\hat{n} \cdot \nabla \varphi = 0$ where $\hat{n}$ represents a unit vector normal to the surface.
- Wall. The flow domain was a wind tunnel with one side open to the atmosphere. As for the rotor blades and the shroud they were modelled as no slip wall in the rotating frame of reference.

3. Coefficients

Among many characteristics of the wind turbines the most important is the torque coefficient. It is commonly defined as

$$C_T = \frac{T}{\frac{1}{4}\rho U^2 D^2 H}. \tag{5}$$

The above definition is valid for both the steady-state and transient flows. For the latter case one should use the time dependent torque $T(t)$ instead of T meaning that in real case we deal with the distribution of C_T as a function of the angular position of the rotor. The torque coefficient (5) is directly related to the efficiency of the rotor.

A typical definition of the efficiency for the steady-state case takes under consideration the wind power $N_w = \dot{m} e_k = \rho U S 2^{-1} U^2$ and the power of the rotor $N = M\,\omega$. The wind power is treated here as the reference power. From the two above definitions we arrive at the following definition of the efficiency

$$\eta_s = \frac{\omega T}{\rho \frac{U^3}{2} HD} = \frac{\omega D C_T}{2U}. \tag{6}$$

In the above definition ω is the angular velocity, ρ represents the density, and U - the reference velocity. D is the diameter of the rotor, and H is the rotor's height. The steady-state efficiency is frequently referred to as the power coefficient C_P. For the transient case, which is typical for the Savonius rotor operation, one should consider the total energy of the wind within the time interval Δt rather than the instantaneous power. This means that the definition (6) takes the following form now

$$\eta_{\mathrm{t}} = \frac{\omega D \int_{t}^{t+\Delta t} C_{\mathrm{T}}(t)dt}{2U\,\Delta\mathrm{t}} = \frac{\omega D \overline{C}_{\mathrm{T}}}{2U} \tag{7}$$

where Δt stands for the time of interest (e.g. one turn). Assuming that the time step $\Delta\tau$ of the transient CFD calculations is constant we can approximate the integral in equation (7) in the following way

$$\int_{t}^{t+\Delta t} C_{\mathrm{T}}(t)\,dt \approx n\Delta\tau \frac{1}{n}\sum_{i=1}^{n} C_{\mathrm{T,i}} = \Delta t\ \overline{C}_{\mathrm{T}} \tag{8}$$

where $\overline{C}_{\mathrm{T}}$ represents the arithmetical average $\overline{C}_{\mathrm{T}} = n^{-1}\sum_{i=1}^{n} C_{\mathrm{T,i}}$ and the time of interest Δt (e.g. one revolution) is expressed as $\Delta t = n\,\Delta\tau$. The total number of time steps is denoted here as n. The last definition (7) has an analogical form as the definition (6).

4. Evaluation and comparison criteria

These criteria allow for direct evaluation and comparison of various designs and solutions. As an example of the existing criterion one can recall enstrophy E^* [15]

$$\mathrm{E}^* = \iiint_V \|\mathbf{\Omega}\|^2\,\mathrm{d}V = \iiint_V \frac{1}{2}\|\vec{\Omega}\|^2\,\mathrm{d}V = \iiint_V \varepsilon^*\,\mathrm{d}V\,, \tag{9}$$

where $\varepsilon^* := 2^{-1}\|\vec{\Omega}\|^2$ may be again treated as a specific enstrophy. Here $\mathbf{\Omega}$ stands for asymmetrical part of the velocity gradient tensor $\nabla\vec{U}$ and $\vec{\Omega}$ is the vorticity vector. The definition of enstrophy plays important role in theory of turbulence. It determines the rate of dissipation of kinetic energy being a global measure of the dissipation rate and vorticity measure. One can imagine even simpler vorticity measure

$$e^* = \iiint_V \sqrt{2\|\mathbf{\Omega}\|^2}\,\mathrm{d}V = \iiint_V \|\vec{\Omega}\|\mathrm{d}V\,. \tag{10}$$

Another vorticity measure is defined by means of the integral of the velocity gradient tensor second invariant Q in the following form

$$Q = \iiint_V \mathrm{tr}\left(\nabla\vec{U}\right)^2\mathrm{d}V = \iiint_V \left(\mathrm{tr}\,\mathbf{D}^2 - \varepsilon^*\right)\mathrm{d}V\,, \tag{11}$$

where $\mathbf{D}$ stands for symmetrical part of the velocity gradient tensor. The above definition is related to enstrophy defined in equation (9)

$$Q = \iiint_V \mathrm{tr}\,\mathbf{D}^2\,\mathrm{d}V - E^*\,. \tag{12}$$

The so called helicity may be treated as a measure of linkage of vortex lines in the flow. The global helicity is defined as

$$H = \iiint_V \vec{U}\cdot\vec{\Omega}\,\mathrm{d}V\,. \tag{13}$$

Typically, it is better to take advantage of absolute helicity $|H|$

$$|H| = \iiint_V \left|\vec{U} \cdot \vec{\Omega}\right| \, \mathrm{d}V \tag{14}$$

to avoid cancelling upon integration over considered volume V. An alternative definition of vorticity measure was given by Truesdell

$$W = \iiint_V \frac{\varepsilon^*}{\mathrm{tr}\,\mathbf{D}^2} \, \mathrm{d}V \,. \tag{15}$$

This measure is combined from two invariants and equals zero for irrotational flow $\left\|\vec{\Omega}\right\| = 0$ and $\mathbf{D} \neq \mathbf{0}$. The measure equals infinity for rotational flow $\left\|\vec{\Omega}\right\| \neq 0$ and $\mathbf{D} = \mathbf{0}$. The former has the largest possible vorticity measure.

Discussed criteria are related to the vortex structures and energy dissipation. These structures are generated by the rotor and may affect the efficiency. There are also different vorticity measure taking advantage of eigenvalues of the velocity gradient tensor.

5. Mesh

The both flow domains, i.e. the rotor and the wind tunnel, were discretised separately. Both domains have an unstructured grid consisting of mostly tetrahedral elements. The total number of elements covering the flow area is about 15 million. There are also special elements around the blades to ensure that flow near a wall is properly resolved. The wall function approach was used to provide near wall boundary conditions for the mean flow. The quality of the grid near the blades may be inspected in terms of y^+ distribution. The average value of y^+ on blades is no greater than 1 for all the time steps.

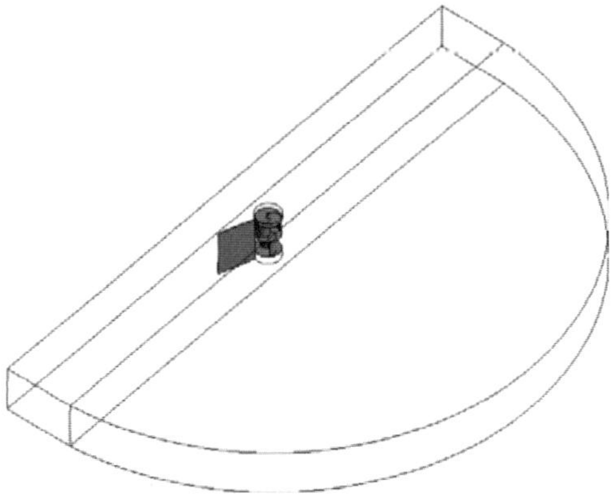

Figure 2. Flow domain.

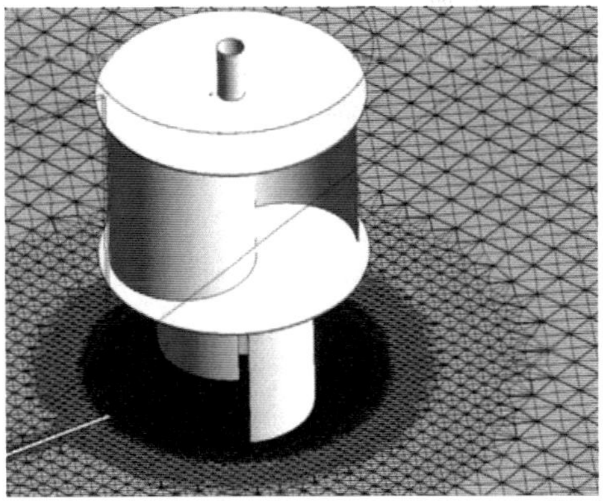

Figure 3. Mesh cross section.

Figure 2 presents global view on the computational domain consists of wind tunnel and cylindrical part of 'far field' surroundings. The mesh can be inspected in figure 3. This figure presents a cross section which is perpendicular to the rotor axis.

6. Results of calculations

Figure 4 presents the torque coefficient C_T distribution as a function of revolution angle. These are results of transient calculations. It has to be pointed out that most of the available data is obtained from steady-state calculations for two dimensions. The present calculations are fully transient and three-dimensional. The most visible difference between present innovative design and literature data [2, 3, 16, 6] is that there are four peaks visible instead of two. This is because the device consists of two rotors rotated by an angle relative to each other. Additionally, there is a stator directing the flow.

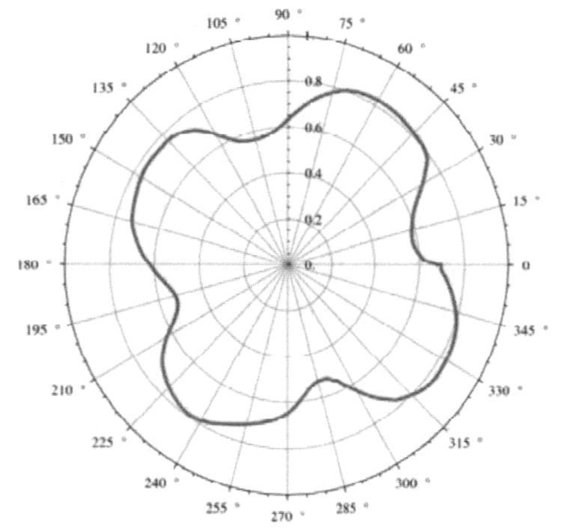

Figure 4. Torque coefficient distribution C_T .

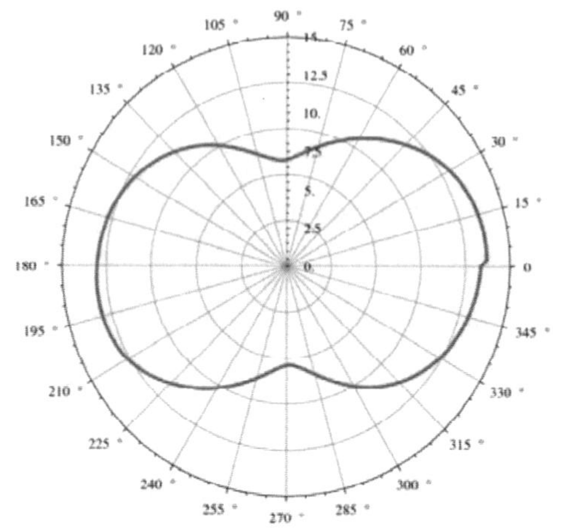

Figure 5. Vorticity measure e^* distribution.

Figure 5 shows distribution of vorticity measure e^* defined by means of equation (10) that can be used to compare various designs indicating the importance of this equation. The time integral of this measure can be used directly to evaluate specific turbine. Furthermore, one can anticipate a relationship between vorticity measure and efficiency of a turbine. It can be expected that the performance of a wind turbine is affected by vortices generated during the rotor revolutions. These vortices are complicated in their nature and dynamically change their configurations and undoubtedly intensities. This may affect pressure distribution around the blades and accordingly the performance of a rotor.

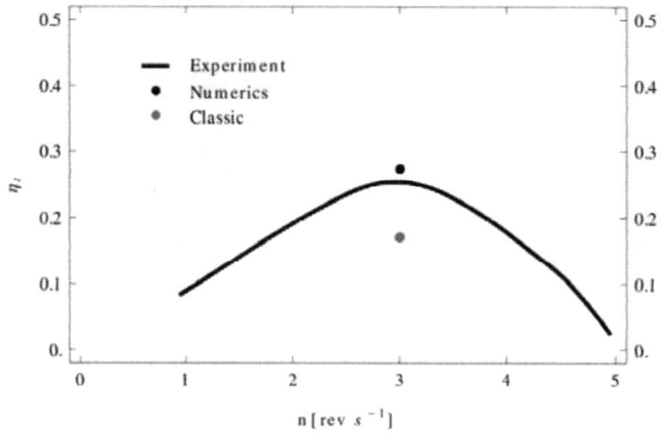

Figure 6. Efficiency comparison η_t .

From the definition (7) it may be deducted that the higher the torque coefficient the more efficient the rotor. The calculated transient efficiency (7) for the innovative design is $\eta_t = 27.3\%$. This value may be easily validated by means of data available in the literature [17] where expected performance of the conventional Savonius rotor for $\lambda = 0.4$ is estimated to be $\approx 17\,\%$. The rotor tip speed ratio λ is defined as

$$\lambda = \frac{\omega D}{2\mathrm{U}}. \tag{16}$$

Figure 6 shows the measured efficiency in the wind tunnel for the innovative design (solid line). Black dot shows the numerical prediction of the transient efficiency whereas the red dot represents literature data [17] for the classic Savonius design. Even for low wind speed $6\,\mathrm{m\,s^{-1}}$ the innovative design if far more efficient.

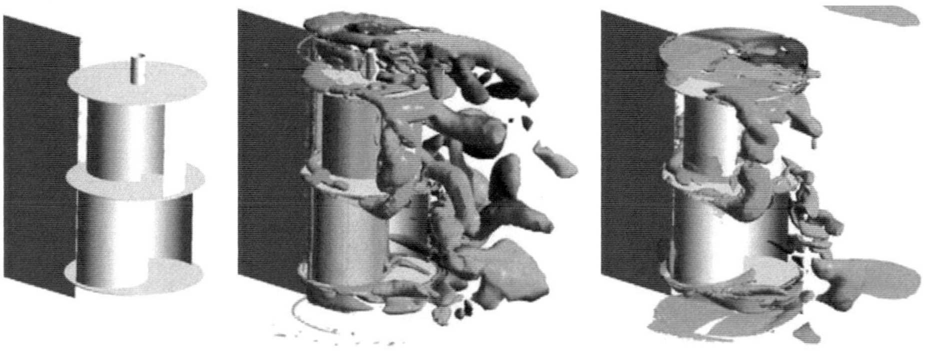

Figure 7. Various vorticity measure distribution.

Figure 7 presents comparison of two example vorticity measure distribution. These are also called vortex cores. In the middle we have $\mathrm{tr}(\nabla\vec{U})^2$ distribution which is a part of the integral of the velocity gradient tensor second invariant Q (equation (11)). On the right it is local absolute helicity $|\vec{U}\cdot\vec{\Omega}|$ being part of global absolute helicity $|H|$ (equation (14)).

7. Conclusions

- The innovative modification of classic Savonius shows higher efficiency. This is true for wind tunnel experiments at least.
- The increased efficiency is due to presence of the stator which directs the air and makes it possible to take better advantage of its energy.
- It is believed that vortex structures are generated by the rotor and affect the efficiency.
- Several vorticity measure have been proposed allowing for evaluation and comparison of various designs.

8. References

[1] Doerffer P, International patent application PCT/PL2012/000125 and National Patent Application P.397052 – 'Aktywny wiatrak o osi poprzecznej do kierunku wiatru'
[2] Blackwell B F, Sheldahl R E and Feltz L V 1977 Wind tunnel performance data for two- and three-bucket Savonius rotor *Sandia Laboratories Report* SAND 76-0131, 105

[3] Kamoji M A, Kedare S B and Prabhu S V 2008 Experimental investigations on single stage, two stage and three stage conventional Savonius rotor *Int J Energ Res* **32** 877–95

[4] Saha U K, Thotla S and Maity D 2008 Optimum design configuration of Savonius rotor through wind tunnel experiments *J Wind Eng Ind Aerod* **96** 1359–75

[5] Fujisawa N and Gotoh F 1992 Visualisation study of the flow in and around a Savonius rotor *Exp Fluids* **12** 407–12

[6] Nakajima M, Iio S and Ikeda T 2008 Performance of double-step Savonius rotor for environmentally friendly hydraulic turbine *Journal of Fluid Science and Technology* **3** 410–19

[7] Menet J 2008 Prediction of the aerodynamics of a new type of vertical axis wind turbine: the reverse bladed rotor *Proc. of the European Wind Energy Conference*, 2008

[8] Swirydczuk J, Doerffer P and Szymaniak M 2011 Unsteady flow through the gap of Savonius turbine rotor *Task Quarterly* **15**(1) 59–70

[9] Afungchui D, Kamoun B, Helali A and Djemaa A B 2010 The unsteady pressure field and the aerodynamic performances of Savonius rotor based on the discrete vortex method *Renew Energ* **35** 307–13

[10] Ogawa T 1984 Theoretical study on the flow about Savonius rotor *J Fluid Eng-T ASME* **106** 85–91

[11] Mohamed M H, Janiga G, Pap E and Thevenin D 2010 Optimisation of Savonius turbines using an obstacle shielding the returning blade *Renew Energ* **35** 2618–26

[12] Kludzinska K and Swirydczuk J 2012 Unsteady flow through the gap of Savonius turbine rotor *Proc. of the European Congress on Computational Methods in Applied Sciences and Engineering (ECCOMAS 2012)* 2012

[13] Kludzinska K and Swirydczuk J 2012 A numerical analysis of the unsteady flow round a Savonius turbine *Proc. of the XX Polish Fluid Mechanics Conf* 2012

[14] Swirydczuk J and Kludzinska K 2012 Improving Savonius rotor performance by shaping its blade edges *Herald of aeroenginebuilding* **2** 86–90

[15] Tesch K 2013 On invariants of fluid mechanics tensors *Task Quarterly* **17**(3–4) 1000–8

[16] Menet J 2004 A double-step Savonius rotor for local production of electricity: A design study *Renew Energ* **29** 1843–62

[17] Le Gourières D 1980 *Energie éolienne. Théorie, conception et calcul pratique des installations* (Eyrolles)

Journal of Physics: Conference Series **530** (2014) 012008 doi:10.1088/1742-6596/530/1/012008

The parametric investigation of influence of DBD actuator on the boundary layer under various Reynolds number

P Procházka[1], **V Uruba**
Institute of Thermomechanics AS CR, v.v.i., Praha, Czech Republic

E-mail: prochap@it.cas.cz

Abstract. In the previous research, the wall-jet-like flow or vortical structures were generated using a single dielectric barrier discharge (DBD) actuator in still air. It has been shown that the actuator can generated so-called vortex train if it is powered by modulated voltage waveform. Now, the plasma DBD actuator is placed inside a rectangular channel and an interaction with boundary layer (BL) is studied. The characteristics of the BL are modifying by varying inlet velocity. The actuator is adjusted to be set in spanwise orientation which means that the generated wall-jet-like flow is oriented in the same meaning as the main flow or in the opposite direction. The interaction of vortical structures generated by actuator with BL will be described quantitatively from mean flow field.

1. Introduction

Recently, the dielectric barrier discharge (DBD) plasma actuator was designed, fabricated and flow field structures generated by this actuator were experimentally investigated during PhD study [1]. This actuator is supplied by very high-voltage high-frequency waveform which makes it easy to control the actuation (via modulation of the waveform) and there are no strict demands on the voltage source. Plasma actuators based on direct current utilization cannot offer above mentioned advantages and thus are not preferable for active flow control over streamlined bodies. The actuator was operated during two different regimes. The first one is steady regime when the plasma discharge induces wall-jet-like flow, whose main feature is that the biggest velocity occurs very close to the surface. The current geometry of actuator and electrical characteristics of power source allowed to gain the maximal velocity inside the wall jet up to 2 $m\cdot s^{-1}$ in the mean flow (peak velocities reached the value of 3-4 $m\cdot s^{-1}$). The power consumption is in order of watts. On the other hand, after the application of rectangular modulation, there are series of vortices generated inside wall jet. The vortex shedding frequency, vortex dimension etc. are related with input voltage properties of the actuator. This unsteady regime seems to be very convenient for purpose to use active flow control very effectively in regard to power consumption and mainly sophisticated control of the flow – e.g. the vortices with different circulation would affect the control volume of the flow in different way – which should be utilized by continuously changing flow conditions of streamlined body.

In [1], a physical model describing the process of vortex street generation by this DBD actuator was presented under various conditions of modulation parameters. The vortex size, the distance between two consequences vortices, the circulation of a vortex, vortex trajectory and convective velocity of the vortex as well as mean flow field were parametrically described in relation on mainly

[1] IT ASCR, Dolejskova 5, 182 00 Praha 8, Czech Republic.

frequency of modulation, duty cycle and voltage magnitude. The equation 1 expresses the relation between convective velocity of the vortex core in dependency on used modulation frequency and duty cycle:

$$u_{conv} = a_1 + a_2 \cdot DC + a_3 \cdot f + a_4 \cdot DC^2 + a_5 \cdot DC \cdot f + a_6 \cdot f^2 \quad (1)$$

where coefficients a_i were obtained by experimental measurement. Now, it is well-known how this actuator works in steady air and the generally valid physical model about vortex generation can be utilized for study of interaction the plasma induce flow with a boundary layer.

Before an investigation of plasma actuation over Glaubert-Goldschmiedt (GG) body will be performed, a simpler case of boundary layer generated inside rectangular channel will be introduced in this article. An experimental approach is proposed to solve this challenging problem due to complex character of this phenomenon. Results will be obviously treated as a comparison of velocity profiles measured by diverse conditions of the flow and quantitative effect of plasma actuation will be presented.

2. Experimental setup

2.1. Actuator description

The plasma DBD actuator is considered as in simple wire configuration which means that it has just two electrodes. The upper electrode is very thin wire (tungsten) and is powered by high-voltage high-frequency waveform. The lower electrode is thin layer of gold and is grounded. There is a layer of dielectric with thickness of only 1 mm. The material of dielectric is silica glass which has extraordinally dielectric strength and thermal expansion. Electrode gap was set to two millimeters.

The plasma actuator was significantly modified according to previous type. The wire electrode was changed to be more thin (now the diameter is only 30 μm) and the wire electrode stretching mechanism had to be completely reconstructed so that the holder of a spiral torsion spring did not affect the free flow. The position of wire electrode is in the center of the actuator. Consecutively, the position of wire electrode due to the channel inlet is the same for both orientation of the plasma induced flow.

2.2. Power source

The power source was designed and manufactured in the Institution of Thermomechanics. The maximum voltage peaks is up to 12 kV and frequency can be adjustable, but is set to optimum value of 16 kHz. It has also so-called shut down function which enables to use amplitude modulation of the voltage waveform. Besides voltage magnitude and the frequency of waveform, the frequency of modulation and duty cycle of one period can be utilized as another control parameters of power feeding.

The parameters related to the modulation are to be fully adjustable during plasma actuator run and are determine to define the characteristics of generated vortices. If no modulation is applied, there is no presence of significant vortical structures inside wall-jet-like flow. The modulation frequencies range from units of Hz to tens of Hz. It was observed that if this parameter is higher than 80 Hz, the vortex train starts to be similar to unmodulated continuous jet (vertical structures are too small and distance between each other is tiny – limitations in spatial resolution of measurement system). Duty cycle determine the rate between the working cycle of the actuator (when plasma is on) and relaxation time (plasma is off). Note that DC of 30% means that the plasma discharge is on for 70% of one period.

2.3. Experimental layout

Since the aim of that study was to study the influence of plasma induced flow to the boundary layer, this actuator was placed inside plexi-glass channel connected to the outlet of the blow-down wind tunnel. The cross-section dimensions were 100 x 250 mm and total length was 3000 m (figure 1). The actuator was mounted in spanwise configuration so that the distance of wire electrode from inlet edge of the channel was 2275 mm for both corresponding direction of induced flow and for opposite direction of induced flow. This distance was used to determine the Reynolds number (see table 1). The wall of the tunnel was smooth without no turbulators or VGs.

Table 1. The Reynolds number based on conventional BL thickness and on horizontal distance inlet-wire.

	Re_x	Re_δ
$V_{ext} = 5\ \mathrm{m \cdot s^{-1}}$	$18 \cdot 10^3$	$750 \cdot 10^3$
$V_{ext} = 10\ \mathrm{m \cdot s^{-1}}$	$34 \cdot 10^3$	$1500 \cdot 10^3$
$V_{ext} = 20\ \mathrm{m \cdot s^{-1}}$	$55 \cdot 10^3$	$3000 \cdot 10^3$

The parameters of the boundary layers were modified by changing velocity inside the channel. These velocities were set to 5, 10 and 20 $\mathrm{m \cdot s^{-1}}$ resulting in conventional boundary layer thickness (at the position of wire electrode) of 54, 51 and 41 mm, respectively. The thickness of BL was evaluated from the base case (the actuator was mounted inside the channel but actuation was off). The Reynolds numbers based on conventional thickness are listed in the table. 1.

The experiment was carried out under three different regimes of the actuator. The first regime - base case - was already introduced. The second regime is continuous wall jet generated by the actuator with condition of maximum applicable voltage magnitude and third regime is unsteady characterized by the varying modulation frequency of 10 or 20 or 50 Hz with the constant duty cycle of 30 %.

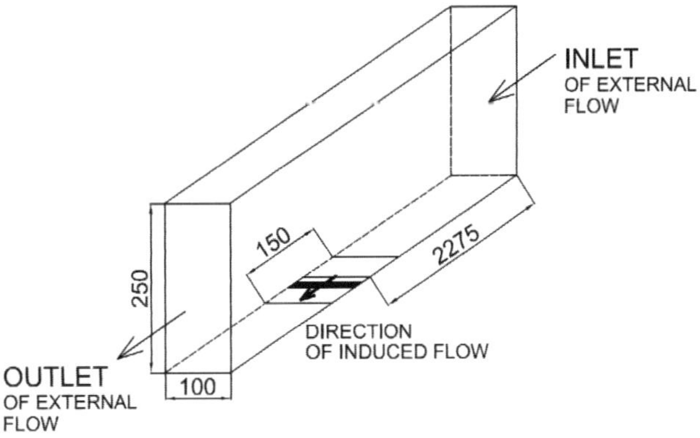

Figure 1. Experimental layout.

2.4. Measurement technique

The standard time-resolved Particle Image Velocimetry (TR-PIV) was used as a measurement technique. PIV composes mainly from a camera and a laser. Laser New Wave Pegasus Nd:YLF is a double-head pulse laser with a cylindrical optics and its wavelength is 527 nm, maximal frequency is 10 kHz and shot energy of one pulse is 10 mJ for 1kHz (10 W per head). Phantom V711 is a CCD camera with maximal resolution 1280 x 800 pixels and corresponding maximal frequency is 6 kHz (3000 double-images per second). The total memory of camera is 8 GB.

This experiment could be handled from two diverse points of view. The first one is to investigate the mean flow field and to compare the statistical characteristics. That is why the lower acquisition frequency was utilized with longer time of data acquisition (100 Hz and 20 sec). This resulted in very good statistical properties. The second point of view, which is not considered in this article, is to study the interaction of induced flow with BL from dynamical point of view where maximal available acquisition frequency will be used.

3. Results

The flow field generated by plasma DBD actuator in still air can be seen at [2]. Further, only the interaction with boundary layer will be presented. Since the resolution of the camera in streamwise direction is not sufficient, the camera was traversing through five distinct positions. The position 0 was set to capture the flow instabilities upstream the wire electrode. Then the positions 1, 2, 3 and 4 were established to sample the flow downstream the actuator. The distance between two consecutive positions was set to allow the image covering. In this paper, there will be presented results from position 0 and position 2 which corresponds to the upstream distance L = -5 mm and downstream distance L = 150 mm, respectively.

Firstly, results based on velocity vector field will be presented. Figure 2 and 3 shows velocity vector map in position 2 with distribution of u-component of velocity. It can be seen that boundary layer with turned actuation on disposes with higher velocities close to the surface. The figure 4 shows the velocity field in the position 0 when the actuation is turned on and the plasma induced flow has opposite direction. A small region of velocity increase upstream of the wire electrode extends. Here, the velocities are increased instead the opposite ionic wind causes the deceleration of the flow. This effect is obviously caused by some instabilities and it should be studied with respect to the dynamics. Further downstream of the wire electrode, the boundary layer velocities are decreased (figure 9).

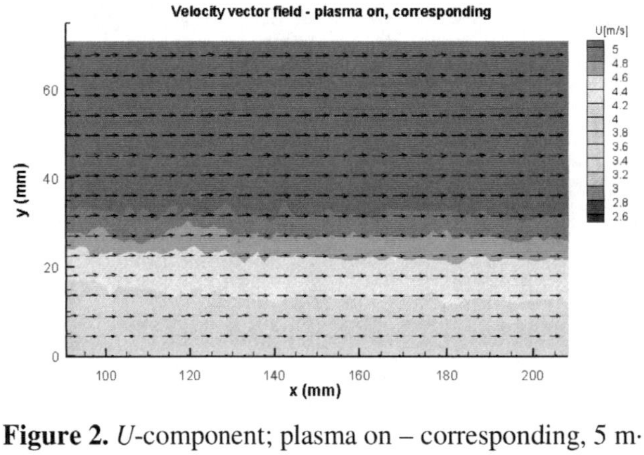

Figure 2. U-component; plasma on – corresponding, 5 m·s^{-1}.

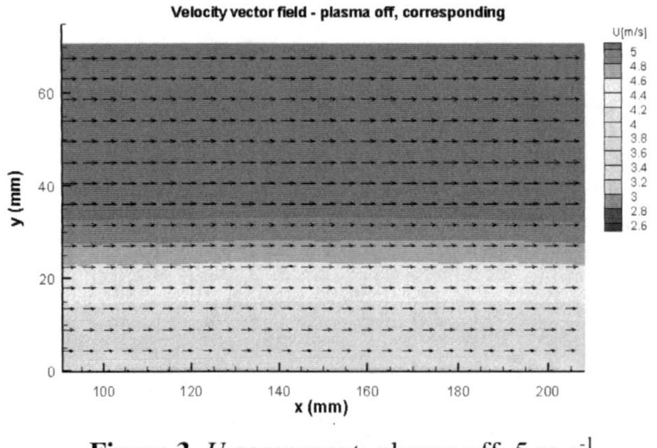

Figure 3. U-component; plasma off, 5 m·s^{-1}.

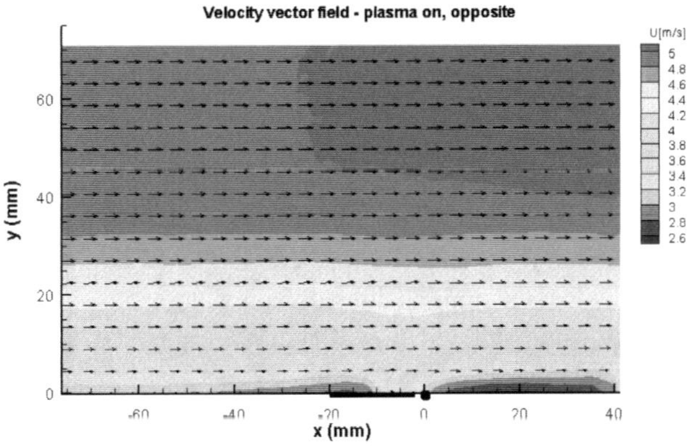

Figure 4. *U*-component; plasma on – opposite, 5 m·s^{-1}.

The effect of plasma actuation is better to see from velocity profiles of the BL that are normalized with respect to the external flow. The base case (when the actuator was mounted in the channel but actuation was off) was compared for the case of corresponding and opposite direction of plasma induced flow. As it was expected, the actuator did not disturb the BL so both profiles look the very same. The figure 5 shows the comparison with the actuation on in steady regime and in unsteady regime (10 Hz and 30% DC) for the external velocity 5 m·s^{-1}. The velocity profile was improved about 8% in bottom part for steady actuation and slightly less for unsteady actuation. Naturally, the effect of plasma actuator became weaker after that the external flow velocity was increased (figure 6 and 7). When the velocity was enhanced to the value of 10 m·s^{-1} and 20 m·s^{-1} the effect was about 6% and 2%, respectively. The comparison of various modulation frequencies is not plotted here, but the most promising results seems for the highest modulation frequency when the vortex train starts to be similar to the continuous wall-jet-like flow which is more effective.

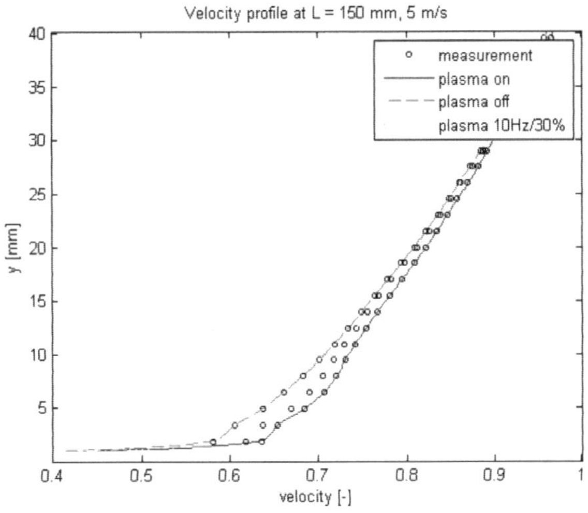

Figure 5. Effect of plasma actuation – steady/unsteady.

The comparison of standard deviation of velocities showed that plasma induced flow could stabilize the BL under specific conditions. The actuation by 5 m·s^{-1} was able to significantly decrease standard deviations in the area of 20 mm from the wall (figure 8) without distinction if the steady or unsteady actuation was used. This stabilization effect became weaker for higher velocities of external flow.

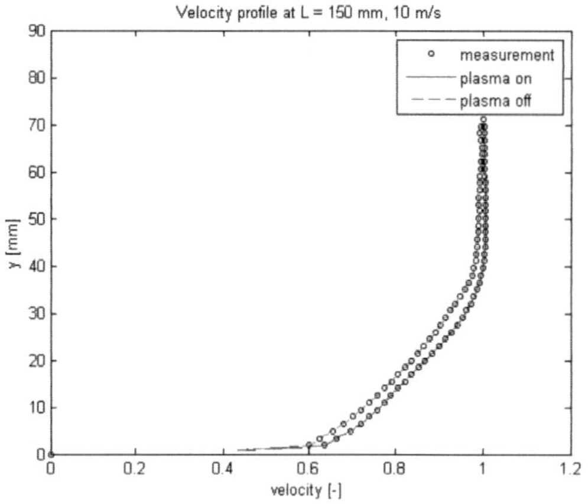

Figure 6. Effect of plasma actuation – steady/unsteady.

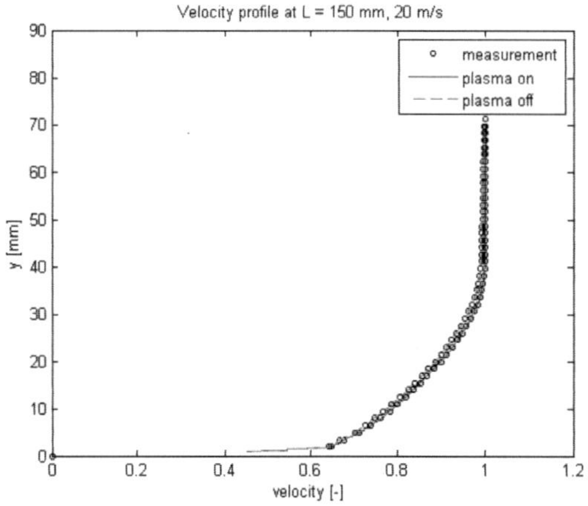

Figure 7. Effect of plasma actuation – steady/unsteady.

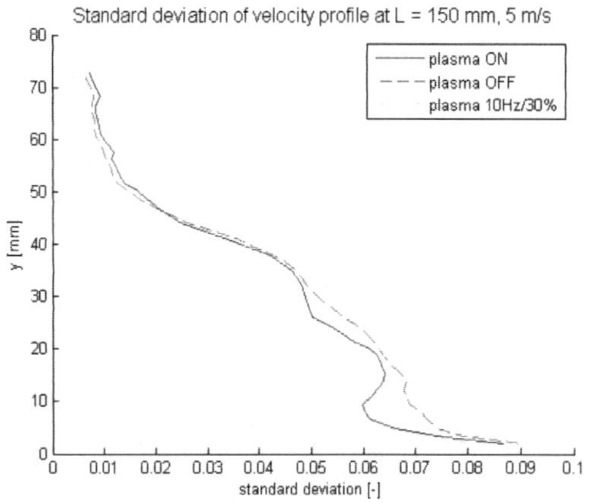

Figure 8. Velocity standard deviations.

Very interesting results were obtained for opposite orientation of the actuator. The figure 9 shows how the velocity profile is decelerated (about 6%) far from wire electrode downstream. And again the deceleration is not so significant for the case of unsteady modulation. The standard deviation distribution shows the destabilization effect of opposite actuation on the BL (figure 10).

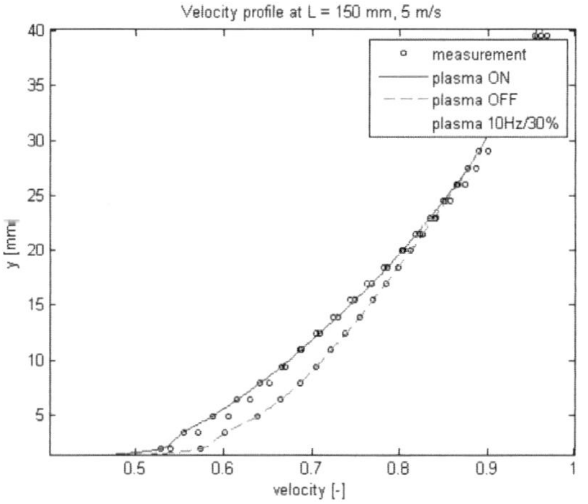

Figure 9. Effect of opposite plasma actuation.

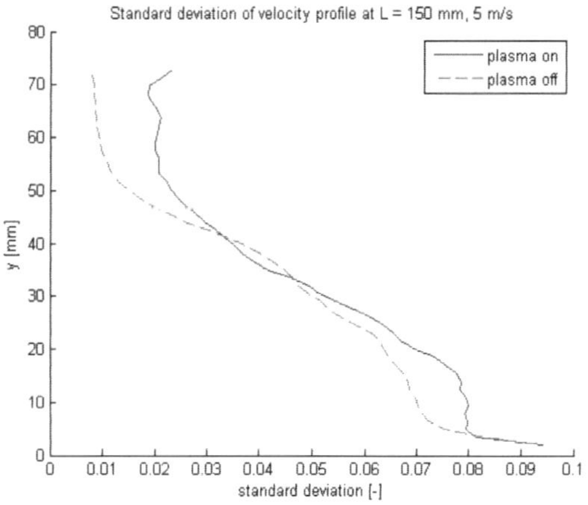

Figure 10. Velocity standard deviations – opposite actuation.

There is also effect upstream of the wire electrode (L = -5 mm) as it can be seen at the figure 11. In this very tiny region, plasma actuation can increase the velocities about 9 % close to the surface. This diagram also shows the case when actuation is in corresponding direction with respect to the external flow (green dotted line) where one can also observe the velocity gain which is caused by suction effect of plasma induced flow. Last picture denotes extensive standard deviation decrease (about three times).

4. Conclusions

The interaction of plasma induced flow with BL in a rectangular channel was investigated via velocity profiles from mean flow fields. Plasma actuation is capable to enhance velocities close to the surface about approx. 10%. Further, analyses of velocity standard deviations showed that plasma actuation can stabilize the region inside the BL. If the induced flow is oriented in opposite direction than mean flow, the velocities are decelerated downstream and are increased in small region upstream.

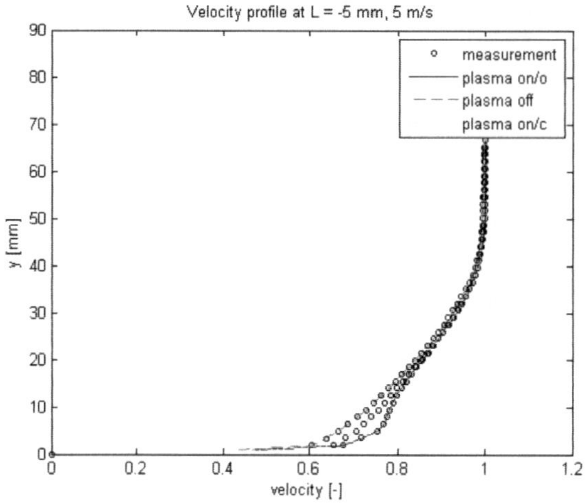

Figure 11. Effect of opposite actuation upstream.

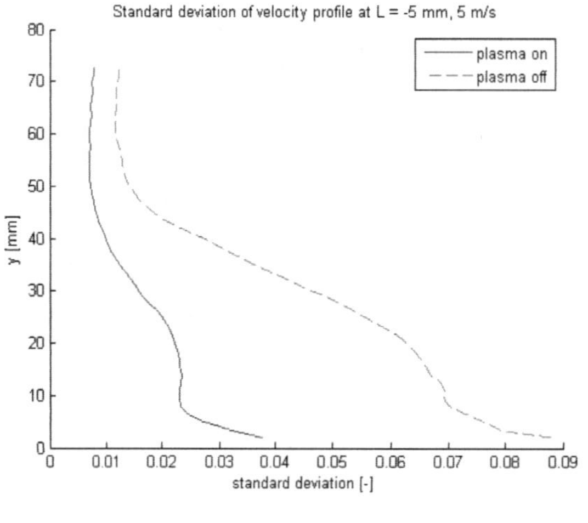

Figure 12. Velocity standard deviations upstream.

5. References

[1] Procházka P 2013 Dissertation thesis. Praha ČVUT

[2] Procházka P, Uruba V, Antos P 2013 *Proceedings in Applied Mathematics and Mechanics* **13** 1 343

[3] Procházka P, Uruba V 2013 *Experimental Fluid Mechanics 2013. Kutná Hora* 577

Acknowledgments

The authors gratefully acknowledge financial support of the Grant Agency of the Czech Republic, project No. GP14-25354P and No. P101/10/1230.

Journal of Physics: Conference Series **530** (2014) 012009 doi:10.1088/1742-6596/530/1/012009

Application of 3D printing technology in aerodynamic study

K Olasek[1] and P Wiklak
Łódź University of Technology, Institute of Turbomachinery, Wólczańska 219/223, 90-924 Łódź, Poland

E-mail: krzysztof.olasek@p.lodz.pl

Abstract. 3D printing, as an additive process, offers much more than traditional machining techniques in terms of achievable complexity of a model shape. That fact was a motivation to adapt discussed technology as a method for creating objects purposed for aerodynamic testing. The following paper provides an overview of various 3D printing techniques. Four models of a standard NACA0018 aerofoil were manufactured in different materials and methods: Multi-Jet Modelling (MJM), Selective Laser Sintering (SLS) and Fused Deposition Modeling (FDM). Various parameters of the models have been included in the analysis: surface roughness, strength, details quality, surface imperfections and irregularities as well as thermal properties.

1. Introduction

Preparation of an appropriate model to conduct experiments is a crucial aspect of every science investigation. Usually, apart from the requirement to achieve a desired level of a model quality, it is also very important to keep in mind the costs and time of its manufacturing. Recently, development of technologies made the production of even very complicated objects simpler, cheaper and less time consuming. One of such techniques that can simplify life of every researcher seems to be a 3D printing (an example of additive manufacturing). This method of fabrication is a process of making a three-dimensional solid objects from a digital model. 3D printing is an additive process, where successive layers of material are laid down in controlled way in order to achieve desired shape. 3D printing is often put in opposition to traditional machining techniques, which mostly rely on the removal of material by methods such as drilling, turning, milling or cutting.

A 3D printer is a limited type of industrial robot that is capable of carrying out an additive process under computer control. The first working 3D printer was created in 1984 by Chuck Hull of 3D Systems Corp. From that point, various additive manufacturing processes, techniques and systems have been developing constantly. The 3D printing has begun to exhibit great applications potential and advantages in the aerospace, construction, architecture, automotive, power engineering, dental and medical industries, biotech (human tissue replacement), education and many other fields providing a cost-effective and time-efficient way to produce low-volume, customized products with complicated geometries and advanced material properties. Objects that are manufactured additively can be used anywhere throughout the product life cycle, from pre-production (i.e. rapid prototyping) to full-scale production (i.e. rapid manufacturing), in addition to tooling applications and post-production customization [1].

[1] To whom any correspondence should be addressed.

Drop in prices of these devices in recent years effects in increase of their accessibility. Traditional models for tests performed in the wind tunnel are fabricated of light wood, aluminum (and other metals) or various plastics. Most of the models are manufactured with computerized numerical control (CNC) milling machines. That is why processing of these materials is not very difficult, but in most cases due to the complicated shape desired several machining processes are required. Due to this fact, operator's skills, appropriate equipment and a lot of time is needed. It happens that for simple testing or classes (study cases) with students it is a much too big expense. Summarising all above, lower prices of 3D printers may become an opportunity for a more practical and accessible knowledge for students and more efficient aerodynamic investigations for researchers. However, not everything is so perfect. In aerodynamic experiment a crucial factor is proper preparation of tested elements surface. Various 3D printing techniques provides different quality and surface finish which is not always acceptable for certain application [2]. The following paper attempts to review chosen properties of rough 3D printouts and analyse them.

2. 3D printing technologies and materials

Aim of the research carried out was to evaluate the utility of most common 3D printing techniques for aerodynamic study. Similar field has been already explored by various research centres (e.g. [3], [4]). In order to perform a reliable comparison of most common 3D printing methods, four models of standard NACA0018 aerofoil have been created in various materials and technologies: Multi-Jet Modelling (MJM), Selective Laser Sintering (SLS) and Fused Deposition Modeling (FDM). Each of them is described briefly in table 1. The most crucial factor defining the 3D printing technology and influencing the properties of final product manufactured is the type of bulk material used. Depending on the fact whether a solid filament is extruded (FDM), fine powder is laser-sintered (SLS) or liquid is solidified by means of UV light curing (MJM), models of various quality and properties are obtained.

Table 1. The overview of the most popular printing techniques and materials.

Type	Technology	Materials
Extrusion	Fused deposition modeling (FDM)	Thermoplastics: PLA, ABS, nylon
Granular	Selective laser sintering (SLS)	Thermoplastics, metal powders, ceramic powders
Liquid	Multi Jet Modeling (MJM)	Acrylic Plastic

Below a description of materials used for printing models discussed in this paper are presented.

ABS - Acrylonitrile Butadiene Styrene is a very popular material used in the industry. Most popular applications are the production of housings for electronic equipment, household appliances, automotive interior parts, sports equipment, and furniture. ABS is known as material to production of LEGO bricks.

Nylon - invented and developed by DuPont Company in 1935 is a synthetic polymer used to create fibres having high tensile strength. It is used, among others, in the production of textiles, ropes and veins as well as gears, bearings etc.

Alumide - The material formed in the sintering process blend of nylon and aluminium dust. It has medium strength and fidelity of detail. It can be polished. Suitable for printouts related to jewellery. Objects printed with alumide are not suitable for recycling.

UV Curable Acrylic Plastic – material used in MJM 3D printing technology. Models prepared from a liquid plastic, which is distributed on the platform by a number of nozzles of the printer, and then UV-cured. This type of material and technology are ideal for printing very fine details, or thin walls.

Table 2 presents the most significant parameters of discussed materials and 3D printing techniques [5], [6], [7].

Table 2. 3D printing materials data.

	Alumide®	PA 2200 (nylon)	UV curable acrylic plastic	ABS
3D printing technology	SLS	SLS	MJM	FDM
accuracy	±0.15mm	±0.15mm	±0.025-0.05mm	- [a]
min. wall thickness	0.8mm	0.7mm	0.3mm	- [a]
density	1.36g/cm³ [b] 0.67g/cm³ [c]	0.93g/cm³ [b] 0.45g/cm³ [c]	1.02g/cm³	1.05 g/cm³
tensile modulus	3800MPa	1700MPa	1108MPa	1627MPa
tensile Strength	48MPa	48MPa	26.2MPa	22MPa
flexural modulus	3600MPa	1500MPa	n/d	1834MPa
flexural strength	72MPa	58MPa	26.6MPa	41MPa
elongation at break	4%	24%	9%	6%
shore D - hardness	76	75	n/d	n/d
thermal properties	172-180°C (melting point) 177°C (heat deflection temp. at 0.45 MPa)	172-180°C (melting point)	46°C (heat distortion temp. at 0.45 MPa) 80°C (heat softening temp)	190-240°C (melting point) 90°C (heat deflection temp. at 0.45 MPa)
recycling	non recyclable	recyclable	most recyclable	recyclable

[a] FDM accuracy depends on filament thickness and layer structure [8], [9], [b] laser-sintered part density, [c] bulk density

3. Manufactured models

Aerofoil with complex structure adapted for aerodynamic study has been designed and printed in 4 different materials described in previous section. Model dimensions are 100mm chord and 170mm span. Due to costs reduction and material saving aerofoil is empty inside. Two ribs with holes improve model's rigidity and enable easy mounting in the test section of the wind tunnel. Additionally two side segments have been manufactured with MJM technique. Segments assembled with the aerofoil extend its span to 300mm (see Figure 1). As only central part of the assembly is used for measurements, side segments were printed in one technology and can be applied with all 4 testing models. Aerofoil wall thickness along its circumference is constant and equals 4mm. Design details are presented in the Figure 2.

Figure 1. Full aerofoil assembly with mounting system for wind tunnel testing - black SLS model & two MJM side segments.

In order to allow static pressure measurements at the aerofoil surface a unique design was proposed. 48 channels with diameter of 1.5mm and 0.4mm are hidden in the shell. 1.5mm diameter channels are distributed in aerofoil's transverse direction along 55% length of its span (ca. 95mm). It gives a very large length to diameter ratio l/d=60. Such a solution is practically impossible to be fabricated using any other manufacturing method than 3D printing. Small measurement holes of diameter equal 0.4mm normal to aerofoil surface are evenly distributed along the aerofoil cross-section circumference.

Apart from pneumatic measurements, designed aerofoil was considered for Particle Image Velocimetry (PIV) experiment, thus, models should be black in order to reduce excessive light reflections. One of the SLS printout as well as the FDM one are originally made of black bulk material and do not require further processing. Acrylic plastic and alumide models could require darkening of the surface for wind tunnel testing if problems with proper quality of PIV measurements occurred. In Figure 1 MJM side segments and mounting shanks painted in black are shown.

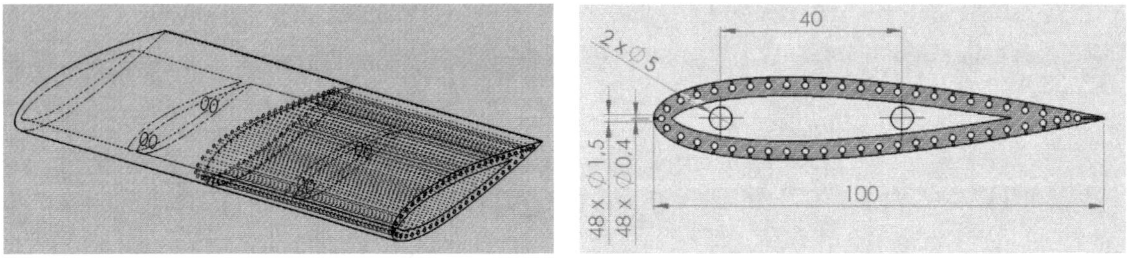

(a) (b)

Figure 2. NACA0018 aerofoil wind tunnel model design (a) isometric view, (b) dimensions.

3D printed models vary in materials used for manufacturing. In the Figure 3 it is clearly visible how printed aerofoils differs from each other. A more specific comparison however, can be made after more detailed analysis of prepared models. Most findings and observations are listed in Table 3.

Table 3. NACA0018 aerofoil models 3D printouts evaluation.

	MJM	SLS	SLS (alumide)	FDM
color/opacity	translucent	black	silver (shiny)	black
surface roughness	low	moderate	moderate	high
surface irregularities direction	transverse	uniform	uniform	longitudinal
ϕ 1.5mm holes quality	most open	blocked	most open	excluded from design
ϕ 0.4mm holes quality	most open	blocked	blocked	excluded from design

4. PIV measurements

One of the experimental techniques used for measurements of printed aerofoil was Particle Image Velocimetry (PIV). Results presented in this paper are of the preliminary character and concern only black SLS model. Rest of the aerofoils are scheduled to be tested later.

PIV results are presented in the form of velocity field around the aerofoil. It gives a full information about the 2D flow velocity components in the proximity of the model. Aim of the experiment carried out was to obtain results for various angle of attack settings. Choice of the measurement plane aligned with the flow and perpendicular to the aerofoil allows to obtain full image of the flow around the aerofoil. Flow structures like separation bubble with reverse flow can be easily detected therefore while increasing the angle of attack. Such a procedure was applied to identify the moment of static stall and find the critical angle.

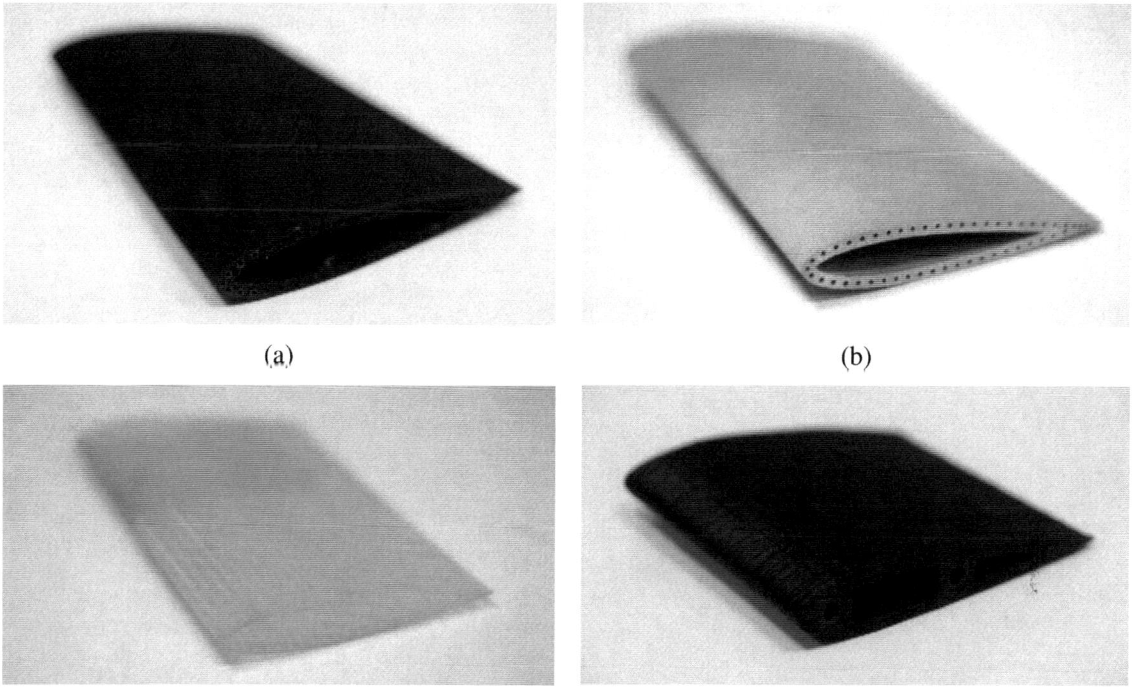

Figure 3. Examples of 3D printed NACA0018 aerofoil models in (a) SLS (black PA2200), (b) SLS (alumide), (c) MJM, (d) FDM technology.

A 2D measurement method was chosen, i.e. each camera pointing at different flow region (with slight overlapping enabling merging results). Cameras were placed in a vertical arrangement observing top and bottom surface of the aerofoil. Laser sheet optics was illuminating observed area from the top creating a plane parallel to the flow. Mirror located at the bottom of test section was used to reflect laser sheet in order to illuminate bottom of the aerofoil.

Experimental setup consisted of the following main components:

- double-pulse Nd:YAG laser from Litron (1200mJ max output, 4ns pulse duration, 1064/532nm wavelength),
- arm with optical system redirecting laser beam to desired position,
- laser sheet optics (cylindrical lens f=-20),
- two PIV cameras Imager pro x 4M from LaVision (2048x2048 pixels),
- Nikkor camera lenses (24-85mm, f2.8),
- mirror (reflectance R=100%) for illuminating bottom part of the aerofoil.

Flow images were captured by two cameras in a double frame mode (giving 4 images per one measurement). Maximum available frame rate of 7.26Hz was used for image acquisition. 100 instantaneous measurements were taken and then averaged to obtain mean velocity fields.

NACA0018 aerofoil model was mounted in the open test section of the subsonic wind tunnel at the Institute of Turbomachinery at the Lodz University of Technology [10]. For the aerofoil chosen (chord $c=0.1$m) and average flow velocity $v=11$m/s Reynolds number $Re\approx75000$. Range of the angle of attack investigated was ±20° with 1° increment. Exemplary results are presented in the Figure 4. Basing on the following velocity fields critical angle has been defined to appear between 15° and 16° angle of attack. The same procedure is planned to be repeated for the rest of 3D printed models. The most crucial observation expected is to determine the influence of printouts surface quality (roughness, imperfections, etc.) on the separation layer formation.

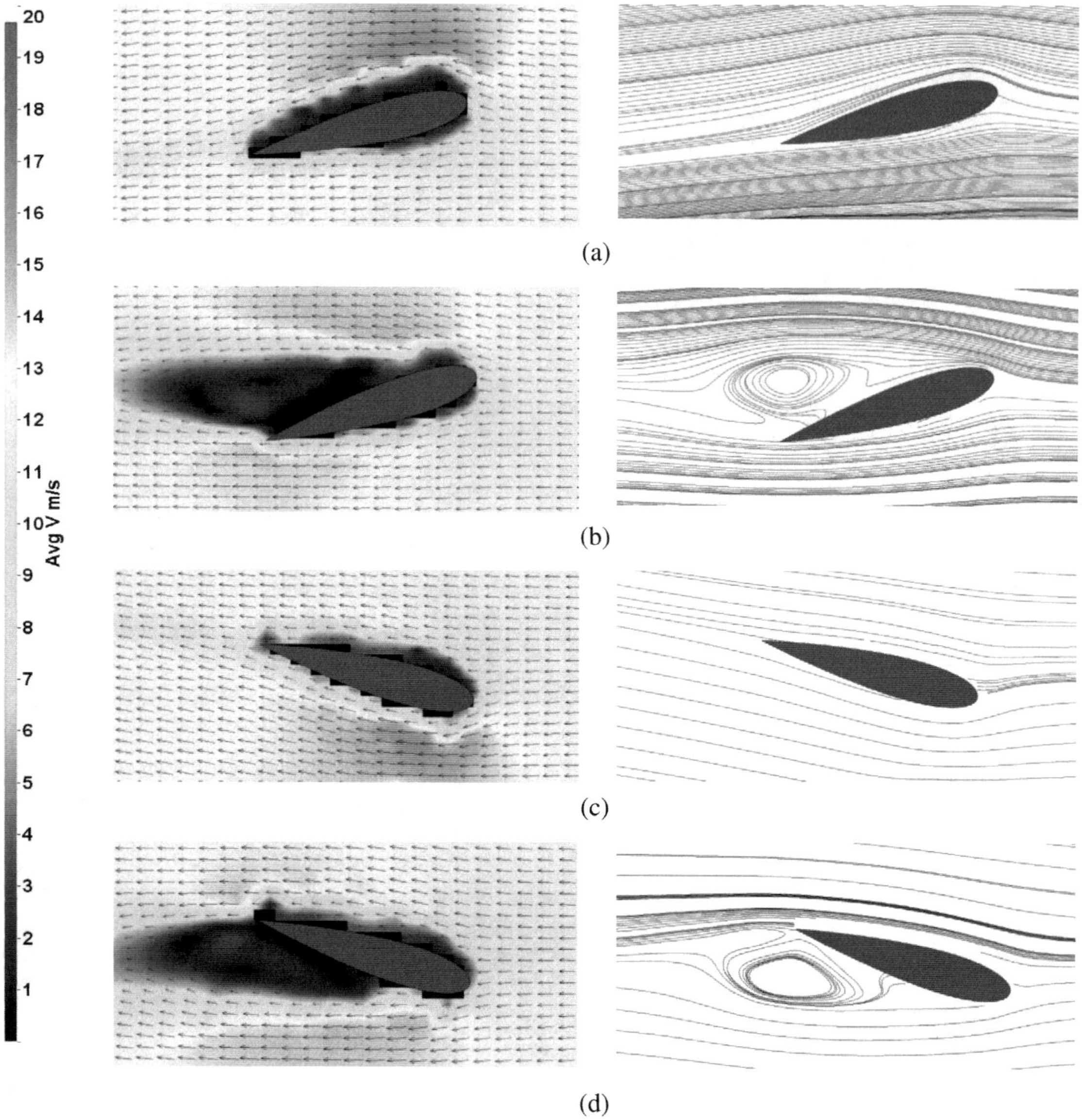

Figure 4. Exemplary PIV results of NACA0018 aerofoil SLS 3D printed model for (a) +15°, (b) +16°, (c) -15°, (d) -16° angle of attack.

5. Further development and perspectives

Results and observations presented in this paper are only the beginning for larger research campaign that includes for instance:

- continuation of PIV measurements,
- performing pneumatic measurements by means of designed system of channels,
- direct lift and drag force measurements with the use of aerodynamic balance,
- lift and drag force measurements basing on PIV measurements [11],
- determination of roughness influence and printouts quality on aerodynamic characteristic of the aerofoils,
- investigation of embossed and engraved artefacts induced by 3D printing technology as a vortex generation modifying the flow,

- determination of influence of the channels inside the aerofoil's shell on flow parameters,
- study aimed at comparison of untreated 3D printed aerofoils and models after surface smoothing process.

6. Conclusions

As for the summary of presented study following observation and conclusions can be drawn:

- 3D printing is a quick, simple and relatively cheap method for production of models for aerodynamic testing.
- As it was presents a wide range of materials and technologies with various features and parameters are available (scope presented in the following paper is very limited as a choice of representative popular technologies only). This enables a potential researcher to choose the right technology fitting exactly his needs and fulfilling the requirements of prepared experiment.
- Thin-walled and tough models have been successfully manufactured.
- It is possible to 3D print holes of very high length to diameter ratio as well as very small objects (like measurement holes of 0.4mm diameter). Still however not all of presented 3D printing technologies give perfect results. Quality strictly depends on method and material chosen.
- Certain 3D printed models are suitable for only limited temperature range. Otherwise they can soften, deflect, deform, etc.

References

[1] Nannan G and Ming C L 2013 Additive manufacturing: technology, applications and research needs *Front. Mech. Eng.* 8(3): (Higher Education Press and Springer-Verlag Berlin Heidelberg) pp 215-43

[2] Daneshmand S, Aghanajafi C and Ahmadi N A 2010 The effect of chromium coating in RP technology for aerofoil manufacturing *Sadhan* a Vol. 35 Part 5 (Indian Academy of Sciences) pp 569–84

[3] Aghanajafi C and Daneshmand S 2010 Integration of three-dimensional printing technology for wind-tunnel model fabrication *Journal of Aircraft* Vol. 47 No. 6 pp 2130-35

[4] Xianghua L et al. 2012 Rapid prototyping of aerodynamics research models *Advanced Materials and Process Technology* pp 2016-25

[5] Alumide® material data sheet 2008 (EOS GmbH – Electro Optical Systems https://www.shapeways.com/rrstatic/material_docs/mds-alumide.pdf)

[6] PA2200 material data sheet 2008 (EOS GmbH – Electro Optical Systems https://www.shapeways.com/rrstatic/material_docs/mds-strongflex.pdf)

[7] FDM material data sheet (Materialise http://manufacturing.materialise.com/sites/default/files/ /public/AMS/Updated%20datasheets/ams_datasheets_fdm.pdf)

[8] Sung-Hoon A et al. 2002 Anisotropic material properties of fused deposition modeling ABS *Rapid Prototyping* Vol. 8 No. 4 (Emeraldin Sight) pp 248–57

[9] Ebel E and Sinnemann T 2014 Fabrication of FDM 3D objects with ABS and PLA and determination of their mechanical properties *RTejournal*

[10] Olasek K and Karczewski M 2012 Multi-phase modernisation of the subsonic wind tunnel oriented towards integration of CFD & experiment *XX Polish Fluid Mechanics Conference*

[11] Olasek K and Karczewski M 2012 Velocity-based lift coefficient calculations *Turbomachinery* no.142

XXI Fluid Mechanics Conference IOP Publishing
Journal of Physics: Conference Series **530** (2014) 012010 doi:10.1088/1742-6596/530/1/012010

Numerical modeling of the pulse wave propagation in large blood vessels based on liquid and wall interaction

K Rup[1] and A Dróżdż[1,2]

[1] Faculty of Mechanical Engineering, Cracow University of Technology, al. Jana Pawła II 37, 31-864 Kraków, Poland
[2] Jagiellonian Centre for Experimental Therapeutics, Jagiellonian University, ul. Bobrzyńskiego 14, 30-348 Kraków, Poland

Abstract. The purpose of this article is to develop a non-linear, one-dimensional model of pulse wave propagation in the arterial cardiovascular system. The model includes partial differential equations resulting from the balance of mass and momentum for the fluid-filled area and the balance equation for the area of the wall and vessels. The considered mathematical model of pulse wave propagation in the thoracic aorta section takes into account the viscous dissipation of fluid energy, realistic values of parameters describing the physicochemical properties of blood and vessel wall. Boundary and initial conditions contain the appropriate information obtained from *in vivo* measurements. As a result of the numerical solution of the mass and momentum balance equations for the blood and the equilibrium equation for the arterial wall area, time-dependent deformation, respective velocity profiles and blood pressure were determined.

1. Introduction

The blood flow in the human body is mostly caused by the deformation of the vessels and other flow conduits. The interaction between the forces acting on fluid and vessel walls often leads to a non-linear pressure drop, wave-flow phenomena and the complex instabilities in the circulation. An existing need to possess a knowledge of these phenomena is some kind of a challenge for experimental research, analytic and numerical calculations [1, 2, 3, 4].

Description of a number of numerical simulation works on aortic blood flow, assuming both rigid and elastically deformable walls as well are shown in [3].

The results of numerical calculation on the self-excited oscillation in the velocity and pressure of the selected sections of the artery, taking into account its local obstruction, is shown in article [4].

In [2] the flow of blood as a Newtonian fluid in the thoracic aorta is analysed, assuming that the wall material is characterized by isotropic linear elasticity. The known dependence of the volume flow rate in time was used in the numerical calculations as the initial condition.

The purpose of this article is to develop a non-linear, one-dimensional model of pulse wave propagation in the arterial cardiovascular system. The main idea of this work involves an analysis of the average blood flow perturbation in the thoracic aorta in human, caused by a cyclically varying pressure in the inlet cross-section.

Arterial segments (human thoracic aorta) are modeled using pipe sections undergoing linear elastic deformations. The model allows to perform an analysis of the performance of a healthy or partially obstructed cardiovascular system.

2. Problem formulation

We will consider an unsteady, laminar blood flow through the thoracic aorta segment in human. Figure 1 shows the considered flow diagram.

Our considerations take into account the following assumptions:

- blood vessel has the shape of a circular cylinder with changing cross-sectional area $S=S(s,t)$, constant wall thickness b and fixed length L,
- the vessel wall undergoes substantial elastic deformations in the radial direction only,
- in a vessel filled with blood pressure $p(s,t)$ and velocity $\upsilon=\upsilon(s,t)$ wave propagation occur,
- the blood filling the vessel is a Newtonian fluid demonstrating a constant viscosity and a slight compressibility.

Given the above assumptions, the considered blood flow in the thoracic aorta is described by the following equations :

$$\frac{d\upsilon}{dt}+\frac{1}{\rho}\frac{\partial p}{\partial s}+g\frac{dz}{ds}+\frac{\lambda}{2D}\upsilon|\upsilon|=0 \tag{1}$$

$$a^2\frac{\partial\upsilon}{\partial s}+\frac{1}{\rho}\frac{dp}{dt}=0 \tag{2}$$

Equations (1) and (2) result from the analysis of the overall mass and momentum balance equations for blood and from the equilibrium equations for an elastic blood vessel wall material.

Symbols in equations (1) and (2) mean:

$$\begin{cases}\dfrac{d\upsilon}{dt}=\dfrac{\partial\upsilon}{\partial t}+\upsilon\dfrac{\partial\upsilon}{\partial s}\\[2mm] \dfrac{dp}{dt}=\dfrac{\partial p}{\partial t}+p\dfrac{\partial p}{\partial s}\end{cases} \tag{3}$$

$$a=\frac{\sqrt{\frac{K}{\rho}}}{\sqrt{1+\frac{K}{E}\frac{D}{b}C}} \tag{4}$$

For the thick blood vessel, constrained at both ends, the parameter C takes the form of [5]:

$$C=\frac{1}{1+\frac{b}{D}}\left[\left(1-\mu^2\right)+2\frac{b}{D}\left(1+\mu\right)\left(1+\frac{b}{D}\right)\right] \tag{5}$$

where μ is the Poisson's ratio for the material of the arterial wall.

The system of equations (1) and (2) is known in the literature for modeling water hammer [5]. To solve this system of equations the method of characteristics is most commonly used [5]. The main idea behind the method of characteristics is an effective reduction of a system of partial differential equations to an equivalent system of ordinary differential equations.

$$\frac{d\upsilon}{dt}+\frac{g}{a}\frac{dH}{dt}-\frac{g}{a}\upsilon\frac{dz}{ds}+\frac{\lambda}{2D}\upsilon|\upsilon|=0 \tag{6}$$

$$\frac{d\upsilon}{dt}-\frac{g}{a}\frac{dH}{dt}+\frac{g}{a}\upsilon\frac{dz}{ds}+\frac{\lambda}{2D}\upsilon|\upsilon|=0 \tag{7}$$

It should be noted that equation (6) is valid only for a C⁺ characteristic, described by a relationship:

$$\frac{ds}{dt} = \upsilon + a \tag{8}$$

while equation (7) is valid along the C⁻ characteristic:

$$\frac{ds}{dt} = \upsilon - a \tag{9}$$

In equations (6) and (7) the value $H(s,t)$ is the piezometric height of the liquid column expressing the pressure changes:

$$p = \rho g (H - z) \tag{10}$$

From equations (8) and (9) it is apparent that the slope of the lines of characteristics depends on the fluid velocity $\upsilon(s,t)$.

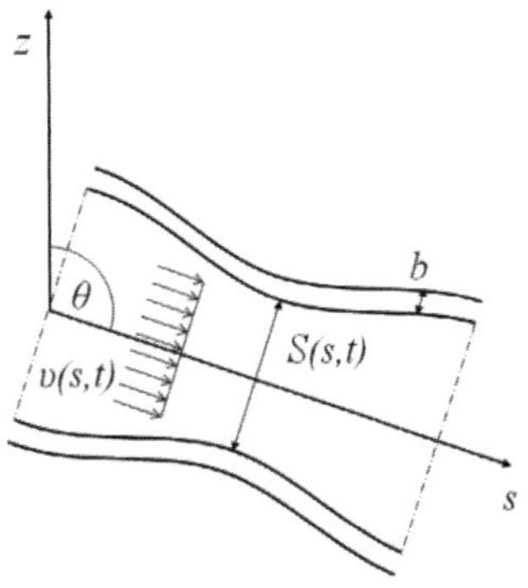

Figure 1. Diagram of the flow in the vessel.

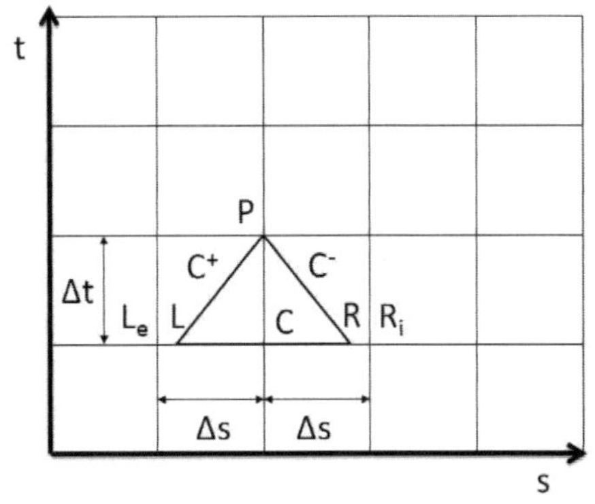

Figure 2. An interpretation of H and υ in the s-t plane.

3. Solving the problem

We will write differential equations (6) and (7) in the form of finite differences. Assuming small values of spatial Δs and time Δt steps, new values of H and υ at the point P of characteristics C^+ and C^- intersection can be obtained, as shown in Figure 2.

It is worth noting that the characteristics C^+ and C^- in Figure 2 are straight line segments due to the small value of the time step Δt. Given the above, equations (6) and (7) will be written in the form of finite differences:

$$\frac{\upsilon_P - \upsilon_L}{\Delta t} + \frac{g}{a}\frac{H_P - H_L}{\Delta t} - \frac{g}{a}\upsilon_L\frac{dz}{ds} + \frac{\lambda}{2D}\upsilon_L|\upsilon_L| = 0 \tag{11}$$

$$\frac{\upsilon_P - \upsilon_R}{\Delta t} - \frac{g}{a}\frac{H_P - H_R}{\Delta t} + \frac{g}{a}\upsilon_R\frac{dz}{ds} + \frac{\lambda}{2D}\upsilon_R|\upsilon_R| = 0 \tag{12}$$

Equations (11) and (12) are valid along the characteristics C^+ and C^- respectively. From the differential equations (11) and (12) we calculate υ_P and H_P for each time-step:

$$\upsilon_P = \frac{1}{2}\left[(\upsilon_L + \upsilon_R) + \frac{g}{a}(H_L - H_R) + \frac{g}{a}\Delta t(\upsilon_L - \upsilon_R)\sin\theta - \frac{\Delta t}{2D}\frac{64\nu}{|\upsilon_L|D}(\upsilon_L|\upsilon_L| + \upsilon_R|\upsilon_R|)\right] \tag{13}$$

Journal of Physics: Conference Series **530** (2014) 012010 doi:10.1088/1742-6596/530/1/012010

$$H_P = \frac{1}{2}\left[\frac{a}{g}(\upsilon_L - \upsilon_R) + (H_L + H_R) + \Delta t(\upsilon_L + \upsilon_R)\sin\theta - \frac{a}{g}\frac{\Delta t}{2D}\frac{64\nu}{|\upsilon_L|D}(\upsilon_L|\upsilon_L| - \upsilon_R|\upsilon_R|)\right] \tag{14}$$

In equations (13) and (14) $\sin\theta = dz/ds$ is positive when a blood vessel rises up the z axis (Figure 1). In order to ensure the stability of solutions, the relationship between the time and spatial step takes the form of inequality:

$$\Delta t \le \frac{\Delta s}{\max|a + \upsilon|} \tag{15}$$

Initial and boundary conditions are based on measurements made *in vivo* with a piezoelectric tonometer (SphygmoCor, AtCor Medical). In particular, the dependence of the blood pressure $p(t)$ at the opening of the aortic valve was obtained. Selected pressure waveforms for persons aged 27 and 78 have been presented in Figure 3. Measured *in vivo* pressure waveforms $p(t)$ differ in the cardiac cycle period and exhibit different levels of intensity changes. The cardiac period is 0.902 *s* for the 78-year old man and 0.998 *s* for the 27-year old man.

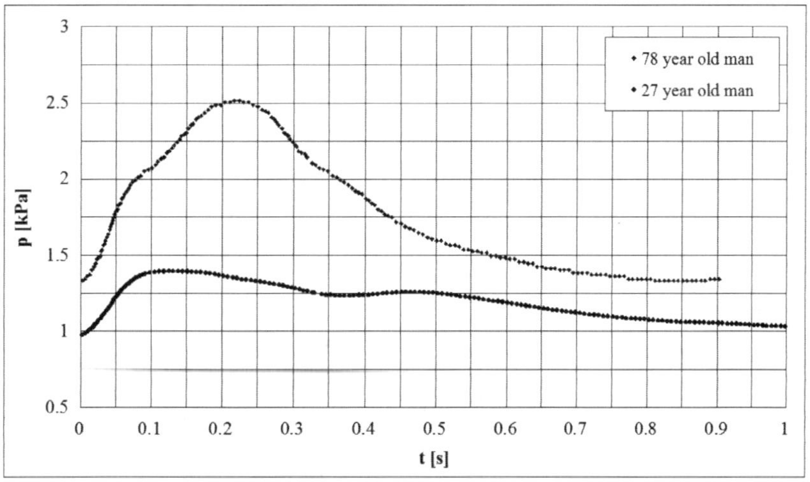

Figure 3. Pressure waveforms included in the calculations obtained with SphygmoCor (AtCor Medical).

For the purposes of the calculation, the pressure changes were converted according to equation (10):

$$H_{P_1}(t) = \frac{p(t)}{\rho g} \tag{16}$$

Taking into account the relation (16) in equation (12), we obtain the initial condition for velocity:

$$\upsilon_{P_1} = \upsilon_2 + \frac{g}{a}\left(H_{P_1}(t) - H_2\right) - \frac{g}{a}\upsilon_2 \sin\theta - \frac{\lambda\Delta t}{2D}\upsilon_2|\upsilon_2| \tag{17}$$

In equation (17) the index 2 corresponds to the index R in equation (12). At the end of the considered segment of the aorta the following condition was assumed:

$$\upsilon_{P_{N+1}} = \upsilon_0\left(1 - \frac{t}{T_C}\right), 0 \le t \le T_C \tag{18}$$

A suitable condition for the pressure at the considered aortic end results from equations (11) and (18):

$$H_{P_{N+1}} = H_N - \frac{a}{g}\left(\upsilon_{P_{N+1}} - \upsilon_N\right) - \upsilon_N \sin\theta - \frac{a}{g}\frac{\lambda \Delta t}{2D}\upsilon_N\left|\upsilon_N\right| \tag{19}$$

In order to perform numerical integration of the equations (1) and (2) with conditions (16)-(19), in accordance with the idea of the method of characteristics, the numerical code in FORTRAN was developed.
The measured changes in the pressure $p(t)$ $(H(t))$ were interpolated using the spline function in the calculation process.

4. Results and discussion

The numerical calculation includes parameter values and physical properties of the thoracic aortic wall at $T = 37°C$ adopted from the literature [6]:

- The blood properties:
  dynamic viscosity $\eta = 0.0035$ N·s/m^2
  density $\rho = 1030$ kg/m^3
  bulk modulus $K = 2.15 \cdot 10^9$ *Pa*
- The aortic wall properties:
  inner diameter $D = 20$ mm
  wall thickness $b = 1.2$ mm
  length $L = 160$ mm
  Poisson's ratio $\mu = 0.5$
  circumferential modulus of elasticity depending on the human age expressed in years:
  $E_\theta(27) = 5.5 \cdot 10^5$ Pa
  $E_\theta(78) = 18.7 \cdot 10^5$ Pa.

Taking into account the assumed values describing the physical properties of blood and the aortic wall, the values of C and a parameters were calculated from equations (4) and (5). They are respectively: C=0.8875, a(27)=6.010 m/s, a(78)=11.078 m/s.
The calculations take also into account the average blood velocity in the aorta which equals υ=0.27 m/s. Also, the vertical arrangement of the vessel was assumed, described by the equation sinθ=-1.0. Considered section of the aorta was divided into 19 equal segments. The length of the time step in the calculation was 0.0014 s and 0.00076 s for the 78-year old and 27-year old human respectively. The results of numerical calculations are presented in graphical form in Figures 4-11.

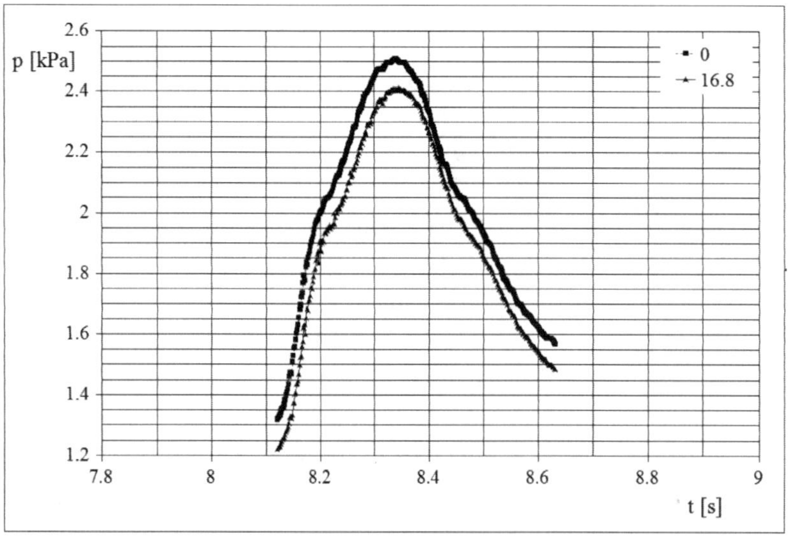

Figure 4. Blood pressure waveforms for the two thoracic aorta segments (0 and 16.8 mm) in the 78-year old man.

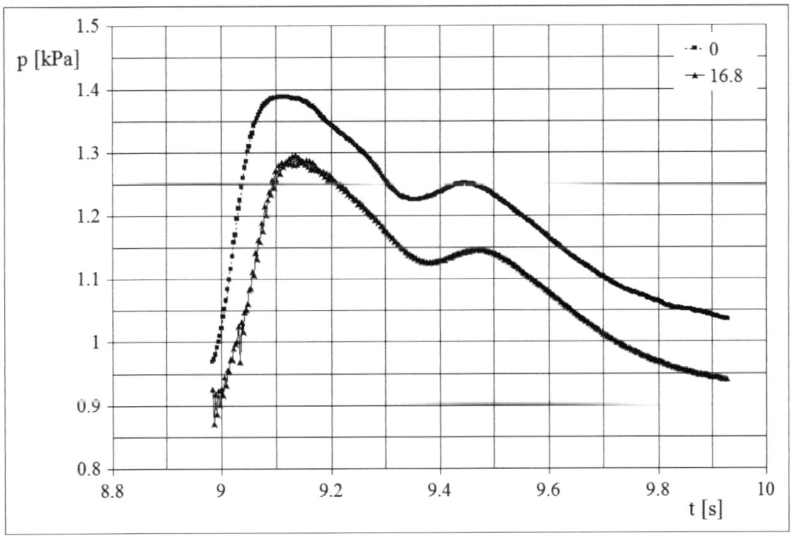

Figure 5. Blood pressure waveforms for the two thoracic aorta segments (0 and 16.8 mm) in the 27-year old man.

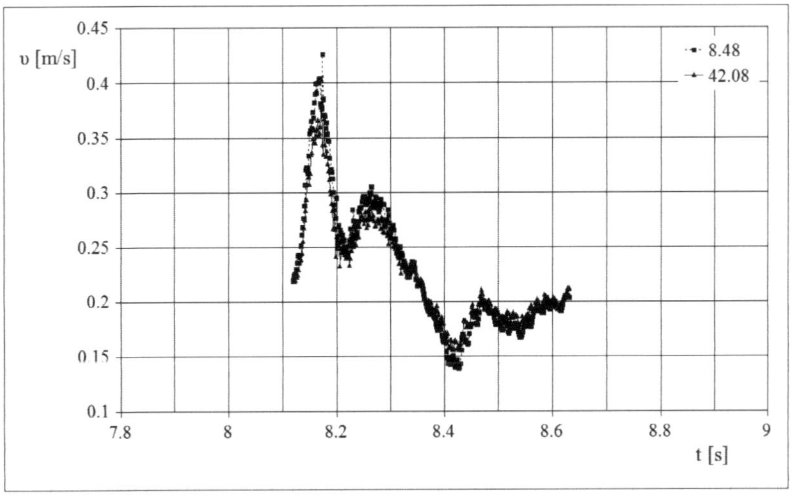

Figure 6. Blood velocity waveforms for the two thoracic aorta segments (8.48 mm and 42.08 mm) in the 78-year old man.

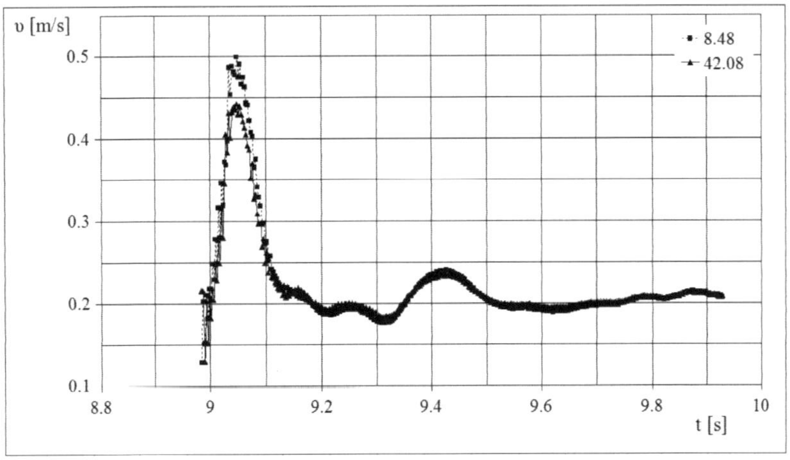

Figure 7. Blood velocity waveforms for the two thoracic aorta segments (8.48 mm and 42.08 mm) in the 27-year old man.

Journal of Physics: Conference Series **530** (2014) 012010 doi:10.1088/1742-6596/530/1/012010

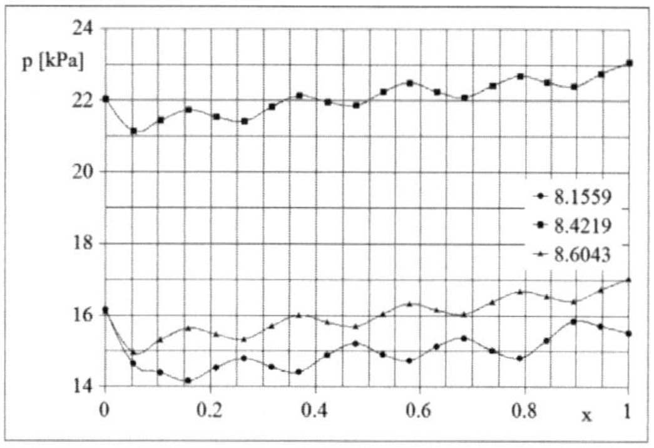

Figure 8. Spatial blood pressure changes at three time points in the tenth cardiac cycle in the 78-year old man.

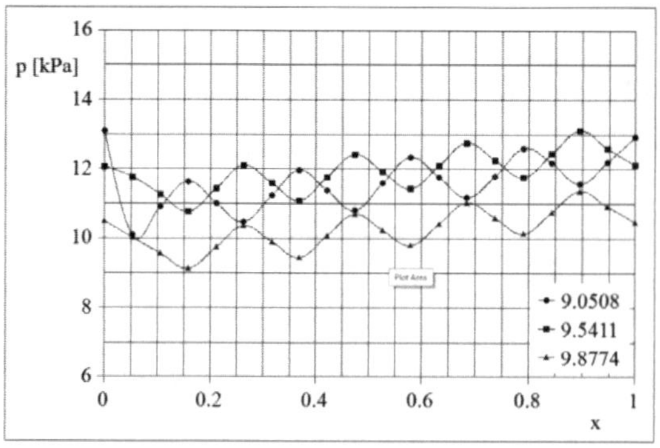

Figure 9. Spatial blood pressure changes at three time points in the tenth cardiac cycle in the 27-year old man.

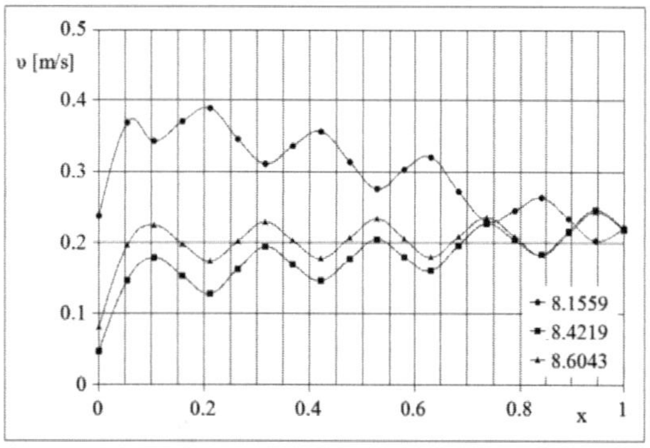

Figure 10. Spatial blood velocity changes at three time points in the tenth cardiac cycle in the 78-year old man.

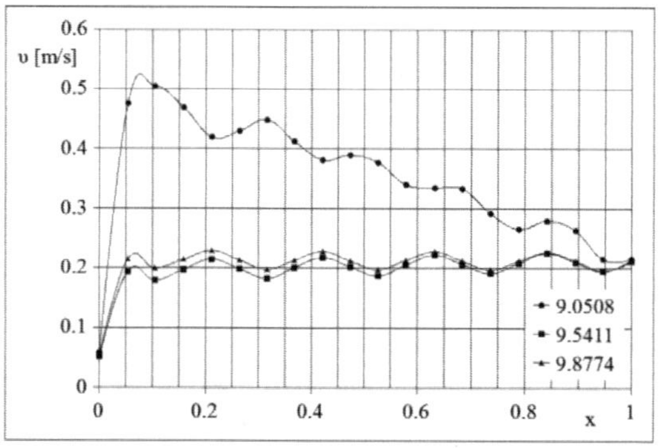

Figure 11. Spatial blood velocity changes at three time points in the tenth cardiac cycle in the 27-year old man.

The time-dependent local changes in blood pressure calculated in the tenth heart cycle for the 78-year old man are shown in Figure 4. Blood pressure plots are assigned to the respective segments along the thoracic aorta. Square-shaped symbols in Figure 4 indicate a change in blood pressure in the inlet cross-section ($x=s/L=0$), triangle-shaped points represent the changes in blood pressure in the second segment ($x=16.8/160=0.105$). It can be easily observed that the predetermined profile of numerical changes in blood pressure in the second aortic segment is considered very similar to the corresponding shape preset in the inlet cross-section, characterized by a suitable, lower values of pressure.

Respective local changes in blood pressure profiles in the case of 27-year old man, also calculated during the tenth cardiac cycle, are shown in Figure 5.

The time-dependent changes in blood velocity calculated in the tenth heart cycle for the 78-year old man are shown in Figure 6. Similar waveforms of blood velocity changes in equally distant segments for a 27-year old man are presented in Figure 7.

From a comparison of the blood velocity waveforms in figures 6 and 7, a greater degree of severity of the flow perturbation is present in the stiffer vessel section in the 78-year old man ($E_0(78) = 18.7 \cdot 10^5$ Pa) compared to the corresponding flow through more elastic part of the aorta in younger man ($E_0(27) = 5.5 \cdot 10^5$ Pa).

Figures 8 and 9 represent the spatial changes in blood pressure through the thoracic aorta segment in 78- and 27-year old man respectively, at selected moments of the tenth cardiac cycle. Due to lower heart rate in the younger man and longer cardiac cycle, the spatial pressure waveforms are selected at different time points.

In figures 10 and 11 changes in blood velocity along the thoracic aorta are presented for 78- and 27-year old man respectively. The diagrams refer to the same moments of the tenth cardiac cycle which are analyzed in the spatial pressure variations in Figures 8 and 9. It can be seen that the blood velocity at the end of the considered aortic segment is achieved about the same value for the three randomly selected moments of the tenth cardiac cycle.

5. Final remarks and conclusions

In this paper we present numerical results of the integration of the one-dimensional differential equations of an unsteady blood flow in the human thoracic aorta. The calculations account for the pressure waveforms in the inlet cross section of the considered aortic segment, measured *in vivo* in two people of different age. The realistic, taken from the literature, values of parameters describing the blood and arterial wall properties are also included. Elastic deformation of the wall caused by pulsatile blood flow was taken into account by varying the value of the modulus of elasticity E, dependent on the human age. The analysis obtained from the numerical results have been limited to the tenth of the cardiac cycle. As a result from the analysis it can be concluded that in cardiac cycle a strong stabilization of the blood flow parameters is found in the considered aorta segment, both in the elderly and the young man. Elastic deformation of the considered aortic wall also stabilizes. The maximum relative changes in the aortic diameter are not greater than $\pm$ 5%.

References

[1] Grotberg J B and Jensen O E 2004 Biofluid Mechanics in Flexible Tubes *Annu. Rev. Fliud Mech.* **36** 121-147

[2] Filipović N, Mijailović S, Tsuda A and Kojić M 2006 An implicit algorithm within the arbitrary Lagrangian-Eulerian formulation for solving incompressible fluid flow with large boundary motions *Comp. Meth. App. Mech. Eng.* **195** 6374-6361

[3] Lin Y, LaiY, Nagaraj A, Kane B, Hamilton A, Greene R, McPerson D and Chandran K 2001 Pulsatiole flow simulation in arterial vascular segments with intravascular ultrasound images *Medical Eng. Phys.* **23** 586-595

[4] Tang D, Yang J and Ku D N 1999 A nonlinear axisymmetricmodel with fluid-wall interactions for steady viscous flow in stenotic elastic tubes *J. Biomech. Eng. Trans.* ASME **121** 494-504

[5] Larock B E, Jeppson R W and Watters G Z 2000 *Hydraulics of Pipeline Systems* CRC Press New York

[6] Black J and Hastings G 1998 *Handbook of Biomedical Properties* Chapman&Hall, London

Acknowledgments

The *in vivo* measurements were made with the support from the resources of the European Regional Development Fund under the Innovative Economy Program (grant coordinated by JCET – UJ, no. POIG.01.01.02-00-069/09).

Journal of Physics: Conference Series **530** (2014) 012011 doi:10.1088/1742-6596/530/1/012011

Two-way FSI modelling of blood flow through CCA accounting on-line medical diagnostics in hypertension

K Czechowicz[1,2], J Badur[1], K Narkiewicz[2]
[1] Energy Conversion Department, Institute of Fluid-Flow Machinery PAS-ci, Gdańsk, Poland
[2] Medical University of Gdańsk, Gdańsk, Poland

E-mail: krzysztof-czechowicz@gumed.edu.pl

Abstract. Flow parameters can induce pathological changes in the arteries. We propose a method to asses those parameters using a 3D computer model of the flow in the Common Carotid Artery. Input data was acquired using an automatic 2D ultrasound wall tracking system. This data has been used to generate a 3D geometry of the artery. The diameter and wall thickness have been assessed individually for every patient, but the artery has been taken as a 75mm straight tube. The Young's modulus for the arterial walls was calculated using the pulse pressure, diastolic (minimal) diameter and wall thickness (IMT). Blood flow was derived from the pressure waveform using a 2-parameter Windkessel model. The blood is assumed to be non-Newtonian. The computational models were generated and calculated using commercial code. The coupling method required the use of Arbitrary Lagrangian-Euler formulation to solve Navier-Stokes and Navier-Lamè equations in a moving domain. The calculations showed that the distention of the walls in the model is not significantly different from the measurements. Results from the model have been used to locate additional risk factors, such as wall shear stress or circumferential stress, that may predict adverse hypertension complications.

1. Introduction

Cardiovascular dieses are becoming the main cause of death and disabilities in the modern world [1]. Treatment for some of the diseases like hypertension is expensive and mostly concentrates on managing the disease [2]. Identifying the early onsets and biological mechanisms leading to the end result is becoming an urging challenge and more detailed research is needed. The focus of this paper is on providing a tool to extract mechanical details of the cardiovascular system, that have not been available up to date. Scientist tried to extract mechanical parameters from the human body since the beginning of modern mechanics. Earliest work regarding the effects of the interaction between blood and the walls of the arteries can be dated to work by Thomas Young from 1808 [3]. Since then many analytical and numerical solutions have been proposed but the computational power of modern computers allows acquiring very detailed information about flow in the arteries.

It is known that the blood vessels react to the flow in many ways. The reactions can be short-term, such as aligning of the cells on the inner most layer of the wall [4], or releasing ATP (Adenosine triphosphate) [5] and long-term such as permanent increase in diameter or vessel wall thickness [6]. The main mediator of these changes is thought to be the Wall Shear Stress (WSS) generated by the pulsate flow of blood, and blood pressure. In a pulsate flow, calculation of the WSS is very problematic and it cannot be measured due to lack of precision of modern medical devices.

Journal of Physics: Conference Series **530** (2014) 012011 doi:10.1088/1742-6596/530/1/012011

Furthermore blood flow is pulsate and as such the profile is not always parabolic, but acts according to the Womersley solution of pulsate flows [7]. Therefore use of computer modelling is the best alternative. The way blood pressure changes within one heart cycle varies between patients, therefore to analyze the problem in detail, a series of patient specific models have to be created.
Most work done in this field focuses on using Magnetic Resonance Imaging or Computated Tomography [8][9].Those imaging methods are expensive, time consuming and may be harmful to the patient, which excludes the use of this method on a significant number of patients. In this paper we present a method of generating patient specific computational models of the human Common Carotid Artery (CCA) based on data acquired during an ultrasound examination of the arteries.

2. Anatomy

Humans have four arteries that supply blood to the head. Two Common Carotid Arteries go through the front of the neck on either side. These are the main suppliers of blood to the brain and other intracranial organs. In the upper neck arteries bifurcate to form internal and external carotid arteries.
The walls of the arteries are build of 3 layers. The inner most is called Intima and it consists of a single layer of endothelial cells. The middle layer is Media and it consists of smooth muscle cells and connective elastic tissue. It is responsible for most of the artery's thickness and mechanical properties. The outer most layer, Adventitia, is mostly built of connective tissue and its main function is supporting the artery. The complex of Intima and Media is the main pressure load bearing part of the artery. Furthermore the interfaces between the artery's lumen and intima, as well as the interface between Media and Adventitia are visible on ultrasound images. Their thickness is often used as a measure of artery thickness and is referred to as Intima Media Thickness (IMT)

3. Methods

We studied a group of 18 people with high normal blood pressure and early detected hypertension. Mean age of our patients was 36 (±10) years. The patients had never been treated for hypertension and had not been diagnosed with any other chronic diseases. All of our measurements were non-invasive and were carried out in supine position in a quiet room after a 15 minute rest. The measurements have been conducted on the right CCA. The exact spot was found by locating the CCA bifurcation and placing the start of it at the distal edge of the ultrasound image.

3.1. Data acquisition

The ultrasound machine used in this study was an Esaote Picus Pro device equipped with a prototype wall tracking system ART.LAB (Esaote, Maastricht, Netherlands). The system provides very high resolution images, with a pixel size of 23µm x 23µm, and a very high frame rate of 600Hz. The wall tracking system provides continuous recordings of the distention of the artery, its diameter and IMT. Data is recorded in 14 lines perpendicular to the probe. Simultaneously other biological signals were recorded. Continuous blood pressure waveform was obtained from the finger using the Finometer device (Finapress Medical Systems, Amsterdam, The Netherlands) linked to a PowerLab (ADInstruments, Dunedin, New Zealand) data acquisition system which also recorded electrocardiograms (ECG). The ECG and blood pressure signals were synchronised automatically with each other. The ECG signal was duplicated and sent to the ART.LAB device and recorded along with the ultrasound data. This was used to synchronise the signals in pre processing.

3.2. Data Preprocessing

Due to the high resolution of the data the length of the ultrasound recordings is limited to 4.7 seconds thus a series of 5-10 measurements were carried out. In order to achieve maximum accuracy ultrasound recordings had to be synchronised with the BP recording. The ECG signal recorded along with the ultrasound dataset was fitted to a window of 10 seconds around the start of the ultrasound measurement.

For each patient recording of one heart beat was selected. The criteria for this selection was the minimum difference between the distention waveforms taken from different ultrasound lines, as well as no interruptions, such as calibration, on the blood pressure recording.

3.2.1. Pressure waveform

Knowing the blood pressure measured in the finger and the distention of the carotid artery it is possible to estimate local pressure in the artery. Arterial pressure is characterized by two main values: Systolic Blood Pressure (SBP) – the maximum pressure in a heartbeat and Diastolic Blood Pressure (DBP) – the minimum blood pressure. Experimentally it is well established [10] that the value of DBP is constant throughout the arterial tree in each heartbeat. The same is true for a value called Mean Arterial Pressure (MAP) calculated as (1)

$$\mathrm{MAP} = \frac{1}{3}\mathrm{SBP} + \frac{2}{3}\mathrm{DBP} \tag{1}$$

The local pressure waveform is estimated by scaling the distention waveform according to (2) where the scaling factor is given by (3) assuming that the wall of the artery to have linear elastic properties.

$$\mathrm{p(t)} = \delta \cdot \mathrm{D(t)} + \mathrm{DBP} \tag{2}$$

$$\delta = \frac{\mathrm{MAP} - \mathrm{DBP}}{\overline{\mathrm{D}}} \tag{3}$$

where $\overline{D}$ is the average distention calculated over the whole heartbeat.

3.2.2. Young's modulus

The arterial wall is considered to be linear elastic and incompressible therefore it has a constant Young's modulus for each patient and Poissons number of $v = 0.5$. For numerical purposes a value of $v = 0.495$ was taken. As for the Young's modulus a value for each patient was calculated as

$$\mathrm{E} = \frac{(1-\upsilon^2)\mathrm{D}}{\mathrm{IMT} \cdot \mathrm{DC}} \tag{4}$$

Where DC is the distensibility coefficient given by

$$\mathrm{DC} = \frac{2\mathrm{D} \cdot \Delta\mathrm{D} + \Delta\mathrm{D}^2}{\mathrm{D}^2 \cdot \Delta\mathrm{p}} \tag{5}$$

Young's modulus defined in such a way takes into account material properties of the entire vessel wall. It is worth mentioning that the value is not representative for the elasticity of both the Media and Intima layers of the artery separately, but it is a surrogate value for the entire vessel including the adventitia layer. Furthermore it is a linear approximation of the real elasticity of the vessel.

3.2.3. Pulse wave velocity

Blood travels through the arterial tree in a pulsate manner. Each pulse travels at a velocity given by the Moens Korteweg [11] equation in a form modified by Bramwell [12]

$$\mathrm{PWV} = \sqrt{\frac{\mathrm{dp} \cdot \mathrm{V}}{\mathrm{d} \cdot \rho}} \tag{6}$$

After introducing the distesibility coefficient (5)

$$\mathrm{PWV} = \frac{1}{\sqrt{\rho \cdot \mathrm{DC}}} \tag{7}$$

The local PWV in the CCA can reach values between 4m/s and 10m/s, which means the pulse needs between 0.0075s and 0.01875s.

3.2.4. Mass Flow

Mass flow of blood was obtained by using a zero dimensional two parameter Windkessel model proposed by Frank in 1899 [13].

$$\dot{m} = \frac{p(t)}{R} + C\frac{dp}{dt} \tag{8}$$

Parameter R represents the mean resistance of the arteries and can be easily calculated when the amount of blood flowing through the artery in every heartbeat is known. Parameter C for this model is obtained by solving equation (8) in a period after the heart has finished contracting when the flow in the artery is expected to be stable.

4. Computer model

Data prepared as described above has been used to create transient computational models of momentum exchange between the fluid (blood) and the solid (arterial walls). The model has been prepared in the ANSYS Workbech enviroment. The fluid domain was solved using the CFX solver and the solid domain was solved using the ANSYS Mechanical solver. Each of these solvers is based on a different discretisation method. CFX uses the finite volume method and ANSYS Mechanical is using the finite element method. Communication between the solvers was provided by ANSYS Workbench.

4.1. Fluid domain

The fluid part of the model was computed by the CFX finite volume solver. The governing equations for this software are the Navier-Stokes equations. These in general consist of three laws of conservation: conservation of mass, momentum and energy (11) [14]. The conservation equations used in the Arbitrary-Lagrangian Eulerian formulation to accommodate for the moving mesh. The formulation is obtained by subtracting the velocity of the mesh from fluid velocity in the convective flux of the standard formulation.

$$\partial_t \begin{Bmatrix} \rho \\ \rho\vec{v} \\ \rho e \end{Bmatrix} + \mathrm{div}\begin{Bmatrix} \rho(\vec{v}-\vec{w}) \\ \rho\vec{v}\otimes(\vec{v}-\vec{w}) \\ \rho e(\vec{v}-\vec{w}) \end{Bmatrix} + \begin{Bmatrix} 0 \\ p\vec{I} \\ p\vec{v} \end{Bmatrix} = \mathrm{div}\begin{Bmatrix} 0 \\ \vec{\tau}^{c} \\ \vec{\tau}^{c}\vec{v} + \vec{q}^{c} \end{Bmatrix} + S \tag{9}$$

This can be written in a different form, reviling the fluxes

$$\partial_t\{\mathcal{U}\} + \mathrm{div}\{\mathcal{F}^{c}\} + \{\mathcal{F}^{e}\} = \mathrm{div}\{\mathcal{F}^{v}\} + \mathbf{S} \tag{10}$$

In this case there is no transfer of energy within the domain, as the temperature is assumed to be constant; therefore the latter equation is always fulfilled. Blood is considered to be incompressible therefore its density is constant. Implementing those changes simplifies the conservation equations (9) to the form of Navier-Stokes equations.

$$\begin{gathered} \nabla\vec{v} = 0 \\ \rho\left(\frac{\partial\vec{v}}{\partial t} + \vec{v}\cdot\nabla\vec{v}\right) + \nabla p = \mu\nabla^2\vec{v} \end{gathered} \tag{11}$$

where $\vec{v}$ is blood velocity, ρ is density, μ is blood viscosity.

The domain has been divided into a structural mesh swept along the length of the vessel with 6 layers of inflating near the arterial wall. The same settings have been used for each patient to increase automatization of the calculation process.

4.2. Solid domain

The solid part of the model has been solved by the ANSYS Mechanical solver. The walls have been modelled by one layer of 3 dimensional elements. The size of the elements has been chosen to be as close to proportions of 1:1:2 as possible, with the elongation in the axial direction. The total number of elements was around 800. This value was proven to be sufficient in several numerical experiments carried out by our team. The governing equations for this solver are the Navier-Lame equation

$$\sigma = 2\mu\epsilon + \lambda\, \mathrm{tr}\epsilon\, \mathrm{I} \tag{12}$$

The distention of the artery is expected to be around 8% of the diameter, reaching up to 14%. The scale of deformations of the walls of the artery required using the finite deformation approach. This means that the elements of the Green's Tensor (13) had to be taken in full form, as its final term is expected to have non negligible values

$$\epsilon_{ij}^{G} = \frac{1}{2}\left(\frac{\partial u_i}{\partial x_j} + \frac{\partial u_j}{\partial x_i} + \frac{\partial u_k}{\partial x_i}\frac{\partial u_k}{\partial x_j}\right) \tag{13}$$

4.3. Fluid Solid Interaction

As mentioned before the solid and fluid domains are solved in separate programs using different methods. Coupling between them has to insure that the forces acting between the domains (14), and the deformation of the domains (15) are equal. We achieved this using the conservation of momentum method.

$$\left(\mathcal{F}^{c} - \mathcal{F}^{e}\right)\cdot \mathbf{n}|_{solid} = \left(\mathcal{F}^{c} - \mathcal{F}^{e}\right)\cdot \mathbf{n}|_{fluid} \tag{14}$$

$$\vec{v}|_{solid} = \vec{v}|_{fluid} \tag{15}$$

where **n** are versors normal to the boundary in solid and fluid.

Movement of the mesh has been described by the use of the Laplace equation (16). It is also necessary that the velocity of the mesh near the wall is equal to the velocity of the wall, to avoid creating a gap (17).

$$\mathrm{div}[(1-\alpha)\mathrm{grad}\,\vec{w}] = 0 \tag{16}$$

$$\vec{w}|_{CS} = \vec{v}|_{fluid} \tag{17}$$

The stiffness coefficient α in equation (16) has been set to increase near the boundaries where thinner elements are located.

4.4. The model

The arteries have been modelled as straight tubes with the diameter and wall thickness measured by the ART.LAB system. The walls at inlet and outlet have been fixed, to prevent the movement of the artery and allow implementation of the boundary conditions to the fluid domain. This means that when the artery is distended nozzles appear at both ends. We found empirically, by testing different lengths of the model, that the influence of the inlet nozzle is observable only within the first 40mm. The outlet nozzle has only local effects in the flow. Therefore the most optimal length of the model is 75mm with only the part between the 40th and 65^{th}mm taken into further analysis.

For the fluid domain we applied an opening type boundary on the inlet side, to allow the fluid to flow both in and out of the domain. The boundary is governed by flow in the axial direction. The outlet boundary has been set as a pressure boundary condition. As mentioned earlier the pulse travels through the arteries with a finite velocity. The model has a time step of 0.005s and, as mentioned in paragraph 3.2.3, the time necessary for the pulse to travel through the artery can be up to 0.019s. To compensate for the delay conditions an offset at the inlet boundary condition given by the time calculated from equation (7) has been implemented.

Blood has been modelled as an incompressible non Newtonian fluid. The viscosity of the blood has been modelled using the power law expression (18). The properties for this model have been taken from literature [15]. According to Neofytou this model takes into account both viscous and two-phase properties of blood.

$$\mu = K\cdot(\dot{\gamma})^{n} \tag{18}$$

Journal of Physics: Conference Series **530** (2014) 012011 doi:10.1088/1742-6596/530/1/012011

where K = 146.7mPas^n and n = 0.7755. Elastic material properties for the vessel wall have been calculated individually for each patient.

5. Results

After each calculation, the results have been inspected manually to verify that the calculated distention waveform is the same as the measured distention waveform used as input. A sample comparison is depicted on Figure 1. The distention is caused by total pressure, a sum of static pressure and dynamic pressure. the latter is a result of the blood velocity. Similarity of the input and output data, shows that the total pressure of blood in the model is the same as that in the artery. Thus the velocity is the same as well.

In all cases an additional high frequency component is visible on the calculated results. This component is a consequence of limitations of the 2-element Windkessel method used to obtain the input velocities. Using a higher order of the Windkessel model would help solve it, however it is impossible to obtain additional parameters needed for them. Nevertheless the scale of this disturbance is much smaller than the scale of the distention.

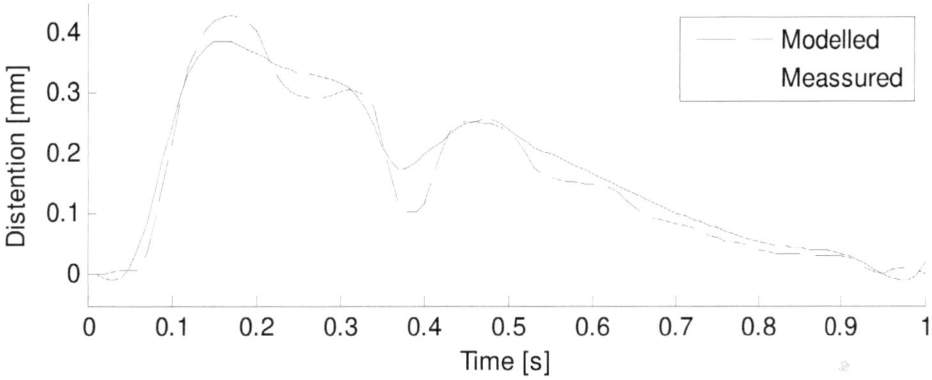

Figure 1. Comparison between the distention of the model versus measurements.

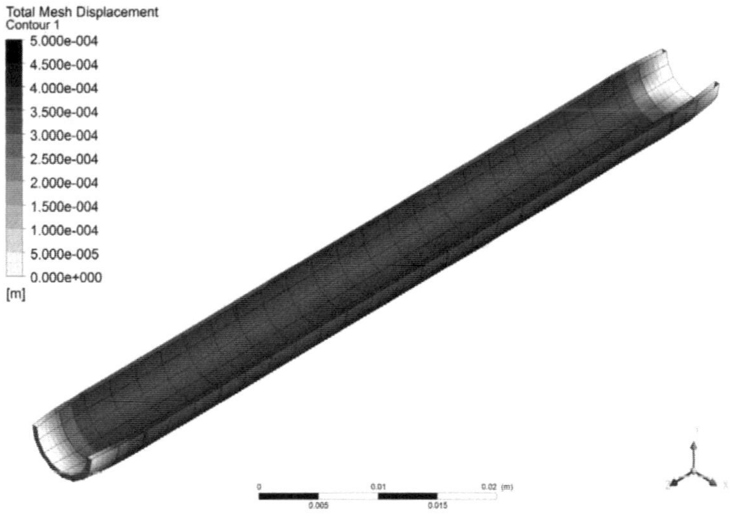

Figure 2. Half of the fully deformed model, presenting total mesh displacement

Journal of Physics: Conference Series **530** (2014) 012011 doi:10.1088/1742-6596/530/1/012011

The deformation of the model can be seen on the images of the full model. Figure 2 shows the full model length with mesh on the surface. The maximal deformation of this model is 0.4 mm. This means that the deformation is over 10%. (the radius of the artery is 3mm).
One of the main results is obtaining transient maps of wall shear stress (Figure 3). High values of the WSS are considered a risk factor in many cardiovascular diseases Identifying those patients will help with administering preventive treatment, to stop development of cardiovascular problems.

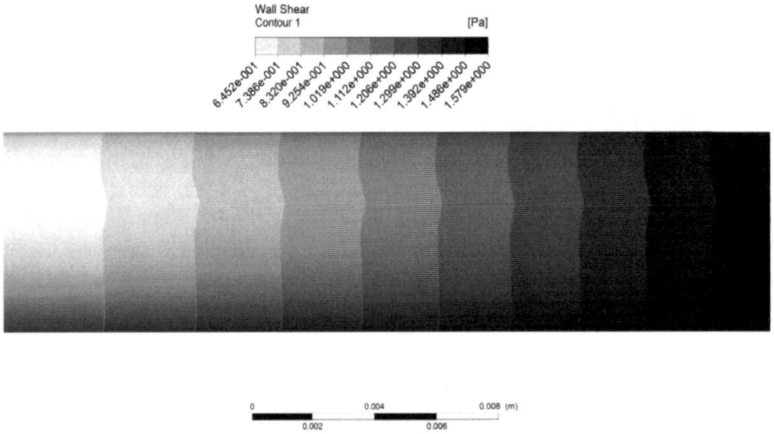

Figure 3. Example Wall Shear Stress map of the model

6. Conclusion

The results of this work show that the approach proposed in this paper is validated and can be used in further medical research on larger groups of patients. High automatization of model preparation reduces labour input needed from the researcher for each patient, making it a viable method for large cohort research.
The method has been developed to work with fully 3 dimensional geometries, therefore it is possible to apply this approach to more complex geometries.

7. Reference

[1] Nichols M, Townsend N, Luengo-Fernandez R, Leal J, Gray A, Scarborough P and Reyner M 2012 *European Cardiovascular Disease Statistics 2012. European Society of Cardiology, Sophia Antipolis* (European Heart Network, European Society of Cardiology, Sophia Antipolis)

[2] Mancia G, et al2013 2013 ESH/ESC Guidelines for the management of arterial hypertension: The Task Force for the management of arterial hypertension of the European Society of Hypertension (ESH) and of the European Society of Cardiology (ESC) *J. Hypertens.* **31** 1281–357

[3] Young T 1808 Hydraulic Investigations, Subservient to an Intended Croonian Lecture on the Motion of the Blood *Philos. Trans. R. Soc. Lond.* **98** 164–86

[4] Topper J N and Gimbrone Jr M A 1999 Blood flow and vascular gene expression: fluid shear stress as a modulator of endothelial phenotype *Mol. Med. Today* **5** 40–6

[5] Bodin P and Burnstock G 2001 Evidence that release of adenosine triphosphate from endothelial cells during increased shear stress is vesicular *J. Cardiovasc. Pharmacol.* **38** 900–8

[6] Boutouyrie P, Bussy C, Lacolley P, Girerd X, Laloux B and Laurent S 1999 Association Between Local Pulse Pressure, Mean Blood Pressure, and Large-Artery Remodeling *Circulation* **100** 1387–93

[7] Womersley J R 1955 Method for the calculation of velocity, rate of flow and viscous drag in arteries when the pressure gradient is known *J. Physiol.* **127** 553–63

[8] Reymond P, Crosetto P, Deparis S, Quarteroni A and Stergiopulos N 2013 Physiological simulation of blood flow in the aorta: Comparison of hemodynamic indices as predicted by 3-D FSI, 3-D rigid wall and 1-D models *Med. Eng. Phys.* **35** 784–91

[9] Moireau P, Xiao N, Astorino M, Figueroa C A, Chapelle D, Taylor C A and Gerbeau J-F 2012 External tissue support and fluid–structure simulation in blood flows *Biomech. Model. Mechanobiol.* **11** 1–18

[10] Pauca A L, Wallenhaupt S L, Kon N D and Tucker W Y 1992 Does radial artery pressure accurately reflect aortic pressure? *CHEST J.* **102** 1193–8

[11] Korteweg D J 1878 Ueber die Fortpflanzungsgeschwindigkeit des Schalles in elastischen Röhren *Ann. Phys. Chem.* **241** 525–42

[12] Bramwell J C and Hill A V 1922 The Velocity of the Pulse Wave in Man *Proc. R. Soc. B Biol. Sci.* **93** 298–306

[13] Sagawa K 1990 Translation of Otto frank's paper "Die Grundform des arteriellen Pulses" zeitschrift für biologie 37: 483–526 (1899) *J. Mol. Cell. Cardiol.* **22** 253–4

[14] Badur J 2005 *Pięć wykładów ze współczesnej termomechaniki płynów* (Gdańsk)

[15] Neofytou P and Tsangaris S 2006 Flow effects of blood constitutive equations in 3D models of vascular anomalies *Int. J. Numer. Methods Fluids* **51** 489–510

Acknowledgments
This research was carried out as part of the TEAM project: „Integrated measurement of arterial function and sympathetic nervous system activity: development, validation and application of a new method" which was realized within the programme Innovative Economy of Foundation for Polish Science, co financed from European Union, Regional Development Fund

Journal of Physics: Conference Series **530** (2014) 012012 doi:10.1088/1742-6596/530/1/012012

Analysis of lymphocytes deformation in microchannel flows using compound drop model

K Tatsumi, K Haizumi, K Sugimoto, K Nakabe
Department of Mechanical Engineering and Science, Kyoto University,
Kyoto daigaku-Katsura, Nishikyo-ku, Kyoto 6158540, JAPAN

E-mail: tatsumi @me.kyoto-u.ac.jp

Abstract. To understand the physical properties of the lymphocytes and develop a numerical model that can predict the motion and deformation of lymphocytes in flows, three-dimensional numerical simulation of the lymphocytes flowing through the contraction region of the microchannel was carried out employing the compound drop model in this study. The present model considers the shear-thinning effect of the cytoplasm, and includes the nucleus by applying another drop inside the cell. The results were compared with the experimental ones measured by recording the lymphocyte deformation using the high-speed camera. The time-dependent characteristics of the deformation degree *DI* of the lymphocyte, and the effects of the flow rate and the nucleus position on the *DI* were discussed. The results showed that the conventional drop model applying the Newtonian fluid properties to the cytoplasm cannot predict the deformation in the contraction region correctly where the nonlinear effects become important. The present model, on the other hand, showed a good agreement with the measured one. Further, the nucleus showed a significant effect on the shape of the lymphocyte. The rigid nucleus supressed the cell deformation and a slight deviation of its position affected the deformation rate of the leading and trailing sides markedly.

1. Introduction

In the past few decades, strong interest has been taken in to develop detail and massive simulations of blood flows in order to understand the physical phenomenon, and applying it to diagnoses and treatments. Further, more attention is paid on single cell analysis using the nano/micro technologies due to its high accuracy, measuring speed, compactness and low cost, a measuring technique which can provide new insights to the molecular biology and also make a large contribution to the fields of medicine by detect clinical diseases in early stages [1-3]. In order to achieve these goals, it is essential to develop a numerical model which can calculate the motion and deformation of a single blood cell in flows intended for blood vessels, medical equipment and micro-devices.
Among the models for blood flow and cell simulation, those for red blood cells (RBCs) have been keenly studied in the past since RBC not only shows large influences on the fluid properties such as viscoelasticity, but also plays an important role in transport phenomena and generation of stresses at the vessel walls. Therefore, various types of models treating the RBCs as bulk fluid and single cells have been presented. On the other hand, not much discussion is made on the motions and modelling for white blood cells (WBC) yet. Since WBC serves a key role in the autoimmune system [4,5], it is strongly related to number of diseases [6], and measurement of the WBC characteristics can be applied

to the diagnosis and treatment. The motion of WBC in the blood vessel is also an important factor to understand the flow and transport phenomena, and also to solve the clinical problems.

Measurements of the lymphocyte deformability have been carried out in the past by several groups mostly using the micropipette aspiration technique [7-12]. This method aspirates the single cell using a pipette of diameter smaller than the cell. By measuring the aspiration speed of the cell and the pressure applied at the upstream of the pipette, the deformability, in other words the elastic properties of the cell can be obtained [13]. Several essential problems, however, still remains in this method which are: the wall friction affects the measurement accuracy; deformation characteristics of the lymphocytes in flow cannot be fully verified; it is time consuming and the number of samples is limited.

In the present study, three-dimensional numerical simulation of lymphocyte deformation in microchannel with a contraction region is carried out for the purpose of modelling the lymphocytes using the compound drop model. The effects of the flow rate and nucleus position on the deformation characteristics are discussed. In addition to the computation, measurements of lymphocytes in microchannel flow were carried out using the high-speed video camera in order to validate the model. By using the microchannel, the shape of each cell can be measurement directly, quickly and continuously. Further, the contraction region will produce an elongational flow, in which the lymphocyte will deform with time and enables to measure and understand the viscoelastic properties of the cell.

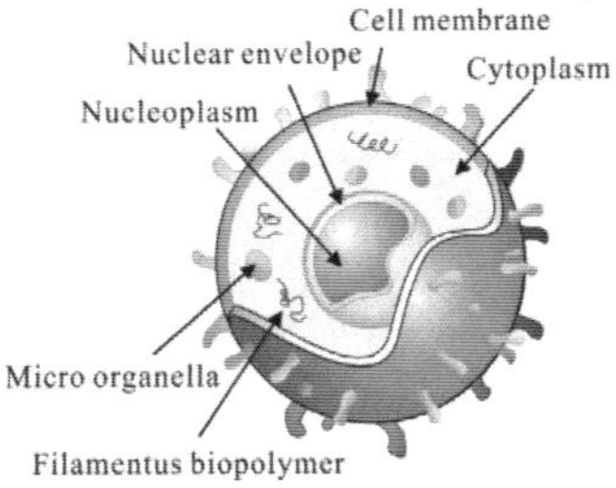

Figure 1. Schematic of the lymphocyte structure.

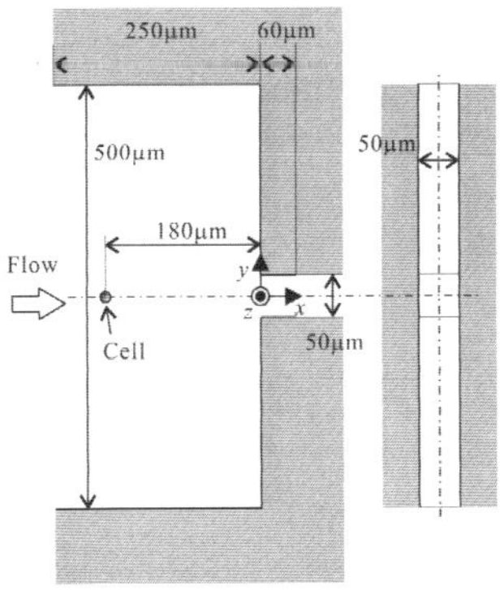

Figure 2. Computational domain.

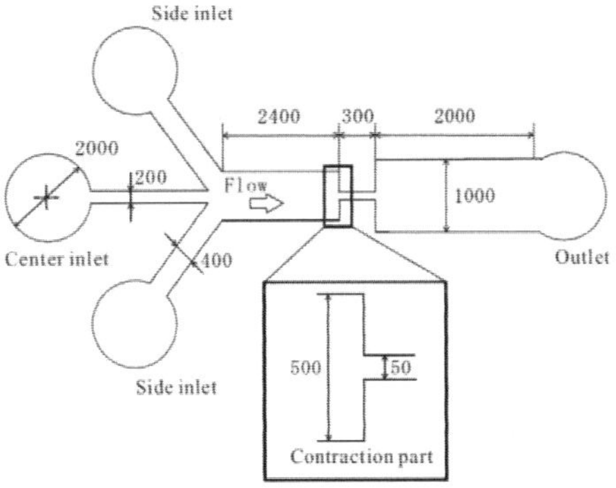

Figure 3. The dimension of the microchannel.

2. Numerical methods

2.1. Lymphocyte model

Figure 1 shows the schematic of the lymphocyte structure. A Lymphocyte consists of membrane, cytoplasm, nuclear membrane and nucleoplasm [12]. Therefore, to model the dynamics of the lymphocytes, one needs to consider the kinetic properties of each factor. Liu et al. treated the lymphocyte as a single drop, the properties of which are equivalent to the bulk properties of the lymphocytes.

The models presented in this study are based on the drop model. The drop model consists of the membrane with zero thickness and the cytoplasm only (the fluid inside). Surface tension is applied to the membrane, and several conditions of the cytoplasm properties are considered in this study. The first one is the model with the Newtonian fluid properties applied to the cytoplasm, and is referred to as ND model hereafter. The second one is shear-thinning drop (STD) model. In this case, the shear-thinning effect of the cytoplasm is considered as Eq. (1) [14].

$$\mu_c = \max\left[\mu_{c0}\left(1+\frac{\dot{\gamma}}{\dot{\gamma}_c}\right)^{-\beta}, 6.25\right] [\text{Pa·s}] \tag{1}$$

μ_{c0} is the zero-shear viscosity. $\dot{\gamma}_c$ and β are the parameters indicating the reduction of the viscosity. Note that a lower limit 6.25Pa·s is applied to μ_c in order to prevent a significant decrease of the viscosity.

The third model is the compound drop with shear-thinning effect (STCD) model. In this case, the nucleus is placed in the cytoplasm by applying another drop of much higher viscosity.

2.2. Numerical Procedure

The governing equations solved in this study to calculate the flow field are the continuity equations and Navier-Stokes equations expressed in unsteady and incompressible form for three-dimensional Cartesian coordinates. Finite volume method is employed in this study to solve these equations. 2nd order central scheme and 1st order upwind schemes are used to discretize the diffusion and convection terms of N-S equation. SIMPLER algorism is applied to couple the continuity and N-S equations to solve the pressure field.

In order to couple and calculate the conjugate problem of the lymphocyte and the flow field, Immersed Boundary method [15] is applied to the boundary of the surface of the drop and the surrounding fluid. In this case, force appearing in each node of the membrane is added as the external force term to the N-S equation which is the method called as the feedback forcing. On the other hand, the velocity of the fluid located adjacent to each node is applied to the membrane nodes assuming that the membrane follows aong the flow field. These velocities determine the motion and deformation of the lymphocyte.

2.3. Numerical Conditions

Figure 2 shows the computational domain. A contraction region with the same dimension with the experiment is calculated. A fully developed flow condition is applied to the inlet, and Neumann condition is applied to the outlet.

The fluid and membrane conditions applied to each model are summarized in Table 1. The tension applied to the membrane is set as 30μN/m referring the results of Yeung et al. [16] and Needham et al. [17] Due to the fact that the cytoplasm has a shear-thinning effect and that during the deformation as the lymphocyte passes the contraction region, the viscosity of the cytoplasm will decrease markedly. In order to match the deformation degree with the experiment in the case of ND model, the viscosity of the cytoplasm is set as 10Pa·s which is higher than those of the measurements.

The viscosity of the cytoplasm in the STCD model should be smaller than that of STD model because the nucleus and its rigidity effect is directly considered. However, since the viscosity of the nucleus itself is not known well, we will apply the same value as the STD model to the cytoplasm in this study. As for the viscosity of the nucleus, the value is on the bases of the measurement results of Kaleridis et al. [18] They have used the value 2.22×10^3Pa·s, however, due to the stability problem of the computation, 1/10 value is applied in the present study. The tension of the nuclear membrane was set based on the assumption that the ratio between the viscosity and the tension of the cytoplasm and the cell membrane is equivalent with the one between the nucleus and the nuclear membrane [9]. It should be noted that although the viscosity of the nucleus and the related nuclear membrane tension are different than those of other references, it was confirmed that the influence on the deformation degree of the lymphocytes was small in the computation.

The working fluid of the channel flow is phosphate buffers saline (PBS) solution. Polyvinylpyrrolidone (PVP) is added to the solvent by 12.5wt% in order to increase the fluid viscosity.

The fluid viscosity is set as μ_m=0.23Pa·s, which matches the one measured in the experiment. The fluid density is ρ=1000kg/m^3. Mass flow rate Q is varied in the range of 0.4~16μL/min. The Reynolds number defined as $Re = \rho U_m H/\mu_m$ is 0.002 when Q=16 μL/min. Here, H is the channel height and U_m is the cross-sectional average velocity of the channel inlet.
The lymphocyte is sphere with the diameter of 15μm at the initial position, which is $(x, y, z) = (-180, 0, 0)$μm. Since the velocity of the lymphocyte increases as it approaches the contraction region, the time step of the iteration procedure was changed depending on the lymphocyte location and was defining as $\Delta t = 0.1\Delta z/u_{WBC,max}$, where $u_{WBC,max}$ is the maximum velocity at each period.

Table 1. Lymphocyte properties of each drop model.

parameters	ND model	STD model	STCD model
Cell membrane tension τ_c [μN/m]	30	30	30
Apparent viscosity [Pa·s]	10	–	–
Zero shear viscosity μ_{c0} [Pa·s]	–	250	250
Shear-thinning parameter β [–]	–	0.5	0.5
Characteristic shear rate $\dot{\gamma}_c$ [1/s]	–	–	0.05
Nuclear membrane tension τ_n [Pa·s]	–	–	300
Nucleoplasm viscosity μ_n [μN/m]	–	–	250

3. Experimental methods

3.1. Microchannel

Figure 3 shows the schematic of the microchannel together with the dimensions. The contraction part is located at the middle. Three inlets were applied to the upstream side to generate a sheath flow which can control the spanwise position of the lymphocyte to be at the channel centreline.
The microchannel was made of PDMS (poly-dimethylsiloxane) and was prepared by using SU-8 (MicroChem Co.) as a casting mold that was fabricated through a lithography process. The PDMS channel removed from the SU-8 was attached to a cover glass. The working fluids were supplied to the inlet of the microchannel by a pressurized pump, the flow of which was confirmed to be stable without any fluctuations.
The motions and shape of the lymphocytes were measured by high speed video camera (Vision Research Co. Inc., Phantom V7.3). Objective lens of ×100 magnification (Olympus Co., LMPLFLN100X) was used, and the depth of field was 1.7μm. The image recoded by the camera had the size of 800×600pixels^2 with the resolution of 0.217μm/pixel.

3.2. Sample preparations

Jurkat cells (ATCC, CRL-2570) is cultured and used as lymphocytes in the experiment. 100μL culture medium, in which the Jurkat cells were suspended, was centrifuged for 2 min at 1000G, and the supernatant was removed. The precipitated cells were collected and washed one time by PBS (Amresco, E404). The collected cell was then suspended in PBS to which PVP (MW=3.6×10^5, Nakalai tesque, K-90) was mixed. PVP 8wt% and 12wt% solutions were used to obtain solutions of different viscosity. By using these two solutions, a wide range of the stresses condition generated in the flow was able to be examined in the experiment. The rheological properties of these fluids were measured by the rheometer (Anton Paar, MCR 301) in advance of the experiment.

3.3. Flow condition

Flow rate were changed in the experiment in the range of 0.2~2.4μL/min. As mentioned in Section 3.2, PVP 8wt% and 12wt% solutions were used to measure the deformation degree in a wide range of stresses. The flow rate of PVP 8wt% and 12wt% solutions were 0.2~2.4 and 0.2~1.88μL/min, respectively.

4. Results and Discussion

4.1. ND model

Figure 4 shows how the lymphocyte deforms as it passes the contraction region in the case of ND model. The upper and lower figures show the projected view of the x–y and the x–z planes. As the lymphocyte approaches the contraction region (x=0), it is stretched in the streamwise direction and contract in the spanwise direction when observed in the x–y plane. Comparing the two planes, one can see that the deformation takes place in a three-dimensional form due to the effects of the top and bottom walls producing stresses which is more dominated by the shear flow than the elongational flow.

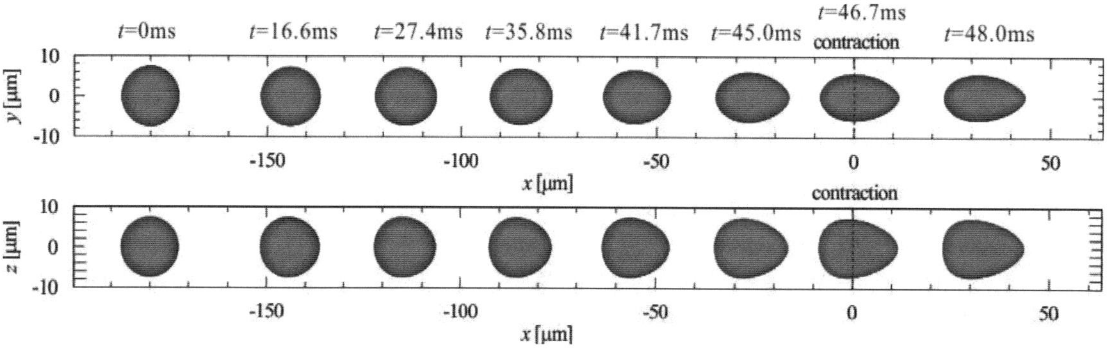

Figure 4. Images of the lymphocyte deformation projected on the x–y and x–z planes.

4.2. Effect of the flow rate on the deformation

The non-linear characteristic of the lymphocyte deformability will be discussed in this section. In this case, the influence of the deformation speed on the deformation characteristic will be considered by changing the flow rate conditions. Figure 5 shows the relation between the flow rate Q and the deformation index DI. DI is defined as $DI = (l_x-l_y)/(l_x+l_y)$, where l_x and l_y are the lengths of the lymphocyte in the x and y directions, respectively. The DI shown in the figure represents the value of the lymphocyte located at the streamwise position of x=0, which is the contraction part.

The square symbols show the experimental results, the error bars of which are calculated from the uncertainty analysis based on the ASME standard. As Q increases, DI increases due to the increase of the flow elongational and shear stresses. The increase rate, however, decreases with Q, and the DI shows an asymptotic behaviour.

The lines show the numerical results of ND, ND+PVP, and STD+PVP models. The ND+PVP and STD+PVP are the cases where the shear-thinning effect of the PVP solution is considered. In the case of ND model, DI remains constant against Q. This is attributed to the fact that the streamwise velocity u_{WBC} and the stresss σ_{ext} both increase proportionally with the flow rate. Since the cell deformation of the ND case mainly has the viscous characteristic, the shape of the cell will be the integrated value of the product of u_{WBC} and σ_{ext}. This will be constant with Q.

In the case of ND+PVP, DI decreases moderately as Q increases. The viscosity of the surrounding fluid decreases as the shear rate of the flow increases due to the shear-thinning effect of the PVP solution. Therefore, the stress applied to the lymphocyte deceases as Q increase. One can easily find that the distributions do not match those of the experiment in the cases of ND and ND+PVP.

On the other hand, the STD+PVP model, which applies the shear-thinning effect to the cytoplasm, shows a good agreement with the experiment. *DI* in this case increases as *Q* increases, and the increase rate decreases as *Q* increases. This shows the validity of the model.

One of the reasons why the *DI* of STD+PVP case is underestimated compared with the experimental ones is believed to be related to the initial shape of the cell. Figure 6 shows the relation between the *DI* and x_{WBC} in the case of Q=10μmL/min. x_{WBC} is the streamwise position of the lymphocyte. The solid and dotted lines with black colour depict the average value and the 95% uncertainty range of the experimental results. At each x_{WBC}, the *DI* of the experiment shows larger value compared with the numerical ones, namely the case of DI_0=0. On the other hand, one can see that the two distributions show a good match qualitatively. Further, the deviation between the two lines observed at the upstream regions tends to remain its width in downstream region. Since the lymphocyte shows viscoelastic characteristics when deforming, the shape of one moment relies not only on the instantaneous force working on the cell, but also on its shape history. The initial shape can, therefore, show a large effect on the *DI* distributions.

To discuss this matter, the initial shape of the lymphocyte was changed from spherical to ellipsoid assuming that certain deformation is initially occurred. To define the ellipsoidal shape we assumed that the cell volume is constant and the lymphocyte will not deform in the height direction. The shape was adjusted by matching the *DI* at x_{WBC}= −83μm of the simulation with the experimental one.

The results are shown by the chain line in the figure as case DI_0=Δ*DI*. It appears that the line of DI_0=Δ*DI* is that having the line of DI_0=0 moved simply in the height direction. Further, the result agrees very well with the experiment particularly in the region located upstream of the contraction part. This result also indicates the validity of the present model for predicting the deformation of lymphocytes.

We can see that deviation still remains at the contraction part and the downstream region. This is believed to be attributed to the elastic properties of the lymphocyte including the influence of the nucleus. The membrane structure shows elasticity and the nucleus will supress the deformation as shown in the next section. In order to evaluate these effects, it is necessary to measure the deformation under constant shear rate, which will be the future work of our study.

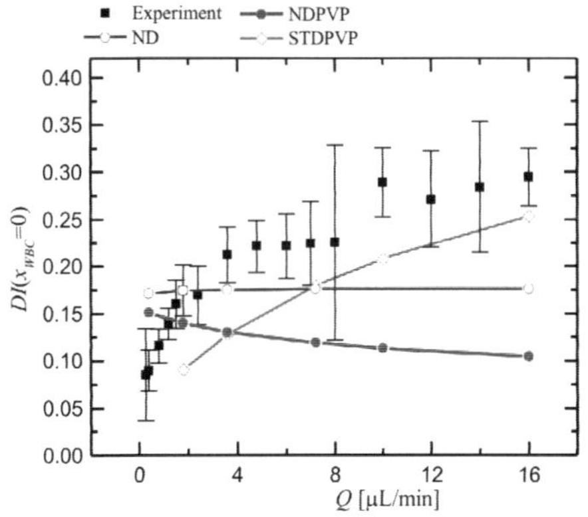

Figure 5. Relationship between deformation index *DI* and flow rate *Q*.

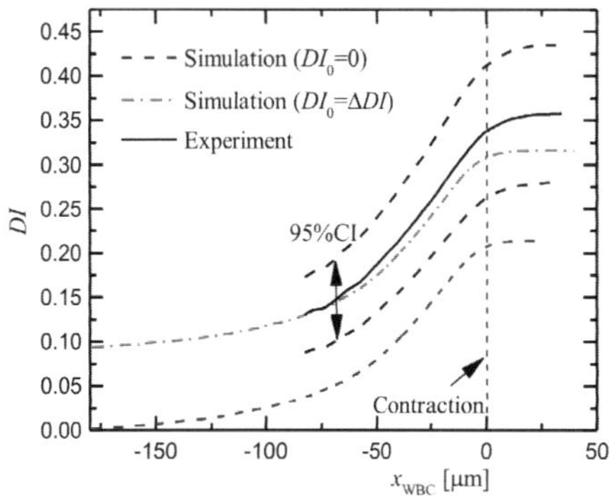

Figure 6. Relation between the deformation index *DI* and streamwise location of the lymphocytes (Q=10μL/min).

4.3. Effect of the nucleus

As shown in figure 6, the uncertainty of the experiment is large. This is believed to be attributed to the dispersion of the cell size, cell position in the flow, the properties of the cells, and nucleus size and position. In this session, we will see how the position of the nucleus in the cell will influence the cell deformation and also the flow generated in the cytoplasm.

Figure 7 shows the velocity vector distributions in the lymphocyte in the cross-sectional plane of z=0. The flow rate condition is Q=1.8μL/min, and the lymphocytes are of those located at x_{WBC}=0μm. The vectors are for velocities relative to the centre of mass of the cell. (a) shows the STCD case, and (b) shows the STD case. In figure 7 (b), one case see that the flow in the cell tends to possess a flow pattern very similar to the one observed in simple elongational flows. On the other hand, in figure (a), flow pattern completely different with the STD case is generated in the cell. Since the viscosity of the nucleus is large, it does not deform as it flows through the contraction region at least under the condition considered in this study. Due to the rigid nucleus, the flow in the cell is interrupted. The relative velocity near the nucleus is zero and increases with distance to the nucleus. Therefore, the cell membrane will easily deform when it is located apart from the nucleus and becomes less deformable in the area adjacent to the nucleus. This leads to the suppression of deformation in the spanwise direction and also the trailing side of the lymphocyte, the area in which the membrane approaches the nucleus. An asymmetric deformation in the streamwise direction is thus observed in the STCD case.

The influence of the nucleus streamwise position on the lymphocyte deformation is shown in figure 8 for. The nucleus positions of each case is shifted Δx_n=0, 1.5 and −1.5μm in the streamwise direction, respectively.

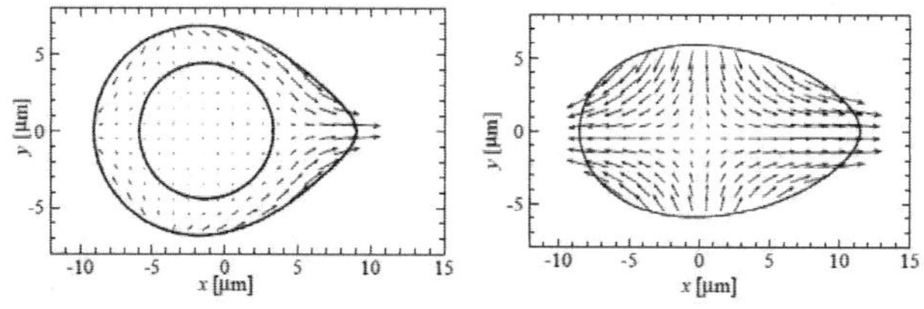

(a) STCD model (b) STD model

Figure 7. Effect of the nucleus on the lymphocyte deformation at x_{WBC}=0μm.

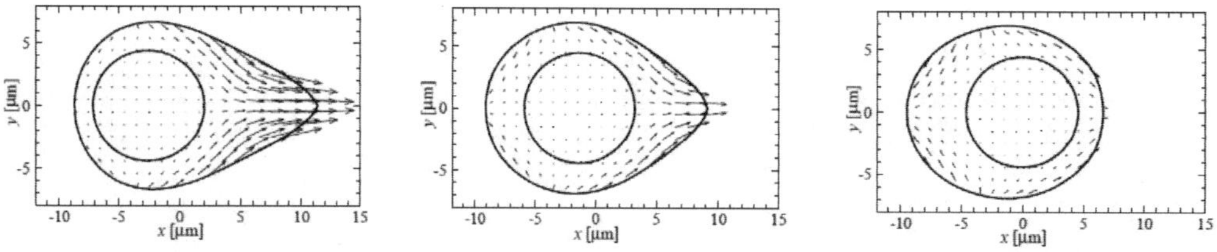

(a) Δx_n= −1.5μm (b) Δx_n= 0μm (c) Δx_n= 1.5μm

Figure 8. Effects of the nucleus position on the lymphocyte deformation.

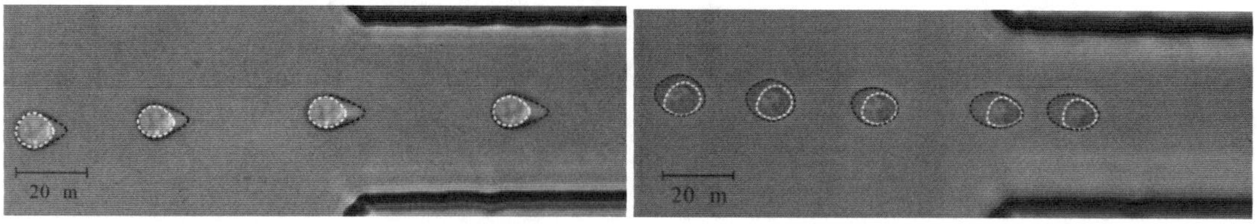

(a) Nucleus located at trailing side (b) Nucleus located at leading side

Figure 9. Snapshots showing the influence of the nucleus position on lymphocyte deformation.

In figure 8 (b), the leading side of the lymphocyte is deformed and stretched in the streamwise direction, while the deformation at the trailing side is small and negligible. In figure (c), when the nucleus is located closer to the trailing side the leading side is largely deformed compared with (b).

When the nucleus is located closer to the leading side as shown in figure (a), the deformation is largely supressed at the leading side while the increase of deformation at the trailing side is moderate.
Figure 9 shows the snapshots of two samples measured in the experiment. The location of the nucleus estimated is depicted by the white dash lines in the photographs. As shown in figure 9, the tendency of the nucleus effect on the deformation shows a reasonable agreement between the computation and experiment. The *DI* variation in relation to the nucleus position was more than 50% in the measurement. This is comparative with the magnitude of the error bar shown in figure 6 indicating that the position of the nucleus is one main factor for the variation of the deformation characteristics.

5. Conclusions

Numerical model of lymphocytes on the basis of drop model was presented. The deformation characteristics of the cell passing the contraction region in a microchannel was calculated and compared with the measurement for validation. Discussing the relationship between the flow rate Q and deformation degree *DI*, the conventional drop model could not predict the increase of *DI* against Q, while the present model applying shear-thinning effects to the cytoplasm showed a reasonable agreement. Further, the discussion on the effects of the nucleus and its position on the deformation showed that the nucleus supresses the deformation particularly in the spanwise direction. A slight deviation of its position changed the shape of the cell markedly showing that the uncertainty of *DI* could be attributed to the nucleus position. This tendency matched well with the experimental data.

6. References

[1] Gregg E C and Steidley K D 1965 *Biophysical Journal* **5** 393-405
[2] Morgan H, Sun T, Holmes D, Gawad S and Green N G 2007 *J. Physics D: Applied Physics* **40** 61-70
[3] Schade-Kampmamm G, Huwiler A, Hebeisen M, Hessler T, and Berardino M D 2008 *Cell Proliferation* **41** 830-840
[4] Mahalingam M, Pozniak A, Mcmanus T J, Vergani D and Peakman M 1995 *Experimental Immunology* **102** 481-486
[5] Wurtzer S, Compain S, Benech H, Hance A J and Clavel F 2005 *J. Virology* **79** 14815-14821
[6] Roda B, Reschinglian P, Zattoni A, Luigi P, Buzzi M, Ricci F and Bontadini A 2008 *Analytical and Bioanalytical Chemistry* **392** 137-145
[7] Schmid-Schönbein G W, Sung K P, Tözeren H, Skalak R and Chien S 1981 *Biophysical J.* **36** 243-256
[8] Tsai M A, Frank R S and Waugh R E 1993 *Biophysical Journal* **65** 2078-2088
[9] Kan H C, Udaykumar H S, Shyy W and Tran-Son-Tay R 1998 *Physics of Fliuds* **10** 760-774
[10] Evans E and Needham D, J. 1987 *Physical Chemistry* **91** 4219-4228
[11] Vigneron P, 1998 *A thesis presented to the graduate school of the university of Florida, University of Florida*
[12] Dong C and Skalak R 1992 *Journal of Theoretical Biology* **158** 173-193
[13] Schmid-Schönbein G W, Shih Y Y and Chien S 1980 *Blood* **56** 866-875
[14] Drury J L and Dembo M 2001 *Biophysical Journal* **81** 3166-3177
[15] Peskin C S 1977 *J. Computational Physics* **25** 220-252
[16] Yeung A and Evans E 1989 *Biophysical Journal* **56** 139-149
[17] Needham D and Hochmuth R M 1992 *Biophysical Journal* **61** 1664-1670
[18] Kaleridis V, Athanassiou G, Deligianni D and Missirlis Y 2010 *Clinical Hemorheology and Microcirculation* **45** 53-65

Acknowledgments

This work was supported by Japan Society for the Promotion of Science (JSPS) KAKENHI Grant Number 24360080.

Local quenching phenomena of a lean premixed flat flame impinging with a pulsating air jet

Y Yahagi[1] and I Makino[2]
[1]Shibaura Institute of technology, 3-7-5 Tyosu, Koto-ku, Tokyo, Japan
[2]Tohoku University, 6-6-4 Aramaki, Aoba-ku, Sendai, Miyagi, Japan

E-mail: yahagi@shibaura-it.ac.jp

Abstract. Local quenching phenomena of a lean methane air premixed flat flame formed horizontally in a wall stagnating flow impinging with a pulsating air jet has been investigated experimentally. The burner system consists of 40mm inverted nozzle burner and a solid wall with 8mm diameter air jet placed in line vertically. The pulsating frequencies set up to 100Hz while the jet intensities generate up to 6 m/s by a loud speaker. Approximately 100mm disk shape flame front is curved by the pulsating air jet and the air jet impacting point is locally quenched. The fuel concentration of quenching start condition increases with increasing the intensity of air jet, because the increased jet intensity linked with the flame strain rate gain. For weak jet intensity range, the quenching hole becomes directly to develop the whole flame extinction. On the other hand, for moderate or strong jet condition, the flame can recover from the local quenching phenomena. In this condition, once the quenching hole creates, but the hole may close by the flame propagation or reigniting process. Then, the whole flame extinction limits are lower than no jet impacting condition depending on the circumstances.

1. Introduction

A part of flamelet locally quenching phenomena namely local quenching occurs in a turbulent combustion. Especially, in the lean combustion, the local quenching possibility increases with increasing lean burn rate. A flame strain due to flame front curvature or velocity gradient, and heat loss to the cold wall are major trigger for the quenching [1]. The local quenching induces unburnt gas emission and also causes for unstable combustion [2]. Those are typical negative factors for the lean combustion with the local quenching. Introducing the local quenching phenomena including that of recovering can be great interest for persons working in the turbulent combustion. However, quenching and recovering mechanisms are very complicated phenomena since those occur in the three-dimensional and are difficult to predict in time and space. Up to now, the local quenching possibility increases with approaching the whole flame extinction condition [3]. Local quenching may not directly trigger to the global extinction depending on the circumstance [1], [2], [4]. Most of case, the locally quenched flame can recover and the continuous flame front recreates. As our focusing points, first question is what the primary trigger for the local quenching is. Second question is how the flame can recover from the local quenching. In the previous study of the wall stagnating turbulent premixed flame, we used 4 stages of turbulence generator for different Lewis number premixed gases such as lean propane air and lean methane air. On one hand, in the weak turbulence condition the whole flame

[1] Yuji Yahagi, Shibaura Institute of technology, 3-7-5 Tyosu, Koto-ku, Tokyo 1358548, Japan

extinction equivalence ratios are higher than the laminar flame extinction, but on the other hand, in the strong turbulence condition those are lower than the laminar [5], [6]. In addition, those phenomena are clearly noticed for the lean methane air flames. The results show an interesting paradox, because the turbulent flame receives additional flame strain effect by the flame having a curvature, so intuitively a turbulent flame more extinguishable than a laminar flame. The question what the mechanism works to extend the whole flame extinction for such kind of turbulent condition. However, in those studies, the discussions have been concluded from averaged data.

In order to elucidate the local quenching phenomena changing in the position at every moment, a time series analysis of detailed flame structures are important issues. As advanced recent studies, three dimensional LES [7] and DNS [8], [9]or high speed laser diagnostics for example PLIF and three dimensional PIV combined technique [10]~[13] adapt for the local structure of turbulent combustion. However, it is quite hard to reproduce the some phenomena as before.

As one of the approach, we proposed a simplified experimental model for the local quenching and recovering phenomena. The important priority matter for the model are to reproduce the

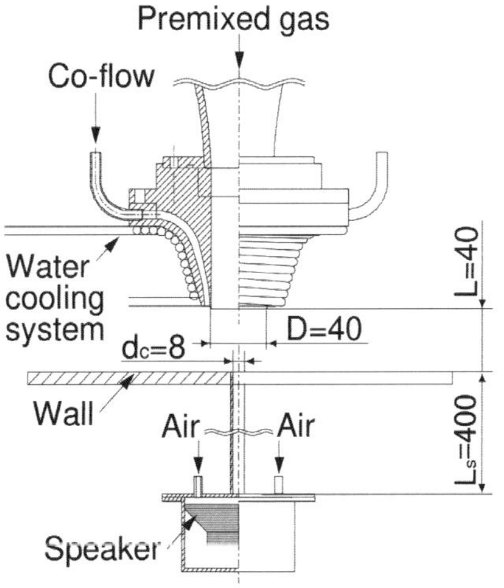

Figure 1. Burner system.

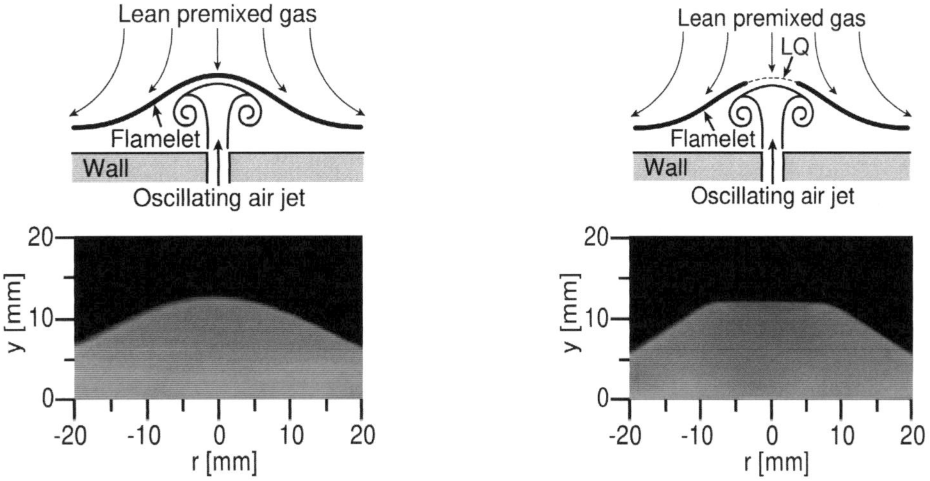

(a) No local quenching flame. (b) Local quenching flame.

Figure 2. Direct photos and flow images .

local quenching position and time. Our proposed model is a lean premixed flat flame impinging with a pulsating air jet perpendicularly in order to re-create simplified quenching phenomena as an experimental model for turbulent combustion.

2. Experimental setup

Figure 1 shows the burner system. The burner system consists by the inverted nozzle burner module in the upper side and the pulsating air jet in the lower side. The lean methane air premixed gas from the inverted nozzle burner having 40mm exit diameter impinges to the 8mm steel flat wall and formed the flat flame in the stagnating flow field. This nozzle has dual nozzle structure which consists a main nozzle for combustible gas and a ring shape secondly air flowing nozzle having 5mm ring width to minimize turbulence due to shear stress. The flat velocity profile methane air lean premixed gas impinges to the flat plate. The mean velocity of the mixture sets 2.0m/s and equivalence ratio (ϕ) range are from 0.72~0.85. At the lower part, the pulsating air jet sets on the center of solid plate. The pulse generates by the loud speaker set up 400mm downstream form the wall surface. Flow structures are obtained using a particle image velocimetry. Fig. 2 show the direct photograph with 1/40s shatter speed and the flows image respectively. In the no air jet condition, a typical stagnating flat shape flame formed parallel to the wall. With the pulsating air jet, the flame has a concave curvature to the unburnt gas due to the jet impinging as shown in Fig. 2 (a). Figure 2 (b) shows the local quenching flame photo and flow image. In this case, the concaved flame top locally quenches. From the quenching hole, the air jet probably impacts to the unburnt gas directly. The wall temperature is measured by K type thermocouple which installed in 3mm deep from the wall surface. The measured temperature is set at 300±50℃ and the measuring point is r=10mm. The pulse intensity (u') is defend as the difference from maximum to minimum velocity of the air jet measured at y=2mm and r=0mm by using a hot wire or PIV. The pulsating jet frequency set up to 100Hz. The local quenching phenomena (LQ) is visualized by laser tomography images. As tracer particles, 1 micro meter diameter of silicone oil droplets with vaporizing temperature 300℃ is used for the premixed gas side, while same diameter of alumina particle for the pulsating air jet side. Then two boundaries such as the premixed flame front and the impinging surface can be identified. The judgment of the LQ is when the premixed side Mie scattering image touches with that of image by the pulsating air jet coming from the lower side. The two dimensional instantaneous velocity is measured by a cross-correlation particle image velocimetry having a 30mJ per pulse at 532nm emission energy Nd YAG laser.

3. Results and discussions

3.1. Flame front movements

Figures 3 (a) and (b) show the high speed tomographic movies for one cycle of the flame motions. The conditions are ϕ =0.8 and f=20Hz, where first image, t=0ms corresponds with the flame front position (y_f) come to minimum in the cycle. The difference between (a) and (b) are the intensity of pulsating air jet (u'). In the no LQ condition shown in Fig. 3 (a), the flame front synchronies with the jet. The air jet forms a mushroom shape flow pattern to impact the flame. The flame front bends and moves as an up and down motion. On the other hand, in the u'=4.6m/s shown in Fig. 3 (b), the air jet inertia force is stronger than u'=1.6m/s shown in Fig.3 (a). Therefore the flame front bending motion and its amplitude are strong. At t=12ms, the air jet penetrates the flame front. In a moment, the air jet attached with the unburnt combustible mixture. The local quenching zone shown as a dotted line spread widely into the flame front. Then, the end of the mushroom shaped impacting jet formed axisymmetric vortex roll (see t=24ms). In the next, at t=36ms, the vortex roll disappeared. The flame seems to be recovering from the local quenching phenomena. In the images from t=36ms to t=0ms, the remaining air from the jet are still attaching with the unburnt mixture. It is hard to judge the flame completely recovers from the local quenching phenomena in this case. However, it is certain that the local quenching never develops to the global extinction. Those kinds of the local quenching and the

Journal of Physics: Conference Series **530** (2014) 012013 doi:10.1088/1742-6596/530/1/012013

recovery phenomena are repeating on every cycle. The quenching area increases with decreasing the equivalence ratio. Finally the flame cannot recover from the local quenching event.

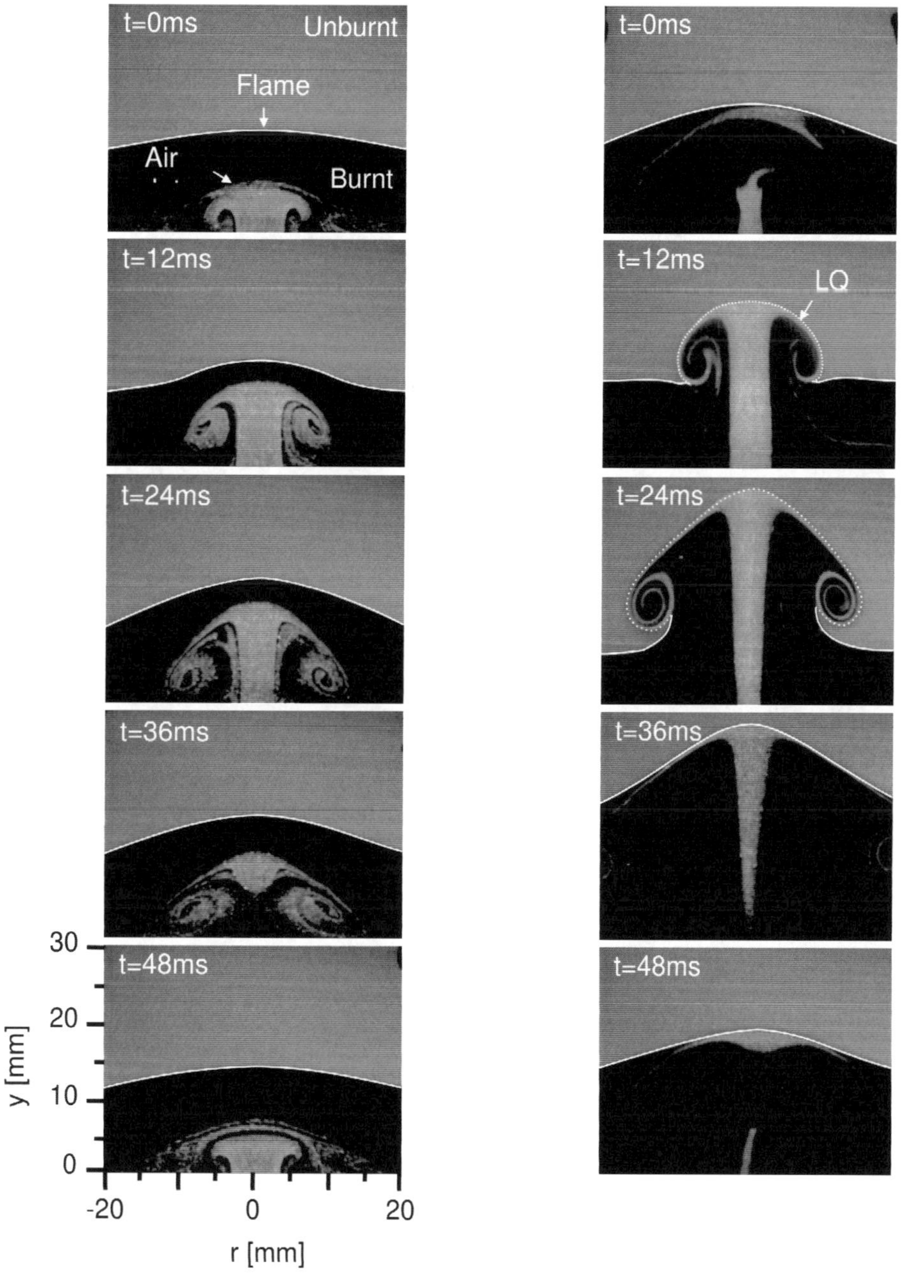

(a) No local quenching mode.

(b) Local quenching and recovering mode.

Figure 3. High speed tomographic movies for one cycle of the regular flame motion and the flame with local quenching and recovering motion (f=20Hz, ϕ =0.8) .

Journal of Physics: Conference Series **530** (2014) 012013 doi:10.1088/1742-6596/530/1/012013

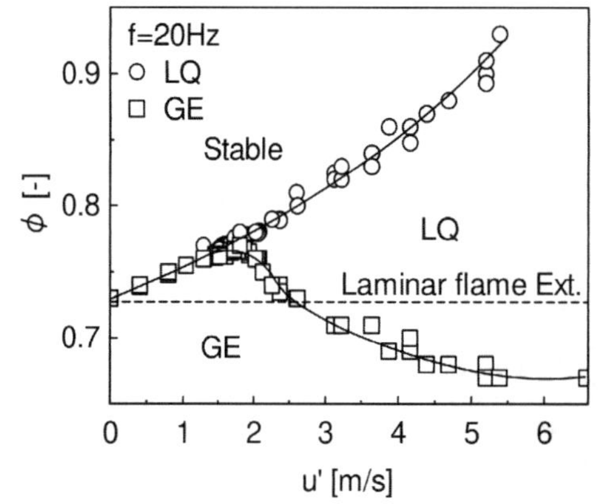

Figure 4. Flames map for stable flame regime, local quenching regime and global extinction regime for f=20Hz. The dotted line indicate the laminar flame extinction limit for this burner.

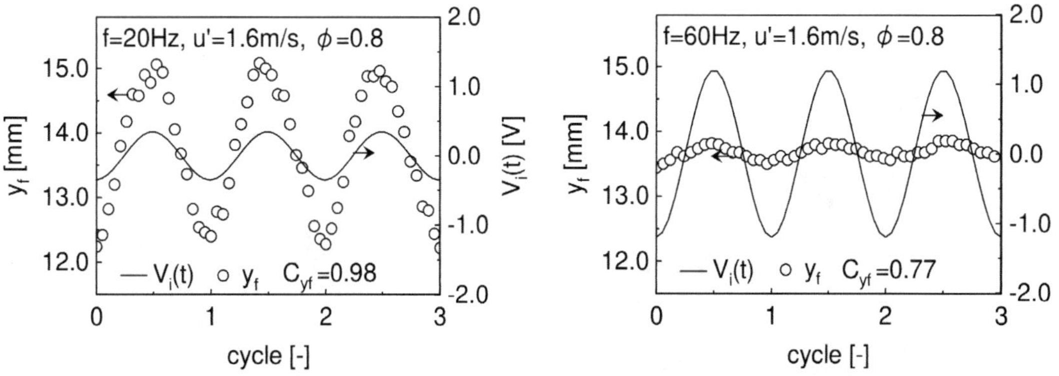

(a) f=20Hz. (b) f=60Hz.

Figure 5. Flame front responsibility for pulsating jet intensity u'=1.6m/s. $V_i(t)$ indicates input voltage of the loud speaker and the flame location (y_f) corresponds as the distance from the exit of air jet to the center of the flame front.

3.2. Global extinction and local quenching

Figure 4 shows the flame regimes of the stable flame, the local quenching and the global extinction. Basically, the local quenching starting equivalence ratio almost lineally increases with increasing the pulsating air jet intensity (u'). For the weak u' condition, the local quenching starting line and the global extinction line overlap in the range of u' up to about 1.5m/s. This indicates that the local quenching becomes directly to develop the global extinction. The flame has no chance to recover from the local quenching in this regime, namely direct global extinction mode. On the other hand, for u'<1.5m/s, the overlapped global extinction line branch off from the local quenching line. From this branch condition, the local quenching flame dose not directly become the global extinction. The flame can recover from the local quenching as described in Fig. 3 (a). The branched global extinction line shows getting low with increasing the u'. And finally the equivalence ratio of global extinction line pass the laminar flame extinction equivalence ratio shown as the dotted line. This means that the disturbance flame has a potential to improve than the laminar flame in some cases.

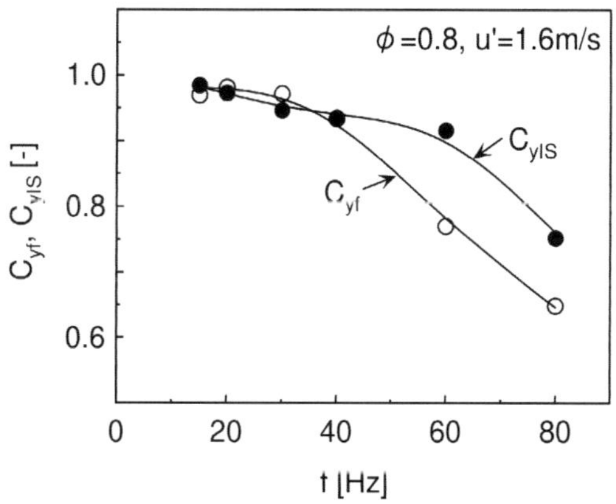

Figure 6. Cross-correlation function for the flame position (C_{yf}) and impinging surface (C_{yIS}).

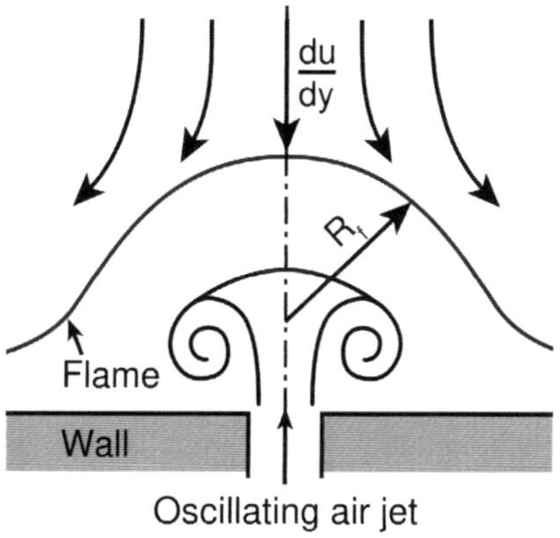

Figure 7. Total stain in the flat flame impinging with the pulsating air jet. R_f is the flame front curvature.

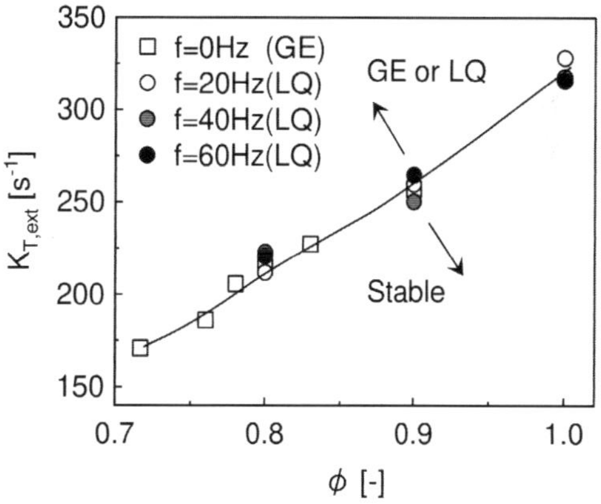

Figure 8. Local quenching or global extinction stating conditions. Ordinate ($K_{T,ext}$) is total strain rate defined as the equation (1).

3.3. Flame front response characteristic for pulsating jet

In this section, we discussed about the flame front response against the pulsating jet. The temperature in the flame zone steeply increases more than 4 times of room temperature. For example, when the burnt gas temperature is 2000K, then the dynamic viscosity just behind the flame is 6 times higher than that of the unburnt gas. It means Reynolds number also steeply decreases in the reaction zone. Our question is how transfer and follow the pulsating jet energy to the flame. Figure 5 shows the flame front movement for input pulse in the no LQ condition. As shown in Fig. 5 (a), for f=20Hz, the flame position at the flow center (y_f) is synchronized with the pulsating jet frequency. In this case, the cross-correlation function for the flame position and input pulse (C_{yf}) marks 0.98. For f=60Hz shown in Fig. 5 (b), C_{yf} is 0.77 and the amplitude of y_f decreases approximately 20% of f=20Hz. In the case of f=60Hz, in order to generate u'=1.6m/s, the input voltage $V_i(t)$ needs more than triple of f=20Hz. This is the cause for the viscous dissipation in the tube.

Figure 6 shows the cross-correlation functions for the flame position (C_{yf}) and the impinging surface (C_{yIS}) versus pulsating air jet frequency (f) at ϕ=0.8 and u'=1.6m/s, where the impinging surface (IS) is defend as the upper side flow and the lower flow impacting surface. In other words, IS corresponds with the boundary of the hot gas and the cold gas. C_{yf} and C_{yIS} drops with increasing f. Both drop rates steeply increase from about 40Hz. This means that the flame cannot perfectly follow the air jet motion.

Journal of Physics: Conference Series **530** (2014) 012013 doi:10.1088/1742-6596/530/1/012013

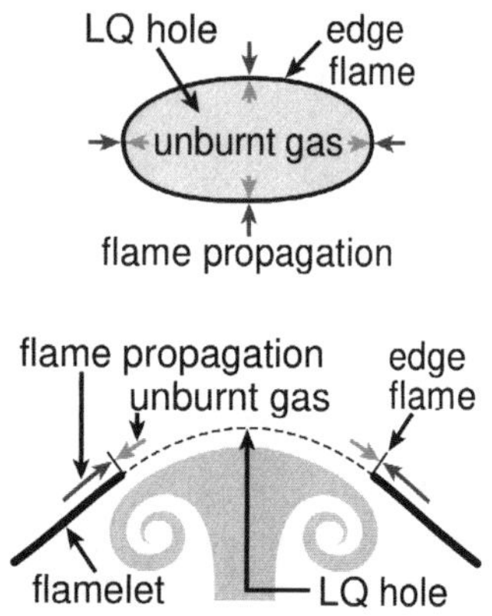

(a) Edge flame propagation mode.

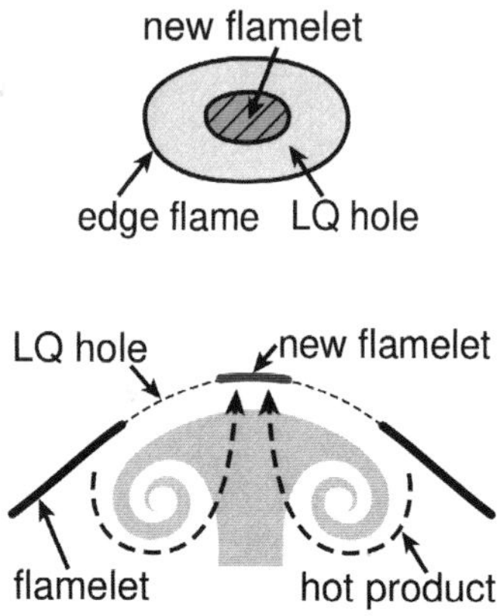

(b) New flame front recreate mode or reigniting mode.

Figure 9. Possible local quenching recovery mode.

3.4. Total strain rate

Important flame strain effects for turbulent flames are a main flow velocity gradient and a flamelet curvature due to the turbulence. In this study, we proposed total strain rate consisting from the sum of strain rate by main premixed gas divergence and flame front curvature by pulsating jet flow as described in Fig. 7. Then, the total strain rate defined as

$$K_T = \frac{du}{dy} + 2\frac{S_{L0}}{R_f} \quad (1)$$

where du/dy is the strain rate due to main flow divergence and $2S_{L0}/R_f$ is that due to flame curvature. Figure 8 shows the LQ quenching starting equivalence ratio for different pulsating jet frequencies. The f=0Hz means no pulsating jet condition which corresponds with laminar stagnating flame global extinction limits. All of data can be plotted on the straight line. This means the LQ is triggered by total strain.

3.5. Possible local quenching and recovering mode

As discussed in previous section, the LQ trigger is mainly a flame strain. However, it is difficult to elucidate the recovering mechanisms from the present our having data. Figure 9 shows two possible recovering mechanisms in present stage. One is an edge flame propagation recovering mode as shown in Fig. 9 (a). The edge flame exists on the boundary between the quenching hole and reacting area. The edge flame has a propagation property to the unburnt gas region. The balance of the edge flame propagation speed and unburnt gas expanding speed determines the LQ hole size. Other mode is a new flame front recreating mode as shown in Fig. 9 (b). In this mode, the hot product is transported by the rolled vortex to the unburnt gas zone. The unburnt gas reignites by the heat transfer or transportation of hot product. The new flame front recreate in the hole and propagate to whole area of the LQ hole.

4. Concluding remarks

There are two distinct local quenching phenomena observed depending on the jet intensity. When the jet intensity is weak, the local quenching becomes directly trigger to the whole flame extinction, namely direct extinction mode. When the jet intensity is moderate or strong, the flame can recover from the local quenching phenomena, namely quenching recovery mode. In the quenching recovery mode, while the quenching start condition increases with increasing the jet intensity almost lineally, the whole extinction fuel concentration decreases with increasing the jet intensity. In the end, the whole extinction fuel concentration for the high jet intensity has lowered than that of flat flame without the pulsating jet. That is the pulsating jet has a potential to prolong the whole flame extinction and to extend lean burn limits. The pulsating jet plays two important work as a negative and a positive. The negative work is a trigger for the local quenching. The positive work is a recovering from the local quenching. We proposed the following possible positive works; to reduce the effect of main strain in here stretch due to flow divergence, to assist the quenched flame edge propagation and to transport high temperature burnt gas to the local quenching part. Those local quenching recovering mechanism may be possible to adapt for a turbulent premixed combustion. The whole flame extinction limits can extend more than that of a laminar flame if the recovering mechanism works well.

Reference

[1] Yahagi Y, Yokoyama T, Nakaza H and Tsunekawa T 2007 Extinction Phenomena of Premixed and Non-premixed combined Flames in an Opposite Turbulent Flow *4th Int. Con. on Flow Dyna.* pp 1-18
[2] Driscoll J F 2008 *Prog. Eng. Comb. Sci.* **34** 91-134
[3] Kawanami T and Yahagi Y 2013 Effects of turbulence and flame propagation on the local quenching *9th Asia-Pacific Conf.on Comb. (Gyeongju)*
[4] Sardi K and Whitelaw J H 1999 *Experiments in Fluids* **27** 199-209
[5] Yahagi Y, Ueda T and Mizomoto M 1992 *JSME International Journal Series* **35** 2 304-309
[6] Yahagi Y, Ueda T and Mizomoto M 1992 *Proc. Comb. Inst.* **24** 537-542
[7] Knudsen E, Kolla H, Hawkes E R and Pitsch H 2013 *Comb. Flame* **160** 2911-2927
[8] Tanaka S, Shimura M, Fukushima N, Tanahashi M and Miyauchi T 2011 *Proc. Comb. Inst.* 3293-3300
[9] Shim Y, Fukushima N, Shimura M, Nada Y, Tanahashi M and Miyauchi T 2013 *Prog. Eng. Comb. Sci.* **34** 1383-1391
[10] Tanahashi M, Taka S, Hirayama T, Minamoto Y and Miyauchi T 2008 Local Burning Velocity Measurements in Turbulent Jet Premixed Flame by Simultaneous CH DPPLIF/OH PLIF and Stereoscopic PIV *14th Int. Symp. on Applications of Laser Techniques to Fluid Mech. (Lisbon)* pp 1-11
[11] Böhm B, Heeger C, Boxx I, Meier W and Dreizler A 2009 *Proc. Comb. Inst.* **32** 1647-1654
[12] Steinberg A M, Boxx I, Arndt C M, Frank J H and Meier W 2011 *Proc. Comb. Inst.* **33** 1663-1672
[13] Kerl J, Lawn C and Beyrau F 2013 *Comb. Flame* **160** 2757-2769

Journal of Physics: Conference Series **530** (2014) 012014 doi:10.1088/1742-6596/530/1/012014

Pressure evolution of ethylene-air explosions in enclosures

C. Movileanu, D. Razus, V. Giurcan and V. Gosa
Ilie Murgulescu Institute of Physical Chemistry, 202 Spl. Independentei, Bucharest, Romania

E-mail: cmovileanu@icf.ro

Abstract. The peak explosion pressure and the maximum rate of pressure rise are important safety parameters for assessing the hazard of a process and for design of vessels able to withstand an explosion or of their vents used as relief devices. Using ethylene-air with various fuel concentrations (4-10 vol% C_2H_4) as test mixture, the propagation of explosion in four closed vessels (a spherical vessel with central ignition and three cylindrical vessels with various L/D ratios, centrally or side ignited) has been studied at various initial pressures between 0.3-2.0 bar. In all cases, the peak pressures and the maximum rates of pressure rise were found to be linear functions on the total initial pressure, at constant fuel concentration. Examining several enclosures, the maximum values of explosion pressures and rates of pressure rise have been found for the spherical vessel. For the same initial conditions, the peak explosion pressure and maximum rates of pressure rise determined in cylindrical vessels decrease with the increase of L/D ratio. Asymmetric ignition, at vessel's bottom, induces important heat losses during flame propagation. This process is characterized by the lowest rates of pressure rise, as compared to propagation of flame ignited in the centre of the same vessel.

1. Introduction

Explosions of fuel-air gaseous mixtures in enclosures are characterized by specific characteristic parameters: the peak explosion pressure, the time to peak explosion pressure, the maximum rate of pressure rise, the propagation speed and the normal burning velocity [1]. The explosions may take place as deflagrations (subsonic propagation speeds) or detonations (larger than sonic propagation speeds). Deflagrations develop usually high explosion pressures (6-9 times the initial pressure), reached in a short time ranging from a 20-50 ms (in small vessels, with volumes below 1 L) to a few seconds (in vessels with volume between 1 L and a few m^3). During such processes the rate of pressure rise varies between 100 and 2000 bar/s, depending on initial composition, pressure and temperature of the flammable mixture, on vessel's form and volume, on ignition source position and energy. Knowledge of characteristic parameters of deflagrations and of factors that influence them allows the assessment of explosion risks for flammable mixtures in various conditions and the formulation of safety recommendations against the damaging effect of such explosions. At the same time, the characteristic parameters of deflagrations are necessary input data for the design of venting systems and for modelling the flame propagation in various conditions. Explosion pressures and maximum rates of pressure rise in closed vessel explosions are influenced by the initial composition, pressure and temperature of the fuel-air mixture (factors which determine the rate of heat release) and by the volume and the shape of the enclosure, the ignition source size, energy and position, the pre-existing or combustion-created turbulence (factors which determine the amount of generated heat as well as the amount of heat losses during flame propagation)[1-3]. Specific propagation features appear when the process takes place in elongated vessels, such as the preheating enclosures for chemical reactors or the pipes connecting fuel tanks. Three stages during flame propagation in long closed vessels have been observed [2]; the flame configuration, propagation speed and aspect of *p(t)* evolution have specific features in each period. The first stage is characterized by a quasi-spherical flame propagation, accompanied by maximal propagation speeds. Here, the flame is accelerated due to the expansion of burned gas and the flame structure is still uniform and homogeneous. At the end of the first stage, the flame front is stretched in axial direction and it is slowed down, as a consequence of decrease in its surface at the contact between flame and vessel's walls. Studies on flame propagation of propane-air mixtures in elongated vessels with L/D ranging between 1 and 19 [3] showed that the pressure increase is almost linear and the propagation speed is quasi constant in the 2nd stage, depending on the equivalence ratio of the flammable mixture and on vessel's geometry. In the last stage, the propagation speed was again decreasing and the flame instability increased so that the amplitude of peak-to-peak pressure oscillations could reach 5 bar. Other specific features of fuel-air deflagrations in tubes with asymmetrical ignition have been outlined by works of Phylaktou et al. [4,5] Fairweather et al. [6], Kindracki et al. [7], Cammarota et al. [8], Xiao et al. [9,10]. Bielert et al. [11] developed a one-dimensional model to simulate the combustion of methane-air flames in long closed pipes (L/D = 12.3 or 18.4), including the heat transfer to walls into simulation. Numerical simulations of fuel-air flames propagating in pipes have been performed also by Bychkov et al. [12] for propane-air flames (using data from [13]) and Bi et al. [14] for methane-air flames (using data from [7]). Other recent publications [10,15,16] examine characteristic aspects of flame propagation in tubes, relevant for the appearance of characteristic tulip flame after the rapid decrease of flame area due to the flame quenching at walls. Flame propagation in long vessels has been investigated also in connection with the process of DDT (Deflagration-to-Detonation Transition) frequently met in such conditions [17].

Ethylene, widely used in the chemical industry for production of valuable products (ethylene oxide, ethylene dichloride or polyethylene), was chosen as test fuel due to fire and explosion hazards associated to its mixtures with air, oxygen or other oxidizers. Explosions of gaseous ethylene-air mixtures in enclosures have been studied in various conditions: in spherical and cylindrical vessels, at initial pressures within 1 - 20 bar and initial temperatures within 20 - 200° C, using mixtures with variable ethylene concentration [4,18-21]. The propagation of C_2H_4-air flames in a cylindrical vessel with L/D = 21.6 and side ignition was studied by Phylaktou [4] who reported the rates of pressure rise and the flame speeds in various stages of flame propagation. Other results on ethylene-air explosions in closed vessels (explosion pressures and severity factors at variable initial pressures (1 - 10 bar), initial temperatures (20 - 200 °C) and C_2H_4 concentration (3 - 30 vol.%)) have been recently obtained in the frame of European Safekinex project [20]. Ethylene-air explosions in cylindrical vessels with L/D between 1.0 and 2.4 and central ignition have been experimentally investigated at ambient temperature and various initial pressures within 0.2 and 1.1 bar [21].

The present paper reports maximum (peak) explosion pressures, explosion times, maximum rates of pressure rise and severity factors reached during asymmetrically ignited explosions of ethylene-air in closed cylindrical vessels with L/D between 2.4 and 20.7 in comparison with centrally ignited explosions in elongated vessels and a spherical vessel. All data are discussed in correlation to the initial pressure and composition of ethylene-air mixtures and with the characteristic features of closed vessels: volume V and aspect ratio L/D.

2. Experimental

The main components of the experimental set-up are the explosion vessels, the cylinder for gaseous mixture storage, the ignition controller and the acquisition data system. The set-up has been described in previous papers [21,22]. Several explosion vessels, with different L/D ratios, have been used in the present measurements. The vessels were made from stainless steel and were equipped with several ports for the gas feed/evacuation valve and for mounting the ionization probes, thermocouples, pressure transducers and ignition electrodes. Each vessel was rated to 40 bar. The mixtures were ignited in vessel's centre or near the bottom (in the case of cylindrical enclosures), at points axially situated along each vessel, using high voltage inductive-capacitive sparks. Their list and characteristic dimensions are given in Table 1.

The transient pressure variation was recorded by means of two piezoelectric pressure transducers (Kistler 601A), in line with Charge Amplifiers Kistler 5001 SN. One pressure transducer was mounted in the centre of the top cover; the other was mounted in the bottom flange, flush with the cylindrical wall, 15 mm far from bottom. The time of flame arrival to the top was monitored by means of 2 ionization probes: one of them was mounted flush with the cylindrical wall, 15 mm far from bottom, the other in a symmetrical position relative to the top cover. The signals from the 2 ionization probes and from the 2 Charge Amplifiers were acquired, at 2500 signals/channel and a maximum of 1 GS/s sampling rate, with a 4-channels Tektronix digital oscilloscope type TDS 2014B connected to a PC.

Table 1. Geometric characteristics of explosion vessels.

Vessels		L / cm	D / cm	V / cm^3	L/D
Spherical	S	-	10.0	524	-
	V1	15.0	10.0	1178	1.5
Cylindrical	V2	20.0	8.1	1030	2.4
	V3	54.7	6.8	1986	8.1
	V4	100.0	4.8	1809	20.7

The following operating procedure has been used: before each test, the combustion vessel was evacuated down to 0.5 mbar, the explosive mixture was admitted and allowed 10 min. to become quiescent, it was ignited and afterwards, evacuated. The initiating spark was used as external trigger of the time base of the oscilloscope.

Ethylene, purity 99.99% (SIAD - Italy) and compressed air, has been used without further purification. Experiments have been made on ethylene-air mixtures with various concentrations within 3.0 - 10.0 vol% at initial pressures within 0.30 - 1.8 bar and ambient initial temperature. Minimum 3 experiments have been performed for each initial condition of explosive mixture.

3. Data evaluation

The computation of maximum explosion pressures from *p(t)* diagrams was made after smoothing the data either by FFT method or/and by using a cubic spline function. The rates of pressure rise were computed by numerical derivation of smoothed *p(t)* data using the Savitzky–Golay method, based on least squares quartic polynomial fitting across a moving window within the data. The method has the advantage of producing a smoothed first derivative without filtering the data. In all cases, we used a 10% smoothing level, since higher values of this level (e.g. 20%) leads to a reduction of the noise accompanied by the signal distortion.

The calculation of adiabatic explosion pressures and flame temperatures of ethylene-air mixtures at various initial pressures has been made with the program COSILAB [23] based on a general algorithm meant to compute the equilibrium composition of products for any fuel–oxidizer–diluent gaseous mixture. The algorithm is based on the thermodynamic criterion of chemical equilibrium: the minimum Gibbs energy, at constant temperature and pressure or the minimum Helmholtz energy, at constant temperature and volume. Fifty-three compounds have been considered as products.

4. Results and Discussion

Typical plots of pressure–time history for explosions propagating in several cylindrical vessels are shown in Figures 1 (central ignition) and 2 (bottom ignition), for the stoichiometric ethylene–air mixture ([C_2H_4] = 6.54 vol %) at ambient initial pressure. In Figure 1, the *p(t)* data measured in the spherical vessel with central ignition, for the same ethylene–air mixture, are overlaid. In centrally ignited explosions (Figure 1), the shortest explosion time (time from ignition to the peak explosion) is reached in the spherical vessel. In this vessel, both the peak explosion pressure and the maximum rate of pressure rise exceed the corresponding indices measured in the cylindrical vessels. A common feature of *p(t)* records from cylindrical vessels is the appearance of a short period of rapid pressure increase, at the beginning of flame propagation. In this stage, the pressure rise of the examined explosions increases according to the cubic law ($\Delta p = k \cdot t^3$)[24] so one can assume that the flame propagates undisturbed according to a spherical symmetry. As the L/D ratio of vessels increases, the duration and height of the rapid pressure increase are lower. In the second stage of both central and the bottom ignitions, the explosions are characterized by lower pressure rise rates and longer explosion times. A set of relevant sets of explosion indices is given in Table 2, where θ_{max} is the time to maximum explosion pressure. In accord to Phylaktou et al. [4] we use the terms "the first rate of pressure rise" and "the second rate of pressure rise" for the pressure rise rate of the first and the second stage of propagation, respectively.

Smooth *p(t)* records have been obtained in centrally ignited explosions (Figure 1). Bottom-ignited explosions in vessels with L/D > 2.4 (Figure 2) are accompanied by pressure oscillations that appear in the last stage of propagation, when flame instability develops. The oscillations are observed both in lean and rich mixtures; their amplitude and frequency depend on the equivalence ratio and initial pressure of flammable mixtures and L/D values of explosion vessels. The peak-to-peak amplitude of oscillations recorded in vessel V4 reached 2.2 bar during the explosion of a 5% C_2H_4-air mixture (φ=0.756) at p_0 = 1 bar. In a shorter vessel, the amplitude of oscillations reached 2.3 bar during the explosion of the stoichiometric C_2H_4-air mixture. In all cases, the pressure oscillations create a more dangerous situation of experiments in comparison with oscillation-free runs.

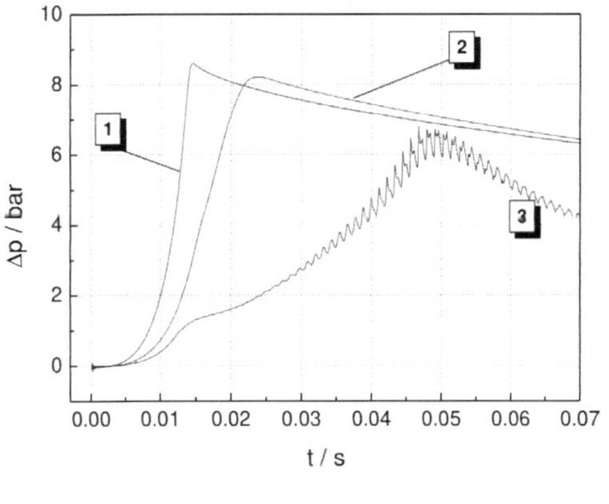

Figure 1. Pressure–time history for explosions of the stoichiometric C_2H_4–air mixture at p_0 = 1 bar in vessels with central ignition: (1) vessel S; (2) vessel V1 (L/D = 1.5); (3) vessel V3 (L/D = 8.1).

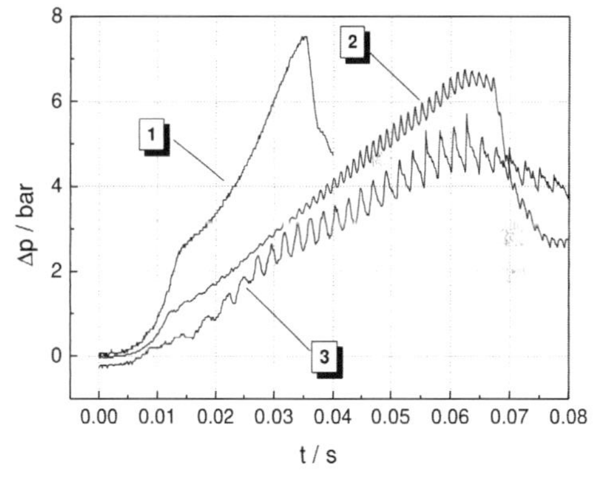

Figure 2. *p*(t) for explosions of the C_2H_4–air stoichiometric mixture at p_0 = 1 bar; vessels with bottom ignition: (1) V2 (L/D = 2.4); (2) V3 (L/D = 8.1); (3) V4 (L/D = 20.7)(offset -0.2 bar).

4.1. The maximum explosion pressures

As in the case of centrally ignited explosions studied in spherical vessel S and in cylindrical vessel V1 using other gaseous fuel-air mixtures (CH_4, C_3H_8, C5H12, C_3H_6, C_4H_6 or C_6H_6 [21,22,25-27]) a linear correlation was found between the peak (maximum) explosion pressure p_{max} and the total initial pressure p_0:

$$p_{\max} = m \cdot p_0 - n \quad (1)$$

The correlation holds for all vessels and all fuel concentrations where oscillation-free explosions occur, in deflagration regime. Data relevant for explosions of three ethylene-air mixtures in vessel V3 are given in Figures 3 and 4. The peak pressures reached in bottom-ignited explosions are lower as compared to the peak pressures reached in centrally ignited explosions, over the whole range of initial pressures.

Table 2. Deflagration indices measured during explosions at p_0 = 1 bar of the stoichiometric (6.54 vol.%) C_2H_4-air mixture in several vessels with central ignition.

Vessel's name	L/D	p_{max} / bar	1st rate of pressure rise $(dp/dt)_{max1}$ / (bar s^{-1})	2nd rate of pressure rise $(dp/dt)_{max2}$ / (bar s^{-1})	θ_{max} / ms
S	-	9.63	2960	-	14
V1	1.5	9.18	820	-	16
V2	2.4	6.16	734	814	20.5
V3	8.1	7.28	268	300	46.8
V4	20.7	6.33	79	105	58.0

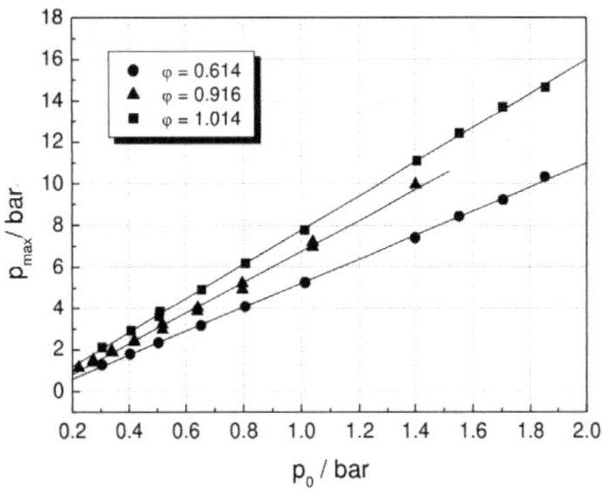

Figure 3. Maximum explosion pressures reached in vessel V3 (L/D = 8.1) with central ignition; C_2H_4–air mixtures with various initial concentrations.

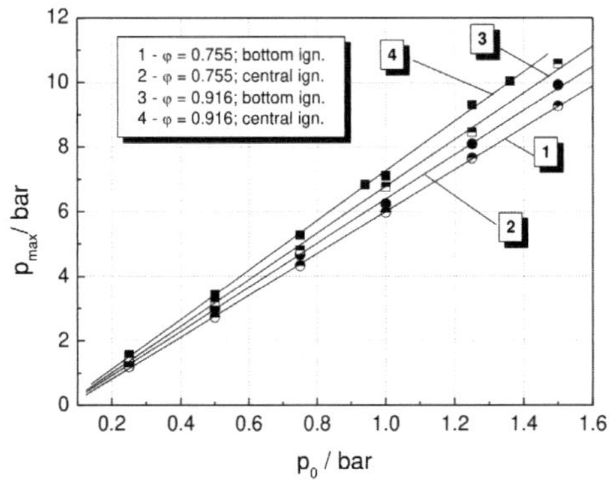

Figure 4. Maximum explosion pressures reached in vessel V3 (L/D = 8.1) with central and with bottom ignition; lean C_2H_4–air mixtures.

The influence of the aspect ratio L/D on peak explosion pressures is shown in Figures 5 and 6, for bottom-ignited explosions; Similar diagrams have been drawn for centrally ignited explosions.

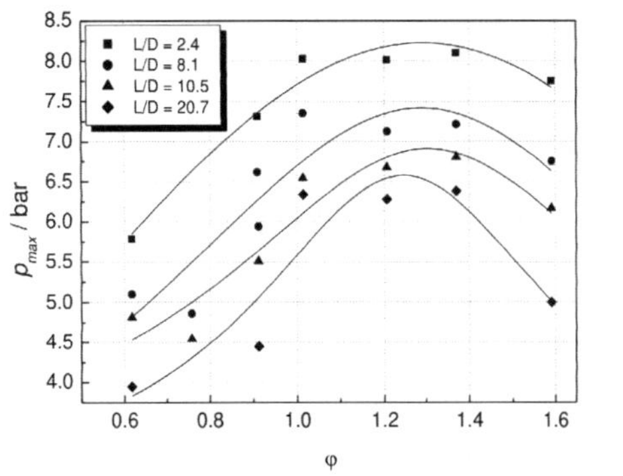

Figure 5. Influence of initial concentration on average peak explosion pressures at p_0 = 1 bar, in vessels with various L/D and bottom ignition.

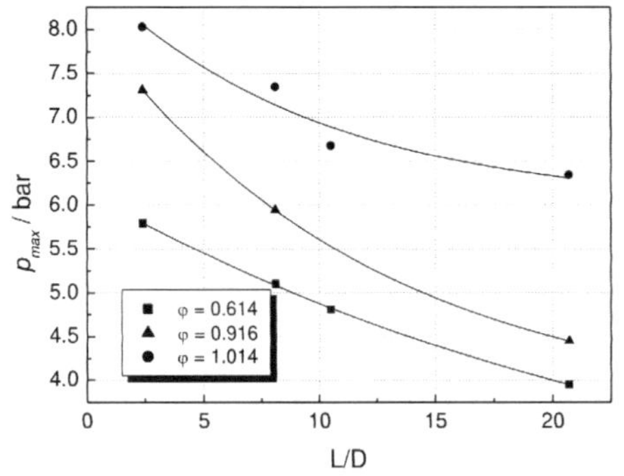

Figure 6. Influence of L/D on average peak explosion pressures of several ethylene-air mixtures at p_0 = 1 bar; bottom ignition.

For the stoichiometric and rich mixtures, characterized by appearance of important pressure oscillations in the 2nd and 3rd stages of propagation, the data scattering in Figures 5 and 6 is larger as compared to lean ethylene-air mixtures, since the peak explosion pressures have been evaluated as average values between the pressure peaks corresponding to pressure oscillations. For all vessels, the maximum explosion pressure is observed in the range of rich fuel-air mixtures ([C_2H_4] = 7-8 vol%). At constant composition, the increase of L/D results in the decrease of peak explosion pressures, due to a longer time of the heat exchange between the hot combustion products and the cold vessel walls.

4.2. The first and second rates of pressure rise

In all experiments, the first stage of flame propagation was oscillation-free and the (dp/dt) = f(t) diagrams were easily evaluated. In the 2nd stage of experiments in vessels with L/D>2.5, where presure oscillations have been observed for all ethylene-air compositions, one could determine instantaneous second rate of pressure rise. Relevant information on the possible damage of explosions in elongated vessels is based, however, on the averaged second rate of pressure rise, computed from smoothed p(t) data. Both the first and the second rates of pressure rise seem to follow linear correlations on initial pressure, p_0. Typical results are plotted in Figures 7 and 8.

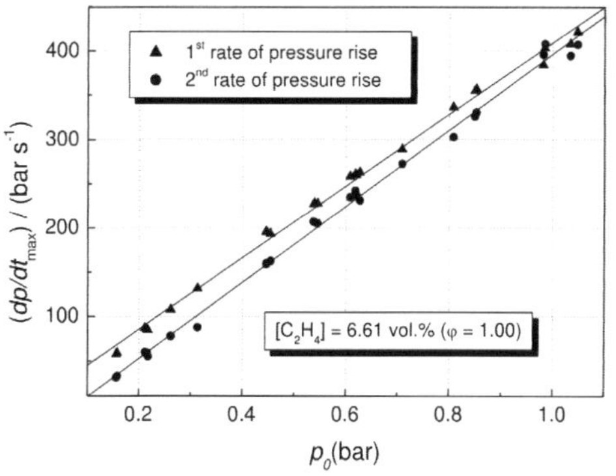

Figure 7. Initial pressure influence on maximum rates of pressure rise measured in vessel V2 (L/D = 2.4); bottom ignition.

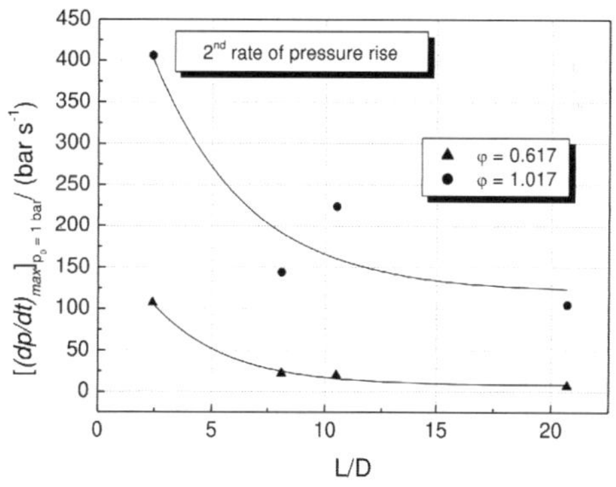

Figure 8. Influence of L/D on maximum rate of pressure rise measured for two ethylene-air mixtures at p_0 = 1 bar; bottom ignition.

The influence of ethylene concentration on first and second rates of pressure rise is seen in Figure 9, where typical results from vessel V4 with bottom ignition have been plotted together with data from literature [4], measured in a horizontal tube with L/D = 21.7 and side ignition. Larger differences between the 1st and the 2nd rate of pressure rise were observed in this vessel, in comparison to other vessels.

The first rates of pressure rise of bottom ignited explosions have been used to calculate the severity factors characteristic for ethylene explosions explosion in elongated vessels, as: $K_{G,L} = \left(\frac{dp}{dt}\right)_{\max} \cdot L$, according to a suggestion from Phylaktou [5]. Here, L (vessel's length), replaced the characteristic dimension $\sqrt[3]{V}$ used in the usual definition of a severity factor $K_G = \left(\frac{dp}{dt}\right)_{\max} \cdot \sqrt[3]{V}$. Representative results are given in Fig. 10. For comparison, $K_{G,L}$ = 184 bar m s^{-1} has been computed from literature data [4] (stoichiometric ethylene-air mixture, vessel with L = 1.64 m, bottom-ignited explosion), a lower value in comparison to our data obtained in vessel V4 with a close aspect ratio, but lower length. Making use of the usual definition of the severity factor, we would obtain K_G = 14.2 bar m s^{-1} for the bottom-ignited explosion of the stoichiometric mixture in vessel V4.

4.3. Examination of heat losses in explosions propagating in elongated vessels with bottom ignition

The energy lost to the walls during explosion propagation in a closed cylindrical vessel has been estimated from the difference between the internal energy of burned gas in the ideal end condition of adiabatic combustion (characterized by p_{max}, ad and $T_{max,\ ad}$) and in the real end condition (characterized by $p_{max,\ real}$ and $T_{max,\ real}$)[3]. According to this, the heat lost to the surface unit of wall, q_{tr}, is: $q_{tr} = \frac{V}{A}\frac{1}{\gamma_{end}-1}\left(p_{\max,ad} - p_{\max,real}\right)$, where V and A are the volume and the inner surface area of the vessel, respectively, and is γ_{end} the adiabatic coefficient of burned gas, at flame temperature.

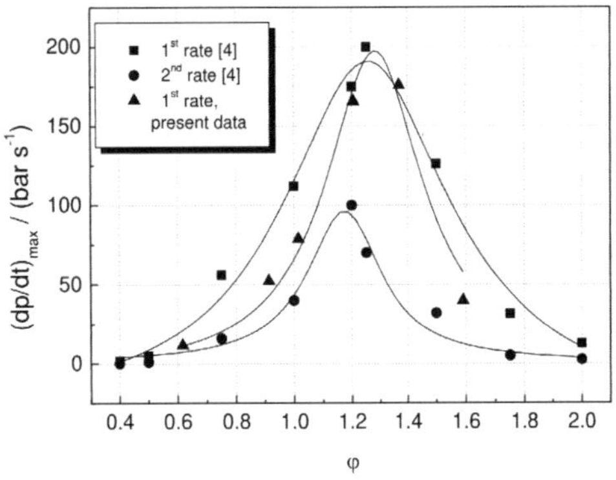

Figure 9. Maximum rates of pressure rise in vessels with bottom ignition, for mixtures with various concentrations at p_0 = 1 bar: (▲) "First" rates, present data, vessel V4 (L/D = 20.7); (•) "First" rates, vessel with L/D = 21.6 [4]; (■) "Second" rates, as above [4].

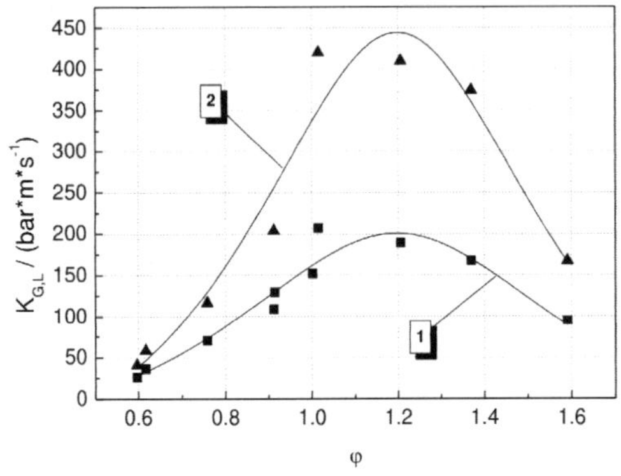

Figure 10. Severity factors of explosions at p_0 = 1 bar in elongated vessels with bottom ignition, for ethylene–air mixtures with various concentrations: (1) vessel V3 (L/D=8.1); (2) vessel V4 (L/D=20.7).

A set of data referring to explosions of ethylene-air mixtures at p_0 = 1 bar, computed for cylindrical vessels V2 - V4, is given in Table 3. The adiabatic coefficient of burned gas at flame temperature, γ_{end}, has been computed by means of burned gas composition obtained from equilibrium computations. The highest amount of transferred heat to the unit surface of walls was observed in vessel V4, characterized by the highest aspect ratio (L/D = 20.7) and the lowest - in vessel V2 (L/D = 2.4). For each case, the heat lost to wall's surface unit was found to depend linearly on the initial pressure.

Table 3. Heat lost to the surface unit of wall, q_{tr} / (J m^{-2}), during ethylene-air explosions at p_0 = 1 bar.

L/D	$[C_2H_4]$ / vol% 6.62	7.77	8.73
2.4	10.92	11.02	11.17
8.1	16.48	16.79	17.76
20.7	18.19	18.95	19.61

5. Conclusions

In elongated vessels, the deflagration indices are strongly influenced by the aspect ratio L/D of the vessel and reach much lower values as compared to short cylindrical vessels or with a sphere.

The important heat losses occurring in such situations explain the found dependencies of peak explosion pressure and of first and second rates of pressure rise against L/D. For all studied systems, linear correlations p_{max} = f(p_0) and $(dp/dt)_{max}$ = f(p_0) have been found. The amount of transferred heat

Journal of Physics: Conference Series **530** (2014) 012014 doi:10.1088/1742-6596/530/1/012014

to the unit surface of walls, computed from the adiabatic and experimental explosion pressures, had highest values for the vessel characterized by the highest aspect ratio (L/D = 20.7). The reported measurements provide useful results concerning explosion evolution in ethylene–air mixtures, in closed vessels with increasing asymmetry and might be useful for scaling explosions in chemical reactors, which are in most cases elongated cylindrical vessels.

References

[1] Hattwig M and Steen H 2004 *Handbook of Explosion Prevention and Protection* (Weinheim, Wiley VCH)
[2] Markstein G H 1970 Non steady flame propagation (Ed. Pergamon Press, Oxford, 1964)
[3] Leyer J C 1970 *Revue Générale de Thermique* vol 98 pp121-138
[4] Phylaktou H, Andrews G and Herath P 1990 *J. Loss Prevent. Process Ind.* vol 3 p 355
[5] Phylaktou H and Andrews G 1991 *Comb. Sci. Technol.* vol 77 pp 27-39
[6] Fairweather M, Hargrave G, Ibrahim S and Walker D 1999 *Comb. Flame* vol 116 pp 504 - 518
[7] Kindracki A, Kobiera G, Rarata P and Wolanski J 2007 *J. Loss Prevent. Process Ind.* vol 20 p 551
[8] Cammarota F, Di Benedetto A, Di Sarli V, Salzano E and Russo G 2009 *J. Loss Prevent. Process Ind.* vol 22 pp 607-613
[9] Xiao H, Wang Q, He X, Sun J, Yao L 2010 *Intern. J. Hydrogen Energy* vol 35 pp 1367-1376
[10] Xiao H, Makarov D, Sun J, Molkov V 2012 *Comb. Flame* vol 159 pp 1523-1538
[11] Bielert U and Sichel M 1998 *Comb. Flame* vol 114 pp 397-419
[12] Bychkov V, Akkerman V, Fru G, Petchenko A and Eriksson L 2007 *Comb. Flame* vol 150 pp 263-276
[13] Clanet C and Searby G 1996 *Comb. Flame* vol 105 pp 225-238
[14] Bi M, Dong C and Zhou Y 2012 *Appl. Thermal Eng.* vol 40 pp 337-342
[15] Dunn-Rankin D 2009 *Tulip Flames: The Shape of Deflagrations in Closed Tubes, in "Combustion Phenomena. Selected Mechanisms of Flame Formation, Propagation, and Extinction"*, Ed Jarosinski J and Veyssiere B (CRC Press)
[16] Zhang P, Du Y, Zhou Y, Qi S, Wu S and Xu J 2013 *J. Loss Prevent. Process Ind.* vol 26 pp 1279-1284
[17] Ciccarelli G and Dorofeev S 2008 *Progr. Energy Comb. Sci.* vol 34(4) pp 499-550
[18] Crescitelli C, Russo G, Tufano V, Napolitano F and Tranchino L 1977 *Combust. Sci. Technol.* vol 15 pp 201-212
[19] Amyotte P R, Patil S and Pegg M J 2002 *Process Saf. Environm. Protection* vol 80 (part B) pp 71-77
[20] Holtappels K 2006 *Report on the experimentally determined explosion limits, explosion pressures and rates of explosion pressure rise, Part 2: ethane, ethylene, ammonia and carbon oxide*, SAFEKINEX project deliverable nr. 9
[21] Movileanu C, Gosa V and Razus D 2013 *J. Hazard. Mater.* vol 235 pp 108-115
[22] Movileanu C, Razus D and Oancea D 2011 Rev. Roumaine Chim. vol 56 pp 11-17
[23] Cosilab, version 3.0.3. (Rotexo-Softpredict-Cosilab GmbH & Co KG, Bad Zwischenhahn, 2012)
[24] Razus D, Oancea D and Movileanu C 2006 *J. Loss Prevent. Process Ind.* vol 19(4) pp 334-342
[25] Razus D, Movileanu C, Brinzea V and Oancea D 2006 *J. Hazard. Mater.* vol 135(1-3) pp 58-65
[26] Razus D, Brinzea V, Mitu M and Oancea D 2009 *J.Hazard. Mater.* vol 165(1-3) pp 1248-1252
[27] Razus D, Brinzea V, Mitu M and Oancea D 2010 *J.Hazard.Mater* vol 174(1-3) pp 548 555

Acknowledgment: This work was supported by a grant of the Romanian National Authority for Scientific Research, CNCS – UEFISCDI, project PN-II-RU-PD-2012-3-0035.

XXI Fluid Mechanics Conference IOP Publishing
Journal of Physics: Conference Series **530** (2014) 012015 doi:10.1088/1742-6596/530/1/012015

Pyrolysis and gasification of single biomass particle - new openFoam solver

K Kwiatkowski[1,2], PJ Zuk[1], M Dudyński[3,4] and K Bajer[1,2,5]

[1] Faculty of Physics, University of Warsaw, Hoża 69, 00-691, Warsaw, Poland
[2] Interdisciplinary Centre for Mathematical and Computational Modelling, Pawińskiego 5a, 02-106, Warsaw, Poland
[3] Faculty of Mathematics, Informatics and Mechanics, University of Warsaw, Banacha 2, 02-097, Warsaw, Poland
[4] Modern Technologies and Filtration, Przybyszewskiego 73/77 lok. 8, 01-824 Warsaw, Poland
[5] University Centre for Environmental Studies, University of Warsaw, Żwirki i Wigury 93, 02-089 Warsaw, Poland

E-mail: `kamil@igf.fuw.edu.pl`

Abstract. We present a new solver *biomassGasificationFoam* that extended the functionalities of the well-supported open-source CFD code OpenFOAM. The main goal of this development is to provide a comprehensive computational environment for a wide range of applications involving reacting gases and solids. The *biomassGasificationFoam* is an integrated solver capable of modelling thermal conversion, including evaporation, pyrolysis, gasification, and combustion, of various solid materials.

In the paper we show that the gas is hotter than the solid except at the centre of the sample, where the temperature of the solid is higher. This effect is expected because the thermal conductivity of the porous matrix of the solid phase is higher than the thermal conductivity of the gases. This effect, which cannot be considered if thermal equilibrium between the gas and solid is assumed, leads to precise description of heat transfer into wood particles.

1. Introduction

The complex nature of gasification, wide range of available feedstocks together with an extensive process parameter space, and various types of reactors and possible reaction paths impede the development of a comprehensive and rigorous description of the process from a numerical perspective.

The process of biomass gasification and its subprocesses are well fitted in the fields that may be simulated within the Computational Fluid Dynamics (CFD) framework, and particularly with open-source CFD code OpenFOAM [1]. The solvers embedded in the standard OpenFOAM distributions for modelling pyrolysis (*fireFoam*) and chemical reactions (*reactingFoam*) or gasification in fluidised bed (*coalChemistryFoam*) are not suitable for modlling gasification of thick biomass particles. Currently available approaches do not consider the mutual influence of the reacting flow and gasified solid and volumetric reactions within porous material.

In this work, we present the new solver *biomassGasificationFoam* and the complementary library *biomassGasificationMedia* [2] developed to simulate a wide range of processes and applications related to the gasification of thin and thick biomass particle and, after further development, to simulate fixed bed gasification.

In this code, we implemented the following:
- volumetric reactions within porous material,
- homogeneous and heterogeneous reactions with the heat of reaction defined directly or derived from enthalpies of formations,
- a flexible definition of biomass and its thermochemical properties,
- customisable kinetic mechanisms of pyrolysis and gasification,
- heat and mass transfer between gases and solids around and within porous medium,
- transient flow within and around porous medium,
- separete energy equations for gas and solid phases for thermal non-equillibrium effects.

2. biomassGasificationFoam solver

2.1. Mathematical model

In figure 1, we present typical considered processes when one or more biomass fragments is pyrolysing or gasifying surrounded by inert or reactive gases. The gas phase is composed of a mixture of gases with different origins: gas that initially filled the reactor, gas supplied to the process, water vapour evaporated while the biomass is drying, and gases produced during thermal conversion of the solid. The potential liquid phase (water and liquid hydrocarbons) is considered vapour. This assumption is well satisfied for most of the considered processes.

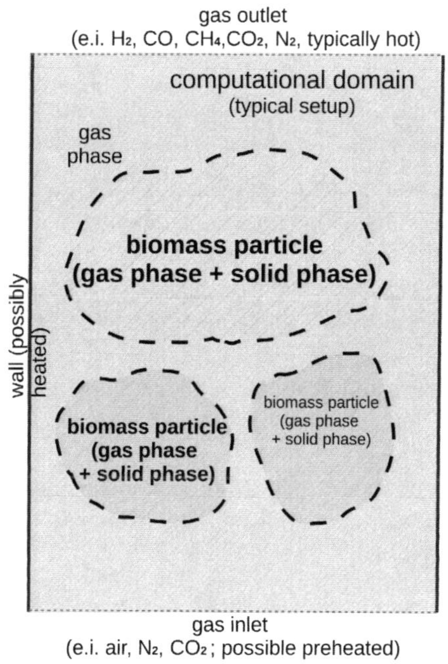

Figure 1. Exemplary computational domain with porous biomass particles.

The other assumptions are as follows:
- In current implementation the movement of porous medium is neglected.
- The moisture is embedded within this porous structure. Moisture released from the solid evaporates immediately.
- Gas flows according to Darcy's Law inside the porous medium.
- Radiative transfer from domain walls to solid surfaces is available .
- Chemical reaction and thermal processes may lead to thermal non-equilibrium.
- The porous medium loses mass via homogeneous and heterogeneous processes.
- The volume fraction of the gas phase γ is determined for the entire domain as $\gamma = \frac{V - V^S}{V}$ ($\gamma = 0$ when only the solid phase is present and as $\gamma = 1$ if there is no solid).

The three-dimensional, time-dependent mathematical model of the thermal conversion of biomass is based on conservation laws that are determined for both gaseous and solid phases. The solid is an isotropic or anisotropic porous medium with a porosity γ, which may change slowny during thermal conversion.

The governing equations for the gaseous phase are as follows: momentum conservation, continuity, species conservation, and energy conservation. The set of species typically involves water vapour, CO, CO_2, O_2, N_2, and CH_4. The solid-phase equations involve conservation of solid components and the energy conservation equation. Details of the governing equations are presented in [3] and [2].

2.2. Volumetric reactions

To allow the volumetric implementation of thermal processes, we propose to introduce porous medium inside the main computational domain as a volumetric porous field, as shown in figure 1. In the current approach the equations inside and outside the porous medium are the same and the boundary has no extra dynamics of its own.

The boundary of the porous material is the surface where the physical parameters present in those equations are discontinuous. In the momentum equation, for example, both the d'Arcy term and the usual viscous term of a Newtonian fluid, as in the Navier-Stokes equation, are formally present on both sides, but the associated coefficients have a jump across the interface.

2.3. Heat transfer

The thermal equilibrium between a gas and solid vanishes when chemical reactions or thermal conversions take place [4]. Our first analysis confirmed that during the thermal conversion of a single thermally thick wood particle, the gases (both outside and inside the particle) are not in equilibrium with the solid [5] and separate energy equations for the gas and solid are required.

The early analysis performed by [6] and repeated recently by [7] suggested that within the microscopic pores of single wood particles, the gases and solid are in thermal equilibrium. Investigations into the miniaturisation of heat exchangers independently led to the same conclusion: the assumption of thermal equilibrium is reasonable for micrometre-sized pores. In such case, the temperatures of the gas (in the pores) and solid phases are always equal and only one energy equation has to be solved within the computational domain. This assumption is common in the numerical modelling of pyrolysis and gasification. However, the above-mentioned analysis did not consider energy fluxes due to chemical reactions.

The withdrawal of the assumption of thermal equilibrium requires that energy transfer between the gas and solid is considered within the entire porous medium according to the source term.

$$\alpha\Sigma(T^G - T^S) \tag{1}$$

where: α - heat transfer coefficient (per surface), Σ - total surface area of pores per solid state volume.

2.4. Implementation

The new set of libraries that we implemented introduces thermal, chemical, and radiation properties of such defined porous field and determines their interactions with the gas phase flowing through and around these fields. Essentially, the libraries handle biomass as a collection of additional scalar and tensor fields within the computational domain, e.g., the mass fraction of cellulose, the thermal conductivity of the porous medium, and the absolute density.

The *biomassGasificationFoam* solver is based on the algorithm used in the unsteady solver *reactingFoam* for reacting flows available in the OpenFOAM standard distribution.

A schematic illustration of the main PIMPLE loop of the *biomassGasificationFoam* solver is provided in figure 2. The main loop contains two inner corrections for solving the pressure-momentum coupling influenced by energy and species distributions. The mass conservation equation is solved and the source terms calculated once per each main loop term.

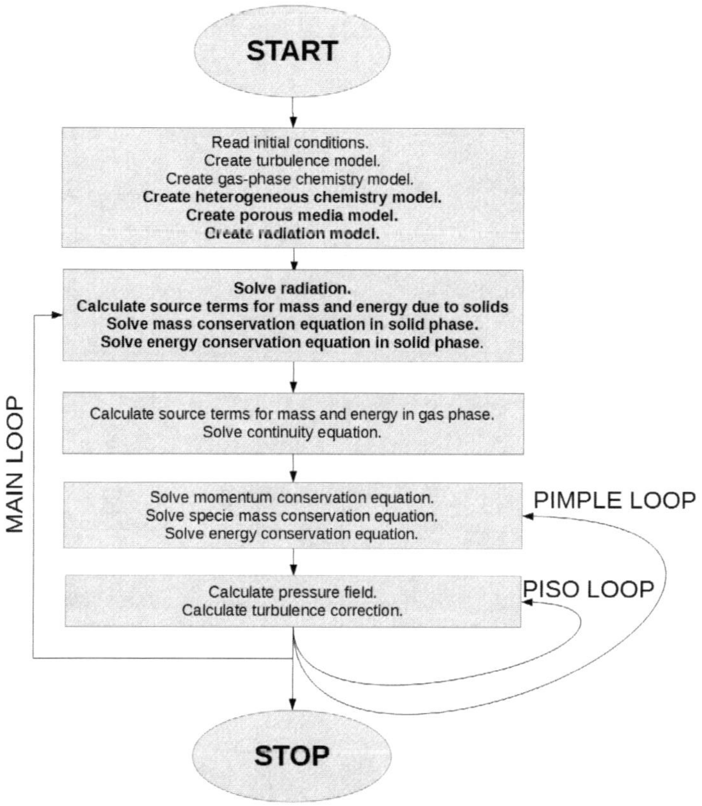

Figure 2. Scheme of the main loop of the *biomass-GasificationFoam* solver based on the PIMPLE algorithm. New development in the solver is in bold.

The model of the thermal conversion of biomass, as presented in figure 3, is constructed from three main elements:

- the definition of a porous medium and its mechanical properties - *porousReactingMedia* library, with the *setPorosity* utility;
- definitions of thermal, chemical, and radiation properties of solids - *thermophysicalModels* libraries;
- class linking the above libraries to the biomass gasification model - *pyrolysisModels*.

3. Validation and results

TGA, through a detailed measurement of mass loss, quantifies the thermal decomposition of wood under well-controlled conditions. The experiment was performed using the Q500 (TA Instrument, New Castle, Delaware, US) thermogravimeter with an initial temperature of 300 K and a constant heating rate of 10 K/min. The experiment was carried out until the temperature reached 1,300 K. Because we provided a nitrogen environment, only the processes of drying and pyrolysis took place. The heating protocol and process environment were introduced through the initial and boundary conditions, which are summarised in table 1.

The solver was validated with the case of pyrolysis of a preprocessed isotropic wood sample with an initial mass of 24 mg. The test calculations were performed with simplified geometry

Journal of Physics: Conference Series **530** (2014) 012015 doi:10.1088/1742-6596/530/1/012015

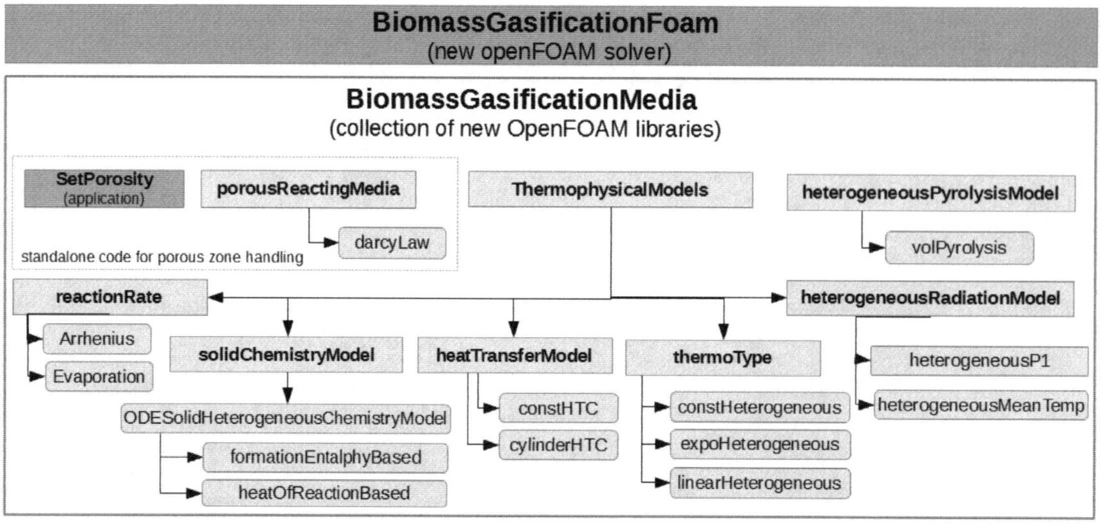

Figure 3. Structure of the developed code: the solver and collection of libraries.

Table 1. Initial and boundary conditions used in the simulations.

	Initial condition	Boundary condition inlet	walls	outlet
Inert species	nitrogen	nitrogen	-	-
Temperature	300 K	300 K + $10 \cdot t$ K/min	300 K + $10 \cdot t$ K/min	zero gradient
Velocity	0 m/s	0.3 m/s 0 m/s	zero gradient	
Pressure	101300 Pa	101300 Pa	-	zero gradient

for the TGA reactor and sample, both of which are modelled as cubes (figure 4). The edge of the reactor is 12 mm. The entire computational domain was composed of 20 x 20 x 20 cells. The porous zone was distributed over 10 x 10 x 10 cells.

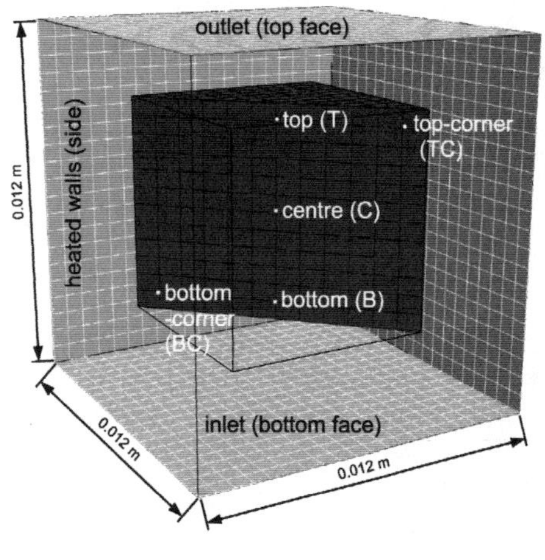

Figure 4. A simplified TGA reactor with cubic geometry and a hexagonal wood sample placed in the centre. In a diagonal plane cross-section of the sample, five points are marked (centre (C), top (T), bottom (B), top-corner (TC), and bottom-corner (BC)) in order to probe results.

In this paper, we use the mechanism of wood biomass pyrolysis, where cellulose, hemicellulose, and lignin decompose into a set of gases: H_2, CO, CO_2, and CH_4 [8]. The heat of pyrolysis was determined from enthalpies of formation of substrates and products. The parameters of kinetics are summarised in table 2.

Table 2. Pyrolysis parameters and kinetics constants used in the simulation.

	Moisture	Cellulose	Hemicellulose	Lignin	Ref.
Initial mass fraction [kg/kg]	0.07	0.359	0.276	0.285	[8]
Pre-exponential factor [1/s]	$1.4 \cdot 10^{14}$	$2.1 \cdot 10^7$	$5.8 \cdot 10^{13}$	$1.06 \cdot 10^3$	
Activation energy [J/mol]	$1.35 \cdot 10^4$	$1.34 \cdot 10^4$	$1.94 \cdot 10^4$	$8.95 \cdot 10^3$	
Minimum temperature [K]	300	450	400	500	
Enthalpy of formation [J/kg]	$-15.9 \cdot 10^6$	$-1.0 \cdot 10^6$	$-8.5 \cdot 10^5$	$-9.2 \cdot 10^5$	[9]

The experimentally determined material properties of wood are typically 'effective', combining the effects of gas and solid phases. Such properties depend strongly on the origin of the wood, the structure of the porosity, and the selected experimental procedures. In the solver presented here, we have introduced separate equations for the solid state that requires parameters for a 'pure solid'. A good example of such a material property is thermal conductivity. Measured effective values vary from 0.1 to 1 W/m/K, but the thermal conductivity of a solid is an order of magnitude higher, approximately 10 W/m/K. The set of material properties are presented in table 3.

Table 3. Physical and material properties of lignocellulose material used in the simulation.

	Moisture	Wood components	Char	Ash	Ref.
Absolute density [kg/m^3]	1000	1480	1957	2500	[10]
Initial void fraction [-]	—	0.55	0.91	-	[11]
Initial pore diameter [m]	—	$1 \cdot 10^6$	$1 \cdot 10^4$	-	[12]
Permeability [m^2]	—	$1 \cdot 10^{-14}$	$1 \cdot 10^{-11}$	-	[7]
Thermal conductivity [W/m/K]	0.58	0.20	0.1^a	0.17	[10]
Heat capacity [J/kg/K]	4200	1380	1100	840	[10]

The comparison of mass loss during evaporation and pyrolysis in the TGA experiment and the corresponding simulation is presented in figure 5. The results are in qualitative agreement, but discrepancies are visible. The main reasons for these discrepancies are the imperfect, oversimplified kinetics of pyrolysis and the variability of wood composition, that cannot be exactly determined.

The subsequent peaks of the inverse of the mass loss rate presented in figure 6 correspond to the processes of evaporation and pyrolysis of hemicellulose, cellulose, and lignin, respectively. All processes stop at approximately 800 K, when only char remains. In reality, as it is shown in figure 5, a small amount of char is still volatilising at higher temperatures. This effect is not included in the selected pyrolysis kinetics. The mass loss rates shown in figure 6 are plotted for five different locations within the wood sample, as we marked in figure 4. The rates of mass loss at the bottom, centre, and top are similar. Only the rates at the corners (top and bottom) are slightly distinguished.

Figure 7 illustrates how the densities (mass) of solid components, embedded moisture, hemicellulose, cellulose, and lignin are decreasing while the density (mass) of char is increasing.

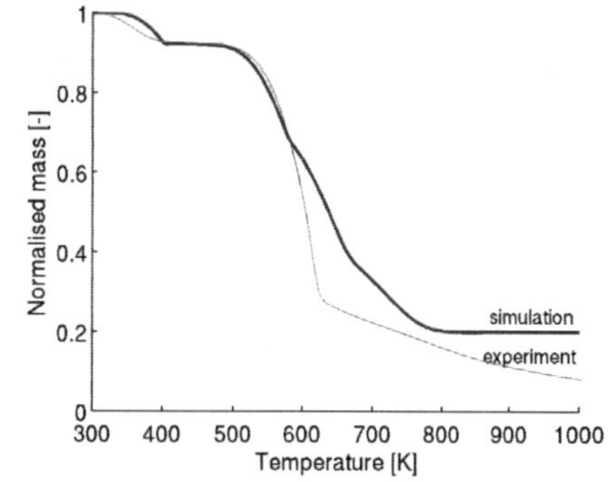

Figure 5. Comparison of the mass loss during the TGA experiment and the corresponding simulation.

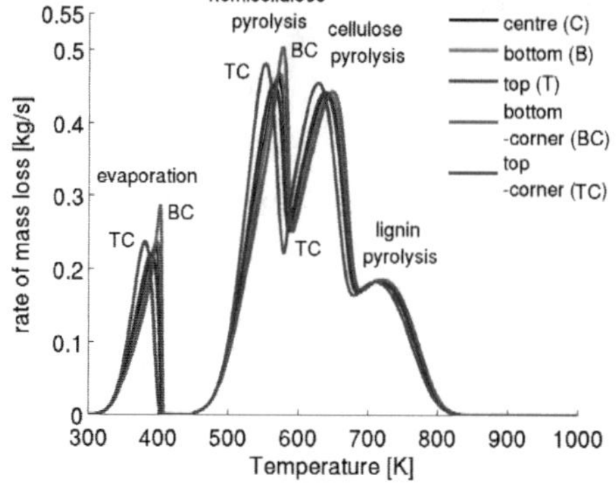

Figure 6. Simulation: rate of mass loss at the five locations within the sample.

Reproducing this well-known sequence of decomposition is possible with separate kinetics constants for individual solid components, which is straightforward in the presented solver.

Figure 8 presents the composition of the produced gases. Because the process takes place in a reactor filled with nitrogen, the presented values are relative. However, the solver with the selected kinetics adequately reproduces the composition of gases and the peaks of the processes.

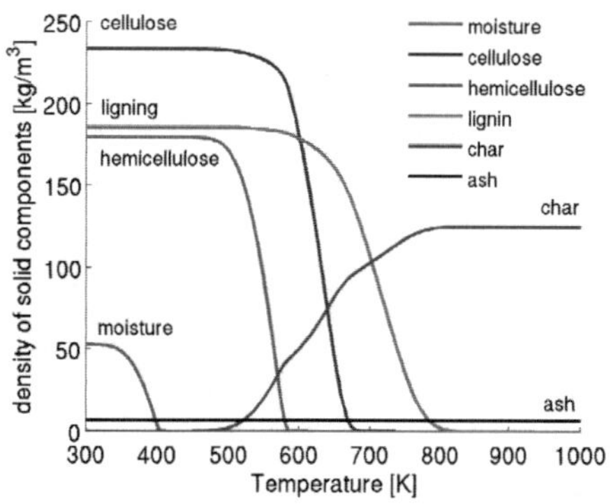

Figure 7. Simulation: changing the densities of the solid components.

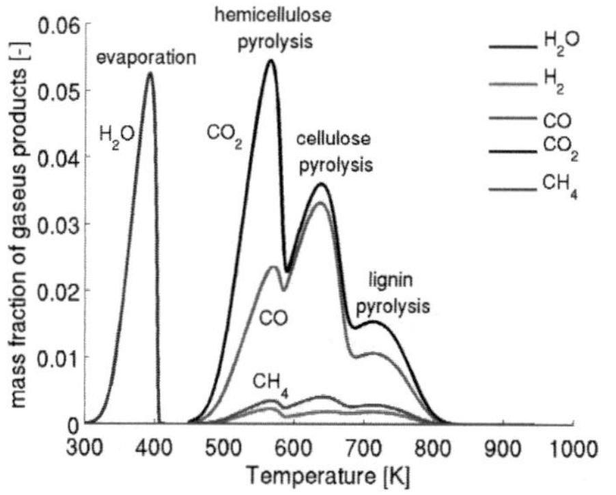

Figure 8. Mass fraction of pyrolytic gases.

Figure 9 illustrates the increase in the temperature of the gas phase at five selected points inside the wood: the centre, the bottom, the top, the bottom-corner, and the top-corner. The differences between points are now small. This result corresponds well with the expectation that temperature gradients should be negligible in such a small sample [13]. The included thermal processes are endothermic and, as noted in figure 9, the gas temperature decreases slightly when these processes occur. As expected, this decrease is most pronounced in the centre of the sample.

In figure 10, we present the difference between the gas- and solid-phase temperatures, plotted for the same points. First, we can see that the temperatures of the gas and solid are close, which is expected for such a small sample. As expected, this difference is negligible when no reaction takes place but increases when thermal processes are present. The subsequent peaks visible in figure 10 are due to the evaporation and pyrolysis of the solid components.

In general, the gas is hotter than the solid except at the centre of the sample (black line in figure 10), where the temperature of the solid is higher. This effect is expected because the thermal conductivity of the porous matrix of the solid phase is higher than the thermal conductivity of the gases. This effect cannot be considered if thermal equilibrium between the gas and solid is assumed.

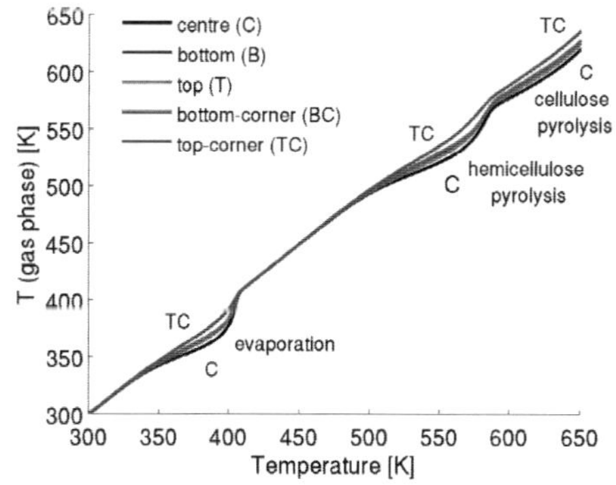

Figure 9. Temperature of gas phase.

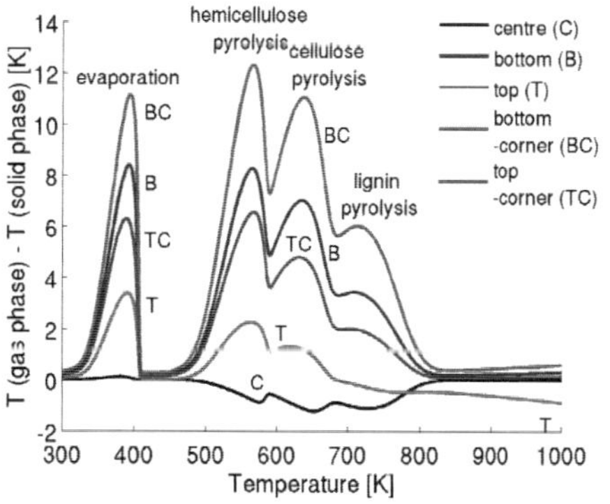

Figure 10. Difference between the temperature of the gas and solid.

4. Conclusion

As demonstrated here, the developed model, with simplified kinetics, reproduces the experimental results of the TGA experiment with an acceptable degree of uncertainty. Moreover, the solver reproduces several expected effects that are difficult for direct measurement and comparison.

5. References

[1] OpenFOAM Foundation 2013 URL `http://www.openfoam.com`
[2] Kwiatkowski K and Zuk P J 2014 biomassGasificationFoam - the openfoam solver for biomass gasification URL `http://www.syngasburner.eu/biomassGasificationFoam`
[3] Kwiatkowski K 2013 *Dynamics of gasification process and combustion of produced gas (in Polish)* Ph.D. thesis University of Warsaw, Faculty of Physics
[4] Kaviany M 2012 *Principles of Heat Transfer in Porous Media* Mechanical Engineering Series (Springer-Verlag, New York)
[5] Kwiatkowski K, Korotko J, Gryglas W, Dudyński M and Bajer K 2012 *Proc. 19th Int. Symp. on Analytical and Applied Pyrolysis, Linz, 21-25 May 2012* p 78
[6] Chan W C R, Kelbon M and Krieger B B 1985 *Fuel* **64** 1505
[7] Dufour A, Ouartassi B, Bounaceur R and Zoulalian A 2011 *Chem Eng Res Des* **89** 2136
[8] Kwiatkowski K, Grecki B, Gryglas W, Korotko J, Dudyński M and Bajer K 2013 *Numer Heat Tr A Appl* **64** 216
[9] Milosavljevic I, Oja V and Suuberg E M 1996 *Ind Eng Chem Res* **35** 653
[10] Porteiro J, Granada E, Collazo J, Patino D and Moran J 2007 *Energ Fuel* **21** 3151
[11] Senneca O 2007 *Fuel P* **88** 87
[12] Bellais M, Davidsson K O, Liliedahl T, Sjöström K and Pettersson J B C 2003 *Fuel* **82** 1541
[13] Di Blasi C 1996 *Ind Eng Chem Res* **35** 37

Acknowledgments

Numerical computations were performed in the Interdisciplinary Centre for Computational and Mathematical Modelling (ICM), University of Warsaw, grant number G34-8.

We thank Marek Kochańczyk for helpful comments on C++.

We thank prof. Marzena Półka from The Main School of Fire Service in Warsaw for performing the thermogravimetry analysis.

CFD modelling of gas flow through a fixed bed of Raschig rings

M Marek
Institute of Thermal Machinery, Częstochowa University of Technology,
al. Armii Krajowej 21, 42-200, Częstochowa, Poland

E-mail: marekm@imc.pcz.czest.pl

Abstract. In this work, direct numerical simulation of incompressible gas flow through a complex geometrical structure of Raschig rings is performed. The bed structure is obtained in a separate simulation in which Raschig rings are added one by one and fall freely into a cylindrical container until mechanical equilibrium is reached. The gas is injected at the top of the container, flows through a packed layer of randomly oriented rings, then leaves the container at the bottom. The flow equations are solved with the use of classical second order solver and projection method for structured regular grid and the complexity of domain geometry is handled by a variant of immersed boundary technique. The model allows to study the characteristics of the flow within the bed, pressure distribution in particular, and can be applied to development and validation of simplified approaches. The simulations are performed for various grid resolutions and the results show good convergence to grid independent solution, especially for lower gas velocities. The obtained dependence of total pressure drop on the inlet velocity is in reasonable agreement with the literature data but the demand for grid resolution significantly increases with the gas velocity.

1. Introduction

A fluid flow through a porous bed, because of highly complex domain geometry, is usually modeled in a simplified fashion using averaged flow equations [1,6]. Such an approach does not require generation of equally complex computational grids and also allows for simulation of full-scale systems taking into account many phenomena essential from the point of view of a given process.

However, for simplified modeling it is necessary to deliver some information about porosity of the bed (or the distribution of porosity when it is not homogeneous), active surface of the bed elements and the influence of the microstructure of the bed on the flow characteristics in a larger scale. Such data can be obtained only in experiments [2,3] or in advanced CFD simulations when the physics of fluid flow in complex geometry is properly captured [4,5].

In a recent publication [7] a method has been presented of numerical generation of a fixed bed structure consisting of cylindrical elements or - after a minor modification – Raschig rings. The idea of the method is to simulate the filling process of a cylindrical container with bed elements added one-by-one, moving due to gravity and reaction forces from the remaining bed elements until eventually mechanical equilibrium is reached. Then a given element is included in the bed structure and remains at rest to the end of the simulation. Structures generated in this way can be examined with respect to porosity distribution to obtain necessary input for averaged flow equations or used directly in CFD simulations as the geometry of the computational domain.

This work is devoted to direct numerical simulation of gas flow through a fixed bed, a cylindrical container filled with Raschig rings, of numerically generated realistic structure. Examination of the flow at the microscopic level can be very useful from the point of view of construction of models focused on larger scales as well as validation of models already formulated on the basis of experimental research. One of the main results obtained here – pressure drop in the bed – is of importance for practitioners designing systems with fixed beds e.g. absorbers.

The plan of the paper is as follows. In Section 2 the flow solver is shortly described. Section 3 presents the results of pressure drop calculations together with grid convergence studies and comparison with reference experimental data. Section 4 closes the paper with summary.

2. Flow solver

In this section, solution of Navier-Stokes equations for incompressible flow is described, together with a variant of Immersed Boundary Method used for handling of complex domain geometry.

2.1. Solution of flow equations (simple geometry)

The flow solver Drifter, in-house code developed at Institute of Thermal Machinery, is a classical second order finite volume code based on staggered MAC grid and projection method [8]. As it was designed for simulation of flow in simple academic geometries, the complex bed geometry is handled by a variant of immersed boundary method [10].

The full computational domain is rectangular box circumscribed on the cylindrical container (see figure 1). Physical dimensions of the box in x and z direction coincide with the diameter of the container and the dimension in y direction depends on the height of the considered bed (possibly including empty buffer zones above and beneath the bed). The number of grid nodes in each direction can be set independently as N_x, N_y, N_z.

Assuming that velocity $\mathbf{u}$ of incompressible flow is expressed as $\mathbf{u}=[u,v,w]$, Navier-Stokes equation can be written as [8,9]:

$$\rho\frac{\partial \mathbf{u}}{\partial t}+\rho\mathbf{u}\cdot\nabla\mathbf{u}=-\nabla p+\nabla\cdot\mu\left(\nabla\mathbf{u}+\nabla\mathbf{u}^T\right)-\rho g\ , \qquad \nabla\cdot\mathbf{u}=0 \tag{1}$$

where ρ - density, p - pressure, μ - dynamic viscosity, g - gravitational acceleration.

Flow equations (1) are solved with projection method in three stages. In the first, the pressure term is omitted – equation (1) takes the form of advection-diffusion equation with a source term – and provisional velocity $\mathbf{u}^*$ is found. In the second stage, Poisson equation is solved for pressure with the right hand side of the equation depending on the divergence of the provisional velocity (this is the most time consuming part of the whole procedure). In the third stage, when the pressure is known, the provisional velocity is corrected and the flow velocity in the next time step is obtained.

The solution is advanced in time with second order Runge-Kutta scheme. The advection term is treated with second order upwind scheme and the diffusion term with the central second order finite difference formula. The Poisson equation is discretised with second order central finite difference in space and the resulting set of equations is solved with preconditioned conjugate gradient method (IPCG). The preconditioner is based on incomplete Cholesky decomposition.

2.2. Handling complex geometry on regular grid

In the literature, there are examples of flow simulations in complex realistic bed structure [4,5]. The methods used, however, require generation of highly complex computational grids fitted to the considered structure (usually simplified e.g. by using spherical bed elements). The approach taken in this work is completely different and enables simulation of flow through the bed on regular structured grid.

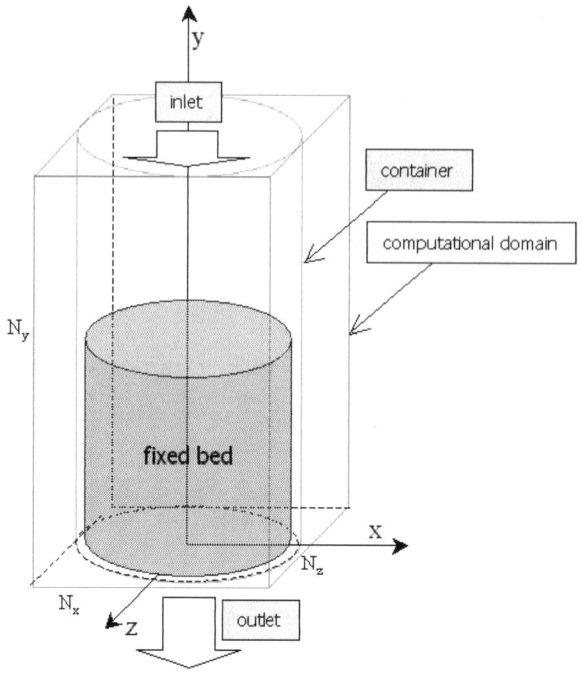

Figure 1. Position of the container with respect to the computational domain.

Immersed boundary method (IBM) was originally formulated by Peskin [10] for simulation of blood flow in heart (dynamically changing geometry) using structured grid. The main concept of the method has been the modelling of "immersed" boundaries by proper source terms in flow equations. In the following years, a large number of variants of the method has been developed. It can be shown [9] that in the case of stationary boundaries (fixed geometry) and no-slip boundary condition, additional source terms are not necessary. The approximation of the appropriate boundary condition and the damping of flow velocity on the immersed boundaries can be realized by modification of provisional velocity in the first stage of projection method:

$$\mathbf{u}^* = \mathbf{u}_0^* b(\varphi) \tag{2}$$

where $\mathbf{u}_0^*$ - provisional velocity calculated as described in 2.1., φ - signed distance of a given point from the immersed boundary. The distance can be easily calculated for a single ring (or container wall) and then the minimum value is taken over all the bed elements. For the points inside the ring's shell the distance is negative (hence the term: "signed" distance). The damping function b should be equal 1 far from immersed boundaries (no damping) and equal 0 inside the ring's shell (full damping). In order to avoid numerical difficulties with discontinuous function, a transition zone is introduced in the definition of function b:

$$b(\varphi) = \begin{cases} 1 & \text{for } \varphi > \Delta h \\ \dfrac{1}{2} + \dfrac{1}{2}\sin\left(\dfrac{\pi(\varphi - \Delta h)}{\sigma h}\right) & \text{for } -\Delta \text{h} < \varphi < \Delta h \\ 0 & \text{for } \varphi < -\Delta h \end{cases} \tag{3}$$

where h is the grid step, Δ, σ - parameters of the damping function describing the shift and the smoothing width. The degree of smoothing should be large enough to filter out the irregularities of provisional velocity near the immersed boundary. However, the smoothing cannot be too large as it widens the area of influence of the boundary on the grid points in the interior of the domain.

The parameters of the damping function has been selected in a test case of a laminar flow through a circular channel (without rings), for which the analytical solution is known and the velocity profile together with the pressure drop may be used as reference data for simulation results.

The simulations were performed for three computational grids differing by number of nodes on the channel cross-section (5×5, 10×10, 20×20 grid points respectively) and for three values of Reynolds number (5, 250 and 1000). Parameters (σ,Δ) were selected to obtain the minimum error of pressure drop value for all considered cases. This resulted in $\Delta=-0.1$ and $\sigma=0.3$.

Beside the change of the provisional velocity, similar damping is applied to corrected velocities (third stage of projection method). Sample distribution of $b(\varphi)$ field for a bed of Raschig rings is shown in figure 2. The distance φ is understood not only as the distance from the nearest ring but also from the container wall.

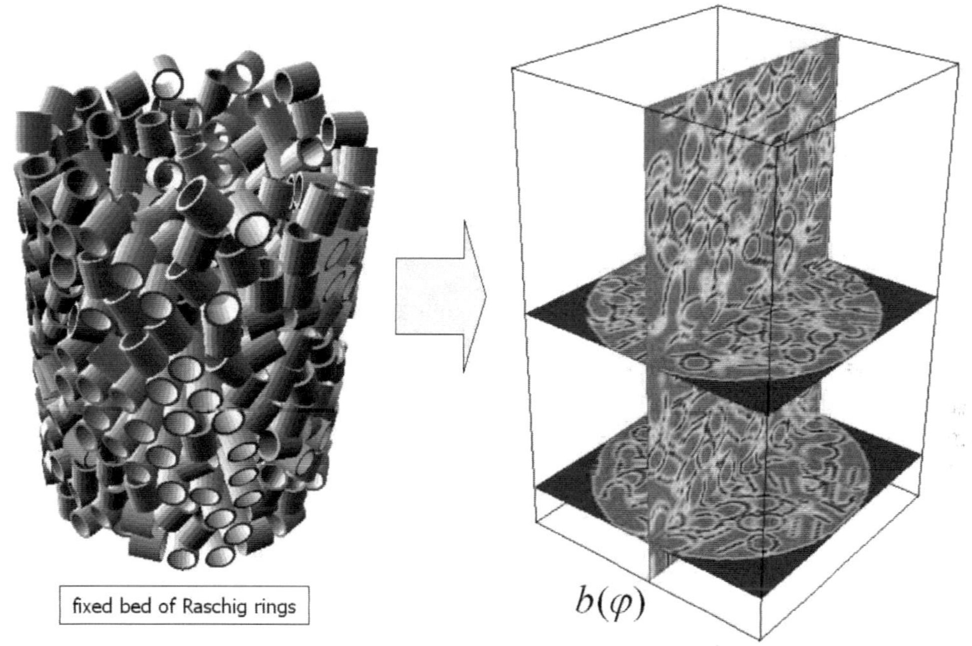

Figure 2. Sample $b(\varphi)$ field for a bed of Raschig rings. Blue colour represents full damping of velocity, red – no damping.

To summarise, a single time step of projection method is realised in the following substeps:

a. Calculation of provisional velocity $\mathbf{u}_0^*$, then modified with the damping function $\mathbf{u}^*=\mathbf{u}_0^*b(\varphi)$.
b. Solution of pressure equation: $\nabla^2 p=\rho\nabla\cdot\mathbf{u}^*/\Delta t$, where Δt - time step.
c. Correction of provisional velocity: $\mathbf{u}^{**}=\mathbf{u}^*-\nabla p\Delta t/\rho$ and the final damping $\mathbf{u}^{n+1}=\mathbf{u}^{**}b(\varphi)$.

3. Simulation of gas flow through the fixed bed

This section presents sample results of simulation of gas flow through a fixed bed of Raschig rings. The gas is injected at the top of the domain with uniform velocity profile. In the considered range of inlet velocity, the flow within the bed remains in the laminar regime. As the flow velocity is damped on the container wall and on the rings' boundaries, this represents no-slip boundary condition. At the outlet (bottom of the domain) zero gradient boundary condition is imposed on the velocity. In the case of pressure Neumann boundary condition is used for the inlet plane and side walls, and Dirichlet boundary condition at the outlet. The computational domain covers the whole container and two empty buffer zones above and below the bed.

Looking at the iso-surface of vorticity (figure 3a), we can see complex structure of the flow inside the bed. Although some longer vortex tubes may be distinguished, the flow is generally irregular which is not surprising bearing in mind the random packing of rings. The distribution of pressure (figure 3b) reveals distinct pressure gradient and local variations of the pressure due to local changes in bed porosity.

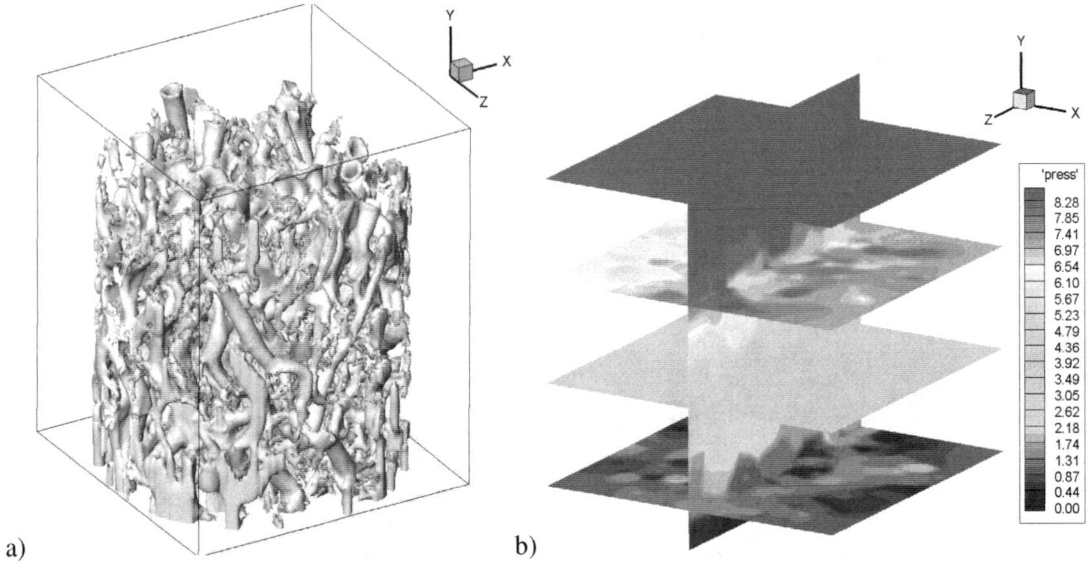

Figure 3. Gas flow through the bed: a) iso-surface of vorticity, b) pressure distribution in cross-sections of the domain.

A good quantitative test is an examination of pressure gradient dependence on the gas velocity v_g at the inlet. In the literature, empirical formulas can be found for pressured drop in the beds consisting of various elements and for different flow conditions. For instance, Billet proposes [11]:

$$\frac{\Delta p}{H} = \xi_0 \frac{a}{\varepsilon^3} \frac{u_V^2}{2} \rho_V \frac{1}{f_s} \tag{4}$$

where: ε - bed porosity, ρ_V - gas density, a - geometric surface are of packing per unit volume,

$$\xi_0 = C_p \left(\frac{64}{\mathrm{Re}_V} + \frac{1.8}{\mathrm{Re}_V^{0.08}} \right) \tag{5}$$

$$\mathrm{Re}_V = \frac{v_g d_p \rho_V}{(1-\varepsilon)\eta_V} f_s, \quad d_p = 6\frac{1-\varepsilon}{a}, \quad \frac{1}{f_s} = 1 + \frac{4}{a d_s} \tag{6}$$

where: η_V - gas viscosity, $C_p = 1.329$ for Raschig rings, d_s - container diameter.

In [12] Maćkowiak suggests another formula (based on a wide range of empirical data):

$$\frac{\Delta p}{H} = \psi_0 (1-\varphi_P) \frac{1-\varepsilon}{\varepsilon^3} \frac{\rho_V v_g^2}{d_p K} \tag{7}$$

where: ϕ_P - proportion of perforated surface area of the bed element (equal 0 for Raschig rings) and K is a coefficient taking into account the influence of the finite diameter of the container (analogous to f_s in Billet formula):

$$\frac{1}{K} = 1 + \frac{2}{3} \frac{1}{1-\varepsilon} \frac{d_p}{d_S} \tag{8}$$

and

$$\psi_0 = \frac{725.6}{\mathrm{Re}_V} + 3.203 \tag{9}$$

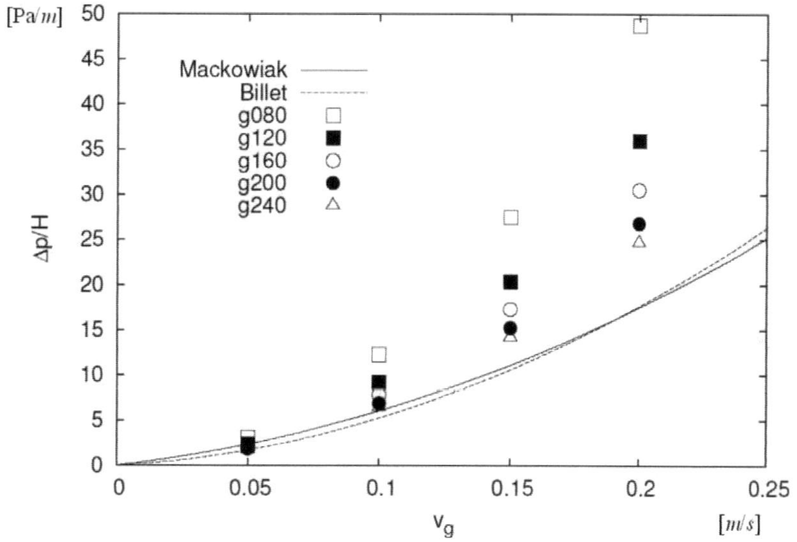

Figure 4. Calculated pressure gradient as the function of gas velocity and grid density. Solid and dashed line represent the values calculated with empirical formulas.

The simulation results for a bed consisting of 1056 rings ($a = 193.6$) is shown in figure 4. The curves represent the values calculated with empirical formulas of Billet and Maćkowiak. It can be seen that in the considered range of gas velocities the difference between them is not large. The simulations were performed for computational grids of various densities to examine the influence of the grid on the results and the convergence of the solution with sufficient level of accuracy. The cases are denoted as g080, g120, g160, g200 and g240, respectively, for grids of $80 \times 80 \times 96$, $120 \times 120 \times 144$, $160 \times 160 \times 192$, $200 \times 200 \times 240$, $240 \times 240 \times 288$ nodes. The transition from one case to the next one resulted in approximately doubling of the computational time required to attain steady state flow and final value of pressure drop.

The obtained results show distinct sensitivity to the number of grid nodes for larger values of gas velocity. This may be related to the change of flow character (regime), separation of boundary layers and no explicit model in the solver to take into account such effects. However, for values of gas velocity lower than 0.15 m/s the convergence of the solution is quite satisfactory and for the finest

Journal of Physics: Conference Series **530** (2014) 012016 doi:10.1088/1742-6596/530/1/012016

mesh (g240) the correction of pressure gradient is not larger than 3%. Comparing the simulation results with empirical formulas one should bear in mind considerable level of uncertainty of the latter ones (reaching even 20%), as typically they are derived as the correlation of a wide range of bed and flow parameters.

4. Summary

In this work, simulation gas flow through a porous bed has been performed by directly solving flow equations in the realistic geometry. Such an approach allows for capturing the physics of the process and construction or verification of various models designed for larger scales. The obtained values of pressure drop within a bed show good convergence with increasing grid density and do not deviate much from empirical formulas derived for the beds of industrial scale.

In future work, the model will be extended with two-phase flow simulation capabilities in order to analyse the structure of liquid flow through a fixed bed which is important in some industrial applications (e.g. absorbers).

5. References

[1] Bazmi M, Hashemabadi S H and Bayat M 2012 CFD simulation and experimental study of liquid flow mal-distribution through the randomly trickle bed reactors *Int. Commun. Heat Mass* **39** 736-743

[2] Zhang W, Thompson K, Reed A H and Beenken L 2006 Relationship between packing structure and porosity in fixed beds of equilateral cylindrical elements *Chem. Eng. Sci.* **61** 8060

[3] Boyer C, Koudila A, Chen P and Dudukowic M P 2005 Study of liquid spreading from a point source in a trickle bed via gamma-ray tomography and CFD simulation *Chem. Eng. Sci.* **60** 6279 – 6288

[4] Jafari A, Zamankhan P, Mousavi S M and Pietarinen K 2008 Modeling and CFD simulation of flow behavior and dispersivity through randomly packed bed reactors *Chem. Eng. J.* **144** 476–482

[5] Freund H, Zeiser T, Huber F and Klemm E 2003 Numerical simulations of single phase reacting flows in randomly packed fixed-bed reactors and experimental validation *Chem. Eng. Sci.* **58** 903 – 910

[6] Asendrych D, Niegodajew P, Drobniak S 2013 CFD modelling of CO_2 capture *Chem. Proc. Eng.* **34**(2) 269-282

[7] Marek M 2013 Numerical generation of a fixed bed structure *Chem. Proc. Eng.* **34** (3) 347-359

[8] Ferziger J H and Peric M 2002 *Computational methods for fluid dynamics* (Berlin: Springer-Verlag)

[9] Tryggvason G, Scardovelli R and Zaleski S 2011 *Direct numerical simulations of gas-liquid multiphase flows* (Cambridge: Cambridge University Press)

[10] Peskin C S 2002 Immersed boundary method *Acta Numer.* **11** 479-517

[11] Billet R 1995 *Packed Towers* (Weinheim: Verlagsgesellschaft mbH)

[12] Maćkowiak J 2012 *Fluid dynamics of packed columns* (Berlin: Springer-Verlag)

Acknowledgments

The work supported by Strategic Research Programme nr 364 SP/E/1/67484/10.

Journal of Physics: Conference Series **530** (2014) 012017 doi:10.1088/1742-6596/530/1/012017

Hydrodynamic lift force generated on the accelerated profile

Tomasz Kozłowski and Henryk Kudela

Department of Numerical Modelling of Flows, Wrocław University of Technology, Wybrzeze Wyspianskiego 27, 50-370 Wroclaw

E-mail: tomasz.kozlowski@pwr.wroc.pl, henryk.kudela@pwr.wroc.pl

Abstract. In the paper we investigate the forces that act on the profile when profile starts suddenly to accelerate. It was proved that that sudden jump of acceleration cause sudden jump of the lift force. The peaks of force jumps do not depend on Reynolds number. To investigate that phenomena it was carry out the numerical simulation using the vortex particle method.

1. Introduction

In present work we examined the influence of the acceleration of the profile on the lift force generation in low Reynolds numbers flows, $Re < 10^4$. This problem applies to the study of non-stationary aerodynamic effects that are typical for flying birds and insects. Birds and insects generate the lift and the thrust force by the motion which is called the flapping motion. Due to fact that flapping motion undergoes changeable acceleration one can observed peaks in the lift force evolution [2, 3]. Similar effect is very clearly visible in sudden starts of the profile [8, 9]. In the initial time at $t = 0$ the flow over the profile is potential and the vorticity around the profile is equal zero. When the viscosity come to the action the large amount of the vorticity on the profile is generated. That is responsible for sudden increase of the lift force. Similar peaks of forces appear when the profile is suddenly accelerated. It is seems that effect is caused by not sufficient production of the vorticity at moment of acceleration. To study this phenomenon we perform the numerical calculations for the accelerating profile. We assume that the velocity of the profile start to growth linearly. The calculations were done also for the case of a sudden start of the profile. It had been investigated how the acceleration of the profile impact on the lift force distribution over the profile. For our study we choose the vortex particle method which allows for easily and direct analysis of vorticity fields evolution around the profile.

2. Vortex in Cell method

The Navier-Stokes equations in primitive variables (u,p) can be transformed to the Helmholtz equations for the evolution of the vorticity. The Helmholtz equations in the non-inertial reference frame have the form

$$\frac{\partial(\omega + 2\Omega_0)}{\partial t} + (\nabla\omega)\cdot\vec{u} = \nu\Delta\omega \tag{1}$$

$$\Delta\psi = -\omega \tag{2}$$

$$\vec{u} = \nabla\times(0,0,\psi) \tag{3}$$

where Ω means the angular velocity of the profile. For our purposes, it was convenient to change the frame of reference in such a way, that profile would be at rest. That cause that boundary condition for the stream function far from the body with specified angular Ω and translational U_0 velocity is

$$\psi_\infty = U_0(y\cos(\alpha) - x\sin(\alpha)) - \frac{\Omega_0}{2}(x^2 + y^2). \quad (4)$$

A detailed description of solution of the Helmholtz equations in the moving reference frame can be found in [1]. In present paper the numerical calculation of the moving profile in fluid were carried out by the Vortex in Cell method (VIC). The velocity of the particles were calculated using the numerical grid. To better fit the numerical grid to the solid boundary we take advantage of the conformal transformation. By the conformal mapping the physical non-uniform flow region (x, y) is replaced by the rectangular one $\xi, \eta)$ where the calculations are carried out very efficiently. The following conformal transformation was applied (figure 1)

$$x + iy = \cosh(\xi + i\eta). \quad (5)$$

In new variables the equations of motion have the form

$$\frac{\partial\omega}{\partial t} + (\nabla\omega)\cdot\vec{u} = \frac{\nu}{J}\Delta\omega, \quad (6)$$

$$\Delta\Psi = -J\omega, \quad (7)$$

where J denotes the Jacobian of the transformation. The velocity field is obtained according to the formula

$$u = \frac{1}{J}\frac{\partial\psi}{\partial\eta}, \quad v = -\frac{1}{J}\frac{\partial\psi}{\partial\xi}. \quad (8)$$

In the VIC method the continuous vorticity field is approximated with the discrete particles distribution. The flow region is covered with the numerical grid $h = \Delta\eta = \Delta\xi$. In every grid node, the particle with $\Gamma_j = \int_A \omega d\xi d\eta$ circulation are placed, where $A = h^2$, and

$$\omega(\xi,\eta) = \sum_p \Gamma_p \delta(\xi - \xi_p)\delta(\eta - \eta_p). \quad (9)$$

The equations of motion were solved using viscous splitting algorithm [5]. First, the inviscid fluid motion equation is solved

$$\frac{\partial\omega}{\partial t} + (\nabla\omega)\cdot\vec{u} = 0. \quad (10)$$

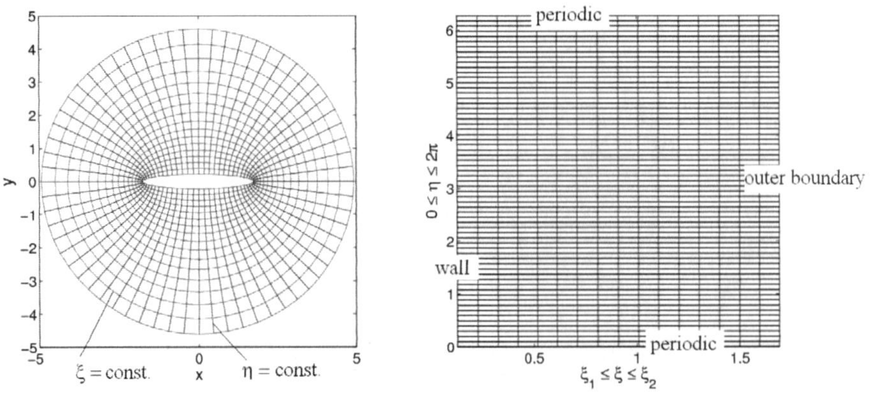

Figure 1. Elliptical grid in the physical domain.

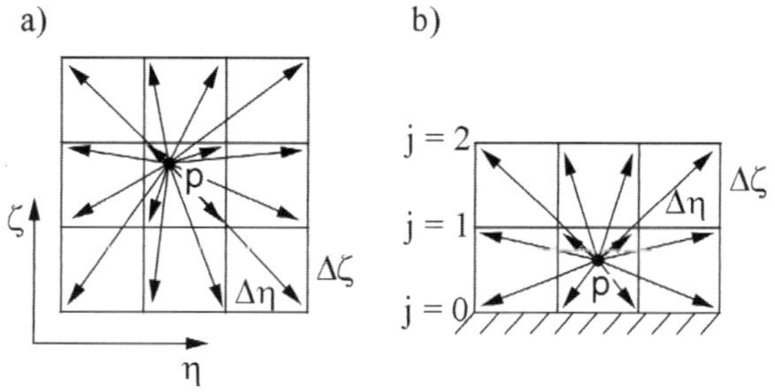

Figure 2. Redistribution of the particle masses onto the neighboring grid nodes, a) for particles laying inside of the computational domain (at least one cell from the wall), b) for the particles in the vicinity of the wall.

From (10) stems that vorticity is constant along the trajectory of the fluid particles. According to Helmholtz theorem [4, 5], vortex particles are moving like material fluid particles. The differential equation (10) is replaced by the infinite set of ordinary equations

$$\frac{d\xi}{dt} = u, \quad \frac{d\eta}{dt} = v, \quad \xi(0, \alpha_1) = \alpha_1, \quad \eta(0, \alpha_2) = \alpha_2. \tag{11}$$

where $\alpha = (\alpha_1, \alpha_2)$ means Lagrangian coordinate of fluid particles. The number of particles is equal to the number of grid nodes. The set of ordinary equations (11) were solved by the four order Runge-Kutta method. The velocity field was obtained by the solution of Poisson equation for the stream function and vorticity on the numerical grid (2). The velocities of the particles between the grid nodes were calculated by the two dimensional bilinear interpolation. In the second step the viscosity is taken into account by the solution of the diffusion equation

$$\frac{\partial \omega}{\partial t} = \frac{\nu}{J}\Delta\omega, \tag{12}$$

$$\omega(\xi, \eta, 0) = \omega_0, \quad \omega|_{wall} = \omega_s. \tag{13}$$

where ω_s denotes the vorticity at the solid boundary. The motion of the body in fluid and vorticity in the flow region generate non-zero tangent velocity component at the wall. The no slip condition is realized by adding the proper amount of vorticity at the wall. In present work we used the Briley formula of fourth order [6]. Nullifying of the normal velocity component at the wall was done by constant stream function at the wall $\psi = \text{const}$. The diffusion equation was solved by alternating direction implicit method (ADI). In our calculations we use the numerical grid, therefore it's necessary to transfer the information from the vortex particles to the mesh nodes, figure 2. The redistribution of the vorticity from fluid particles to the mesh nodes was performed according to the formula

$$\omega_{ij} = \frac{1}{h^2}\sum_p \Gamma_p \varphi_h(\xi)\varphi_h(\eta). \tag{14}$$

where $\varphi_h(\cdot)$ denotes the interpolation kernel. The redistribution process plays a fundamental role for the accuracy of the vortex particle methods. In the current study we use the Z-splines interpolation functions [7] have a very good interpolation properties, especially when applied to vortex particle methods.

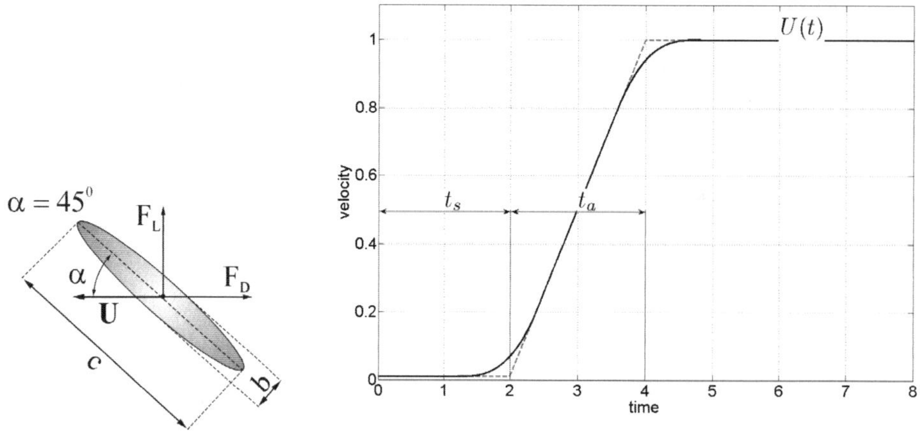

Figure 3. Computational parameters. Investigated profile on the left and change profile velocity.

3. Formulation of the problem and computation details

The numerical calculations were performed for a elliptic profile with relative thickness $e = b/c = 1/25$ and the angle of attack $\alpha = 45^0$. The flow configuration and parameters are shown on the left side of figure 3. The velocity of the profile U is plotted on the right side of the figure. It is well known that for potential flow over the plate inclined to the direction of the incident flow a stagnation points appear on the upper and lower surface of the profile. Such a flow pattern cause the flow around the trailing edge from the lower to the upper surface. Theoretically an infinitely high velocity at the trailing edge cause an infinitely high negative pressure. By adding the circulation to the profile the rear stagnation point can be moved to the sharp trailing edge. In this way one can eliminate the flow around the trailing edge. In vortex particle method we begin the calculations from potential flow solution. In order to eliminate the flow over the trailing edge that cause also the generation of the large amount of the vorticity we decided to create the vorticity background around the profile. It was done by assuming at the beginning the small velocity around the profile $U = 0.01U_0$. The incremental velocity U lasted a short interval time t_s. After the time t_s the linearly growth of the velocity $U(t)$ to the nominal speed $U(t_a) = U_0$ occured. To obtain the smooth acceleration dU/dt, the curve $U(t)$ was smoothed in points ts and $t_s + t_a$ by polynomial of third order. The Reynolds number was defined based on the chord of the profile c: $Re = U_0 c/\nu$ and the numerical calculations were performed in the range of Reynolds numbers 100 to 1000. The physical flow domain was transformed onto rectangular computational domain by the mapping (5), figure 1. The calculations were performed on a uniform grid of size: 512×512. The resolution of the numerical grid was chosen in such way that the further increasing of the mesh resolution does not affect the calculation results.

4. Computational results

The problem of moving profile in fluid belong to the category of external problems. In such a problems it is necessary to set the nominal velocity of the profile U_0. It is often done by the change of the profile velocity suddenly at the initial time: $U = 0$ for $t = 0$ and $U = U_0$ for $t > 0$. The sudden jump in the velocity of the profile from zero to the nominal velocity cause infinite increase of the lift force. In figure 4 it was shown the variation lift coefficient with time. Our numerical results was compared to the numerical results of [8]
The phenomenon of the high increase of the lift force at the initial time has been reported

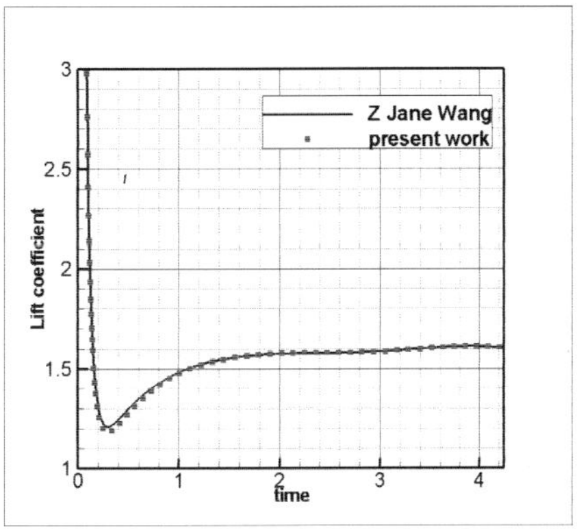

Figure 4. Lift coefficient versus time for sudden start of the profile, $Re = 1000$, angle of attack $\alpha = 45^0$.

repeatedly[8, 9]. In order to better investigate the this effect we performed a series of numerical tests where the profile was accelerated in different ways. To create the vorticity background around the profile during the time t_s we assume small velocity of the profile: $U = 0.01U_0$. On figure 5 (on the left) the lift force for the three different start time t_s is presented. The end of acceleration time t_a was the same as well as the Reynolds number, $Re = 300$. We slightly smooth the acceleration curve around points t_s and t_a. The presented test shows that the effect of jump in the lift force is not an numerical artifact and is connected with sudden acceleration of the profile. This effect is also independent of the length of the start time t_s and of the existence of the background vorticity around the profile.
On the figure 6 the sequence of the stream function and vorticity field for the accelerated profile that correspond to the line $t_s = 4$ in figure 5 (line with number 3), is presented. The figure 6a correspond to the potential flow. It is visible the rear stagnation point on the upper surface of the ellipse. In the frame t = 3.7 (figure 6b) due to production of the background vorticity, one can see that the stagnation point was moved to the trailing edge. In the frame t = 3.8, after start of acceleration one can observe that stagnation point appears on the upper surface. Farther, this point was moved along the plate in upward direction. It correspond the sudden

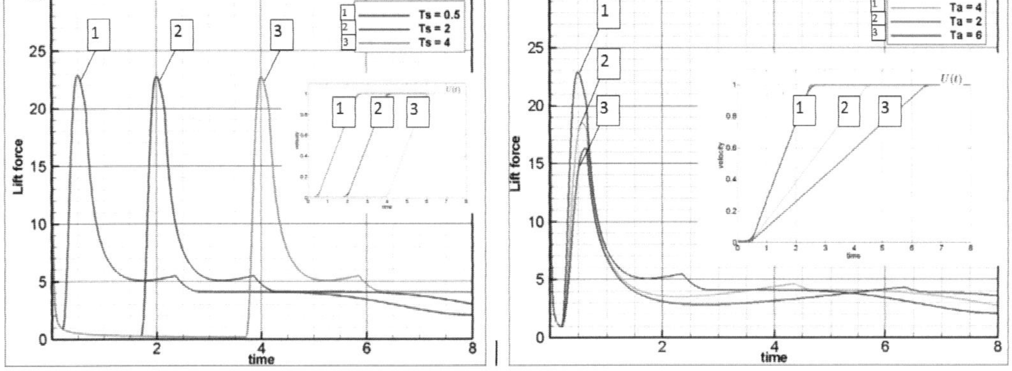

Figure 5. The effect of sudden start of the profile. On the figure the lift force for different start time t_s on the left and with different acceleration time is presented.

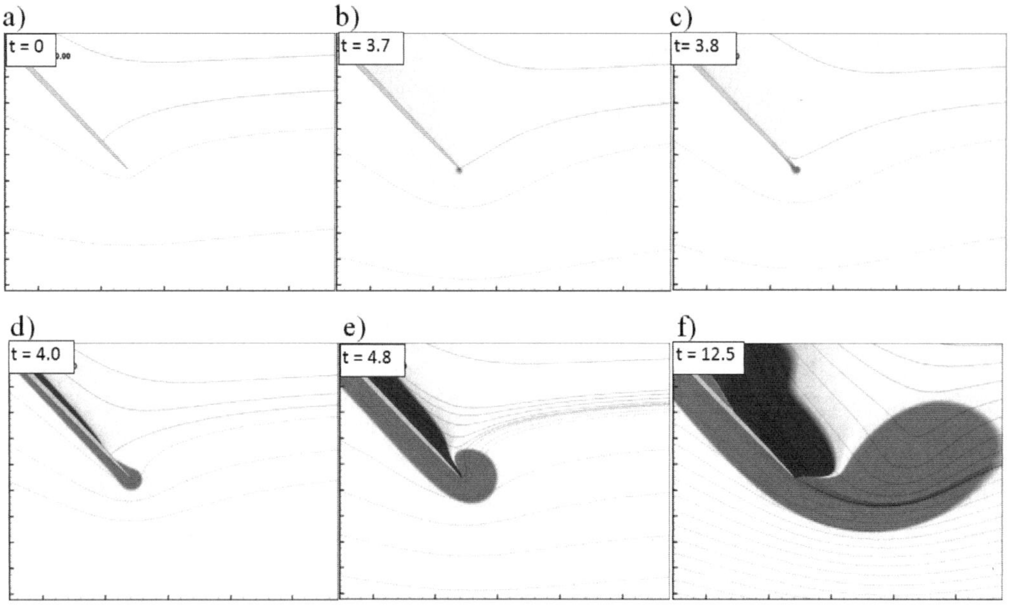

Figure 6. Vorticity contour and stream function over suddenly started profile.

increase of the lift force in figure 5. We believe that the effect of a sudden increase in lift force is related to the shift of the stagnation point on the upper edge of the profile.
We also check the influence the duration of acceleration time t_a for the lift force. The calculations were done by selecting the $t_a = 2, 4$ and 6. figure 5 (on the right) shows plots of the calculated lift force. As we expected the greater velocity of acceleration causes the greater increase in the lift force. On the left of figure 7 the lift force for different way of smoothness of the $U(t)$ curve is presented. The curve with number 4 presents the acceleration of the profile without smoothing. One can see also that if the slope of the $U(t)$ is change more smoothly the increase with the lift force is reduced. On the right of figure 7 we presented the lift force of the accelerated profile for different Reynolds number. The peak in the lift force is the same for all examined Reynolds numbers. The increase in the lift force is the result of the inertia caused by the acceleration of

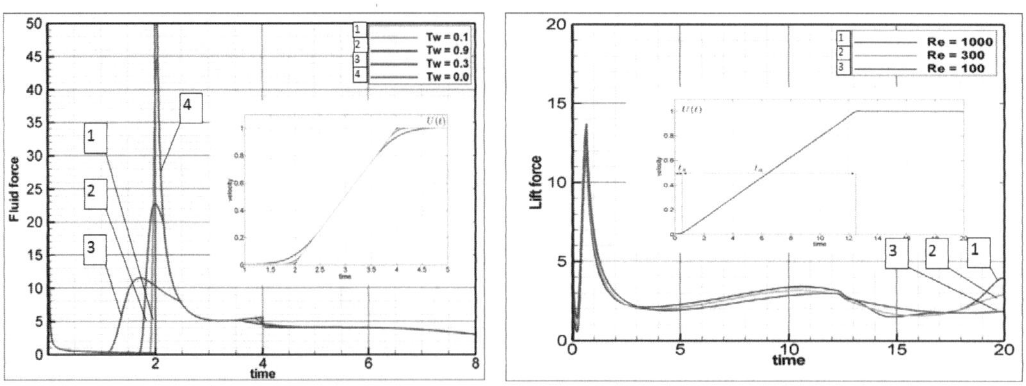

Figure 7. The lift force observed for sudden start of the profile. Influence of the Reynolds number on the left and influence of the Reynolds number on the right.

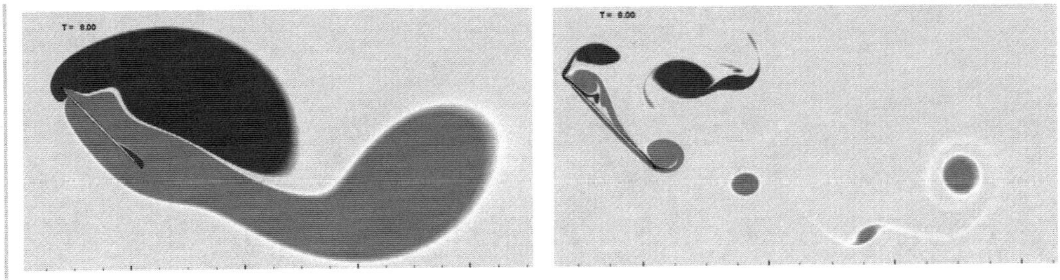

Figure 8. Vorticity contour around profile. On the left: $Re = 100$ and on the right: $Re = 1000$.

the profile. It worth notices that enlarge in the Reynolds number has effect on the reduction of the scale of the vortices around the body (see figure 8).

5. Conclusions

In the paper the effect of the acceleration of the profile was presented. It was found that acceleration caused sudden increase of the lift force. The effect of strong increase of the lift force is related with shift of the stagnation point from the trailing edge to the upper side of the profile. It has been shown that the peaks of the lift force do not depend on the Reynolds number. For the numerical simulation the Vortex in Cell method was used. The method is very well suited to study of the unsteady movement of body in the fluid, which is dominated by the vortex structures and it's interactions.

6. References

[1] Gustafson K E Leben R Freymuth P Society for Industrial and Applied Mathematics **143** 1991
[2] Wang Z J Birch J M Dickinson M H J. Exp. Biol. **207** 2004
[3] Wu J C *Theory for aerodynamic force and moment in viscous flows* AIAA **19** 1981
[4] Wu J Z Ma H Y Zhou M D Springer 2005
[5] Cottet G-H Koumoutsakos P D Cambridge University Press, 2000
[6] Koumoutsakos P Leonard A and Pepin F J. Comput. Phys. **113** 1994
[7] Becerra Sagredo Research Report No. 10 Zurich Switzerland 2003
[8] Wang Z J J. Fluid Mech. **410** 2000
[9] D'Alessio S J D Dennis S C R Nguyen P J. Eng. Math **35** 1999

An asymmetrical λ-foot of condensing steam flow in the IMP PAN nozzle

S Kornet[1,2*] **and J Badur**[1]

[1]Energy Conversion Department, Institute of Fluid-Flow Machinery PAS-ci, Gdańsk, Poland
[2] Conjoint Doctoral School at the Faculty of Mechanical Engineering, Gdansk University of Technology, Gdańsk, Poland

E-mail: sebastian.kornet@imp.gda.pl, jb@imp.gda.pl

Abstract. In the present paper we have focused on the precise prediction of the spontaneous condensation phenomena in wet steam flow. Novelty of our approach lies on modelling both the moment of initiation of a phase transition, as well as the moment of its reverse progress - called here re-vaporization of the condensate phase. The practical issue is to elaborate of a model of spontaneous condensation/vaporization of water steam flow under low-pressure conditions by using methodology of non-equilibrium thermodynamics [2]. The basic tests including comparison with an experimental data have been performed using the IMP PAN nozzle with the de Laval geometry [1]. Having observed the finishing of a foggy flow within the shock wave, according to Puzyrewski's postulate [1], we would like to analyse the topography of the shock wave pattern in the IMP PAN symmetric nozzle. This phenomenon, independently from a type of compressible fluid, has been observed to be the result of interaction between a normal shock wave and the boundary layer – it has been known as a λ-foot structure [3]. The asymmetry of the shock structure is measured by optical system and visible since the foggy flow can be easily observed. Our paper is a trial towards to an explanation of this problem.

1. Introduction

The first time the de Laval nozzle has been used in steam turbines in 1888. Nowadays nozzles are commonly used to acceleration different fluids in various technical devices.

In the low-pressure part of steam turbine, the state path usually crosses the saturation line in penultimate stages. At least last two stages of this part of turbines operate in two–phase region. The liquid phase in this region in mainly created in the process of homogeneous and heterogonous condensation. Several observations confirm however, that condensation often occurs earlier than it is predicted by theory i.e. before the Wilson line. It is because the nucleation can start at some soluble and insoluble impurities, particle of dust, chemical compounds or corrosion products. Droplets of condensed phase can move with velocity different than gas phase velocity and can seriously damage blades. Erosion of rotor blade leading edge is mainly due to droplets that form behind the trailing edge of guide vanes. The flow in the low-pressure part of steam turbine is complicated and still requires thorough experimental and numerical analysis [5, 6].

Shock wave is the result of interaction between a normal shock wave and the boundary layer, which produces a λ-foot structure. It is well-known and experimentally confirmed that for low Mach numbers the λ-feet are of the same size at both nozzle walls. However, at higher velocities (Ma > 1.4) the λ-feet generated on the upper and lower walls of a symmetric nozzle become different in size. Flow separation inside propulsive nozzles generates unsteady and asymmetrical shock structures and induce fluctuating side loads which can damage the structure. It has also been observed that the tendency towards asymmetry depends on the nozzle divergence angle [3, 7, 8, 9]. The most of the experimental data used for validation of the numerical models are relatively old, and do not include a precise investigation of the shock wave behaviour in the wet steam regime.

The present work is concerned with investigation into coupled phenomena occurring in the supersonic section of the de Laval nozzle, characterized by the presence of shock the flow of condensing steam.

2. Conditions of experiment

The experiment was carried out in a symmetric nozzle of rectangular cross-section. The nozzle dimensions and variation of cross section area along the nozzle axis are shown in figure 1. The nozzle inlet conditions were varied within a rather narrow range by throttling or by injecting water into superheated steam. The exhaust pressure was varied by flow throttling in the pipeline leading steam out to the condenser. Thus it was possible to change the location of the shock wave within the supersonic section of the nozzle. The pressure measurements were made on both side walls of the channel. Initial parameters were close to the saturation line, in order to obtain a condensation [4].

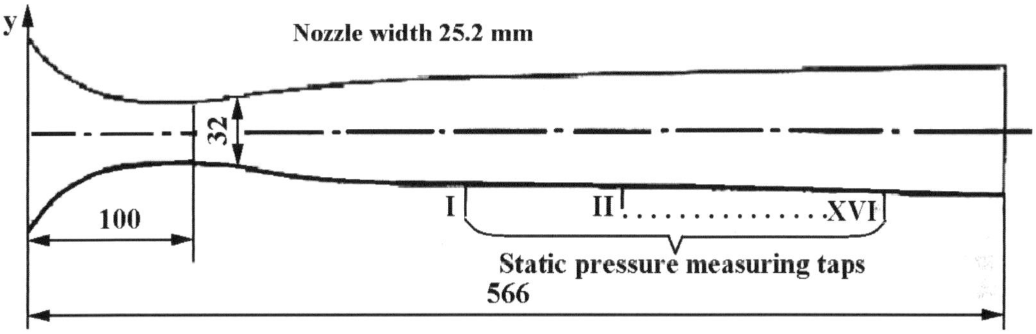

Figure 1. The shape and dimensions of the IMP PAN nozzle [4].

3. Conditions of numerical analysis

Modelling conditions have been taken from description of experiment [4] that was originally conducted at Institute of Fluid-Flow Machinery site. Inlet pressure was $p_{inlet} = 2.26$ bar, inlet temperature $T_{inlet} = 502.15\,\mathrm{K}$. Pressure at the point of intersection of the isentropic expansion line and saturation line was $p_{sat} = 0.55\,\mathrm{bar}$. These input conditions correspond to the V-th set of experimental conditions [4]. Nozzle geometry, was prepared within preprocessing utility Gambit version 2.4.6. The mesh was generated very precisely to eliminate any incidental asymmetry. Mesh for full geometry of the nozzle (figure 2) has consisted of 380000 hexahedral finite volumes.

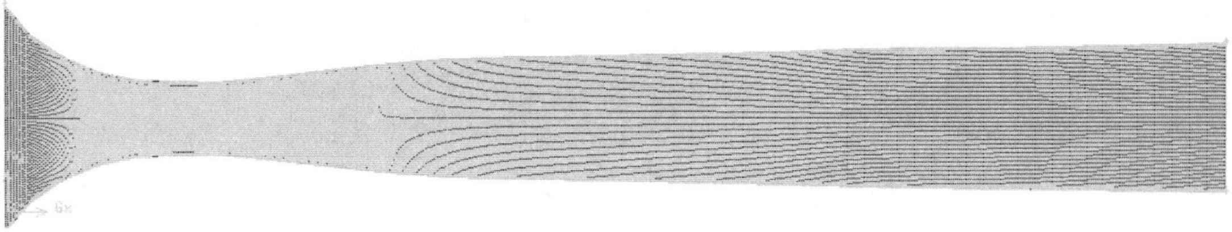

Figure 2. FVM discretization of the de Laval nozzle.

4. Governing equations of CFD and assumptions for a employed model.

Wet steam governing equations are based on the following balance of liquid-vapour mixture:

- balance of mass – ρ:

$$\partial_t(\rho) + \text{div}(\rho\,\mathbf{v}) = 0 \tag{1}$$

- balance of momentum – $\rho\,\mathbf{v}$:

$$\partial_t(\rho\,\mathbf{v}) + \text{div}(\rho\,\mathbf{v}\otimes\mathbf{v} + p\mathbf{I}) = \text{div}(\tau^c) + \rho\,\mathbf{b} \tag{2}$$

- balance of energy – e:

$$\partial_t(\rho e) + \text{div}(\rho e\mathbf{v} + p\mathbf{v}) = \text{div}(\tau^c\mathbf{v} + \mathbf{q}^c) + \rho\,\mathbf{b}\cdot\mathbf{v} \tag{3}$$

- turbulence energy evolution – k:

$$\partial_t(\rho k) + \text{div}(\rho k\mathbf{v}) = \text{div}(\mathbf{J}_k) + \rho S_k \tag{4}$$

- turbulence dissipation evolution – ε:

$$\partial_t(\rho\varepsilon) + \text{div}(\rho\varepsilon\,\mathbf{v}) = \text{div}(\mathbf{J}_\varepsilon) + \rho S_\varepsilon \tag{5}$$

- evolution of wetness fraction – x:

$$\partial_t(\rho x) + \text{div}(\rho x\,\mathbf{v}) = \text{div}(\mathbf{J}_x) + \rho S_x \tag{6}$$

- evolution of number of droplets – a:

$$\partial_t(\rho a) + \text{div}(\rho a\,\mathbf{v}) = \text{div}(\mathbf{J}_a) + \rho S_a \tag{7}$$

An additional balance equations for k, ε, x and a are related to the evolution of turbulence in the condensing flow, or the condensation evolution under turbulent flow condition. Both are fully non-equilibrium phenomena, that should be combined by the sources S, and thermodynamical forces which constitute fluxes $\mathbf{J}$. The above set of nine equations is to be integrated within every finite volume [5, 10,11,]. The following assumptions are made in this model:

- the velocity slip between the droplets and gaseous-phase is negligible,
- the interactions between droplets are neglected,
- the mass fraction of the condensed phase is small: $\beta < 0.2$ (wetness factor),

- since droplet sizes are typically very small (from approximately 0.1 microns to approximately 100 microns), it is assumed that the volume of the condensed liquid phase is negligible [10, 11].

5. Results and discussion

For a given de Laval nozzle geometry authors have conducted calculations for wet steam model. Our numerical investigation (figure 3a) have shown that for low Mach numbers the feet are of the same size. However, at higher velocities (Ma > 1.4), the λ-feet generated on the upper and lower walls of a symmetric nozzle become different in size (figure 3b). λ-feet is result of interaction between a normal shock wave and the boundary layer.

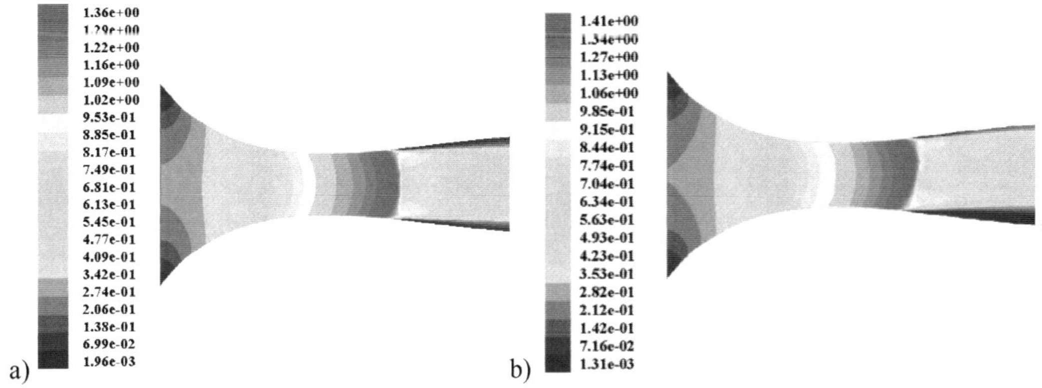

Figure 3. Shock wave: a) symmetric pattern for low Mach number, b) asymmetric pattern for Mach number higher than 1.4.

The shock/boundary layer interaction produces two separated flows: 1 - with reattachment downstream for upper wall (this is a restricted shock separation) and 2 - without reattachment downstream for lower wall (this is a free shock separation). The interaction between the two oblique shocks (C1 and C3 in figure 4a) forms a Mach disc (MD). The λ-shock structures C1 – C2 and C3 – C4 are jointed to Mach disc by two triple points T1 and T2. Above the critical pressure ratio $p_{inlet}(p_{outlet})^{-1}$ shock wave becomes symmetrical with respect to the nozzle axis again, what we can see in figure 4b.

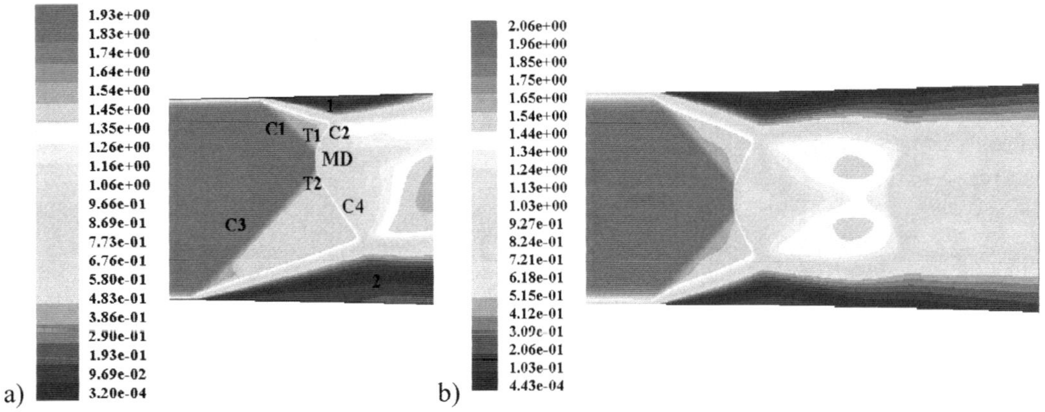

Figure 4. Shock wave: a) asymmetrical shock pattern for Ma > 1.4, b) symmetrical shock pattern above the critical pressure ratio.

Visual observations (in the experiment) were carried out either with direct lighting, or marking use of the Töpler optical system to improve the possibility of indentifying shock wave. In the figure 5a we can see shock wave photo taken using this optical system and in figure 5b we can see the same shape and localization of shock for CFD simulation for the same flow conditions.

a) 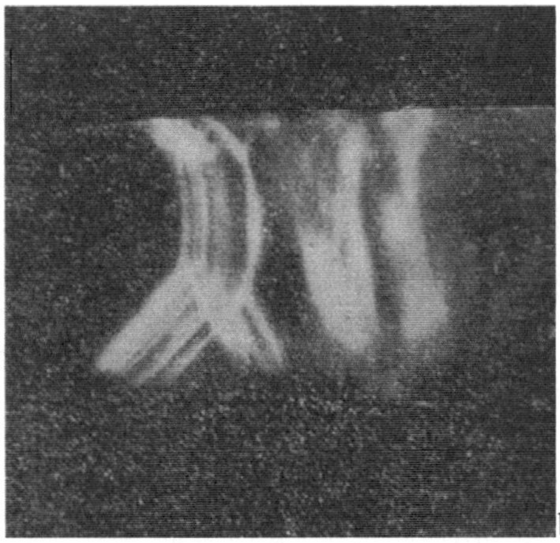 b) 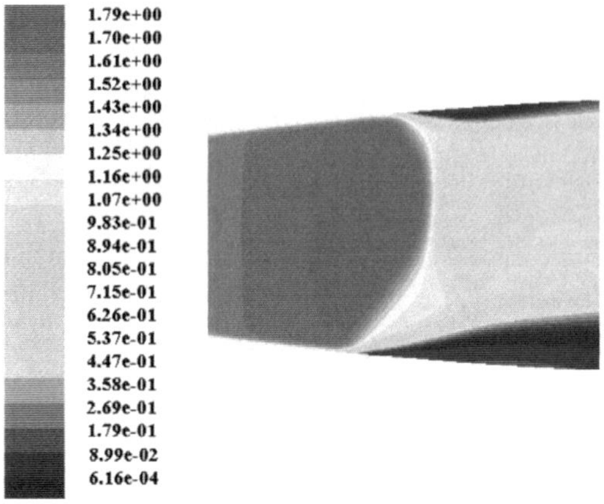

Figure 5. The same localization and shape of shock in the nozzle a) experiment, b) CFD simulation.

Figure 6 shows a comparison of the results of the experiment with nine-equation non-equilibrium model of wet steam for the V-th set of experimental conditions. This model of condensation provides a good agreement with experiment. Figure 7 presents expansion in the „i-s" diagrams for chosen flow conditions obtained by numerical analysis. In the figure 8 we can see this expansion in wet steam region. Phenomena such as: shock, condensation and evaporation cause that the expansion in wet steam region is very complicated.

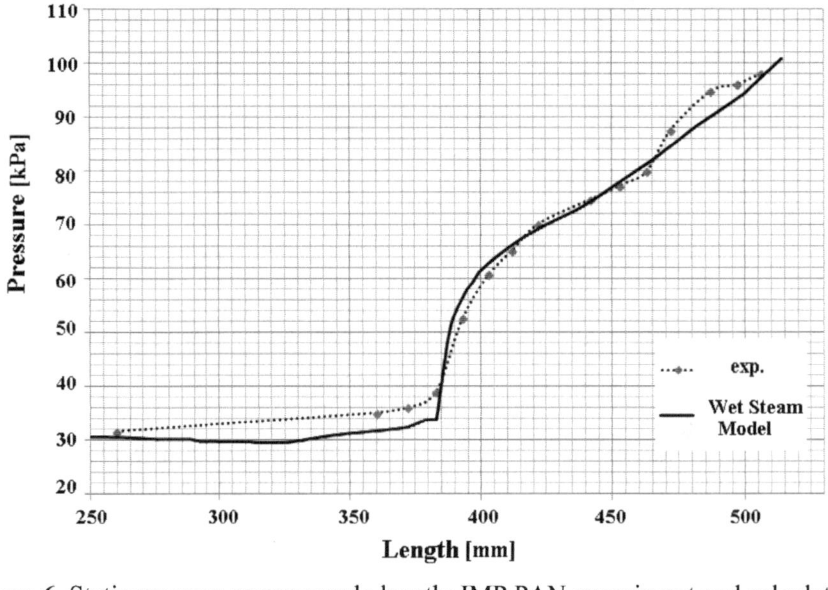

Figure 6. Static pressure versus nozzle length: IMP PAN experiment and calculations.

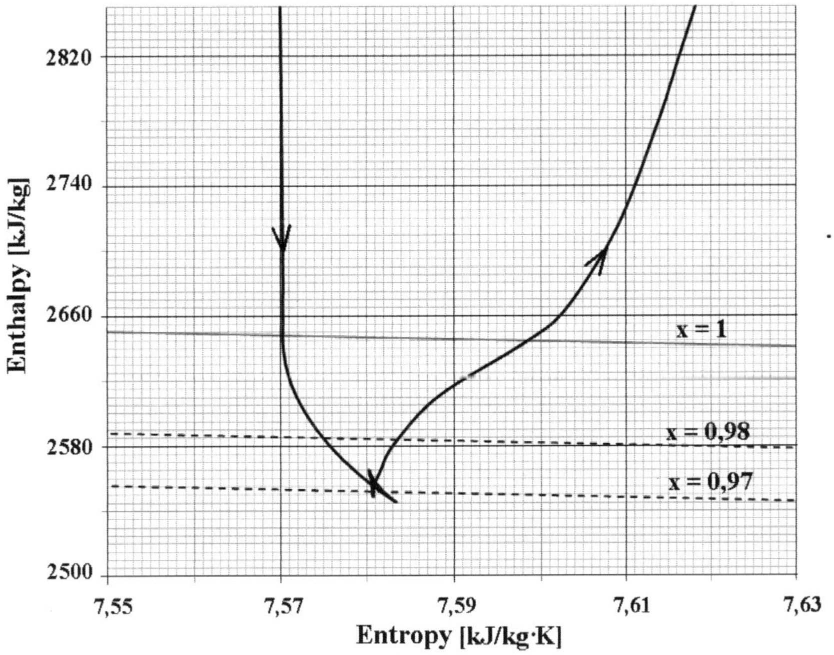

Figure 7. Change of steam entropy during nozzle expansion.

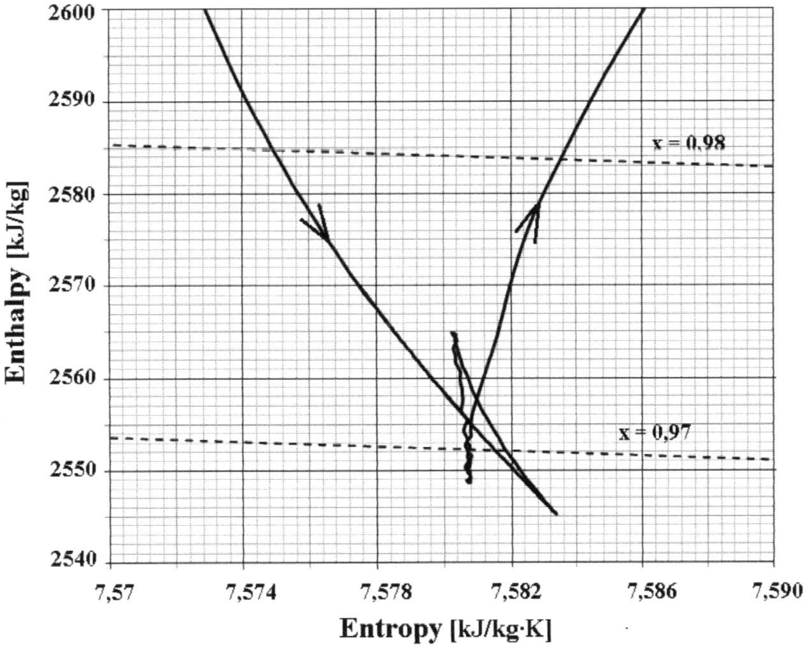

Figure 8. Change of steam entropy during nozzle expansion – a detail.

Figure 9 shows wetness for calculation of CFD simulation for the V-th set of experimental conditions. Results of our analysis (condensation and shock wave) is the same like in Puzyrewski's experiment: *For V-th set of experimental conditions, there were only very slight traces of mist ahead of the shock wave, disappearing in the shock*. Due to the increase of steam conditions through the wave, there is a complete evaporation of the condensed water drops.

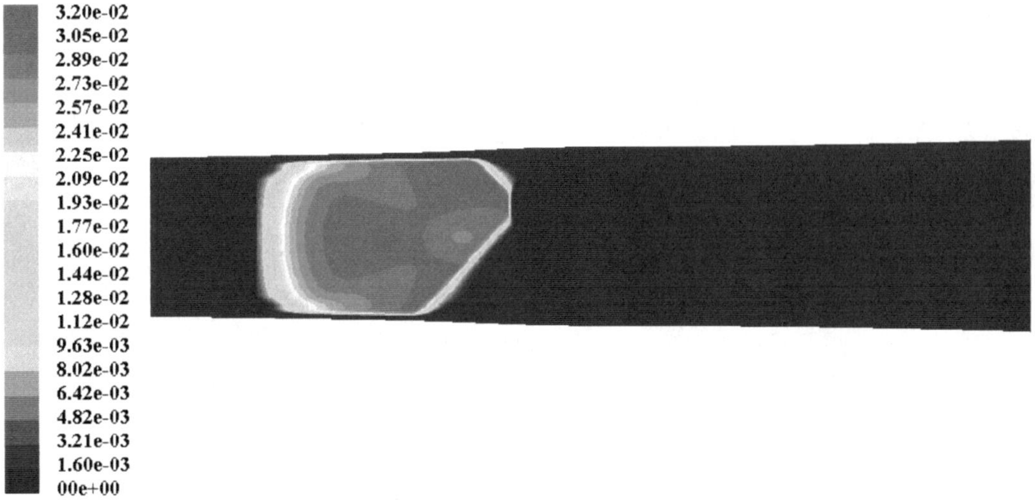

Figure 9. Frog appearing and mist disappearing in the shock for a CFD analysis.

6. Conclusions

Models of nine-equation non-equilibrium have been tested and compared with experimental data. This Wet-Steam model applied to de Laval nozzle gives satisfactory results that agree well with experimental data. Presented data indicates correctness and utilization of employed model for condensing flows in more complex geometries.

Our numerical investigation have shown that for low Mach numbers the feet are of the same size. However, at certain Mach values (Ma > 1.4), the shock pattern becomes asymmetric. λ-feet generated on the upper and lower walls of a symmetric planar nozzle become different in size for Ma > 1.4. This phenomena is depending on interaction between a normal shock wave and the boundary layer.

7. References

[1] Puzyrewski R 1969 Theoretical and experimental studies on formation and growth of water drops in LP steam turbines *Transactions of IFFM* **42-44** 289-303

[2] Bilicki Z, Badur J 2003 A thermodynamically consistent relaxation model for a turbulent, binary mixture undergoing phase transition *Journal of Non-Equilibrium Thermodynamics* **28** 145-172

[3] Namieśnik K, Doerffer P 2004 Numerical simulation of shock wave patterns in supersonic divergent symmetric nozzles *TASK QUARTERLY* **9** 53–63

[4] Puzyrewski R, Gardzielwicz A, Bagińska M 1973 Shock wave in condensing steam flow through a Laval nozzle, *Archives of Mechanics* **25, 3** 393-409

[5] Zakrzewski W 2011 Walidacja modelu pary mokrej przez porównanie z eksperymentem, *Współczesne technologie i konwersja energi*i, Wydawca Wydział Mechaniczny

[6] Dykas S, Majkut M, Smołka K, Strozin M 2013 Reserach on steam condensation flow in nozzle with shock wave *Journal of Power Technologies* vol 93 No 5

[7] Bourgoing A, Reijasse Ph *2005* Experimental analysis of unsteady separated flows in a supersonic planar nozzle *Shock Waves* **14 (4)** 251-258

[8] Sellam M, Fournier G, Chpoun A, Reijasse Ph 2014 Numerical investigation of overexpanded nozzle flow *Shock Wave* **24** 33-39

Journal of Physics: Conference Series **530** (2014) 012018 doi:10.1088/1742-6596/530/1/012018

[9] Verma S B, Manisanka C 2014 Origin of flow asymmetry in planar nozzle with separation *Shock Wave* **24** 191-209

[10] Badur J, Karcz M, Lemański M, Zakrzewski W, Jesionek K *2011* Remarks on steam condensation modeling related to steam turbine of large output *Proc. of microCAD "International Scientific Conference"*, Miscolc

[11] Karcz M, Zakrzewski W, Lemański M, Badur J 2011 Zagadnienia modelowanie 3D spontanicznej kondensacji w części niskoprężnej turbiny parowej 200MW *Mat. X Konf. Nauk.-Tech. „Elektrownie Cieplne Eksploatacja-Modernizacje-Remonty"*, Słok k.Bełchatowa 253-262

Journal of Physics: Conference Series **530** (2014) 012019 doi:10.1088/1742-6596/530/1/012019

Large-eddy simulations of 3D Taylor-Green vortex: comparison of Smoothed Particle Hydrodynamics, Lattice Boltzmann and Finite Volume methods

A. Kajzer, J. Pozorski and K. Szewc

Institute of Fluid-Flow Machinery, Polish Academy of Sciences, Gdańsk, Poland

E-mail: `adkajzer@gmail.com`, `jp@imp.gda.pl`, `kamil.szewc@imp.gda.pl`

Abstract. In the paper we present Large-eddy simulation (LES) results of 3D Taylor-Green vortex obtained by the three different computational approaches: Smoothed Particle Hydrodynamics (SPH), Lattice Boltzmann Method (LBM) and Finite Volume Method (FVM). The Smagorinsky model was chosen as a subgrid-scale closure in LES for all considered methods and a selection of spatial resolutions have been investigated. The SPH and LBM computations have been carried out with the use of the in-house codes executed on GPU and compared, for validation purposes, with the FVM results obtained using the open-source CFD software OpenFOAM. A comparative study in terms of one-point statistics and turbulent energy spectra shows a good agreement of LES results for all methods. An analysis of the GPU code efficiency and implementation difficulties has been made. It is shown that both SPH and LBM may offer a significant advantage over mesh-based CFD methods.

1. Introduction

In recent years, Smoothed Particle Hydrodynamics (SPH) [7] and the Lattice Boltzmann Method (LBM) [9] become more and more popular approaches for incompressible flow simulations. One of the advantages of SPH is its meshless character and suitability to solve multiphase flows with interfaces [11]. The LBM offers easy implementation, computational efficiency and ability to treat complex geometries on regular grids [5]. Another reason for the growing use of both approaches is their adaptivity for massively parallel computations on Graphics Processing Units (GPU). Thus, it is worth making research on turbulence models that would widen the field of applicability of SPH and LBM for high-Re and multiphase flows.

2. Smoothed Particle Hydrodynamics

SPH is a meshfree method [7] that solves the Navier-Stokes equations in the Lagrangian setting of notional fluid particles indentified by their locations $\mathbf{r}(t)$ and velocities $\mathbf{u}(t)$:

$$\frac{d\mathbf{r}}{dt} = \mathbf{u}, \tag{1}$$

$$\frac{d\mathbf{u}}{dt} = -\frac{1}{\rho}\nabla p + \nu\Delta\mathbf{u} + \mathbf{g}, \tag{2}$$

$$\frac{d\rho}{dt} = -\rho\nabla\cdot\mathbf{u}. \tag{3}$$

It consists in a spatial discretisation in terms of moving fluid volumes (particles) and associates with them some special interpolating functions (further called kernel functions or, shortly, kernels). A scalar field A is interpolated at position $\mathbf{r}$ as follows:

$$\hat{A}(\mathbf{r}) = \int_{\Omega} A(\mathbf{r}')W(\mathbf{r}-\mathbf{r}', h)d\mathbf{r}', \tag{4}$$

where Ω is the simulation domain, h is the smoothing length and $W(\mathbf{r}, h)$ is the kernel function. Function W should have the following features:

- $W(\mathbf{r}, h) = W(-\mathbf{r}, h)$,
- $\lim_{h\to 0} W(\mathbf{r}, h) = \delta(\mathbf{r})$, where $\delta(\mathbf{r})$ is the Dirac delta function,
- $\int_{\Omega} W(\mathbf{r}, h)d\mathbf{r} = 1$.

The flow field gradients can be easily obtained from integration by parts using the kernel symmetry:

$$\widehat{\nabla A}(\mathbf{r}) = \int_{\Omega} A(\mathbf{r}')\nabla W(\mathbf{r}-\mathbf{r}', h)d\mathbf{r}'. \tag{5}$$

In the discretised form, the integrals are expressed as sums within the kernel range (kernels of compact support are used to decrease the computational effort):

$$\hat{A}(\mathbf{r}_a) \approx \langle A\rangle_a = \langle A\rangle(\mathbf{r}_a) = \sum_b A(\mathbf{r}_a)W(\mathbf{r}_a-\mathbf{r}_b, h)\Omega_b, \tag{6}$$

$$\widehat{\nabla A}(\mathbf{r}_a) \approx \langle \nabla A\rangle_a = \langle \nabla A\rangle(\mathbf{r}_a) = \sum_b A(\mathbf{r}_a)\nabla W(\mathbf{r}_a-\mathbf{r}_b, h)\Omega_b, \tag{7}$$

where a and b are particles indexes and Ω_b is the volume of particle b. In this paper we present results of weakly compressible SPH (WCSPH) [10] where the pressure field is related to density according to the following equation of state:

$$p = \frac{c_s^2 \rho_0}{\gamma}\left(\left(\frac{\rho}{\rho_0}\right)^{\gamma} - 1\right), \tag{8}$$

where ρ_0 is the density of fluid, γ is a constant (here equal to 7) and c_s denotes the speed of sound. In the SPH formalism eqs. (2) and (3) can take different forms. For eq. (3), due to good stability properties, we decided to choose the form proposed by Hu and Adams [6]:

$$\rho_a = m_a\Theta_a = m_a \sum_b W(\mathbf{r}_a - \mathbf{r}_b, h), \tag{9}$$

where Θ_a denotes the inverse of particle a volume. In eq. (2) the terms corresponding to pressure forces and viscous forces are approximated respectively as follows:

$$\left\langle \frac{\nabla p}{\rho} \right\rangle_a = \frac{1}{m_a}\sum_b \left(\frac{p_a}{\Theta_a^2} + \frac{p_b}{\Theta_b^2}\right)\nabla W(\mathbf{r}_{ab}, h), \tag{10}$$

$$\langle \nu \Delta \mathbf{u}\rangle_a = \frac{1}{m_a}\sum_b \left(\frac{\mu}{\Theta_a^2} + \frac{\mu}{\Theta_b^2}\right)\frac{\mathbf{r}_{ab}\cdot\nabla W(\mathbf{r}_{ab}, h)}{r_{ab}^2 + \eta^2}\mathbf{u}_{ab}, \tag{11}$$

where:

- $\mathbf{r}_{ab} = \mathbf{r}_a - \mathbf{r}_b$ - the distance between particles a and b,
- $\mathbf{u}_{ab} = \mathbf{u}_a - \mathbf{u}_b$, - the relative velocity of particles a and b,
- μ - the dynamic viscosity,
- $\eta = 0.01h$ - a coefficient preventing from singularity.

3. Lattice Boltzmann Method

The LBM is based on the kinetic theory of gases. The discretised Boltzmann equation is solved (instead of the Navier-Stokes equations) for the probability density function of fluid particles moving at specified speeds in specified directions. The spatial discretisation is done on a regular square grid; moreover, only a finite number of directions are allowed for particle velocity [9].

In this paper we present results obtained with a code implementing the D3Q15 lattice, i.e., 3 dimensions and 15 allowed directions. The discretised Boltzmann equation takes the following form:

$$f_\alpha(\mathbf{x} + \mathbf{e}_\alpha \Delta t, t + \Delta t) - f_\alpha(\mathbf{x}, t) = -\frac{1}{\tau}\left(f_\alpha(\mathbf{x}, t) - f_\alpha^{eq}(\mathbf{x}, t)\right), \tag{12}$$

where:

- α - the index of velocity direction ($\alpha = 0, 1, ..., 14$),
- τ - the nondimensional relaxation time,
- $\mathbf{x}$ - position in space,
- $\mathbf{e}_\alpha$ - the lattice velocity in direction α,
- Δt - the time step.

The equilibrium distribution f^{eq} is calculated as follows:

$$f_\alpha^{eq} = w_\alpha \rho \left[1 + \frac{3}{c^2}\mathbf{e}_\alpha \cdot \mathbf{u} + \frac{9}{2c^4}(\mathbf{e}_\alpha \cdot \mathbf{u})^2 + \frac{3}{2c^2}\mathbf{u} \cdot \mathbf{u}\right], \tag{13}$$

where:

- $\rho = \sum_\alpha f_\alpha$ - the fluid density,
- $\mathbf{u} = \rho^{-1} \sum_\alpha \mathbf{e}_\alpha f_\alpha$ - the macroscopic velocity,
- w_α - a weighting coefficient (depending on direction α),
- $c = \Delta x / \Delta t$ - the lattice speed.

The discretised Boltzmann equation is solved in two steps, called respectively the collision step and the propagation step:

$$\widetilde{f_\alpha}(\mathbf{x}, t) = f_\alpha(\mathbf{x}, t) - \frac{1}{\tau}(f_\alpha(\mathbf{x}, t) - f_\alpha^{eq}(\mathbf{x}, t)), \tag{14}$$

$$f_\alpha(\mathbf{x} + \mathbf{e}_\alpha \Delta t, t + \Delta t) = \widetilde{f_\alpha}(\mathbf{x}, t). \tag{15}$$

In our implementation $\Delta x = \Delta t = 1$, what results in $c = 1$. The speed of sound $c_s = 1/\sqrt{3}$. The pressure field is obtained from the following equation of state: $p = c_s^2 \rho$. It can be shown that $\mathbf{u}$ and p satisfy the Navier-Stokes equations with the kinematic viscosity $\nu = (\tau - 0.5)c_s^2$ with error $\mathcal{O}(\mathrm{Ma}^2)$.

4. Large-eddy simulation in SPH and LBM

LES is a method of turbulence modelling that consists in spatial filtering of flow fields [8] that can be implemented as a local increase of kinematic viscosity ν:

$$\nu_{\text{eff}} = \nu + \nu_{\text{turb}}, \tag{16}$$

where ν_{eff} is the effective local viscosity and ν_{turb} denotes a turbulent (or sub-grid scale, SGS) viscosity that in the Smagorinsky model is computed as

$$\nu_{\text{turb}} = (C_S \Delta)^2 |S|, \tag{17}$$

where:

- Δ - the filter size,
- $|S| = \sqrt{2S_{ij}S_{ij}}$, where $S_{ij} = \frac{1}{2}(\frac{\partial u_i}{\partial x_j} + \frac{\partial u_j}{\partial x_i})$ is the strain rate tensor,
- C_S - a constant, most often equal to 0.17.

4.1. LES implementation in LBM

We present now the application of the Smagorinsky SGS model to LBM [4, 12]. In the lattice Boltzmann method, the kinematic viscosity of fluid ν is uniquely linked with the relaxation time τ. The effective relaxation time is determined as:

$$\tau_{\text{eff}} = 3(\nu + \nu_{\text{turb}}) + \frac{1}{2}. \tag{18}$$

It can be shown that

$$\nu_{\text{turb}} = (C_S\Delta)^2|S| = \frac{1}{6}\left(\sqrt{\tau^2 + \frac{18(C_S\Delta)^2 P}{\rho}} - \tau\right), \tag{19}$$

where $P = (2P_{ij}P_{ij})^{\frac{1}{2}}$ with $P_{ij} = \sum_\alpha e_{\alpha i}e_{\alpha j}(f_\alpha - f_\alpha^{eq})$ being the stress tensor. Combining eqs. (18) and (19), we obtain:

$$\tau_{\text{eff}} = \frac{\tau}{2}\left(1 + \sqrt{1 + \frac{18(C_S\Delta)^2 P}{\rho\tau^2}}\right). \tag{20}$$

In our implementation of LES in the LBM we set $\Delta = \Delta x$.

4.2. LES implementation in SPH

In SPH the strain rate tensor needed in eq. (17) is approximated in the same way as other flow fields, i.e., with the use of kernel functions:

$$\langle S_{ij}\rangle_a = \frac{1}{2}m_a\left(\sum_b \frac{u_{ab,i}\nabla_j W(\mathbf{u}_{ab}, h)}{\rho_b} + \sum_b \frac{u_{ab,j}\nabla_i W(\mathbf{u}_{ab}, h)}{\rho_b}\right). \tag{21}$$

Consequently, eq. (11) will now take the following form:

$$\langle \nu\Delta\mathbf{u}\rangle_a = \frac{1}{m_a}\sum_b\left(\frac{\rho_a\nu_{\text{eff},a}}{\Theta_a^2} + \frac{\rho_b\nu_{\text{eff},b}}{\Theta_b^2}\right)\frac{\mathbf{r}_{ab}\cdot\nabla W(\mathbf{r}_{ab}, h)}{r_{ab}^2 + \eta^2}\mathbf{u}_{ab}. \tag{22}$$

The LES filter size is taken equal to the kernel smoothing length in SPH, i.e. $\Delta = h$.

5. Test case and results

As the test case the Taylor-Green Vortex (TGV) was chosen [3]. The computational domain is a periodic cube with edge length $L = 1$ m and the initial velocity amplitude is $U_0 = 1$ m/s. The initial conditions for velocity are defined as follows:

$$\begin{aligned}
u_x(x, y, z, 0) &= U_0\cos(2\pi x)\sin(2\pi y)\sin(2\pi z),\\
u_y(x, y, z, 0) &= -\frac{U_0}{2}\sin(2\pi x)\cos(2\pi y)\sin(2\pi z),\\
u_z(x, y, z, 0) &= -\frac{U_0}{2}\sin(2\pi x)\sin(2\pi y)\cos(2\pi z).
\end{aligned}$$

The TGV flow, due to subsequent nonlinear interactions, develops in time to isotropic turbulence. Recently, the SPH simulation of the 2D TGV at Re=100 was performed by Adami et al. [2] using a modified WCSPH approach.

For LBM simulation the initial pressure (that uniquely determines density) was obtained by solving the Poisson pressure equation with the right-hand side involving gradients of the initial velocity. This step allowed us to avoid transient density oscillations that would occur when initializing the simulation with a constant pressure distribution. In SPH and FVM approaches the initial density was constant. Each method was tested at the Reynolds numbers $Re = 3 \times 10^3$ and $Re = 10^4$ ($Re = U_0 L/\nu$) and at two spatial resolutions: 64^3 and 128^3. Additionally, the LBM simulations at $Re = 3 \times 10^4$ and $Re = 10^5$ at resolution 128^3 were made.

In order to compare the methods in local, i.e. non-statistical sense, temporal evolution of velocity components at position $\mathbf{x} = (L/3, L/5, L/3)$ was investigated at $Re = 10^4$. The results presented in Figure 1 indicate that the TGV flow tends to the isotropic state beyond $t^+ = 4$, say. Figure 2 shows the temporal evolution of the total kinetic energy of the decaying TGV flow; a good agreement among the three methods is noticed.

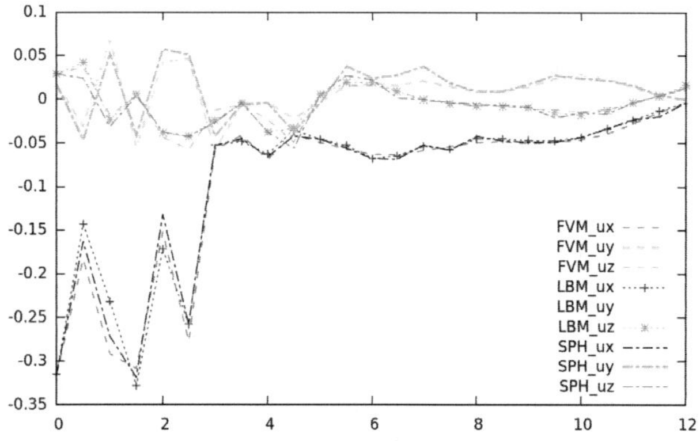

Figure 1. The temporal evolution of normalized velocity components $u_i(t^+)/U_0$ at position $\mathbf{x} = (L/3, L/5, L/3)$. Results for $Re = 10^4$ at resolution 128^3.

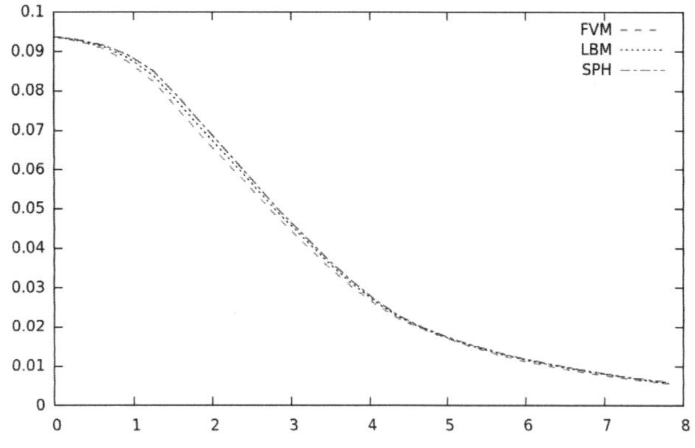

Figure 2. Temporal evolution of total kinetic energy $k(t^+)$. Results for $Re = 10^4$ at resolution 128^3.

For the assesement of the methods, the kinetic energy density spectra

$$E(\kappa) = -\frac{\kappa}{2}\left(\frac{dF_{11}}{d\kappa} + 2\frac{dF_{22}}{d\kappa}\right), \tag{23}$$

were calculated from the 1D longitudinal and transversal spectra, F_{11} and F_{22}, where in general

$$F_{ij}(\kappa) = \frac{1}{\pi}\int_{-\infty}^{\infty} Q_{ij}(\mathbf{r}) \exp(-i\kappa r) dr \tag{24}$$

and $\mathbf{r} = r\mathbf{x}^0$ with $\mathbf{x}^0$ being the x-axis unit vector. The 1D spectra are computed out of the two-point velocity correlation Q_{ij}. In our problem, the ensemble averaging is replaced by the space integral over all the locations $\mathbf{x} \in \Omega$ (because of the flow periodicity, also $(\mathbf{x}+\mathbf{r}) \in \Omega$):

$$Q_{ij}(\mathbf{r}) = \frac{1}{|\Omega|} \int_{\Omega} u_i(\mathbf{x}) u_j(\mathbf{x}+\mathbf{r}) d\Omega. \tag{25}$$

The TGV flow case is deterministic and such an averaging corresponds to what is usually done in the direct numerical simulation (DNS) of isotropic turbulence. The TGV becomes ultimately isotropic in this sense, as ilustrated by the maps of vorticity magnitude (Figure 3). However, there is no large-scale forcing (unlike the DNS), so the flow is unsteady and decaying.

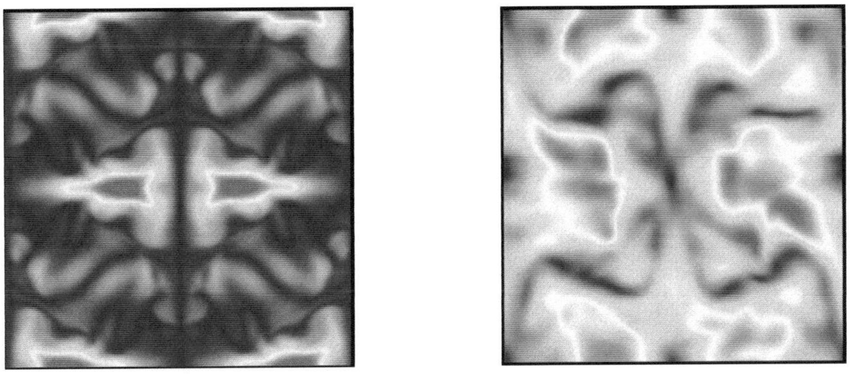

Figure 3. The vorticity magnitude at $t^+ = 3.125$ (left) and $t^+ = 12.5$ (right) in x–z plane at $y = L/2$. Results obtained in the LBM approach at $Re = 10^4$.

In Figures 4 and 5 we present the energy spectra (normalised by the total kinetic energy k) at $Re = 10^4$ and at a dimensionless time $t^+ = 6.25$, where $t^+ = tU_0/L$. The graphs look noisy because the spectra are obtained by means of numerical differentiation. Nevertheless, the characteristic $\kappa^{-\frac{5}{3}}$ slope in the inertial range can be clearly distinguished. The minimal size of eddies recovered by the spectra corresponds to approximately $2.3\Delta x$ and $3.5\Delta x$ for resolutions of 64^3 and 128^3, respectively. A similar agreement of results at $Re = 3 \times 10^3$ was obtained. A comparison of energy spectra at $Re = 3 \times 10^3, 10^4, 3 \times 10^4$ and 10^5 at resolution 128^3 obtained by means of LBM is shown in Figure 6. As the Reynolds number grows the inertial range widens and shorter wavelengths occur, what corresponds with theoretical predictions.

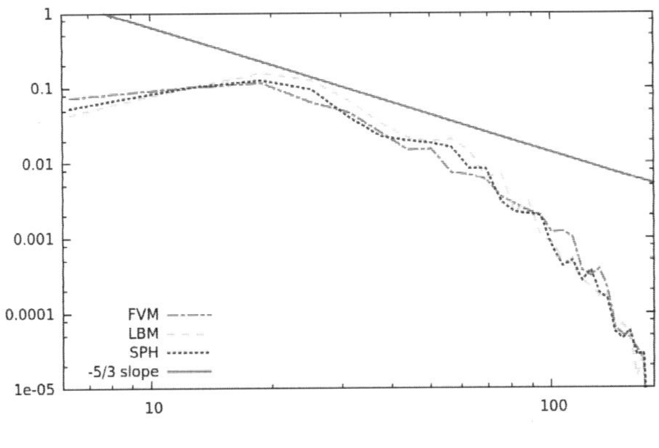

Figure 4. The comparison of kinetic energy spectra $E(\kappa)/k$ at $Re = 10^4$ and resolution 64^3 at dimensionless time $t^+ = 6.25$.

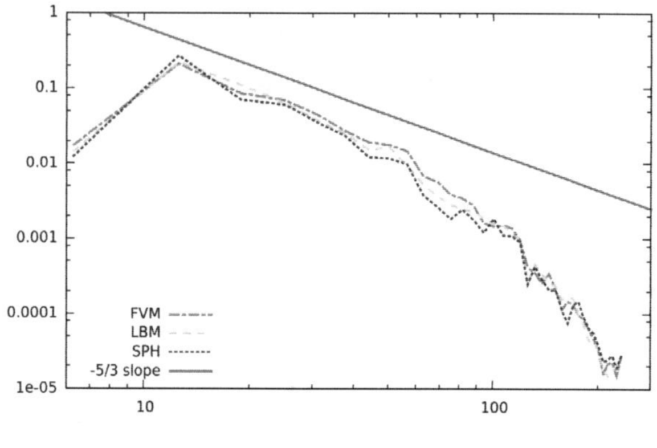

Figure 5. The comparison of kinetic energy spectra $E(\kappa)/k$ at $Re = 10^4$ and resolution 128^3 at dimensionless time $t^+ = 6.25$.

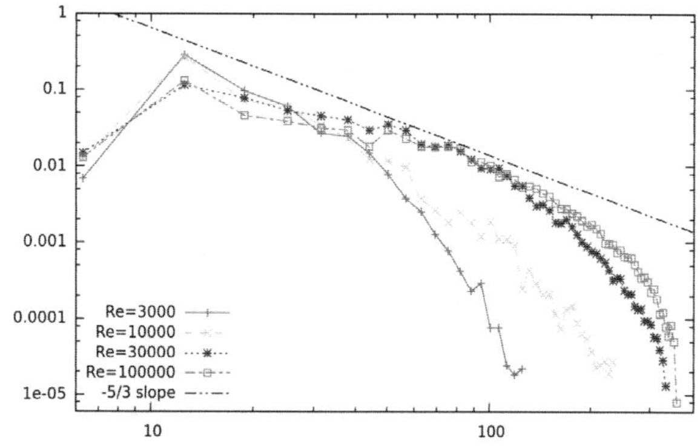

Figure 6. The kinetic energy density spectra at dimensionless time $t^+ = 6.25$ at different Reynolds number values obtained by means of LBM.

6. Computational efficiency and GPU implementation

The explicit and local character of both LBM and SPH makes them perfect for parallel implementation. The in-house codes were written in C++ with the use of NVIDIA-CUDA API and executed on NVIDIA GeForce GTX 780 GPU. The LBM is the fastest among the presented methods. It is also well suited for GPU implementation - the code was written without use of any external libraries neither any user-defined structures, only CUDA built-in array pointers have been used. The SPH method is slower than LBM and its implementation is a bit more difficult. To optimise the execution time it is necessary to use the linked-list structure as well as sorting and hashing algorithms that reduce the computational complexity from $\mathcal{O}(n^2)$ to $\mathcal{O}(n)$. In our code we use the THRUST library that is CUDA implementation of C++ STL (Standard Template Library). The OpenFOAM FVM simulation [1] was executed on 2 cores of Intel Core2Duo CPU and is obviously significantly slower than GPU codes despite the largest timestep used. Figure 7 presents the execution time needed to obtain 1 second of physical time of TGV simulation.

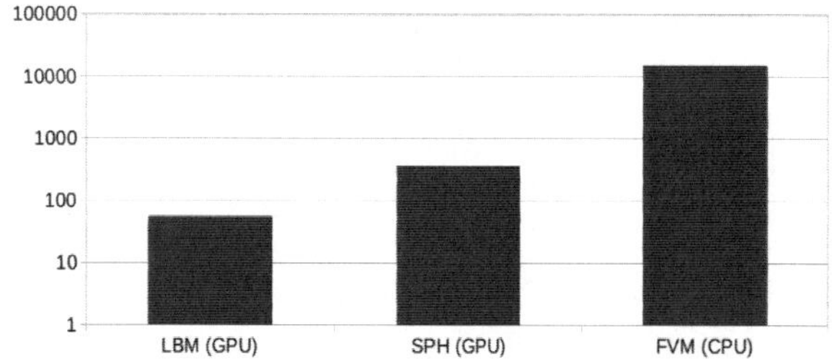

Figure 7. The comparison of the methods efficiency: the wall-clock time (in seconds) needed to obtain 1 s of simulation at resolution 128^3. Timesteps in LBM, SPH and FVM approaches were 0.0003125 s, 0.0005 s and 0.004 s, respectively.

7. Conclusions and perspectives

In the present work, we have analysed the Taylor-Green vortex as a benchmark to validate different computational approaches. The chosen flow case is of interest because of its affinity to isotropic turbulence and LES studies. We have retrieved the TGV flow properties known from the literature in terms of temporal evolution of local velocity components as well as one-point statistical quantities and the kinetic energy spectra. Results are in good correspondence with theoretical predictions as the Reynolds number varies. It has been shown that spatial resolution has significant influence on the results obtained by means of investigated approaches. From this point of view the great advantage of SPH and LBM is the computational complexity, that scales as $\mathcal{O}(n)$ with the problem size.

Promising results presented in this paper and very high efficiency of the codes run on GPU encourage us to further investigation of the LBM and SPH methods. We are planning to test more complex phenomena with these approaches, especially turbulence and two-phase dispersed flows are of interest.

8. References

[1] www.openfoam.com, 2014
[2] Adami S, Hu X Y and Adams N A 2013 A transport-velocity formulation for smoothed particle hydrodynamics *J. Comp. Phys.* **241** 292–307
[3] Aubard G, Volpiani P S, Gloerfelt X and Robinet J C 2013 Comparison of subgrid-scale viscosity models and selective filtering strategy for large-eddy simulations *Flow Turbul. Combust.* **91** 497–518
[4] Chang S C, Yang Y T, Chen C K and Chen W L 2013 Application of the lattice Boltzmann method combined with large-eddy simulations to turbulent convective heat transfer *Int. J. Heat Mass Transfer* **66** 338–48
[5] Grucelski A and Pozorski J 2013 Lattice Boltzmann simulations of flow past an obstacle and in simple porous media *Comp. & Fluids* **71** 406–16
[6] Hu X Y and Adams N A 2006 A multi-phase SPH method for macroscopic and mesoscopic flows *J. Comp. Phys.* **213** 844-861
[7] Monaghan J J 2012 Smoothed Particle Hydrodynamics and its diverse applications *Annu. Rev. Fluid Mech.* **44** 323–46
[8] Sagaut P and Grohens R 1999 Discrete filters for large eddy simulation *Int. J. For Numer. Meth. In Fluids* **31** 1195–220
[9] Succi S 2001 *The Lattice Boltzmann Method for Fluid Dynamics and Beyond* (Oxford: Clarendon Press)
[10] Szewc K, Pozorski J and Minier J P 2011 Analysis of the incompressibility constraint in the Smoothed Particle Hydrodynamics method *Int. J. Numer. Meth. Engng* **92** 343–69
[11] Szewc K, Pozorski J and Minier J P 2013 Simulations of single bubbles rising through viscous liquids using Smoothed Particle Hydrodynamics *Int. J. Multiphase Flow* **50** 98–105
[12] Yu X, Tang X, Wang W et al. 2010 A lattice Boltzmann model coupled with a Large Eddy Simulation model for flows around a groyne *Int. J. Sediment Res.* **25** 271–82

Acknowledgments

This research was supported by the National Science Centre (NCN, Cracow, Poland) through the research project 2011/03/B/ST8/05677.

Journal of Physics: Conference Series **530** (2014) 012020 doi:10.1088/1742-6596/530/1/012020

Vorticity generation in compressible multiphase flows

A Ballil, A F Nowakowski[1], S Jolgam and F C G A Nicolleau
Sheffield Fluid Mechanics Group, Department of Mechanical Engineering, University of Sheffield, Mappin Street, Sheffield S1 3JD, United Kingdom

E-mail: a.f.nowakowski@sheffield.ac.uk

Abstract. The simulations of flows in inhomogeneous media of various physical regimes leading to shock-bubble interactions were performed using a developed numerical code based on a multi-component flow model. The numerical method which considers interfaces represented by contact discontinuities as numerically diffused zones, has been applied to simulate compressible two-phase flows. The approach takes advantage of the inherent numerical diffusion present in solutions. The mathematical formulation of the presented method is obtained after an averaging process of the single phase Navier-Stokes equations and contains the non-conservative equations and non-conservative terms that exist in the model to fulfill the interface condition. The finite volume Godunov-type computational technique, equipped with an approximate Riemann solver for calculating fluxes, is applied to simulate flows in two space dimensions. The approach accounts for pressure non-equilibrium. It resolves interfaces separating compressible fluids and captures the baroclinic source of vorticity generation. A numerically challenging shock bubble interaction problem is investigated to evaluate the effect of the Atwood number and shock wave intensity (various Mach numbers) on the interface evolution and vorticity generation.

1. Introduction

Many engineering processes and systems involve the dynamics of high speed compressible multiphase flows. Good examples of such flows with significant density variations, are the supersonic combustion in aircraft engines [1] and the inertial confinement fusion (ICF) [2], [3]. The flow pattern is diverse and the dynamical interactions of phases separated by a free surface significantly complicate the mathematical description of the inherently multi-component flow problem. The propagation of a shock wave through this type of inhomogeneous media leads to baroclinic vorticity generation and the development of the Richtmyer-Meshkov instability (RMI) [4]. This hydrodynamic instability grows nonlinearly and subsequently shows transition to turbulence without any wall effects. Vorticity generation is proportional to the misalignment between the density gradient at the fluid interface and induced upstream-directed pressure gradients related to a shock wave.

Numerous experiments have been carried out to investigate the initial stages of this process. The interaction of a planar shock wave with a single spherical or cylindrical bubble has been a typical experimental setup chosen to study the interface perturbation and deformation. The studies have been performed using shadowgraph diagnostic techniques as in the work of Haas & Sturtevant [5] and the series of experiments of Layes *et al*, see for example [6]. The other experiments applied modern laser technologies [7]. All these experiments allowed for better understanding of the mechanism of the

[1] To whom any correspondence should be addressed.

shock-bubble interaction and provided the reference data for computational studies, which have been performed using different mathematical models and numerical methods.

For the purpose of this study a diffuse interface approach has been developed to mathematically represent and numerically simulate flows with interfaces separating components of various density ratios. The method considers interfaces as numerically diffused zones, taking advantage of inherent numerical diffusion. This is in contrast to other approaches that consider interfaces between flow components as sharp non-smeared discontinuities, see for example [8]. The formulation is based on the model known, in its one-dimensional form, as the six-equation model of Saurel *et al* [9]. It was first derived by Kapila *et al* [10] from the generic model of Baer and Nunziato [11]. A brief introduction of the model as well as of the numerical method is provided in the next sections. This is followed by the general description of the mechanism of vorticity generation, the numerical results and conclusions.

2. Multi-component flow model

The original six equation model formerly derived by [10] in one dimensional space is now considered in two dimensional space. It consists of a statistical volume fraction equation, two continuity equations, mixture momentum equation in x and y direction and two energy equations. This model is characterized by two pressures attributed to each phase and a single velocity. It is therefore considered as a pressure non-equilibrium model, which requires further relaxation. Therefore, its solution discussed in the following section is based on a predictor-corrector, Godunov-type method and a diffuse-interface approach for the material interfaces. The model for compressible flows without mass and heat transfer can be reduced to:

$$\begin{aligned}
&\frac{\partial \alpha_1}{\partial t} + u\frac{\partial \alpha_1}{\partial x} + v\frac{\partial \alpha_1}{\partial y} = \mu(p_1 - p_2),\\
&\frac{\partial \alpha_1\rho_1}{\partial t} + \frac{\partial \alpha_1\rho_1 u}{\partial x} + \frac{\partial \alpha_1\rho_1 v}{\partial y} = 0,\\
&\frac{\partial \alpha_2\rho_2}{\partial t} + \frac{\partial \alpha_2\rho_2 u}{\partial x} + \frac{\partial \alpha_2\rho_2 v}{\partial y} = 0,\\
&\frac{\partial \rho u}{\partial t} + \frac{\partial[\rho u^2 + (\alpha_1 p_1 + \alpha_2 p_2)]}{\partial x} + \frac{\partial \rho uv}{\partial y} = 0,\\
&\frac{\partial \rho v}{\partial t} + \frac{\partial \rho uv}{\partial x} + \frac{\partial[\rho v^2 + (\alpha_1 p_1 + \alpha_2 p_2)]}{\partial y} = 0,\\
&\frac{\partial \alpha_1\rho_1 e_1}{\partial t} + \frac{\partial \alpha_1\rho_1 e_1 u}{\partial x} + \frac{\partial \alpha_1\rho_1 e_1 v}{\partial y} + \alpha_1 p_1\frac{\partial u}{\partial x} + \alpha_1 p_1\frac{\partial v}{\partial y} = -p_i\mu(p_1 - p_2),\\
&\frac{\partial \alpha_2\rho_2 e_2}{\partial t} + \frac{\partial \alpha_2\rho_2 e_2 u}{\partial x} + \frac{\partial \alpha_2\rho_2 e_2 v}{\partial y} + \alpha_2 p_2\frac{\partial u}{\partial x} + \alpha_2 p_2\frac{\partial v}{\partial y} = p_i\mu(p_1 - p_2),
\end{aligned} \tag{1}$$

where α_k, ρ_k, p_k, e_k, ρ, u, v and p are the volume fraction, the density, the pressure, the specific internal energy of the phase k, the mixture density, mixture velocity components and mixture pressure, respectively. The variable μ is a homogenization parameter controlling the rate at which the pressure tends towards equilibrium. An additional equation has been derived by [9] and considered in the numerical solution of the model (1) to eliminate the errors in the values of the thermodynamic variables due to the numerical approximation of the two non-conservative internal energy equations in the presence of shocks. This equation is obtained by summing up the two internal energy equations with mass and momentum equations and takes the following final form:

$$\frac{\partial \rho E}{\partial t} + \frac{\partial u(\rho E + \alpha_1 p_1 + \alpha_2 p_2)}{\partial x} + \frac{\partial v(\rho E + \alpha_1 p_1 + \alpha_2 p_2)}{\partial y} = 0, \tag{2}$$

where the mixture total energy is given by:

$$E=\frac{\alpha_1\rho_1}{\rho}e_1+\frac{\alpha_2\rho_2}{\rho}e_2+\frac{1}{2}u^2+\frac{1}{2}v^2, \tag{3}$$

The stiffened gas equation of state relates the thermodynamic variables and has the following form:

$$p_k=(\gamma_k-1)\rho_k e_k-\gamma_k\pi_k, \tag{4}$$

where γ_k and π_k are respectively the heat capacity ratio and the pressure constant for the phase k.The stiffened equation of state for the mixture can be written in the following compact form:

$$p=\frac{\rho e-\sum_k^2(\frac{\alpha_k\gamma_k\pi_k}{\gamma_k-1})}{\sum_k^2(\frac{\alpha_k}{\gamma_k-1})}. \tag{5}$$

3. Numerical method

The model (1) is hyperbolic but non-conservative, i.e. it cannot be written in the divergence form. Therefore the standard numerical methods developed for solving the hyperbolic conservation laws are not applied directly. In order to solve the model (1) the usual numerical procedure is to decompose the governing equations into hyperbolic and relaxation parts and solve them in succession of operations. The hyperbolic part of the model (1) can be rewritten as:

$$\begin{aligned}
&\frac{\partial\alpha_1}{\partial t}+u\,\frac{\partial\alpha_1}{\partial x}+v\frac{\partial\alpha_1}{\partial y}=0,\\
&\frac{\partial\alpha_1\rho_1e_1}{\partial t}+\frac{\partial\alpha_1\rho_1e_1u}{\partial x}+\frac{\partial\alpha_1\rho_1e_1v}{\partial y}+\alpha_1p_1\frac{\partial u}{\partial x}+\alpha_1p_1\frac{\partial v}{\partial y}=0,\\
&\frac{\partial\alpha_2\rho_2e_2}{\partial t}+\frac{\partial\alpha_2\rho_2e_2u}{\partial x}+\frac{\partial\alpha_2\rho_2e_2v}{\partial y}+\alpha_2p_2\frac{\partial u}{\partial x}+\alpha_2p_2\frac{\partial v}{\partial y}=0,\\
&\frac{\partial U}{\partial t}+\frac{\partial F(U)}{\partial x}+\frac{\partial G(U)}{\partial y}=0,
\end{aligned} \tag{6}$$

where the conserved variables U and the fluxes $F(U)$ and $G(U)$ are given in a vector form as:

$$U=\begin{bmatrix}\alpha_1\rho_1\\ \alpha_2\rho_2\\ \rho u\\ \rho v\\ \rho E\end{bmatrix},\quad F(U)=\begin{bmatrix}\alpha_1\rho_1u\\ \alpha_2\rho_2u\\ \rho u^2+p\\ \rho uv\\ u(\rho E+p)\end{bmatrix}\quad\text{and}\quad G(U)=\begin{bmatrix}\alpha_1\rho_1v\\ \alpha_2\rho_2v\\ \rho uv\\ \rho v^2+p\\ v(\rho E+p)\end{bmatrix}. \tag{7}$$

The elements in the last row of the above vectors in (7) represent the additional equation of the total mixture energy (2). The hyperbolic part (6) is solved using a Godunov-type scheme. The explicit first-order Godunov scheme can be written as:

$$\begin{aligned}
U_{i,j}^{n+1}=U_{i,j}^{n}&-\frac{\Delta t}{\Delta x}[F(U^*(U_{i,j}^n,U_{i+1,j}^n))-F(U^*(U_{i-1,j}^n,U_{i,j}^n))]\\
&-\frac{\Delta t}{\Delta y}[G(U^*(U_{i,j}^n,U_{i,j+1}^n))-G(U^*(U_{i,j-1}^n,U_{i,j}^n))].
\end{aligned} \tag{8}$$

Equation (8) is applied to the mass, momentum and total energy equations, where the superscript * in (8) refers to the state at cell boundaries. The discretisation of the non-conservative equation for the volume fraction is:

$$\alpha_{1i,j}^{n+1} = \alpha_{1i,j}^{n} - \frac{\Delta t}{\Delta x} u_{i,j}[(u\alpha_1)_{i+1/2,j}^{*} - (u\alpha_1)_{i-1/2,j}^{*} - \alpha_{1i,j}^{n}(u_{i+1/2,j}^{*} - u_{i-1/2,j}^{*})] - \frac{\Delta t}{\Delta y} v_{i,j}[(v\alpha_1)_{i,j+1/2}^{*} - (v\alpha_1)_{i,j-1/2}^{*} - \alpha_{1i,j}^{n}(v_{i,j+1/2}^{*} - v_{i,j-1/2}^{*})]. \tag{9}$$

The non-conservative equations for the internal energy are discretised as follows:

$$(\alpha\rho e)_{ki,j}^{n+1} = (\alpha\rho e)_{ki,j}^{n} - \frac{\Delta t}{\Delta x} u_{i,j}[(\alpha\rho eu)_{ki+1/2,j}^{*} - (\alpha\rho eu)_{ki-1/2,j}^{*} - (\alpha p)_{ki,j}^{n}(u_{i+1/2,j}^{*} - u_{i-1/2,j}^{*})] - \frac{\Delta t}{\Delta y} v_{i,j}[(\alpha\rho ev)_{ki,j+1/2}^{*} - (\alpha\rho ev)_{ki,j-1/2}^{*} - (\alpha p)_{ki,j}^{n}(v_{i,j+1/2}^{*} - v_{i,j-1/2}^{*})] \tag{10}$$

In the present work the second order accuracy is achieved by using the MUSCL scheme. The HLL approximate Riemann solver is employed to calculate the numerical fluxes.

When the hyperbolic part of the solution algorithm is accomplished, the pressure is modified by relaxation solver based on the instantaneous pressure relaxation assumption, where the variable μ for this purpose is considered to be infinite. This part of the solution requires an iterative procedure to solve the ordinary differential equations containing the source terms of the original model (1):

$$\frac{\partial}{\partial t}\begin{bmatrix} \alpha_1 \\ \alpha_1\rho_1 \\ \alpha_2\rho_2 \\ \rho u \\ \rho v \\ \alpha_1\rho_1 e_1 \\ \alpha_2\rho_2 e_2 \end{bmatrix} = \begin{bmatrix} \mu(p_1 - p_2) \\ 0 \\ 0 \\ 0 \\ 0 \\ -\mu p_i(p_1 - p_2) \\ \mu p_i(p_1 - p_2) \end{bmatrix} \tag{11}$$

The new value of pressure resulting in the updated value of the volume fraction and the mixture energy ρe calculated from (2) are used to obtain the mixture pressure from the mixture equation of state (5). The new pressure can be now used to update the internal energies for each constituent with the help of the corresponding equations of state.

4. Mechanism of vorticity generation

The acceleration of a gas bubble by a plane shock wave causes a number of complex phenomena such as bubble compression and acceleration, interface instabilities and vorticity generation. The main source of vorticity generation in such cases is the misalignment of the pressure and the density gradients, which is known as the baroclinic vorticity deposition. Figure 1 shows a schematic diagram of the possible scenarios when a planar shock wave propagating initially in air passes through a heavier or lighter cylindrical gas bubble. To explain the evolution of the vorticity field, it is convenient to consider Euler equations written in a vorticity transport form. In 2D form the equation is:

$$\frac{D\omega}{Dt} = -\omega(\nabla\cdot U) + \frac{1}{\rho^2}(\nabla\rho\times\nabla p) \tag{12}$$

The only term of importance in the early stages of a shock-bubble interaction for which ω is initially equal to zero is the baroclinic torque ($\nabla\rho \times \nabla p$). The misalignment of the local pressure and density gradients leads to the non-zero source term in equation (12). The baroclinic torque is the largest if the pressure gradient is perpendicular to the density gradient. Whereas, at the most upstream and downstream poles of the gas bubble, the baroclinic torque is equal to zero, owing to the collinearity of density and pressure gradients. The curvature of the shock wave front (the refracted shock wave) has also been used to build a theory behind the vorticity generation This mechanism is based on the conservation of the tangential velocity and the angular momentum across the shock wave as the compression only affects the motion normal to the shock surface (for details see [12] and [13]).

Journal of Physics: Conference Series **530** (2014) 012020 doi:10.1088/1742-6596/530/1/012020

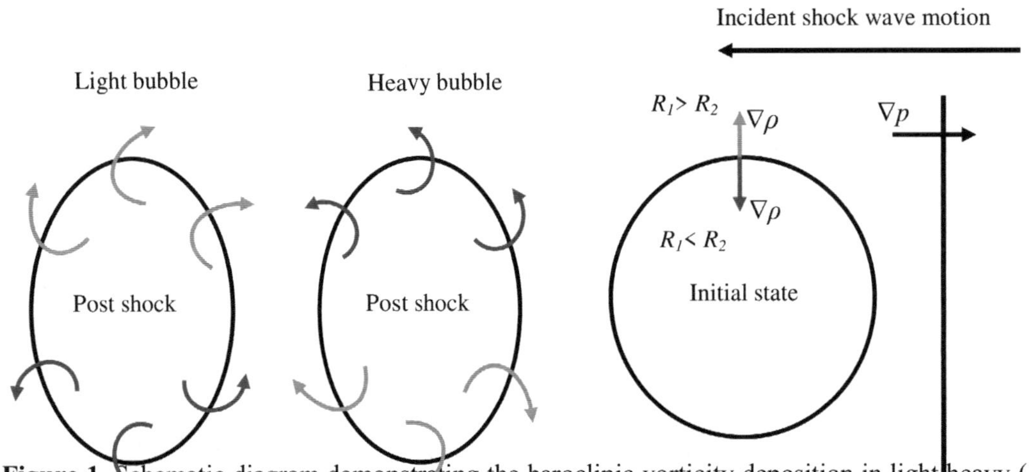

Figure 1. Schematic diagram demonstrating the baroclinic vorticity deposition in light/heavy ($R_1 < R_2$) and heavy/light ($R_1 > R_2$) shock-bubble interaction scenarios. R_1 is the acoustic impedance of the air and R_2 refers to the acoustic impedance of the bubble.

5. Numerical results

Numerical simulations of different scenarios of the interaction of a planar shock wave propagating in the ambient air and encountering a single circular bubble are conducted. Three air/bubble arrangements that represent heavy/light, equal density and light/heavy are considered. The configurations include air/krypton (Kr), air/nitrogen (N_2) and air/helium (He), which represent negative, close to zero and positive Atwood number A respectively. The Atwood number is expressed in terms of density ratio as follows:

$$A = \frac{(\rho_b - \rho_s)}{(\rho_b + \rho_s)} \tag{13}$$

where, ρ_b and ρ_s are the densities of the gas filling the bubble and the surrounding medium respectively. The simulations consider shock waves of Mach numbers (M = 1.5, 1.7 and 2.0). The physical domain and the initial conditions are illustrated in figure 2 and tables 1 and 2.

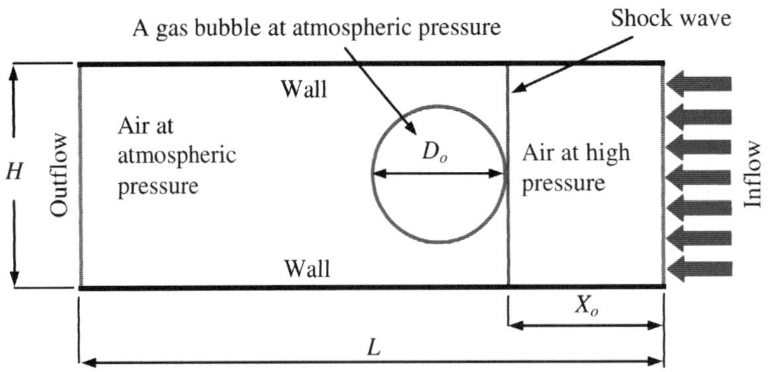

Figure 2. Schematic diagram of the initial state of shock-bubble interaction simulations.

Table 1. Initial conditions of air and different gas bubbles in the pre-shock chamber.

Physical Property	**Air**	**He**	**N_2**	**Kr**
Density, kg/m^3	1.29	0.167	1.25	3.506
Pressure, Pa	101325	101325	101325	101325
Horizontal velocity, m/s	0.0	0.0	0.0	0.0
Vertical velocity, m/s	0.0	0.0	0.0	0.0
Sound speed, m/s	332	1005	337	220
Heat capacity ratio γ	1.4	1.67	1.4	1.67

Table 2. Initial conditions of air in the post-shock chamber.

Physical Property	$M = 1.5$	$M = 1.7$	$M = 2$
Density, kg/m^3	2.4021	2.835	3.44
Pressure, Pa	249091	324747	455963
Horizontal velocity, m/s	230.28	374.55	414.5
Heat capacity ratio, γ	1.4	1.4	1.4

The characteristic dimensions of the domain as shown in figure 2 are: $L = 30$ cm, $H = 8$ cm, $X_o = 5$ cm and $D_o = 4$ cm. The initial and boundary conditions of these computational cases reflect the experimental setup of Layes and Le Métayer [14]. The present computations were performed using 1800×480 computational cells. The Schlieren images of the structures obtained in the experiment and computational simulations are in good agreement. Although it has to be remembered that the experiments involves a spherical bubble and the computational results are restricted to the cylindrical bubble. The results of the interface evolution as a function of time for the different air/gas bubble arrangements are shown in figure 3 for the $M = 1.5$.

The left plots in figure 3 show the evolution of the Kr bubble interface. The initial Atwood number is 0.462 and this case represents light/heavy arrangement. After the planar shock wave impacts the Kr bubble, a refracted shock propagates inside the bubble, the outer parts of the incident shock propagate outside the bubble and a reflected shock propagates back towards the right side. The acoustic impedance inside the bubble is higher than the acoustic impedance outside ($R_2 > R_1$). This causes the refracted shock wave, inside the bubble to travel more slowly than the incident shock outside the bubble, figure 3 at time 47 μs. A sheet of positive vorticity along the upper surface and another sheet of negative vorticity along the lower surface correspond to the interface formed behind the shock wave. By the time 166 μs the vortices on the top and bottom of the bubbles become more visible and a penetrating jet is formed at the downstream pole of the Kr bubble. As the time goes the vortical flow dominates the interface deformation process.

The middle plots in figure 3 present the interaction of a planar shock wave with N_2 bubble. The Atwood number in this case is close to zero and the difference in the acoustic impedance is small. In this case the speed difference of the refracted shock wave and the incident shock wave is small as it can be seen in figure 3. The N_2 bubble undergoes a compression process, and the vorticity production in this case is negligible.

The right plots in figure 3 show the snapshots of the interaction of a shock wave with the He bubble. The case represents an opposite scenario to the case of the Kr bubble. In this experiment the Atwood number is equal to -0.77 which is representative to the heavy/light arrangement. The shock wave hits the right interface of the He bubble causing a refracted shock wave to propagate inside the bubble faster than the incident shock wave outside the bubble, as a result of the acoustic impedance misalignment (in this scenario $R_2 < R_1$). In this scenario a rarefaction wave reflects and propagates back towards the right side. The bubble interface deforms and a penetrating jet acting along the line of symmetry in the main flow direction is formed. The upstream pole of the bubble moves towards the downstream one and the bubble interface evolves into a kidney shape.The penetrating high velocity jet along the flow direction moves through the bubble forming symmetric patterns, figure 3. When the bubble deforms the associated flow field is subsequently split into two rings of vorticity.

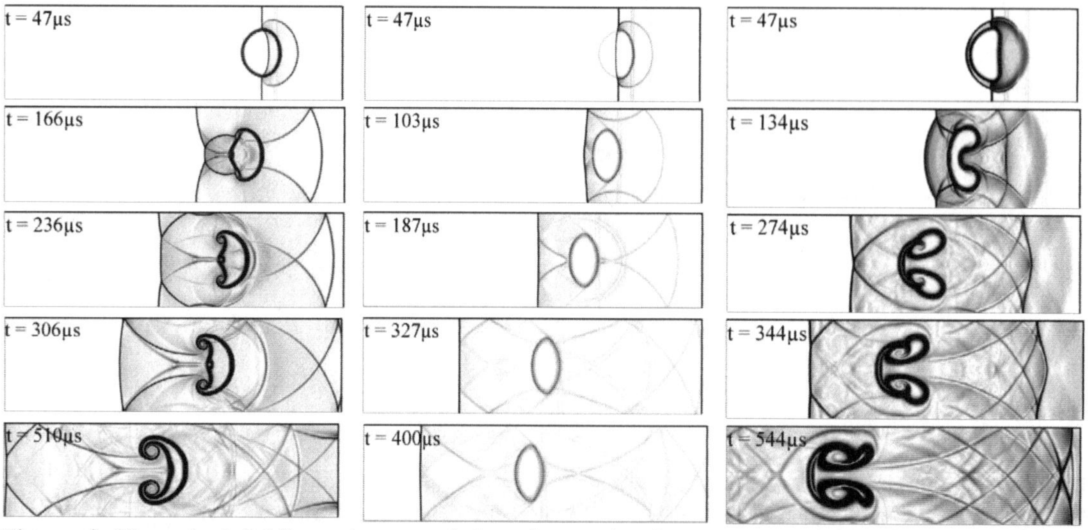

Figure 3. Numerical Schlieren images of the mixture density contours representing the interface evolution with respect to time for air/Kr (left), air/N_2 (centre) and air/He (right) at M = 1.5.

The intensity of the vortex rings in the cases of He and Kr bubbles is expressed by the total circulation Γ. The calculated values of the circulation in the symmetrical half of the computational domain with respect to time are represented in figure 4 for all Mach numbers considered. The values of the circulation show that: if the Mach number increases the values of the circulation also increase and their growth rate during the shock-bubble interaction process is increased too. In the case of the relatively high absolute values of the Atwood number the vorticity generation becomes higher as well.

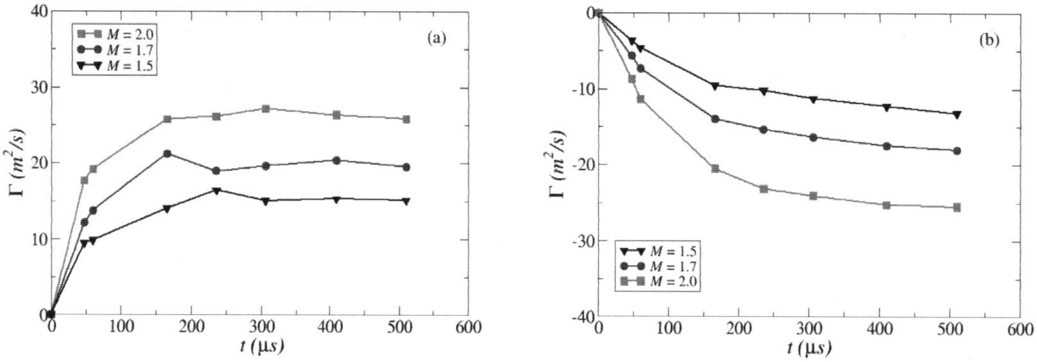

Figure 4. Circulation time evolution of air/He (a) and air/Kr (b) constitutions for different shock wave celerities.

6. Conclusions

Numerical simulations were performed to capture vorticity generation and the interface evolution in multi-component compressible flows. The study considered the effect of Atwood and Mach numbers on vorticity generation resulting from the interaction of a planar shock wave with a single cylindrical bubble. The constant Mach number comparison demonstrated that an increase of the Atwood number leads to a higher vorticity generation and its effect on the interface evolution becomes more pronounced. The constant Atwood number comparison proved that the flow with increasing Mach number produces higher circulation and vorticity generation.

7. References

[1] Yang J, Kubota T and Zukoski E E 1993 Applications of shock-induced mixing to supersonic combustion *AIAA Journal* **31** pp 854-862

[2] Lindl J D, McCrory R L and Campbell E M 1992 Progress toward ignition and burn propagation in inertial confinement fusion *Physics Today* **45** pp 32-40

[3] Aglitskiy Y; Velikovich A L; Karasik M; Metzler N; Zalesak S T; Schmitt A J; Phillips L; Gardner J H; Serlin V; Weaver J L and Obenschain S P 2010 Basic hydrodynamics of Richtmyer-Meshkov-type growth and oscillations in the inertial confinement fusion-relevant conditions *Philosophical Transactions of the Royal Society A: Mathematical,Physical and Engineering Sciences* **368** pp 1739-1768

[4] Brouillette M 2002 The Richtmyer-Meshkov instability *Annual Review of Fluid Mechanics* **181** 41

[5] Haas J and Sturtevant B 1987 Interaction of weak shock waves with cylindrical and spherical gas inhomogeneities *Journal of Fluid Mechanics* **181** pp 41-76

[6] Layes G, Jourdan G and Houas L 2009 Experimental study on a plane shock wave accelerating a gas bubble *Physics of Fluids* **21** 074102

[7] Ranjan D, Niederhaus J, Motl B, Anderson M, Oakley J and Bonazza R 2007 Experimental investigation of primary and secondary features in high Mach number shock-bubble interaction *Physical Review Letters* **98** 024502

[8] Terashima H and Tryggvason G 2009 A front-tracking/ghost-fluid method for fluid interfaces in compressible flows. *Journal of Computational Physics* **228** pp 4012–4037

[9] Saurel R, Petitpas F and Berry R A 2009 Simple and efficient relaxation methods for interfaces separating compressible fluids, cavitating flows and shocks in multiphase mixtures *Journal of Computational Physics* **228** pp 1678–1712

[10] Kapila A K, Menikoff R, Bdzil J B, Son S F and Stewart D S 2001 Two-phase modeling of deflagration to detonation transition in granular materials: Reduced equations *Physics of Fluids* **13** pp 3002-3024

[11] Baer M K and Nunziato J W 1986 A two-phase mixture theory for the deflagration-to-detonation transition (DDT) in reactive granular materials *International Journal of Multiphase Flow* **12** 861-888

[12] Kevlahan N K -R 1997 The vorticity jump across a shock in a non-uniform flow *Journal of Fluid Mechanics* **341** pp 371-384

[13] Huete Ruiz de Lira C, Velikovich A L and Wouchuk J G 2011 Analytical linear theory for the interaction of a planar shock wave with a two or three-dimensional random isotropic density field *Physical Review E* **83** 056320

[14] Layes G and Le Métayer O 2007 Quantitative numerical and experimental studies of the shock accelerated heterogeneous bubbles motion *Physics of Fluids* **19** 042105

Journal of Physics: Conference Series **530** (2014) 012021 doi:10.1088/1742-6596/530/1/012021

Numerical study of the vortex tube reconnection using vortex particle method on many graphics cards

Henryk Kudela and Andrzej Kosior

Wroclaw University of Technology, Wybrzeze Wyspianskiego 27, 50-370 Wroclaw, Poland

E-mail: henryk.kudela@pwr.wroc.pl, andrzej.kosior@pwr.wroc.pl

Abstract. Vortex Particle Methods are one of the most convenient ways of tracking the vorticity evolution. In the article we presented numerical recreation of the real life experiment concerning head-on collision of two vortex rings. In the experiment the evolution and reconnection of the vortex structures is tracked with passive markers (paint particles) which in viscous fluid does not follow the evolution of vorticity field. In numerical computations we showed the difference between vorticity evolution and movement of passive markers. The agreement with the experiment was very good. Due to problems with very long time of computations on a single processor the Vortex–in–Cell method was implemented on the multicore architecture of the graphics cards (GPUs). Vortex Particle Methods are very well suited for parallel computations. As there are myriads of particles in the flow and for each of them the same equations of motion have to be solved the SIMD architecture used in GPUs seems to be perfect. The main disadvantage in this case is the small amount of the RAM memory. To overcome this problem we created a multiGPU implementation of the VIC method. Some remarks on parallel computing are given in the article.

1. Introduction

Understanding the dynamics and mutual interaction of various types of vortical motions is the key ingredient in clarifying and controlling fluid motions. One of the most fundamental 3D vortical interactions is related to vortex tube reconnection. Reconnection is regarded as a one fundamental mechanism that increases the complexity of the flow what is important in the study of the turbulence phenomena [14].

In most of the real life experiments the vorticity evolution is tracked using some kind of passive markers (like paint particles). But in viscous fluid the particles does not follow the vorticity exactly [4]. In this situations the numerical simulation can be used.

As a numerical method we chose the 3D Vortex–in–Cell (VIC) method [1]. In this method particles carry information about vorticity. It is well known that the velocity may be calculated from the vorticity distribution. Next the vortex particles are displaced according to the local velocity field. Particles intensities are then interpolated back to the grid nodes. To simulate the effect of the viscosity the viscous splitting algorithm was used.

Numerical solution of the 3D Navier – Stokes equations for high Reynolds number, using any method, is a very time consuming process. It may be noticed that recently there is not much increase in the computational power of a single processor. We are forced to use multicore architectures and parallel computation.

The VIC method is very well suited for parallel computation. The displacement and redistribution processes, which have to be done at each time step, have a local character and the computations for each particle can be done independently. So the whole set of particles can be divided into independent groups and operations over these groups can be done concurrently.

Graphics Processing Units (GPU) that were developed for video games provide cheap and easily accessible hardware for scientific calculations. It is build of hundreds of simple streaming processors which altogether give a great computational power. GPUs are relatively cheap and commonly available but in order to take advantage of its potential one needs to use proper numerical algorithms.

In the next chapter the equations of the motion and description of the vortex particle method were be given. In chapter 3 we discussed problems connected to the parallel computations. In chapter 4 we presented a numerical example of a head-on vortex rings collision. The last chapter are conclusions.

2. Equations of the motion and description of the vortex particle method

Equations of incompressible and viscous fluid motion have the following form:

$$\frac{\partial \vec{u}}{\partial t} + (\vec{u} \cdot \nabla)\vec{u} = -\frac{1}{\rho}\nabla p + \nu \Delta \vec{u}, \tag{1}$$

$$\nabla \cdot \vec{u} = 0. \tag{2}$$

where $\vec{u} = (u, v, w)$ is velocity vector, ρ is fluid density, p is pressure, ν is kinematic viscosity. Equation (1) can be transformed to the Helmholtz equation for vorticity evolution [18]:

$$\frac{\partial \vec{\omega}}{\partial t} + (\vec{u} \cdot \nabla)\vec{\omega} = (\vec{\omega} \cdot \nabla)\vec{u} + \nu \Delta \vec{\omega}, \tag{3}$$

where $\vec{\omega} = \nabla \times \vec{u}$.

Generally in all vortex particle methods the viscous splitting algorithm is used [3]. The solution is obtained in two steps: first, the inviscid - Euler equation is solved.

$$\frac{\partial \vec{\omega}}{\partial t} + (\vec{u} \cdot \nabla)\vec{\omega} = (\vec{\omega} \cdot \nabla)\vec{u}. \tag{4}$$

Next, the viscosity effect is simulated by solving the diffusion equation

$$\frac{\partial \vec{\omega}}{\partial t} = \nu \Delta \vec{\omega}, \tag{5}$$

$$\vec{\omega}(\vec{x}, 0) = \vec{\omega}_I, \tag{6}$$

where $\vec{\omega}_I$ is vorticity distribution obtained after inviscid step.

For the solution of the equations (5) and (6) one can use any suitable method like the Particle Strength Exchange (PSE) method [1] or the Finite Difference method. In the present paper we solved equations (5) and (6) using the implicit finite difference scheme.

In inviscid flow (4), according to the third Helmholtz theorem [18], vorticity lines move as the material particles of fluids. Thus the movement of the vortex particles can be described by the infinite set of ordinary differential equations:

$$\frac{d\vec{x}_p}{dt} = \vec{u}(\vec{x}_p, t), \qquad \vec{x}(0, \vec{c}) = \vec{c}. \tag{7}$$

where $\vec{c} = (c_1, c_2, c_3)$ means the Lagrange coordinates of fluid particles. Solution of the equation (7) gives the particle-trajectory mapping $\Phi(\cdot, t) : R^3 \to R^3; \vec{c} \to \Phi(\vec{c}, t) = \vec{x} \in R^3$ that is one–to–one and onto. Incompressibility implies that

$$det(\nabla_c \Phi(\vec{c}, t)) = 1. \tag{8}$$

The velocity can be express through vorticity distribution using the Biot–Savart law

$$\vec{u}(\vec{x}, t) = \int \vec{K}(\vec{x} - \vec{x}')\vec{\omega}(\vec{x}', t)d\vec{x}' = (\vec{K} * \vec{\omega})(\vec{x}, t), \tag{9}$$

where $*$ denotes convolution and $\vec{K}$ is a 3×3 matrix kernel

$$\vec{K}(\vec{x}) = \frac{1}{4\pi}\frac{1}{|\vec{x}|^3}\begin{bmatrix} 0 & x_3 & -x_2 \\ -x_3 & 0 & x_1 \\ x_2 & -x_1 & 0 \end{bmatrix}, \qquad |\vec{x}| = \sqrt{x_1^2 + x_2^2 + x_3^2}. \tag{10}$$

Equations (9) are the fundamental formulas for direct vorticity methods [12]. By covering the domain of the flow with a numerical mesh $(N_x \times N_y \times N_z)$ with equidistant spacing h, the i-th component of the intensity vector particle $\vec{\alpha}$ is defined by the expression:

$$\alpha_i = \int_{V_p} \omega_i(x_1, x_2, x_3)\mathrm{d}\vec{x} \approx h^3\omega_i(\vec{x}_p), \quad \vec{x}_p \in V_p, \quad |V_p| = h^3, \tag{11}$$

where V_p is the volume of the cell with index p. The velocity field is found by summing over all of the particles

$$\frac{d\vec{x}_p}{dt} = \vec{u}(\vec{x}_p, t) = \textstyle\sum_{k=1}^{N_p} \vec{K}(\vec{x}_p - \vec{x}_k)\vec{\alpha}_k(t), \tag{12}$$

$$\frac{d\vec{\alpha}_p}{dt} = (\vec{\alpha}_p(t) \cdot \nabla)\,\vec{u}(\vec{x}_p, t). \tag{13}$$

The disadvantage of the direct vortex particle method given by formula (12) is that the method is very computational time consuming. It stems from the fact that one should take into account all of the mutual interactions between particles that are in the flow. In practical calculation one should also introduce regularization in the formulas (12) removing the singularity of the kernel $\vec{K}$ [16, 13, 12]. To overcome those difficulties we replaced the Biot-Savart law for velocity calculation with the grid method.

From incompressibility (2) stems the existence of the vector potential $\vec{A}$, such that:

$$\vec{u} = \nabla \times \vec{A}. \tag{14}$$

The vector potential $\vec{A}$ defines the velocity field with accuracy to some potential field $\nabla\phi$. Adding $\vec{A} + \nabla\phi$ we do not change the the left side of (14). It is convenient to assume that $\mathrm{div}\vec{A} = 0$. The vorticity is related to $\vec{A}$ as follows

$$\nabla \times \nabla \times \vec{A} = \vec{\omega} = \nabla(\nabla \cdot \vec{A}) - \Delta\vec{A}. \tag{15}$$

So assuming that $\nabla \cdot \vec{A} = 0$ the vector potential can be obtained by solving three Poisson equations

$$\Delta A_i = -\omega_i, \qquad i = 1, 2, 3. \tag{16}$$

The Poisson equation can be solved effectively using the numerical gird and finite difference method. The velocity in grid nodes was obtained also by the finite difference using the formula (14).

Solving equation (12) by Runge-Kutta method we need the velocity of the particles that are out of the grid nodes. This is obtained using the interpolation of the velocity onto that new particle positions. Using grid for solution of equations (12) and (16) significantly accelerates ($\sim$ 1000 times [2]) the calculations. For the solution of the algebraic systems obtained form discretisation of the Poisson equations (16) we chose the Multigrid method that is very well suited for parallelization. Multigrid method used in a sequential architecture can be as fast as the Fast Poisson Solvers (FPS) [17]. The disadvantage of the FPS is that it is a sequential algorithm and can not take the full advantage of the parallel architecture. Our parallel implementation of the Multigrid method gave a speed-up up of 10-12 times comparing to FPS running on a single thread of the processor [7]. The system of equations (12) was solved by the Runge–Kutta method of the 4th order. After solution of the equations (12) obtained intensity of the particles was redistributed onto the grid nodes.

We did the redistribution of the intensities of particles onto grid nodes in each time step, before the solution of the Poisson equations (16).

It was done using an interpolation:

$$\omega_j = \sum_p \tilde{\alpha}_{p_n} \varphi \left(\frac{\vec{x}_j - \tilde{\vec{x}}_p}{h} \right) h^{-3}, \tag{17}$$

where j is the index of the numerical mesh node, p is the index of a particle.

Let us assume that $x \in R$. In this article, we used the following interpolation kernel [1]

$$\varphi(x) = \begin{cases} (2 - 5x^2 + 3|x|^3)/2 & \text{for } 0 \le |x| \le 1, \\ (2 - |x|)^2(1 - |x|)/2 & \text{for } 1 \le |x| \le 2, \\ 0 & \text{for } 2 \le |x|. \end{cases} \tag{18}$$

For the 3D case, $\varphi = \varphi(x)\varphi(y)\varphi(z)$.

If φ satisfies the following moment condition [1]:

$$\sum_j (x - x_j)^l \varphi \left(\frac{x - x_j}{h} \right) = 0, \qquad 1 \le |l| \le m - 1, \tag{19}$$

then

$$f(x) = O\left(h^m\right), \tag{20}$$

and the redistribution process will be of the order m. It means that the polynomial functions up to the order m will be exactly represented by this interpolation.

Kernel (18) used in this work is of order $m = 3$ [1].

3. Remarks on parallel computing on GPUs

As can be seen from the description of the VIC method, most calculations are related to the vortex particles displacement and redistribution. These processes are independent of one another and can therefore be done in parallel. For our calculations we selected Graphics Processing Units (GPUs - also called as a device in [15]). The computer in which GPU is mounted is called as a host. The graphics cards have a great computing power-to-cost ratio, but require quite a different approach to programming.

Details of implementation of the VIC method on a single GPU can be found in the authors' paper [7]. Rewriting the code to use the graphics card in calculations allowed for a speed-up of about 46 times. This result was obtained on a GTX480 GPU compared to a single thread of i7 960 @ 3.2GHz CPU.

Program running on a single GPU has its limitations. The most important is the amount of the RAM memory available on a single device. In our vortex particle method implementation, a single GPU used at that time (GTX480 with 1.5GB of RAM) allowed for a numerical mesh not larger than 128x128x128 nodes. To overcome this limitation we had to write a program capable of using multiple graphics cards. This required distribution and exchange of the data between devices.

Our final implementation of the VIC method was supposed to run on the computer cluster - that means a group of computers (called „nodes") used as servers connected to each other through local area network (LAN). Each of the nodes is a stand alone computer and can work on its own. Our target hardware configuration was that each node has more than one GPU (in supercomputing centres one can find nodes with even up to 8 GPUs).

To take advantage of the GPUDirect we introduced the so called hybrid MPI-OpenMP parallel programming (also called multilevel parallel programming) in our code. OpenMP is a standard for parallel programming on systems with shared-memory (in opposite to MPI which is dedicated to systems with distributed memory). The main difference between these two standards is the fact that in MPI each process makes its own copy of the data that it needs, and in OpenMP processes may read or modify memory spaces created by other processes (variables can be set as a shared or private). The main idea in hybrid MPI-OpenMP programming is that OpenMP assures communication between the processes running on the same node and MPI transfers the data between the nodes. OpenMP threads can use the GPUDirect for the communication between GPUs on the same node.

4. Collision of two vortex rings ($Re_\Gamma = 1000$)

This test shows the head–on collision of two vortex rings. The experimental results were published by T.T. Lim and T.B. Nickels in [11] and also on the web side of Lim [10] http://serve.me.nus.edu.sg/limtt/.

Distribution of vorticity in the cross section of the ring was assumed as:

$$\omega(r) = \omega_0 e^{-\frac{r^2}{a^2}} \tag{21}$$

where $\omega_0 = 20$ is the maximal value of vorticity in the tubes cross-section at initial time, $a = 0.5$ is a vortex tube radius. The radius of the ring was $R = 1.0$.

Based on this the total circulation Γ and the Reynolds number were calculated as

$$\Gamma = \int_0^a \omega(r) 2\pi r \, dr, \qquad Re_\Gamma = \frac{\Gamma}{\nu} \tag{22}$$

where $\nu = 0.01$ is kinematic viscosity coefficient. Calculated Reynolds number for this test was $Re_\Gamma = 1000$.

The results in the form of an isosurface plot can be seen in the figure 1 and the visualisation with passive markers is presented in figure 2.

When two rings approach each other the velocities induced by them cause both rings to grow in diameter and become flat. At first one can notice that the vorticity distributions around the axis of symmetry create a thin sheet that was called by Lim as a membrane (frame $t = 2$ in figure 1). As was reported in [11] (see also movies obtained in experiments by Lim) when size of the rings are approximately four times their initial diameter, a symmetrical instability in the form of azimuthal waviness begin to developed (frame $t = 3$ in figure 1, see also [10]). That instability lead to the creation, on the circumference, four bulges of vorticity. Our results are related to the experimental data presented on the web side of Lim for Reynolds number

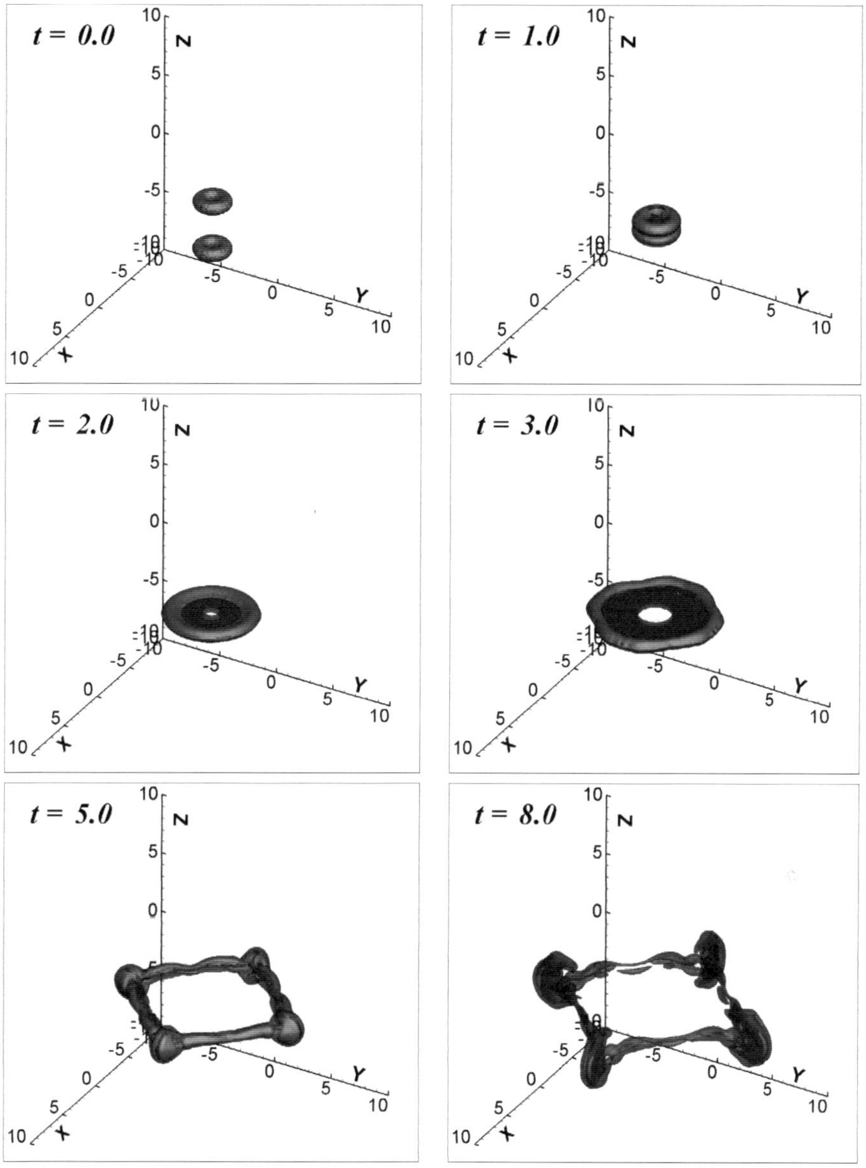

Figure 1. Isosurface plot $|\omega| = 0.05\omega_0$ (blue) and $|\omega| = 0.2\omega_0$ (yellow - look like grey when covered with blue) of collision of two vortex rings. $Re_\Gamma = 1000$.

smaller then 1000. In Lim's experimental results the number of the bulges is five times higher than in our computations. It is hard to compare Lim's results to our computations exactly. We conducted computations on a finite size cube $20 \times 20 \times 20$ with the periodic boundary conditions. Lim related the Reynolds number to the diameter of the hole by which the rings were produced and not to the circulation of the rings as we did. Smaller number of bulges on circumference of vortex structure and rectangular shape of the structure in the end of our computation may results from that. Also the size of our computational domain and periodic boundary conditions impact on our results.

It is known that vorticity is not carried by the fluid particles in the viscous flow and that the

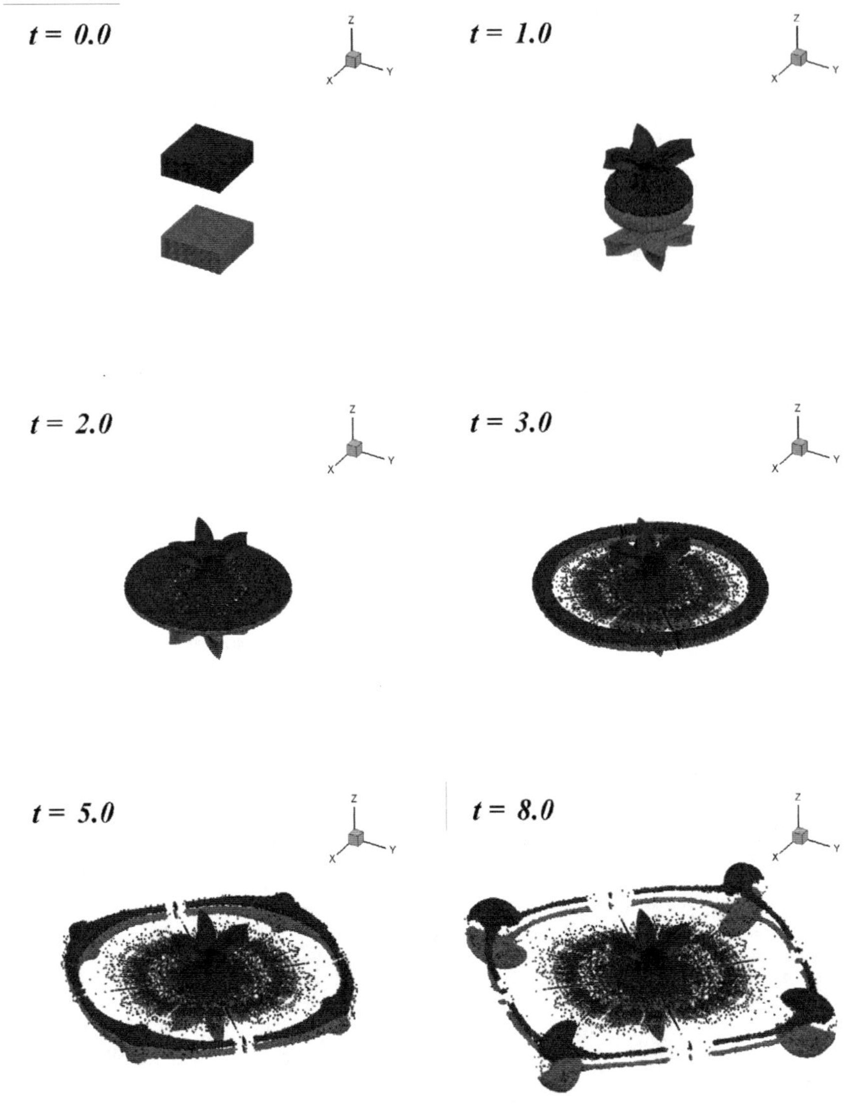

Figure 2. Collision of two vortex rings visualized by the passive marker. On the frame $t = 0$ the vortex rings are merged inside of the cuboids.

vorticity is not the prefect mean for visualization of the flow. Due to this, in order to better compare our results to the real flow visualization we simulated movement of the passive markers carried by the fluid. The vortex rings were merged in boxes of passive markers (colored with the initial position: red - lower ring, blue - upper ring). The size of the cuboids was equal to the diameter of the vortex rings and their thickness. Inside of each box we put a grid of $100 \times 100 \times 50$ passive markers. It is clear that some passive markers laid outside of the rings. Our numerical results were presents in figure 2. One can see that in frame $t = 1$ a tail created by the particle initially placed outside of the rings is formed. In the frame $t = 3$ one can see formation of the membrane. The structure that can be seen in the frame $t = 5$ very much resembles the experimental results by Lim presented in [10]. From the frame $t = 8$ it is clear that reconnection did not take place. The particles did not mix. The reconnection phenomena that would appear in the vorticity bulges could be responsible for the creation of the small rings on a circumference as was reported in [11].

5. Closing remarks

We showed that our implementation of vortex particle method is capable of reproducing result obtained in experiments. Because vorticity is not carried by the fluid particles in the viscous flow numerical simulation seems to give better understanding of the processes behind the reconnection phenomena. In this paper we showed comparison with the real experiment with the use of isosurfaces of vorticity and passive markers.

Nowadays it is not difficult to notice that the computational power of a single processor has stopped rising. Parallel architectures need to be used to deliver the means to speed up computation. Developing programs on GPUs is an interesting alternative to using the CPU. Thanks to hundreds of streaming processors working in parallel we can get the results faster. The processors also quite cheap and easily accessible.

It is obvious that if one wants to have a good resolution of the physical phenomena, one has to use a fine numerical mesh in computations. That requires greater memory and computational time. To overcome this problems, one can use many graphics cards. Introducing of multi GPU computation is our aim in the nearest future. Properly used GPUs (memory management, parallel algorithms, etc.) allows programs to be executed much faster at relatively low cost.

References

[1] G.-H. Cottet, P. D. Koumoutsakos, *Vortex Methods: Theory and Practice*, Cambridge University Press, (2000)

[2] G.-H. Cottet, P. Poncet, *Advances in direct numerical simulations of 3D wall-bounded flows by Vortex-in-Cell methods*, Journal of Computational Physics **169**, 136–158 (2004)

[3] H. Holden, K. H. Karlsen, K.-A. Lie, W. H. Risebro, *Splitting Methods for Partial Differential Equations with Rough Solutions*, European Mathematical Society, (2010)

[4] S. Kida, M. Takaoka, *Breakdown of frozen motion of vorticity field and vorticity reconnection*, Journal of Physical Sociecty of Japan, **60**(7), 2184–2196 (1991)

[5] S. Kida, M. Takaoka, *Vortex Reconnection*, Annual Review of Fluid Mechanics **26**, 169–189 (1994)

[6] S. Kida, M. Takaoka, F. Hussain, *Collision of two vortex rings*, Journal of Fluid Mechanics **230**, 583–646 (1991)

[7] A. Kosior, H. Kudela, *Parallel computations on GPU in 3D using the vortex particle method*, Computers & Fluids **80**, 423–428 (2013)

[8] H. Kudela, P. Regucki, *Vorticity particle method for simulation of 3D flow*, in Computational Science - ICCS 2004 ed by M. Bubak, G. D. van Albada, P. M. A. Sloot and J. J. Dongarra, 356–363 (2004)

[9] T. T. Lim, *A note on the leapfrogging between two coaxial vortex rings at low Reynolds numbers*, Phys. Fluids **9**, 239–241 (1997)

[10] T.T. Lim webpage, `http://serve.me.nus.edu.sg/limtt/`

[11] T. T. Lim, T. B. Nickels, *Instability and Reconnection in the Head-On Collision of Two Vortex Rings*, Nature **357**, 225–227 (1992)

[12] A. J. Majda, A. L. Bertozzi, *Vorticity and Incompressible Flow*, Cambridge University Press, (2002)

[13] E. Meiburg, *Three-dimensional vortex dynamics simulations*, in Fluid Vortices ed by S. I. Green, Kluwer Academic Publishers, 651–685 (1995)

[14] V. M. Melander, F. Hussain, *Cross-linking of two antiparallel vortex tubes*, Phys. Fluids A **1**, 633–636 (1989)

[15] *NVIDIA CUDA C Programming Guide*, `http://www.nvidia.com/cuda`

[16] E. G. Puckett, *Vortex Methods: An Introduction and Survey of Selected Research Topics*, in Incompressible Computational Fluid Dynamics Trends and Advances ed by M. D. Gunzburger, R. A. Nicoladies, Cambridge University Press, 335–407 (1993)

[17] U. Trottenberg, C.W. Oosterlee, A. Schuller, *Multigrid*, Academic Press, (2001)

[18] J. Z. Wu, H. Y. Ma, M. D. Zhou, *Vorticity and Vortex Dynamics*, Springer, (2006)

[19] N. J. Zabusky, M. V. Melander, *Three-dimensional Vortex Tube Reconnection: Morphology For Orthogonally - Offset Tubes*, Physica D **37**, 555–562 (1989)

DNS benchmark solution of the fully developed turbulent channel flow with heat transfer

M Jaszczur
Faculty of Energy and Fuels
AGH University of Science and Technology, Krakow, Poland

E-mail: marek.jaszczur@agh.edu.pl

Abstract. In the present paper direct numerical simulation (DNS) of the fully developed turbulent non-isothermal flow has been study for Re_τ=150 and for Pr=1.0. The focus is on the role of the thermal boundary condition type on the results. Various types of thermal boundary conditions presented in literature has been considered in this work: isoflux wall boundary conditions, symmetrical isofluxes wall boundary conditions and isothermal b.c. also with combination with adiabatic or isothermal second wall. Turbulence statistics for the fluid flow and thermal field as well turbulence structures are presented and compared. Numerical analysis assuming both zero and non-zero temperature fluctuations at the wall and zero and non-zero temperature gradient in the channel centre shows that thermal structures may differ depend on case and region. Results shows that the type of thermal boundary conditions significantly influence temperature fluctuations while the mean temperature is not affected. Difference in temperature fluctuation generate the difference in turbulent heat fluxes. Presented results are prepared in the form of the benchmark solution data and will be available in the digital form on the website http://home.agh.edu.pl/jaszczur.

1. Introduction

The turbulent boundary layer is for the long time of interest from the engineering and scientist point of view. In many practical applications the transport of heat in boundary layer of turbulent flows is of great importance. Heat exchangers, combustion chambers, gas turbines and recently fuel cells or high temperature nuclear reactors are a few examples. A large number of numerical and experimental study has been carried out to understand the heat and momentum transport. To solve non-isothermal turbulent fluid flow using Reynolds averaged equations for the fluid flow and for the energy similar turbulence closure problem occurs. In the literature many approaches has been proposed to solve the non-isothermal turbulence flows. Most study employed averaged transport equations with high orders moments models. More recently large number of numerical computations have employed Large Eddy Simulations (LES) technique in which subgrid scale models need to be used. In case of using any mathematical model a validation of implemented approach and approximations are highly required in order to justify that turbulence models used for momentum and for energy transport performing well in analysed case. In the last decade Direct Numerical Simulations (DNS) technique which employ no simplifications of the governing equations and in consequence use no model, became very important research tool to study transport mechanisms in wall bounded turbulent flows [1]. This technique with a continuous progress in supercomputers has been extended to the non-isothermal [2] and multiphase flows [9]. One of the key properties in understanding heat transport between solid walls and the fluid is the how temperature and velocity are fluctuating very close to the wall or even at the solid walls. The typical approach to study the phenomena is to consider fully developed hydrodynamically and thermally flow.

For the channel flow computation, the boundary conditions for the momentum transport in almost all DNS computations presented in literature are similar. The non-slip boundary condition at the solid walls (sometimes on the second top wall there are slip conditions or the free surface instead solid wall if applicable) and deviation from this standard boundary condition configuration is rather unique. The situation for the cases with thermal boundary conditions is different and different groups of researchers apply different conditions, resulting in different thermal field presented. The first results, published by Kim and Moin [2], were attained for heat generated in the flow internally and at the same time removed at the two constant temperature walls. This procedure allowed them to maintain the temperature at the same level. Kasagi *et al.* [3] performed their computations with both walls heated at constant heat flux whereas Lyons *et al.* [4] heated one wall while letting the other cool at the same heat rate. In general, two main types of thermal conditions are applied with different combination and implementation (conducting wall, or non-conducting wall are implemented) i.e. constant temperature or constant heat flux. How thermal condition are implemented may also lead to final results.

The equations derived by Kasagi *et al.* [3] are used for the isoflux boundary condition on the wall. When they are used, the mean temperature increases linearly in the streamwise direction. After the energy equations are transformed into a dimensionless form, the isothermal boundary condition is used. This procedure results in zero temperature fluctuations on the wall.

Numerous experiments have shown that thermal boundary conditions may differ reasonably from case to case, resulting in a difference in the turbulent heat transfer [5-7]. Kasagi *et al.* [3] have shown that in the experiments where air is the working fluid the wall temperature fluctuations are usually negligible, but the experiments with the water can be performed with isothermal boundary conditions if a thick copper plate is used as a heater or with isoflux boundary conditions if the heated wall is thin (Khabakhpasheva [8]).

In the present work the heat transport in a turbulent flow has been study solving numerically three dimensional unsteady Navier-Stokes and energy equations directly without using of any additional models. Various configurations for the wall thermal boundary conditions are considered and compared: (H1) symmetrical heat flux, (H2) asymmetrical heat flux, (I) isothermal walls, (HA) heat flux at bottom wall and adiabatic top wall, (IA) isothermal bottom wall and adiabatic top wall. Presented results are prepared in the form of the benchmark solution data to validate turbulence models as well as numerical procedures and will be available in the digital form on the Author website.

2. Mathematical model

The geometry of the problem along with the dimensions and coordinate system is shown in Figure 1.

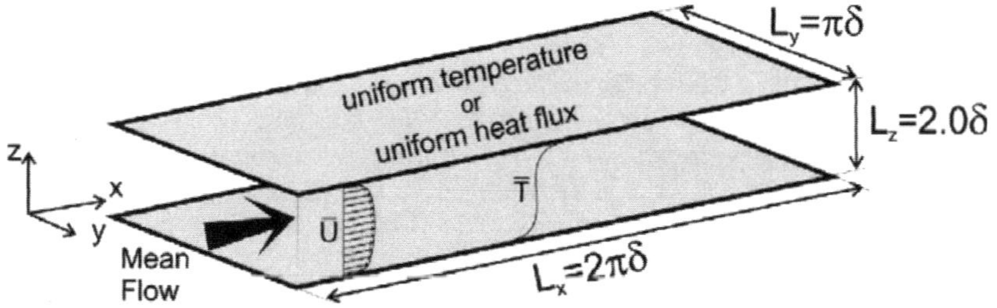

Figure 1. Computational domain

The elementary governing equations for the fluid eqs. (2.2-2.3), with equations for the thermal energy of the fluid (2.4) need to be solved in discretized form. It is assumed that temperature is treated as a passive scalar and have no influence on the fluid flow nor properties. To accommodate the governing equations, a normalization procedure has been applied. In the present case the coordinates and flow variables are normalized by the channel half height δ, the kinematic viscosity ν, the friction velocity u_τ, and the friction temperature T_τ.

$$u_i^+ = \frac{u_i}{u_\tau} \qquad x_i^+ = \frac{x_i^* u_\tau}{\nu} \qquad t^+ = \frac{t u_\tau^2}{\nu} \qquad T^+ = \frac{T}{T_\tau} \tag{2.1}$$

where

$u_\tau = \sqrt{\frac{\tau_w}{\rho}}$ is the shear (or friction) velocity, τ_w is the mean shear stress at the wall and ρ is the fluid density, $T_\tau = \frac{q_w}{\rho c u_\tau}$ is the shear (or friction) temperature and q_w is the mean heat flux at the wall.

The governing equations dimensionless form are:

$$\frac{\partial u_k^+}{\partial x_k^*} = 0 \tag{2.2}$$

$$\frac{\partial u_i^+}{\partial t^+} + u_k^+ \frac{\partial u_i^+}{\partial x_k^*} = -\frac{\partial p^+}{\partial x_i^*} + \frac{1}{Re_\tau} \frac{\partial^2 u_i^+}{\partial x_k^* \partial x_k^*} \quad \text{for } i\text{=1,2 and 3} \tag{2.3}$$

$$\frac{\partial T^+}{\partial t^+} + u_k^+ \frac{\partial T^+}{\partial x_k^*} = \frac{1}{Re_\tau Pr} \frac{\partial^2 T^+}{\partial x_k^* \partial x_k^*} + s \tag{2.4}$$

where:

$Re_\tau = \frac{u_\tau \delta}{\nu}$ is the fluid flow Reynolds number

$Pr = \frac{\nu}{\alpha}$ is the fluid molecular Prandtl number

s is the source term dependent on boundary condition type

When the heat flux q_w is applied, reference temperature θ should be introduced. It represents the difference between the fluid temperature and the mean wall temperature. This allows flow field quantities to be directly compared with thermal field ones. Please note that all the figures (except instantaneousfields) will show temperature θ and it statistics.

$$\theta^+ = \overline{T}_{wall}^+ - T^+ \tag{2.5}$$

In the present turbulent channel flow, assuming the flow is Newtonian and incompressible. To force the fluid motion, a constant pressure-gradient is imposed along the streamwise direction (x). The flow is: heated by a uniform heat-flux from both walls and cooled with a negative source for case (H1); heated by a uniform heat-flux q from bottom wall and cooled by a uniform heat-flux from top $-q$ wall (the same amplitude but opposite direction) for case (H2); maintain walls at constant but different temperatures for isothermal case (I); heated by a uniform heat-flux q from bottom wall and cooled with a negative source while the top wall is set adiabatic for case (HA) and last case where bottom wall is maintain at constant temperature while the top wall is set adiabatic (IA). In any case periodic boundary conditions are imposed in both the streamwise and spanwise directions for the velocity and for the temperature. To solve the governing equations (2.2-2.4) a finite-volume method on a staggered grid was employed together with a two-step predictor-corrector method for incompressible flow. A second-order Adams-Bashforth scheme is used for time-advancement, with the time-step determined by the Courant criterion. The convective and diffusive terms in all the equations are discretized using a second-order central scheme. The Poisson equation is solved by applying a Fourier Solver in the periodic directions and a TDMA solver for the remaining direction [9].

Preliminary calculations were performed for the different grid resolution and different size of the computational domain to ensure adequacy of the results. As a basic domain the size $2\pi\delta\times\pi\delta\times2\delta$

(942×471×300 in wall units) and the grid resolution 128×128×64 is used but the higher grid resolutions up to 256×256×256 or 512×512×192 were also tested. The use of larger computational domain doesn't produce significant difference in results but required proportional increase in the number of grid points to reproduce the same near-wall structures. It was verify using 2-point velocity or temperature correlations that this domain is sufficient in both cases and allow to obtain good quality flow and thermal fields. The turbulent statistics for the velocity and temperature are compared at different grid resolutions and what is the main purpose of this work for different boundary conditions type configurations. First and second order statistics were calculated using the following procedure. When the statistically steady state was developed (hydrodynamically, and thermally) then the saving of the raw data for post-processing statistical analysis was begun. A database of velocity and temperature at each point was created. This database was post-processed. In order to obtain good statistics it has been found that about 200 uncorrelated fields are required to average (in time and in two homogenous directions) over non-dimensional time t^{+}=60000.

3. Results

Type of the boundary conditions significantly influence the temperature fluctuations in the near wall region. Before the turbulence statistics will be presented the instantaneous temperature field is shown. Figure 2 show temperature fields in the x-y and x-z plane for z^{+}=3.6 and for y^{+}=235 respectively. It should be notice that for this case only the fluid temperature (denote here as T) is not referenced to the wall temperature. This mean that high value corresponds to the high fluid temperature. In the other plots normalised and calculated with eq. (2.5) the wall temperature value θ^+ has been used for which the high value of T^{+} corresponds to low fluid temperature θ^+. In Figure 2 the temperature field is shown at: (a)symmetric-isoflux-H1, (b)isoflux-H2 and (c)isothermal-I walls.

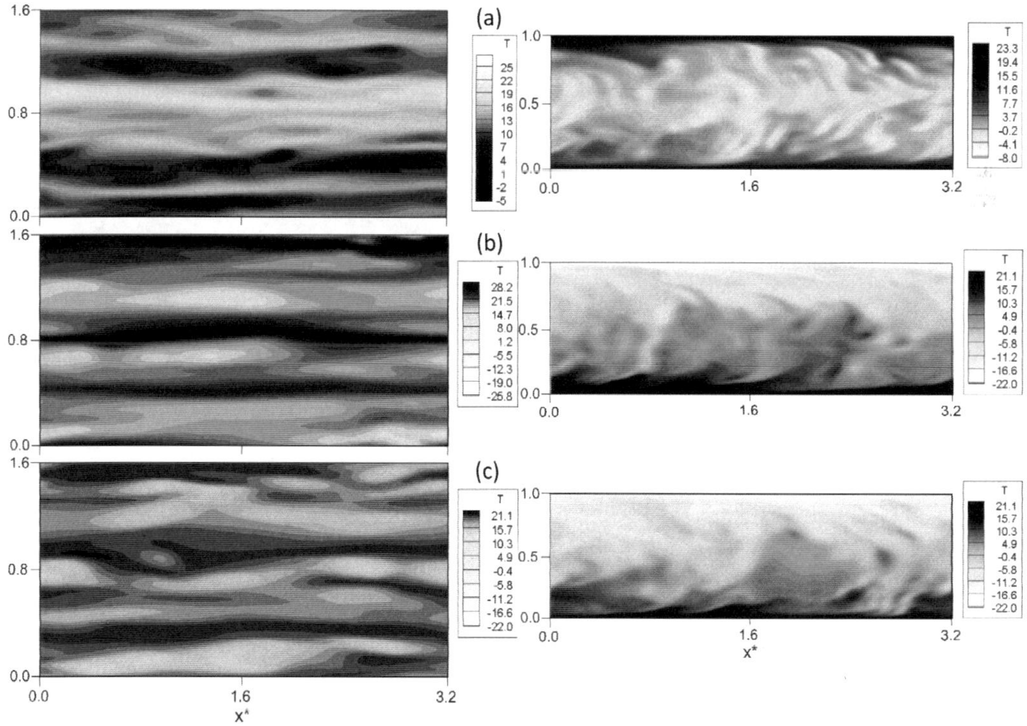

Figure 2. Instantaneous temperature fields at x-y plane at z^{+}=3.6 (left) or x-z plane at y^{+}=235 (right) for (a)symmetric-isoflux, (b)isoflux, and (c)isothermal boundary conditions

All results indicate the existence of the large scale thermal structures. For the isothermal cases (I,IA) large temperature gradient at viscous layer in wall normal direction is detected and the same time for x-y plane uniform temperature field is seen. This can be explain by the stabilizing influence of constant temperature wall. For isothermal cases better similarity between momentum transport and thermal transport is observed [10]. The mean velocity and the mean turbulence intensities normalized by the wall shear velocity are presented in Figure 3 for the different grids. Results are also compared with literature data from Kawamura *et al.*[11]. Profiles are in good agreement and the existing small discrepancy can be explained by different numerical approach used for both simulations.

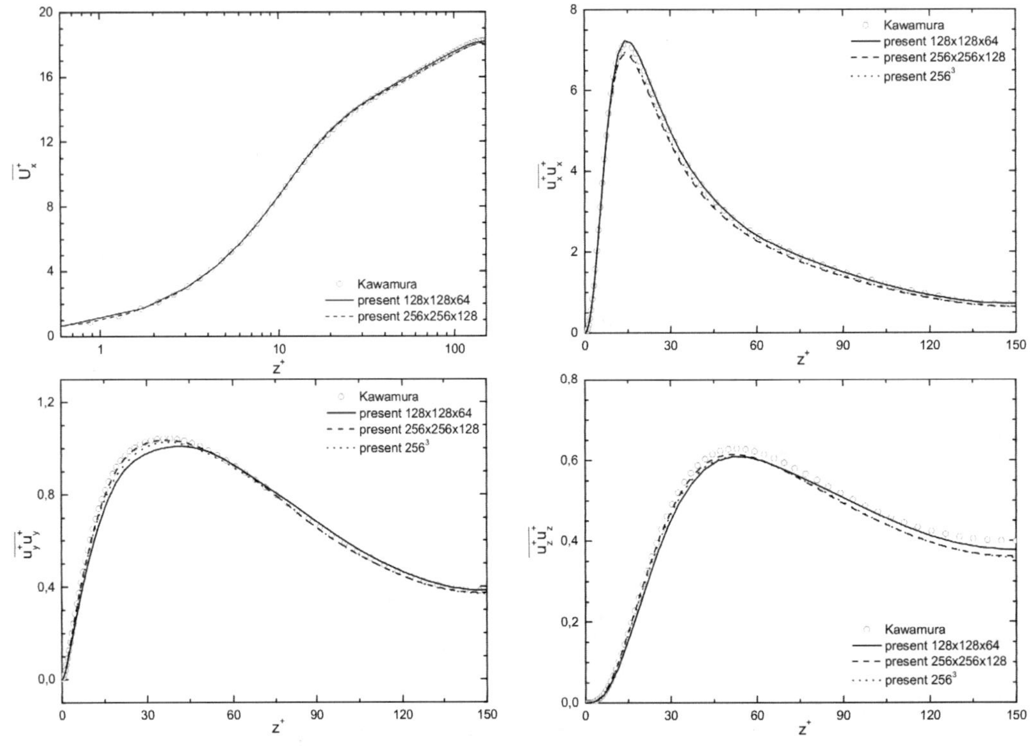

Figure 3. Profiles of the mean velocity and mean velocity fluctuations components, comparison with literature data Kawamura *et al.*[11]

The mean temperature normalized by the shear temperature and reference to the wall temperature is show in Figure 4. Results are presented for all the thermal boundary conditions considered in this work. Results indicate that wall thermal conditions do not influence mean temperature profile. The only difference in the temperature is observed in the near channel centre region where due to existence different boundary conditions at the top wall temperature is increasing, decreasing or remain constant. Figure 4 also shows the results for temperature fluctuations for isothermal cases and for isoflux cases. It can be seen that isothermal and isoflux cases up to z^+ gives similar results in each groups. In the outer region results due to different conditions at the second wall increases or decreases. Because temperature fluctuations for isothermal wall is zero similarly as for the velocity thus for that case similarity between thermal and velocity field is easy recognised. For the isoflux cases temperature fluctuations remains at constant level up to $z^+=2$ and is almost equal to the maximum fluctuation level observed for isothermal case. The maximum for the temperature fluctuation in both general boundary conditions types are located at the same distance but the fluctuations level for the isoflux cases is higher. In Figure 4 bottom plots shown the turbulent heat flux axial component $\overline{u_x^+\theta^+}$ versus the distance from the wall. In the near wall region turbulent heat flux axial component $\overline{u_x^+\theta^+}$ is much higher for the isoflux cases than for isothermal cases [12,13]. At larger distance from the wall this component is always higher for isoflux case. Only in the channel centre region this component became smaller and tends to zero at centre because the flux changes direction.

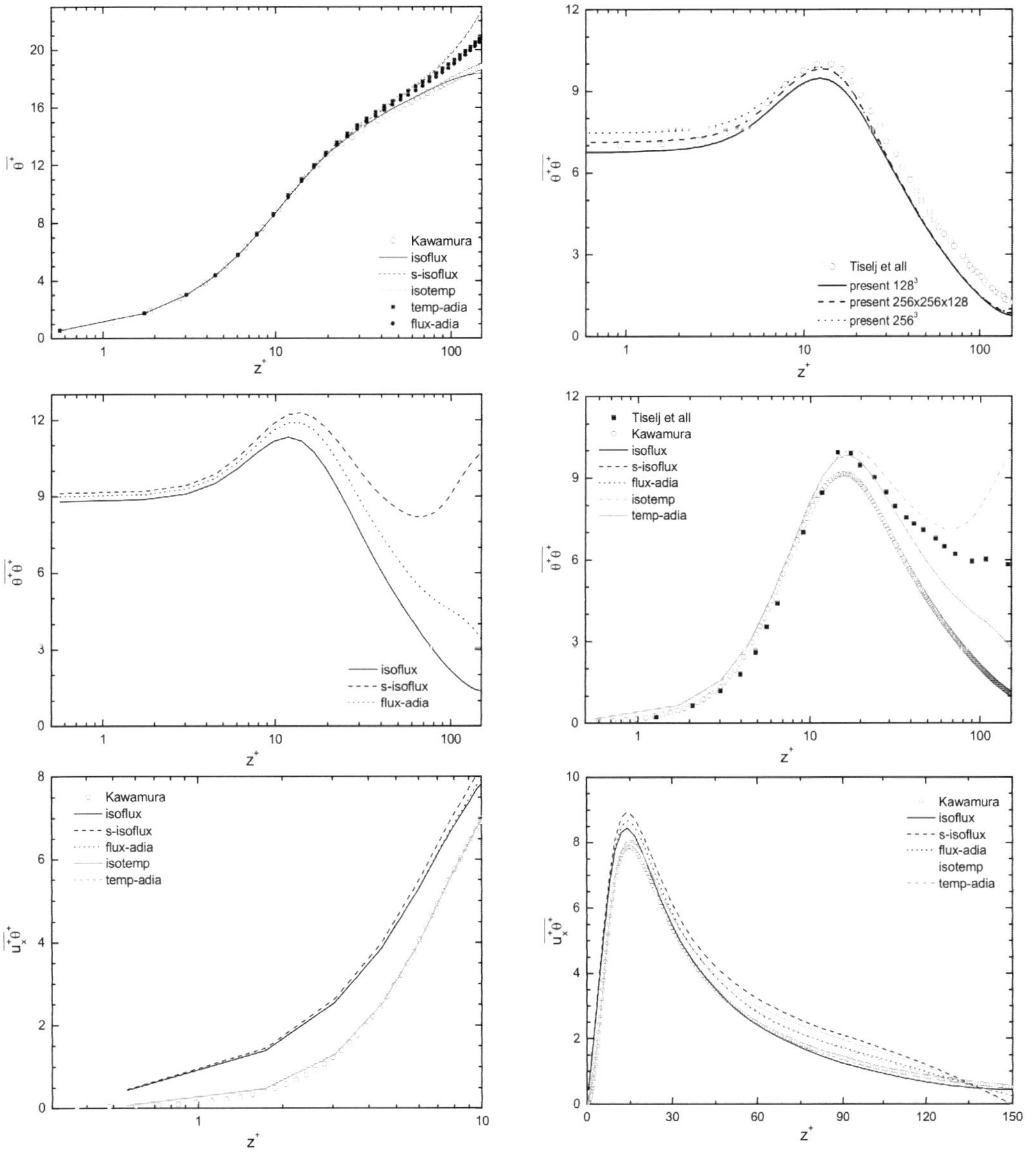

Figure 4. Profiles of the mean temperature $\overline{\theta^+}$, temperature fluctuations $\overline{\theta^+\theta^+}$, and the streamwise velocity-temperature correlations $\overline{u_x^+\theta^+}$ for the different thermal boundary conditions, comparison with literature data Kawamura *et al.*[11]

The turbulent heat flux wall normal component is presented in Figure 5. It can be seen that this component in near wall region do not depend on thermal boundary conditions. In outer region maximum is observed at two different locations depends on thermal conditions on second wall. For the cases where top wall is adiabatic this components has almost constant value in large part of the channel. In the channel centre three different values for the $\overline{u_z^+\theta^+}$ are seen (zero, maximum and average).

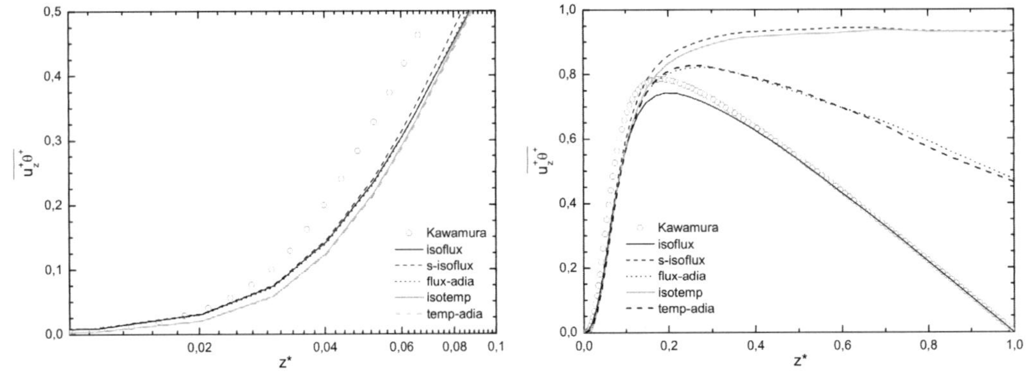

Figure 5. Profiles of the mean wall-normal velocity-temperature correlations $\overline{u_z^+\theta^+}$ for the different thermal boundary conditions, comparison with literature data Kawamura *et al.*[11]

4. Conclusion

In the present paper DNS of the fully non-isothermal channel flow has been done for the Reynolds number Re_τ=150 and for the molecular Prandtl number Pr=1. Various configurations of the thermal boundary conditions on top and bottom walls were taken into account including isothermal which not allow wall temperature fluctuations and isoflux where wall temperature fluctuations is non zero. Results for some cases are also compare with the existing literature data.

It has been found that mean temperature profile in near wall region is identical for all six configuration considered here and independent of the boundary condition type. In the channel centre region difference in the mean temperature profiles are observed cause by different thermal condition on the second wall. For example for isoflux mean temperature gradient is zero by definition while in this region for isothermal case temperature increases linearly. Although mean temperature profiles are identical close to the wall observed turbulent large scale structures are different. This behaviour is indirect captured with the first order statistics. For the cases with isoflux type wall boundary conditions mean temperature fluctuations in near wall region is relatively high (also on the wall) and get almost constant value while for isothermal cases temperature fluctuations on the wall is zero. Analysis of the instantaneous results shows that flow and thermal field are better correlated for isothermal cases. Much larger temperature fluctuations for isoflux cases has also impact on the axial turbulent heat flux which is larger not only in near wall region but also in outer region. Close look on the turbulent heat fluxes components shows that axial component in viscous sublayer is much larger for isoflux cases than for isothermal ones while the difference in turbulent wall normal heat flux component is very small in this region. Although the results of this study are limited to the relatively low Reynolds number it can be assumed that similar difference in turbulent heat transport phenomena, generated in the viscous sublayer will also appear for higher Reynolds numbers.

Presented results uses no model for momentum nor for heat transfer and were compare with literature data and verify with different grid resolution and size of domain. All key statistics for flow and thermal fields are available in digital form on Author's home page. Results for higher grid resolution are also presented. No significant discrepancy other than cause by different boundary condition types has been found between results. Existing small difference in results are cause by the different methodology and numerical methods used in DNS codes. Domain decomposition for parallelization or even way of computing statistics may have impact on the final results. Only in the limit when computational grid became very large all DNS computations (even with very different approaches) converged to the same solution. Before this limit is reached small difference also in Direct Numerical Simulation case are inevitable.

Acknowledgements
The present work was supported by the Polish Ministry of Science (Grant AGH No. 11.11.210.198)

5. References

[1] Kim J, Moin P and Moser R D 1987 Turbulence statistics in fully-developed channel flow at low Reynolds number *Journal of Fluid Mechanics* **177** 133–166

[2] Kim J and Moin P 1989 Transport of Passive Scalars in a Turbulent Channel Flow *Turbulent Shear Flows **VI*** 85-96

[3] Kasagi N, Tomita Y and Kuroda A 1992 Direct Numerical Simulation of Passive Scalar Field in a Turbulent Channel Flow *J. Heat Transfer* **114** 598-607

[4] Lyons S L, Hanratty T M and McLaughlin J B 1991 Direct numerical simulation of passive heat transfer in a turbulent channel flow *Int. J. Heat and Mass Transfer* **34** 1149-1161

[5] Kawamura H, Abe H and Matsuo Y 2004 Very large-scale structures observed in DNS of turbulent channel flow with passive scalar transport, *15th Australasian Fluid Mechanics Conference*, The University of Sydney, Sydney, Australia

[6] Tiselj I, Bergant R, Mavko B, Bajsic I and Hetsroni G 2001 DNS of turbulent heat transfer in channel flow with heat conduction in the solid wall *Journal of Heat Transfer* **123** 849-857

[7] Lu D M and Hetsroni G 1995 Direct numerical simulation of a turbulent open channel flow with passive heat transfer *Int. J. Heat and Mass Transfer* **38** 3241–3251

[8] Khabakhpasheva Y M 1986 Experimental Investigation of Turbulent Momentum and Heat Transfer in the Proximity of the Wall *Proc. 8th Int. Heat Transfer Conf.* CA, USA

[9] Jaszczur M 2011 Numerical analysis of a fully developed non-isothermal particle-laden turbulent channel flow *Archives of Mechanics* **63** 77-91

[10] Antonia R A, Abe H and Kawamura H 2009 Analogy between velocity and scalar fields in a turbulent channel flow *Journal of Fluid Mechanics* **628** 241–268

[11] Kawamura H, Abe H and Shingai K 2000 DNS of Turbulence and Heat Transport in a Channel Flow with Different Reynolds and Prandtl Numbers and Boundary Conditions *Turbulence, Heat and Mass Transfer* **3** 15-32

[12] Bergant R and Tiselj I 2006 On the role of the smallest scales of a passive scalar field in a near-wall turbulent flow *Heat and Mass Transfer* **42** 411-426

[13] Na Y and Hanratty T J 2000 Limiting Behavior of Turbulent Scalar Transport Close to a Wall *Int. J. Heat and Mass Transfer* **43** 1749–1758

Experimental research on coarse water formation in steam condensing flow on a transition through the shock wave

S Dykas, M Majkut, M Strozik and K Smołka
Silesian University of Technology, Institute of Power Engineering and Turbomachinery, 44-100 Gliwice, Konarskiego 18, Poland

E-mail: slawomir.dykas@polsl.pl

Abstract. In this work the condensation phenomena in steam turbine were discussed. The motivation for presented research was an interaction of liquid phase with shock waves existing in the transonic flow. The paper presents the experimental results of the steam condensing transonic flow in Laval nozzles. For the tests the geometries of the half arc nozzles were used. The behaviour of shock waves in the wet steam region was investigated. Due to the high back pressure, in the divergent part of the nozzle the shock wave was induced and interacting with the nozzle walls caused instability in the flow.

1. Introduction

Condensation, i.e. the phase change of steam to liquid water occurs in a number of engineering applications. Condensation of steam is in many applications desired, such as in heat exchangers or steam condensing turbine. The condensation process in steam turbines takes place in the last stages of the low-pressure part by very high steam velocity, in transonic flow conditions. Condensation of steam occurs when the temperature drops below the saturation temperature at a given pressure, i.e. the saturation line. Latent heat is released from liquid water as the phase change occurs, this heat has to be taken over by surrounding vapour phase. Since the phase change affects pressure, temperature, velocity and heat transfer a lot of work has been done in experimental investigation [1, 2, 3, 4, 5] and computational modelling [6, 7, 8, 9] of condensation.

The process of condensation can be divided into primary (homogeneous/heterogeneous) condensation and surface condensation. Numerous methods exist to determine primary condensation that occurs in the steam flow [10, 11], while the modelling of condensation onto a surface is by no means trivial.

Primary condensation is the phenomena where droplets grow in size in supercooled vapour in a free flow. The liquid phase consists of the small droplets (0.1–0.001 μm) forming a fog. Supercooling is the difference between the saturation temperature at the given pressure and vapour phase temperature. If small foreign particles exist in the steam, these will serve as condensation nuclei. The heterogeneous condensation mainly consists of droplets growth process only. Saturation lines of many admixtures (NaCl, HCl, NaOH, and other) dissolved in the steam are located higher than that for pure steam [12], therefore, droplets of these admixtures can act as condensation nuclei, thus causing the condensation front to shift toward the point of less supercooling.

The homogeneous condensation takes place in the absence of foreign particles, where due to the rapid expansion of the flow the nucleation process occurs, i.e. the formation of critical droplets, on which the further droplets growth is continued [1, 13]. When pure steam expands in the turbine flow path, its

Journal of Physics: Conference Series **530** (2014) 012023 doi:10.1088/1742-6596/530/1/012023

condensation in the flow occurs with relatively high supercooling (by approximately 30-35°C) and the front of initial condensation is located within the so called Wilson's zone, which corresponds to an equilibrium steam wetness of 2.5-3.5%.
As water vapour comes into contact with a surface at a temperature below the saturation temperature for the corresponding partial pressure of the water vapour, droplets start to form on the surface. As the water condensate on the cooled surface, the latent heat is released onto the surface. The thickness of the water film depends also on the surface roughness. The condensation on the steam turbine blade surface can occur at a very small level of supercooling and even with some steam superheating, +0.3 ÷ –0.3°C. The water film thickness for a curvature radius of the surface roughness of 1 μm is in the range of 0.1-0.001 μm. The above values of the film thickness are of the same order of magnitude as the radius of the droplets formed in the primary condensation front region.

2. Motivation

As it was already mentioned, the liquid phase formation occurs by a very high steam velocity, even supersonic. Usually, for the nominal load of the steam turbine the primary condensation takes place in the stator of the low-pressure part last stage. For the stator blades the transition from sub- to supersonic flow conditions and primary condensation onset correspond to the area between the upper blade trailing edge and the bottom blade suction (convex) side surface (figure 1). The oblique shock wave set comprises the so-called fishtail shock starting downstream of the trailing edge of the blade and the shock wave reflected from the suction side.

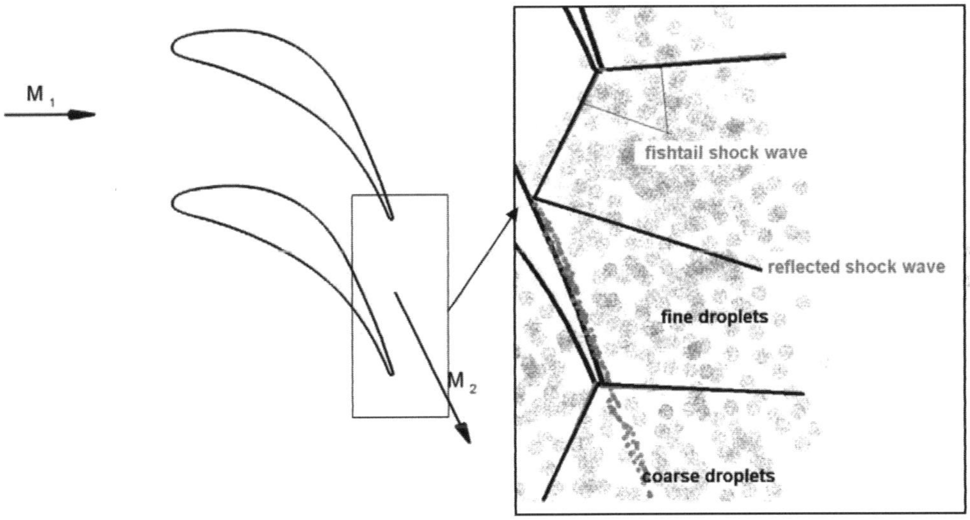

Figure 1. Sketch of the shock waves configuration in blade-to-blade channel for the turbine stage stator (M – Mach number).

The liquid film forms on the suction side surface and leaves the blade together with flowing steam, forming the coarse water droplets behind blade.
Very important issue is the behavior of the fine droplets on the transition through the shock waves. On the shock wave transition, we deal with the rapid increase of the pressure and temperature and decrease of velocity. The fine droplets have a slightly higher temperature and pressure and move with a little smaller velocity than the surrounding vapour phase, therefore, it is expected that the part of the droplets evaporate on the shock wave and they start colliding between each other in the result of the rapid slowing down.
Generally, the interaction of the shock wave with a solid surface causes, in the most cases, separation of the boundary layer [14]. These phenomena have a significant (typically negative) impact on the

flow, especially when there is a liquid film on the solid surface. The interaction of the liquid phase, fine or/and coarse droplets, with the shock waves in supersonic flow affects the unsteadiness. The numerical modelling of this phenomenon is very difficult and requires to use a two-fluid model [10] including both mentioned condensation processes, also model of water droplets collision and coagulation. Additionally, for the elaboration and validation of such model the experimental data are necessary. The main motivation of presented work is to experimentally investigate the interaction of the shock wave with liquid phase in steam condensing transonic flow. To this end the flow through the Laval nozzle was taken into account and the shock wave was generated by increasing the pressure at the nozzle outlet.

3. Experimental facility

The experimental facility is a part of the small steam condensing power station that is located in the Institute of Power Engineering and Turbomachinery of the Silesian University of Technology. The steam tunnel facility (figure 2) was designed in order to perform the experiments for steam condensing flows in nozzles or linear cascades [15]. The superheated steam of the very stable parameters is supplied from the 1MWt boiler. The maximum steam mass flow rate is about 3 kg/s. The parameters ahead of the test section are controlled by means of the control valve and desuperheater providing the steam with parameters corresponding to the conditions prevailing in low-pressure turbine stages. The total inlet pressure can vary in the range of 70-150 kPa(a) and the total temperature between 70-120°C. The pressure behind the test section is controlled by the throttle valve assuring the corresponding value of the static back pressure.

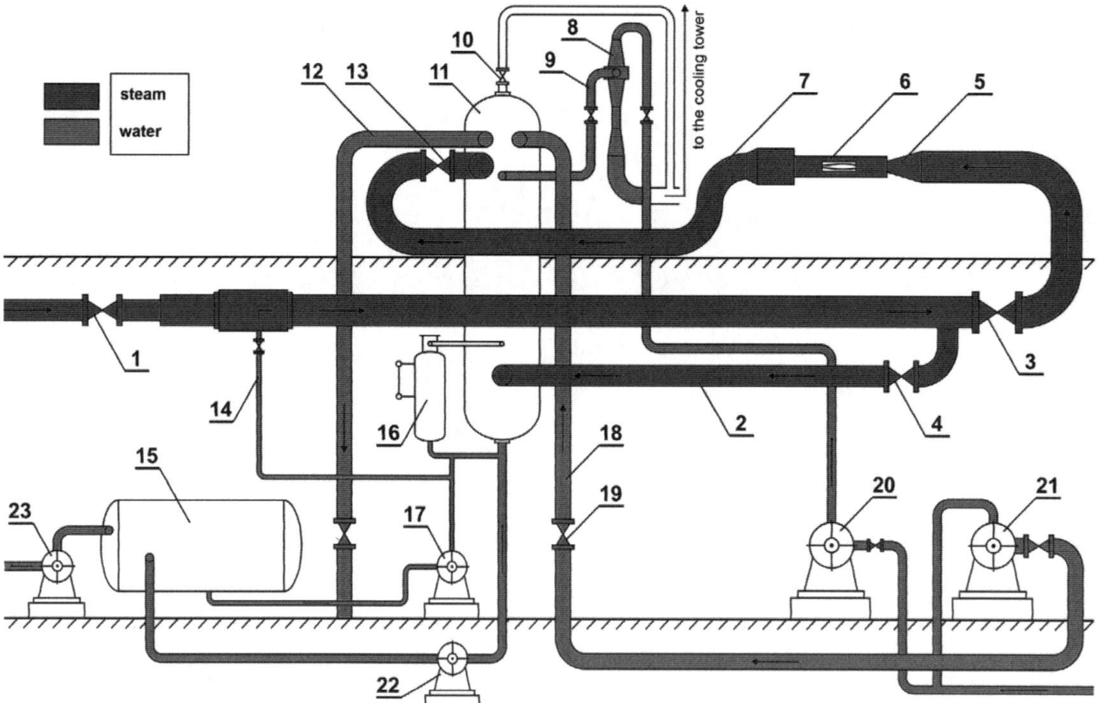

1) Control valve 2) By-pass 3) Stop gate valve 4) Stop gate valve at by-pass 5) Inlet nozzle 6) Test section 7) Outlet elbow 8) Water injector 9) Pipe 10) Safety valve 11) Condenser 12) Suction line 13) Throttle valve 14) Desuperheater 15) Condensate tank. 16) Control system of condensate level. 17) Condensate pump 18) Discharge line 19) Stop valve 20) Water injector pump 21) Cooling water pump 22) Condensate pump 23) Pump

Figure 2. Steam tunnel with auxiliary devices.

4. Experimental results

The experimental measurements were performed for the geometry of the half Laval nozzle. The applied half nozzle is an arc nozzle with lower critical throat height of H* = 27.5 mm and wall curvature radius of R = 525 mm (figure 3). The width of the nozzles amounted to 110 mm. The static pressure measurement in the nozzles was carried out on a distance of 200 mm downstream from the nozzle throat along the line in the middle of the nozzle width. The distance between pressure taps was 10 mm. Each measurement series lasted 30seconds and the static pressure was measured with the frequency of 400 Hz, what gave the total value of 12000 samples. The accuracy of the applied pressure transducers (Honeywell 243PC15M) was ±100 Pa. This accuracy was achieved by individually calibration of the pressure transducers within the range of 0÷100 kPa(a).

The measurement uncertainty within one series was calculated as a difference between maximum and minimum value plus accuracy of the sensor. The length of the pressure tubes connecting the pressure transducers with the taps was about 200 mm, what is too big value in order to capture the flow unsteadiness in high accuracy.

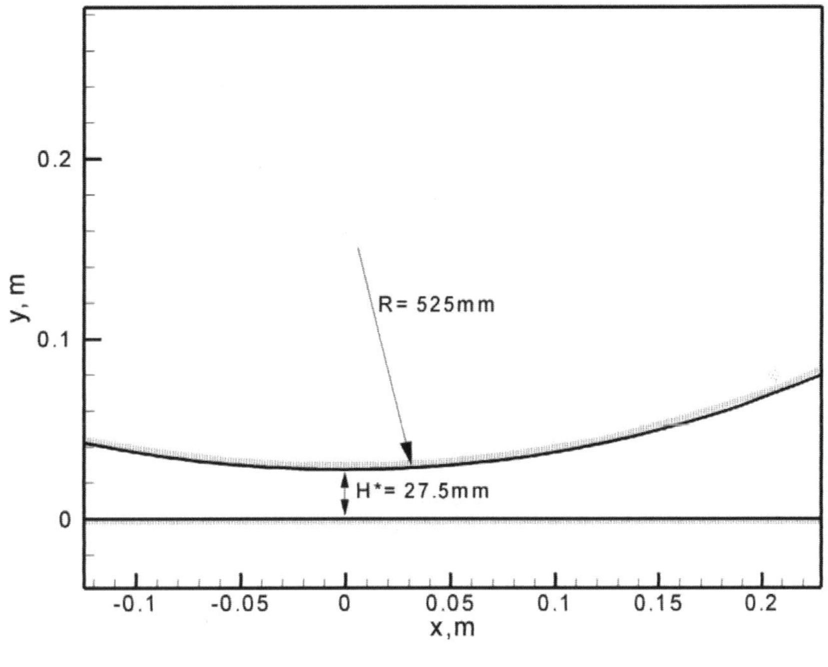

Figure 3. Geometry of the half arc Laval nozzle.

For nozzle at the inlet the value of total pressure was $112500^{\pm 250}$ Pa(a) and temperature of $111^{\pm 0.25}$ °C. The back pressure value amounted to 50000 Pa(a).

Figure 4 shows the observed water membrane created on the shock wave. This phenomenon was visible by naked eye, without additional flow visualisation techniques.

In order to explain the appearance of water film on the shock wave the sketch presented in the figure 5 may be useful. Figure 5 depicts the possible configuration of the shock waves in the nozzle caused by too high back pressure.

It can be conclude that the one possible reason for creating the observed water film is the rapid retardation of the fine droplets on the transition through the shock wave. It contributes to the collision and coagulation of the fine droplets, finally creating the coarse droplets behind the shock wave.

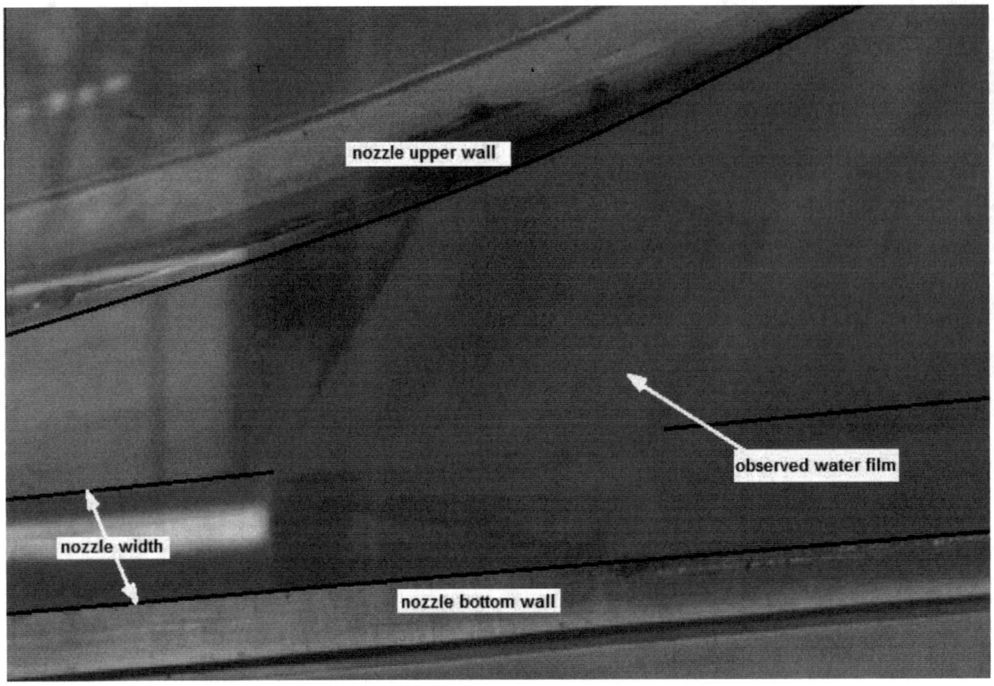

Figure 4. Photo of the observed water film on the shock wave.

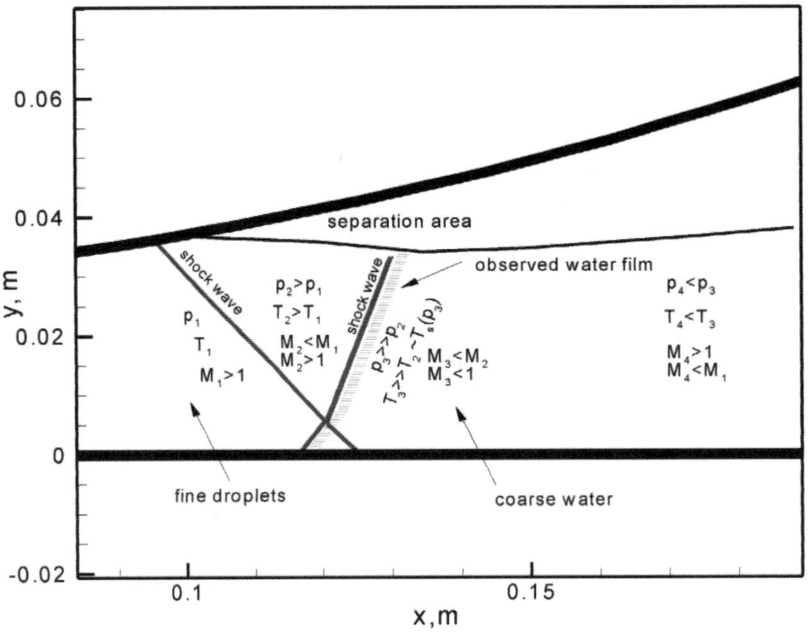

Figure 5. Sketch of the shock waves configuration caused by high back pressure in supersonic part of the Laval nozzle (*M – Mach number, p – static pressure, T – static temperature*).

Figure 6 presents the values of measured static pressure together with measurement uncertainty along the bottom wall of the nozzle. It is visible that in the vicinity of the shock wave the measurement uncertainty rapidly increases. It is caused by the interaction of the shock wave with the water film on the bottom and top walls and chaotic movement of the coarse water behind the shock.

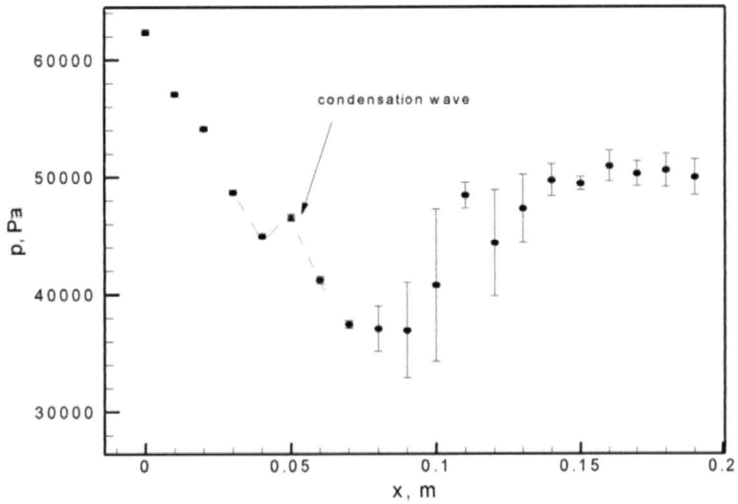

Figure 6. Measured static pressure distribution along the bottom wall of the nozzle *(p – static pressure)*.

In spite of too long pressure tubes that were applied in experiment the dominated frequency of the unsteadiness appearing in the shock wave vicinity was a little above 100 Hz (figure 7). The value of this frequency depends on the flow conditions as well as the nozzle geometry (expansion rate $\dot{P} = -\frac{1}{p}c\frac{dp}{dx}$, where c is the speed of sound).

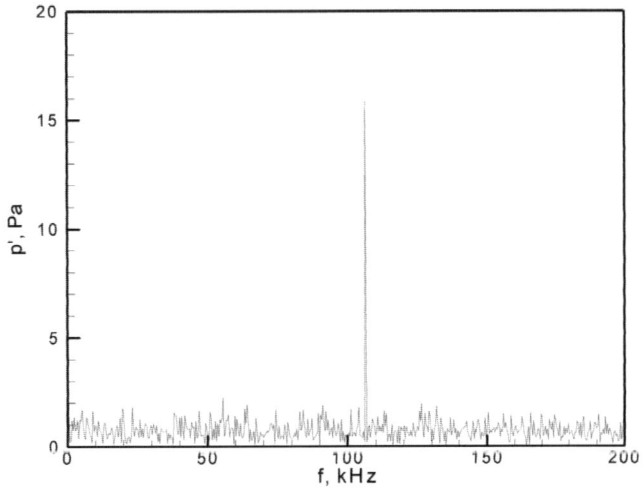

Figure 7. FFT analysis of the static pressure fluctuation on the bottom wall for a position $x = 0.1\ m$ *(p' – pressure fluctuation)*.

5. Conclusions

The paper presents the experimental results of a steam condensing flow in the Laval nozzles. The effect of condensation phenomena and the shock wave behaviour in the wet steam region were investigated.

The obtained experimental results led to the following conclusions:

- On a strong shock wave the water film (water membrane) was observed by naked eye that was created due to the strong retardation of the fine droplets created in the results of primary condensation.
- Behind the water film observed on the shock wave the coarse water droplets are created that were formed in the result of collision and coagulation of the fine droplets. The coarse water moves in the chaotic way.
- The experiments of the shock wave behaviour in the wet steam region showed a coarse water formation in detachment zone, i.e. in the place of the shock wave interaction with the solid wall.
- The coarse water formation and its chaotic movements cause the flow unsteadiness with the frequency of hundreds hertz, this value depends mainly on the expansion rate (nozzle geometry and boundary conditions at the nozzle inlet).

The future experimental work will concentrate on a detailed explanation of the phenomenon of coarse water formation in separation region caused by shock wave. This phenomenon can be crucial for further research in nozzle but in particular in linear cascades. Also, obtained experimental data will be helpful in the future work on a numerical model, its validation.

6. References

[1] Gyarmathy G 1960 *Grundlagen einer Theorie der Nassdampfturbine* (Zürich: Juris Verlag)

[2] Krol T 1971 Results of optical measurement of diameters of drops formed due to condensation of steam in a Laval nozzle *Trans. Inst. Fluid Flow Machinery (Poland)* **57** pp 19–30

[3] Moses C A and Stein G D 1978 On the growth of steam droplets formed in a Laval nozzle using both static pressure and light scattering measurements *ASME* **100** pp 311–22

[4] Schnerr G H 1986 *Homogene Kondensation in stationaeren transsonischen Stroemungen durch Lavalduesen und um Profile* (Habilitationsschrift, Fakultaet fuer Maschinenbau, Universitaet Karlsruhe)

[5] White A J, Young J B and Walters P T 1996 Experimental validation of condensing flow theory for a stationary cascade of steam turbine blade *Philosophical Transaction of Royal Society London A* **354** pp 59–88

[6] Bakhtar F, White A J and Mashmoushy H 2005 Theoretical treatments of two-dimensional two-phase flows of steam and comparison with cascade measurements *Proc. IMechE, Part C: J. Mechanical Engineering Science* **219(C12)** pp 1335–55

[7] Bakhtar F and Zamri M Y 2011 On the performance of a cascade of improved turbine nozzle blades in nucleating steam - Part 3: Theoretical analysis *Proc. IMechE Part C: J. Mechanical Engineering Science* **225** pp 1649–71

[8] Dykas S, Wróblewski W and Łukowicz H 2007 Prediction of losses in the flow through the last stage of low-pressure steam turbine *International Journal for Numerical Methods in Fluids* **53 (6)** pp 933–45

[9] Young J B 1982 Spontaneous condensation of steam in supersonic nozzles *Physicochemical Hydrodynamics (PCH)* **3** pp 57–82

[10] Dykas S and Wróblewski W 2013 Two-fluid model for prediction of wet steam transonic flow *International Journal of Heat and Mass Transfer* **60** pp 88-94

[11] Halama J, Fort J and Seifert M 2010 Numerical solution of wet steam flow with a priori droplet size distribution *Proceedings of Conference "Topical Problems of Fluid Mechanics" (Prague)* pp 63–66

[12] Stastny M and Sejna M 1999 Numerical analysis of hetero-homogeneous condensation of the steam flowing in turbine cascade *ImechE Conference Transactions Turbomachinery Vol. B (London)*
[13] Frenkel J 1946 *Kinetic Theory of Liquids* (New York: Oxford University Press)
[14] Doerffer P and Dykas S 2005 Numerical analysis of shock induced separation delay by air humidity *Journal of thermal sciencc* **14(2)** pp 120 25
[15] Smołka K, Strozik M, Majkut M and Dykas S 2011 Możliwości badań eksperymentalnych przepływu pary mokrej w układzie mini-siłowni kondensacyjnej *Prace Naukowe. Konferencje* **27** pp 233–40

Acknowledgements
The presented work was supported by the National Science Centre founds within the project with no. 2011/03/B/ST8/06125

Study on Flow behind Backward-Facing Step in a Narrow Channel

V Uruba[1]
Institute of Thermomechanics AS CR, v.v.i., Praha, Czech Republic

E-mail: uruba@it.cas.cz

Abstract. Both statistics and dynamics of the flow-field behind a backward-facing step in a narrow channel is studied experimentally using time-resolved PIV technique. The channel width is 4 step heights, Reynolds number based on the step height was about 35 thousands. Secondary flow effect is detected clearly in the region behind the step resulting in strong 3D flow structure just behind the step. The low-frequency quasi-periodical structures appearing in the region behind the step and in the region of the flow reattachment close to the channel bottom are studied using the Oscillation Pattern Decomposition (OPD) technique. Strouhal numbers of those phenomena are very low, order of 0.01. Typical time-mean structures as well as dynamical structures are to be presented.

1. Introduction

The backward-facing step flow has been established as a benchmark configuration for separated flow studies in fluid mechanics.

The flow over a backward-facing step is a very simple as for its geometry but the flow structure is extremely complex both in space and time. The flow over the backward facing step is considered to be 2D in sense of statistical characteristics as a rule, meaning that the ratio of the step height and channel width is very high, say more than 20. Then the flow-field in the middle third channel width is pretty 2D, with constant statistical characteristics along the span. Detailed study of the flow in a narrow channel, when the ration of channel width over step height is lower than 10 is very limited within public resources. The flow-field structure model in the recirculation region behind a step in narrow channel was suggested in [5]. The flow visualization on the channel bottom just behind the step showed a kidney-shaped region, suggesting existence of recirculation vortices impinging perpendicularly to the bottom – see figure 1. Here the channel bottom is shown with result of the surface visualization using painting, published in [10]. Flow direction is from left in x direction, step edge position is $x = 0$ .The red lines correspond to vanishing streamwise mean velocity component in the measuring plane (more details see section 3.1).

The case has been studied in our lab in more details. The first publications [6,8] cover the mean flow structure, those results have been summarized in [3]. Then, the dynamical behavior of the flow-field has been studied. Information on secondary structures dynamics in prismatic channel of rectangular cross-section is in [9]. Then some studies on flow in channel with backward facing step

[1] IT ASCR, Dolejskova 5, 182 00 Praha 8, Czech Republic.

have been carried out. Analysis of velocity field dynamical behavior in a few planes perpendicular to the mean flow has been presented in [4].

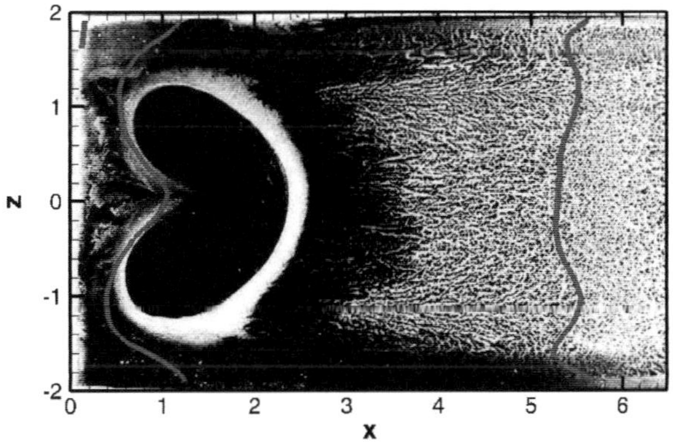

Figure 1. Surface visualization on the channel bottom.

2. Experimental Setup and Used Methods

Experimental setup and methods are to be introduced first.

2.1. Experimental Setup

The existing blow-down facility has been used for the experiments. The wind tunnel has rectangular cross-section with filled corners within the contraction (to suppress corner vortices), honeycomb and a system of damping screens followed by contraction with contraction ratio 16. The area of the test section input is 0.25 m in height and 0.1 m in width. The time mean velocity departures from homogeneity in planes perpendicular to the tunnel axis are of order tenth of per cent with the exception of corners, where corner vortex starters could be detected. The channel downstream the backward facing step was 1 m in length. The step is placed close to the channel inlet.

In the channel inlet the boundary layer thickness on the walls in the step position was about 3 mm, the step height was $h = 25$ mm. Velocity in the inlet on the step was $U_e = 20$ m/s and natural turbulence level was about 0.1 % in the working section input. The Reynolds number defined using the step height was 34 400. All presented dimensions are related to the step height and velocities to the inlet velocity U_e.

Schematic view of the experimental setup and measurement areas is given in figure 2.

The Cartesian coordinate system is defined, x is the streamwise direction, y is spanwise perpendicular to the channel bottom and z is spanwise direction parallel to the step edge. All coordinates values are non-dimensioned using the step height. The system origin was chosen on the step edge.

Two measuring domains are defined. The first 2D domain is the *measuring plane* placed behind the step oriented parallel to the channel bottom in the distance of about 1 mm above it ($y = -0.96$). It is used mainly for studying the flow dynamics behind the step and in the reattachment region.

The second 3D domain is the *measuring cuboid* covering one half of the volume behind the step. The cuboid is filled by 20 measuring planes and mean flow characteristics are interpolated to the 3D domain. This is used only for representation of time-mean data. The remaining half of the mean-flow-field is supposed to be symmetrical.

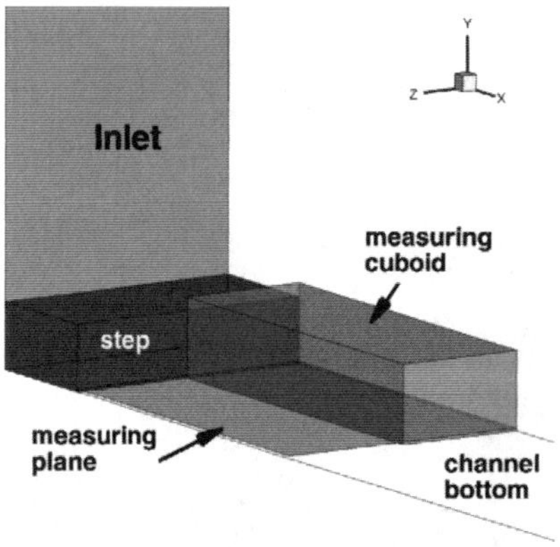

Figure 2. Experimental setup and measuring areas.

Inlet section is shown in blue, measuring cuboid in green, measuring plane in red and the step is grey.

2.2. Measuring Technique
The time-resolved PIV method was used for the experiments. The measuring system DANTEC consists of a double-pulse laser with cylindrical optics and CCD camera. The software Dynamics Studio 3.4 was used for velocity-fields evaluation. Laser New Wave Pegasus Nd:YLF, double head, wavelength 527 nm, maximal frequency 10 kHz, a shot energy is 10 mJ for 1 kHz (corresponding power 10 W per head). Camera Phantom V711 has maximal resolution 1280 x 800 pixels and corresponding maximal frequency 3000 double-snaps per second. The presented experiments are intended to cover low-frequency dynamics offering good statistics. Thus, for the measurements, the frequency 100 Hz and 4000 double-snaps in sequence corresponding to 40 s of record for mean evaluation was acquired. As the seeding source the SAFEX system has been used producing oil droplets of about 1 μm in size. Details are given in references [2-9].

2.3. Analysis Methods
The mean velocity field has been evaluated first to obtain the flow statistics and remove the mean flow. The dynamical behavior is characterized by the turbulent kinetic energy defined in usual way as a half of sum of individual velocity components variances. The dynamical behavior was studied using the Oscillation Pattern Decomposition (OPD) method, which is described in details in [5,7].

The OPD method evaluates the basis representing oscillating modes of the extended dynamical system, which are characterized by a single frequency and damping representing cyclostationary elementary processes involved in the phenomenon notwithstanding that hidden. The structure can be typically a wave or travelling structures propagating in space, or pulsating pattern represented by complex vector field.

Categorizing of the resulting vector field is performed with help of a classical approach presented in [1].

3. Results
The selected results of experiments are to be presented in following paragraphs.

First, statistical characteristics of the flow quantities distributions will be presented, then dynamical features of selected regions are to be shown.

3.1. Statistics
Statistical characteristics were evaluated in all inspected regions, i.e. in the 3D domain – measuring cuboid and in the 2D region – measuring plane. In all figures the kidney shape (in black) and $U = 0$

lines (in red) are shown. Please note that the kidney is located on the bottom ($y=-1$), while lines of vanishing mean streamwise velocity component are in the measuring plane ($y=-0.96$), so in the perspective view the alignment of the curves is affected.

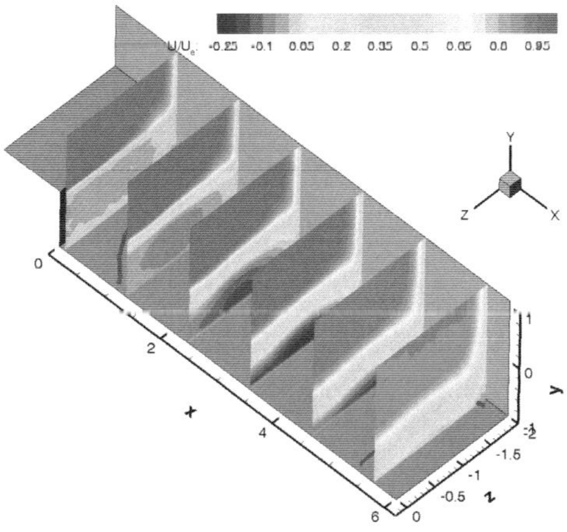

Figure 3. Streamwise velocity component distribution.

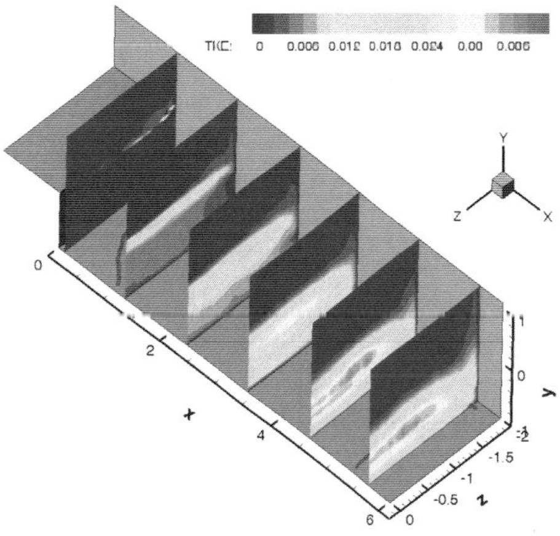

Figure 4. Kinetic energy distribution.

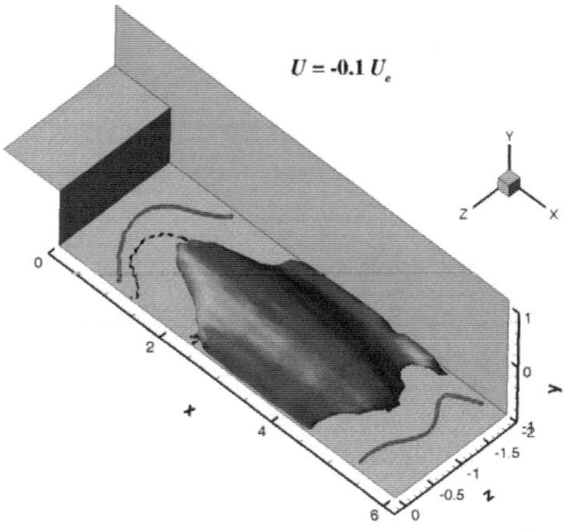

Figure 5. Iso-surface $U=-0.1U_e$.

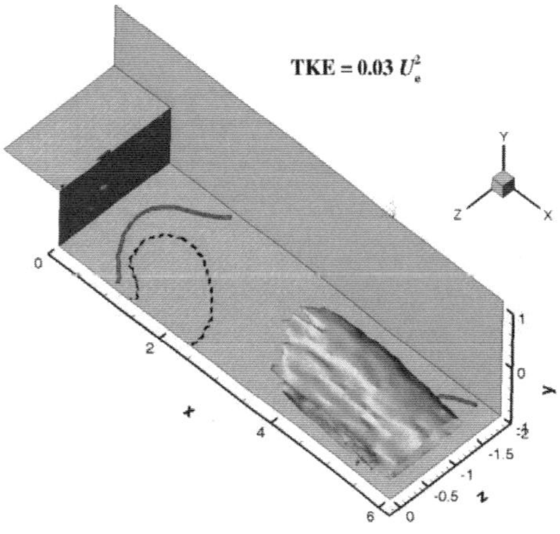

Figure 6. Iso-surface $\text{TKE}=0.03U_e^2$.

In figures 3 and 4 distributions of streamwise mean velocity component U and turbulent kinetic energy are shown in 6 planes $z=const$. Both velocities and kinetic energy are non-dimensioned using the velocity in the channel inlet U_e.

Position of the recirculation region (in blue) could be recognized clearly with local maximum of back-flow velocity in position $z=-1$.

Just behind the step the kinetic energy maximum is located within the free shear layer in the step edge height ($y=0$) falling down in the streamwise direction. In the reattachment region $x=5.5$ its position is about half step bellow $y=-0.5$.

Then, the iso-surfaces of constant mean streamwise and spanwise vertical components and shown in figures 5 and 6. The back-flow region shape is clearly visible in figure 5, in figure 6 the high kinetic energy region is shown located above the reattachment region.

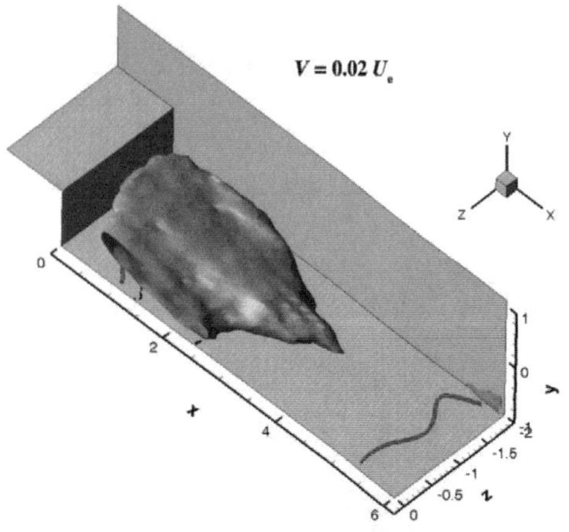

Figure 7. Iso-surface $V = 0.02U_e$.

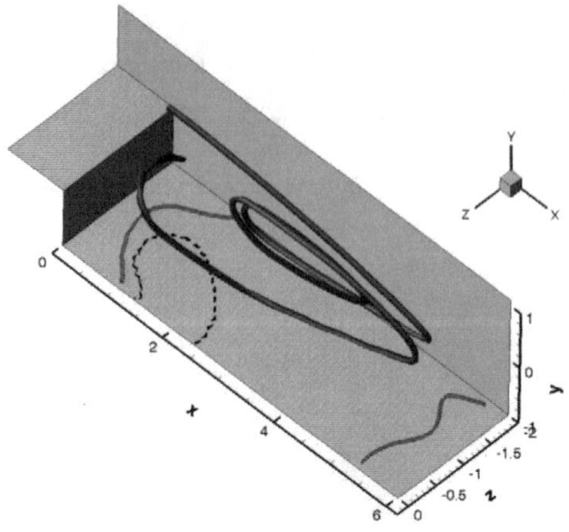

Figure 8. Vector-line of the time-mean velocity field.

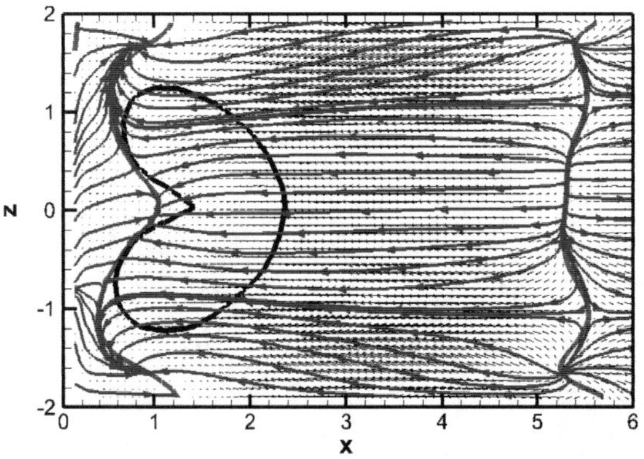

Figure 9. The mean velocity field in the measuring plane.

In the recirculation region the vertical spanwise mean velocity component seems to be of great importance. In figure 7 there is iso-surface of $V = 0.02U_e$. The spiky shape streamwise oriented is located just above the kidney pattern on the channel bottom.

Topology of the recirculation region is characterized by the vector-line of the time-mean velocity field shown in figure 8, the line originated in the channel corner just above the step.

The flow pattern in the measuring plane above the channel bottom is analyzed as well. In figure 9 the vector lines are added in blue, in red the contour of the kidney structure is given. The black lines represent the position of vanishing longitudinal mean velocity component U, dividing the area into parts with forward ($U > 0$) and backward ($U < 0$) mean flow. Note that between the black lines in the figure middle the back-flow region is located. Forward flow direction close to the step for $0 < x < 1$ is connected with appearance the tertiary vortex, while secondary vortex generates back-flow close to the bottom.

For the purpose of dynamical behaviour analysis the measuring plane has been divided into 2 parts. The first region is located just behind the step, where the kidney structure appears $x \in (0;2.5)$, the second is the flow reattachment region $x \in (4.5;6.5)$.

3.2. Dynamics of the Region just behind the Step

The region behind the step is studied first. The dominant 3 modes are those with biggest e-folding time and/or periodicity value. For the given case, the modes 1, 2 and 3 could be considered as dominant. Topology of the dominant modes is shown in Figures 10, 11 and 12. The modes topology is represented by vector fields of real and imaginary parts of the spatial mode. To visualize the structures involved (especially vortices), vector-lines (in blue) are added.

The dominant modes are all oscillating with travelling character. Topology is characterized by attracting focuses representing vortices.

For the OPD mode 1, the real and imaginary parts are in figure 10, the cores of vortices are marked by circles in red with numbers. The mode parameters, namely corresponding Strouhal number, frequency, e-folding time and periodicity was respectively: $Sr_1 = 0.0094$, $f_1 = 8.18$ Hz, $\tau_{e1} = 37.7$ ms, $p_1 = 0.308$.

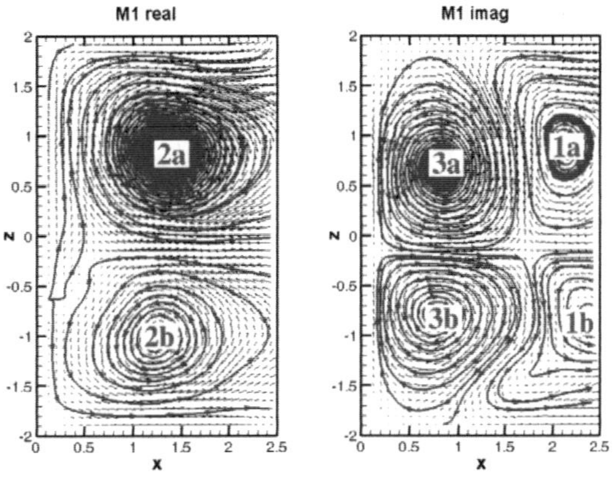

Figure 10. The OPD mode 1 in the flow behind the step.

In figure 10 a pair of contra-rotating vortices with alternative orientation travelling against the step periodically is captured. The vortices are marked “a” and “b”. The process could start by vortices in positions 1a, 1b (in the imaginary part). The vortex pair moves towards the step reaching in time corresponding to a quarter of the period position 2a, 2b (see real part). In the time half period (i.e. 61 ms) the vortices reach positions 3a, 3b, however theirs orientation is still the same. In imaginary part of figure 10 the situation of the vortices in position 3 is related to preceding vortex pair with opposite orientation. Thus, the OPD mode 1 represents the ***train of contra-rotating vortex pairs*** oriented in y direction with distance about 1.5 in z direction moving towards the step by the mean velocity of about 0.0146 of the inlet velocity U_e. The vortex pairs appear periodically in trains changing theirs orientation on the regular basis during the single mode appearance.

The OPD mode 2 is shown in figure 11. The mode relevant parameters: $Sr_2 = 0.0079$, $f_2 = 6.85$ Hz, $\tau_{e2} = 24.6$ ms, $p_2 = 0.168$. The topology shows two vortices of the same orientation, 1a and 1b, moving to the position 2a and 2b, the vortex 2c appears in the same time a quarter of the period in the channel center. Then all 3 vortices of the same orientation merge into a single vortex 3 in half period time 73 ms. The last stage of the development is represented by the vortex 4 located just behind the step in the channel middle. The process is repeated periodically, however the opposite oriented vortices are involved with the shift equal to a half process period. The vortices cores (a and b) mean velocity is about 0.0122 of the U_e. Thus, the OPD mode 2 represents ***coalescence of the co-rotating vortices into a single vortex*** of the same orientation.

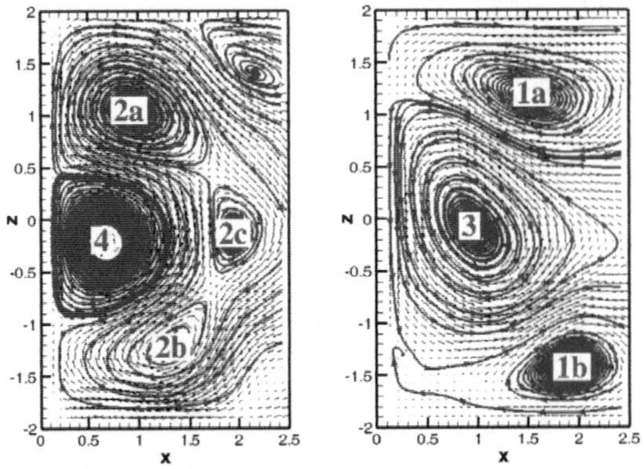

Figure 11. The OPD mode 2 in the flow behind the step.

The last dominant OPD mode 3 is shown in Figure 12. The mode relevant parameters: $Sr_3 = 0.0129$, $f_3 = 11.14$ Hz $\tau_{e3} = 23.3$ ms, $p_3 = 0.259$. The mode could be interpreted as splitting of the single vortex 1 located on the channel axis into two smaller vortices 2a and 2b moving towards the step corners by mean velocity of about 0.0133 U_e. The next stages in a quarter period distances are 3a, 3b and finally 4a, 4b. The vortices cores are just filling the kidney-shaped area, shown in green. Thus, the OPD mode 3 represents ***splitting of the central vortex into two co-rotating vortices***, all vortices of the same orientation.

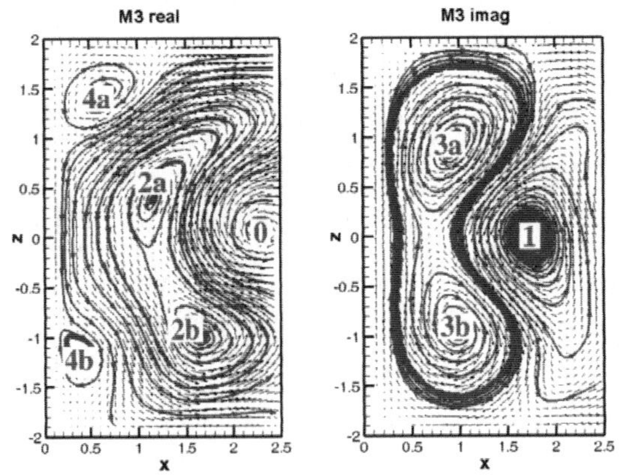

Figure 12. The OPD mode 3 in the flow behind the step.

The OPD mode 3 is the most probably directly connected with development of the kidney structure on the channel bottom, especially its downstream contour (on the right-hand side). The contour close to the step is connected with the zero U line (confirm figure 9).

3.3. Dynamics of Reattachment Region

Dynamics of the reattachment region is completely different in comparison with the region just behind the step. The typical patterns are the nodes in streamwise orientation of alternatively repealing and attracting character. A few saddle points and focuses could be recognized.

As an example the OPD mode 2 is shown in figure 13. The mode relevant parameters: $Sr_2 = 0.0088$, $f_2 = 7.61$ Hz $\tau_{e3} = 15.8$ ms, $p_2 = 0.120$. This mode involves attracting nodes (red arrows) repealing nodes (green arrows) as well as a saddle point (green square). The mode dynamics could be interpreted as ***transversely moving nodes*** in *z* direction though the region.

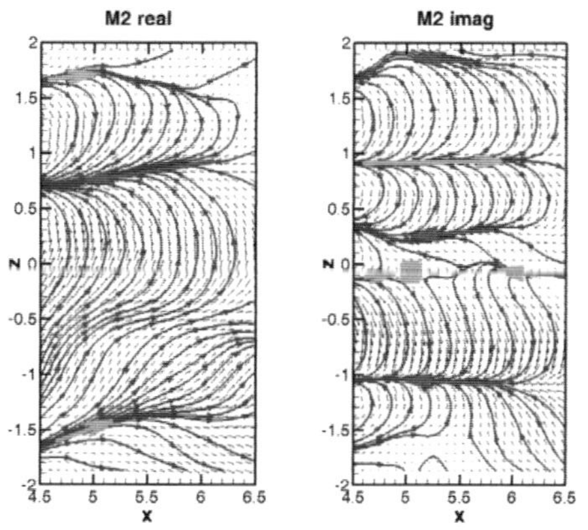

Figure 13. The OPD mode 2 in the reattachment region.

However, the OPD modes in the reattachment region are typically decaying with a small periodicity value.

4. Conclusions

The flow-field statistics and low-frequency dynamics of the flow behind the backward-facing step in narrow channel have been analyzed in the presented paper.

The results suggest presence of several families of vortical structures with particular dynamics involving contra-rotating vortex pairs train, vortex splitting and coalescence and travelling nodes.

5. References

[1] Helman J, Hesselink L 1989 *IEEE Computer* **22**, 8 27
[2] Uruba V 2014 *EPJ Web of Conferences* **67** 02120
[3] Uruba V 2013 *Colloquium FLUID DYNAMICS 2013* 35 http://www.it.cas.cz/cs/kolokvium-dynamika-tekutin-2013?q=en/node/5039
[4] Uruba V 2013 *EPJ Web of Conferences* **45** UNSP 01108
[5] Uruba V 2012 *EPJ Web of Conferences* **25** 01095
[6] Uruba V 2012 *Colloquium Fluid Dynamics 2012* 33 http://www.it.cas.cz/cs/kolokvium-dynamika-tekutin-2012?q=en/node/4886
[7] Uruba V 2012 *XX Fluid Mechanics Conference XX. Gliwice* 152
[8] Uruba V, Jonáš P 2012 *Proceedings in Applied Mathematics and Mechanics* **12** 1 501
[9] Uruba V, Hladík O, Jonáš P 2011 *Journal of Physics Conference Series* **318** 062021
[10] Uruba V, Jonáš P, Mazur O 2007 *International Journal of Heat and Fluid Flow* **28** 4 665

Acknowledgements

This work was supported by the Grant Agency of the Czech Republic, project No. P101/10/1230.

Journal of Physics: Conference Series **530** (2014) 012025 doi:10.1088/1742-6596/530/1/012025

Hot – wire study on the axisymmetric free jet dependence on the nozzle shape.

A Pawlowska, P Domagala and M Wysocki
Institute of Thermal Machinery, Czestochowa University of Technology,
Al. AK 21, 42-200 Czestochowa, Poland

E-mail: pawlowska@imc.pcz.czest.pl

Abstract. The paper presents the results of experimental research in axisymmetric free jet performed in order to verify the previous numerical calculations. During the study the influence of the initial conditions at the outlet of the nozzle on the formation of self-sustained oscillations in free jet was determined. The variables affecting the initial conditions were: the shape of nozzle tip (free – slip and no - slip), contraction ratio (C = 128 and 144) and Reynold's number (Re = 5 000 – 40 000). Measurements were taken using the CTA probes. An analysis of the distributions of mean velocity at the nozzle exit shows that a laminar boundary layer was obtained, values of integral boundary layer parameters were not significantly affected by the type of nozzle tip. Some differences were observed in the turbulence intensity distributions in the exit plane and along the jet axis, where characteristic plateau could be noted for case with the free – slip nozzle tip, which indicates the presence of coherent structures in the flow. These results suggest that for almost the same initial conditions obtained for plane and profiled nozzle tip the modification of organized structures in the far flow field was obtained. For physical explanation of these phenomena it was necessary to perform spectral and spatio – temporal correlation analysis (not discussed in the present paper).

1. Introduction

The paper presents the results of research devoted to the experimental verification of the self-sustained oscillations in axi – symmetric jet, which were found in numerical experiment of Boguslawski et al. [1]. Preliminary experiments and LES calculations reported in the above paper allowed to formulate the idea that in the free axisymmetric jet there is a possibility to obtain the regime of self-sustained oscillations, provided that the boundary layer at the nozzle exit is sufficiently thin and the perturbation level sufficiently low. In particular it was found that the thickness of the exit shear layer is of great importance for the existence of regime of self-sustained oscillations.

The main problems with the comparison of experimental and numerical investigations in free jets is the identity of initial conditions, which in the flow considered should be a matter of extreme care. As it was stated in the recent review paper by Ball et al.[2] the round jet is extremely sensitive to inlet conditions, which change not only the near flow field but this influence may propagate even to the far field. As it was also noticed by Nathan [3] "...jet never forgets...", that is attributed to the underlying structure of turbulent inlet motion, which is carried throughout the entire flow field. The reason is that even small changes of inlet conditions cause considerable modification of large – scale vortices, it concerns in particular their shape and convection velocities, as it was proved by many researchers e.g. by Verzicco and Orlandi [4] and Romano [5]. That is why not only experimental values of Re number and turbulence intensity Tu but also the D / δ^{**} had to match numerical calculations, where δ^{**} is the momentum thickness of initial shear layer, while D is the diameter of jet nozzle outlet.

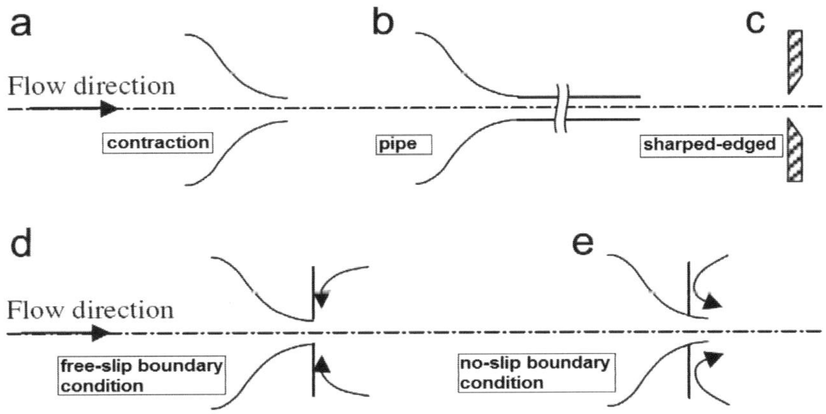

Figure 1. The sketch of possible inlet conditionf for round jet, *Ball et al. [2].*

As far as the initial flow geometry is concerned two aspects are important, the first is the shape of nozzle, which affects the initial velocity profiles, the second aspect is the shape of the nozzle end (tip). Ball et al.[2] distinguish three shapes of jet nozzle, the first ('figure 1a') is the classical contoured nozzle widely used in aerodynamic research, the second ('figure 1b') is the pipe flow and the third one ('figure 1c') is the sharp edge orifice. In the present research the contoured nozzle was selected because in this case only a sufficiently low level of turbulence could be achieved.

The additional problem related to comparison of numerical and experimental initial conditions results from the fact, that in LES calculations, which allowed to establish the idea of self-sustained regime, the moderate degree of coflow was used, it was necessary because periodic boundary conditions were applied. The numerical tests performed by da Silva and Metais [6] suggest that the coflow up to 5 % of mean velocity does not influence the results of computations, but it was decided to verify experimentally whether this numerical "trick" may exert some influence upon the behaviour of the free jet. The case without a coflow corresponds to "free – slip boundary condition" ('figure 1d'), while the case with coflow is mimicked by "no – slip b.c." ('figure 1e'). Both types of boundary conditions were investigated during the present research as it will be discussed later.

2. Experimental Setup and Signal Processing

The sketch of the test rig is presented at 'figure 2', in fact two rigs with different contraction ratios were investigated but both rigs were of identical design. The air is supplied at the bottom, the fan, filter, ducts and acoustic silencers are located outside the measuring area and are not shown at the sketch. The air first passes through the heater, where electric heaters are filled with iron pellets which are designed to increase the heat capacity and to make the flow more uniform. Then the air enters the settling chamber where 6 wire gauzes are located, which makes the flow more uniform and reduces the turbulence level at the outlet.

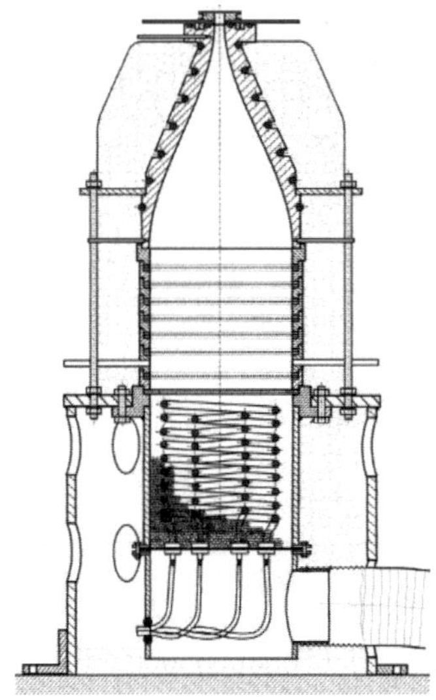

Figure 2. The sketch of the test rig.

Table 1 summarizes the main flow parameters applied in the experiment. One should notice the large contraction ratio of nozzles, which is much larger than in most experiments described in literature. The rig allows to perform measurements at the range of Reynolds numbers from 5 000 to 40 000, the Reynolds number is based on the outlet nozzle diameter. The values of shape parameter H confirm that the shear layer is laminar, the values of δ^{**} integral parameter are given in millimetres.

Two different values of contraction ratio C were used (see table 1), these two designs are supposed to provide different values of initial turbulence intensity at the nozzle outlet. The wire gauze surrounding the jet facility was used to prevent the disturbing influence of draughts and possible convective motions in the room.

Table 1. Inlet parameters for both experiments.

	C	δ^{**}[mm]	H	D/δ^{**}
Exp.No.1	128	0.080 – 0.199	2.50 – 2.66	75 - 188
Exp.No.2	144	0.088 – 0.193	2.56 – 2. 68	78 - 171

As far as the influence of the shape of the nozzle tip is concerned, it was decided to investigate both free - slip and no-slip initial nozzle shapes shown at 'figure 3'. The no – slip nozzle tip ('figure 3a') corresponds to the case without coflow, while the free – slip ('figure 3b') imitates the case with coflow at the nozzle edge.

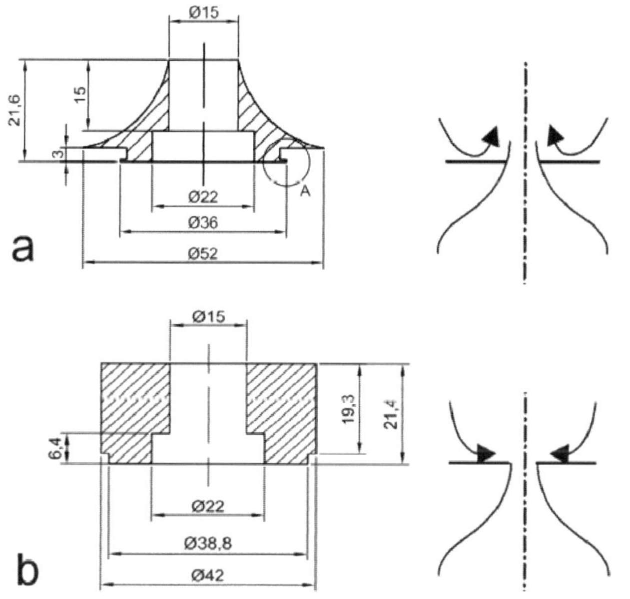

Figure 3. The sketch of free – slip (a) and no – slip (b) nozzle tips.

The thickness of the exit shear layer was among the important flow parameters, so the contoured nozzle was combined with cylindrical extensions. The CTA single and "X" wire probes with wire length to diameter ratio 250 were used, the accuracy of probe positioning was 0.01 [mm]. The sketch of CTA equipment together with traversing mechanism is shown at 'figure 4'. For CTA measurements dual-channel acquisition was carried out, the first channel was connected directly to the CTA bridge and contained both DC and AC signals, while the second channel contained the AC signal only. Data from first channel were used to calculate the average velocity and from the second one to calculate the turbulence intensity.

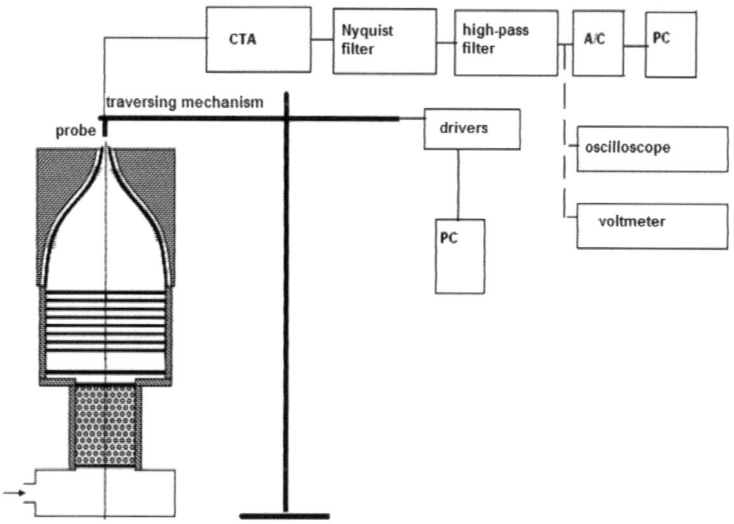

Figure 4. The sketch of CTA equipment.

3. Discussion of results

Velocity profiles measured at the nozzle exit are presented at 'figure 5' in universal coordinates for all values of Reynolds numbers used in the experiment. The results obtained close to the wall were corrected for wall proximity, one may notice that all the points follow the linear velocity distribution $U+ = y+$, the corresponding values of shape parameter H collected in table 2 confirm that we have here an undisturbed laminar shear layer and that the type of nozzle tip does not influence the integral parameters of the exit shear layer.

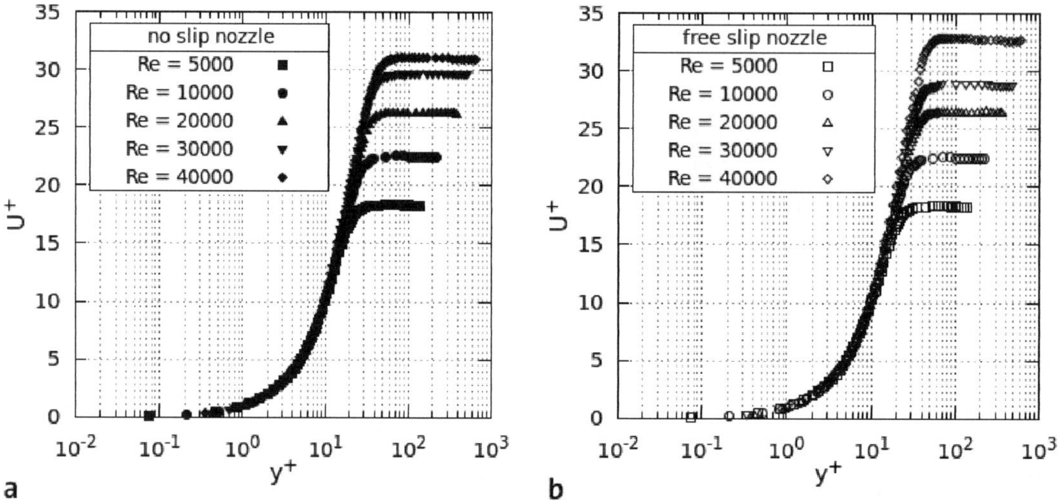

Figure 5. Velocity profiles in exit shear layers for various Re and C as well as for free – slip (a) and no – slip (b) nozzle tips.

Table 2. Values of shape parameter for various Re and C as well as for free – slip (a) and no – slip (b) nozzle tips.

		Re = 5 000	Re = 10 000	Re = 20 000
Exp.No.1	free - slip	2.652	2.363	2.630
	no - slip	2.664	2.594	2.617
Exp.No.2	no - slip	2.591	2.567	2.617

'Figure 6' shows the comparison of turbulence intensity measured at the nozzle exit with the free – slip nozzle tip for all Reynolds numbers, turbulence intensity Tu is presented versus the r / D relative radial coordinate, the value r / D = 0 corresponds to the jet axis, the value r / D = 0.5 corresponds to nozzle edge. The distributions of Tu reveal that for the lowest values of Reynolds number there is almost no increase of Tu in the shear layer, in fact Tu level is identical as in the potential core, the increase of Reynolds number brings about the higher level of Tu in the shear layer, one may also notice that the increase of Re shifts the maximum Tu towards the outer region of the shear layer. Very similar are the results for the no - slip nozzle tip (not shown here), but contrary to the previous case the increase of Re shifts the maximum Tu towards the inner region of the shear layer.

Journal of Physics: Conference Series **530** (2014) 012025 doi:10.1088/1742-6596/530/1/012025

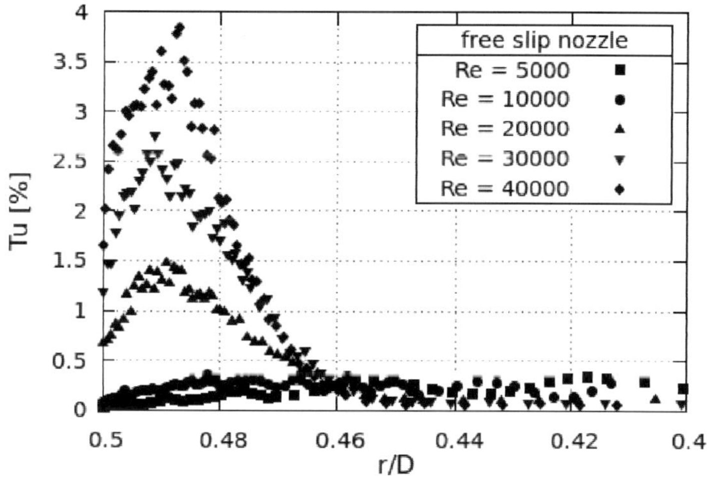

Figure 6. Turbulence intensity Tu profiles in exit shear layers for various Re and for the free – slip nozzle tip.

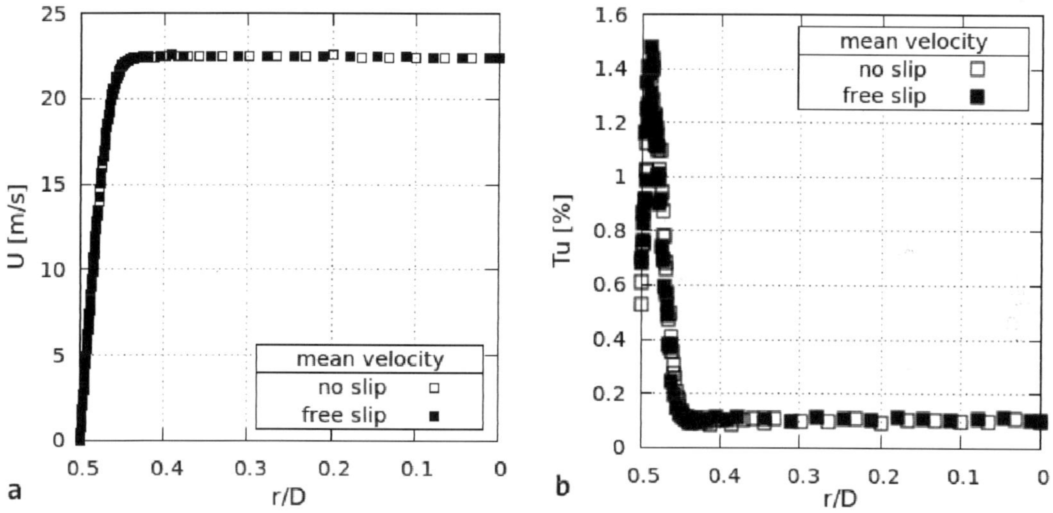

Figure 7. Sample comparison of mean velocity profiles (a) and turbulence intensity Tu (b) at the exit shear layer for Re = 20 000 and for free – slip and no – slip nozzle tip.

'Figure 7' shows the comparison of mean velocity and turbulence intensity at the nozzle exit for both free – slip and no – slip nozzle tips for sample value of Re = 20 000 and for C = 128 (Exp. No.1). One may easily notice that both these distributions are in fact identical, one may conclude therefore that the shape of nozzle tip does not affect the jet initial mean velocity and Tu profiles.
'Figure 8' presents the downstream evolution of turbulence intensity at the jet axis for the sample value Re = 20 000, one may notice a substantial difference in Tu profiles, in particular the plateau visible for plane tip becomes much more pronounced for the profiled nozzle tip. This plateau is related to the presence of coherent structures, that means that no – slip nozzle tip creates more favourable conditions for the development of large scale structures. This comparison is to some extent surprising, because it suggests that even if initial conditions are the same for free - slip and no - slip conditions the amplification of organized motion may be substantially different. In order to explain these findings the

spectral and spatio - temporal correlations was performed, which allowed to identify the development of large scale vortices.

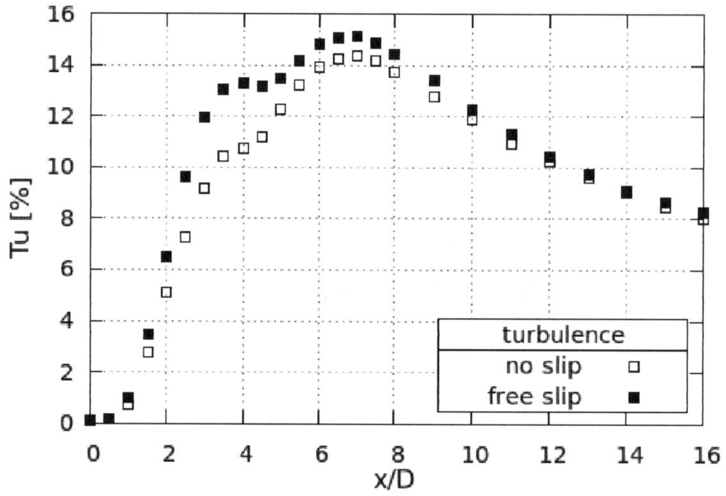

Figure 8. Sample comparison of turbulence intensity profiles along the jet axis for Re = 20 000 and for free – slip and no – slip nozzle tip.

The next part of the research was devoted to analysis of the influence of nozzle contraction ratio, which was performed by comparative tests of rigs No. 1 and 2. The higher contraction ratio applied in the test rig No. 2 allowed to expect the lower turbulence level at the jet exit. However, the data presented at 'figure 9' do not support this expectation. 'Figure 9a' presents a comparison of Tu measured at the exit plane for both test rigs, 'figure 9b' presents the mean turbulence intensity at the potential core, for test rigs No. 1 and 2. One may notice that for all Reynolds numbers the mean value of Tu in potential core and maximum value in the shear layer are identical, with the exception of the smallest Re number 5 000, where the increase in the nozzle contraction caused the visible increase of mean turbulence intensity.

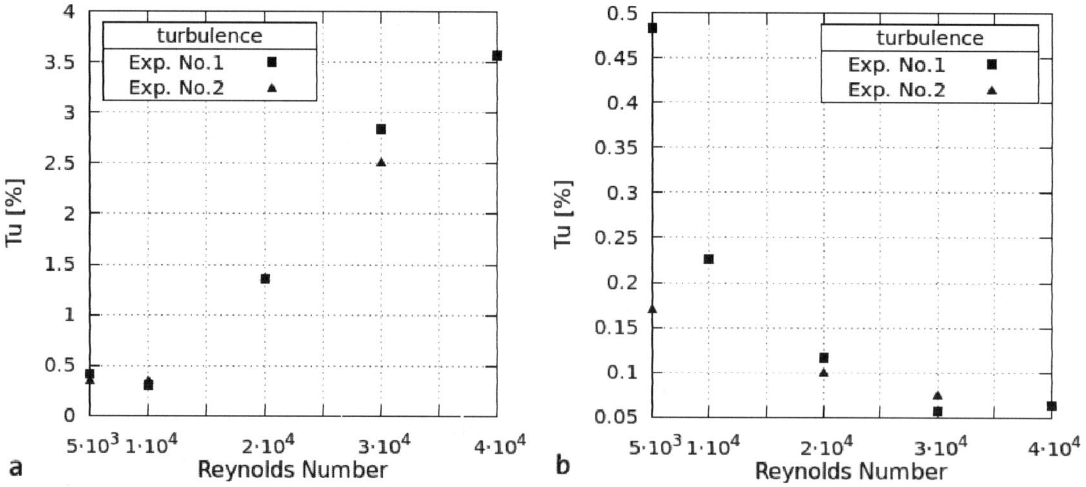

Figure 9. Evolution of mean (a) and maximum value (b) of turbulence intensity at the exit plane versus Re number for two test rigs with different nozzle contraction ratio.

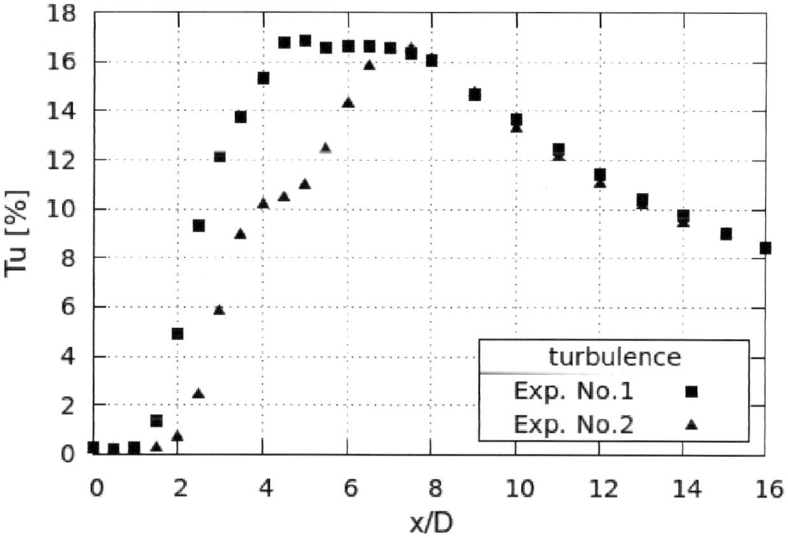

Figure 10. Downstream evolution of turbulence intensity at the jet axis for two nozzle contraction ratios applied for Re = 10 000.

However, the data presented at 'figure 10' reveal, that despite identical Tu mean and maximum values at the jet exit a substantial difference in Tu downstream distribution along the jet axis was observed for Re = 10 000. These results suggest the need to apply more advanced characteristics of the jet initial conditions, in particular the spectral contents of turbulence and the analysis of turbulence scales will be necessary.

4. Conclusions

The investigations confirmed the sensitivity of free jets to initial conditions for a broad range of relative boundary layers thicknesses and Re, the importance of the shape of cylindrical nozzle tips has also been confirmed. For physical explanation of these phenomena it was necessary to perform spectral and spatio – temporal correlation analysis.

5. References

[1] Bogusławski A, Tyliszczak A, Drobniak S and Asendrych D 2013 Self-sustained oscillations in a homogeneous-density round jet *J. Turb.* **14:4** pp 25-52
[2] Ball CG, Fellouah H and Pollard A 2012 The Flow Field of a Turbulent Round Jet *Prog. Aerosp. Sci.* **50** pp 1-26
[3] Nathan G J, Mi J, Alwahabi Z T, Newbold G J R and Nobes D S 2006 Impacts of a jet's exit flow pattern on mixing and combustion performance *Prog. Energ. Combust.***32** pp 496-538
[4] Verzicco R and Orlandi P 1995 Mixedness in the formation of vortex rings *Phys. Fluids* **7** p 1513
[5] Romano G P 2002 The effect of boundary conditions by the side of the nozzle of a low Reynolds number jet *Exp. Fluids* **33(2)** pp 323 - 333
[6] da Silva C and Metais O, Vortex control of bifurcating jets: A numerical study *Phys. Fluids* **14(11)** p 3801.

Acknowledgements
The investigations presented in this paper have been performed with funding from Polish National Science Centre within the grant DEC 2011/03/B/ST8/06401.

Journal of Physics: Conference Series **530** (2014) 012026 doi:10.1088/1742-6596/530/1/012026

Determination of thermodynamic gas parameters in branched pipes in internal combustion engines

W Mitianiec
Cracow University of Technology, Mechanical Faculty
Al. Jana Pawla II 37, 31-864 Krakow, Poland

E-mail: wmitanie@usk.pk.edu.pl

Abstract. The paper presents theoretical and experimental results of calculation of basic gas parameters in the branched pipes. These parameters are required in one-dimensional computer models for prediction of non-steady gas flow in complicated multi-cylinder engine ducts. The gas flow near the junction is described with assumption of compressed and unsteady flow. Mathematical equations describing the gas flow are given in the paper on basis of mass, energy balance in the junction, pressure drop between pipes and conservation of energy in the section of supplied pipe. Equation systems enable to solve values of pressure, gas velocity, sound speed, density and concentration of gas components in every pipe connected to the joint. The different cases of the flow area are considered. The obtained parameters at the junction outflow are needed as initial values for calculation of unsteady gas flow in the outflow pipes. Verification of the method was conducted experimentally and pressure loss coefficients are given in the paper. Additionally by using Fluent program with high mesh density of the T-pipe junction the thermodynamic parameters (pressure, velocity, temperature) are compared with those obtained from 0D model. The model enables calculation the thermodynamic parameters of inflow and outflow systems in multi-cylinder IC engines in computer program.

1. Introduction

Numerical calculation in 0D and 1D models of gas flow in multi-cylinder reciprocating internal combustion engines is widely used for prediction of thermodynamic parameters of gases in the branched ducts of inlet and outlet systems [1]. These ducts are connected to the branch of "Y" and "T" type, where the gas is transferred from supplier pipes to the outflow pipes in varying of changing in the time gas properties. In the engine ducts the gas parameters are a function of temperature and chemical composition. Taking into account the dynamics of the gases with incident and reflected pressure waves the filling and emptying of the cylinder depends on the flow in the branches particularly in multi-cylinder engines. The concept of calculations, based on the balance of mass, energy, and the values of the pressure loss between the pipes oriented relative to each other, has to determine the thermodynamic parameters of gas and the mass fraction of each component in the branch node. Although the model refers to the three branched ducts, and test results are considered to "T" type of branched pipes, one can apply it to the layout consisting of more pipes. This allows a definition of incoming or outgoing pressure waveforms in the ducts from the valves or the ports of individual cylinders in internal combustion engines. Unsteady gas flow in the engine ducts influences on the mass filling ratio of the gas in each cylinder. In the simulation of any engine, where its performance is related to the mass trapped in the cylinder and pressure wave motion through the branched pipes was considered by Blair [2]. The incident and reflected pressure waves define gas velocity, density and mass flow rate. The non-isentropic analysis is required for determination all gas parameters in the branched pipes where unsteady gas flow takes place for computation of the intake or outflow engine systems. Calculation of the gas flow in the parallel branched pipes with different resistance was considered by Yoo *at al* [3]with taking into account κ-ε turbulent model.

2. Model Flow 3-pipe branch

The mass rate of gas flowing from the individual pipes to the node in the branched ducts depends on the different pressure in pipes and the node. Schematic of a typical combination of three pipes located relative to each other by an angle θ is shown in figure 1. It was assumed that the gas flow is unsteady and non-isentropic resulting from the existence of friction (pressure loss) in the node and the pipes. A positive direction of gas flow was assumed where the gas is directed towards the node A, which corresponds to the gas pressure wave. It is known the flow direction and gas parameters in the cross-section S in the duct 3, which is located Δx away from section 3. These parameters are also known for the remaining two pipes, and they are as a result of solution of wave motion of the gas in the pipes. For this reason, a positive pressure wave amplitude P_R is known in every branch, which has moved to the node. Most of the duct connections in combustion engines have "T" shape form considered in the work and it is shown in figure 2. We should determine the parameters of gas in sections 1, 2 and 3 for different gas flow cases.

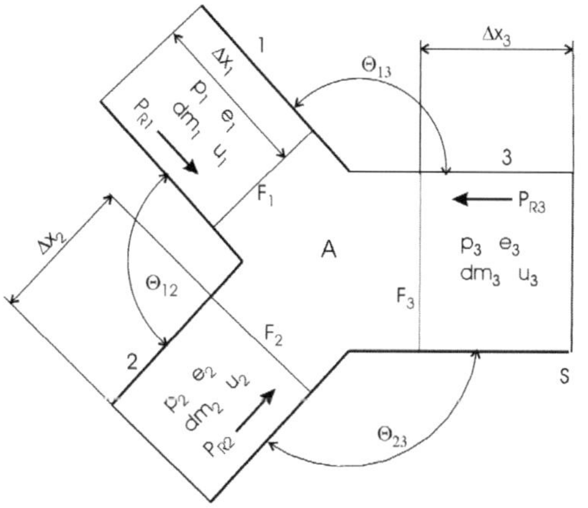

Figure 1. Overall diagram of mass flow of gas through 3 branched pipes.

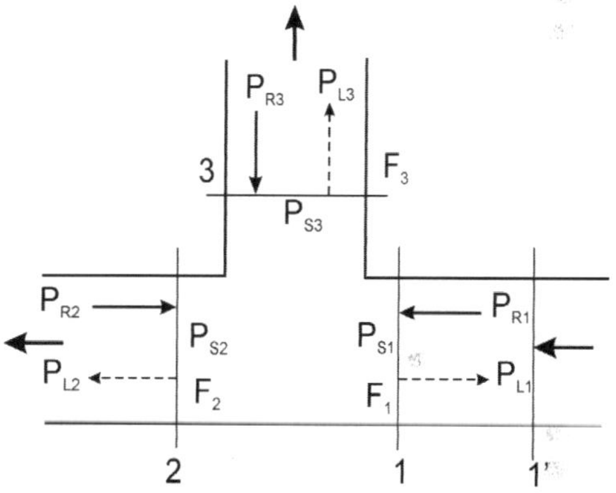

Figure 2. Diagram of pressure impulse of gas in 3 branched pipes of type "T".

The gas in the pipes of internal combustion engines consists of different chemical composition. The gas composition is determined by mass ratio of the individual species and their individual gas constant *R*. In each branch the gas flows into the node or flows out from the node with charge mass *dm*, velocity u_j, pressure p_j and specific energy e_j. The incident and reflected pressure waves and other gas parameters at unsteady flow in the pipes can be solved by using different calculation techniques for example: Rieman, QUB, Lax-Wendroff or Harten-Lax-Leer methods [1]. The relative gas pressure *P* at each point of the system is defined by formula:

$$P = \left(\frac{p}{p_0} \right)^{\frac{\kappa - 1}{2\kappa}} \tag{1}$$

where p — absolute pressure in the system, p_0 — reference pressure and κ — ratio of the specific heats at a given temperature *T*.

The pressure wave arising at the end of each duct P_S is a function of the impulse of wave in negative direction P_L and the impulse of wave in positive direction P_R derived by Benson [4]:

$$P_{Sj} = P_{Lj} + P_{Rj} - 1 \tag{2}$$

wherein *j* is the sequence number of the pipes.

In contrast, the gas particle velocity reaching the node A is:

$$u_j = c_j \cdot a_{0j} \cdot (P_{Lj} - P_{Rj}) \tag{3}$$

where $a_{0j} = \sqrt{\kappa R T_{0j}}$ is the local speed of sound at reference conditions and variable c_j is a function of the specific heats ratio and calculated from the following dependence:

$$c_j = \frac{2 \cdot \kappa_j}{\kappa_j - 1} \tag{4}$$

For gas flow in the ducts of internal combustion engines it can be assumed that for every branched pipe near the node reference temperature T_{0j} is the same ($T_0 \approx 298K$).

3. The mathematical model of the flow

In many previous methods there assumed that the node A (figure 1) the thermodynamic state and gas dynamic state is the same for all waves running across the pipes, and the pressure in the pipes near the node is the same [2,4].

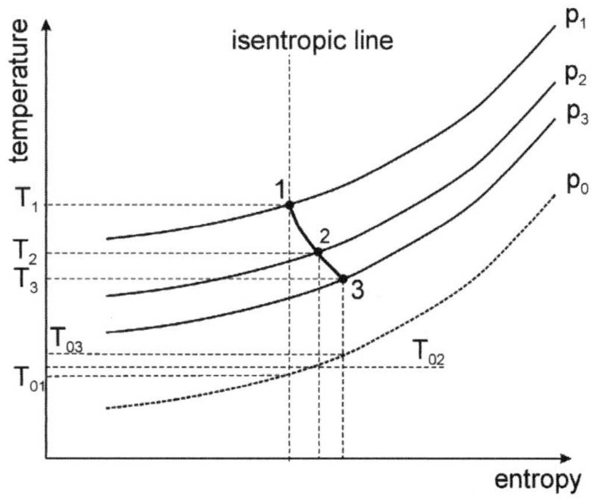

Figure 3. Non-isentropic gas flow from one supplying pipe to two receiving ducts.

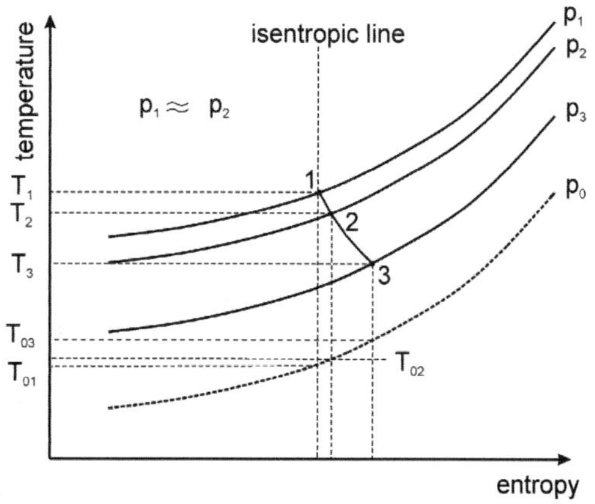

Figure 4. Non-isentropic gas flow from two supplying pipes to one receiving duct.

As a result of experimental studies conducted by Bingham [5], and the author [6,7] the mass flow from one pipe to the other depends on the pressure drop between the ducts. The actual flow in the branch is non-isentropic as shown in figure 3 at a flow from one pipe to the other two, and in figure 4, when two pipes feeding the one pipe. At the isentropic flow the quantities T_{01}, T_{02} and T_{03} are respectively the reference temperatures at pressure which is usually the ambient pressure p_0. When one pipes feeds the other two the difference between pressure p_2 and p_3 pressure is small (figure 3). Therefore, it can be considered that the reference temperature T_{02} and T_{03} are equal (figure 4). When the gas is supplying a single pipe, for example the pipe 3 with two pipes 1 and 2, the pressure in the pipes 1 and 2 differ slightly ($p_1 \approx p_2$), which also means that $T_{01} \approx T_{02}$ (figure 4). The pressure drop between the pipe 1 and the other pipes k is expressed by the given equations:

$$p_0 \cdot \left((P_{S1})^{1/c_1} - (P_{Sk})^{1/c_k} \right) = C_{L1k} \rho_k u_k^2 \tag{5}$$

where the loss factor determined by the flow [5] amounts:

$$C_L = 1.6 - \frac{1.6\Theta}{167} \tag{6}$$

The flow loss coefficient C_L in equation (6) for an angle $\Theta > 167°$ between two ducts is set to zero. For an angle of 90° the flow loss coefficient is 0.737. The assumption of pressure drop eliminates an implementation of the complex system of momentum conservation equations to the calculations.

Journal of Physics: Conference Series **530** (2014) 012026 doi:10.1088/1742-6596/530/1/012026

In order to find the amplitude of the pressure wave reflected from the node in each branched pipe the law of conservation of mass, law of conservation of energy and pressure drop equation (4) between the pipes are applied. At the branching occurring in the intake system one can be omitted a consideration of change of residual exhaust gas ratio γ, gas constant R and the ratio of specific heats κ. The gas composition and thermodynamic parameters of the gas in the intake system are almost identical along the pipes. For the conservation of flow continuity the following condition must be met:

$$\sum_{j=1}^{k} \dot{m}_j = 0 \tag{7}$$

Gas pressure at the ends of the pipes near the node is determined from the formulas (4). For the three branching pipes, there are two equations of the pressure drop. In contrast, the energy flow equation for the three pipes can be written as follows:

$$\dot{m}_1 \cdot e_{01} + \dot{m}_2 \cdot e_{02} + \dot{m}_3 \cdot e_{03} = 0 \tag{8}$$

where e_{01}, e_{02} and e_{03} are the accumulation energies of fluid (gas enthalpy increased by the specific kinetic energy). This corresponds to a gas state on the line of the pressure p_0 in Figure 3 and 4. The accumulation energy of gas can be written as follows:

$$e_{0j} = \left(c_p\right)_j \cdot T_j + \frac{u_j^2}{2} = \frac{\kappa_j}{\kappa_j - 1} \cdot R_j \cdot T_j + \frac{u_j^2}{2} = \frac{1}{2} \cdot \left(c_j \cdot a_j^2 + u_j^2\right) \tag{9}$$

where a_j — gas sound speed.

Speed of sound a_j can be expressed using the pressure wave resulting P_{Sj} as:

$$a_j = a_{0j} \cdot P_{Sj} = a_{0j} \cdot \left(P_{Lj} + P_{Rj} - 1\right) \tag{10}$$

After substitution of equations (3) and (10) to (9) the accumulated energy is a function of reflected and positive (incidence) pressure waves:

$$e_{0j} = c_j a_{0j}^2 \left[\left(P_{Lj} + P_{Rj} - 1\right)^2 + c_j \left(P_{Lj} - P_{Rj}\right)^2\right] \tag{11}$$

The density of the gaseous medium obtained from gaseous state and sound speed eq. takes the form:

$$\rho_j = \frac{\kappa_j \cdot p_0}{a_{0j}^2} \left(P_{Lj} + P_{Rj} - 1\right)^{\frac{2}{\kappa_j - 1}} \tag{12}$$

The flow rate in the duct j with cross section area F_j can be expressed by the equations (3) and (9) as:

$$\dot{m}_j = \frac{2\kappa_j^2}{\kappa_j - 1} \cdot \frac{p_0}{a_{0j}} \left(P_{Lj} + P_{Rj} - 1\right)^{\frac{2}{\kappa_j - 1}} \cdot \left(P_{Lj} - P_{Rj}\right) \cdot F_j \tag{13}$$

Finally, one can write the appropriate systems of equations for the two types of gas flow:

a) gas flow from the first to the second and third duct,

b) gas flow from the first and second duct to the third duct.

Due to the similar arrangement of the equation there are given only equations for the first case:

1. Pressure drops between the pipes 1-2, 1-3 from eq. (5):
2. The equation of mass conservation in the node:

$$\sum_{j=1}^{3} \frac{2\kappa_j}{\kappa_j - 1} \cdot \frac{p_0}{a_{0j}} \left(P_{Lj} - P_{Rj}\right) \cdot \left(P_{Lj} + P_{Rj} - 1\right)^{\frac{2}{\kappa - 1}} \cdot F_j = 0 \tag{14}$$

3 The equation of energy conservation in the node as a function of known values of positive pressure waves P_{R1}, P_{R2}, P_{R3} and additionally relative speed sound a_{01} which is based on equations 11 and 13:

$$\sum_{j=1}^{3} m_j e_{0j} = f\left(P_{R1}, P_{R2}, P_{R3}, a_{01}\right) \tag{15}$$

4. The equation of conservation of energy in the supply section 1 – 1':

$$\frac{a_{1'}^2}{\kappa - 1} + \frac{\left(\rho u^2\right)_{1'}}{2} = \frac{a_1^2}{\kappa - 1} + \frac{\left(\rho u^2\right)_1}{2} \tag{16}$$

Journal of Physics: Conference Series **530** (2014) 012026 doi:10.1088/1742-6596/530/1/012026

For disposal there are five equations for each case and five unknown values: three pulses of the reflected wave P_{Lj} and two reference sound speeds: a_{02} and a_{03}. This nonlinear equations system can be solved for example by using the Newton-Raphson method.
Below there is presented one example of the air flow in the branched ducts as shown in figure 2 with the basic data for three different configurations of the branched ducts:
- cross section areas of the ducts $F_1 = 4\ \text{cm}^2$, $F_2 = 3\ \text{cm}^2$, $F_3 = 1{,}5\ \text{cm}^2$
- positive pressure impulses: $P_{R1} = 1{,}06$, $P_{R2} = 0{,}98$, $P_{R3} = 0{,}98$.

The supplier pipe was the pipe No 1 and two others (No 2 and 3) were being supplied by the branch node. As a result of solution of the equation system (5, 14 – 16) the required gas parameters were obtained in the pipes near the branch node for the given data and the results are presented in table 1.

Table 1. Calculation results of gas flow in branched ducts (figure 2)

θ_{12} °	$\theta_{2:}$	θ_{23} [-]	P_{S2} [-]	P_{S3} [-]	u_1 m/s	u_2 m/s	u_3 m/s	$\dot{m}_1$ kg/s	$\dot{m}_2$ kg/s	$\dot{m}_3$ kg/s	a_{01} m/s
120	120	1,043	1,0321	1,0321	126,8	-120	-119,8	0.0775	-0,0511	-0,0264	340,4
90	150	1,045	1,0293	1,0341	122,6	-115,7	-123,6	0.0758	-0.0483	-0.0275	341,5
180	90	1,039	1,0385	1,0245	133,9	-129,8	-106,6	0,0802	-0,0576	-0,0226	338,5

The reference acoustic velocities in the pipes being supplied have almost the same value because of a small difference of gas temperature and density.

4. The spatial gas flow in the three branched ducts

The presented model of gas flow through the branching pipes refers to unsteady flow, but determination of the gas parameters is held on an assumption of stationary gas motion in a very small period of time. In this paper the analysis of the gaseous medium is limited to the branching "T“ ”ype, in which the theoretical flow loss coefficients calculated according to the equation (5) are respectively $C_{L12} = 0$, $C_{L13} = 0.738$ and $C_{L23} = 0.738$. In order to determine the nature of the flow the spatial gas flow model was created using the finite element method in the package CFD software ANSYS FLUENT Ver.13. The geometrical model was created in CAD system and next it was translated to the step model. Such universal CAD model was imported to ICEM CFD preprocessor, where the mesh was created with definition of surface and volume zones.

Figure 5. Mesh model of “T” pipes branching.

The grid of model shown in figure 5 contains 134425 hexahedral cells and 141000 nodes with high mesh density near the walls and in the pipes connection due to the turbulence flow. The total length of the model in both directions was 0.1 m and diameter of the main pipe was 4 cm and 3 cm of the perpendicular pipe. For simplicity, it was assumed that the flowing gas is the air treated as an ideal gas with RNG k-ε turbulence model. Elongation of the pipes did not influence on the calculation results. The air flow was analyzed at assumption of supplying of one branching pipe (2) by two pipes (1 and 3), wherein the inlet and outlet boundary parameters were as follows at reference pressure 1.013 Mpa:

inlet 1 : $p_1 = 6800$ Pa, $u_1 = 80$ m/s, T_1=500 K,
inlet 3 : $p_3 = 8000$ Pa, $u_3 = 40$ m/s, T_3=450 K
outlet 2: $p_2 = 0$ Pa.

Formulations of the boundary conditions of the gas flow in the branched pipes were considered by Chae *et al* [9]. Waveforms of pressure, velocity and density of the gas obtained from solution of the spatial flow model in the x-z plane cross section are shown in figures 6 – 9, respectively. The purpose of the analysis of this model was to determine the gas pressure difference across sections near the node, and checking the validity of the formulas (5). The analysis confirmed the assumption that in the case of gas flow from two ducts to a single duct, the static pressure (figure 5) is almost equal (the supply lines in figure 4).

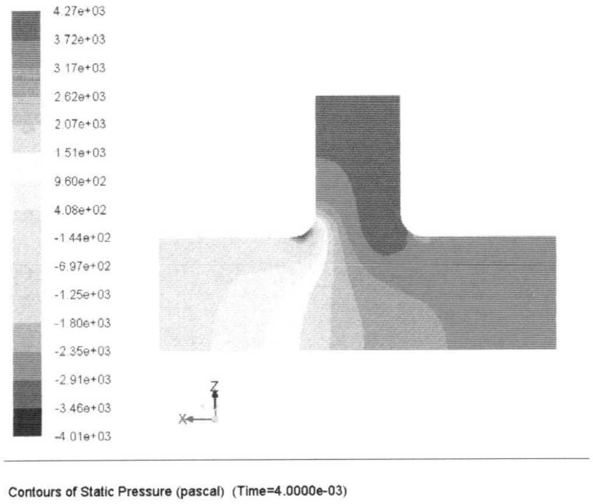

Figure 6. Static pressure in "T" branch type of 3D model.

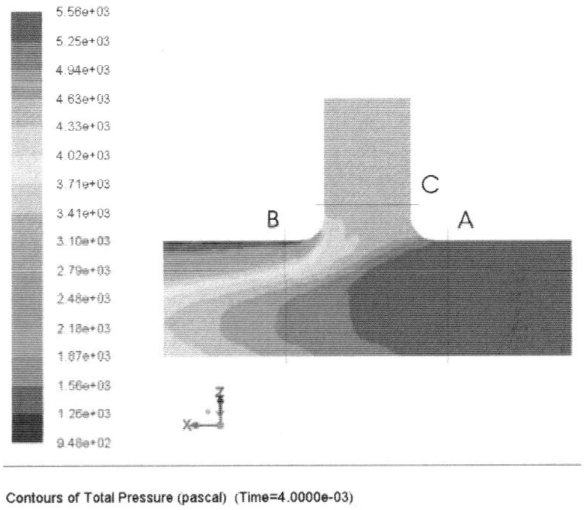

Figure 7. Total pressure in "T" branch type of 3D model.

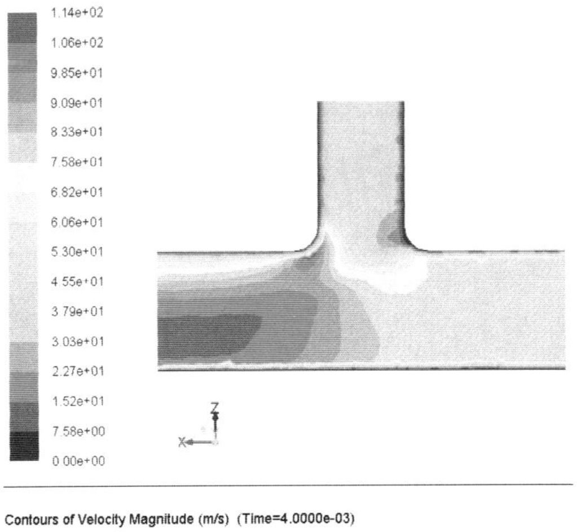

Figure 8. Contours of velocity magnitude (m/s) in x-z cross section.

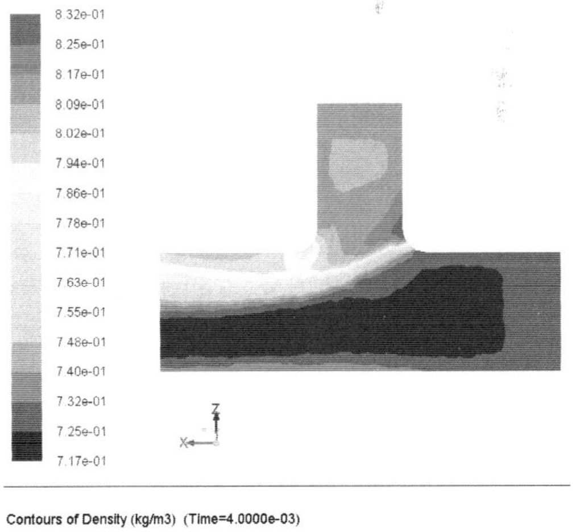

Figure 9. Contours of gas density at x-z cross section

During unsteady air flow high differences of velocity and density occur (figure 8 and 9). Differences in the density of the gas (figure 9) and specific energy (figure 11) have the rapid change of values in the node as a result of different temperature of the medium in inlet 1 and 3 (figure 10). The static temperature of the gas (figure 10) and total specific energy (figure 11) have the same distribution profile. In the numerical calculation of transient motion of the gas one assumes the average length of the cells equal to 2 cm, and therefore for benchmarking the distance 2 cm from the node was indicated in figure 7 by the letters A, B and C.

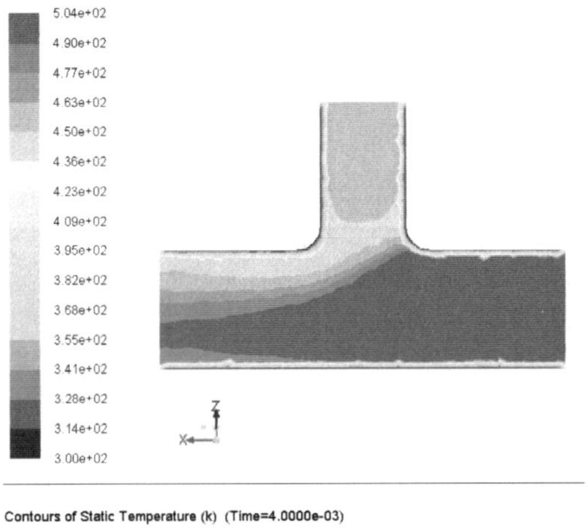

Figure 10. Static temperature (K) at x-z cross section.

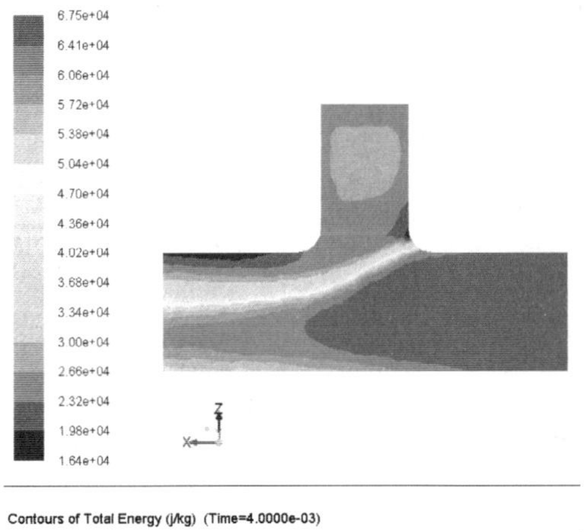

Figure 11. Contours of total specific energy (J/kg) in x-z cross section.

The table 2 shows the average pressure, velocity and density in these conductors. The loss factors determined on the spatial air motion analysis deviate significantly from the coefficients determined by the formula (6) because of non-constant thermodynamic parameters in the sections A, B and C obtained from simulation.

Table 2. Gas parameters in „T" branch of pipes

Parameter (mean values)	A	B	C
static pressure p [Pa]	3650	0	4250
density ρ [kg/m^3]	0.73	0.75	0.81
velocity u [m/s]	75,0	97.3	40
Loss coefficient in p. B according to (5)	C_{L12}=0,514		C_{L32}=0,598

5. Experimental determination of pressure losses

Pressure drops in the junction depend on the angles between the pipes and direction of gas flow. The experimental work has been carried out with the use of air in order to determine the loss coefficients C_L for the various directions of gas motion. Air velocity and pressure measurement have been conducted on the volume intake system with "T" branched pipes in steady conditions by using of thermo-anemometers and pressure gauges. Variation of pressure drop between pipes 1 and 2 ($\theta_{12} = 180°$) with the air inflow to the cylinder is presented in the figure 12. The pressure drop is the function of dynamic pressure of gas in the duct being supplied. The pressure difference between lines 1 and 2 is linearly dependent on the total value ρu^2. The existence of a pressure difference has been confirmed experimentally for the same gas flow direction where C_L≈0.6. The pressure drop between pipes 3 and 2 (flow from the duct 3 to 2) is also expressed as a linear function of the parameter ρu^2 (figure 13) and the pressure loss coefficient is 0.64 and according to the Blair formula [2] this factor should be 0.737 and is almost close to the value achieved experimentally. The experiments done by Daisuke et al [8] enabled to define the pressure loss coefficient in a polynomial function of the dynamic pressure $\rho u^2/2$. At the air flow from the cylinder connected with the duct 2 to the ducts 1 and 3 with the open ends to the atmosphere, the pressure loss coefficient between the duct 2 and 1 was determined with value C_{L21}≈ 0.598. The pressure loss coefficient C_{L23} between the pipe 2 and 3 in reverse flow direction is about 0.64 and these values differ to the value determined in [5] about 13%.

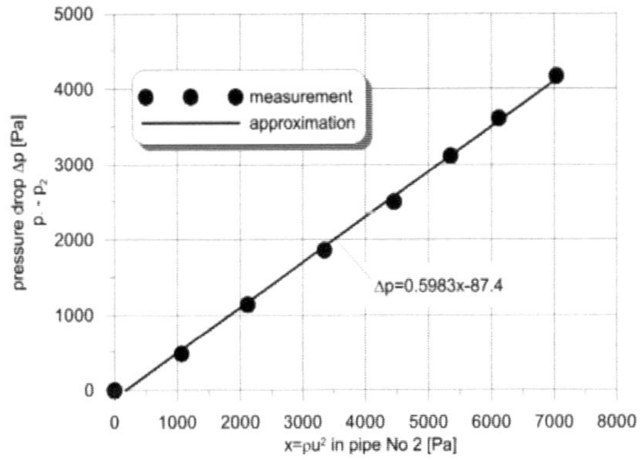

Figure 12. Air pressure drop between pipe No 1 and 2 (pipes No 1 and 3 open).

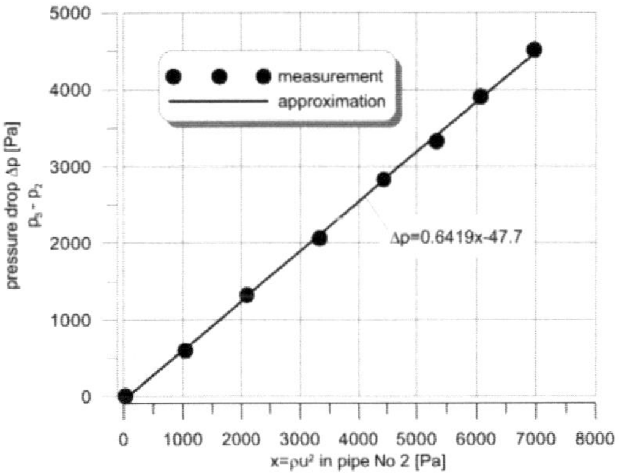

Figure 13. Air pressure drop between pipe No 3 and 2 (flow into pipe No 2).

6. Conclusions

The given mathematical model and carried out the analysis of spatial motion of gas in the manifold was developed independently by the author on the basis of the work [2].

1. During charge flow through manifolds the unequal pressures in the pipes reaching the node takes place. Two pipes supplying the gas into the branched node have the same pressure. The mathematical model enables gas flow calculation in the branched pipes in the combustion engines.
2 Solution of the equations (5, 14-16) and knowledge of the nature of the flow due to the pressure loss coefficient allows a determination the charge thermodynamic parameters in individual ducts. The use of CFD method to analyze the gas flow through the branch "T" of three ducts confirmed the assumption of the analytical method of one-dimensional gas motion but with correction of flow coefficient.
3 If the flow from two pipes into the third pipe takes place at the same time, the pressure in the supplying pipes is almost identical. In the case of gas flow from one pipe to two pipes the pressures in the pipes being supplied are different.
4. The pressure losses obtained from the experiments and simulations between pipes set at angle 180° are not equal zero as it comes from formula (6). Further experimental test are foreseen for modification of the formula on the coefficient of pressure losses (6).

7. References

[1] Chen C and Veshagh A 1992 A comparison between alternative methods for gas flow and performance prediction of internal combustion engines *SAE 921734* Warrendale
[2] Blair G 1999 Design and simulation of four stroke engines *SAE R-161* Warrendale
[3] Yoo G Choi H and Kim C 2012 Characteristics of turbulent flow distribution in branch piping system DOI: 10.1007/s11771-012-1397-3 *Springer-Verlag* Heidelberg
[4] Benson R 1982 The thermodynamics and gas dynamics of internal combustion engines **Vol. 1** and **2** Clarendon Press Oxford
[5] Bingham J and Blair G 1985 An improved branched pipe model for multi-cylinder automotive engine calculations *Proc IMechE* **Vol 199** No D1 Belfast 65-77
[6] Mitianiec W 1999 Improvements of crankcase delivery ratio in two-stroke engines with volume intake system *SAE 1999-01-1334* Warrendale
[7] Mitianiec W 1994 Zawory membranowe i układy pojemnościowe Ossolineum Kraków
[8] Daisuke S Chongho Y and Toshiharu K 2013 Pressure Change in tee branch pipe in oscillatory flow *Hindavi Pub Cor AdvMechEng* **Vol 2013** ID 257283
[9] Chae K S Lee K T Hwang C J and Lee D J 2006 Formulation and validation of boundary conditions at a branched junction for nonlinear waves *J. of Sound and Vib.* **2006**|**295**|1-2|13-27

Journal of Physics: Conference Series **530** (2014) 012027 doi:10.1088/1742-6596/530/1/012027

PERFORMANCE OF A MICRO-UAV LIFTING SYSTEM BUILT WITH THE USAGE OF RAPID PROTOTYPING METHODS

R T Dalewski[1], K Gumowski[1], T Barczak[1] and J Godek[2]
[1]IAAM, Warsaw University of Technology, ul. Nowowiejska 24, 00-665 Warsaw, Poland
[2] IMP, Warsaw University of Technology, ul. Narbutta 85, 02-524, Warsaw, Poland

E-mail: rdalewski@meil.pw.edu.pl

Abstract. This article presents results of the aerodynamic testing of a micro unmanned aerial vehicle rotor efficiency. The rotors were prepared as a set of two rotors in a counter-rotating ducted drive. Prototypes of the drives were made using two rapid prototyping techniques - FDM – fused deposition modelling method and SLS - selective laser sintering. Rotors were made then treated by introducing additional finishing cyanoacrylate coating and abrasive processing. Main differences between those models were observed in fan shape, porosity, surface roughness and mechanical properties – stiffness. An influence of these factors was observed on an aerodynamic efficiency. For the obtained prototypes both simulations and experimental testing were conducted with thrust, power, torque measurements, as well as the measurement of velocity and pressure distribution at the outlet of the duct. The results show the possibility of using rapid prototyping techniques to produce prototypes of drives operating in the low and medium Reynolds numbers (6000-60000), and the aerodynamic shape relevant factors affecting the preparation and performance of such drives. In addition, simulation studies were performed using the Fluent environment where experimental results were confronted with the results of simulation studies.

1. Introduction

A growing scope of applications for MAV's encourages further development in low and ultra-low Reynolds number aerodynamics [1], [2] and [3]. This field may be characterized by very low lift-to-drag ratio (15-35 for Re in the range from 5000 to 40000). Presented work relates to low Reynolds number drive manufactured by means of rapid prototyping methods designed for rough terrain purpose. The idea for a ducted counter-rotating drive came as an idea for the most ergonomic drive for the purpose to which the designed mechanism is addressed, namely autonomous flight in a rough terrain (shrubs, trees etc.). On the same basis requirements for dimensions were defined as a compromise between weight and performance. Taking these aspects into account, basic assumptions were fulfilled by the construction presented below.

[1] To whom any correspondence should be addressed.

Journal of Physics: Conference Series **530** (2014) 012027 doi:10.1088/1742-6596/530/1/012027

2. Design

The system was designed with the usage of software based on calculations according to Blade Element Method developed by Glauert [4], and further adopted to VTOL flight by McCormick [5] and profile characteristics estimated in Xfoil. Primary design was later verified by Fluent modelling taking into account effect of a duct. Samples were then produced using two rapid prototyping methods – FDM (fused deposition modelling) and SLS (selective laser sintering). Rough samples were then tested and after the test finished using hand abrasive polishing. Samples were then tested again. Different designs were tested according to this manner. This study was performed to find the best design out of the tested concepts.

A basic NACA 3404 with extended trailing edge gap was adopted for blade geometry (figure 1). Pitch angle and chord distribution along the radial coordinate are presented in figure 2 and 3. Duct and hub geometry is sketched in figure 4.

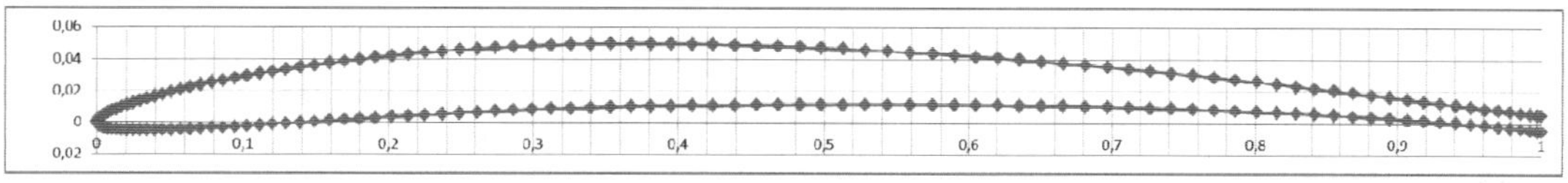

Figure 1. Naca 3404 with a broader trailing edge gap (1% c).

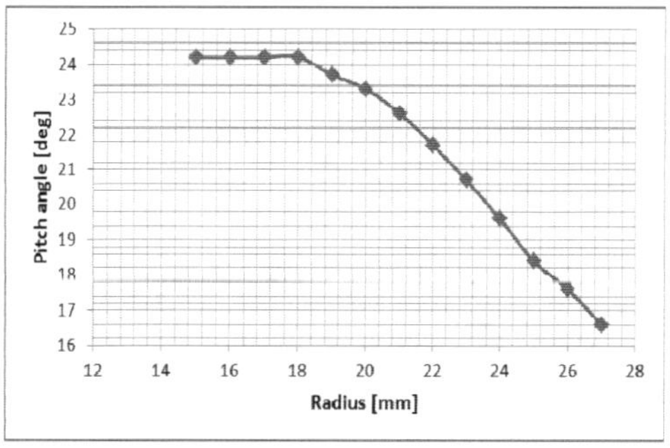

Figure 2. Pitch angle distribution along the span.

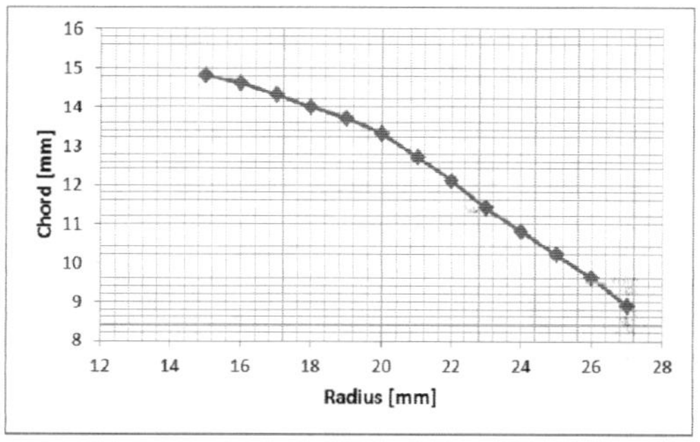

Figure 3. Chord length distribution along the span.

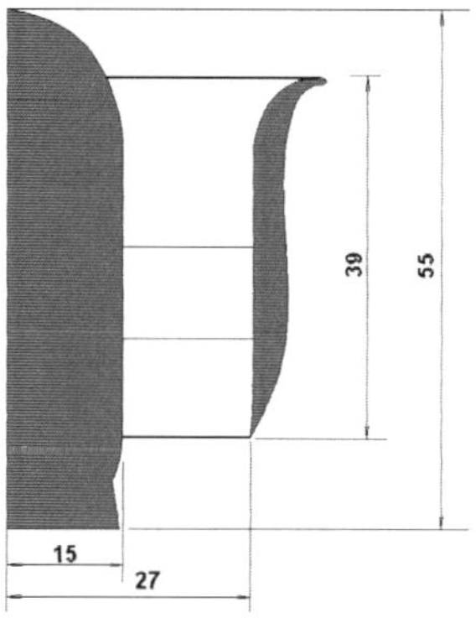

Figure 4. Duct and hub geometry.

3. Production process output

Manufactured samples are presented in figure 5 (rough samples) and figure 6 (comparisons before and after finishing). Rough samples produced by SLS method (Polyamide material was used) characterize good consistency, material is rough with low porosity. Blades are not very stiff and during testing flutter occurs for significant range of RPMs. FDM made rough samples (base material ABS) picture clear artefacts of manufacturing process – single fibres are clearly seen, for thin parts of blade single layer of fibres gives significant transparency. Samples are quite stiff and do not suffer from aeroelastic problems.

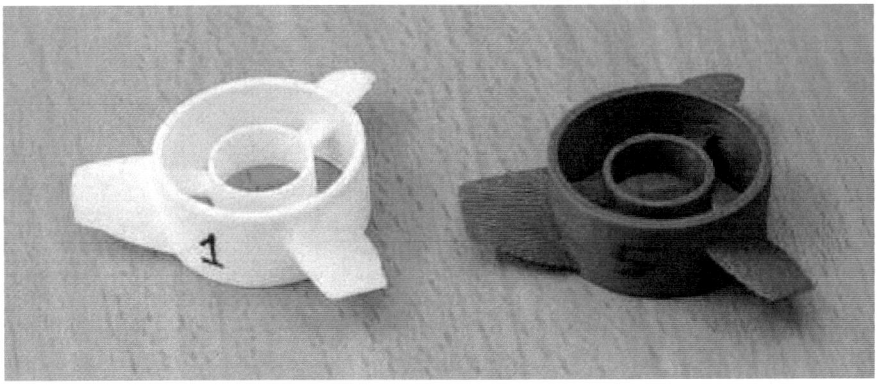

Figure 5. FDM (ABS) rotor, right and SLS (PA) rotor, left after manufacturing.

To increase aerodynamic performance samples were finished with the usage of: firstly cyanoacrylic coating, secondly abrasive treatment of the coating. Both processes were hand-made. Significance of the second phase is vital for the final effect. For SLS samples low abrasive properties of polyamide were very useful because rough sample was a very good base for the process and the surplus coating was eliminated without significant destruction of base material, therefore the intended shape was not significantly corrupted. In the FDM samples, on the other hand, base material (ABS) was vulnerable to abrasive treatment, therefore final geometry depends very much on worker skills and the recurrence properties of each blade may be questioned. SLS properties were finally improved also because of improved stiffness of the blade, therefore aeoroelastic behaviour of the blades was insignificant for the results of the test. Final mass of FDM rotor after finishing was 1.45g, and SLS 1.15 gram.

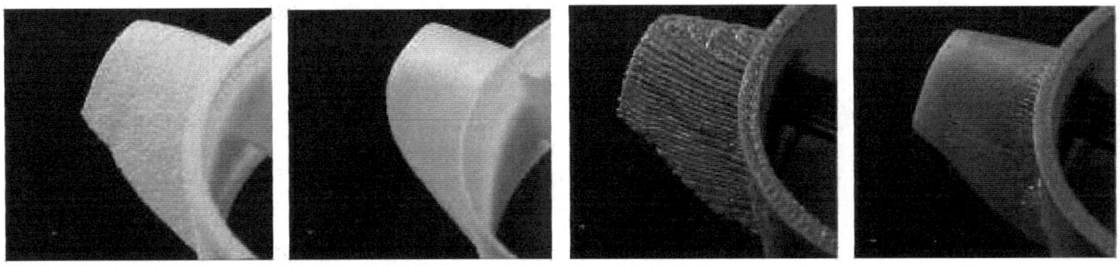

Figure 6. Raw samples roughness and porosity (SLS on the left and FDM on the right).

Effects of surface treatment can be observed in following pictures (figures 7 to 12). Figure 7 presents roughness of rough SLS sample, figures 8 and 9 present roughness of respectively treated upper and lower surface of the blade. Shape of first diagram reflects non-structured composition of the rough sample reflecting grains of sintered base material. The following two pictures show small scale roughness of the surfaces after treatment. No significant differences between upper and lower surface are observed.

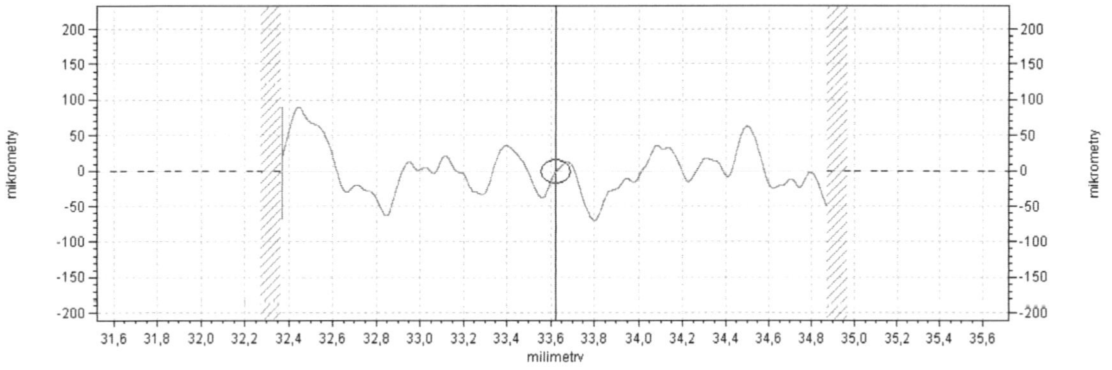

Figure 7. Raw samples roughness (SLS).

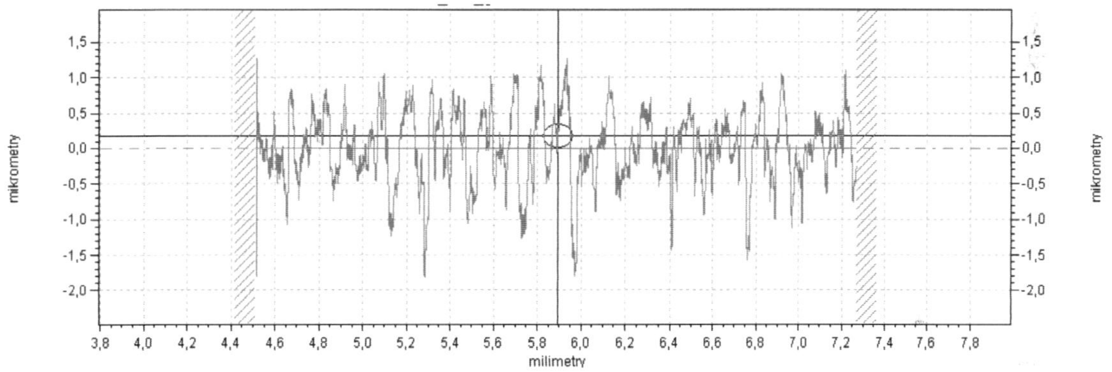

Figure 8. SLS surface-treated samples roughness (upper surface).

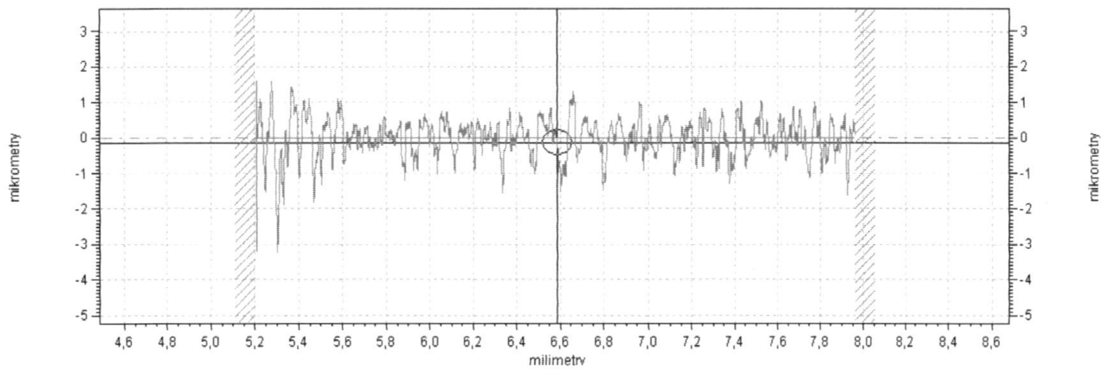

Figure 9. SLS surface-treated samples roughness (lower surface).

Effects of surface treatment of the FDM samples can be observed in figures 10 to 12. Figure 10 presents roughness of rough FDM sample – fibres are clearly seen, figures 11 and 12 present roughness of respectively treated upper and lower surface of the blade. Important increase of lower blade surface roughness is caused by deep crevasses – artefacts of initial structure.

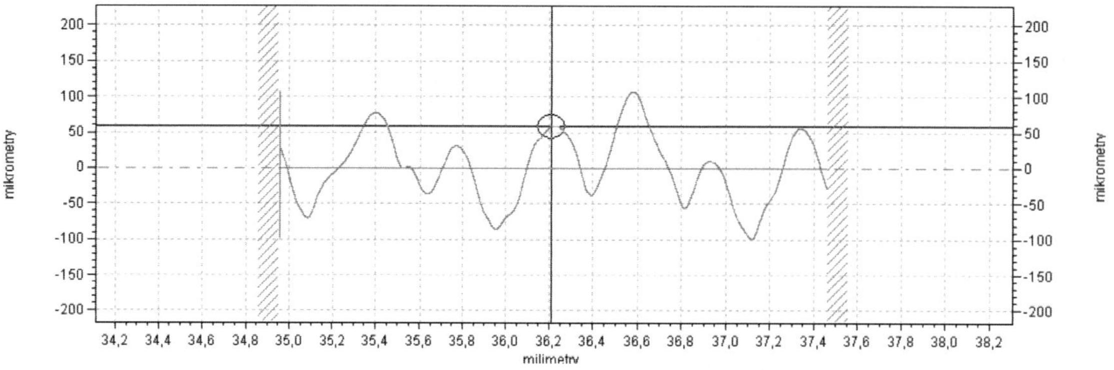

Figure 10. Raw samples roughness (FDM).

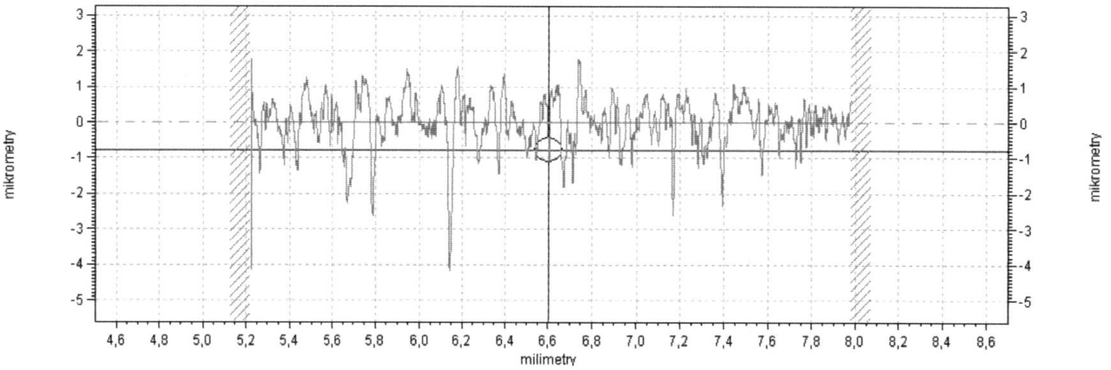

Figure 11. FDM surface-treated samples roughness (upper surface).

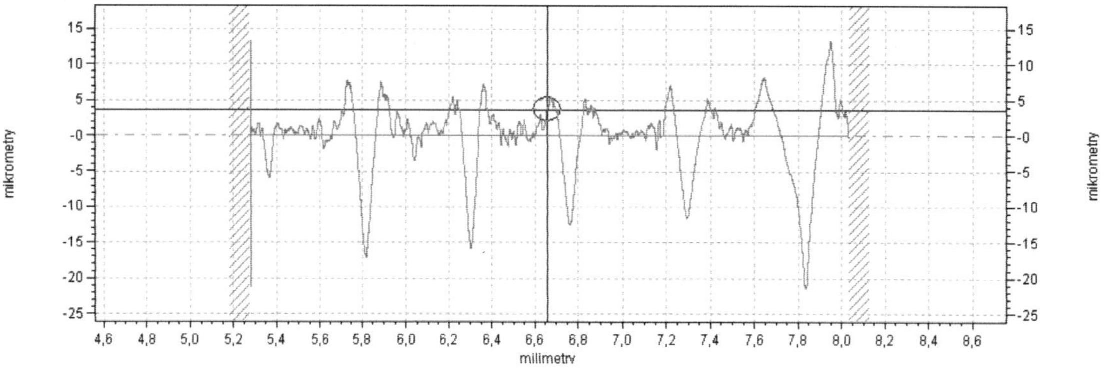

Figure 12. FDM surface-treated samples roughness (lower surface).

Roughness results were summed up in table 1. Surface treatment decreased roughness by two orders from approximately 35 μm (Ra) to 0.5 μm (Ra). It is likely that further decrement of roughness may further influence the result but in this course of investigation resulting performance was satisfactory. Further investigation will also cover further improvement of surface quality, although rather ?? decrement of aerodynamic performance may be expected due to increased likelihood of laminar separation occurrence.

Journal of Physics: Conference Series **530** (2014) 012027 doi:10.1088/1742-6596/530/1/012027

Table 1. Roughness measurements results.

Sample	Ra (μm)
SLS rough	16.53 (+8.93/-4.78)
SLS finished upper/lower	0.39 (+0.04/-0.06)
FDM rough	35.73 (+3.67/-4.55)
FDM finished upper	0.55 (+0.03/-0.03)
FDM finished lower	3.01 (+0.4/-0.4)

4. Simulations

After the design stage, Fluent simulations were performed as well as corrections to the preliminary design were implemented. Pressure-based Navier-Stokes model with transient SST turbulence modelling, on transient solver with rotating parts was used.

Figure 13. Meshed duct and the domain.

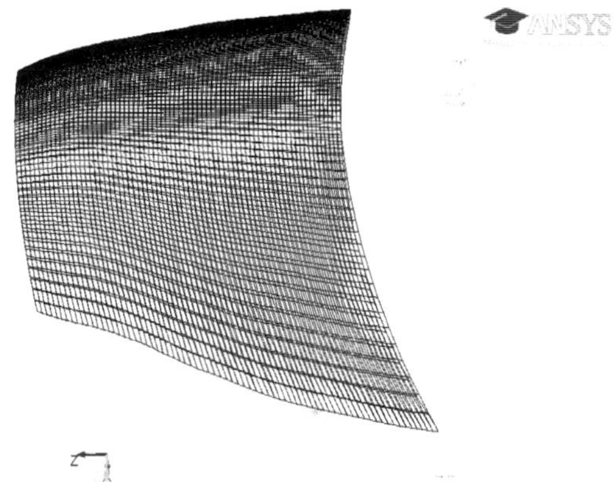

Figure 14. Meshed blade and the domain.

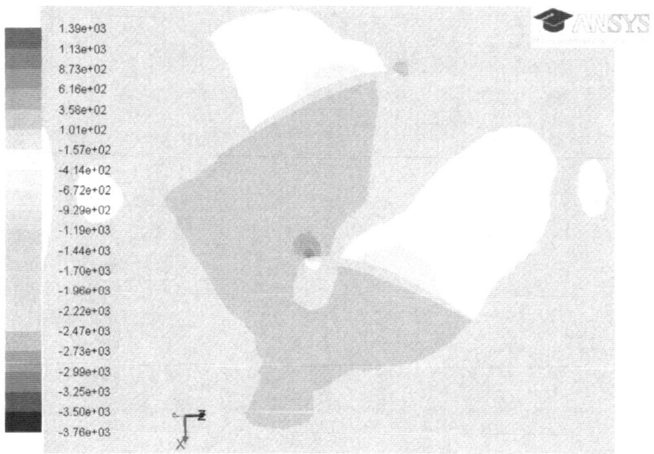

Figure 15. Pressure contours in passing-by configuration of blades (19 kRPM).

Fluent results at this stage were performed on a mesh with approximately 1.6 mln nodes. Although layered mesh was generated around the blades to solve boundary layer more accurately, it needs further improvement of quality. A special mesh treatment is planned for further investigation. Some properties of rough and smooth profiles may be controlled by transition properties, and to some extent by shape deformation (3404ZZ profile) in Xfoil [6] (figure 16 and 17), although it is a problematic issue and significant shape deformation (such as in rough FDM samples) should be further investigated.

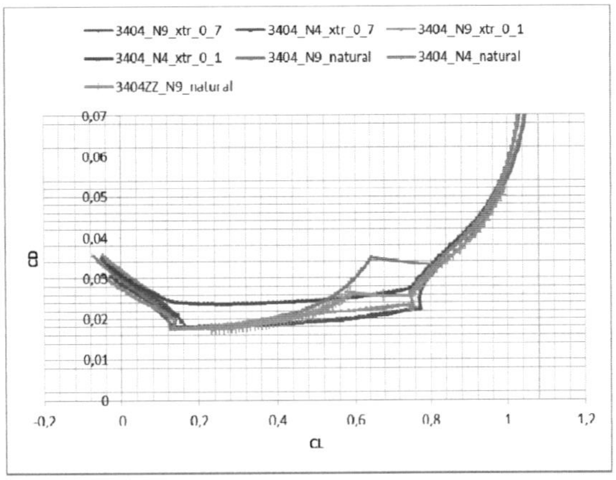

Figure 16. Xfoil – generated polars graphs for different transition scenarios.

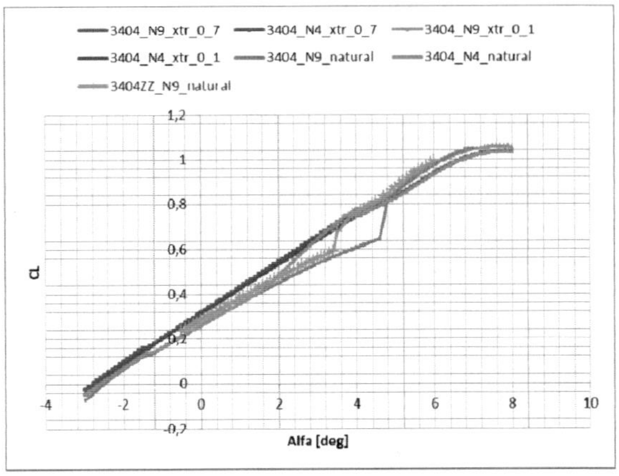

Figure 17. Xfoil – generated CL(α) graphs for different transition scenarios.

5. Testing

Tests were performed on a dedicated test stand presented in figure 18. During the course of design process many configurations were tested (rotors, motors and regulators). Results presented in table 2 were obtained for final set of motor, regulator and rotors.

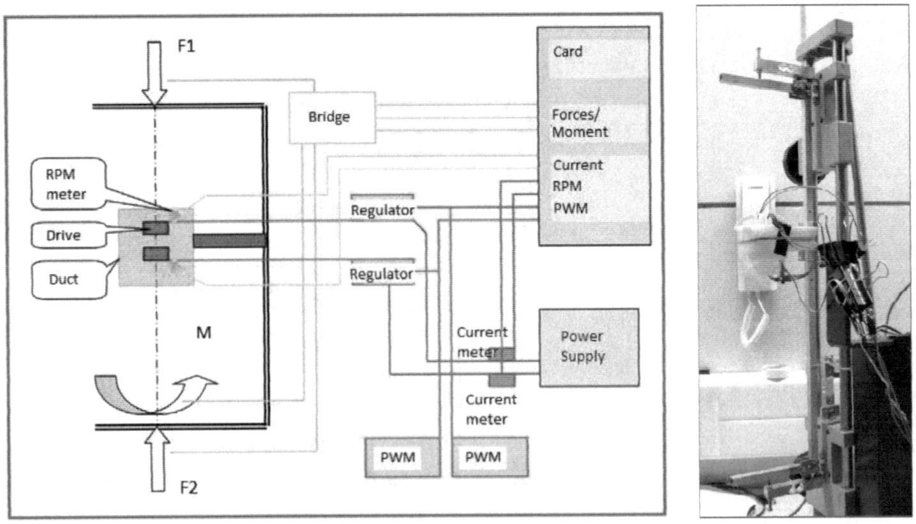

Figure 18. Test stand arrangement.

Journal of Physics: Conference Series **530** (2014) 012027 doi:10.1088/1742-6596/530/1/012027

6. Results and discussion

Final tests were presented in table 2. Rough samples present significantly lower thrust and definitely lower thrust@power-supplied performance. Two reasons seem to be of greatest significance – overall and surface shape and porosity. FDM samples' results, where porosity is clearly visible, are lower comparing to SLS samples' results (although roughness is similar). Finished samples results are significantly better (over 50% increase in thrust, similar power consumption), where SLS samples are slightly better than FDM samples. This effect is most probably because surface geometry was less rough and overall geometry reproduction was better because of low-abrasive properties of polyamide.

Table 2. Thrust force in gram (@power-supplied) results. Force in G (gram).

	19 kRPM	21 KRPM	22.5 kRPM
SLS rough	28 (@ 12W)	36 (@ 14.7W)	40.7 (@ 16.9W)
SLS finished	42 (@ 13.2W)	53 (@ 15.5W)	60.5 (@ 17.6W)
FDM rough	23 (@ 13.6W)	32 (@ 15.2W)	37 (@ 17.4W)
FDM finished	41 (@ 13.6W)	51 (@ 16.8W)	58 (@ 18.9W)
BEM	40.5 (@ 13.5W)	49.5 (@ 18.2W)	56.9 (@ 22.4W)
Fluent	44 (@ 16W)	-	68 (@ 24W)

Presented study showed the need for further investigation for dependency between roughness and performance, taking into account transition control and detailed study of shape reproduction.

7. References

[1] Schafroth D M 2010 *Aerodynamics, Modelling and Control of an Autonomous Micro Helicopter* (Zürich: DISS. ETH NO. 18901)

[2] Kunz P J 2003 *Aerodynamics and Design for Ultra-Low Reynolds Number Flight* (Stanford: PhD Dissertation)

[3] Kunz P J and Kroo I 2001 Analysis and Design of Airfoils for Use at Ultra-Low Reynolds Numbers, Fixed and Flapping Wing Aerodynamics for Micro Air Vehicle Applications, edited by T. J. Mueller, Vol. 195, *Progress in Aeronautics and Astronautics, AIAA*, Reston, VA, 2001, Chap. 3.

[4] Glauert H 1948 *The Elements of Aerofoil and Airscrew Theory* (London: Cambridge University Press).

[5] McCormick and Barnes W 1999 *Aerodynamics of V/STOL Flight* (New York: Dover Publications Inc.).

[6] Drela M 1989 *XFOIL: an Analysis and Design System for Low Re Airfoils* (Springler-Verlag Lec. Notes in Eng. 54)

Acknowledgements

This work was supported by The National Center for Research and Development in course of a Lider Program. Grant no. LIDER/04/143/L-3/11/NCBR/2012.

Journal of Physics: Conference Series **530** (2014) 012028 doi:10.1088/1742-6596/530/1/012028

Dynamic contact of droplet with superhydrophobic surface in conditions favour icing

M Remer[1], G Sobieraj[1], K Gumowski[1] , J Rokicki[1],
M Psarski[2], J Marczak[2], G Celichowski[2]
[1] Institute of Aeronautics and Applied Mechanics, Warsaw University of Technology, Nowowiejska 24, 00-665 Warsaw, Poland
[2] Department of Materials Technology and Chemistry, University of Lodz, ul.Pomorska 163, 90-236 Lodz, Poland

E-mail: mremer@meil.pw.edu.pl

Abstract. Flight like droplet impact with superhydrophobic substrate in conditions favour icing is discussed in this work. Test stand with fast camera and equipment eligible to obtain temperatures and humidity at different ranges, lead to results which can prove, that superhydrophobic surface might be good ice repellent substrate. The influence of air humidity on droplet freezing was confirmed.

1. Introduction

The fluid mechanics of drop impact with surfaces is of importance in a variety of different fields. Scientist has been researching the droplet collision with dry or wet substrate since nineteenth century. The precursor of such investigations was Worthington [1]. It is relevant to many industrial applications and has many undesired effects. Droplet collisions have place in agriculture and daily life (processes like spray cooling, coating, painting, ink-jet printing, etc). Erosion at steam turbine blades caused by high-speed impact is one of the undesired effects. Wind-shields, cloths and roofs are examples of things in which repelling water properties have to be enhanced. Recently, application to aircraft anti-icing has been considered. An aircraft during flight has to wrestle with varied flight conditions, which depend on humidity, temperature and pressure. The work of Bobiñski [2] has shown that water was not freezing, when experiment has been done in volume filled with waterless nitrogen (the low humidity, around 8%). That is why, this work is focused on results more likewise occurring during flight. Low temperatures and changes of humidity favorable to the ice formation. Such phenomenon is extremely dangerous for all kind of aircrafts and huge amount of money are spent to overcome it. Rapid development of material engineering introduced new type of materials, such as Graphene, textured surfaces coated with different chemical compound, which can be used as a coating of an aircraft wings. Understanding how does the collision of the droplet with aircraft surface in varying conditions may lead to overcome or reduce ice formation during flight.

The main objectives of this thesis were:

- To investigate the droplet behaviour during the impact with surface in flight like conditions - temperature in range from +25°C to -30°C for varying conditions, depending on surface hydrophobicity, humidity and varying temperature.

- To investigate the droplet behaviour during impact with varying phases of ice formation
- To investigate the supercooled droplet behaviour impact

2. Water properties. Hydrophobicity.

One of the most important factor relevant to droplet impingement is surface upon which it falls. Therefore, the ultimate goal would be to engineer surface in a proper way for certain application, by enhancing or decreasing surface wettability. Recently, the subject of droplet impinging surface is more and more popular because it became possible to produce surfaces with regular, specially designed patterns [3]. What determines whether the surface is hydrophobic or hydrophilic and what does it mean the surface is superhydrophobic? To explain the difference, first it has to be described the nature of surface tension. If we consider a droplet on a perfectly flat, dry and smooth solid surface it may remain as a drop of finite area or it may spread over the surface. Spreading appears if the energy, which is gained in forming unit area, of the solid-liquid interface exceeds that required to form unit area of the liquid-air surface. Hence, spreading will occur if inequality1is fulfilled. In case it is not, the drop has finite size and equilibrium contact angle exists. In terms of surface tension one can write

$$\sum Fx = \gamma_{SA} - \gamma_{SL} - \gamma_{LA}\cos(\Theta_E) = 0 \ (1), \quad \cos(\Theta_E) = \frac{\gamma_{SA} - \gamma_{SL}}{\gamma_{LA}} \ (2)$$

Rearranging this balance formula yields (1), which describes the droplet's contact angle and related surfaces free energy. Contact angle is measured as the angle (2), determined by vertex at the three-phase line, where the droplet meets the solid. In the formulas above: Θ_E is the contact angle, γ_{SL} denotes the solid/liquid interfacial free energy, γ_{SA} is the solid surface free energy, γ_{LA} is the liquid surface free energy, Fx is the surface tension. For common smooth surfaces like steel and aluminium plates droplet's contact angle Θ_E is close to 90°. However, there exist surfaces which exhibit other wetting properties. When droplet's contact angle is lower than 90° the surface is called hydrophilic (Fig. 1a). Surfaces with contact angle higher than 90° are hydrophobic (Fig. 1b). Superhydrophobic surfaces can be characterised by contact angle higher than 150° (Fig. 1c). Any contamination of surface inhibits wetting resulting in higher contact angle in comparison to clean surface. Contact angle can be also affected by the environmental conditions like temperature, relative humidity and surface roughness. More detailed description is given by Cassie [4].

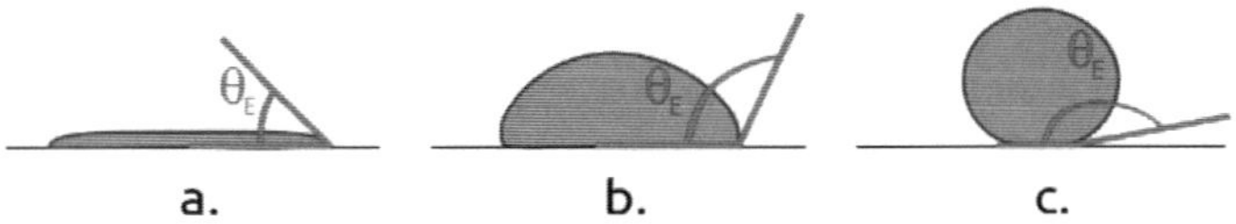

Figure 1. Surface nomenclature – a)hydrophilic, b)hydrophobic, c)superhydrophobic.

Another important issue devoted to hydrophobic properties is named contact angle hysteresis. Contact angle hysteresis (CAH) is one of the most important and classic elements of wetting of liquid droplets in systems from centimetre to micrometer scales. If the three-phase contact line is in actual motion, the contact angle produced is called a "dynamic" contact angle. In particular, the contact angles formed by expanding and contracting the liquid are referred to as the advancing contact angle θ_A and the receding contact angle θ_R, respectively (Fig. 3).

Water has an abnormally high surface tension and surface enthalpy with an abnormally tightly packed surface compared to bulk liquid water. Water molecules at the liquid-gas surface have lost potential hydrogen bonds directed at the gas phase and are pulled towards the underlying bulk liquid water by the remaining stronger hydrogen bonds [5]. Energy is required to increase the surface area (removing a molecule from a well hydrogen bonded interior bulk water to the lesser hydrogen bonded surface), so it is minimized and held under tension. As the forces between the water molecules are several and relatively large on a per-mass basis, compared to those between most other molecules, and the water molecules are very small, the surface tension is large. Lowering the temperature greatly increases the hydrogen bonding in the bulk causing increased surface tension (see Fig. 2).

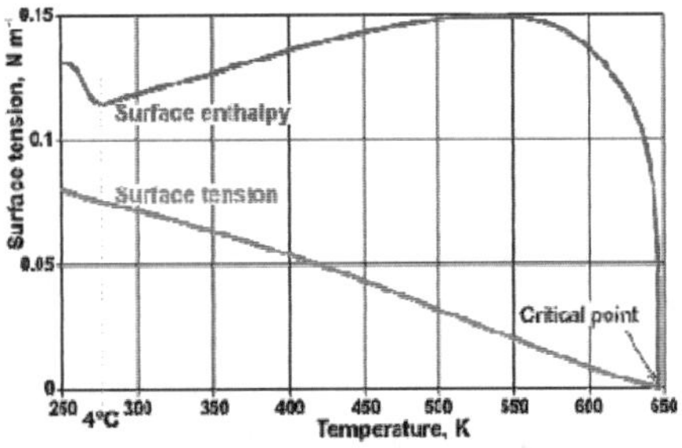

Figure 2. Surface tension and surface enthalpy of water at temperature [5].

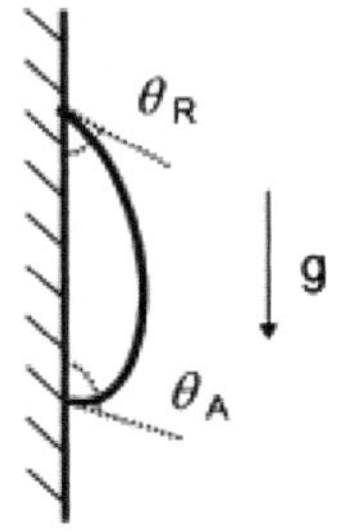

Figure 3. A drop on a vertical surface, stuck at the advancing angle θ_A and the receding angle θ_R [6].

3. Experiment stand.

The stand is designed to enable experiments devoted to dynamic droplet contact with different surfaces in conditions favour freezing. Physical properties of droplet are recorded with fast camera Photron FASTCAM SA5. Test chamber is lighted by two LED lamp Cree XM-L T6 1000 lumen. It is equipped with thermocouples which enable measurement of the temperature of sample and test area. There is also pyrometer which enables additional measurement of surface temperature. The construction of stand enables obtaining varying values of velocity, droplet diameter, surface and surrounding temperatures. Surface can be cooled down with the use of Peltier Cell and water cooling. Testing space can be cooled by means of heat exchanger connected to liquid nitrogen container. Whole circuit can be also filled with vapour free nitrogen to obtain vapour free domain or to reduce humidity inside the chamber. Ambient temperature and probe temperature are measured with DT-8891E thermometer and humidity inside the test space is measured by A1H Rotronic Hygromer hygrometer. To get live image from a test area, Logitech Quickcam 9000 Pro was installed. Whole additional electronic devices and sensors are supplied by 3 Korad KA3005P DC power suppliers.

Measurement range :

- Test surface diameter 150 mm
- Impact velocity ~ 3-5 m/s
- Droplet diameter ~ 0.5 – 2 mm
- Surface temperature ~ -30^{0}C ~ +25^{0}C
- Ambient temperature ~ -30^{0}C ~ +25^{0}C

4. Results

The experiments were carried out to answer question if superhydrophobic materials can prevent icing. In first part of this work concerned room temperature conditions and it was a referring point when we tested dynamic contact with surfaces under conditions that favour icing.

Two types of surface were used:

- Polished steel (hydrophilic)
- Silicon wafer with microstructure (superhydrophobic)

Polished steel was used a reference point, since it is a common used substrate, which is subjected to ice accretion conditions. Silicone substrate was sample with regular, unequivocally determined micro-structure (Fig.5) was performed. On basis of the results given by Varanasi et al. [7] height of posts was set to 10μm. It was necessary to provide possibly the largest range of post-to-width ratio for which Cassie or metastable Cassie states exist. During experiments sample no. 1 was used, which has measured 162^0 wetting angle (static contact angle, which means it was measured in static way). The solid and the dotted lines denote theoretical prediction for Cassie and Wenzel models. Solid circles represent measured values of the WCA – Water contact angle (Fig.4). Micro scale geometry is given by a, b, h, where denote the post width, spacing, and height respectively). The microstructured silicon substrates were hydrophobized in wet chemical process, following the procedure described by Psarski et al [8].

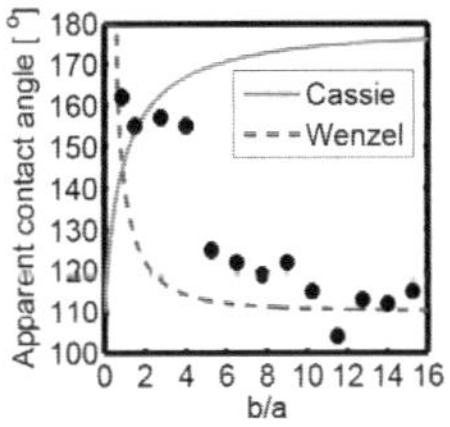

Figure 4. Static contact angle of 5μL water droplet as a function of b/a ratio (spacing/post width).

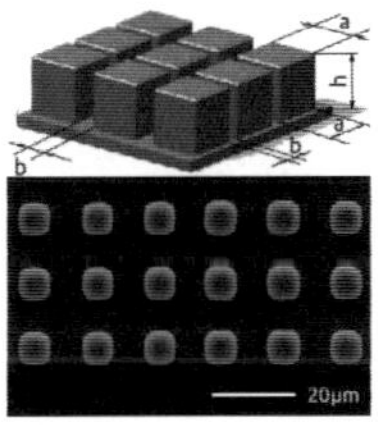

Figure 5. The model and the SEM image of hydrophobic sample with post array b/a = 0.875.

4.1.1. Room temperature results

The first part of the research concerned room temperature conditions. Experiments allowed to distinguish (in hydrophobic and other cases) different phases of motion and give a reference point to results obtained for temperatures below 0°C. Both for polished steel and superhydrophobic test cases experiment were carried out in ambient conditions for test volume which means :

- Temperature – 26°C
- Relative humidity – 30%

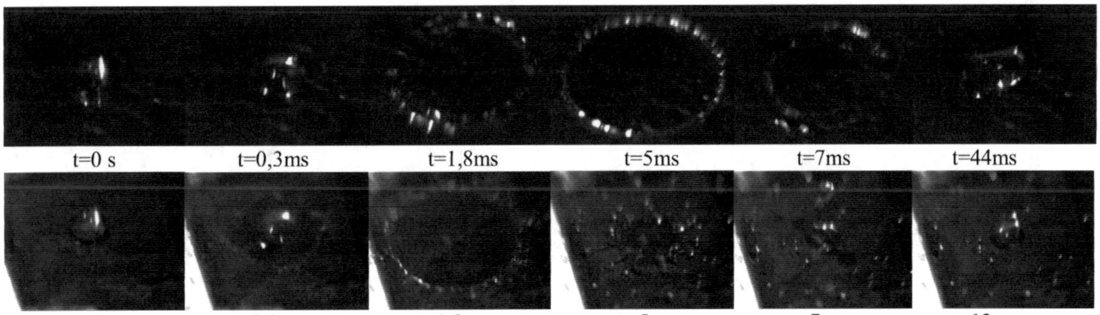

Figure 6. Time elapsed images of water droplet impingement polished steel substrate (top) and for superhydrophobic surface (bottom).

This part of experiments is a reference point for further work with samples exposed on negative temperature. Obtained results show differences in time to reach final diameter for steel (see Fig. 6 top) and superhydrophobic (see Fig. 6 bottom) substrates. It comes from bigger retraction speed at superhydrophobic surface which is directly related to contact angle hysteresis. Another important fact is for such impact velocities may occur some partial bouncing or defragmentation which can be substantial while considering conditions favour icing.

Interesting is fact that for superhydrophobic surface time to reach maximum diameter is shorter than for steel. Table 1 shows exact values of time differences. Such results are different as compared

Journal of Physics: Conference Series **530** (2014) 012028 doi:10.1088/1742-6596/530/1/012028

to e.g. work of Bobiński [2]. Due to the fact such results were not in scope of this work, there was no thorough discussion about, but it is good example for further investigation.

Table 1. Time differences to reach diameter

	Steel (ms)	Superhydrophobic (ms)
Time to reach max diameter	5	1,8
Time to reach final diameter	44	13

4.1.2. Negative temperature surface results

This part of the research concerned droplet impacting superhydrophobic surfaces with temperature lower than water freezing point. Anti-icing properties of hydrophobic substrates were confirmed by Wang et al. [9]. It was stated that increasing surface hydrophobicity leads to retardation of ice accumulation (heat transferring capability and water-cooling rate decrease). Important to mention is that, the pure, deionized water was used. The first set of experiments was carried out both for steel and for superhydrophobic surfaces. It is focused on differences for both substrates when icing appears in equal level of humidity (20-24%) and for temperatures: -1°, -3 °, -7 °, -11 °, - 13 ° , -15 ° C

Such temperatures were obtained by setting constant values of voltage and current at Peltier Cell. The ambient temperature was, room temperature 23 °C. To achieve repeatability the sample was blown each time with waterless nitrogen. This was also the way to maintain relative humidity at the same level. The specimen was subjected to water impingement when both temperature and humidity were stable. To remove water from probe between experiments, humid free nitrogen was used. The humidity and temperature just after blow off have increased rapidly, but for experiments it was set as stationary. The last part of results contains additionally carried out experiments with e.g. substrate cooled down for 180sec before impingement or substrate covered by water film of ice. Such test can be better approximation of phenomena appearing on e.g. airfoil during flight.

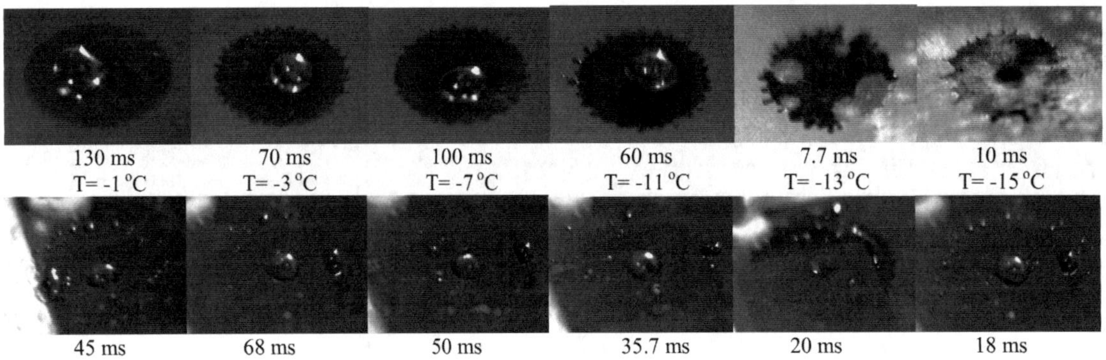

Figure 7. Time elapsed images of water droplet impingement polished steel substrate (top) and for superhydrophobic surface (bottom).

The second part of work devoted to water (at ambient temperature) contact with cold surfaces shows results for both steel and superhydrophobic substrates. It might be easily said that for obtained conditions, relative humidity around 20-25% and temperatures 0 ~ -25°C, superhydrophobic surface shows perspective results for anti icing application. Results obtained for this part of work acknowledge work of other scientist, but also imply additional issues such as changes of humidity, which may lead to qualitatively different results. For example in work of Bobiński [2], droplet has not froze both on steel and superhydrophobic substrate, but there whole area was fulfilled with non vapor nitrogen (RH ~ 8%).

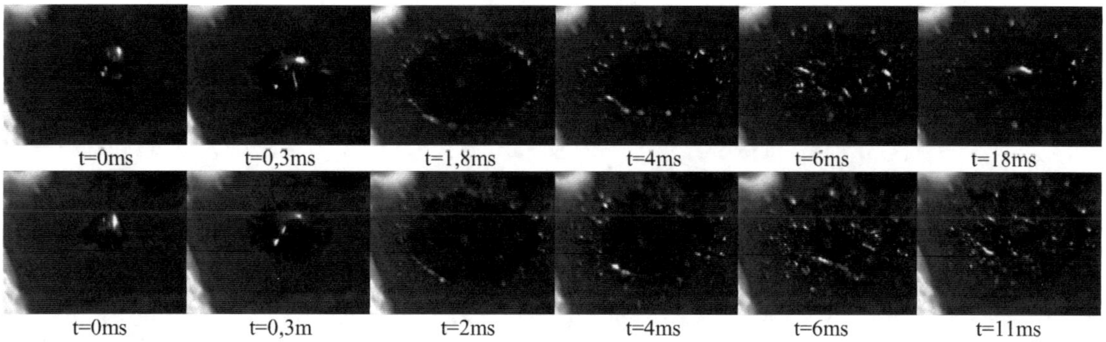

Figure 8. Impingement of droplet on superhydrophobic surface for -15 °C (top) and -19 °C (bottom).

In this work droplet does not freeze until reach temperature around -11 ~ -13 °C (see Fig.7). Such results came up across during experiments and may by directly connected saturation pressure which changes for different humidity and temperatures and of course thermodynamic properties of water, steel and silica. Direct comparison steel with silica demonstrates that for ranges till -15°C water does not freeze during impingement on superhydrophobic substrate. Time to reach final diameter is relatively 2 times shorter for superhydrophobic than for steel (until reaching temperature that freezing occur for steel). Comparing results it can be said that for "smooth" superhydrophobic substrate frost formation is not so rapid and intense as for steel. Interesting fact is that for -19°C at superhydrophobic surface water splashes, reaches maximum diameter and retracts, but much slower than for -15°C (see Fig.8). Water slows rapidly and pretends to freeze but it does not during impingement. Additional "flight like" test cases show how does the water behaves on cold water and ice. Droplet impinged at around 0°C degree water causes wavy structures and some partial defragmentation, but while droplet hit glassy ice at around 0 ~ -1°C it freeze almost immediately. Similar experiment carried out for superhydrophobic substrate partially covered with ice showed that water also froze on surface of ice and retracts at hydrophobic surface. Results from this part of work lead to conclusion that superhydrophobic smooth surface is eligible to be used as a proper anti icing coating. Such results obtained for ambient temperature of water and cold surface may be good approximation what actually happen during icing.

4.1.3. Supercooled droplet result.
As it was mentioned in before, ice repellant in an aircraft application is one the most attractive (and possibly lucrative) application. If we consider superhydrophobic coatings for airfoils it has to be taken into account droplets, which are supercooled. For the experiment, whole system was cooled down by heat exchanger connected to vessel containing liquid nitrogen. Temperature was determined (using thermocouples) in several points of the system. Additionally to tested surface two thermocouples were added – one inside chamber (connected to hygrometer), second one was placed in tube in vicinity of syringe. The obtained temperature was around −10°C. The height of casting was 1.2m to satisfy the conditions for cooling water droplet below 0°C. Startup time of air circulation at heat exchanger was used to obtain different relative humidity values inside circuit. In this section of work test cases were carried out focusing not only on supercooled droplet, but also to provide more possible conditions of icing occurring in real life conditions. Several test cases in range of relative humidity 30% ~ 80%, temperatures of substrate -6 °C ~ -21°C, (temperature around syringe -5 °C ~ -10 °C and temperature inside chamber -7 °C ~ -30 °C were carried out. All test cases were realized in steady conditions, with air circulation switched off. Last part of work is focused on conditions which are highly probable to occur during flight. Changing humidity, supercooled water and various temperatures are variables which can change when e.g. plane flight across clouds. For all obtained results there is no visible ice accretion.

Journal of Physics: Conference Series **530** (2014) 012028 doi:10.1088/1742-6596/530/1/012028

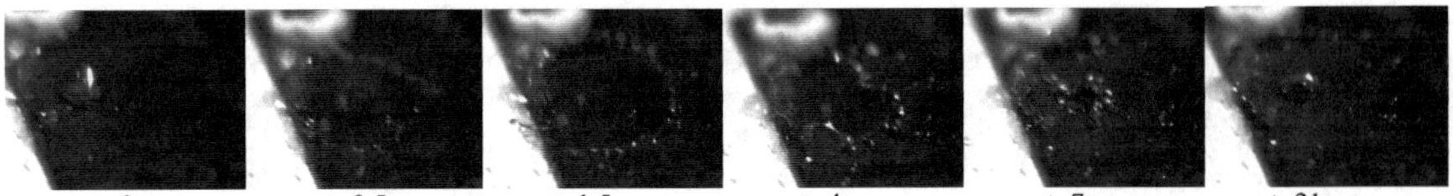

t=0ms t=0,5ms t=1,5ms t=4ms t=7ms t=21ms

Figure 9. Impingement of supercooled droplet on superhydrophobic substrate. 35% Relative Humidity, -16^{0}C temperature of substrate, temperature in chamber -16^{0}C.

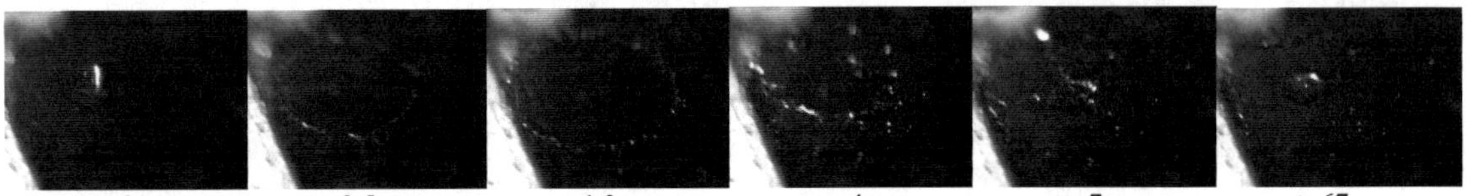

t=0ms t=0,5ms t=1,3ms t=4ms t=7ms t=67ms

Figure 10. Impingement of supercooled droplet on superhydrophobic substrate. 60% Relative Humidity, -21°C temperature of substrate, temperature in chamber -30°C.

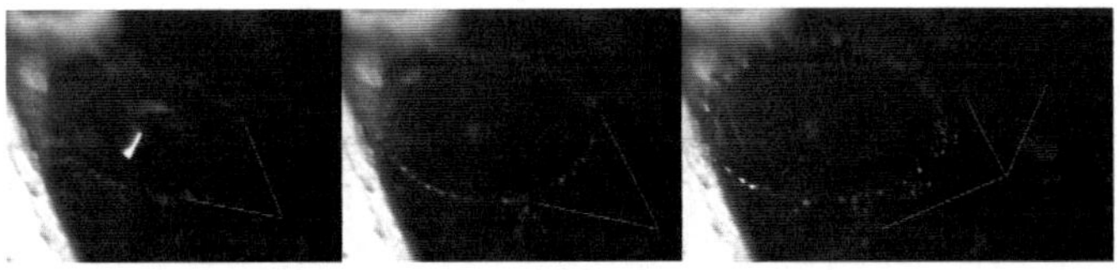

Figure 11. Axis symmetric jetting.

The very first test was carried out at equal temperature of substrate and air (-16°C) and relative humidity 35% (see Fig.9). Higher relative humidity and lower temperatures did not lead to icing on superhydrophobic substrate. Three presented examples have similar solutions for droplet spreading/splashing and then retraction scenarios. Despite the high humidity e.g. 80% there is no frost observed on superhydrophobic substrate which might lead or accelerate icing. For three different values of humidity and different temperatures time to reach final diameter for supercooled droplet is not varying much.

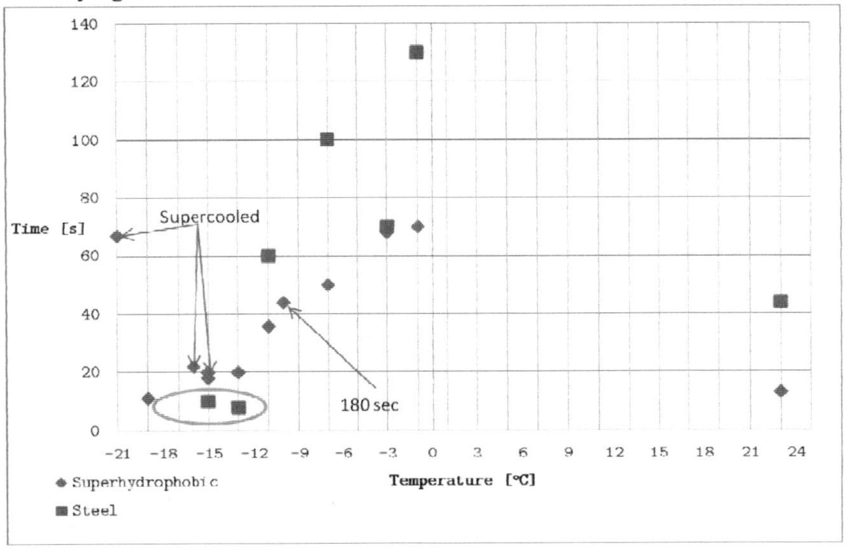

Figure 12. Time to reach final diameter for given temperatures.

Interesting is fact that there might occur some partial jetting of small amount of water in quite different shape (see Fig.11) than known from bibliography. Figure 12 shows times to reach final diameter for all experiments mentioned in work. One of the most interesting fact is for examples with supercooled water and higher relative humidity droplet is not freezing e.g. for -21°C, RH 60% , while for -19°C , RH 20% retraction phase and velocity is slowed down and it tends to freeze. That fact may come from lower surface tension for higher temperature and so called Mpemba effect. At this point of experiments it can be assumed that : for warmer water evaporation reduces mass to be frozen, reduction of water density might suppress the convection currents that cool lower part of the liquid mass and thus colder droplet does not tend to freeze. It can be also concluded that it takes two times longer to reach final diameter for steel than for superhydrophobic substrates, of course not for samples which freeze immediately (see Fig.12, green circle). For lower temperatures time to reach final diameter is smaller, which can be directly connected to dynamic viscosity which is much higher below 0 °C , which effects with no defragmentation and no bouncing. Results mentioned in this conclusion are strictly related to contact angle hysteresis. It is the property which plays important role while considering velocity of retraction. For substrate kept in cold for longer period of time (180 s) before impingement there is no change in time to reach final diameter. That means that frost formation has no meaning for this range of temperatures. Quite interesting is experiment for -21°C which, despite high relative humidity – 60% (see Fig.10) and low temperature in chamber -30°C does not freeze and also takes more time to reach final diameter than other examples.

5. Conclusions

The present paper dealt with investigation of droplet impact on surfaces with different wetting properties and in the icing conditions. The first part consisted of validation of wetting properties of the surfaces in the ambient temperature environment. Second part of experiment related to negative temperature of surface proved ice repellent properties and characteristics temperatures to freeze water at steel substrate were obtained. Interesting is fact that with lowering temperatures time scale is changing, but droplet does not freeze. The main part of the research, presented in this paper, considered the passive anti-icing application of hydrophobic surfaces for various temperatures and humidity. For even high humidity (85%) and low temperatures (-21°C), water does not freeze during impingement on "smooth" silica, superhydrophobic substrate. It seems therefore that, within the range of investigated parameters, the crucial factor for controlling the droplet freezing process is the existence of water vapour in the system and thermodynamic processes which are on going inside and on the surface of water droplet. Some additional interesting behaviours of water were observed during test cases and it is reasonable to carry on further experiments.

References

[1] Worthington A 1876 *Proceedings of the Royal Society of London* **25** p 261-272

[2] Bobinski T 2014 Droplet impact in icing conditions – the influence of ambient air humidity *Arch. of Mech.* **66** 2

[3] Varanasi K, Deng T, Hsu M and Bhate N 2009 Hierarchical Superhydrophobic Surfaces Resist Water Droplet Impact *Tech. Proceedings of the 2009 NSTI Nanotechnology Conference and Expo*

[4] Cassie A B D 1948 *Discussions of the Faraday Society 3* **1116**

[5] Hacker P T 1951 Experimental values of the surface tension of supercooled water *National Advisory Committee for Aeronautics Technical note* **2510**

[6] Eral H B, Mannetje D J C M and Oh J M 2012 *Contact angle hysteresis: a review of fundamentals and applications*

[7] Psarski M, et al 2012 Hydrophobization of epoxy nanocomposite surface with 1H,1H,2H,2H-perfluorooctyltrichlorosilane for superhydrophobic properties *Central European J. of Phy.* **5** 10 p 1197-1201

[8] Wang H, Tang L, Wu X, Dai W and Qui Y 2007 *Applied Surface Science* **253** 22 p 8818–8824

Journal of Physics: Conference Series **530** (2014) 012029 doi:10.1088/1742-6596/530/1/012029

An Experimental Validation of a Turbulence Model for Air Flow in a Mining Chamber

M. Branny[1*], M. Karch[1], W. Wodziak[1], M. Jaszczur[2], R. Nowak[2], J.S. Szmyd[2]
[1]Faculty of Mining and Geoengineering
[2]Faculty of Energy and Fuels
AGH University of Science and Technology, Krakow, Poland

e-mail: branny@agh.edu.pl

Abstract. In copper mines, excavation chambers are ventilated by jet fans. A fan is installed at the inlet of the dead-end chamber, which is usually 20-30m long. The effectiveness of ventilation depends on the stream range generated by the fan. The velocity field generated by the supply air stream is fully three-dimensional and the flow is turbulent. Currently, the parameters of 3D air flows are determined using the CFD approach. This paper presents the results of experimental testing and numerical simulations of airflow in a laboratory model of a blind channel, aired by a forced ventilation system. The aim of the investigation is qualitative and quantitative verification of computer modelling data. The analysed layout is a geometrically re-scaled and simplified model of a real object. The geometrical scale of the physical model is 1:10. The model walls are smooth, the channel cross-section is rectangular. Measurements were performed for the average airflow velocity in the inlet duct equal 35.4m/s, which gives a Reynolds number of about 180 000. The components of the velocity vector were measured using the Particle Image Velocimetry approach. The numerical procedures presented in this paper use two turbulence models: the standard *k-ε* model and the Reynolds Stress model. The experimental results have been compared against the results of numerical simulations. In the investigated domain of flow - extending from the air inlet to the blind wall of the chamber - we can distinguish two zones with recirculating flows. The first, reaching a distance of about 1m from the inlet is characterized by intense mixing of air. A second vortex is formed into a distance greater than 1m from the inlet. Such an image of the velocity field results from both the measurements and calculations. Based on this study, we can conclude that the RSM model provides better predictions than the standard *k-ε* model. Good qualitative agreement is achieved between Reynolds Stress model predictions and measured components of the velocity.

1. Introduction

Mine headings in board and pillar excavation system are typically ventilated with the use of jet fans. Polish copper mining widely used such ventilation systems. Most headings are 20-30m long and fans are typically installed at the inlet to the heading. The ventilation efficiency depends on the range of air jet generated by the fan. The air jet penetration distance is influenced by a number of factors, among them such as inlet air flow rate, the flow velocity or the intensity of turbulence. In real flow the velocity field generated by the supply air stream is fully three-dimensional and the flow is unsteady and fully turbulent. Currently the parameters of 3D air flows are determined using the CFD (Computational Fluid Dynamics) approach [1]-[5]. The validation test of CFD simulation results for

Journal of Physics: Conference Series **530** (2014) 012029 doi:10.1088/1742-6596/530/1/012029

the air flow in the mining face area was done in [7]-[10]. The flow field, investigated in this paper differs significantly from those analyzed in the mentioned papers where the connections between the ventilation ducts has been arranged different way.

This paper presents the result of experimental and numerical predictions of airflow in a laboratory model of a blind heading ventilated by a jet fan. The heading geometry considered in the study is a re-scaled and simplified in reference to the real object. The walls of the model representing chamber and main gallery are smooth and the channels cross-sections are rectangular. The fan generating the air jet is located beyond the measurement section. The measurements were performed for the steady and isothermal flow conditions. In that case, an equal Reynolds number in the model and in the real object assures that the flow criteria are similar. Air was used as the experimental fluid in the laboratory stand, thus the equality of the Prandtl number was fulfilled.

The components of the velocity vector were measured using the Particle Image Velocimetry (PIV) approach. The numerical procedures presented in this paper use two turbulence models: the standard *k-ε* model and the Reynolds Stress Model (RSM). Numerical calculations are supported by the Ansys Fluent software . The main objective of this research is the qualitative and quantitative verification of modelling results.

2. Experimental set-up

Experimental measurements were conducted in a laboratory settings (shown schematically in Fig.1), incorporating two intersecting rectangular channels with a cross-section of 0.4 x 0.2 m. The blind chamber is 3.35 m length. Air is supplied with a 3.0 m long tube , where the tube diameter is 0.076 m. Measurement is done on a purpose-built 1:10 scale model of the real object. In order to obtain the axisymmetric flow velocity profile at the inlet to the chamber a 1.35 m jet-stabilised section was provided to stabilise the air jet. The stabiliser comprises a diffuser, a confuser and a 0.2 m diameter tube. The distance between the axis of the air-supply duct to the upper and side walls of the blind chamber is 0.08 m. The air inlet is at the distance of 2.0 m from the chamber's blind wall, which is a typical chamber length in copper mines, re-scaled accordingly.

The measuring PIV system was subjected to testing and calibration procedures under flow condition to yield the number of registered flow images required to ensure accurate statistics (around 1000) and the size of the optimal interrogation window for the cross-correlation method (32x32 pixels). The Davis software was used to analyse the images to evaluate the velocity vector components.

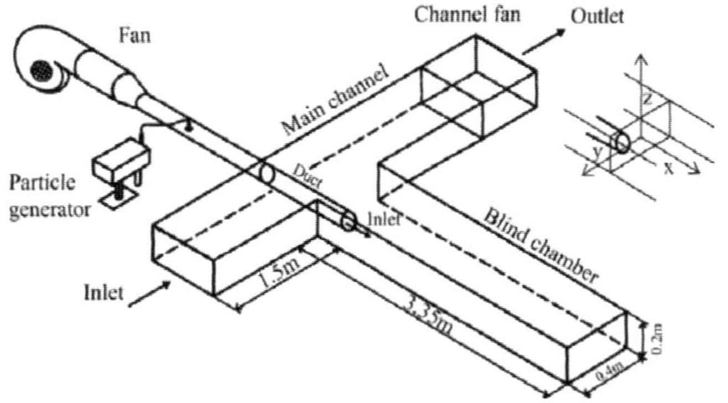

Figure 1. Schematic diagram of the measurement set-up.

3. Mathematical model

Classical modeling of turbulence is based on the Reynolds concept, which for incompressible and Newtonian fluids yields the following equations [6]

$$\frac{\partial U_i}{\partial x_i} = 0 \tag{1}$$

$$\frac{\partial U_i}{\partial t} + U_j \frac{\partial U_i}{\partial x_j} = -\frac{1}{\rho}\frac{\partial P}{\partial x_i} + \frac{\partial}{\partial x_j}\left(\nu \frac{\partial U_i}{\partial x_j} - \overline{u_i u_j} \right) \tag{2}$$

$$-\overline{u_i u_j} = \nu_t \left(\frac{\partial U_i}{\partial x_j} + \frac{\partial U_j}{\partial x_i} \right) - \frac{2}{3}\rho k \delta_{ij} \tag{3}$$

where: $\nu_t = C_\mu \frac{k^2}{\varepsilon}$ is the turbulent viscosity, k is kinetic energy, ε is the dissipation rate of k, C_μ is the constant, δ_{ij} the Kronecker delta, U_i and u_i the mean and fluctuating components of velocity, ρ the density, P the pressure, and ν the molecular viscosity.

In this study, two models of turbulence were tested: the standard k-ε model and the Reynolds stress model. The standard k-ε model is the most widely used turbulent model for solving industrial problems, particularly in mining for solving ventilation problems. This is a semi-empirical model based on the Boussinesq's conception of turbulent viscosity and on two transport equations for turbulent kinetic energy and its dissipation rate (equations (4) and (5)).

$$\frac{\partial k}{\partial t} + U_i \frac{\partial k}{\partial x_i} = \frac{\partial}{\partial x_i}\left(\left(\nu + \frac{\nu_t}{\sigma_k} \right) \frac{\partial k}{\partial x_i} \right) - \overline{u_i u_j} \frac{\partial U_i}{\partial x_j} - \varepsilon \tag{4}$$

$$\frac{\partial \varepsilon}{\partial t} + U_i \frac{\partial \varepsilon}{\partial x_i} = \frac{\partial}{\partial x_i}\left(\left(\nu + \frac{\nu_t}{\sigma_\varepsilon} \right) \frac{\partial \varepsilon}{\partial x_i} \right) - C_1 \frac{\varepsilon}{k} \overline{u_i u_j} \frac{\partial U_i}{\partial x_j} - C_2 \frac{\varepsilon^2}{k} \tag{5}$$

where C_1, C_2, σ_k, σ_ε are constants [6].

The Reynolds Stress Model (RSM) closes two systems of equations (1), (2) by solving transport equations for Reynolds stresses, together with an equation for the dissipation rate. The RSM is expected to give most accurate predictions for complex turbulent flows. A wall function was used in the near-wall region.

3.1. Mesh size

A structured, non-uniform mesh was generated for the computational domain. Three different size meshes of about 1,200,000, 2,800,000 and 5,600,000 cells were examined. The most visible effects of mesh refinement may be observed in cross-sections further apart from the inlet opening, especially in lower chamber region and the results for two finer meshes differ slightly in comparison with the coarse one. Thus, the mesh consisting of 5,600,000 cells was used in further calculations.

4. Numerical procedure

In the numerical procedure, the boundary conditions at the inlet were expressed by the experimental velocity profile of the air jet entering the test section in the laboratory model (see Fig 2).

The mean velocity at the inlet to the blind chamber was U=35,4m/s and the corresponding Reynolds number was approximately Re=180 000.

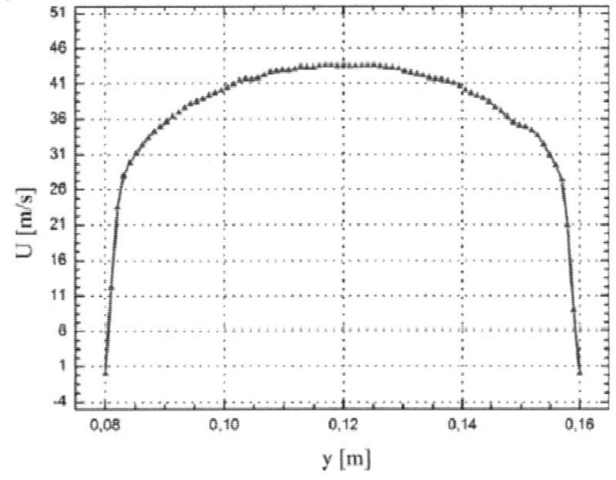

Figure 2. Velocity profile at the inlet to the blind chamber.

The kinetic energy of turbulence, based on the velocity fluctuation measured in the axis of the inlet, was equal to 5.9m^2/s^2. The calculated turbulence intensity was 5.6% while the length scale was 0,00532m.

The velocity field in the inlet cross-section of the main duct was assumed to be uniform. Measured average velocity was 2,1m/s, the intensity of turbulence 5% and the length scale 0.0187 m.

Boundary conditions on walls were assumed in the form of wall functions. The distance between the nodes nearest to the walls should be such that it would fall in the lower range of validity of the logarithmic law for mean velocity. In the calculation procedure this distance was controlled by the parameter y^* :

$$y^* = \frac{\rho C^{1/4} k_P^{1/2} y_P}{\mu} \tag{6}$$

where: C- model constant ; k_P- turbulence kinetic energy at point P; y_P- distance from point P to the wall. In calculations this parameter was in the range between 40 and 80. Boundary conditions in the outlet opening emulate the fully developed flow in duct.

5. Comparison of experimental and numerical data

The velocity field in the horizontal plane located on the level of the air-supply duct is shown in Fig 3. Within the investigated area, two zones with recirculating flow can be distinguished. In the first zone, extended to about 1.0 m from the inlet, there is intensive air mixing. At distances further away from the inlet than 1.0m, the second vortex tends to form. The vorticity magnitudes in these two regions have the opposite sign. In contrast to measurement results and RSM simulations, the calculations relying on the k-ε model reveal a third zone with recirculating flow in the vicinity of the chamber's blind wall.

Fig. 3 (bottom) shows the velocity field contours in the horizontal plane on the level of the axis of inlet opening. To better highlight the characteristic flow features, the coloured velocity scale is restricted to the interval (-1.8 m/s, 2.2 m/s). Velocities beyond that range are represented by extreme colours in the attached colour scale. The inlet is situated near the right upper chamber corner. The air jet entering the system tends to flow towards the left wall of the chamber. Starting from the distance of 1.05m from the inlet, the air jet begins to flow towards the blind wall, along the left-hand side wall of the chamber. The virtual stream of air flowing into system, determined with the use of the k-ε model reaches the left side wall of the chamber approximately 0.8m from the inlet. More accurate values of this parameter,

Journal of Physics: Conference Series **530** (2014) 012029 doi:10.1088/1742-6596/530/1/012029

approaching the accuracy levels available in experiments, were obtained with the use of the RSM model.

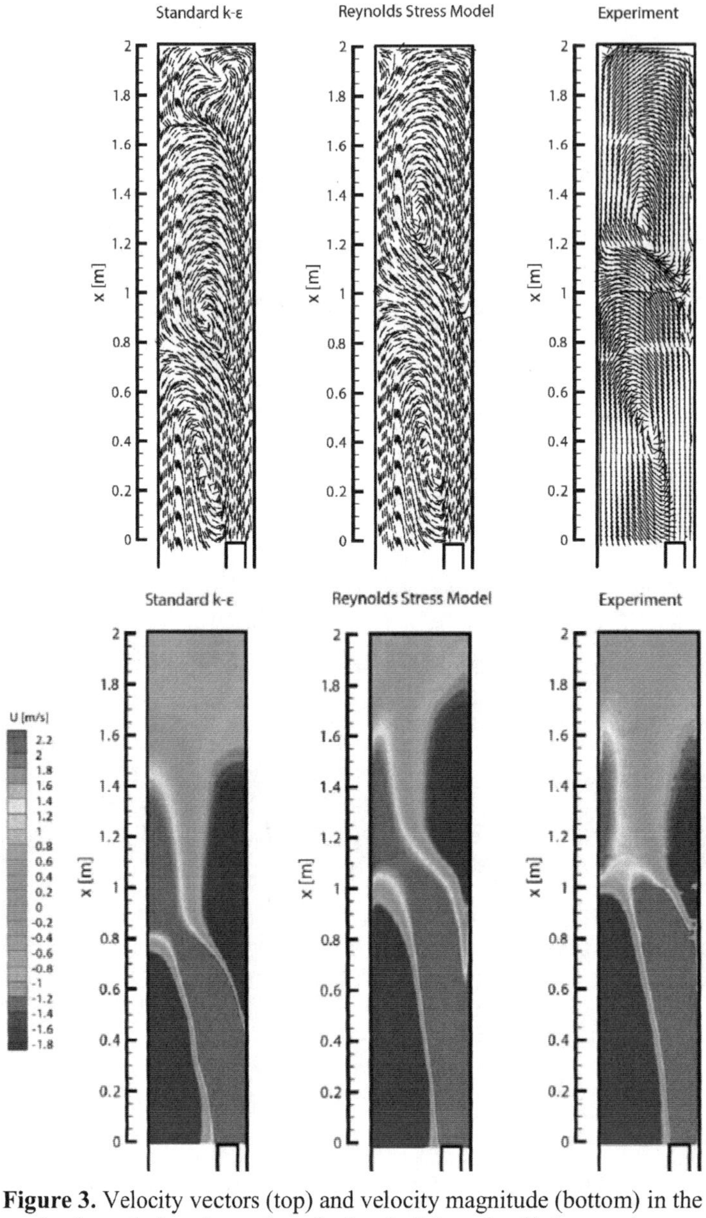

Figure 3. Velocity vectors (top) and velocity magnitude (bottom) in the horizontal plane located on the level of the axis of inlet opening

In the latter case the inlet jet tends to adhere to the left-side wall 1.0m from the inlet. Such a velocity field pattern is obtained both from measurements and simulations using the RMS model and agrees well with the results of investigations conducted in the copper mine chambers.

The experimental and numerical data are compared quantitatively in Fig.4 and Fig.5. These figures show the distributions of two velocity vector components along the selected lines. Horizontal lines are situated on the level of the axis of the air supply duct and 5 cm above the chamber bottom, the distance from the inlet being 0.5 m, 1.0 m and 1.5 m.

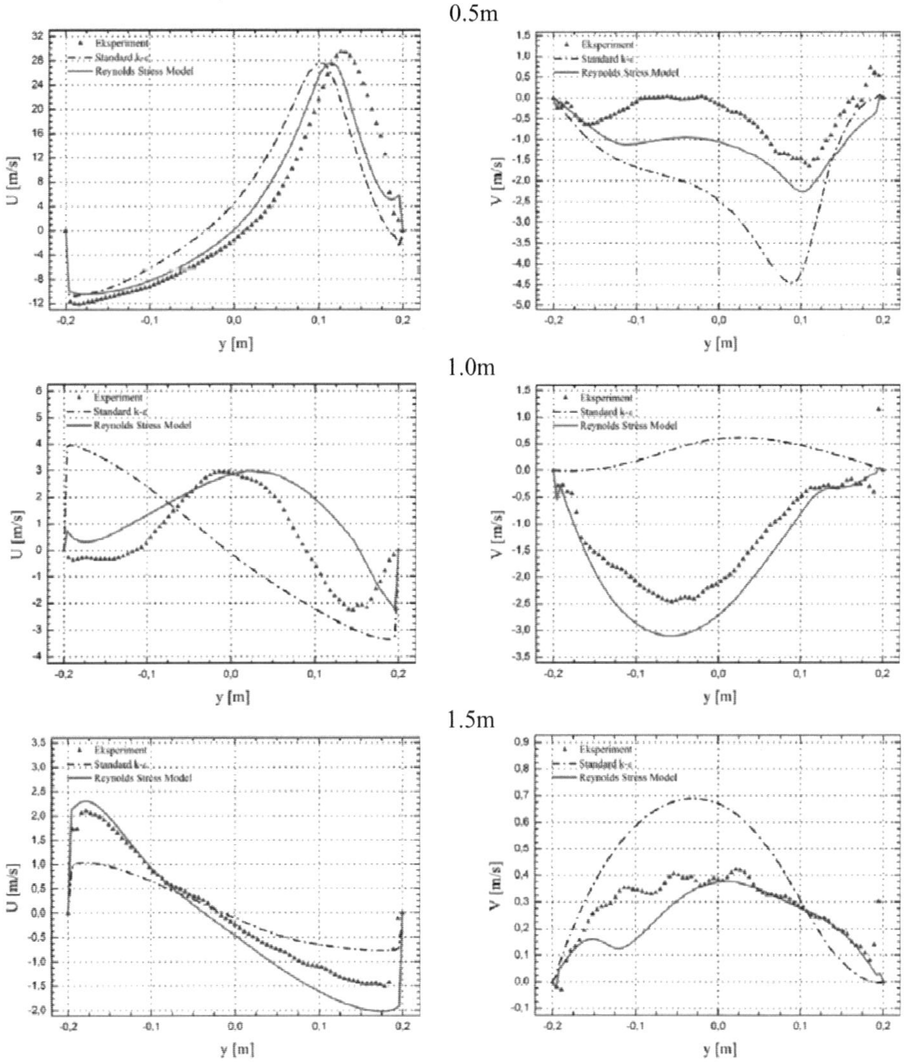

Figure 4. Distribution of the longitudinal U and streamwise V velocity component along horizontal lines on the level of the axis of inlet opening.

The RSM model provides better predictions than the standard k-ε model. In the horizontal cross-section on the level of the axis of supply duct, the longitudinal and streamwise components of the velocity vector obtained using the RSM model agree well with the experimental data (Fig.4). Considerable differences between the experimental and numerical simulation results are registered for the lower chamber regions (Fig.5), particularly for the distance of 1 m from the inlet. These discrepancies are revealed in the region separating two zones with recirculating flow (Fig.3). Despite some discrepancies between the experimental and calculation data, it is still reasonable to conclude that the RSM model captures the real flows with sufficient accuracy for the needs in ventilation problems.

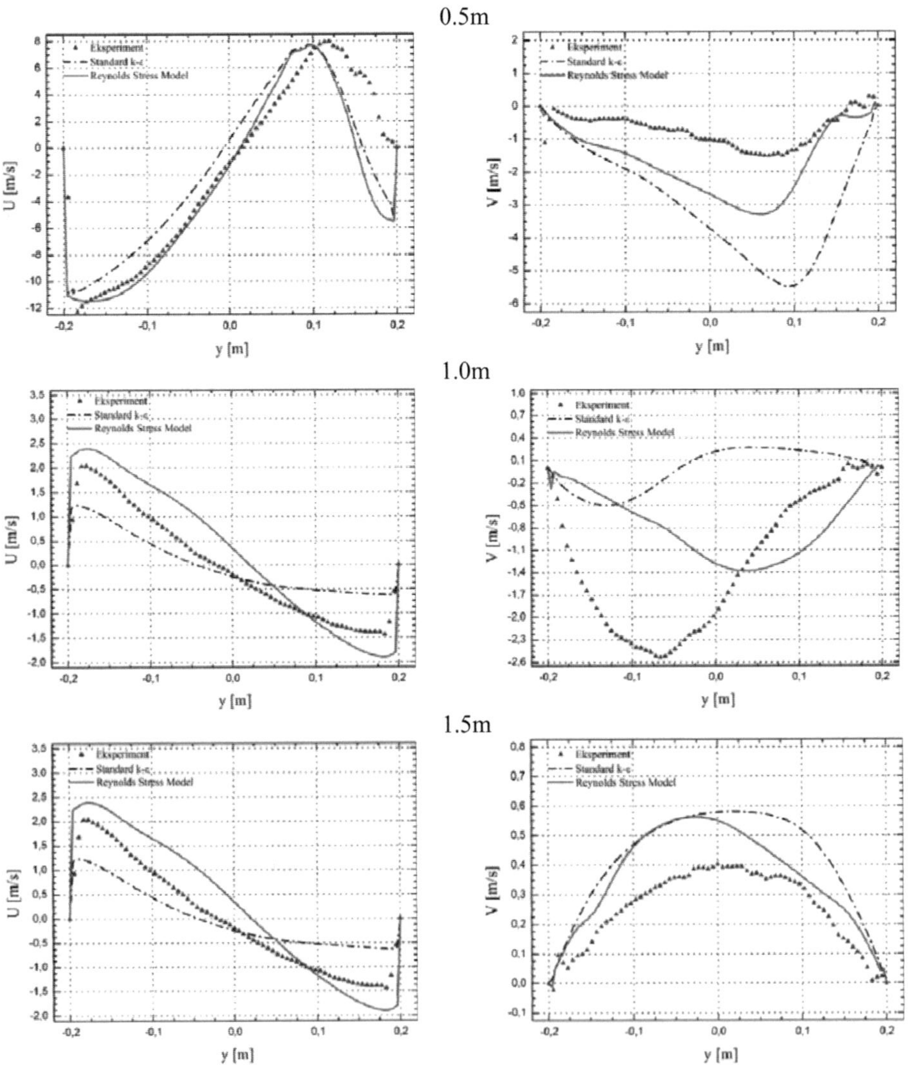

Figure 5. Distribution of the longitudinal U and streamwise V velocity component along horizontal lines 5 cm above the bottom

6. Conclusion

Two zones with recirculating flow can be distinguished due to the shape of the velocity field created by the inlet air jet. The first zone extends approximately 1m from the inlet. Another vortex rotating in the opposite direction is formed further from the inlet. Such a velocity field pattern is obtained from measurements and qualitative agrees well with numerical data obtained using the RSM model. The RSM model provides better predictions than the standard k-ε model. The largest quantitative differences between the calculated (RSM model) and measured velocity vector components are registered in the lower section of the blind chamber, particularly in the zone separating the recirculating flow regions. Despite these discrepancies, it is reasonable to conclude that computer simulations offer a good insight into ventilation processes in long blind headings aired with jet fans and that the RSM model captures the real flow features with sufficient accuracy for the needs of mine ventilation practices.

Acknowledgements
The research was supported by the Polish Ministry of Science and Higher Education (project No. 4653/B/T02/2010/39)

7. References

[1] Aminossadati, S.M., Hooman K.,2008: Numerical simulation of ventilation air flow in underground mine workings. 12th U.S./North American Mine Ventilation Symposium, 253-259
[2] Krawczyk J., 2007: Single and multiple-dimensional models of unsteady air and gas flows in underground mines (in polish), Archives of Mining Sciences, Monograph, 2
[3] Kuan, B., Yang, W., Schwarz, M.P., 2007: Dilute gas-solid two-phase flows in a curved 90° duct bend: CFD simulation with experimental validation. Chemical Engineering Science, 62, 2068-2088
[4] Nakayama, H., Hirota, M., Shinoda, K., Koide, S., 2005: Flow Characteristics in a Counter-Flow Type T-junction. Thermal Science and Engineering, 13, 17-2
[5] Mossad R., Yang W., Schwarrz M.P., 2009:Numerical prediction of air flow in a sharp 900 elbow, Seventh International Conference on CFD in the Minerals and Process Industries, CSIRO, Melbourne,Australia, 1-5
[6] Rodi,W., 1979: Turbulence Models and their Application in Hydraulics. Delft: IAHR
[7] Skotniczny P., 2013: Three Dimensional Numerical Simulation of Mass Exchange Between Longwall Headings and Goafs, in the Presence of Methane Drainage in U-Type Ventilated Lonwall, Archives of Mining Sciences, 58, 3
[8] Szmyd J., Branny M., Karch M., Wodziak W., Jaszczur M., Nowak R., 2013: Experimental and numerical analysis of the air flow in T-shape channel flow, Archives of Mining Sciences, 58,2, 333–348
[9] Wala A.M., Vytla S., Taylor C.D., Huang G., 2007: Mine face ventilation: a comparison of CFD results against benchmark experiments for CFD code validation, Mining Engineering, 59, 10-17
[10] Wala, A.M., Vytla, A., Huang G. & Taylor C.D. 2008: Study on the effects of scrubber operation on the face ventilation. 12th U.S. North American Mine Ventilation Symposium 2008 Reno, USA

Journal of Physics: Conference Series **530** (2014) 012030 doi:10.1088/1742-6596/530/1/012030

Experimental analysis of the flow structure in the laboratory model of SOFC fuel cell channels

M. Jaszczur[1], G. Pucillo[1,2], R. Nowak[1], L. Magistri[2]
[1]AGH University of Science and Technology, Faculty of Energy and Fuels Department of Fundamental Research in Energy Engineering, Poland
[2]University of Genova, Faculty of Engineering, Italy

E-mail: marek.jaszczur@agh.edu.pl

Abstract. In the presented paper a flow structure in the gas channel of planar SOFC fuel cell is presented. The model taken for analysis was constructed based on the channel geometry manufactured by SOFC Power company. The shape of a channel was rectangular filled with large number of obstacles which role is to divide the flow into segments with possibly homogenous velocity distribution. The model itself was constructed from Plexiglas and the reactant gases flow was modelled by water motion. To investigate and visualize the flow structures a PIV technique was applied. Three different flow rates were taken for investigations and the flow uniformity and time dependence was studied.

1. Introduction

Fuel cells are devices that generate electricity by a chemical reaction. In general, hydrogen is normally used as the basic fuel, which reacts electrochemically with oxygen producing electricity, heat and, as a by product, water. Fuel cells are usually classified by the electrolyte employed in the cell and the main five types are: Alkaline Fuel Cell, Molten Carbonate Fuel Cell, Phosphoric Acid Fuel Cell, Polymer Electrolyte Fuel Cell and Solid Oxide Fuel Cell (SOFC).

A second grouping can be done by looking at the operating temperature for each of the fuel cells. The high-temperature fuel cells operate at temperatures approx. 600±1000°C and two different types have

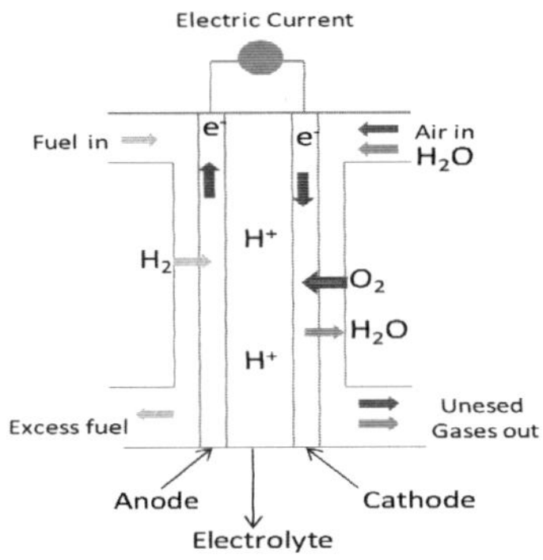

Figure 1. Transport processes within a SOFC

Journal of Physics: Conference Series **530** (2014) 012030 doi:10.1088/1742-6596/530/1/012030

been developed, the Molten Carbonate Fuel Cell and the Solid Oxide Fuel Cell. In this paper, an experimental model of gas distributing channels for planar Solid Oxide Fuel Cell has been tested.
In a planar SOFC, cell components are configured as flat plates which are connected in electrical series. Cell itself consists of electrodes, anode (negative) and cathode (positive), where the reactions which produce the electricity takes place. Between them, there is an electrolyte, which allows the transport of electrically charged particles from one electrode to the other, and a catalyst, which enables and speeds up the reaction.
Technical details of the Solid Oxide fuel cell are given below:

- Operating temperature: 800±1000 °C
- Anode reaction: $H_2+O^2 \rightarrow H_2O+2e^-$
- Cathode reaction: $\frac{1}{2}O_2+2e^- \rightarrow O^?$
- Applications: combined heat and power for stationary decentralised systems and for transportation (trains, boats, etc.)
- Realised Power: small power plants 100-250 kW
- Charge Carrier in the Electrolyte: O^{2-}

In general SOFC cells employ a solid oxide material as electrolyte and are, thus, more stable than the molten carbonate fuel cells as no leakage problems due to a liquid electrolyte can occur. The SOFC cell is a straightforward two-phase gas-solid system so it has no problems with water management, flooding of the catalyst layer, or slow oxygen reduction kinetics. On the other hand it is difficult to find suitable materials which have the necessary thermal and stability properties for operating at high temperatures. An important advantage of SOFCs is the internal reforming. Due to the high temperature of the exhaust gases a combination with other power generation systems (e.g., gas turbines) is also possible which can provide a high overall electrical efficiency (up to 70% in a combined cycle system).

A cell configuration must be designed to have the desired electrical and electrochemical performance, along with required thermal management and mechanical/structural integrity to meet operating requirements of specified power generation applications. The requirements of electrical performance means that the design must minimise ohmic losses in the stack. Thus, the current path in the components must be designed to be as short as possible. Electrochemical performance means that the design must provide the full open circuit voltages and minimal polarisation losses. The gases must be able to quickly reach in the reaction sites to reduce mass transport limitation. The requirement of thermal management means that the cell must provide means for more uniform temperature distribution during operation. Finally, the mechanical/structural integrity means that any planar SOFC stack must be designed to have adequate mechanical strength for assembly and handling.
Different designs for solid oxide fuel cells have been developed over the years. Flat plates have an easier stack configuration while tubular designs have a smaller sealing problem. Due to the high power density of SOFCs compact designs are feasible. The tubular design has been developed by Westinghouse. Tubular designs have a self sealing structure which improves thermal stability and eliminates the need for good thermal-resistant sealants.

The planar design is more efficient and cheaper than the tubular as the current path is shorter and it is easier to stack than the tubular design.
In a SOFC cell there are different plate structures that produce different flow field as, for example, squared spot, serpentine spirals, cascade (this structure uses sealed-off channel that make blockages by water droplets above all it induces a large pressure difference in the cell) and series-parallel, with geometries of various complexity levels. Depending on the concentration of the reactant different flow designs may be advantageous.

The square spot structure in not optimal for an equal distribution of the gases as the reactant gas can flow through the cell by any path available. The advantage of this flow is the fact that the pressure drop is minimal.

In the serpentine configuration (difficult to fabricate if the machining is done manually) reactant gases are obliged to follow the path; therefore a large pressure drops over the gas path may occur and the flow is also affected by impingement, recirculation and flow separation. For this reason it is important to determine the critical Reynolds number necessary for the onset of instability.

J. Martin *et al.* [1] investigated the fluid dynamic phenomena for a model of a U-shaped fuel cell channel and they compared two perpendicular flow planes in terms of absolute and averaged velocity values as well as Reynolds stress correlations. Generally, the flow undergoes a transition to a different regime when two recirculation zones, which originally develop in the U-bend region, merge into one separation region. This transition corresponds to generation of additional vortices in the secondary flow plane. The analytic foundations for secondary flows in curved pipes were given by Dean [2] who established that the onset of secondary vortices in curved pipes is characterized by ratio of Reynolds number to the square of the product of the inertia and centrifugal forces. Hawthorne [3] showed that the secondary flow could occur in a perfectly inviscid flow as a result of a non uniform distribution of velocity at the entrance.

The studies of Martin reveal that at Re<381 the flow in the bend was time independent in both the stream wise and cross-flow planes and was downstream from the bend. At Re>436 the flow in the bend was clearly time dependent in both the stream wise and cross-flow planes and was unstable downstream from the bend. The secondary flow at the midpoint of the turn appeared to contain four major vortices. Besides the ratio of secondary to primary motion was found to decrease after the transition occurred. This suggests that enhanced mixing in the cathode flow channel of fuel cell could be achieved by designing a serpentine channel such that transition occurs around the Reynolds number corresponding on design point operation.

The studies of flow behaviour [4],[5] in cell flow channels were also made in commercial parallel bipolar plate and also in this case the results reveal a non-homogeneous flow distribution across the bipolar plate produce a limited performance of the fuel cell energy conversation.

It is particularly important for the cells with direct internal reforming where the cooperating endothermic reforming reaction, exothermic water gas shift and hydrogen conversation reactions are among those responsible for the rise in the temperature difference along the reaction channels. This phenomenon was studied by Ho *et al.* [6], who observed that fuel cooled by the reforming reaction at the inlet to the cell in gradually warmed up as a consequence of the two above reactions. Moreover, Laurencin *et al.* [7] concluded that the "Nusselt number depends strongly in the geometry of the gas channels"; so the channel design is also responsible for the relationship between the conductive and convective heat transfer in the cell. It is in these grounds that R. Nowak and J. Szmyd [8] tested three kinds of radial flow interconnect configurations for a SOFC, to uncover the possibilities for maintaining the uniform velocity distribution in the channel which directly translates into temperature difference and thermal stresses which can occurs in SOFC cell. They concluded that the most promising geometry for the fuel channel keeps almost constant width along its entire length. In this configuration the velocity field was the most uniform and the smallest velocity gradient along its length were observed.

2. Experimental setup

The main components of the experimental setup are shown in Fig. 2. The flow system consisted of a reservoir with a pump and a flow channel. The PIV system was deployed to acquire data.

During the measurement process the channel was laid on a flat surface with distilled water supplied from the tip of the plate. To excite the tracer particles, which have been added to the fluid, a double cavity Nd: Yag laser has been used and was positioned at a distance of about 90 cm from the axis of the cell model. The cell model has been illuminated through lateral surface of one side and then of the other side. The images of particles motion have been recorded through upper surface of the model, by the CCD camera, with a 50 mm f#8 Nikon lens, placed perpendicularly to the light sheet. This arrangement does not introduce any perspective distortion in the images and eliminates any possible direct reflection of the laser on the CCD camera. Data sets have been recorded for a field of view of 300 mm x 190 mm. The flow rate of water was controlled by the flow controller, and the particles for PIV were added and mixed with the water in the reservoir.

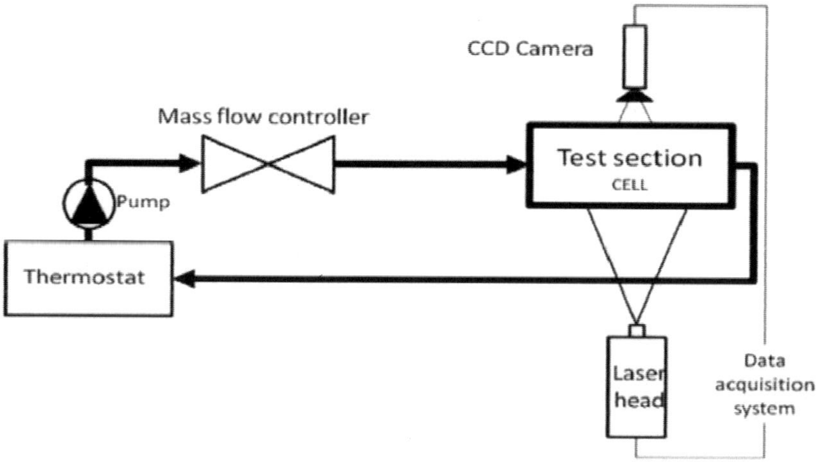

Figure 2. Schematic of the experimental setup

In this paper a rescaled model based on the geometry of the realistic SOFC fuel cell has been tested with four different flow rates. The exact shape of a channel is presented in Figure 3a. The rough surface area, filled with 32 rectangular cross section (1cm x 3 cm) obstacles, was of 378 cm^2 and the thickness of the cell was of 2 cm. The inner diameter of the water supporting tube was approximately 10 mm. Due to the transparency requirements of PIV measurements, the channel was made from a Plexiglas material to allow the investigated flow area to be visible. Because the melting temperature of Plexiglas is about 170°C, it was impossible to carry out the experiment at the high temperatures that characterize SOFC operation. The measurement process was carried out in room temperature and the water motion mixed tracer particles was used as an experimental fluid.

(a) 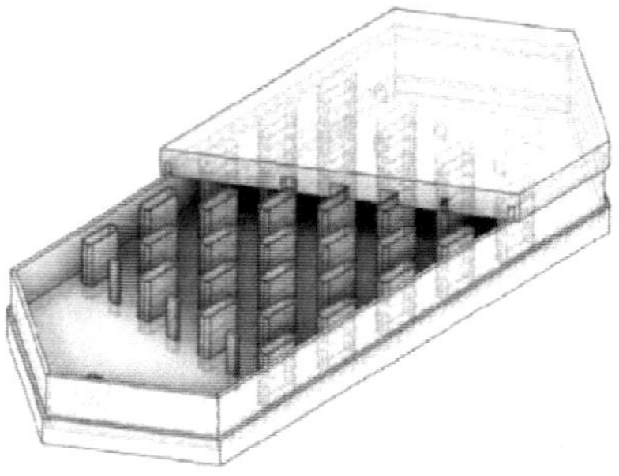 (b) 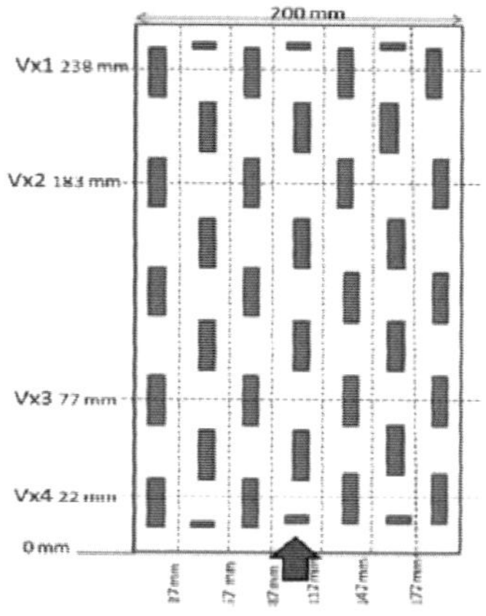

Figure 3. (a) Plexiglass model of SOFC channel, (b) Location of the analysed sections

Journal of Physics: Conference Series **530** (2014) 012030 doi:10.1088/1742-6596/530/1/012030

3. Measurement setup

It was chosen to investigate the flow field inside the channel with the optical measurement technique called Particle Image Velocimetry (PIV). PIV is one of the most recent techniques which allow to determine the instantaneous velocity fields by measuring the displacements of numerous fine particles that accurately follow the motion of the fluid. The PIV method is getting more and more popular in experimental investigations due to its two main advantages: it not interfere with the flow and as a result returns a global measure of the flow field, unlike all the other measurement techniques (except Global Doppler Velocimetry) that allow to perform point measurements.

The standard planar PIV system consists of a double-pulsed laser, light-sheet-forming optics, particle seeding, a monoscopic (single-lens) camera, image digitization hardware and a computer for data storage and analysis.

Double-pulsed illumination is the current norm in PIV because of the ability in using more than two laser pulses. The intensity of illumination required to form visible images of micrometer-sized particles usually requires the use of pulsed solid-state laser sources emitting pulses with energy between 5 and 500 mJ. The energy needed is closely coupled with the scattering properties of the particles.

The flow should be seeded with particles that are small enough to follow the fluid acceleration but large enough to scatter light energy sufficient to form bright images, usually a few tens micrometers in a liquid. The best PIV results are obtained by using artificially seeded particles, thereby permitting careful control of the uniformity of the particle concentration and the optical and fluid mechanical properties of the particles. Particles offer important advantages as markers of the fluid. They produce a stronger optical image than dyed fluid; they are small, of order 0.1-50 μm; they do not diffuse and a solid particles do not deform. It is best to use spherical particles because their images are independent of particle rotation, and barring camera aberrations, the centred of the particle image corresponds to the centre of mass of the particle. The disadvantage of particle markers is that they must slip with respect to the fluid in order to generate the drag force needed to follow the fluid acceleration.

In PIV it is important to strive to achieve a uniform dispersion of scattering particles throughout all regions of the flow that pass though the measurement domain, so as to avoid preferentially sampling selected regions of the velocity field and thereby biasing measurements toward more highly seeded regions.

In PIV the tracer particles are illuminated, by the sheet of laser light, in a plane of measurement in two successive moments of time. The reflected light from the particles is recorded by the sensor of a digital camera , getting two separate frames which will be further processed; the laser must therefore be sufficiently powerful to be able to " impress " the sensor of the digital camera with the small light scattered by small tracer particles , but also must be able to produce a light pulse of short duration, in order to " freeze" the motion of the particles during the exposure and to prevent playback of the image trails . The delay time between two light pulses is a function of the flow velocity: it must be sufficient to identify the displacement of the particles between the two digital images which will act on the software. If the interval is too short, the shift is too small and the noise, due to the "noise" and to the imperfect correlation between the pairs of images, become preponderant compared to the actual displacement.

4. Experimental Results

The measurements of flow velocity distribution in the presented geometry have been performed for four different flow rates ($2{,}44 \cdot 10^{-5} m^3/s$; $2{,}32 \cdot 10^{-5} m^3/s$; $3{=}8{,}91 \cdot 10^{-6} m^3/s$; $4{,}98 \cdot 10^{-6} m^3/s$) which correspond to Reynolds numbers (calculated in the section of the inlet tube) of Re=3101, 2948, 1132 and 632. All the statistical calculations were carried out using DaVis 7.2 software. For each flow rate a set of 300 images was recorded. In Figure 4 an average velocity distribution for a case of $2{,}32 \cdot 10^{-5} m^3/s$

and 4,98·$10^{-6}m^3/s$, corresponding to Reynolds numbers of 2948 and 632 is presented. The average field distribution for the other cases are quite similar to the presented ones.

While analysing the obtained results, it is immediately evident that using lower flow rates, the fluid is less affected by the negative influence of obstacles. The regions with much lower velocities occur mostly in the layers close to the side walls. The side walls caused the flow velocity to decrease mostly in the low flow rate where the general velocity gradient is small.

For the Flow 4 case a good velocity distribution almost all over the channel is obtained with the exception of the lower part near the entrance of the jet. Near the bottom of the channel, a low velocity zone is observed in the flow field, mostly close to the middle and right section of the channel. This low-velocity zone is present in all different flow rates but, as can be verified from the average field distribution, in the higher Reynolds number, is more pronounced. Also the distribution of the flow (for all flow rates) in this part of the channel is evidently not homogenous. It's interesting to note that although the cell structure is assumed to be symmetrical, the measurements results are not completely symmetrical. It has been verified that for all investigated cases, the flow moves with the higher velocities through the left lateral channels. In this part of the channel the velocities are in the range of 0,020-0,023 m/s for higher flow rate and 0,0035-0,0045 for the lower one.

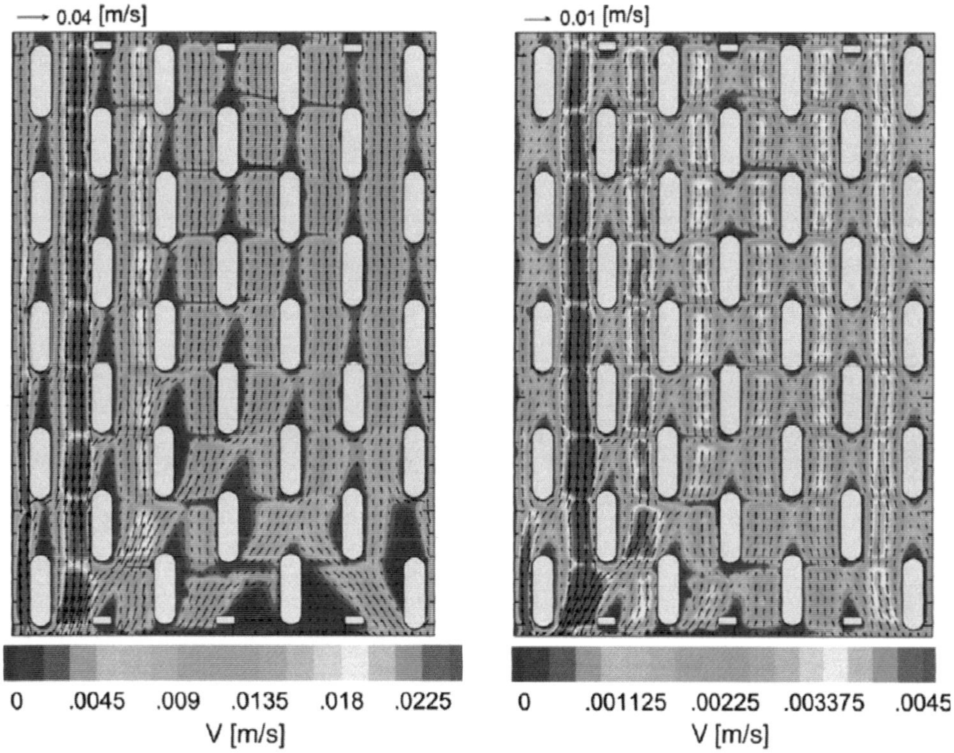

Figure 4. Velocity distribution in the investigated channel; (left) Re=2948, (right) Re=632.

Moreover, the visualized flow is characterized by large velocity gradients in the left part of the channel. It was pointed out that the decrease in velocity is higher for higher flow rates with a gradient of about 0,018 m/s and a gradient of 0,0034 m/s for the lowest flow rate. The higher velocity reached with the case Re=632 about the lower reached with the case Re=2948. On the other hand, inside the channel, the flow is better controlled and is more homogenous with the lower flow rate.

To carry out a more detailed flow structure analysis, the test section was divided transversally by four horizontal sections where the velocity field profiles were calculated. The graphs in Figure 6 show the trend of the velocity, corresponding to the different flow rates, in various stretch of each horizontal section of the channel; obviously the velocity in all sections increased along with the flow rate.

In this case again, the sections most problematic are those nearest the entrance of the jet and in particular in the stretch of width's cell between 70 mm and 120 mm (see in Figure 5b). Curious is the considerable lowering of the velocity to approximately half of each stroke considered; probably the flow is affected by the presence of the obstacle placed in the section above that which is considered. As it has been discussed previously, even the plots indicate the maximum of the fluid velocity is located in the left side of the channel for all different flow rates.

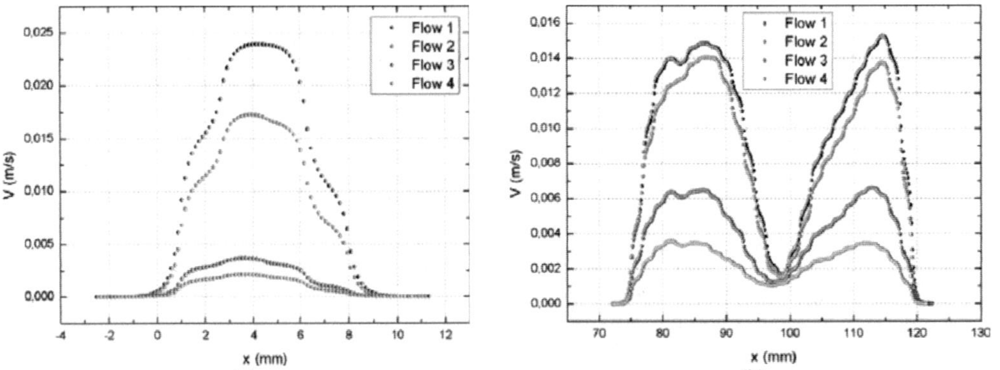

(a) (b)

Figure 5. Comparison of velocity profiles between various Flow rates – Section Vx2.

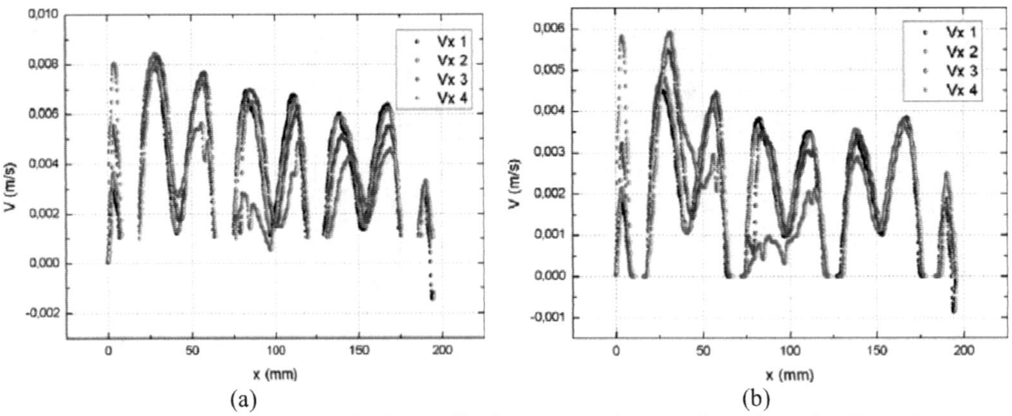

(a) (b)

Figure 6. Comparison of velocity profiles between various sections along the channel; (a) Re=1132, (b) Re=632

The graphs in Figure 6 show, at constant flow rate, the trend of the velocity in four different horizontal sections of the cell (the location of each section is presented in figure 3b). The velocities development for the three sections positioned far from the inlet is very similar, this means that there are not high losses during the crossing of the cell. More problematic and with an anomalous trend (especially at half the width of the channel, where the entrance of the jest is positioned) is instead the trend of the velocity in the lowest section, where probably the flow is heavily influenced by the presence of three small obstacles erected between the four standard dimensions obstacles. This is especially visible in the highest flow rates where the amplitude of the oscillations increases. Moreover, the velocity difference between the inlet and the outlet of the channel grows for higher flow rates.

As it has been showed in Figure 6, the existence of flow back in the section at the entrance of the channel can also be observed. This phenomena gradually disappears in the above sections and it is less pronounced in the low rates. The reason is probably the influence of the proximity of the inlet of the fluid and the presence of the three small obstacles placed in the bottom section. Very suspicious is a remarkable reversal of the velocity (present in all flow rates) at the highest section, near the right side wall of the channel. Additional measurements have to be performed to verify this phenomenon.

5. Conclusions

In the presented paper an experimental investigation of the flow in the model of SOFC cell gas channel have been performed. A velocity vector fields for two selected flow rates and the velocity plots in various locations of the channel have been shown. The results showed that the flow in such geometry was better controlled for lower flow rates. It was also found that the velocity distribution itself was not fully homogenous. Especially, significantly larger velocities were observed in the regions next to the channel left wall. Additionally an undesirable disturbances of the flow field were caused by the small obstacles located just in the inlet to the channel. However to get a better understanding of the obtained results, additional analysis should be performed in the future works.

Acknowledgements
The present work was supported by the Polish Ministry of Science (Grant AGH No. 11.11.210.198)

References

[1] Martin J., Oshkai P., Djilali N., 2005, "Flow Structures in a U-Shaped Fuel cell Flow Channel: Quantitative Visualization Using Particle Image Velocimetry", Journal of Fuel Cell Science and Technology, 2, 70-80

[2] Dean, W.R., 1928, "Fluid Motion in a Curved Channel", Proc. R. Soc. London, Ser. A, 121, 402-420

[3] Hawthorne, W.R., 1951, "Secondary Circulation in Fluid Flow", Proc. R. Soc. London, Ser. A, 206, 374-387

[4] Nishino T., Szmyd J., 2010, "Numerical analysis of a cell-based indirect internal reforming tubular SOFC operating with biogas", Journal of Fuel Cell Science and Technology, 7 , 051004-1–051004-8

[5] Nowak R., Szmyd J, 2010, "An experimental analysis of transport phenomena in a models of circular-planar SOFC", 19th Polish National Fluid Dynamics Conference, 5-9. September, Poznań, Poland

[6] Ho T.X. Kosinski P., Hoffmann A.C., Vik A., 2008, "Numerical study of an SOFC with direct internal reforming using charge diffusion-based model", Proceeding of 8th European SOFC Forum Lucerne, 1-13

[7] Laucerin J., Morel B., Bulter Y., Lefebvre-Joud F., 2006, "Thermomechanical model of solid oxide fuel cell fed with methane", Fuel Cells, 6, 64-70

[8] Nowak R., Szmyd J.S., 2011, "An experimental analysis of the flow structure in varius configurations of a circular-planar SOFC fuel channel", Journal of theoretical and applied mechanics, 49, 541-564

Journal of Physics: Conference Series **530** (2014) 012031 doi:10.1088/1742-6596/530/1/012031

3D Computations of Flow Field in a Guide Vane Blading Designed by Means of 2D Model for a Low Head Hydraulic Turbine

Z Krzemianowski[1,3], R Puzyrewski[2]
[1] Department of Hydraulic Machinery, the Szewalski Institute of Fluid-Flow Machinery, Polish Academy of Sciences, Gdansk, 80-231, Poland
[2] Faculty of Mechanical Engineering, the Gdansk University of Technology, Gdansk, 80-233, Poland

E-mail: krzemian@imp.gda.pl, rpuzyrew@pg.gda.pl

Abstract. The paper presents the main parameters of the flow field behind the guide vane cascade designed by means of 2D inverse problem and following check by means of 3D commercial program ANSYS/Fluent applied for a direct problem. This approach of using different models reflects the contemporary design procedure for non-standardized turbomachinery stage. Depending on the model, the set of conservation equation to be solved differs, although the physical background remains the same. The example of computations for guide vane cascade for a low head hydraulic turbine is presented.

1. Introduction

The chain of models applicable for a turbomachinery blading consists with 1D, 2D and 3D models as it is shown in figure 1. The complexity rises up with the number of dimensions taken into account.

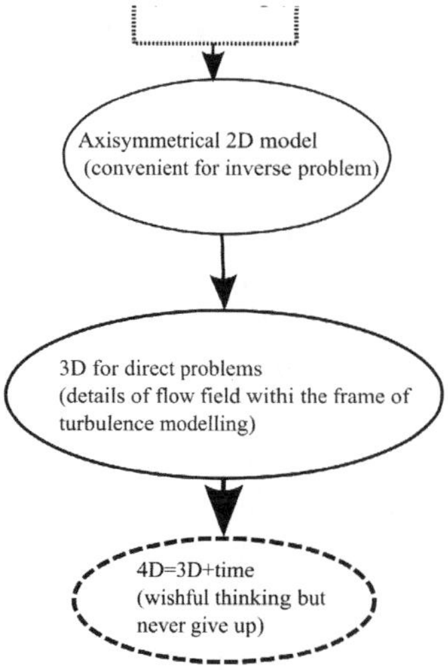

Figure 1. The chain of models for a turbomachinery designing process.

[3] Author to whom any correspondence should be addressed.

The simplest 1D model serves generally for preliminary prediction of flow field behaviour for a given blading. It predicts the main kinematics parameters assuring the maximum efficiency. This was illustrated for low head hydraulic turbine stage in [1].
A 2D model is particularly convenient for the so-called inverse problem when the shapes of blade can be determined. The 2D model was originated by the publication of Lorenz [2] and later developed by Wu [3]. For the inverse problem 2D model was adopted by Puzyrewski, Krzemianowski [4], [5], [6], [7] and others e.g. [8].
A 3D model is recently recognized as the most powerful and enables to determine the details of the flow field including such parameters as the loss coefficients and outlet angles. These parameters are crucial for the evaluation of the blading quality expressed by the efficiency.

2. The comparison of the governing equations for the 2D and 3D models

As it was mentioned above 2D model is convenient for the inverse problem, which enables predicting the shape of blading according to the assumed boundary conditions. The boundary conditions should be determined due to the formal character of equations to be solved. Therefore, it seems reasonable to confront the forms of basic governing equations. The main assumptions for the 2D and 3D models are listed in table 1.

Table 1. Comparison of the parameters for the 2D and 3D models.

Parameters	2D model	3D model
Geometry	• No shapes of blades are known • Given the expected shapes of axisymmetric stream surfaces • Given inlet and outlet planes between inner and outer borders of meridional cross-section • Given expected thickness of blades i.e. blockage factor	• Geometry of flow domain is known. Shapes of blades are given between inner and outer borders of a meridional cross-section • Given expected blades
Coordinates	• Curvilinear axisymmetric coordinates ($x^{(1)}, x^{(2)}, x^{(3)}$) • Assumed stationary flow	• Cartesian coordinates (x, y, z) • Assumed stationary flow
Inlet boundary conditions	• Distribution of flow parameters in front of a blading is assumed	• Distribution of velocity components ahead of a blading area is assumed
Outlet boundary conditions	• Outlet flow parameters results from computation	• Pressure at the outlet of computational domain is assumed
Dissipation effects	• Distribution of dissipation effects is introduced by means of an expected loss coefficient	• Dissipation effects are computed within the frame of an applied turbulence model

It is noteworthy that for 2D inverse problem wishful shapes of stream surfaces distributed in meridional cross-section as shown in figure 2 are assumed. In figure 2 three systems of coordinates are marked, namely: Cartesian (x, y, z), Cylindrical (r, φ, z) and Curvilinear ($x^{(1)}, x^{(2)}, x^{(3)}$). The Curvilinear system is based on the assumed stream surfaces.
The shapes of stream surfaces are chosen in the form of an assumed function as follows:

$$r = f\left(x^{(1)}, x^{(3)}\right) \tag{1}$$

It is easy to note that function $f(x^{(1)}, x^{(3)})$ does not contain the circumferential coordinate $x^{(2)}$.
In a 2D model the intensity of dissipation process has to be introduced on realistic level in advance. For a 3D direct model the geometry of the blading channels is given as created by means of 2D model. It is free to choose the turbulent model which closes the set of governing equation for 3D computation. As a consequence the intensity of dissipation in 3D model is computed.

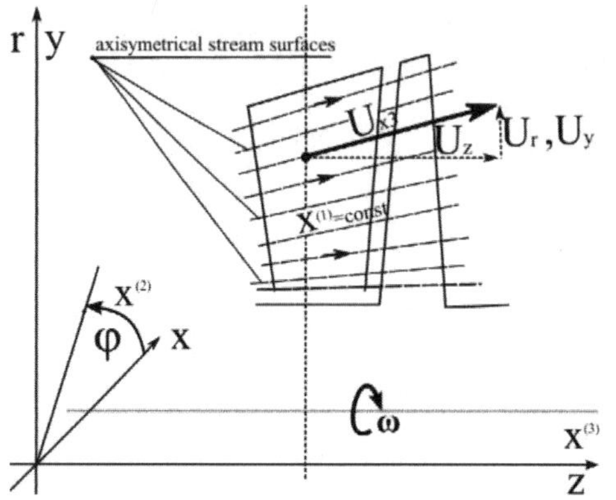

Figure 2. Meridional scheme and Cartesian, Cylindrical and Curvilinear coordinate systems of a turbine channel.

- **Mass conservation equation**
  - ✓ **3D model**:

$$div\vec{U} = 0 \tag{2}$$

  - ✓ **2D model**:

$$div\vec{U} = 0 \tag{3}$$

turns into algebraic form [5]

$$\left(1 - \tau\left(x^{(1)}, x^{(3)}\right)\right) \rho U_{x^{(3)}} \frac{f \frac{\partial f}{\partial x^{(1)}}}{\sqrt{1+\left(\frac{\partial f}{\partial x^{(3)}}\right)^2}} = m\left(x^{(1)}\right) \tag{4}$$

where ρ is density, $m\left(x^{(1)}\right)$ is mass flow rate distributed at each of a streamline.

- **Momentum conservation equation**
  - ✓ **3D model** [9]:

$$\rho\vec{U} grad\vec{U} = -grad\ p + div\ \bar{\bar{\tau}} \tag{5}$$

where tensor $\bar{\bar{\tau}}$ is given as follows:

$$\bar{\bar{\tau}} = \mu\left(grad\vec{U} + grad\vec{U}^T\right) - \frac{2}{3}\mu\ div\vec{U}\ \bar{\bar{I}} \tag{6}$$

where $\bar{\bar{I}}$ is unit tensor, μ is dynamic viscosity.

  - ✓ **2D model**:

where body force $\vec{F}$ replaces the differential part $div\ \bar{\bar{\tau}}$:

$$\rho\vec{U}\ grad\vec{U} = -grad\ p + \vec{F} \tag{7}$$

and it may be rearranged into the following forms in all three curvilinear ($x^{(1)}, x^{(2)}, x^{(3)}$) directions:

$$-\frac{\rho U_{x^{(2)}}^2}{f} + \frac{\rho U_{x^{(3)}}^2 \frac{\partial^2 f}{\partial x^{(3)2}}}{1+\left(\frac{\partial f}{\partial x^{(3)}}\right)^2} = -\frac{\partial p}{\partial x^{(1)}} \frac{1+\left(\frac{\partial f}{\partial x^{(3)}}\right)^2}{\frac{\partial f}{\partial x^{(1)}}} + \frac{\partial p}{\partial x^{(3)}} \frac{\partial f}{\partial x^{(3)}} + \rho F_{x^{(1)}} \tag{8}$$

$$\frac{\rho U_{x^{(3)}}}{f\sqrt{1+\left(\frac{\partial f}{\partial x^{(3)}}\right)^2}} \frac{\partial}{\partial x^{(3)}}\left(f\ U_{x^{(3)}}\right) = \rho F_{x^{(2)}} \tag{9}$$

$$\frac{\rho U_{x^{(3)}}}{\sqrt{1+\left(\frac{\partial f}{\partial x^{(3)}}\right)^2}} \left(\frac{\partial U_{x^{(3)}}}{\partial x^{(3)}} - \frac{U_{x^{(3)}}}{1+\left(\frac{\partial f}{\partial x^{(3)}}\right)^2} \frac{\partial f}{\partial x^{(3)}} \frac{\partial^2 f}{\partial x^{(3)2}}\right) = \frac{\partial p}{\partial x^{(1)}} \frac{\frac{\partial f}{\partial x^{(3)}}}{\frac{\partial f}{\partial x^{(1)}}} \sqrt{1+\left(\frac{\partial f}{\partial x^{(3)}}\right)^2} - \frac{\partial p}{\partial x^{(3)}} \sqrt{1+\left(\frac{\partial f}{\partial x^{(3)}}\right)^2} + \rho F_{x^{(3)}} \tag{10}$$

For 2D model components of body force $\vec{F}\left(F_{x^{(1)}}, F_{x^{(2)}}, F_{x^{(3)}}\right)$ become the unknowns beside pressure p and velocity components $\vec{U}\left(U_{x^{(1)}} \equiv 0, U_{x^{(2)}}, U_{x^{(3)}}\right)$ ($U_{x^{(2)}}$ is tangential and $U_{x^{(3)}}$ is meridional velocity components). As it is shown in [4] the set of the above presented equations may be reduced to hyperbolic type for pressure p. Potential force is neglected because of horizontal flow.

- **Energy conservation equation**
  - ✓ **3D model** [9]:

$$\rho div(\vec{U}h) = \bar{\bar{\tau}} : grad\, \vec{U} \tag{11}$$

where $h = \frac{p}{\rho} + e$ (e is an internal energy).

  - ✓ **2D model**:

$$\frac{U^2_{x^{(2)}} + U^2_{x^{(3)}}}{2} + e + \frac{p}{\rho} + \Pi = const \tag{12}$$

where Π is gravity potential $\left(\Pi = gx^{(3)}\right)$, g is gravitational acceleration.

- **Closing relations**
  - ✓ **3D model** [9]:

$$\frac{\partial}{\partial x_i}(\rho k u_i) = \frac{\partial}{\partial x_j}\left(\Gamma_k \frac{\partial k}{\partial x_j}\right) + G_k - Y_k + S_k \tag{13}$$

$$\frac{\partial}{\partial x_i}(\rho \omega u_i) = \frac{\partial}{\partial x_j}\left(\Gamma_\omega \frac{\partial \omega}{\partial x_j}\right) + G_\omega - Y_\omega + S_\omega + D_\omega \tag{14}$$

Left hand sides present the convection terms. First terms of right hand sides in the above equations represent diffusion effect. The rest terms are sources ones.

  - ✓ **2D model**:

The internal energy rises along the trajectory with time $t_2 > t_1$

$$\Delta e = e_{t_2} - e_{t_1} = \varsigma\left(x^{(1)}, x^{(3)}\right) \frac{U^2_{x^{(2)}} + U^2_{x^{(3)}}}{2} > 0 \tag{15}$$

in a proportion to loss coefficient $\varsigma\left(x^{(1)}, x^{(3)}\right)$.

Body force component can be expressed as follows:

$$F_{x^{(1)}} = \delta\left(x^{(1)}, x^{(3)}\right)\left(Q_0\left(p, \rho, U_{x^{(2)}}, U_{x^{(3)}}, S, \dots\right) + Q_1\left(p, \rho, U_{x^{(2)}}, U_{x^{(3)}}, S, \dots\right)\frac{\partial p}{\partial x^{(3)}} + Q_2\left(p, \rho, U_{x^{(2)}}, U_{x^{(3)}}, S, \dots\right)\frac{\partial U_{x^{(2)}}}{\partial x^{(3)}}\right) \tag{16}$$

where the shape factor $S = \frac{\partial x^{(2)}}{\partial x^{(1)}}$ can be evaluated for skeleton surface (see details in [5]).

The parameter $\delta\left(x^{(1)}, x^{(3)}\right)$ is the third one beside $\varsigma\left(x^{(1)}, x^{(3)}\right)$ and $\tau\left(x^{(1)}, x^{(3)}\right)$ required to close conservation equations. The following shapes of functions were used in 2D computation:

$$\tau\left(x^{(1)}, x^{(3)}\right) = t_1\left(\overline{x^{(3)}}\right)^{t_2}\left(1 - \overline{x^{(3)}}\right)^{t_3}\left(1 - t_4 x^{(1)}\right) \tag{17}$$

$$\varsigma\left(x^{(1)}, x^{(3)}\right) = \left(w_0 + w_1 x^{(1)}\right)^{n_1}\left(w_2 + w_3\overline{x^{(3)}}\right)^{n_2} \tag{18}$$

$$\delta\left(x^{(1)}, x^{(3)}\right) = 0 \tag{19}$$

where $\overline{x^{(3)}}$ is normalized axial coordinate and coefficients t_1, t_2, ... to be determined according to the intuition and experience of the designer. It means the necessity of some experience possessing. The value $\delta\left(x^{(1)}, x^{(3)}\right) = 0$ means that the body force vector is tangential to stream surface $x^{(1)} = const.$ This assumption can be omitted using method of successive approximation as presented in [5].

3. The example of 3D computation

A computation performed by the 2D model on the basis of assumed meridional view allows getting the skeleton surface of blading as shown in figure 3. The details of such computation have been presented in [6].

Journal of Physics: Conference Series **530** (2014) 012031 doi:10.1088/1742-6596/530/1/012031

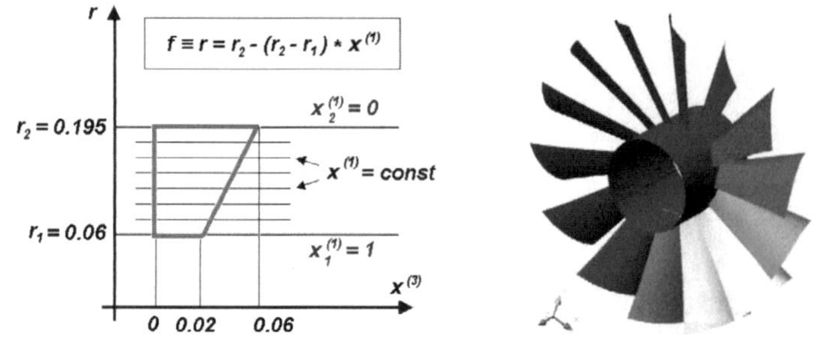

Figure 3. Meridional (left) and isometric (right) views of the guide vanes (skeleton) used to computations

The open problem is the distribution of the thickness blade along the skeleton surface. Figure 4 presents the Göttingen 428 profile with skeleton line in the middle. Along the skeleton line the thickness can be distributed in a way defined by the distribution parameter (thickness coefficient). Along the height of blades the Göttingen 428 profile can be scaled in proportion to the length of chord.

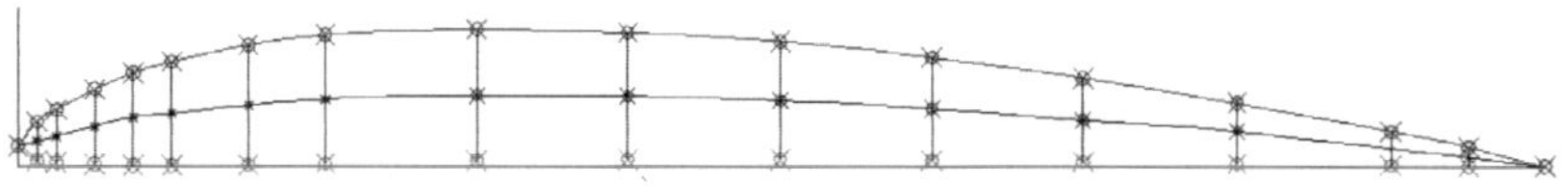

Figure 4. View of the thickness distribution for Göttingen 428.

Five cases of thickness distribution along an assumed constant skeleton were adopted (0.00, 0.25, 0.50, 0.75 and 1.00) to 3D computations as shown in figure 5. The profile with thickness coefficient 0.50 means that thickness was distributed symmetrically along skeleton (mean line). The pressure side of profile with thickness coefficient 0.00 covers the suction side with thickness coefficient 1.00.

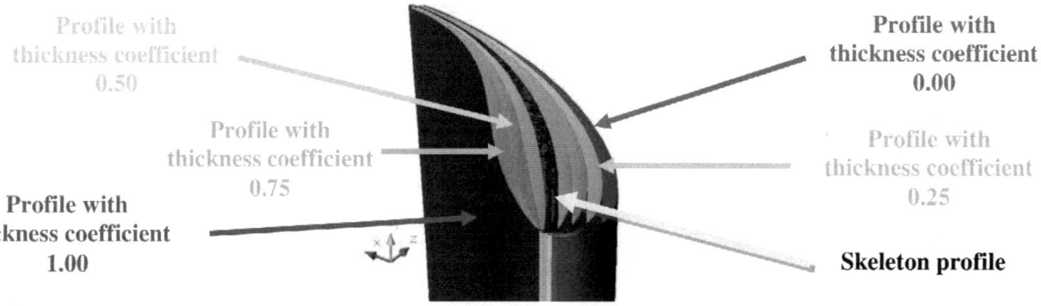

Figure 5. View of the 5 profiles used to 3D computations with Göttingen 428 thickness distribution.

It should be noted that 2D model is sensitive only to the thickness of blading (blockage factor) but irrelevant to the distribution parameter since no influence of circumferential direction is taken into account in this model.

Therefore the aim of 3D model computation was to establish the influence of distribution parameters upon the flow field behind the designed guide vane blade. Two factors are the matter of discussion: (1) the influence of finite number of blades and (2) the influence of thickness coefficient along the skeleton surface.

For a 3D computation the computational grid was created (example of surface mesh is shown in figure 6). The domain of calculation is much higher than domain of blading. It spreads ahead in about ~2*d* distance (*d* is the diameter of blade).

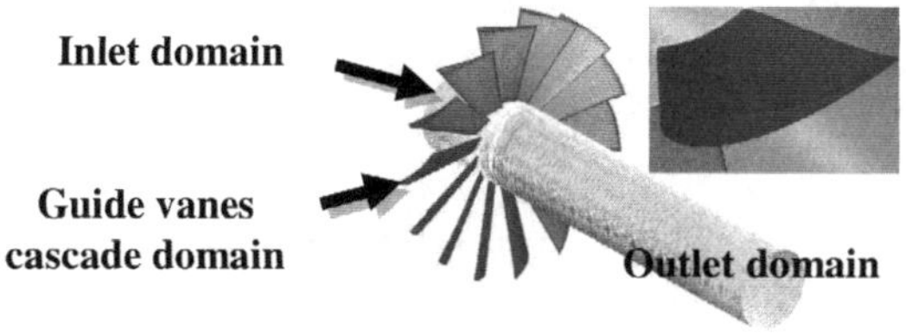

Figure 6. View of the guide vanes surface computational mesh used to 3D computations

Boundary conditions applied for 3D model computation were taken from 2D model where the skeleton surface was determined. They were assumed as follows: at the inlet to computation domain uniform

Journal of Physics: Conference Series **530** (2014) 012031 doi:10.1088/1742-6596/530/1/012031

distribution of axial velocity, mass flow rate: 235 kg/s, water density: 999.1 kg/m^3, dynamic viscosity: 0.00116 Pa s.

The examples of computation for are presented in the following figures. The only most characteristic figures are chosen for this presentation. The axial velocity at the front of cascade for different distribution parameter is shown in figure 7. Not too high difference one can distinguish between curves in figure 7 although the non-uniformity is noticeable (this will be commented below).

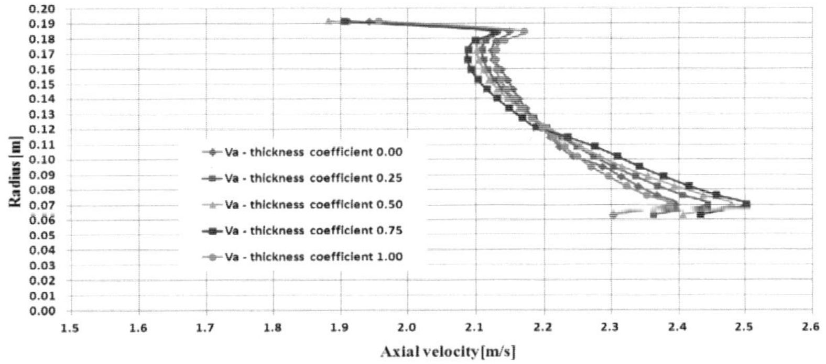

Figure 7. Radius vs. axial velocity at the inlet of guide vanes cascade for 5 different thickness distributions used for 3D computations.

In figure 8 axial velocities at the outlet are presented. The higher blockage at the inner diameter causes the higher axial component of velocity at the root of blading in front and backside of guide vane blading. In spite of the fact that ahead of the guide vane distribution of flow components has been inserted as constant, redistribution occurred as a result of upstream influence of blading. The conclusion can be drawn for 2D model – the axial velocity ought to have characteristic inclination visible in figure 7.

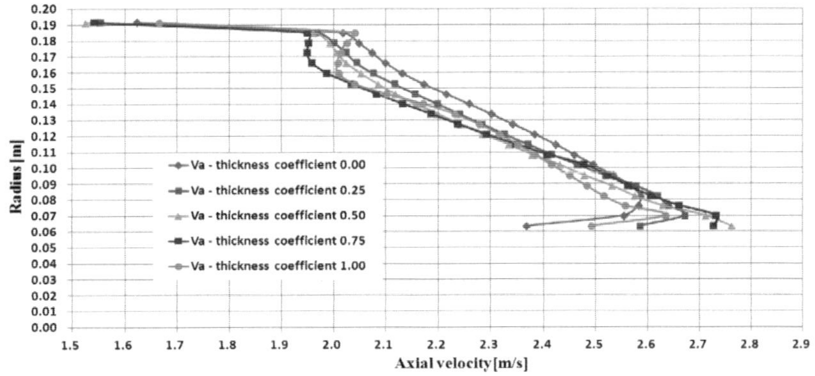

Figure 8. Radius vs. axial velocity at the outlet of guide vanes cascade for 5 different thickness distributions used for 3D computations.

The same conclusion can be drawn from figure 9 for the pressure distribution ahead of blading. For the same blockage factor pressure configuration along the height of blading depends on thickness distribution coefficient. The mean value of pressure ahead of cascade is shown in figure 10. Pressure distribution behind guide vanes cascade is shown in figure 11.

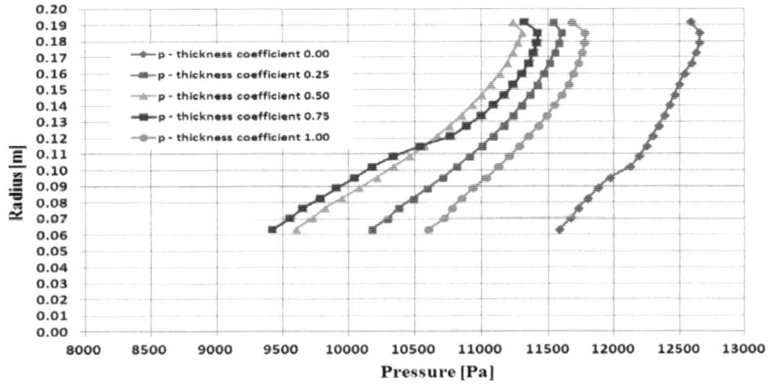

Figure 9. Radius vs. pressure at the inlet of guide vanes cascade for 5 different thickness distributions used for 3D computations.

One can notice deeper expansion at the root of blading. It is also worth noting that the influence of thickness distribution coefficient is not very high upon the pressure distribution behind the cascade. The reason for this fact comes from applied boundary condition for the pressure introduced at the end of flow domain (behind the cascade).

The deeper static pressure at the root is the reason for redistribution of axial velocity between inlet to computation domain and inlet to the guide vane cascade. As it is shown in figure 7 the axial velocity at the root is higher than at the outer diameter. But at the inlet to computation domain uniform velocity was introduced. In front of the computational domain redistribution of mass flow rate occurred by inducing radial component of velocity. Figure 12 presents the radial component of velocity at the inlet to cascade generated by a deeper expansion at the root of blading.

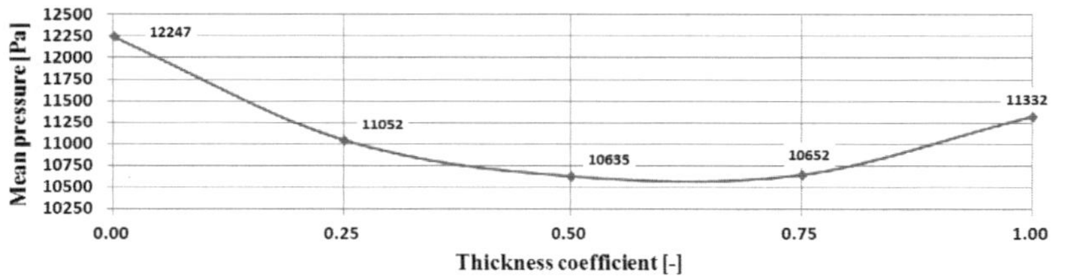

Figure 10. Mean pressure vs. thickness coefficient at the inlet of guide vanes cascade for 5 different thickness distributions used for 3D computations.

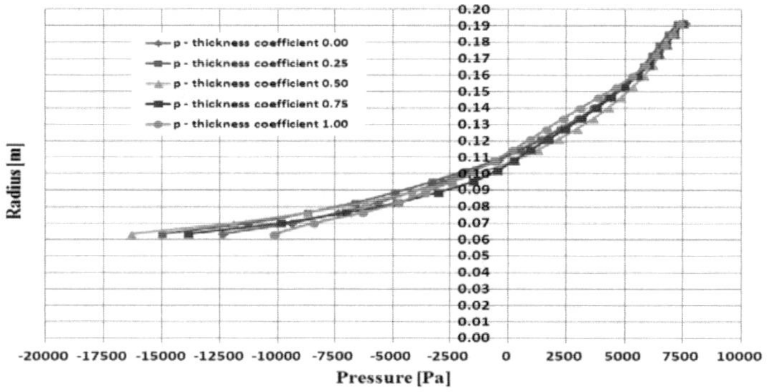

Figure 11. Radius vs. pressure at the outlet of guide vanes cascade for 5 different thickness distributions used for 3D computations.

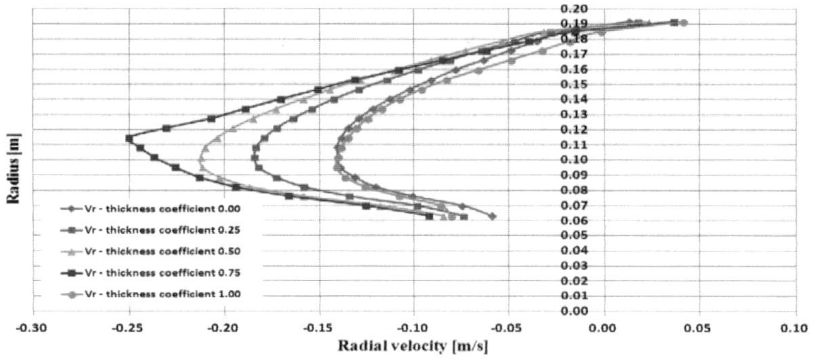

Figure 12. Radius vs. radial velocity at the inlet of guide vanes cascade for 5 different thickness distributions used for 3D computations.

4. Reference to 2D computational results

The main difference between 2D and 3D models consist in the fact of formal difference of the type of differential equations. The governing equations of 2D model are of hyperbolic type where the information spreads along the characteristics [10]. 3D model represented by governing equations has the ability of spreadig the information in the whole domain of computation. This is the essential reason, which influences the results of computation. Due to this fact 2D model is not able to predict the change of parameters ahead of blading. The boundary conditions for both models ought to be formulated according to the character of governing equations.

The aim of guide vane blading is to generate angular momentum at the entrance of a runner at the lowest losses. This means the generation of suitable outlet angle at the exit of blading. The value of this angle results already in 2D model of computing. It is very important to know the correction of exit angle between 2D and 3D models. On the base of presented results the correction is shown in figure 13.

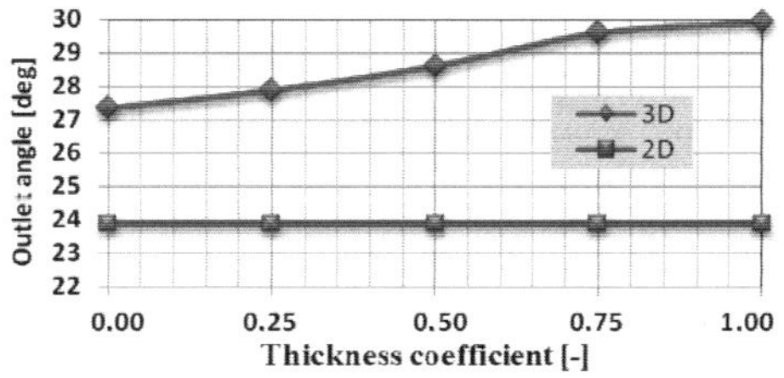

Figure 13. Outlet angle vs. thickness distribution at the inlet of guide vanes cascade for 5 different thickness distributions used for 3D computations.

Having in mind such a value of difference shown in figure 13 the designer ought to aim when performing 2D inverse computation at lower value of outlet angle from blading. The finite number of blading in 3D computation causes the increase of outlet angle.

5. Conclusions

The design procedure of a non-standardized low head hydraulic turbine stage can be supported by the chain of typical flow models like 1D, 2D and 3D. Each of the models plays a specific role. Here the main accent was paid to the relation between 3D and 2D models:

1. 2D inverse problem serves for the design of cascade blading.
2. For the formulation of proper boundary condition for 2D inverse computation it is important to take into account the character of parameters distribution in front of cascade.
3. Non-uniformity of velocity and pressure distribution ahead of guide vanes cascade appears in a 3D direct problem computation.
4. The most important conclusion from presented comparison concerns the difference of the outlet angles between 2D and 3D models due to the finite number of blades in guide vanes cascade.

6. References

[1] Puzyrewski R and Krzemianowski Z 2013 *Impact of Blade Quality on Maximum Efficiency of Low Head Hydraulic Turbine* (J. of Mechanics and Mechanical Engineering vol 17 no 3, Lodz-Poland: Lodz University of Technology Publishing) pp 269-84

[2] Lorenz H 1905 *Theorie und Berechnung der Vollturbinen und Kreisel Pumpen* (Zeitschrift des Vereines Deutscher Ingenieure band 49 no 41, 14 Okt 1905) pp 1670-1675

[3] Wu C H 1952 *A General Theory of Three-Dimensional Flow in Subsonic and Supersonic Turbomachines of Axial Radial and Mixed-Flow Types* (Tran. of ASME Nov 1952) pp 1363-80

[4] Puzyrewski R 1998 *14 Lectures on Theory of Turbomachinery Stage – Two dimensional Model (2D) (in Polish)* (Gdansk-Poland: Gdansk University of Technology Publishing) 1998

[5] Puzyrewski R 2005 *Two Dimensional Inverse Method of Turbomachinery Stage Design Developments in Mechanical Engineering* (Gdansk-Poland: Gdansk University of Technology Publishing) pp 57-72 2005

[6] Puzyrewski R and Krzemianowski Z 2014 *2D Model of Guide Vane for Low Head Hydraulic Turbine: Analytical and Numerical Solution of Inverse Problem* (J. of Mechanics Engineering and Automation David Publishing vol 4 no 3) pp 195-202

[7] Krzemianowski Z 2003 *Blade Shape Design Method for the Reversible Hydraulic Turbines Ph.D Thesis (in Polish)* (Gdansk University of Technology) 2003

[8] Tesch K 2002 *Design of turbomachinery profile by means of model hydraulic turbine stage Ph.D Thesis (in Polish)* (Gdansk University of Technology) 2002

[9] ANSYS/Fluent 12.1 *Manual*

[10] Korn G A and Korn T M 1983 *Mathematical Handbook* (Warsaw: McGraw-Hill Book Company Copyright for the Polish edition by PWN Warsaw) 1983

Acknowledgements

The work was supported by National Science Committee under grant no. 6694/B/T02/2011/40 for the Szewalski Institute of Fluid-Flow Machinery, Polish Academy of Sciences in Gdansk (Poland).

Journal of Physics: Conference Series **530** (2014) 012032 doi:10.1088/1742-6596/530/1/012032

Modelling of flow with cavitation in centrifugal pump

D Homa[1], **W Wróblewski**
Institute of Power Engineering and Turbomachinery, Silesian University of Technology, Gliwice Konarskiego Street 18, Poland

E-mail: dorota.homa@polsl.pl

Abstract. The paper concerns flow modelling in centrifugal pump with special consideration of cavitation phenomena. Cavitation occurs when local pressure drops below the saturation pressure according to the temperature of the flow. Vapour bubbles are created and then they flow through the areas with higher pressure. The bubbles collapse rapidly generating pressure wave, noise and vibration. Working under cavitation condition is very dangerous to a pump and can significantly shorten its lifetime. The investigated centrifugal pump consists of three two-flow rotors and stators working on a single shaft. The modelling process started with grid independence study. When the grid was chosen, the pump performance curve was obtained using the single phase fluid model. Next, using the results from pump performance curve calculations, the cavitation characteristic was obtained. The constant capacity was held when the pressure at the inlet was reduced. The two - phase model was used with Zwart cavitation model. The results indicate that the pump work in safe range of parameters. The analysis also provides wide range of information about the areas of vapour appearance. The most endangered regions are leading edges of rotor. When pressure at the inlet drops to about one third of pressure that calculations started from the cavitation cloud appears in whole rotor. The intense of vapour bubbles creation is greater near the shroud of the pump, rather than near the hub. As cavitation is strongly unsteady phenomena, the transient calculations were performed to check if the results are close to those obtained using the steady state type. The differences are not significant.

1. Introduction

Computational fluid dynamics provides range of possibilities to predict the flow parameters and to observe the phenomena that can occur under different work conditions. Cavitation can occur in pump flow when the conditions on suction side change (increasing hydraulic losses) or when a pump works with capacity much more higher than the capacity in Best Efficiency Point (BEP). Working under cavitation condition leads to serious damage of blades and walls. In paper the simulation of flow with cavitation in centrifugal pump is described.

The investigated pump consists of three two – flow rotors and stators. They are installed on a single shaft. The pump works on steam turbine condenser cooling water. In 2013 the modernization of pump flow system took place. Three new rotors with enlarged outflow diameter were made and installed so that the cooling water flow through the condenser increased. After the modernization the characteristics of pump changed. The numerical analysis of flow was performed to obtain pump

[1] To whom any correspondence should be addressed.

performance and efficiency curve and H(NPSH) curve. To include cavitation in the analysis the cavitation model must have been defined.

2. The cavitation model

To simulate flow with cavitation it is necessary to choose appropriate mathematical model. The multiphase model is assumed to be homogenous fluid. In the analysis the cavitation model is based on Rayleigh – Plesset equation describing the growth of a vapour bubble in flow as follows:

$$R_B \frac{\mathrm{d}^2 R_B}{\mathrm{d}t^2} + \frac{3}{2}\left(\frac{\mathrm{d}R_B}{\mathrm{d}t}\right)^2 + \frac{2\sigma}{\rho_l R_B} = \frac{p_v - p}{\rho_l} \quad (1)$$

where R_B stands for the bubble radius, p_v is pressure in the bubble (assumed to be equal to the vaporization pressure), p is the pressure in liquid, σ represents surface tension between the liquid and vapor and ρ_l is the liquid density. The surface tension as well as second order terms is neglected in the used model. Therefore the equation (1) is reduced to:

$$\frac{\mathrm{d}R_B}{\mathrm{d}t} = \sqrt{\frac{2}{3}\frac{p_v - p}{\rho_l}} \quad (2)$$

Rate of change of bubble mass is (assuming vaporization):

$$\frac{\mathrm{d}m_B}{\mathrm{d}t} = 4\pi R_B^2 \rho_g \sqrt{\frac{2}{3}\frac{p_v - p}{\rho_l}} \quad (3)$$

The vapour volume fraction r_g is defined as:

$$r_g = V_B N_B = \frac{4}{3}\pi R_B^3 N_B \quad (4)$$

where N_B is bubbles per unit volume. According to Zwart model [1] the total interphase mass transfer rate per unit volume is expressed as:

$$\dot{m}_{\mathrm{lg}} = N_B \frac{\mathrm{d}m_B}{\mathrm{d}t} \quad (5)$$

In case of condensation:

$$\dot{m}_{\mathrm{lg}} = F \frac{3 r_g \rho_g}{R_B} \sqrt{\frac{2}{3}\frac{|p_v - p|}{\rho_l}} \operatorname{sgn}(p_v - p) \quad (6)$$

In case of vaporization:

$$\dot{m}_{\mathrm{lg}} = F \frac{3 r_{nuc}(1 - r_g)\rho_g}{R_{nuc}} \sqrt{\frac{2}{3}\frac{|p_v - p|}{\rho_l}} \operatorname{sgn}(p_v - p) \quad (7)$$

where F is an empirical factor which is different for condensation and vaporization because they occur at different rates (condensation is usually slower than vaporization). R_B in equation (6) is replaced by nucleation site radius R_{nuc} and r_{nuc} stands for the volume fraction of the nucleation sites. It was assumed that F_{vap}=50; F_{cond}=0.01; R_{nuc}=1 µm; r_{nuc}=0.0005 [2].

3. The grid independence study

As the rotors of investigated pump work in parallel only flow through one of them was taken to the analysis. Moreover, the rotor is two – flow, so the symmetry in a plane perpendicular to the axis of rotation could be assumed. The flow was divided into 7 parts, as the rotor consists of 7 blades. Analogically, the flow in stator was split into 10 parts. Next the grid independence study was performed. The fluid was assumed to be single phase (liquid). However this assumption leaded to unphysical results near the leading edge of the rotor blade. The dropped below 0 Pa absolute pressure.

It was decided to use the two – phase homogeneous fluid model with cavitation described in chapter 2. It provided the possibility of vapour bubbles creation in the areas with pressure below the vaporization pressure. The data taken to the calculations are described in table 1.

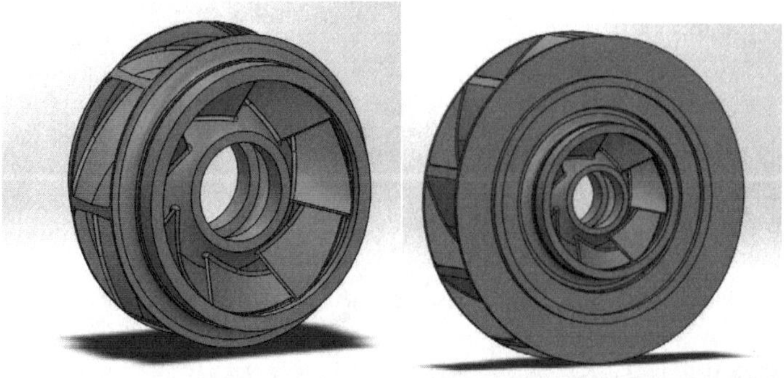

Figure 1. The two–flow rotor (left) and rotor and stator (right).

Table 1. Calculation data to perform the grid independence study.

Angular velocity	985 rev/min
Reference pressure	0 Pa
Heat transfer model	isothermal
Fluid temperature	25°C
Fluid density	997 kg m^{-3}
Vaporization pressure	3574 Pa
Turbulence model	SST
Boundary condition inlet	Static frame total pressure
Turbulence	Medium (Intensity = 5%)
Boundary condition outlet	Mass flow
Mass flow rate	23.08 kg/s

Five grids were studied. For each of them the work parameters were monitored. The pressure distribution on rotor was also compared. The pressure distribution at span 0.5 obtained for each grid is shown in figure 2. The grid with symbol S4 was chosen.

Table 2. Number of nodes for grids taken to the grid independence study.

Grid symbol	Number of nodes
S1	693204
S2	967267
S3	1266659
S4	2059135
S5	3568498

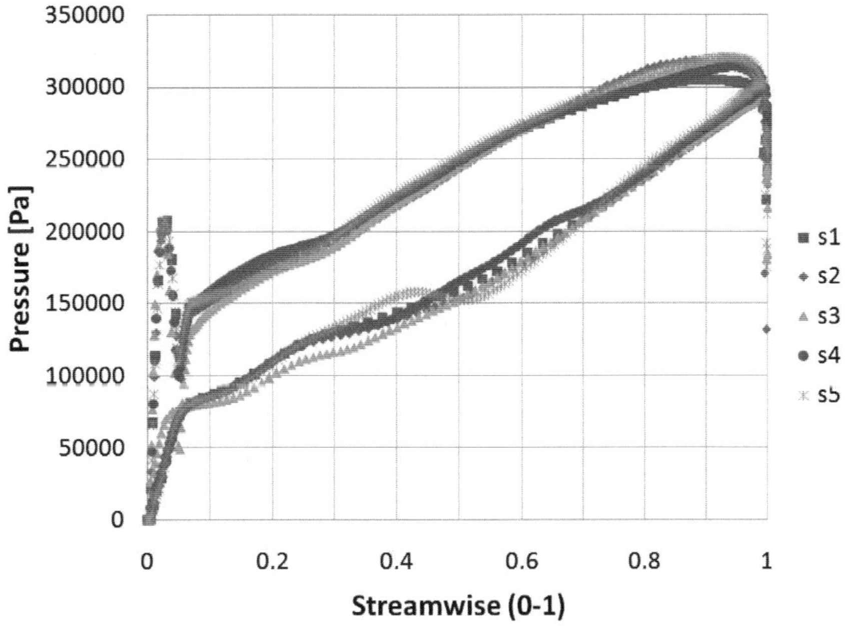

Figure 2. The pressure distribution on the rotor blade (span = 0.5).

4. Pump performance curve

To predict new operating point after the modernization of pump flow system it was vital to obtain pump performance curve H(Q). The series of calculations were made. The flow rate was changed from 5000 to 9000 m^3/h with step 500 m^3/h. The intersection of pump performance curve and system curve takes place at the flow rate about 7600 m^3/h. The efficiency curve is relative flat; in range of flow rate from 6500 to 9000 m^3/h it varies from 89.5 to 92.7 %. The highest efficiency occurs when flow rate is equal to 8000 m^3/h.

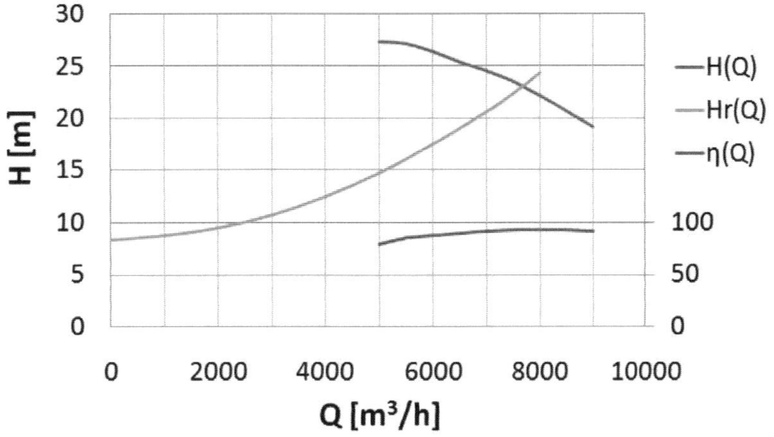

Figure 3. The pump performance curve.

5. Modelling cavitation in the flow

The suction conditions are crucial to avoid cavitation occurrence in a pump. Net positive suction head (NPSH) is a value in hydraulic circuit that indicates the difference between pressure at the inlet (p_s) and vaporization pressure (p_v) [3].

$$NPSH = \frac{p_s - p_v}{\gamma} + \frac{c_s^2}{2g} \tag{8}$$

Where γ stands for specific gravity and c_s is the velocity at the inlet. To preserve pump from working in cavitation condition the NPSH value must be greater than the Required NPSH (NPSHR), usually specified by the manufacturer. NPSHR is defined on the basis of measurements. One of the symptoms of cavitation is a change in pump characteristics. It means that if a pump works with constant flow rate and cavitation occurs, the head of pressure drops significantly. To determine the critical value of NPSH corresponding to the initial cavitation, the H(NPSH) curve is plotted.

5.1. Simulation to obtain H(NPSH) curve

The H(NPSH) curve was plotted at the flow rate equal to 8000 m^3/h. The value of NPSH was gradually decreased by lowering the inlet pressure. The Head of pressure was monitored. The data used to calculations are the same as specified in table 1. The H(NPSH) curve is shown in figure 4.

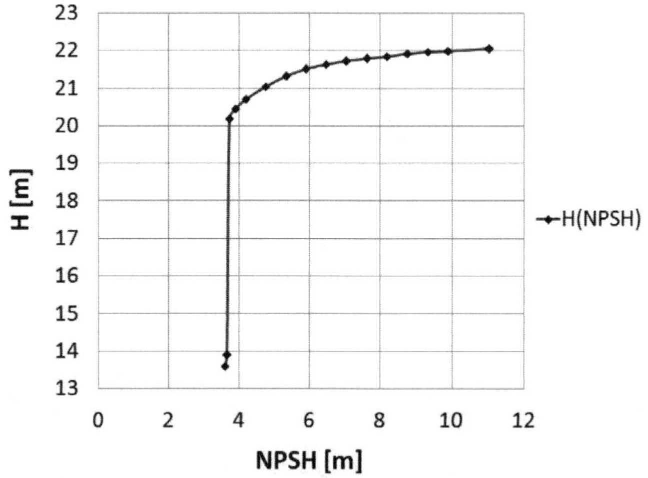

Figure 4. The plotted H(NPSH) curve.

The critical value of NPSH is defined as NPSH for which the Head drops to 97% of its initial value. The Required NPSH is greater than critical NPSH, usually about 10-30% [4]. Initial value of NPSH in the described case was 11,03 m and critical NPSH is about 5,35 m. It indicates that the pump works in a safe range of parameters. Lowering the inlet pressure causes Head to decline, but the changes are smooth. The rapid drop of Head occurs when inlet pressure is decreased to about one third of initial pressure.

5.2. Areas of vapor appearance in rotor

The lowest pressure in a rotor occurs near the leading edge of the blade. In this area the vapour bubbles may appear even during the work with NPSH significantly higher than NPSHR. In figure 5. the area of vapour bubbles appearance is compared for four values of inlet pressure. The cloud of vapour bubbles firstly occurs near the leading edge of the rotor blade and then gradually enlarges as the inlet pressure is decreased. The cloud was located only in the rotor. In the stator there was no vapour in the whole range of inlet pressure. When inlet pressure reaches the value of one third of the initial inlet pressure the cavitation cloud covers whole area beyond rotor blades. However the vapour distribution through the blade span is not uniform. In figure 6 the area of vapour occurrence at lowest inlet pressure is compared for different values of blade span. Most of the cavitation cloud gathers near the shroud of the rotor (span greater than 0.5). This distribution tallies with available data on cavitation damage in centrifugal pump [5, 6]. The most endangered place is the blade surface near the shroud. The results show that the used model correctly predict this phenomenon in rotor.

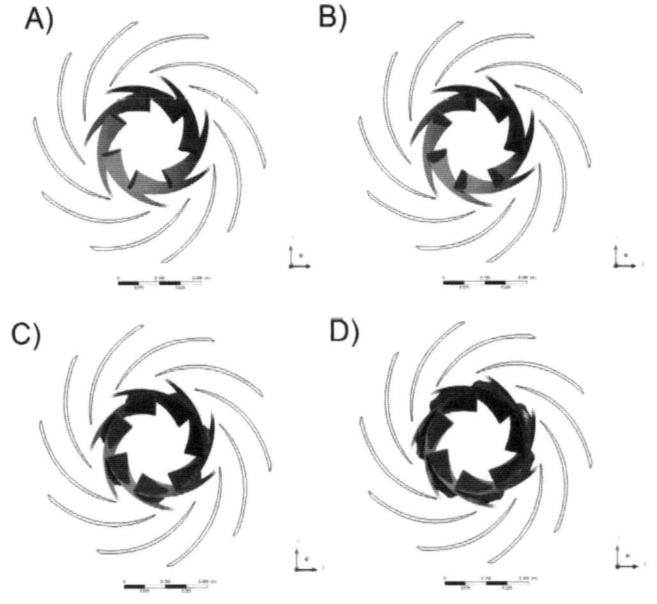

Figure 5. Areas of vapour bubbles appearance in the pump rotor. A) Inlet pressure 110 kPa B) Inlet pressure 55 kPa C) Inlet pressure 40 kPa D) Inlet pressure 39 kPa.

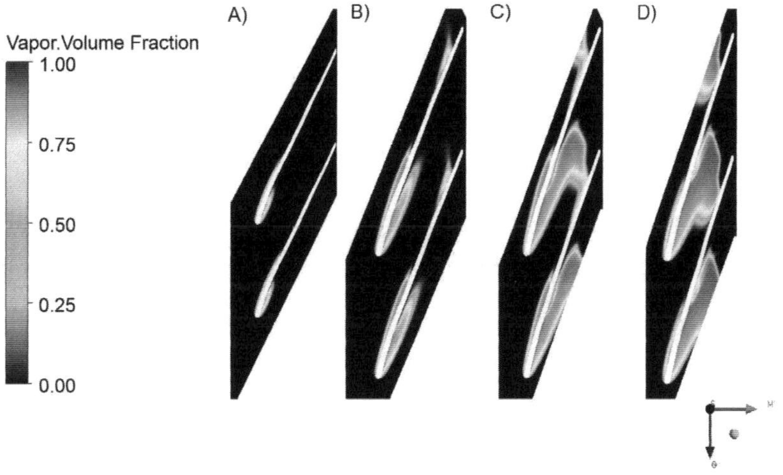

Figure 6. Areas of vapour bubbles appearance in the pump rotor. A) Span 0.2 B) Span 0.5 C) Span 0.7 D) Span 0.9.

5.3. Transient calculations

Cavitation is strongly unsteady phenomena. The process of vaporization and condensation of the bubbles proceed in very turbulent way. The results shown above were obtained with the assumptions of steady state type of calculations. It definitely shortened the time of calculations. The next step was to recalculate a few of the cases above and check if the results from transient calculations are similar to those obtained with the assumption of steady state. The chosen points to recalculations was the points characterized by the inlet pressure: 110 kPa (100% initial inlet pressure), 55 kPa (50% initial pressure), 40 kPa (36% initial pressure) and 39 kPa (35% initial pressure). The results are shown below in figure 7. The values of pressure in both rotor and stator are changing periodically and as a consequence the values of H are changing as well. The values of H shown below are calculated as the average in one period. The differences of H values are specified in table 3.

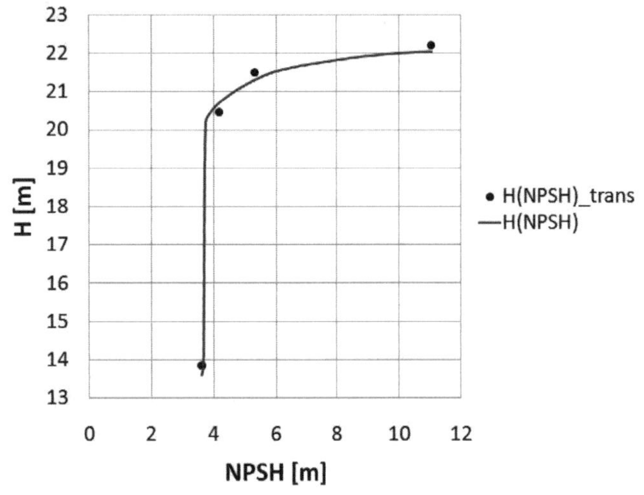

Figure 7. Transient results compared to steady state results.

Table 3. H values for different type of calculations.

NPSH, m	H steady state, m	H transient, m	Relative difference, %
11.03	22.05	22.20	0.67
5.33	21.32	21.50	0.85
4.19	20.73	20.45	1.33
3.62	13.60	13.86	1.90

After analysing differences between transient and steady state results it was affirmed that characteristic H(NPSH) on the basis of transient results would be very close to the curve obtained after steady state calculations. The differences are below 2%. In this case to predict cavitation occurrence in a pump it is sufficient to use steady state type of calculations.

To study pressure changes in time the monitor points were added in both rotor (point A) and stator (point B). The point in the rotor was located near the trailing edge. The point in the stator was far from the area of vapour appearance. In the figures 8 and 9 the values of pressure in both points are plotted in condition of flow without and with cavitation.

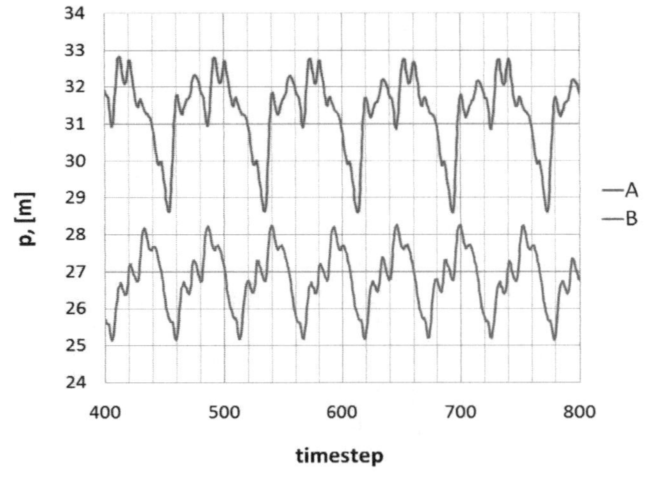

Figure 8. The changes of pressure values in point A and B in condition of non – cavitating flow.

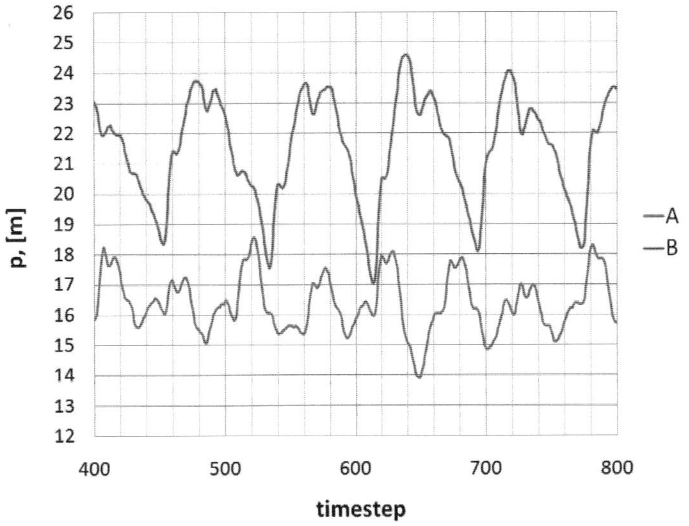

Figure 9. The changes of pressure values in point A and B in condition of cavitating flow.

Frequency of changes in both points was related to the rotational velocity and geometry of the stage. In case of point A this frequency was about 170 Hz, in case of point B – about 115 Hz. However, in case of point A during the flow with cavitation the additional frequency was observed. This frequency was equal to 96.5 Hz and most probably resulted from cavitation model that was used during the calculations.

6. Conclusions

The performed analysis of the flow in a centrifugal pump enabled to predict new working conditions after the modernization of pump flow system and define pump performance curve. For modelling characteristic of pump the single phase fluid model was sufficient, but it gave unphysical results of pressure in the areas where bubbles of vapour should appear. To study cavitation phenomena it was necessary to use two-phase fluid model with appropriate model of cavitation. In the analysis the H(NPSH) curve was also plotted. It indicated that there is no risk of cavitation during work with specified flow rate, because operating NPSH is almost two times greater than critical NPSH corresponding to the initial cavitation. In the case of developed cavitation the location of vapor cloud in the rotor is described. Most of the cloud gathered near the shroud of the rotor. Transient type of calculations was also performed. The results do not vary significantly from the results attained by steady state type of calculations.

7. Literature

[1] Zwart P J, Gerber A G and Belamri T, 2004 *Proceedings of Fifth ICMF* Yokohama, Japan
[2] Bakir F, Rey R, Gerber A and Hutchinson B 2004 *Int J Rotating Machinery* **10** 15-25
[3] Troskolanski A, 1968 *Pompy wirowe* WNT 405-423
[4] Korczak A and Rokita J, 1985 *Pompy i układy pompowe.* Gliwice 105
[5] Brennen C E ,1994 *Hydrodynamic of Pumps*
[6] Janczak M and Plutecki W, 2011 *Mechanik* **8-9** 738-743

Journal of Physics: Conference Series **530** (2014) 012033 doi:10.1088/1742-6596/530/1/012033

Investigation of the tip-leakage losses in turbine axial stages

A Szymański, S Dykas, W Wróblewski and S Rulik
Institute of Power Engineering and Turbomachinery, Silesian University of Technology, 44-100 Gliwice, ul. Konarskiego 18, Poland

E-mail: artur.szymanski@polsl.pl

Abstract. In turbomachinery, an influence of a tip-leakage flow on overall blade loss is crucial and its reduction is still worth striving for. In this paper a numerical analysis of the flow in tip seal of high-rotating gas turbine engine has been made. This analysis is a part of experimental research for testing the commercially used different tip seals solutions. Described test rig is predicted to be an universal tool for developing and examining different configurations of turbine blade tips. Presented numerical analysis is used to predict physical phenomena that may affect the rotor blade performance. In the numerical investigation the commercial Ansys CFX software was employed. The most important parameters were: mass flow rate at the inlet and outlet of the test bench, pressure and velocity distribution and the air temperature growth above the rotor. Also, an influence of test rig inlet and outlet geometry on flow uniformity was investigated. During the analysis the attention was focused also on minimizing the turbulence intensity in outlet area, that could cause significant difficulties in flow and stable work of the machine – generated eddies contributes to lower the mass flow rate.

1. Introduction

In gas turbines applied the leakage flow has a crucial impact on the efficiency. Seals isolate a region of a higher pressure from one of a lower pressure when relative rotational motion occurs, to prevent the leakage. There are many types of seals, for instance gas annular seals reduce leakage well, but it takes certain amount of air from compressor, reducing performance of entire machine. Although new designs of film riding and contacting seals that fundamentally reduce leakage [1, 2], the labyrinth seal constantly is a prime turbine sealing solution because of low price, low maintenance, minimal rub particulate contamination and high temperature capability. However, operating at tight clearances and high pressure ratios, tend to develop considerable force imbalance and dynamic phenomenon, like rotor vibrations. Honeycomb seals were first applied in 1989 in High Pressure Oxygen Turbo Pump in Space Shuttle main engine [3]. Application of honeycomb seals improves rotordynamic performance and force balance, reducing vibrations provides high resistance to flow in peripheral as well as axial directions. Works found in literature give weight the importance of friction factor coefficient evaluation for anticipation of the seal rotor-dynamic characteristics. Majority of researches, both experimental and computational, have focused their attention on stationary configurations – where the relative motion of rotor and stator has been omitted. Stationary test-rigs are much more workable than rotational ones, but an overlooking of a rotational velocity component may affect the test results significantly. Experimental and theoretical predictions on seal configuration performed by Childs and Kleynhans [4], found six parameters influencing rotor dynamics coefficients – associated with friction factor: seal inlet pressure, pressure ratio across the seal, seal clearance, seal inlet fluid rotation, rotor speed and honeycomb cell width. Experimental analysis performed on static set rig by Al-Qutub et al.

[5], found out a dependency on friction factor on clearance and Reynolds number only. Rotational test rigs give full outlook on the phenomena which occur in actual sealing. Results from preliminary CFD calculations are presented in this paper. In the first section test rig and used CFD model is described. This is followed by review of used non-dimensional parameters to describe results. Finally, conclusions and future work ideas are given.

2. Description of designed test rig

Numerical campaign exploiting steady state RANS analysis has been performed to support the design process of the test facility. Presented test rig setup consists of Roots type vacuum pump with a capacity of 600 m^3/h connected directly to the pressure tank of 3 m^3 volume. Tank is connected to the test rig by a pipeline. The valve between the rig and the pressure tank allows to adjust the pressure at outlet. A scheme of the installation is shown in Figure 1.

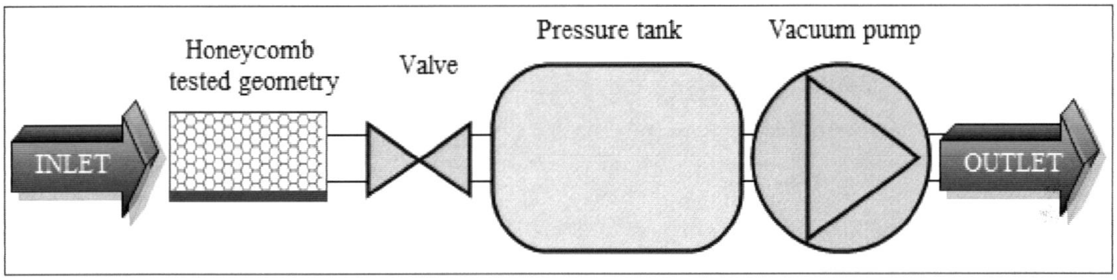

Figure 1. Test rig scheme

Described test rig was designed to investigate various types of sealing system, but mostly "honeycomb on labyrinth" or "smooth on labyrinth". Due to its construction, the test rig allows to inquire different types of sealing, by a large variety of inlet and outlet geometry configurations. Iconic parameters of rotational test bench are gathered in Table 1. This work shows CFD calculation for "smooth on labyrinth" case. Aims of analysis are to check impact of rotor surface roughness, and rotational velocity both on mass flow rate through the labyrinth seal and velocity components. Results are believed to support designing decisions.

Table 1. Parameters of designed test rig.

Maximum mass flow rate (air)	0.2 kg/s
Pressure ratio, up to	0.5
Rotational speed, up to	24000 rpm
Seal outer diameter (rotor), up to	290 mm
Working temperature	Ambient

3. CFD model

In this article a stationary analysis of flow through sealing system "labyrinth on smooth" has been done. For CFD calculations commercial software Ansys CFX has been used. Investigated labyrinth sealing consists of two oblique fins, and is typical in gas turbines appliances. In this case radial dimension of cavity is 0.5 mm, and length (at top) of one fin is 1.45 mm. Analyzed geometry is shown in Figure 2.

Figure 2. Geometry of described labyrinth seal test rig.

Figure 3. Geometry of 60° fluid above rotor.

For discretization the volume of flow in rotor a structural mesh has been used. To obtain both good mesh quality and low time of calculations a 60° geometry has been generated. This procedure let us minimalize number of mesh elements, while accuracy of calculations remains high. The geometry of fluid above the seal is shown in Figure 3. The mesh of rotor – area above the labyrinth seal consists of 2.6M nodes. The mesh was done with good wall layers discretization – parameter $y^+ \approx 1$. Described mesh is presented in Figure 4.

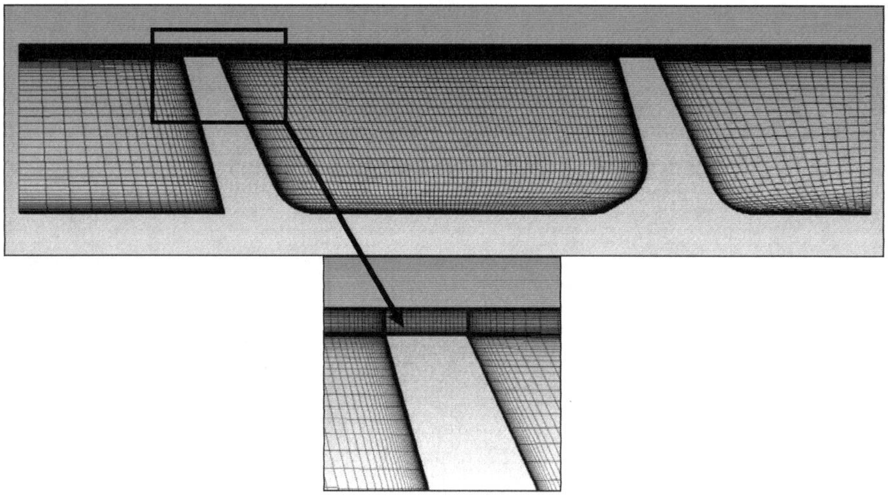

Figure 4. Structural mesh of the rotor domain.

Reference analysis was performed with stationary domain of rotor, in the rest of cases rotational speed varies. Moreover, the in reference case walls of rotor and stator were assumed as smooth, in the rest of cases rotor roughness varies. In each case walls were adiabatic. At sides of rotor – perpendicular to inlet and outlet, a rotational periodicity interface was chosen. Convergence criteria were set to 10^{-5} of RMS, with mass imbalance at rotor domain under 0.1%. Both advection scheme and turbulence numeric were set to high resolution. Boundary conditions are presented in Tab. 2.

Table 2. Boundary conditions used in CFX model

GLOBAL	Turbulence model	SST
	Gas model	Air ideal gas
	Heat transfer	Total energy
	Viscous work term	Yes
	Reference pressure	0 bar
	Timescale	Local
INLET	Total pressure	1 bar
	Total temperature	293 K
	Turbulence intensively	5 %
OUTLET	Total pressure	0.9 bar
	Pressure averaging	Yes

4. CFD results

Results are presented as dimensionless values. Relative mass flow rate was calculated as m_{RPM}/m_{RPM0}, where m_{RPM} – is mass flow rate for given *RPM* , m_{RPM0} – mass flow rate for reference case (i.e. for 0 *RPM*). In each case relative mass flow rate was shown as a function of circumferential rotor velocity, to axial-flow velocity – u/c_{ax}. An arrow depicted in Figure 5 shows plane where the data for further calculation are taken from. Axial-flow velocity is calculated as mass flow averaged velocity X component, in rotational cases as velocity in stationary frame X component.

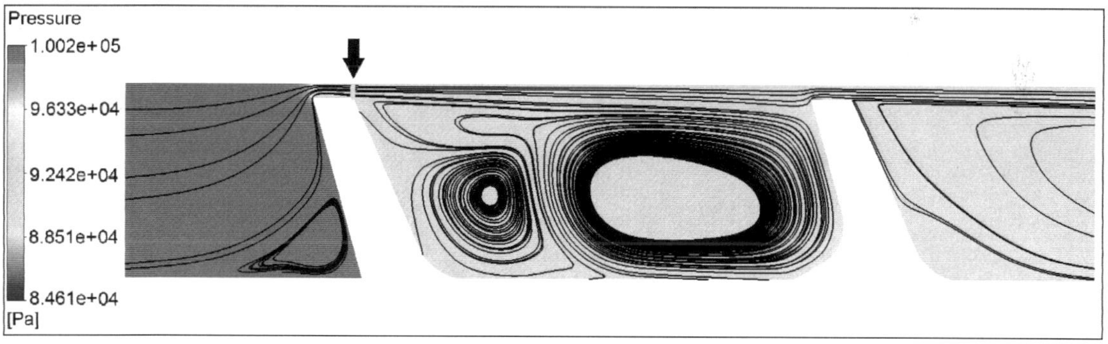

Figure 5. Streamlines superimposed on contours of static pressure for reference, smooth walls, non-rotating case.

4.1. Results of rotational speed influence on relative mass flow rate

In this case only efect of rotational speed on mass flow rate was inqired. Rotor walls were set as smooth. For described geometry certain influence was found. Relative mass flow was described as m_{RPM}/m_{0RPM}, where m_{RPM} – is mass flow for various rotational velocity, and m_{0RPM} is result for stationary case. Relative mass flow decreases in function of rotational speed. The lowest mass flow coresponds to the highest rotational speed. The dependecy is shown in Figure 6. Due to limitation of roational speed – 24000 *RPM* – of the test rig, maximum u/c_{ax} is ca 4. This value corrsponds to relative mass flow rate at 0.88.

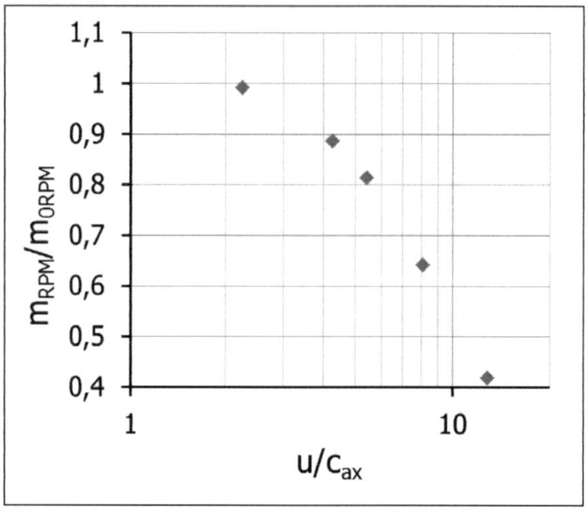

Figure 6. Rotational speed impact on discharge characteristic.

4.2. Results of surface roughness influence on relative mass flow rate

Default wall roughness in CFX is set to smooth – it means that wall condition has no influence on flow. In actual devices roughness has crucial impact both on flow and construction parameters. Roughness is considered to be determinant of part performance. As a consequence, constructors establish an upper limit on roughness. In this work three different values of roughness were set on rotating wall – sedentary wall remain smooth to investigate influence only the rotor roughness. This analysis was done to establish lower limit of surface roughness – to minimalize mass flow through the seal. Values of roughness corresponding to different types of material finishing, are presented in Table 3.

Table 3. Examined values of roughness.

Type of material finishing	**Roughness value**
Fine machining/grinding	2.5 μm
Thorough machining	10 μm
Coarse machining	50 μm

Results of roughness influence investigation are shown in the Fig. 7. Relative mass flow was described as $m_{RPM,Ra}/m_{0RPM,Ra}$, where $m_{RPM,Ra}$ – is mass flow for various rotational velocity and the same surface roughness, and $m_{0RPM,Ra}$ is result for stationary case with the same roughness as rotational case. For higher values of relative circumferential velocity, impact of surface roughness is stronger. The highest value of roughness Ra = 50 μm corresponds with rapid drop of mass flow rate. The difference in relative mass flow rate between for highest rotational velocity and reference case is 20.1 %, which means that mass flow rate, and axial velocity drops, while u/c_{ax} rises. In table 4 the values of relative mass flow for different rotational velocities and surface roughness are shown. For higher velocities, the difference between smooth and rough wall is significant, and is equal to 8 % in smooth vs. 50 μm case for 24000 rpm. This proves that roughness has strong influence on flow regime in seal, and to reduce leakage should be as high as possible. Also another conclusion can be made, that roughness shouldn't be omitted in CFD analysis.

Table 4. Rotational speed and surface roughness dependency on rel. mass flow

	50 μm	10 μm	2.5 μm	Smooth
Rotational velocity (rpm)	m/m_0	m/m_0	m/m_0	m/m_0
0	1	1	1	1
7500	1	1	1	1
15000	0.959	0.983	0.987	0.991
20000	0.887	0.927	0.939	0.942
24000	0.799	0.864	0.877	0.882

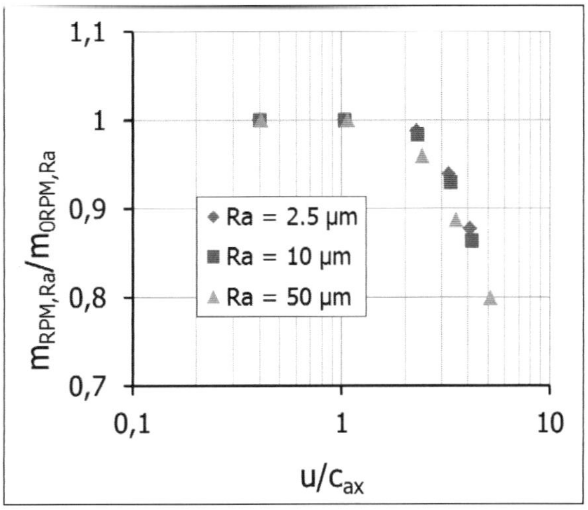

Figure 7. Roughness and rotational speed imact on relative mass flow.

4.3 Results of inlet angle influence on relative mass flow rate

Default angle of inlet flow in CFX is set to 90°. Mentioned angle is measured between inlet velocity direction and circumferential direction, as shown in Figure 8.

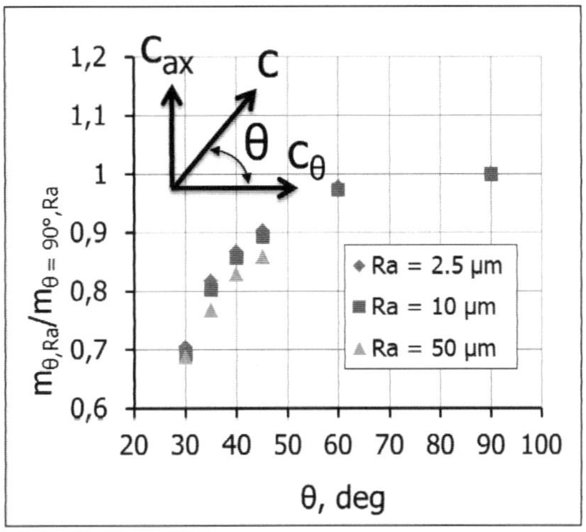

Figure 8. Impact of inlet velocity angle on discharge characteristic for 20000 rpm.

Results of roughness inlet velocity investigation are shown in the Fig. 8. Relative mass flow was described as $m_\theta/m_{\theta=90°}$, where m_θ – is mass flow for various inlet velocity angle the same surface roughness as reference value, and $m_{\theta=90°}$ is result for reference case with flow angle 90°. Every point was calculated for the same rotational velocity 20000 rpm. Both for high values of θ angle – 60° and low – 30° impact of surface roughness was slight. For mentioned angles difference of relative mass flow rate between 2.5 μm and 50 μm roughness case is up to 1.5 percent points. The highest difference was observed for inlet angles 35° – 45°, and was equal up to 5 %. This proves that inlet velocity angle has strong influence on flow regime in seal, and to reduce leakage should be as high as possible. Also another conclusion can be made, that roughness has the highest impact for inlet velocity angles ca. 45°.

5. Conclusions and future work

Numerical calculations were performed and analyzed for the impact both of rotational speed and surface roughness on the discharge characteristics of labyrinth seal geometry. The CFD analysis confirmed that onset of drop in mass flow through the leakage depends on critical value of characteristic velocity, $u/c_{ax} = 1$ – as reported previously by Waschka [2]. For designed geometry influence of rotational velocity up to 24000 rpm on relative mass flow rate was up to 12% reduction, in case with smooth walls. When wall roughness is considered, relative mass flow rate drops by 20,1 % - for maximum rotational velocity and roughness. This work provided information, that surface roughness can be very important aspect in designing labyrinth seals. Also seal inlet fluid rotation has an impact on seal discharge. For presented geometry rel. mass flow rate drops 31% for case with roughness 50 μm and inlet angle 30°. Although, roughness has strong influence only for exact range of inlet angle values. For extreme values of this angle (namely – lower than 30° and higher than 60°), its influence was slight. It shows, that for the presented seal geometry, there is strong possibility to improve its work by simply modifying surface roughness and applying pre-swirl nozzles at inlet.

The future work will cover the more expanded CFD calculations, including transient – time dependent models, comparison presented results with both other turbulence models and different boundary conditions – pressure ratios (to get higher flow velocity and Reynolds numbers), flow pre-swirl on inlet for higher rotational velocities and surface roughness. Another task is compare results and finally validation presented data with experiment done on rotational and stationary test rigs.

6. References

[1] Paolillo R 2006 Rotating Seal Rig Experiments: Test Results and Analysis Modeling, *ASME Turbo Expo 2006: Power for Land, Sea and Air* **3** pp 1551-1559

[2] Waschka W 1992 Influence of high rotational speeds on the heat transfer and discharge coefficients in labyrinth seals, *ASME Paper 90-GT-330 Journal of Turbomachinery* **114** pp 462-468

[3] Scharrer J 1989 Discussion in paper, annular honeycomb seals: test results for leakage and rotordynamic coefficients; comparisons to labyrinth and smooth configurations. *ASME J. Tribol.* **111** pp 300-301

[4] Childs D W and Kleynhans G F 1992 Experimental rotordynamic and leakage results for short (l/d) honeycomb and smooth annular pressure seals. *IMechE Proceedings, Fifth international Conference on Vibrations in Rotating Machinery* **11** pp 305-311

[5] Al-Qutub A M, Elrod D and Coleman H W 2000 A new friction factor model and entrance loss coefficient for honeycomb annular gas seals. *ASME J Tribol 200*;**122** pp 622-627

The boundary condition at the valve for numerical modelling of transient pipe flow with fluid structure interaction

S Henclik

Department of Hydraulic Machinery, Institute of Fluid-Flow Machinery, Polish Academy of Sciences, 80-231 Gdansk, ul. J. Fiszera 14, Poland

E-mail: shen@imp.gda.pl

Abstract. Transient flows in pipes (water hammer = WH) do appear in various situations and the accompanying pressure waves may involve serious perturbations in system functioning. To model these effects properly in the case of elastic pipe the dynamic fluid–structure interaction (FSI) should be taken into account. Fluid-structure couplings appear in various manners and the junction coupling is considered to be the strongest. This effect can be especially significant if the pipe can move as a whole body, which is possible when all its supports are not rigid. In the current paper a similar effect is numerically modelled. The pipe is fixed rigidly, but the valve at the end has a spring-dashpot mounting system, thus its motion is possible when WH is excited by the valve closuring. The boundary condition at the moving valve is modelled as a differential equation of motion. The valve hydraulic characteristics during closuring period are assumed by a time dependence of its loss factor. Preliminary numerical tests of that algorithm were done with an own computer program and it was found that the proper valve fixing system may produce significant lowering of WH pressures.

1. Introduction

Water hammer (WH) is a phenomenon produced by sudden change in steady pipe flow conditions due to valves operation, hydraulic machinery load variation or other reasons [1,2]. When the pipe is elastic the dynamic fluid-structure interaction (FSI) influences the flow and should be taken into account [3,4,5]. Three main fluid-structure coupling mechanisms are pointed in literature. The friction between the pipe wall and the liquid is the weakest one. The Poisson effect is responsible for the pipe-wall longitudinal stress wave generated by pressure-induced circumferential stresses and known as the precursor wave. The strongest FSI effect is the junction coupling which appears at pipe ends, bends, valves and other places where strong forces between the pipe and the liquid exist. This effect can be especially important if the pipe can move as a whole structure on the supports [6]. Such a motion forced by shock of the foundation can be even the source of WH event.

In general it could be expected that elastic pipe supports should influence in lowering of transient pressures due to the energy transfer from the liquid to the structure. This conclusion is however not unambiguous as reported by scientists [7] who pointed that elastic pipe supports and the junction coupling effect may also produce increase in the transient pressure magnitude. Thus it is important to determine the limitations for the supports parameters for which the pressure reduction is possible. In

Journal of Physics: Conference Series **530** (2014) 012034 doi:10.1088/1742-6596/530/1/012034

the current paper the pipe supports system will not be the main goal, however a similar problem is examined. Valves operation is known to be one of the possible, and frequently used in the analyses, source of WH event. It is a known and utilized behaviour that setting the proper time dependence of the valve closuring process may highly reduce the transient pressures (usually elongation of the closuring time). In various scenarios, like check valves, such possibility may not exist, thus another solution can be valuable. An idea of a valve fixed in the pipe with a spring-dashpot system seems to be a natural concept, if the physics of WH phenomenon is considered. Such a possibility is mentioned in [1] but only some general remarks with no detailed solution nor conclusions are presented there. In the current paper the boundary condition (BC) at the valve is formulated as a differential equation of the valve motion and its solution for numerical application is found and preliminary tested.

2. Mathematical model

In the current study the four equations (4E) mathematical model [4,5,8] of WH-FSI is used. In this model one-dimensional (1D) liquid flow is assumed and only the longitudinal pipe motion is taken into account. The 4E model is reported [5] to give a good description of many real hydraulic systems. Moreover, the 1D treatment is justified, as the longitudinal effects at the valve are predominant.

2.1. Assumptions and governing equations

The standard assumptions of the WH-FSI 4E model are used herein. The pipe of the length L, inner diameter D and wall-thickness e is straight, slender (D/L<<1), thin-walled (e/D<<1) and prismatic of circular cross section. Its material of Young modulus E and the density ρ_s is linearly elastic, which produces the standard relation for the celerity of longitudinal elastic waves in it:

$$c_s = \left(\frac{E}{\rho_s}\right)^{1/2} \tag{1}$$

The liquid of bulk modulus K is weakly compressible, linearly elastic and has the density ρ. The density changes are small which is the consequence of the low pressure assumption (*p*/K<<1). The flow velocity *v* is of little relativity to the celerity c of the pressure wave in the liquid (*v*<<c) given in:

$$c = \left(\frac{K}{\rho}\right)^{1/2} \left(1 + (1-\nu^2)\frac{KD}{Ee}\right)^{-1/2} \tag{2}$$

In the equation above ν is the Poisson coefficient and the second right-side factor is responsible for the influence of pipe wall elasticity. The cavitation is assumed not to be present. The liquid - pipe-wall friction shear stresses τ_s are assumed to be given with the quasi-steady formula (λ is the Darcy-Weisbach friction factor and *w* is the pipe cross-section velocity):

$$\tau_s = \frac{\lambda\rho}{8}(v-w)|v-w| \tag{3}$$

The 4E model is described with four hyperbolic partial differential equations of the first order [3,4,5]. Two of them govern the 1D liquid flow:

$$\frac{\partial v}{\partial t} + \frac{1}{\rho}\frac{\partial p}{\partial x} = -g\sin\alpha - \frac{4\tau_s}{\rho D} \tag{4}$$

$$\frac{\partial v}{\partial x} + \frac{1}{\rho c^2}\frac{\partial p}{\partial t} = 2\nu\frac{\partial w}{\partial x} \tag{5}$$

The other two equations govern the pipe longitudinal motion:

$$\frac{\partial w}{\partial t}-\frac{1}{\rho_s}\frac{\partial \sigma}{\partial x}=-g\sin\alpha+\frac{\tau_s}{e\rho_s} \tag{6}$$

$$\frac{\partial w}{\partial x}-\frac{1}{\rho_s c_s^2}\frac{\partial \sigma}{\partial t}=-\frac{\nu D}{2Ee}\frac{\partial p}{\partial t} \tag{7}$$

In the equations above x and t are the position and time coordinates, g (=9.81 m/s^2) is the gravitational acceleration and α is the angle of the pipe inclination against the horizon. The pipe longitudinal stresses are $\sigma=Q/A_s$ (Q is the longitudinal force and pipe cross section area is $A_s=\pi De$). For the solution of that model the method of characteristics (MOC) is frequently used [3,5,8,9] and will be also applied in the current approach.

2.2. Method of characteristics

The standard MOC transformation of the governing equations produces two pairs of compatibility equations (CE) governing of two elastic waves. The WH wave is defined with a pair (+/-) of equations C0 presented below in the form proposed by the author [10,11]:

$$\frac{\mathrm{d}}{\mathrm{d}t}(v+Sw)\pm\frac{\mathrm{d}}{\mathrm{d}t}(r-\tilde{S}q)=-(1+S)g\sin\alpha-(1-R)\frac{4\tau_s}{\rho D} \tag{8}$$

They are valid for positive (+) or negative (-) characteristic slope and $x(t)$ dependence given with the relation $\mathrm{d}x/\mathrm{d}t=\pm c_0$. The WH celerity c_0 is slightly changed relative to (2) due to the Poisson coupling:

$$c_0=cA^{-1/2} \tag{9}$$

The precursor wave is governed by the following pair of equations C1:

$$\frac{\mathrm{d}}{\mathrm{d}t}(Rv-w)\pm\frac{\mathrm{d}}{\mathrm{d}t}(\tilde{R}r+q)=(1-R)g\sin\alpha-(1+S)\frac{\tau_s}{\rho_s e} \tag{10}$$

They are valid respectively for the characteristics $\mathrm{d}x/\mathrm{d}t=\pm c_1$ and the wave celerity c_1 is:

$$c_1=c_s A^{1/2} \tag{11}$$

In the above equations the unified variables were introduced. The normalized pressure [m/s] is defined as $r=p/(\rho c_0)$ and normalized stresses [m/s] as $q=\sigma/(\rho_s c_1)$. The above MOC transformation is valid for most practical cases, but in fact the condition $1-\gamma+\chi>0$ is necessary. The parameters γ and χ are:

$$\gamma=(K\rho_s)/(E\rho) \tag{12}$$

$$\chi=(KD)/(Ee) \tag{13}$$

The Poisson coupling constants S, R and the parameter A are defined with the following formulas:

$$S=4\gamma\nu\left[(1-\gamma+\chi)+\sqrt{(1-\gamma+\chi)^2+4\gamma\chi\nu^2}\right]^{-1} \tag{14}$$

$$R=(S\chi)/(4\gamma) \tag{15}$$

$$A=1+S\chi\nu(2\gamma-S\chi\nu)^{-1} \tag{16}$$

For water in steel pipe and D/e=20 these parameters are A=1.016, S=0.042, R=0.027. The "tilde" parameters have the following definition ($\eta=c_0/c_1$):

$$\tilde{R}=R\eta \tag{17}$$

Journal of Physics: Conference Series **530** (2014) 012034 doi:10.1088/1742-6596/530/1/012034

$$\tilde{S}=S/\eta \tag{18}$$

2.3. The solution scheme

In the equations (8) and (10) two types of fluid-structure couplings can be identified. The first one occurs due to the friction term on their right side. But the friction is weak, thus it can be taken into account iteratively and each of the waves C0 and C1 can be considered to be independent in its own "wave variables" kept in the parenthesis on the left side. The real physical quantities – pressure, stresses and velocities - are finally calculated as the combination of the proper wave variables producing superposition of WH and PC waves due to the Poisson coupling. To solve each of the waves a standard scheme is used. Integrating all the CE within the same time step Δt along their characteristic produces two pairs of finite difference equations (FDE) that can be solved marching in time on the properly designed space-time grid.

The pipeline may consist of a number of straight pipe reaches joint at junctions where proper BC exists. At such boundaries only half of the FDE are valid for each sub-pipe and to find the solutions additional equations of BC have to be formulated and taken into account. The basic BC for the liquid is the continuity equation which can be formulated in the following form (L=left, R=right):

$$v_{\mathrm{L}}-w_{\mathrm{L}}=v_{\mathrm{R}}-w_{\mathrm{R}} \tag{19}$$

The structure velocity w is in fact the same for both edges of the junction, which is rigid in itself, thus the above condition can be simplified to the equality of liquid velocities. At the close-end the BC is:

$$v-w=0 \tag{20}$$

The pressure balance at the junction is the result of minor losses defined with the coefficient ζ:

$$p_{\mathrm{L}}=p_{\mathrm{R}}+0.5\zeta\rho(v-w)^2 \tag{21}$$

If the junction is fixed rigidly to the foundation the BC for the structure is w=0 and should be put to the above equations. However, if it is fixed with elastic mounting or is unfixed, the differential equation of motion should be formulated and solved concurrently with the CE.

3. Boundary condition at the valve

For a rigidly fixed valve the BC problem is limited to resolving the valve hydraulic characteristics during closuring process. In the current study however, the major importance is devoted to the valve fixing system and its influence on WH run parameters.

3.1. Piping physical model

For the current analyses a pipeline physical model is composed of a pressure tank at the beginning, a straight pipe fixed to the foundation with a number of supports and a valve at the end. The flow is driven by the constant pressure of the tank. The valve is being closed with a certain time dependence to excite a water hammer. The valve is fixed to the pipe with an elastic mounting, which is modelled as a spring-dashpot system (Kelvin-Voigt model). The scheme of the pipeline is presented in figure 1.

3.2. Valve characteristics

In the equation of pressure balance at the valve it is assumed the outside pressure is zero, thus:

$$p=0.5\zeta(t)\rho(v-w)^2 \tag{22}$$

The valve characteristics are determined as the dependence of its pressure losses coefficient ζ against the valve closuring rate ϕ, $\zeta=f(\phi)$. In fact ϕ is a conventional parameter and it can be the closing angle (for a globe valve) or just a relative value equal 0 for a complete opening and 1 for a complete closure.

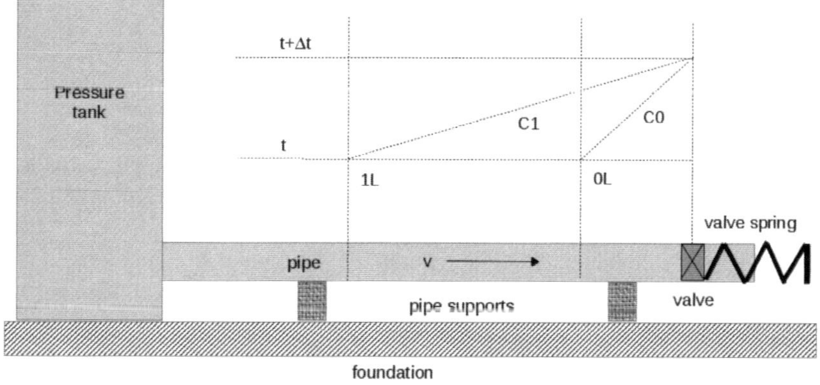

Figure 1. A pipeline with a valve at the end fixed with a spring.

To model the time dependence of the closuring process the rate ϕ is a function of time $\phi=g(t)$, for which the time τ of the total closure is especially important, but the detail shape $g(t)$ could be also valid. For the valve completely closed the losses coefficient becomes infinite so the condition (22) has to be changed to (20). The characteristic $f(\phi)$ can be defined in an analytical way [8], but the most general could be a numerical description, as usually such characteristics are the result of experiments [12]. In the author's approach the valve characteristic is defined numerically in several (N+1) equally spaced points of closing rate ϕ and interpolation is used between them. Lin-lin interpolation is applied at the first interval (0..1/N), lin-hyp at the last one (1-1/N..1) and lin-log within the inner intervals.

3.3. Equation of motion

We can assume various types of support systems for the pipeline. A quite general situation is when the valve of mass m is fixed to the pipe with a spring-dashpot mounting and the pipe itself is also fixed to the foundation with elastic supports. If $y(t)$ is the valve movement and $z(t)$ is the motion of the pipe-end the following equation was formulated for the valve ($A_c=\pi D^2/4$ is a liquid cross-section area):

$$m\ddot{y}+b(\dot{y}-\dot{z})+k(y-z)=-b_s(\dot{y}-\dot{z})+p\,A_c \tag{23}$$

In the above equation k and b are respectively the stiffness and damping coefficients of the valve spring and b_s models the friction between the valve and the pipe. Another equation is valid for the pipe-end. Now m_p is the mass of that junction, b_p and k_p are the coefficients of the pipe-end spring.

$$m_p\ddot{z}+b_p\dot{z}+k_p z=-Q-b_s(\dot{z}-\dot{y})-b(\dot{z}-\dot{y})-k(z-y) \tag{24}$$

Other mounting configurations can be also considered, but for the current analyses let us focus on the upper one simplified with the assumption of pipe-end rigidly fixed to the foundation. Thus $z(t)\equiv 0$ and the equation (23) can be presented in a new form. In fact, we have to measure the pressure at the valve relative to the initial, steady-state one as we measure the motion $y(t)$ relative to the initial, steady-state valve position. Dividing also the equation of motion by ($A_s\rho_s c_1$) and using the "divided" mass, stiffness and damping coefficients (with hats) we can write it in the following way, in which r_0 is the initial, steady-state pressure (normalized) at the valve:

$$\hat{m}\ddot{y}+(\hat{b}+\hat{b}_s)\dot{y}+\hat{k}y=\tilde{R}S^{-1}(r-r_0) \tag{25}$$

To transform the differential equation to the finite form the Newmark method is used to the left side which allows to express it as a function of the valve velocity u=dy/dt in a new time instant. To transform the right side of the equation the two left FDE, C0 and C1 are used. Out of the four system variables v, w, r, q the pipe-end velocity is zero (w=0), thus r and q can be calculated as a function of v. But liquid velocity v and the valve velocity u are related with a formula similar to (22) (u has to be used instead of w), which allows to determine v ($v \geq u$) with:

$$v = u + \left(2c_0 / \zeta(t)\right)^{1/2} \sqrt{r} \tag{26}$$

Thus, the pressure r can be expressed as a function of u and putting it to (25) transformed with the Newmark method, the valve velocity u can be found. Then, all the system variables r, v, q can be calculated and the BC is solved. For the current approach an assumption is made that the BC at the valve is valid at a fixed point in space being the initial valve position, in spite of the valve motion. But this motion is small, so the assumption is considered to be acceptable.

4. Numerical results

The above algorithms were implemented in an own computer program and preliminary numerical tests of the proposed method in various scenarios were done.

4.1. Assumptions for computations

For the numerical analyses a steel pipe of the length L=50m, inner diameter D=50mm and pipe-wall thickness e=2.5mm is assumed. It is fixed rigidly at 0, 14, 28, 42 [m] from the pressure tank and at the end. The flow is driven by constant pressure of 1.014 MPa supplied by the tank. The valve at the end is partially closed to keep the velocity at the steady state at v_0=0.5m/s and to avoid cavitation during WH. The valve characteristic is defined with the following function $f(\phi)$ = (0.1, 0.2, 0.5, 1.2, 3, 10, 30, 100, 400, 1800, ∞). The closuring time is 6 ms and the function $g(t)$ is linear. The mass of the valve mechanism was assumed at m=0.5kg. The Darcy-Weisbach friction factor is constant λ=0.03. To focus on the effects produced by the valve fixing system the Poisson phenomenon is neglected (ν=0). The damping coefficient b of the valve dashpot is estimated with a formula valid for simple oscillator:

$$b = 2\xi\sqrt{kM} \tag{27}$$

M (≅98 kg) is the total mass of water in the pipe (and the valve) and ξ is a conventional damping parameter corresponding to a damping ratio of simple oscillator.

4.2. Results

In the following figures valve displacements and pressure records at a distance of 1m before the valve are presented. In figure 2 the results for zero damping and various stiffness k of the spring (7.5N/mm, 25N/mm or 75 N/mm) are plotted. For these parameters one can observe that the WH pressure magnitudes decrease with lowering the stiffness of the valve spring. For rigidly fixed valve (classic WH) the pressure amplitude is 0.67MPa. In figure 3 the influence of damping is presented. For higher damping the transient decays faster due to energy dissipation in the valve damper.

4.3. Analyses and conclusions

The analyses of numerical results has allowed to conclude the presented numerical method works correctly – the analytical solutions for a simplified model of the scenarios presented in figure 2 gave similar results. Some behaviours however may still require explanation. Such a specific effect in the current results are the very small and short peaks in the pressure records appearing twice a basic WH period ($f_{WH}=c_0/4L \cong 6.7$Hz). They could be physical effects as they are smaller for massless valve. It is possible however they are of numerical origin being, e.g. the result of approximation that the point of the valve BC does not move.

The preliminary analyses of the system behaviour allows to conclude that lowering of transient pressures is possible for the proper valve fixing system. For the presented scenarios decreasing of the mounting stiffness produces WH pressure reduction but larger magnitudes of the valve motion. It was also observed within other results that pressure increase (relative to classic WH) for specific valve mounting parameters may happen. All these effects depend also on the damping properties of the valve fixing system. In general, energy dissipation produce reduction of the transient pressures and magnitudes of the valve motion, however this may not happen for the first peak in the pressure record which is visible in figure 3. All these effects are going to be tested more thoroughly.

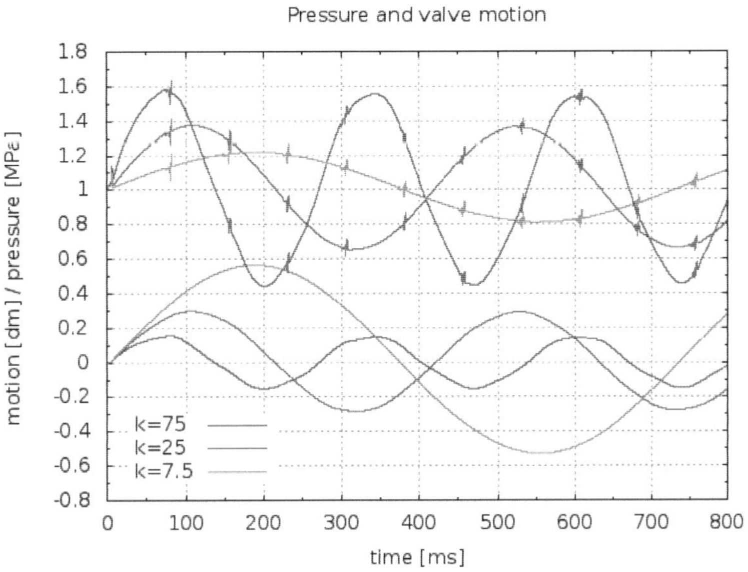

Figure 2. Pressures and valve motion for various stiffness and zero damping of the valve spring.

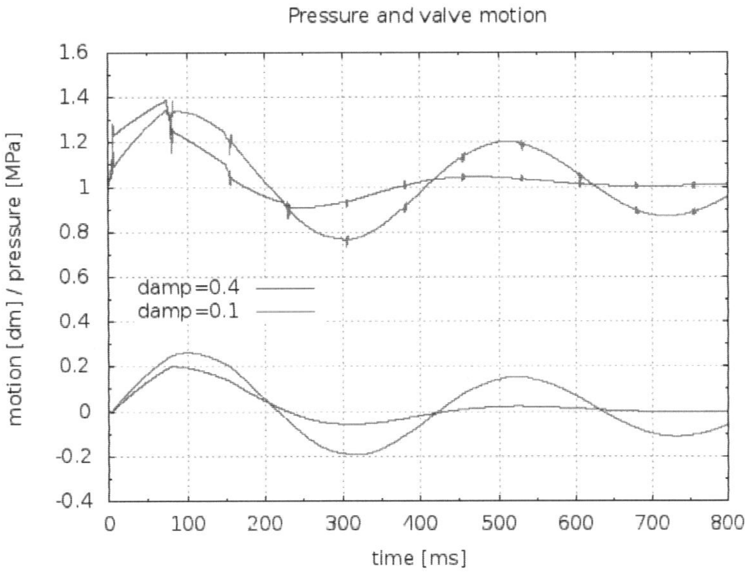

Figure 3. Pressures and the valve motion for damping parameter ξ=0.1 or ξ=0.4 (k=25N/mm).

5. Summary

In the paper a solution for numerical application of the BC at the valve during WH-FSI is presented. The 4E model of WH-FSI is applied and shortly discussed. The numerical approach of the solution of this model is used in a form proposed by the author. Two issues of the BC problem are pointed. The valve closuring hydraulic characteristics are modelled in a numerical way but the main problem analysed and solved within the BC is the idea of fixing the valve in the pipe with viscoelastic mounting. Though this proposal may seem to be a natural concept if the physics of WH is considered, the author does not know other works where this problem has been examined. The proper BC was formulated, solved and implemented in an own computer program. The preliminary numerical tests were done and it was found that WH pressure magnitudes can be significantly reduced in comparison with the traditional design of rigidly fixed valve. The idea of fixing the valve with viscoelastic mounting is expected to be applicable in practice and may be especially valuable in systems where rapid valve closuring may happen (e.g. check valves) and transient pressure variation is undesired. Further studies within this subject are intended with the use of the developed software. The job will be also extended towards examination of the influence of viscoelastic pipe supports, including 3D effects.

References

[1] Wylie E B and Streeter V L 1993 *Fluid transients in systems* (Prentice-Hall Inc. NJ)

[2] Adamkowski A 2013 *Fluid transients in closed conduits* (in Polish *Przeplywy nieustalone cieczy w przewodach zamknietych)* Publishing House of Inst. Fluid-Flow Machinery Gdansk

[3] Wiggert D C, Hatfield F J and Stuckenbruck S 1987 Analysis of liquid and structural transients in piping by the method of characteristics *ASME J. Fluids Eng.* **109** 161-165

[4] Tijsseling A S and Lavooij C.S.W 1990 Waterhammer with fluid-structure interaction *Applied Scientific Research* **47** 273-285

[5] Wiggert D and Tijsseling A 2001 Fluid transients and fluid-strucure interaction in flexible liquid-filled piping *Applied Mech. Review* **54** 455-481

[6] Heinsbroek A G 1997 Fluid – structure interaction in non-rigid pipeline systems *Nucl. Eng. Design* **172** 123-135

[7] Wiggert D, Otwell R S and Hatfield F J 1985 The effect of elbow restraint on pressure transients *Trans. of ASME* **107** 402-406

[8] Tijsseling A S 2003 Exact solution of linear hyperbolic four-equation system in axial liquid-pipe vibration *J. Fluids Struct.* **18** 179-196

[9] Zhang L, Tijsseling A S and Vardy A E 1999 FSI analysis of liquid filled pipes *J. Sound Vib.* **224** 69-99

[10] Henclik S 2010 Mathematical model and numerical computations of transient pipe flows with fluid-structure interaction *Trans. Inst. Fluid-Flow Mach.* **122** 77-94 (Gdansk)

[11] Adamkowski A, Henclik S and Lewandowski M 2010 Experimental and numerical results of the influence of dynamic Poisson effect on transient pipe flow parameters *IOP Conf. Series Earth and Environ. Science, doi:10.1088/1755-1315/12/1/012041* (25 IAHR Symp. on Hydraulic Machinery and Systems – Timisoara)

[12] Thorley A R D 2004 *Fluid transients in pipeline systems* (Professional Eng. Publ. London)

Acknowledgments

The work presented in the paper is supported by the research project No N N504 478839 sponsored by the Ministry of Science and Higher Education of the Government of Poland.

Navier number and transition to turbulence

P Ziółkowski[1,2,3] **and J Badur**[1,4]

[1] Energy Conversion Department, Institute Fluid-Flow Machinery PAS-ci, Fiszera 14, 80-231 Gdańsk, Poland
[2] Conjoint Doctoral School at the Faculty of Mechanical Engineering, Gdansk University of Technology, Gdańsk, Poland

Abstract. The motivation behind this article is to explain a role of the Navier number (*Na* - dimensionless slip-length) in prediction of closures for laminar to turbulent transition undergoing via eddies detachment from the slip layer in nano-cannals. Additionally the role of the Navier number *Na* in universal modeling of phenomenon of enhanced mass flow rate reported in micro- and nano-channels has been explained. The *Na* number should be treated as a ratio of internal viscous to external viscous momentum transport and therefore this notion cannot be extended onto whole friction resistance phenomena. Our proposal for unique expressing of a critical point in turbulence transition is that on a plane of f_D - *Re* one needs two coordinates. The second critical coordinate has been discovered by Stanton and Pannell and is known to be the Stanton-Pannell number *StPa*. Finally dependence of the Stanton-Pannell number *StPa* on Navier number *Na* and Reynolds number *Re* is presented.

1. Introduction

In the MEMS and NEMS literature, it is assumed that the external friction between solid and liquid surfaces cannot be further neglected [3]. It involves the velocity slip, the slip length l_s and the Navier number (*Na* - dimensionless slip-length), which is important in the surface momentum transfer, coupled with an interfacial transport [2].

Generally, from the point of view of slip flow enhancement, two kinds of the "$\ln f_{\mathrm{D}} - \ln Re$" diagrams should be considered. In the first one a type of fluid is fixed and different material of the wall surface is considered – in the second one, conversely, the material of wall is fixed and the sort of fluid is changed [12]. In both cases we have with a whole manifold of the external viscosity ν, the lengths of slip l_s and the Navier numbers *Na*. Unfortunately, this key distinguish for microflows is not clearly exposed in the literature – even in the recent works [7,19] a problem of different wall frictions for different fluids and wall solids is omitted.

In this article we study a consistency of the Navier-Stokes model of fluid friction on the slip boundary layer form the point of view of a possible prediction of enhanced mass flow rate. Our subject of investigation coming from a fact that in the literature there is no of unified approach to modeling of the slip length since its role is not clarified in classical fluid dynamics with the slip boundary conditions. However it appears, studding more precisely pioneering papers of Navier (1827);

[3] E-mail: pawel.ziolkowski@imp.gda.pl
[4] E-mail: jb@imp.gda.pl

Helmholtz and Piotrowski (1860); Maxwell (1879), one could find that there exist also some additional developments of a slip-boundary concept, which can be overlapped by a one concept of the Navier number.

1.1. Friction and Navier number Na

The external friction between different continua is a physical reason of the breakdown of conventional macroscopic ideas at small scales [2,3,9,15]. Indeed, there exist many physical reasons for the slip over solid surfaces, among which are: molecular slip small dipole moment of polar liquids, and trapping the naturally present gas bubbles on the solid surface [16,24].

The slip velocity cannot be arbitrary – it should undergo of the laws of external friction between three-dimensional fluid continua touching a solid surface. According to early Newton's hypothesis [11], the external friction between bodies depends mainly from three mechanisms: a) the pressure within the contacting layer, b) the relative velocity of the contacting bodies, and 3) square of their relative velocity, nowadays identified with a relative kinetic energy. Leonard Euler was a scientist who proposed two kinds of the external friction – the first, called "static", that dominates for small relative velocity and the second one, called "dynamic", is proper for fully developed motion. This line of reasoning even nowadays is to be observe in the literature and is called "rate dependent external friction" [18,25,27]. The concept of "external friction" between laminas of different bodies and the "internal friction" between lamina of the same body has been introduced by Coulomb [4], who has made a first repetitive experiments having the aim to verify his own laws of friction.

But it was Navier [23], who standing on the data of Girard's capillary experiment, has proposed to use both: an idea of dissipative kinetic energy of external friction and the idea of external friction force between fluid and a surface of a fixed solid body, defined by him to be: $\mathbf{f}_f = \nu \mathbf{v}_s$. He called the coefficient of the external friction ν as "la résistance provenant du glissement" – the external viscosity. The tangential component of the fluid velocity $\mathbf{v}$ on surface has been called the "slip velocity". The magnitude of the slip velocity is governed by the value of external friction coefficient – if $\nu \to \infty$ we have the no-slip condition, for $\nu = 0$ we have an ideal slip – the boundary condition which correspond to the Euler model of an internally perfect fluid [5,6].

In contrast, in some special applications, for example, in microfluidic and nanofluidic devices, where the surface-to-volume ratio is huge, the slip velocity behaviour is more typical, and the "slip" hydrodynamic boundary condition is usually used. Regardless the slip physical mechanism, the degree of slip is normally quantified through the "slip length" l_s, or a dimensionless slip length (Navier number) Na [21,28], namely:

$$l_s = \frac{\mu}{\nu} \quad ; \qquad Na = \frac{l_s}{a} \tag{1}$$

being the ratio of the internal viscosity μ and the external viscosity ν. This last coefficient depends form, both, a kind of fluid and a contacting solid. The characteristic dimension of a channel a usually is identified with a radius of tube. The value of Na can be: $0 < Na < \infty$, for instance; $Na = 50$ for water flowing through a carbon nanotube [22]. The slip length l_s can be identified as the distance from the liquid to the surface within the solid phase, where the extrapolated flow velocity vanishes.

In spite of a numerous controversy concerning recognition of the possibility of slippnes, let us recall shortly the main measurements of the external viscosity coefficient, called also sometime the "external friction coefficient". If the linear (Newtonian) bulk shear viscosity of fluid μ is known, then the problem of measurement of ν can be reduced to determination of the slip length $l_s = \mu / \nu$. The first technical closure for ν between water and glass has been proposed yet by [23] in a proportion to the water density ρ:

$$\nu = \rho \times 0{,}0023 \quad [\text{mm}] \tag{2}$$

and was rigorously checked in work by Helmholtz & von Piotrowski [4]. The first closure for the external friction between rarefied air and glass has been prepared by Knudt & Warburg [4] in the following form:

$$l_s = \frac{\mu}{\nu} = 0.7122\, l = 0.7122\, l_0 \frac{760}{p} \qquad \text{in [mm]} \tag{3}$$

where l, l_0 are the mean free paths of molecule in the actual pressure and atmospheric pressure, respectively. At the same time, a systematic experimental study reporting the external viscosity in as well liquids as gases has been apparently reported by Wlademen who was also an inventor for technical devices important for measurements of ν or l_s [17]. Many years after, Schnell [25], has measured the flow rate of water in glass capillaries of radius of the order 100 μm .
Trying to control the external friction by any "interfacial lubrication", Schnell treated the capillaries with dimethyldichlorosilane, making its surface hydrophobic one. Then larger flow rates has been obtained and they were interpreted as a lowering of friction and increasing of the slip length at the wall. Schnell's slip length data was consistent with the analytical solution of Navier that predict the enhancement of the mass flow rate within the capillary tube. Historically, many of the pioneering investigations of non-continuum flows were conducted by researchers in the rarefied gas community who were primarily interested in low-pressure applications. Apart from a numerous experimental data still now there are no benchmarks experiments for identifying of the length-slip – the best candidates are the Couette and Poiseuille flows [7,20].

2. Dimensionless coefficients at a bulk and surface

Traditionally, the question of dimensionless numbers in fluid dynamics appears when the set of governing equations (mass, momentum, angular momentum, entropy and total energy) is undergoing of a sort of purely mathematical analysis in a co-called dimensionless form. The fundamental dimensionless number related with the integral form of equation of mass balance is the Reynolds number:

$$\mathrm{Re} = \dot{m}/(\pi d\, \mu) = \dot{m} d_n /(A\mu) \tag{4}$$

which is understood as a dimensionless mass flow rate $\dot{m}$. Here, in a case of rectangular cross section we use d_n - hydraulic diameter. Since for many flows viscosity μ weakly depends on the temperature and pressure, and since $\dot{m}_{in} = \dot{m}_{out}$, the Reynolds number can be interpreted as an integral (total) flow parameter. One should remember that in above definition the mass flow rate is defined via normal component of velocity $v_n = \mathbf{v} \cdot \mathbf{n}$, which ordinary is less than the velocity length $c = |\mathbf{v}| > v_n$. For open canals, where $\dot{m}$ cannot be determined, Re usually means dimensionless inflow velocity u_∞ [6].

2.1. Navier flow enhancement

For much longer tubes, for a slow (laminar) flow, yet another type of integral characteristics for different length and diameter of glass tubes has been experimentally found by Poiseuille [16] as a linear formula:

$$\dot{m}_{no-slip} = \dot{m}_{Poiseuille} = \frac{\pi \rho^2 d^4}{128 \mu L}(p_{in} - p_{out}) \sim \Delta p \tag{5}$$

It is a celebrated referential formula for no-slip resistance of flow in a capillary tube. Since in this formula only one coefficient μ is responsible for fluid resistance, then during the next century (refer with equation (5)) and Poiseuille-like experiments was a base for experimental executing the value of the internal viscous friction. In hydraulics it is known as a $\Delta pd^4 / L$ law.

The original slip Navier's solution [23] up to year 2000, was treated, in the literature, as wrong law $\Delta pd^3 / L$. However, it is also a linear in Δp, but only with a little bit grater coefficient of discharge such that:

$$\dot{m} = \dot{m}_{Navier} = \frac{\pi a^2 \rho}{\nu} \frac{\Delta p}{L} \frac{a}{2} \left(1 + \frac{\nu}{\mu} \frac{a}{2}\right)^{-1} = \frac{\pi a^3 \rho^2}{2\mu} \frac{\Delta p}{L} \left(1 + \frac{1}{2} \frac{a}{l_s}\right)^{-1} \tag{6}$$

which leads to the following flow enhancement (where $a = d$):

$$\eta_{\text{enh}} = \frac{\dot{m}_{Navier}}{\dot{m}_{Poiseuille}} = (1 + \frac{l_s}{8a})\left(1 + \frac{a}{2l_s}\right)^{-1} = (1 + Na/8)(1 + Na^{-1}/2)^{-1} . \tag{7}$$

It practically means that a notion of flow enhancement can't be discovered by Navier solely, since the referential no-slip rate of mass flow has been finding latter by Poiseuille. Nevertheless, the expression for $\dot{m}_{Navier}$ is a first case of integral characteristics where the slip length explicitly appears. In order to best appreciate the effect of a wall friction Hagen preformed experiments in long capillary tubes with pressure-driven flows of water of different temperature. He fined the following integral characteristics [11]:

$$\alpha_1 \dot{m}^2 + \alpha_2 L \dot{m} = a^4 \Delta p \tag{8}$$

where α_1 is a temperature-dependent constant related with some *vis-viva* flow resistance and α_2 is a temperature-independent constant. Perhaps because the concept of "*vis viva* wall friction" was treated to be mistaken, the credit to the discovery of a linear part has been often being given to Poiseuille and Hagen and is known as the "Hagen-Poiseuille law".

Quite similar linear characteristics for pressure driven flow in a long tube have been find by Helmholtz & von Piotrowski [4]:

$$\eta_{\text{enh}} = \frac{\dot{m}_{slip}}{\dot{m}_{Poiseuille}} = \left(1 + 4\frac{l_s}{a}\right) = 1 + 4Na \quad . \tag{9}$$

2.2. *Navier and Knudsen number at gases*

Extending this line of reasoning to taking into account an acceleration of fluid within a tube and the exact form of continuity equation, in work [1] have been find, a integral form of solution of one-dimensional Navier-Stokes equation with the appropriate Navier slip boundary condition. This characteristic written down in terms of "mass flow rate – drop of pressure" is as follows:

$$\dot{m} = \frac{H^3 W}{24 \mu L R T_{out}} \left[p_{in}^2 - p_{out}^2 + 12 Na_{out} p_{out} (p_{in} - p_{out})\right] \sim (\Delta p)^2 \tag{10}$$

where H, W, L are height, width and length of the channel, respectively, R is the specific gas constant, T_{out} is the outlet gas temperature, p_{in} and p_{out} are the pressures at the inlet and outlet of the channel, and the Navier number calculated on the outlet is expressed by Maxwell's clousure $Na_{out} = (2 - f) / f \, Kn_{out} = l_s / H$.

Ewart et al [14] have obtained the mass flow rate through a tube of diameter d by solving the Navier-Stokes equations with the second order slip condition in the following form:

$$\dot{m} = \frac{\pi d^4 p_m}{128 \mu R T L}(p_{in} - p_{out})\left(1 + 8Na_m + 32A_2 \frac{p_m}{p_{in} - p_{out}} \ln \frac{p_{in}}{p_{out}} Kn_m^2\right) \quad (11)$$

where Kn_m and $Na_m = A_1 Kn_m$ are the mean Knudsen and Navier number based on the mean pressure $p_m = 0.5(p_{in} + p_{out})$. Accordingly to the previous remarks, the second order part described by second slip coefficient A_2 cannot be expressed as a "square of Navier number" - it is rather described by a second Navier number. We should remember that the Navier number always is connected with a ratio of internal and external viscosities (refer with equetion 1). [13] have expressed the flow enhancement as:

$$\eta_{\text{enh}} = \frac{\dot{m}_{slip-2order}}{\dot{m}_{Poiseuille}} = 1 + 8Na_m + 16A_2 \frac{\mathcal{P}+1}{\mathcal{P} \;\; 1} \ln \mathcal{P} \, Kn_m^2 \quad (12)$$

where $\mathcal{P} = p_{in} / p_{out}$ is the inlet-outlet pressure ratio. This compact form of the flow enhancement is a proper for comparisons with the appropriate measured values [8,14]. For another integral flow characteristics that explicitly involve the slip length (or the dimensionless slip length) see: [2,10, 16,17].

3. Transition to turbulence

Stanton and Pannell, [26], making measurements of water flow within a capillary pipe, have proposed a change of the paradigm in approach to description of flow characteristics. They have courage to describe a results of measurements in a quite new way – they resigned with the celebrated pressure-discharge characteristics "$\Delta p - \dot{m}$", deciding to presents their own results as a diagram "dimensionless wall stress – dimensionless mass flow rate". Dimensionless wall stress, called the Stanton-Pannell friction factor, has been defined as a wall stress divided by the *vis viva* of a flow: $f_{SP} = \tau_w / (\rho U^2)$. Dimensionless mass flow rate they described as the Reynolds number to be: $Re = 4\dot{m}/(\pi d \mu) = \rho U d / \mu$. In this "$f_{SP} - Re$" chart the analytical solutions of one-dimensional Navier Stokes equation can easily be presented:

$$f_{SP} = \frac{\tau_w}{\rho U^2} = \begin{cases} \dfrac{4\mu}{a\rho U} = \dfrac{8}{Re} & \text{– no - slip solution} \\ \dfrac{4\mu}{a\rho U(1 + 4l_s / a)} = \dfrac{8}{Re(1 + 4Na)} & \text{– slip solution} \end{cases} \quad (13)$$

In contemporary literature instant of *vis viva* we use the kinetic energy which is two-time grater quantity, and such obtained dimensionless wall stress is called the Fanning friction factor : $f_F = 2 f_{SP}$. Frequently, according to Darcy, we apply four time grater coefficient, called the Darcy friction factor $f_D = 4 f_F = 8 f_{SP}$ [20], it leads to celebrated in the literature: $f_D = 64 / Re$.

In the laminar regime, the classical no-slip solution Hagen-Poiseuille $f_D = 64 / Re$ and the slip Helmholtz-Piotrowski solution $f_D = 64 / Re(1 + 4Na)$ are plotted. In the turbulent regime for smooth pipes: and the no-slip Blasius solution $f_D = 0.316 Re^{-0.25}$ is plotted as continuous lines. Some experimental results [19] for flow of gas within peek-coated, fused-silica microchannal is shown also. The friction factor is reduced due to slip at the walls as $f_{SP\text{-slip}} / f_{SP\text{-no-slip}} = 1/(1 + 4Na)$ [10]. For comparisons, slip analytical solution with $Na = l_s / a = Kn$ within the laminar regime is drowned by a dotted line. Unfortunately, up to now there is no analytical solution (like the Blasius one) for turbulent flow with slip.

Journal of Physics: Conference Series **530** (2014) 012035 doi:10.1088/1742-6596/530/1/012035

Due to unusual and rapid development and improvement of specific measurements techniques for the microfluids field, frequently we are ask about any refinement of analytical methodology in expressing the new results. Especially we are asking; do the classical "$f_D - Re$" diagram is sufficient for universal expressing of the specify phenomena of microflows [20].

One of a key element of $f_D - Re$ universality is appearance of a one critical point exactly in this same point on the diagram for every flow. Let recall, that this point appears also on "$\dot{m} - \Delta p$ diagram, thus in the classical literature by Du Buat, Prony, Darcy, Eytelwein there are exist some qualitative information about "transition" from a quiet form of flow into a loud one. Probably the first quantitative information about this point has been given and measured by Hagen [11] who used some kind of dimensionless critical number $Ha = (2gh)^{0.5}$ where h being the pressure head. Later Reynolds has considered another candidates for a universal number – finally expressing a criterion for laminar-turbulence transmission as: $Re_{cr} = 2250$.

For unique expressing of a critical point on a plane of $f_D - Re$ we need two coordinates. The second critical coordinate has been discovered by Stanton and Pannell [26], which have been written the following celebrated condition:

$$Re = \frac{\rho U_{cr} d}{\mu} \text{ is constant} \sim 2300 \qquad StPa = \frac{\rho d^3}{\mu^2} \frac{dp}{dx}_{|cr} \text{ is constant} \sim 0.004 \,. \tag{14}$$

The Stanton-Pannell number depends only on a drop of pressure and it can give a comfortable information about of type of flow yet in a moment of an experiment design. This number is connected with another dimensionless factor called the Poiseuille number $Po = f_D \cdot Re$ to be:

$$StPa = \frac{1}{2} Po\, Re = \frac{1}{2} f_D (Re)^2 \,. \tag{15}$$

From equation (13) follows that the critical Darcy friction factor always is : $f_D = 0.32$.

Enhancement of the Poiseuille number due to slip is $Po_{slip} / Po_{no-slip} = 1/(1+4Na)$. This result is consistent with [19] where for square channel they have obtained: $f_D \cdot Re = 56.9/(1+7.88Kn)$. Notice that, due to the fact that phenomenon of the laminas-to-turbulent transition do not occurs simultaneously in the whole channel, in practice, we are speaking on a "transition zone" or of the "transition regime" [7,9].

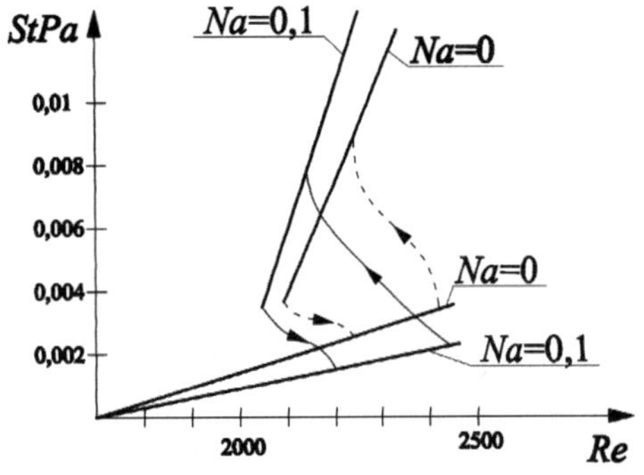

Figure 1. Dependence of the Stanton-Pannell velocity profile (the Stanton-Pannell number $StPa$) on a dimensionless mass flow rate (Reynolds number Re).

Do the $f_D - Re$ curve is enough for explaining universality of microflow phenomena ? Probably not. For instance, note that, a real non-measurable wall stress τ_w [treated as surface friction force per unit area] is replacement by means an analytical solution to the measurable fall of pressure dp/dx. Thus, practically, the Darcy friction factor f_D, has the same information what keep the Stanton-Pannell's number: $StPa$. Dependence of the Stanton-Pannell number $StPa$ on Navier number Na and Reynolds number Re is presented on figure 1. This figure is obtained on the basis Morini and Celata measurements [7,19,20].

4. Conclusions

Reassuming, the special role of Reynolds number coming from Reynolds' discovery that for geometrically similar glass tubes there was a critical mass velocity U_{cr} when stream-line or laminar in character flow at low values of the velocity, suddenly change in eddying or turbulent flow at high velocities. The special role of the Stanton-Pannell number follows from the discovery of the "critical drop of pressure". Both Reynolds and Stanton and Pannell have discovered, on making a series of observations, that for different geometries in the moment of changing of flow character from the laminar [stream-line] to the turbulent [eddying] that two numbers have still the value $Re_{cr} = 2300$ and $StPa_{cr} = 0.004$ (on Figure 1). This fact that there exists two dimensionless numbers proper for description of arbitrary flows within a frame of a one integral law of "flow resistance" nowadays is fundamental and is a base of paradigmatic view on mathematical modeling.

Unfortunately, an obstacle for a further progress of turbulence understanding is a logical mistake, which lays to identification the "wall stress" with a "surface friction force". Therefore, a reader can be astonished finding the main aim of many experimental works to be: "The object is to furnish evidence conforming the existence the laws of the surface friction of fluids." It is a mysterious "similarity of motion" anticipated by a numerous researchers. After discovery of Stanton and Pannell, the scientific word was absolutely astonish – because instant of numerous laws of pipe flows, separately proposed by Du Buat, Prony, Darcy, and other hydraulics engineers, there is only one for every sort of fluid and type and geometry of pipe. And, against the numerous 3D models – the notion of "wall friction" plays in this diagram the fundamental role. From this moment, the enigma of turbulence was identified with external eddies produced during a wall friction phenomena. In other words, if our turbulence model is based on a notion of turbulent wall stress then we must to consider an additional internal transport of momentum by a flux called the "Reynolds stress tensor". However, if our turbulence model is based on turbulent surface friction force concept when we should take into account the "*vis viva*" .

In turn, Stanton and Pannell were the first experimentalists who have believed in a theoretical anticipation of different profiles of fluid velocity at a pipe cross section, therefore they try to measure two kinds of velocities – the first one is the classical da Vinci mass velocity, measured by weight of the total discharge of fluid passing through the pipe in a given time, and the second one is the maximum velocity at the axis of the pipe, estimated by measuring the difference of pressure between that in a small Pitot tube facing the current and placement in the axis of the pipe and that in a small hole in the wall of the pipe. As we know from a basic consideration, this pressure difference is $\frac{1}{2}\rho u_{max}^2$ and from this relation the maximum velocity [speed] can be calculated. Everyone were strongly astonish looking on a plot of a velocity ratio $\varphi_{SP} = U / u_{max}$ has been made for different falls of pressure, expressed by a logarithm of dimensionless [Reynolds] number. Prior studies available in the literature, have no pay attention on the question of the Stanton-Pannell critical number during laminar to turbulent transition [12,7], but should be stated that nanochanels notion of Reynolds number and Stanton-Pannell number is meaningless and should be replaced by the Navier number. Additionally, a role of Navier number leas on expressing flow enhancement.

5. References

[1] Arkilic E B, Schmidt M A, Breuer K S 1997 Gaseous slip flow in long microchannels *J. Micro Electro Mechanical Syst.* **6** 167–178.

[2] Barber R W, Emerson D R 2006 Challenges in modeling gas-phase flow in microchannels: from slip to transition *Heat Transfer Engineering* **27** 3-12

[3] Beskok A, Karniadakis G E 1999 A model for flows in channels pipes and ducts at micro and nano scales *Microscale Thermophys. Eng.* **31** 43–77.
[4] Badur J, Karcz M, Lemański M, Nastałek L 2011 Foundations of the Navier-Stokes boundary conditions in fluid mechanics *Transactions of IFFM* **123** 3-55
[5] Badur J, Ziółkowski P, Ziółkowski P J 2014 Do we need the Navier number? *5th International Conference on "Heat Transfer and Fluid Flow in Microscale"* Marsylia: Campus St. Charles, France, 1-6.
[6] Badur J, Ziółkowski P 2013 Further remarks on the surface vis impressa caused by a fluid-solid contact *in Proceedings of the 12th Joint European Thermodynamics Conference, JETC 2013.* Brescia, Italy: Eds. M. Pilotelli and G.P. Beretta 581-586.
[7] Celata C P, Cumo M, McPhall S J, Tesfagbir L, Zummo G 2007 Experimental study on compressible flow in microtubes *Int. J. Heat Fluid Flow* **28** 28-36
[8] Colin S, Lalonde P, Caen R 2004 Validation of a second-order slip flow model in rectangular microchannels *Heat Trans. Eng* **25** 23-30
[9] Colin S 2005 Rarefaction and compressibility effects on steady and transient gas flows in microchannels *Microfluid Nanofluid* **1** 268-279
[10] Colin S 2006 *Single-phase gas flow in microchannels* [in:] S. Kandlikar et al, Heat transfer and fluid flow in minichanals and microchannals, Elsevier, pp 9-86
[11] Darrigol O 2002 Between hydrodynamics and elasticity theory: the first five births of the Navier-Stokes equation *Arch. Hist. Exact Sci.* **56** 95-150
[12] Demsis A, Prabhu S V, Agrawal A 2010 Influence of wall conditions on friction factor for flow of gases under slip condition *Experimental Thermal and Fluid Science* **34** 1448-1455
[13] Ewart T, Perrier P, Graur I, Méolans J G 2006 Mass flow rate measurements in gas micro flows *Experiments in Fluids* **41** 487-498
[14] Ewart T, Perrier P, Graur I, Méolans J G 2007 Tangential momentum accomodation in microtube *Microfluid Nanofluid* **3** 689-695
[15] de Gennes P, Brochard-Wyart F, Quere D 2004 *Capillarity and Wetting Phenomena* (New York: Springer)
[16] Karniadakis G E, Beskok A, Aluru N 2005 *Microflows and Nanoflows: Fundamentals and Simulation* (New York: Springer)
[17] Lauga L, Stone H A 2003 Effective slip in pressure-driven Stokes flow *J. Fluid Mech.* **489** 55–77
[18] Mooney M 1931 Explicit formulas for slip and fluidity *Jour. of Rheology* **2** 210-222
[19] Morini G L, M. Lorenzi M, Salvigni S, Spiga M 2009 Analysis of laminar to turbulent transition for isothermal gas flows in microchanells *Microfluid Nanofluid* **7** 181-190
[20] Morini G L, Yang Y, Chalabi H, Lorenzini M 2011 A critical review of the measurement techniques for the analysis of gas microflow through microchannels *Exp. Thermal Fluid Sci.* **35** 849-893
[21] Morris D L, Hannon L, Garcia A L 1992 Slip length in a dilute gas *Phys. Rev A* **46** 5279- 5282
[22] Myers T G 2011 Why are slip lengths so large in carbon nanotubes? *Microfluid Nanofluid* **10** 1141-1145
[23] Navier C L M H 1822 *Mémoires l'Acad. Royale des Sciences de l'Institut de France* **2** 375-393
[24] Priezjev N V, Troian S M 2004 Molecular origin and dynamic behavior of slip in sheared polymer films *Phys. Rev. Lett.* doi; 92.018.302
[25] Schnell E 1956 Slippage of water over nonwettable surfaces *J. Appl. Phys.* **271** 1149–1152
[26] Stanton T E, Pannell J R 1914 Similarity of motion in relation to the surface friction in fluids *Phil. Trans. Royal Soc. London* **214** 199-221
[27] Thomson P A, Troian S M 1997 A general boundary condition for liquid flow at solid surface *Nature* **389** 360-362
[28] To Q D, Bercegeay C, Lauriat G, Léonard D, Bonnet G 2010 *Microfluid Nanofluid* **8** 417–422

Study of unsteady natural convection induced by absorption of radiation based on a three-waveband attenuation model

T Hattori, J C Patterson and C Lei
School of Civil Engineering, The University of Sydney, Sydney, Australia

E-mail: tae.hattori@sydney.edu.au

Abstract. The present study considers unsteady natural convection induced by the absorption of radiation for possible applications in the water quality management for the shallow regions of lakes and reservoirs. The direct absorption of the incoming radiation by the water body forms a stable thermal stratification, whilst residual radiation reaching the bottom bathymetry is re-emitted as a boundary flux, forming an unstable thermal stratification, which is a potential source for a Rayleigh-Benard type instability. The bottom boundary layer instability drives intermittent vertical convection in the form of rising plumes. The plume rise is, however, limited by the stable thermal stratification due to the direct absorption, which is controlled by the attenuation coefficient of water. The attenuation coefficient is therefore an important parameter in determining the plume rise and the associated vertical mixing. The wavelength dependency of the attenuation coefficient of water is accounted for by using a three-waveband model. A theoretical prediction is made for the plume rise distance, which represents the region of vigorous mixing. Two-dimensional numerical simulation provides verification for the accuracy of the theoretical prediction.

1. Introduction

Natural convection has a significant impact on transport and mixing processes in the shallow regions of lakes and reservoirs [1, 2]. The radiative thermal forcing from the sun induces both horizontal exchange flow and vertical mixing flow in the water reservoirs. The direct absorption of the incoming radiation by a water body follows Beer's law and decays exponentially with depth, and therefore forms a stable thermal stratification, whilst residual radiation reaching the bottom bathymetry is absorbed and then re-emitted as a boundary flux, forming an unstable thermal stratification in that region [3], which is a potential source for a Rayleigh-Benard type instability. A steady horizontal exchange flow is induced by the differential heating of water columns with different local depths, whilst the bottom boundary layer instability drives intermittent vertical convection in the form of rising plumes. The present study focuses on the latter flow mechanism in isolation, and therefore uses a horizontal fluid layer with the spatial variation of the bathymetry being discounted. Further, the present study considers an idealised and controlled situation in laboratory-scale experiments, rather than a full-scale field situation, and assumes a constant radiation intensity at the water surface with a 90° incident angle, and the heat exchange between the water surface and the ambient being negligible.

As expected from the studies of laminar plume behaviour in a quiescent environment with a stable thermal stratification [4, 5], the plume rise is limited by the stable thermal stratification which forms due to the direct absorption of the incoming radiation. The stable stratification is governed by the decay rate of the radiation intensity, which is controlled by the attenuation coefficient of water. The attenuation coefficient is therefore an important parameter in determining the plume rise and the associated vertical mixing. While the attenuation coefficient depends on the turbidity of water and the wavelength of the radiation source [6], most of the existing and closely related literature assumes a single bulk value for this parameter. However, this simple assumption does not correctly model the absorption of the long wave radiation, which contains a significant portion of the total radiative energy. In the present study, a three-waveband attenuation model is developed based on a blackbody radiation distribution and the spectrum of the attenuation coefficient for water available in the literature [7]. A theoretical prediction is then made for the plume rise distance, based on a one-dimensional analytical solution for the temperature increase. The theoretical prediction is validated by two-dimensional numerical simulation of the flow.

2. Problem formulation

The flow under consideration is governed by the incompressible Navier-Stokes equations with the Boussinesq approximation and a volumetric heating source term in the energy equation, representing the direct absorption of the incoming radiation. The quantities in the governing equations are normalised by the following scales:

Length scale: $x_i, h \sim \frac{1}{\eta_0}$,

Time scale: $t \sim \frac{1}{\kappa\eta_0{}^2}$,

Temperature scale: $T \sim \frac{I_0}{\rho_0 C_p \kappa \eta_0}$,

Pressure scale: $p \sim \frac{g\beta I_0}{C_p \kappa \eta_0{}^2}$,

Velocity scale: $u_i \sim \kappa\eta_0$,

where u_i $(= u, v, w)$ is the velocity component in the x_i $(= x, y, z)$ direction, respectively. η_0 is the characteristic attenuation coefficient. The present study uses $\eta_0 = 10m^{-1}$, which is a typical bulk value for the PAR (photosynthetic active range) of the spectrum measured in lakes [6]. It is noted that η_0 is only used for the normalisation, hence the choice of η_0 does not affect the results and discussions in this paper. I_0 is the total surface intensity. g is the gravitational acceleration, β is the thermal expansion coefficient, ρ_0 is the density, C_p is the specific heat and κ is the thermal diffusivity. The thermal properties of water are assumed to be constant. The governing equations are then given in non-dimensional forms as follows:

$$\frac{\partial u_i}{\partial t} + u_j \frac{\partial u_i}{\partial x_j} = -RaPr\frac{\partial p}{\partial x_i} + Pr\nabla^2 u_i + RaPr\delta_{i,2}T, \tag{2.1}$$

$$\frac{\partial T}{\partial t} + u_i \frac{\partial T}{\partial x_i} = \nabla^2 T + \sum_{n=1}^{N} \frac{\eta_n}{\eta_0}\frac{I_n}{I_0} e^{\frac{\eta_n}{\eta_0}y}, \tag{2.2}$$

$$\frac{\partial u_i}{\partial x_i} = 0. \tag{2.3}$$

With the spectrum of the attenuation coefficient of water divided into N wavebands, η_n and I_n represent the average attenuation coefficient and average surface intensity over the nth waveband, respectively, i.e. $I_0 = \sum_{n=1}^{N} I_n$. Ra is the Rayleigh number $\left(= g\beta I_0/\rho_0 C_p \upsilon \kappa^2 \eta_0{}^4\right)$ and Pr is the Prandtl number $(= \upsilon/\kappa)$ fixed at $Pr = 7$ for water, where υ is the kinematic viscosity. $\delta_{i,2}$ is the Kronecker delta.

The present study assumes the initial condition of the flow to be stationary and iso-thermal. The radiation is instantaneously applied at $t = 0$, and maintained thereafter. The fluid layer is bounded by a stress-free, adiabatic top $(y = 0)$ and a rigid bottom $(y = -h)$ with a fixed flux condition associated with the re-emission of the residual radiation as mentioned earlier, which is given as:

$$\frac{\partial T}{\partial y} = -\sum_{n=1}^{N} \frac{I_n}{I_0} e^{-\frac{\eta_n}{\eta_0}h}. \tag{2.4}$$

3. Three-waveband attenuation model and plume-driven mixing

The spectrum of the attenuation coefficient of water obtained from the literature [7, 8] is shown in figure 1(a). The smallest attenuation coefficient is found in the visible range ($250 - 700nm$) of the spectrum, whilst longer waves are increasingly attenuated. Otanicar *et al.* [8] argues that the accuracy of their measurement is reduced for the visible range where the largest variation from the data of Hale and Querry [7] is found. The present study therefore uses the data of Hale and Querry [7]. In figure 1(b), the blackbody radiation distribution is shown for the theatre spot light ($3200K$) used as a radiation source in the laboratory-scale experiment [9] and the solar radiation ($5800K$). For the colour temperature of $5800K$, a large portion of the intensity is carried by the visible, less attenuated waves, whilst for the colour temperature of $3200K$, a large portion of the intensity is carried by the infra-red waves with increased attenuation. To determine the effect of the wavelength dependence on the attenuation coefficient, for each colour temperature, the spectrum of the attenuation coefficient is divided into $N = 50$ wavebands of equal radiation intensity, therefore the variation of the radiation intensity in a water body follows $I(y)/I_0 = \sum_{n=1}^{N=50} I_n/I_0\, e^{\eta_n y/\eta_0}$ with $I_n/I_0 = 0.02$. Further increasing the number of wavebands has a negligible effect on the intensity distribution. The least square fit of a three-waveband model, $I(y)/I_0 = \sum_{n=1}^{N=3} I_n/I_0\, e^{\eta_n y/\eta_0}$, to the $N = 50$ intensity distribution is performed to determine the values of I_n/I_0 and η_n/η_0, which are shown in table 1 for each colour temperature.

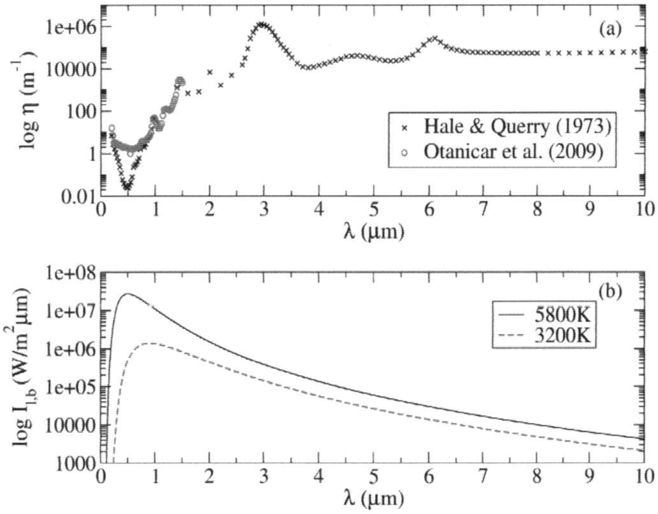

Figure 1. (a) Spectrum of attenuation coefficient [7, 8]; and (b) blackbody radiation distribution for the colour temperature of $3200K$ and $5800K$.

Table 1. Three-waveband models obtained for two colour temperatures.

Colour temperature	3200K			5800K		
	$n = 1$	$n = 2$	$n = 3$	$n = 1$	$n = 2$	$n = 3$
I_n/I_0	0.544	0.194	0.252	0.202	0.126	0.66
η_n/η_0	110.0	5.25	0.24	103.0	4.1	0.1

During the initial stage of the flow development, the flow is quiescent and the temperature variation is therefore purely diffusive with equation (2.2) reduced to:

$$\frac{\partial T}{\partial t} = \frac{\partial^2 T}{\partial y^2} + \sum_{n=1}^{N} \frac{\eta_n}{\eta_0}\frac{I_n}{I_0} e^{\frac{\eta_n}{\eta_0} y}. \tag{3.1}$$

Equation (3.1) and the associated boundary conditions ($\partial T/\partial y|_{y=0} = 0$ and equation (2.4)) is solved analytically using the method of eigenfunction expansion to obtain:

$$T(y,t) = \sum_{n=1}^{N} \frac{I_n}{I_0}\left\{\frac{t}{h} + \frac{h e^{-\frac{\eta_n}{\eta_0}h}}{3} + e^{-\frac{\eta_n}{\eta_0}h}\left[\frac{(y+h)^2}{2h} - (y+h)\right]\right.$$

$$\left. + \sum_{i=1}^{\infty} \frac{2h}{i^2\pi^2}\left[\frac{(-1)^i - e^{-\frac{\eta_n}{\eta_0}h}}{1 + \frac{i^2\pi^2}{\frac{\eta_n^2}{\eta_0^2}h^2}}\left(1 - e^{-\frac{i^2\pi^2 t}{h^2}}\right) - e^{-\frac{\eta_n}{\eta_0}h - \frac{i^2\pi^2 t}{h^2}}\right]\cos\left(\frac{i\pi(y+h)}{h}\right)\right\}. \tag{3.2}$$

In figure 2, equation (3.2) is plotted for $N = 50$ wavebands of equal intensity and the three-waveband model shown in table 1 for $h = 0.5$ and $h = 2.0$ at two different times for the colour temperatures of $3200K$ and $5800K$ respectively. Variations between the $N = 50$ waveband solution and the three-waveband solution are found to be marginal, therefore the three-waveband model is used in the subsequent analysis to account for the wavelength dependency of the attenuation coefficient. The formation of the stably stratified boundary layer below the surface ($y = 0$) is evidently associated with the absorption of long wave radiation, and is more substantial for the lower colour temperature of the radiation source. The formation of a potentially unstable boundary layer at the bottom ($y = -h$) is also dependent on the colour temperature; increased short wave radiation at the higher colour temperature penetrating to the bottom imposes an increased boundary flux.

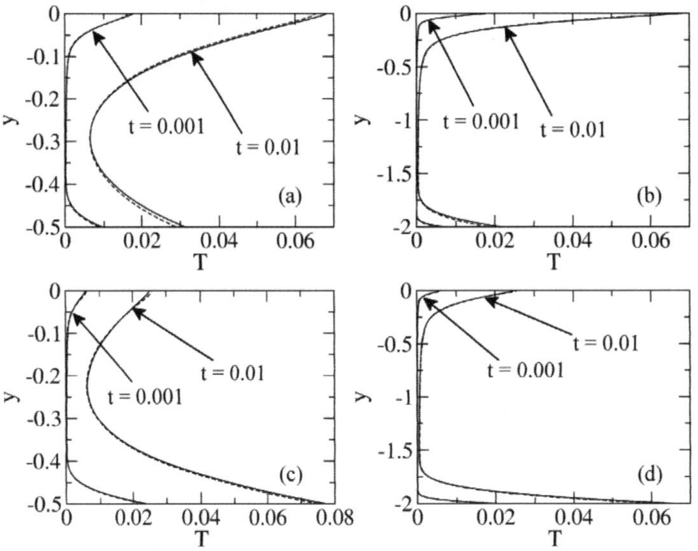

Figure 2. Temperature solution (3.2) for $N = 50$ wavebands of equal intensity (solid line) and the three-waveband model in table 1 (dashed line) at $t = 0.001$ and $t = 0.01$ for (a) $3200K$, $h = 0.5$; (b) $3200K$, $h = 2.0$; (c) $5800K$, $h = 0.5$; and (d) $5800K$, $h = 2.0$.

After the bottom boundary layer grows to a sufficient thickness, it becomes unstable to a Rayleigh-Benard type instability. Any small disturbance then grows exponentially with time, eventually leading to the formation of rising plumes [10]. As discussed earlier, the stable thermal stratification in the surface boundary layer limits the plume rise. In the present study, the term 'mixing depth' is used to quantify the plume rise distance from the bottom boundary since this length scale represents the region of vigorous mixing. The plume rise is considered to be determined by the location of neutral buoyancy (y_{nb}), which is obtained as follows:

$$T(y_{nb},t) = \frac{\int_{-h}^{y_{nb}} T(y,t)dy}{y_{nb}+h}, \tag{3.3}$$

where $T(y,t)$ is the analytical solution (3.2) and $l_m \equiv y_{nb} + h$ is the mixing depth. Equation (3.3) is solved numerically to obtain y_{nb} and hence l_m ($l_m = l_m(t)$ for a given h).

4. Numerical validation

The validation of the above theoretical prediction of the mixing depth is conducted via a two-dimensional numerical simulation of an idealised laboratory-scale experiment using a horizontal fluid layer subject to constant and uniform radiation at the water surface supplied by a theatre spot light ($3200K$) with the Rayleigh number in the range $10^7 \lesssim Ra \lesssim 10^9$. The numerical simulation uses the *SnS* code [11, 12], which is a non-staggered, Cartesian mesh, finite volume code based on a fractional step method [13, 14], with the Adams-Bashforth and Crank-Nicolson time discretisation schemes being used for the advection and diffusion terms, respectively. The spatial discretisation of the diffusion terms uses the second-order central differencing, whilst the spatial discretisation of the advection terms uses the second-order central differencing with the ULTRA (Universal Limiter for Tight Resolution and Accuracy) flux limiter [15]. A variable time step is used to maintain the CFL (Courant-Friedrichs-Lewy) number within the range 0.15~0.3; the time step is updated at every time step based on the CFL number.

The horizontal extent of the computational domain is $0 \le x \le 10$, and the vertical extent is $-h \le y \le 0$ with h varied in the range $0.5 \le h \le 2.0$ at a 0.5 interval. The boundary conditions imposed at the horizontal boundaries ($y = -h$ and $y = 0$) are as aforementioned, while the periodic condition with $f_{x=0} \equiv f_{x=10}$, where f is any flow quantity, is applied at the vertical boundaries. 2000 grid points are uniformly distributed in the horizontal (x) direction. In the vertical (y) direction, the grid size near the horizontal boundaries is $4 \times 10^{-4} < \Delta y < 5 \times 10^{-4}$ with 1%~3% linear stretching applied, giving at least 30 grid points within the bottom boundary layer after the flow becomes unsteady; the largest grid size is of $8 \times 10^{-3} < \Delta y < 9 \times 10^{-3}$ in the centre of the domain; and the total grid points is 200, 360, 540 and 700 for $h = 0.5$, 1.0, 1.5 and 2.0, respectively. A grid dependence test has shown that further increasing the number of grid points has only marginal effect on the solution accuracy.

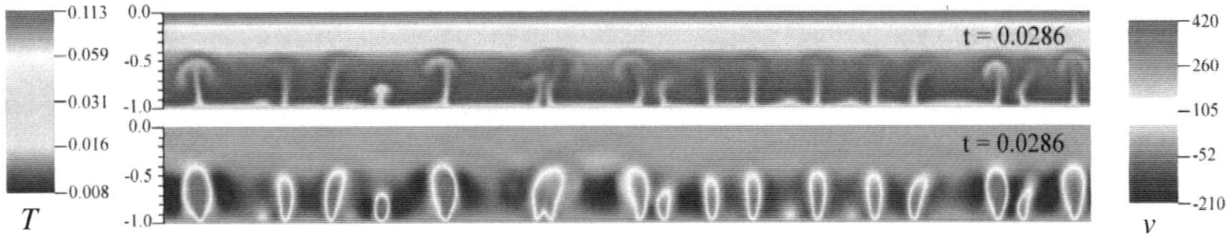

Figure 3. Instantaneous temperature field (top) and vertical velocity field (bottom) obtained for the case of $Ra = 10^8$ and $h = 1.0$ at $t = 0.0286$.

A typical flow structure obtained from the numerical simulation is shown in figure 3. The figure contains instantaneous temperature and vertical velocity fields for the case of $Ra = 10^8$ and $h = 1.0$ at $t = 0.0286$. The temperature field shows the formation of rising plumes from the bottom boundary layer and the stably stratified layer below the surface limiting the plume rise. The vertical velocity field also shows that the plume-driven convection is limited in the lower part of the flow domain and the upper part remains quiescent.

In figure 4(a), the theoretical prediction for the mixing depth, $l_m \equiv y_{nb} + h$, obtained by solving equation (3.3), is plotted over time for h in the range, $0.5 \le h \le 2.0$. It is shown that the mixing depth undergoes a rapid reduction with time and approaches a constant value for a given h, i.e. quasi-steady state. In figure 4(b), the vertical distribution of v_{dev}, where v_{dev} is the standard deviation of the vertical velocity over a horizontal line, $y = a\ (-h \leq a \leq 0)$, is shown for Ra in the range, $10^7 \leq Ra \leq 10^9$, and h in the range, $0.5 \leq h \leq 2.0$, at different times. The abscissa is normalised by the maximum value over the depth. The ordinate is translated by h to fix the location of the bottom boundary for different cases and is then normalised by the theoretical l_m plotted in figure 4(a). It is shown in this figure that v_{dev} rapidly reduces as $(y + h)/l_m \to 1$ and is negligible for $(y + h)/l_m > 1$ in all the cases. The accuracy of the theoretical prediction based on the location of neutral buoyancy in equation (3.3) is therefore verified.

5. Conclusions

Unsteady natural convection induced by the absorption of radiation has been investigated in the present study for possible applications in the water quality management for the shallow regions of lakes and reservoirs. An unstable thermal stratification which forms due to the absorption and re-emission of residual radiation by the bottom bathymetry triggers convective instability, which then drives intermittent vertical convection in the form of rising plumes. The plume rise is, however, limited by a stable thermal stratification which forms due to the direct absorption of the incoming radiation and is controlled by the attenuation coefficient of water. The attenuation coefficient is therefore an important parameter in determining the plume rise and the associated vertical mixing. The wavelength dependency of the attenuation coefficient of water has been accounted for by using a three-waveband model for a given colour temperature of the radiation source. A theoretical prediction has been made for the plume rise distance, i.e. 'mixing depth', based on a one-dimensional analytical solution for the temperature increase. The corresponding two-dimensional numerical simulation has provided verification for the accuracy of the theoretical prediction.

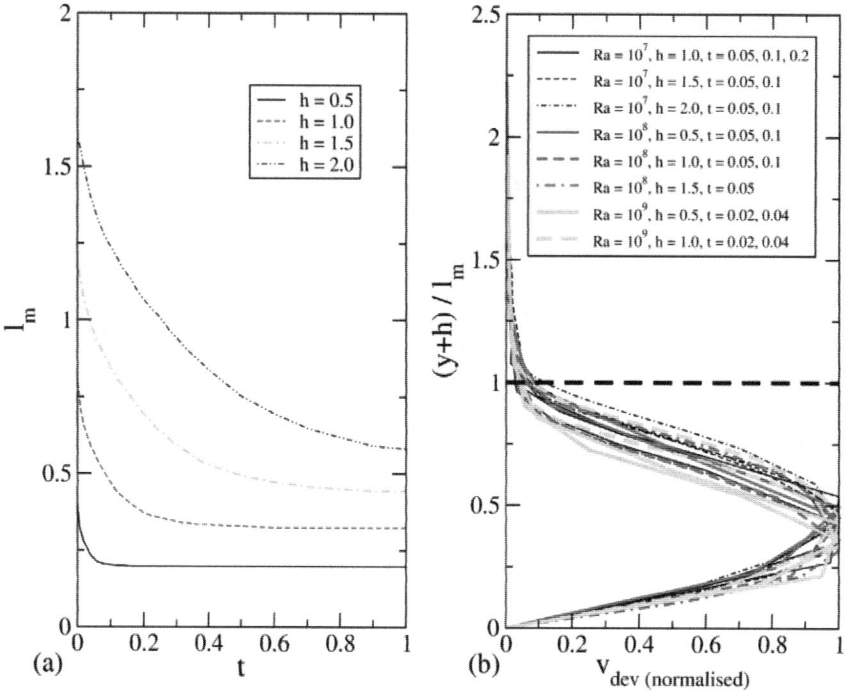

Figure 4. Numerical validation of the theoretical prediction of the mixing depth based on the location of neutral buoyancy in equation (3.3). (a) Theoretical prediction for the mixing depth, $l_m = l_m(t)$. (b) The vertical distribution of v_{dev} for Ra in the range, $10^7 \leq Ra \leq 10^9$, and h in the range, $0.5 \leq h \leq 2.0$, at different times.

6. References

[1] Adams E E and Wells S A 1984 *J. Hydraul. Eng.* **110** 773
[2] Monismith S G, Imberger J and Morison M L 1990 *Limnol. Oceanogr.* 1676
[3] Farrow D E and Patterson J C 1993 *Int. J. Heat Mass Trans.* **36** 89
[4] Morton B R 1967 *Phys. Fluids* **10** 2120
[5] Wirtz R A and Chiu C M 1974 *Int. J. Heat Mass Trans.* **17** 323
[6] Kirk J T O 1994 *Light and photosynthesis in aquatic ecosystems* (Cambridge: Cambridge University Press)
[7] Hale G M and Querry M R 1973 *Appl. Opt.* **12** 555
[8] Otanicar T P, Phelan P E and Golden J S 2009 *Solar Energy* **83** 969
[9] Lei C and Patterson J C 2002 *Exp. Fluids* **32** 590
[10] Lei C and Patterson J C 2003 *J. Fluid Mech.* **480** 161
[11] Armfield S W, Norris S E, Morgan P and Street R 2002 A parallel non-staggered Navier-Stokes solver implemented on a workstation cluster *Proc. the International Conference on Computational Fluid Dynamics (Sydney, Australia, July 2002)* pp 30-45
[12] Norris S E 2000 *A parallel Navier-Stokes solver for natural convection and free surface flow* (PhD thesis, University of Sydney)
[13] Armfield S W 1994 *J. Comput. Phys.* **114** 176
[14] Armfield S W and Street R 2000 *ANZIAM Journal* **42** C134
[15] Leonard B P and Mokhtari S 1990 *Int. J. Numer. Methods Eng.* **30** 729

Acknowledgments

This research was financially supported by the Australian Research Council (DP120104849) and The University of Sydney.

Numerical investigation of the Taylor-Couette and Batchelor flows with heat transfer: physics and numerical modelling

K Kiełczewski[1]**, E Tuliszka-Sznitko**[1] **and P Bontoux**[2]
[1]Institute of Thermal Engineering, Poznań University of Technology, ul. Piotrowo 3, 60-965 Poznań, Poland
[2]M2P2 Laboratory, UMR 7340 CNRS, Aix-Marseille Université, Ecole Centrale Marseille, 13451 Marseille Cedex 20, France

E-mail: ewa.tuliszka-sznitko@put.poznan.pl

Abstract. In the paper the authors present the results obtained during a numerical investigation (Direct Numerical Simulation/Spectral Vanishing Viscosity method - DNS/SVV) of a flow with heat transfer in rotating cavities (i.e. the flow between two concentric disks and two concentric cylinders). These model flows are useful from numerical and experimental point of view among others because of the simplicity of their geometry. Simultaneously, the flows in rotating cavities appear in numerous industrial installations and machines in the field of mechanics and chemistry, e.g., in ventilation installations, desalination tanks and waste water tanks, in cooling system, in gas turbines and axial compressors. In the paper attention is focused on the laminar-turbulent region in the configuration of the large aspect ratio i.e. Taylor-Couette flow (a Batchelor flow case of small aspect ratio $\Gamma = 0.04$ is also presented for comparison). The main purpose of computations is to investigate the influence of different parameters (the aspect ratio, the end-wall boundary conditions and temperature gradient) on the flow structure and flow characteristics. For the non-isothermal flow cases the Nusselt number distributions along cylinders are presented and are correlated with the flow structures. The λ_2 method has been used for visualization.

1.Introduction

The flows in rotating cavities, i.e. the flows between two concentric cylinders and two concentric disks, are the simplest possible wall flows which show most of the phenomena the understanding of which enables us to reconstruct the 3D flows in more complex industrial configurations, [1]. These flows are particularly suitable for investigating the phenomena which occur in gas turbines, ventilations, chemical mixing equipment in geophysics and astrophysics. The flow in rotating cavity is characterized by four main parameters: aspect ratio $\Gamma = 2h/(R_2 - R_1)$, curvature parameter $Rm = (R_2 + R_1)/(R_2 - R_1)$ or radius ratio $\eta = (Rm-1)/(Rm+1) = R_1/R_2$, Reynolds number $\mathrm{Re} = (R_2 - R_1)R_1\Omega/\nu$ and thermal Rossby number $B = \beta\Delta T$. The aspect ratio, curvature parameter, Reynolds number and thermal Rossby number create a large parameter space very rich in physical phenomena. For configurations of the small aspect ratio Γ attention was focused on the flows between rotating and stationary disks, in which the flow is pumped radially outward along rotor and recirculates along stator. When two boundary layers are merged, i.e. when the gap between disks is very thin ($\Gamma < 0.02$, so called torsional Couette flow) different types of finite-size turbulent structures (turbulent spots, turbulent spirals) appear with the gradually increasing Reynolds number, which leads finally to turbulent flow. In this flow case the influence of the inner and outer cylinders boundary conditions on the flow structure is small. When the gap between disks is more than the thickness of two disks' boundary layers (the boundary layers are separated by the inviscid rotating core, $1 > \Gamma > 0.02$) the flow is named Batchelor. In the Batchelor flow the stationary disk boundary layer is more unstable than the rotating one; the transition process stars with the occurance of circular waves in the stator boundary layer that propagate inward [2]. At higher values of the Reynolds number, another unstable mode appears [2], [3], [4], which creates a spiral pattern (spirals co-exists with the circular waves for a range of Reynolds numbers). The heat transfer in the transitional and turbulent rotor/stator cavity was also discussed in many papers [5], [6], [7], [8].

With the increasing aspect ratio the impact of the inner and outer cylinder boundary conditions on the flow structure gradually increases. The flow between two concentric disks and two concentric cylinders of aspect ratio $\Gamma \geq 1$ is named Taylor-Couette flow. In its simplest form, the Taylor-Couette flow arises from the shear flow between a rotating inner cylinder and a stationary outer cylinder (the stable, laminar flow for this configuration is known as cylindrical Couette flow). Beyond a certain critical Reynolds number, the Couette flow becomes unstable, which results in the appearance of pairs of counter-rotating, axisymmetric vortices (TVF Taylor vortex flow) that fill in the annulus (first bifurcation, figure 1a). Each pair of vortices has an axial wavelength, figure 1a. For higher Reynolds numbers a supercritical Hopf bifurcation leads to wavy vortex flow, referred here as WVF, the movement of which consists of azimuthally traveling along the Taylor-Couette vortices waves. However, the flow state for wavy vortex flow depends strongly on how the final rotational speed of the inner cylinder is achieved: impulsively or quasi-statically. With further increase of Reynolds number the modulated waves state (MWV), the chaotic and turbulent Taylor vortex flows appear. It was shown in [9] that the spatial and temporal behavior of modulated waves are directly connected to the outflow jet of the Taylor-Couette vortices.

Experimental measurement and flow visualization have traditionally been the main source of information about transitional and turbulent Taylor–Couette flows, [10], [11]. However, recently the Taylor-Couette flow is also intensively investigated by DNS and LES methods, [12], [13], [14]. In many numerical investigations the assumption of infinitely long annuli, with the periodicity condition in axial direction, has been undertaken. But, the presence of end-walls (even in the case of large height of cylinders) destroys the translation invariance in z direction, which exists in the theoretical infinitely long annuli. In experimental investigations, the presence of the end-walls strongly influences the flow structures by constraining the axial motion near the end-walls and by causing Ekman pumping. For the flow cases with the rotating inner cylinder, the outer cylinder fixed and the end-walls rotating with the inner cylinder, centrifugal viscous pumping causes an outflow movement of the fluid at the end-walls. For the end-walls attached to the stationary outer cylinder, the inflow movement at the end-walls occurs. Asymmetric boundary conditions at the end-walls (stationary top and rotating bottom disks, for example) were also investigated [15], but these flow cases are not fully documented. The importance of the top and bottom disks boundary conditions on selecting the flow pattern (including problem of anomalous modes) was investigated among others in [15], [16].

The heat transfer in Taylor-Couette flow has also been an object of great interest for scientists and engineers, [11], [17] and [18]. Investigations have shown, among others, that above the critical threshold, the transport of fluid from one cylinder to the other, enhances the heat transfer. In [19] the authors have found the second, less pronounced, transition altering the evolution of heat transfer. This thermal transition coincides with the shortest wavelength in axial direction, [20]. The review of papers on the Taylor-Couette flow with heat transfer was done recently in [21]. In [21] the authors stated that, although the problem of the heat transfer in the Taylor-Couette flow is well documented, there are still areas which need further research, for example, fluid flows in configurations of large cylindrical gaps.

In the present paper the authors show transitional and turbulent results obtained numerically (using the DNS/SVV method) in the rotating cavity of the aspect ratios from the range $\Gamma = 0.04 - 20.0$. In the paper, the authors propose to research the influence of the presence of the end-walls (symmetric and asymmetric boundary conditions) together with the investigation of the influence of temperature variation on the Taylor-Couette flow. The authors propose to study the phenomena in Taylor-Couette flow for multy-parameter space which should help to answer the fundamental question concerning transitional structures, their origins, their role in creating stress and transporting energy, and the relationship between momentum and heat transport. This knowledge can help to control the boundary layer flow in more complex configurations. The results presented in the paper are the authors' first step towards such formulated scientific purpose. The presented Taylor-Couette calculations are obtained for relatively low Reynolds numbers, however, computations are still being continued in order to obtain a higher Re. In order to perform calculations in conditions that would be similar to the experimental ones the Reynolds number is gradually increased by $\Delta \mathrm{Re} = 20.0$ and computations are performed for large global time units up to 2000.0, i.e. in terms of Ω^{-1}. The authors are going to confirm experimentally at least a part of obtained results in the frame of EUHit project.

The paper is organized in the following way: The mathematical and numerical approach used in investigations are described in section 2. The selected results obtained for the Batchelor flows (discussed in detail in [8]) are shortly presented in section 3.1. Different stages of the Taylor-Couette laminar-turbulent transition flow obtained for $\Gamma = 11.75$ are presented in section 3.2. In section 3.3 the exemplary Taylor-Couette flow structures obtained for the symmetric and asymmetric end-wall boundary conditions are presented. For the flow case of $\Gamma = 20.0$ the results with heat transfer are shown and distributions of the local Nusselt number along the inner and outer cylinders are analyzed, section 3.4. Conclusion remarks are given in section 4.

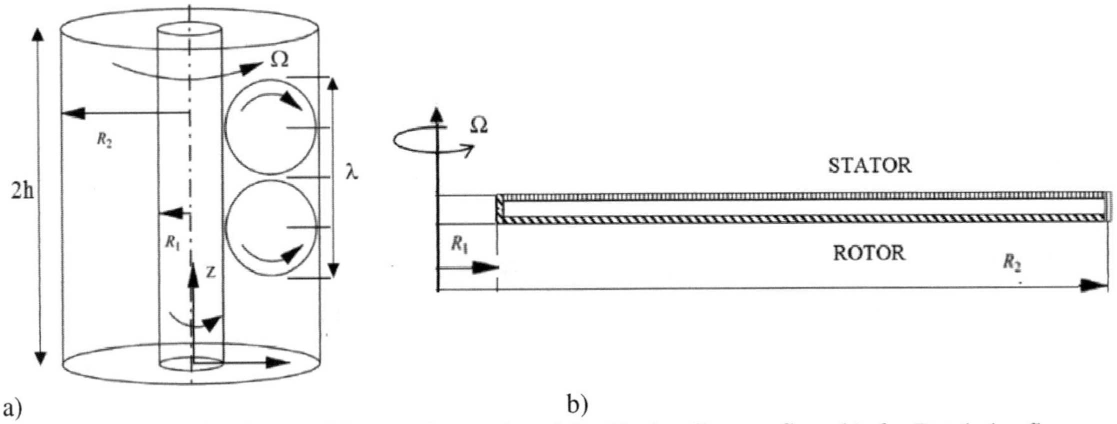

a) b)

Figure 1. Schematic picture of the rotating cavity, a) for Taylor-Couette flow, b) for Batchelor flow.

2. Mathematical and numerical approach

In the paper the authors investigate the isothermal and non-isothermal fluid flows in the rotating cavity of the inner and outer cylinder radius R_1 and R_2, respectively (figure 1a and 1b). The inter-disks spacing is denoted by $2h$. The rotor (the inner cylinder and in some flow cases also the bottom disk) rotates at uniform angular velocity $\boldsymbol{\Omega} = \Omega \boldsymbol{e}_z$, where $\boldsymbol{e}_z$ is the unit vector on the z axis. The flow is described by the Navier-Stokes, continuity and energy equations written in a cylindrical coordinate system (R, φ, Z). The equations are written with respect to rotating frame of reference. The Boussinesq approximation is used to take into account the buoyancy effects induced by the involved body forces. The dimensionless axial and radial coordinates are: $z = Z/h,\ z \in [-1, 1]$, $r = (2R - (R_2 + R_1))/(R_2 - R_1),\ r \in [-1, 1]$. The velocity components and time are normalized by ΩR_2 and Ω^{-1}, respectively. The dimensionless components of the velocity vector in radial, azimuthal and axial directions are denoted by $u = U/\Omega R_2,\ v = V/\Omega R_2,\ w = W/\Omega R_2$. In the non-isothermal flow cases the dimensionless temperature is defined in the following way: $\Theta=(T-T_1)/(T_2-T_1)$, where T_1 is the temperature of the cooled surface (the inner cylinder and the top disk, for example), and T_2 is the temperature of the warmed surface (the bottom or both disks and the outer cylinder). The Prandtl number Pr = 0.71 and the thermal Rossby number $B=\beta(T_2-T_1)$ are constant (β is thermal expansion coefficient). The no-slip boundary conditions are applied to all rigid walls ($u=w=0$). For the azimuthal velocity component the boundary conditions are: v=0 on the rotating inner cylinder and on the rotating disk, and $v=-(Rm+r)/(Rm+1)$ on the stationary outer cylinder and on the stationary disks. The time scheme is second-order semi-implicit, which combines an implicit treatment of the diffusive term and an explicit Adams–Bashforth scheme for the non-linear convective terms. The spatial scheme is based on a pseudo-spectral Chebyshev-Fourier-Galerkin collocation approximation. All presented results are obtained with the use of DNS/SVV method (using meshes of 2-30 million collocation points), [7], [8], [23].

3. Results

3.1 Batchelor flows

The authors have performed broad numerical investigations of the flow between rotor and stator with and without heat transfer for different aspect ratios $\Gamma=0.04-0.5$ (Batchelor flows), different curvature parameters, and for different Reynolds numbers (including fully turbulent flows). The results were published, among others, in [8]. These investigations delivered many correlational formulas, which can be useful for scientists and engineers. The exemplary temperature field obtained for Re=163257, $\Gamma=0.04$, *Rm*=1.8, *B*=0.1 is presented in figure 2. The flow is pumped radially outward along the rotating bottom disk and recirculates along the stationary top disk (the inner cylinder is attached to the rotor and the outer one to the stator, the rotor and the outer cylinder are heated). In the considered flow case the stator boundary layer is fully turbulent, whereas the rotor one is turbulent in the area near the outer cylinder (*r* = 0.0 - 1.0). The axial distributions of the main Reynolds stress tensor components normalized with the wall units (obtained for radius *r* = 0.0 and different Re, stator boundary layer) as a function of z^+ are presented in figure 3 together with the wall asymptotes and [22] numerical results. The definition of Reynolds number in figure 3 is based on R_2, $\mathrm{Re}=R_2^2\Omega/\nu$ and $L=1/\Gamma$. The influence of the cylinders boundary conditions on the flow structure for aspect ratio 0.04 is small, however, the computations performed by the authors for higher Γ have shown that with an increasing aspect ratio this influence increases significantly.

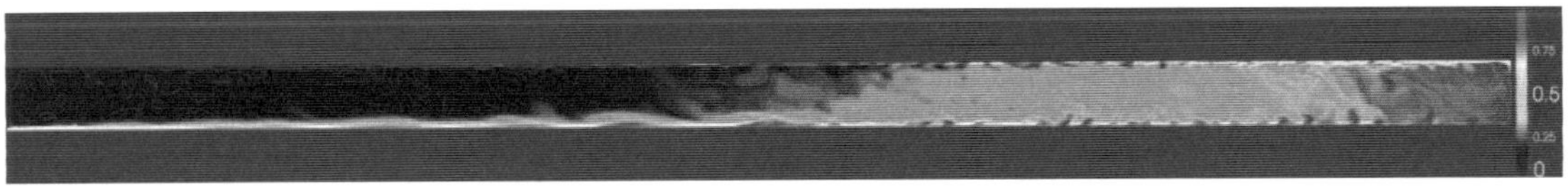

Figure 2. The meridian temperature field obtained for Re=163257, *L*=0.04, *Rm*=1.8, *B*=0.1.

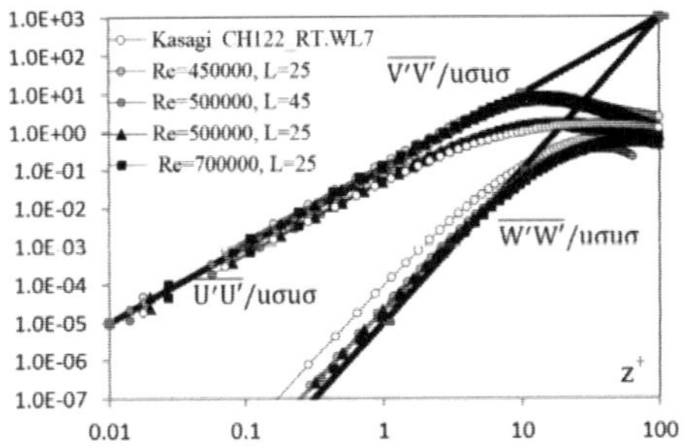

Figure 3. The axial profiles of the three main Reynolds stress tensor components normalized with wall units versus the wall coordinate obtained for different Re and $L=1/\Gamma$. Definition of Re is based on R_2, $\mathrm{Re}=R_2^2\Omega/\nu$. *Rm*=1.8, *B*=0.1.

3.2 Different stages of Taylor-Couette vortices

Figure 4 shows the exemplary 3D flow structures obtained for higher aspect ratio equal $\Gamma = 11.75$ and for the radius ratio 0.9 (*Rm*=19, isothermal flow case); both end-walls are attached to the outer stationary cylinder. In figure 4 the iso-surfaces of instantaneous λ_2 are presented. The comparison of the results obtained for different Reynolds numbers shows that for Re smaller than 80 only Ekman vortices at the end-walls are present in the cavity (flow is pumped radially inward along end-walls). At higher Re the Taylor vortices gradually appear and the Ekman vortices are squeezed. The vortex flows obtained for Re = 139, 248, 342 and 540 are presented in figure 4a, 4b, 4c and 4d, respectively. For Re=139 the stationary Taylor vortex flow is observed (figure 4a) but for higher Re the modulated waves state occurs (figure 4b, 4c and 4d). The flow is dominated by the outflow jet, which is driven directly by the centrifugal force and throws fluid outward from the rotating inner cylinder. The inflow jet is broader than the outflow jet and at the same time is weaker than the outflow jet (the outflow jet narrows with the increasing Re). Figure 5 presents time histories of the azimuthal velocity component obtained at the monitoring points (r=-0.871, z=0.718, $\varphi = 0.0$). The instability characteristic obtained for Re=199.8 (figure 5a) shows that the rotation change initiates a disturbance characterized by a wave packet which is rapidly damped (reaching a steady state) and then (t >350) grows exponentially. At Reynolds number Re=248, the disturbances are oscillatory from the beginning of the time history with constant amplitude.

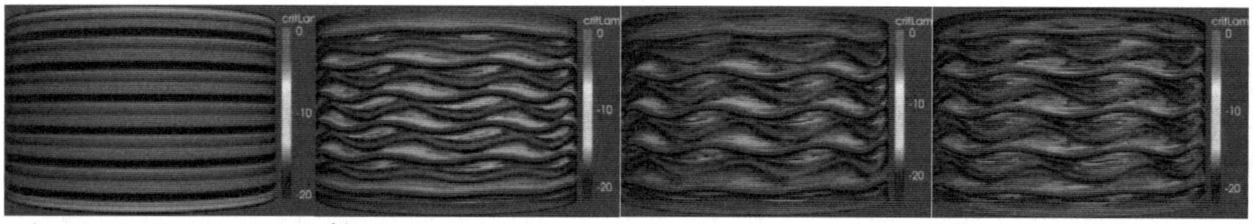

a) b) c) d)

Figure 4. The iso-surfaces of instantaneous λ_2 obtained for Reynolds numbers: a) Re=139, b) 248, c) 342, d) 540, $\Gamma = 11.75$, Rm=19.

Table 1. Geometrical and physical parameters

	Γ	Rm	$\eta = R_1 / R_2$
1	20	3	0.5
2	12	19	0.9
3	11.75	19	0.9
4	4	4.2	0.615
5	2.35	3	0.5
6	1.88	2.2	0.375
7	0.68	2.2	0.375

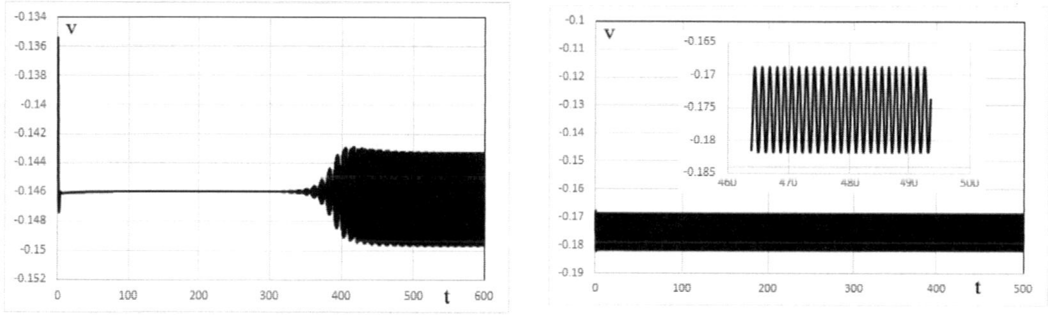

a) b)

Figure 5. Instability characteristics obtained at the point (r=-0.871, z=0.718, $\varphi = 0.0$), the azimuthal velocity component as a function of time, a) Re=199.8, b) Re=248.4.

3.3 Influence of the end-walls

The authors consider two types of boundary conditions at the disks: 1) configurations with both end-walls attached to the stationary outer cylinder, and 2) with the top disk attached to the outer stationary cylinder and the bottom one attached to the inner rotating cylinder. These two types of the boundary conditions have been applied to all flow cases listed in table 1. The influence of the end-wall boundary conditions on the Taylor-Couette flow structure is shown in figure 6, where the meridian flow and the iso-lines of the azimuthal velocity component obtained with the symmetric and asymmetric boundary conditions are presented. In figure 6 two flow cases are analyzed: 1) $\Gamma = 4$, Re=530, 2) $\Gamma = 1.88$, Re=633. In the flow cases with the symmetric boundary conditions the onset of instability process gives rise to a primary mode with an even number of cells. The direction of the rotation at both disks is consistent with the Ekman vortex, so there is an inward flow towards the rotating inner cylinder. The symmetry of the flow in the midplane is fully preserved. When the bottom disk is attached to the rotating cylinder and the top one to the stationary outer cylinder there is an inward flow towards the rotating inner cylinder at the stationary top disk and an outward flow towards the stationary outer cylinder. In these flow cases there is an odd number of vortices. The cavity with the asymmetric boundary conditions was investigated experimentally in [24] where the authors observed interaction between three and five cells, and in [15] where interaction between one and three cells was observed. The steady states with the odd number of cells and so called anomalous modes are discussed in [15], [16], [25].

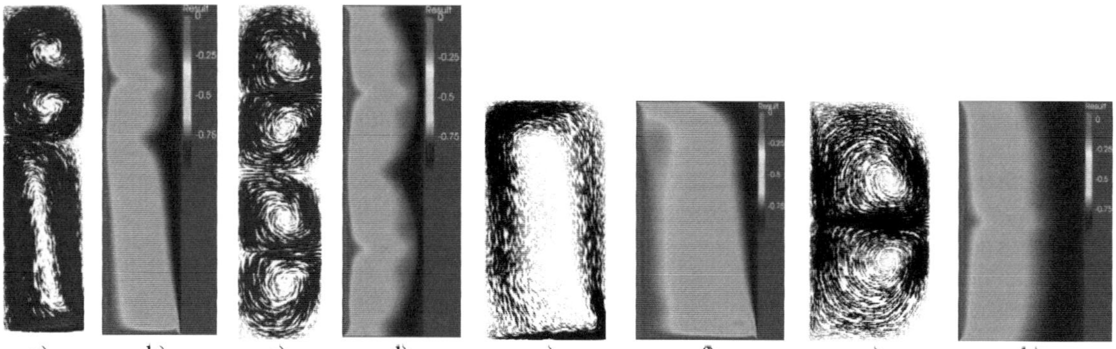

a) b) c) d) e) f) g) h)

Figure 6. The instantaneous velocity field in meridian section and the iso-lines of the azimuthal velocity component obtained for: a and b) $\Gamma = 4$, Re=530, asymmetric boundary conditions at the disks, c and d) $\Gamma = 4$, Re=530, symmetric boundary conditions at the disks, e and f)

$\Gamma = 1.88$, Re=633, asymmetric boundary conditions, g and h) $\Gamma = 1.88$, Re=633, symmetric boundary conditions.

3.4 Taylor-Couette flow with heat transfer
Figure 7 presents the results obtained by the authors for non-isothermal fluid flow in the cavity of aspect ratio $\Gamma = 20.0$, radius ratio $\eta = 0.5$ and Reynolds numbers Re=500.0 and 300.0 (the outer stationary cylinder and two stationary end-walls are heated). The iso-lines of temperature in figure 7a show the presence of 24 vortices (shorter wavelength in axial direction in comparison to the isothermal flow case). The distributions of the local Nusselt number obtained for Re=500.0 and 300.0 along the outer heated cylinder and inner one are presented in figure 7b and 7c, respectively. The maximum values of local Nusselt numbers along the outer cylinder are observed in the areas of outflow jets (figure 7c), whereas the maximum values along the inner cylinder are observed in the areas of inflow jets. The peaks of Nu along the outer cylinder are sharper than the peaks along the inner cylinder. This results from the fact that the outflow jets are thinner than inflow jets. The similar analysis was done in [18].

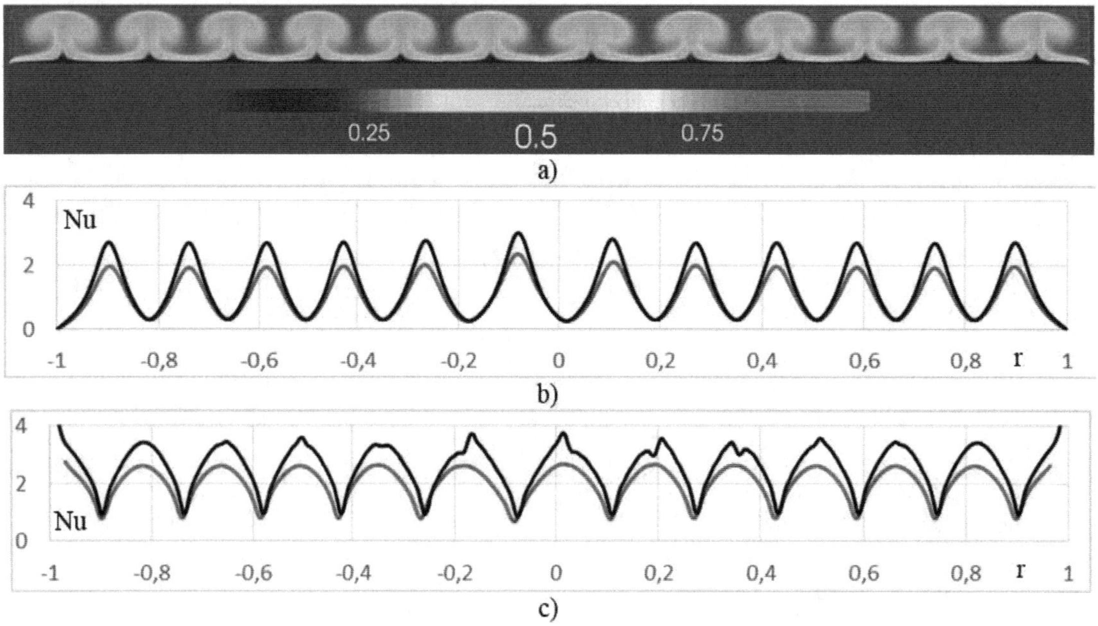

Figure 7. Numerical results obtained for $\Gamma = 20.0$, radius ratio $\eta = 0.5$ and Re=500.0 (black line) and 300.0 (red line), a) the temperature field in meridian section, b) the distributions of the local Nusselt number along the outer stationary cylinder, c) the distributions of the local Nusselt number along the inner rotating cylinder.

4. Conclusions
In the paper the authors presented the results obtained numerically (DNS/SVV) during the investigations of the flow with the heat transfer in rotating cavity of the aspect ratio from the range 0.04-20.0, for different curvature parameters Rm, different Reynolds numbers, for symmetric and asymmetric boundary conditions at the end-walls and for different thermal conditions. Attention was focused on the transitional results obtained for the Taylor-Couette flow. The computations performed for symmetric and asymmetric boundary conditions at the disks ($\Gamma = 12, 11.75, 4.0, 2.35, 1.88$) showed that the asymmetric boundary conditions break the mirror-plane flow symmetry, which is preserved in symmetric boundary conditions cases. For configuration of $\Gamma = 4.0$ the authors observed

4 Taylor-Couette vortices for the symmetric and 3 vortices for asymmetric end-walls conditions. For configurations of $\Gamma = 1.88$ the authors obtained 2 and 1 vortices respectively for the symmetric and asymmetric conditions. For non-isothermal flow case ($\Gamma = 20.0$, the symmetric boundary conditions at the disks) the authors observed shorter wavelength in axial direction in comparison to the isothermal flow case. The local Nusselt number distributions along the inner and outer cylinders were presented and discussed in the light of the flow structure. The computations will be continued for higher Reynolds numbers and also for modulated in time boundary conditions.

Acknowledgment
Computations were performed in the Poznań Supercomputing and Networking Center, which is gratefully acknowledged.

References
[1] Hanjalic K and Launder B 2011 Modelling Turbulence in Engineering and the Environment *Cambridge*
[2] L. Schouveiler L, Le Gal P, Chauve M and Takeda Y 1999 *Exp. Fluids* **26** 179
[3] Serre E, Tuliszka-Sznitko E and Bontoux P 2004 *Phys. Fluids* **16** 688-707
[4] Serre E, Crespo del Arco E and Bontoux P 2001 *J. Fluid Mech.* **434** 65
[5] Randriamampianina A, Bontoux P and Roux B 1987 *Int. J. Heat Mass. Transfer* **30** 1275-1292
[6] Poncet S and Schiestel R 2007 *Int. J. of Heat and Mass Transfer* **50** 1528
[7] Tuliszka-Sznitko E, Majchrowski W and Kiełczewski K 2012 *Int. J. Heat and Fluid Flow* **35** 52
[8] Tuliszka-Sznitko E and Kiełczewski K 2014 CRC Press Taylor and Francis Group 369
[9] Coughlin K and Marcus P 1996 *Phys. Rev. Lett.* **587** 373
[10] Egbers Ch and Pfister G 2000 Physics of rotating fluids *Springer*
[11] Harlander U, Wright G and Egbers Ch 2012 *Geophysical Research Abstracts* **14** EGU2012-5368
[12] Czarny O, Serre E and Bontoux P 2002 *Theor. and Comp. Fluid Dynam.* **16** 5
[13] Dong S 2007 *J. Fluid Mech.* **587** 373
[14] Brauckmann H and Eckhardt B 2013 *J. Fluid Mech.* **718** 398
[15] Mullin T and Blohm C 2001 *Phys. Fluids* **13** 136
[16] Benjamin T and Mullin T 1981 *Proc. R. Soc. Lond.* A **377** 221
[17] Poncet S, Haddadi S and Viazzo S 2010 *Int. J. Heat and Fluid Flow* **32** 128
[18] Ghayoub N, Bertin Y and Saulnier J 1992 Caractérisation des transferts de chaleur dans un espace annulaire avec cylindre intérieur tournant: écoulement de type CouetteeTaylor in SFT 223
[19] Becker K and Kaye J 1962 *J. Heat Transfer* **84** 106
[20] Peres I 1995 Contribution à l'analyse de l'ecoulement et des transferts convectifs dans un espace annulaire lisse ou encoché par voie de simulations numeriques Université de Poitiers Thèse de doctorat
[21] Fenot M, Bertin Y, Dorignac E and Lalizel G 2011 *Int. J. Thermal Sciences* **50** 1138
[22] Kasagi N 1999-2003 Micro Gas Turbine/Solid Oxide Fuel Cell Hybrid Cycles for Distributed Energy System The University of Tokyo
[23] Kiełczewski K and Tuliszka-Sznitko E 2013 *Arch. Mech.* **65** 527
[24] Cliffe K and Mullin T 1986 AERE Harwell Report No TP.1179
[25] Cliffe K, Kobine J and Mullin T 1992 *Proc. R. Soc. London* A**439** 341
[26] Youd A 2005 Bifurcations in Forced Taylor–Couette Flow, PhD Thesis Newcastle University

Journal of Physics: Conference Series **530** (2014) 012038 doi:10.1088/1742-6596/530/1/012038

Heat transfer characteristic of an impingement cooling system with different nozzle geometry

K Marzec[1] and A Kucaba-Pietal[2]
[1] Rzeszow University of Technology, MTU Aero Engines, Rzeszow, Poland
[2] Rzeszow University of Technology, Rzeszow, Poland

E-mail: [1]k.marzec@prz.edu.pl, [2]anpietal@prz.edu.pl

Abstract. The influence of different geometries of the nozzles of an impingement cooling array of six jets directed to the flat surface on the flow mechanism and the heat transfer were investigated numerically. Basically the setup consisted of a cylindrical plenum with an inline array of impingement jets. Simulation were performed using Computational Fluid Dynamics (CFD) code Ansys CFX. The $k - \omega$ shear stress transport (SST) turbulence model was used in calculations. The physical model was simplified by using the steady state three-dimensional analysis and incompressible and viscous flow of the fluid. The study focused on an usage of different nozzles shapes in the cooling system for constant inlet flow parameters and boundary conditions. The numerical analysis of the different mesh density resulted in good convergence of the GCI index, what excluded mesh size dependency. The obtained results indicate, that the usage of various types of nozzles results in different values of the heat transfer coefficient and the Nusselt number in the affected area.

1. Introduction

Literature includes many investigations of arrays of impingement cooling jets. A variety of geometrical or physical parameters of cooling systems has been studied. The majority of the studies were experimental ones [1-7]. Many numerical simulations of the impingement cooling affecting different types of the jets [8-11] or different CFD turbulence models [12] are presented in the literature. Many studies have been made to analyze the different dimensionless y/D parameter or different Reynolds number values. Many correlations were introduced to calculate the heat transfer coefficient or the Nusselt number in the stagnation region, applicable for the adequate Reynolds number and nozzle height range [12].

An impingement cooling system is an array of jets of high velocity fluid which is made to strike a target surface. An impinging jets can be classified as a submerged jet or a free jet. If the fluid issuing from the jet is of the same density and nature as that of the surrounding fluid then the jet is called the submerged. If the fluid issuing from the nozzle is of a different density and nature than that of the surrounding fluid then the jet is called free.

Further classification between jets can be made for confined and un-confined jets. In the case of confined, the jet remains bounded between two surfaces during its flow. There is less entrainment of air from the atmosphere. Un-confined jets are free to expand after they imping on the target surface.

[1] Corresponding author: k.marzec@poczta.onet.pl , phone +(48)-731-706-731

A jet impingement heat transfer cooling systems plays a significant role in many industrial applications. This is due to the fact, that the impinged jet provides higher rate of heat and mass transfer. Such applications as drying of paper, cooling of photovoltaic cells, de-icing of aircraft wings, annealing of metals, heat exchangers in automotive and aeronautical applications, cooling in grinding process or cooling of gas turbine blades and other moving parts of small distances between them, are a few examples of jet impingement cooling.
A flow field is characterized by a jet zone, a stagnation zone and a wall jet zone, as indicated in Figure 1.

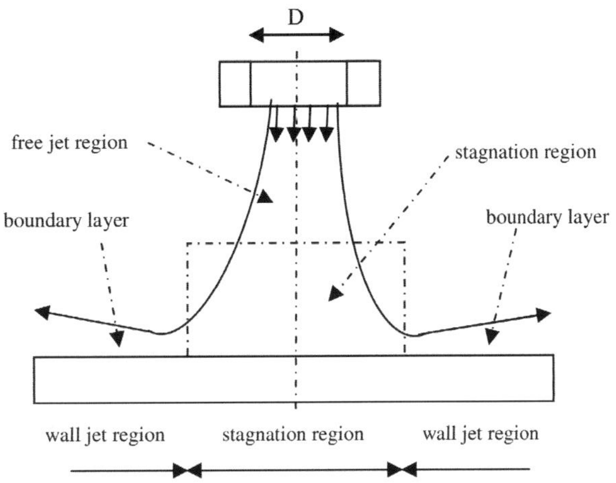

Figure 1. Flow field characteristics.

2. Numerical Setup

In this section, an overview on the numerical setup investigations is presented. The numerical grid density is taken into consideration.

2.1. Boundary and Initial Conditions

The schematic geometry of the system under consideration is shown in Figure 2. Both: the left hand side and the right hand side of the system is open to allow the fluid to flow. Fluid is free to expand after imping target surface. The velocity of the flow $V=(u,w)$ at the inlet of the supply tube u=14m/s was prescribed to obtain the Reynolds number Re=4800 at the impingement nozzle. Thereby, the Reynolds number was based on the mean velocity at the nozzle and the jet diameter D. The air density was ρ=1.17 kg/m^3, thermal conductivity was k=0.025 W/mK and the dynamic viscosity was μ=1.8 10^{-5} Pas. The fluid temperature at the inlet was T_{jet}=20 C^o. The wall, the jet impinged onto was heated with the heat flux q_w=5000W/m^2. The outlet boundaries of the calculation domain were represented by opening pressure boundary conditions which permitted the fluid to both enter or leave the domain. To simplify the analysis, the steady-state fluid flow is considered and it was assumed that the fluid physical properties are constant and the effect of the gravity and radiation is neglected. The flow field is three-dimensional. The roughness of the tube which contains the flowing fluid is 30 μm.

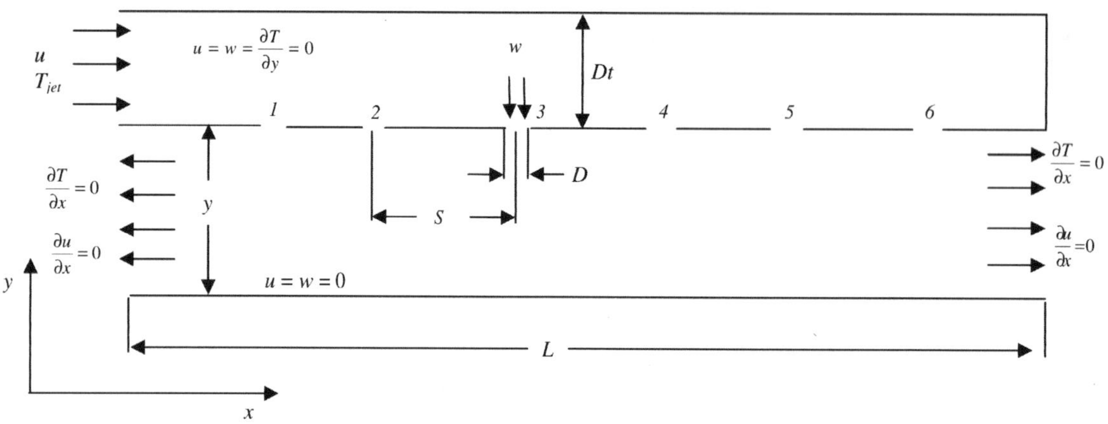

Figure 2. A schematic 2D diagram of geometry of the considered problem.

2.2. Computational domain

The chosen computational domain corresponds to an array of six cooling impingement jets directed to a flat surface. In this study four different nozzle shapes were taken into consideration (see Figure 3). The nozzle exit-plate distance y/D=8 and nozzle pitch-diameter radio S/D=8 were constant for all simulations. For the chamfer and countersunk nozzle type, the smaller diameter d=0.8mm and the convergence ratio D/d=2 were set. The tube inlet diameter was D_T=5mm. The total length and width of the target plate was L=56mm. This ensured that the outlet boundaries were away from the section of interest where the jet impinged onto the target plate. Air was supplied to the system by the supply-tube only from the left side. The right side of the tube was closed.

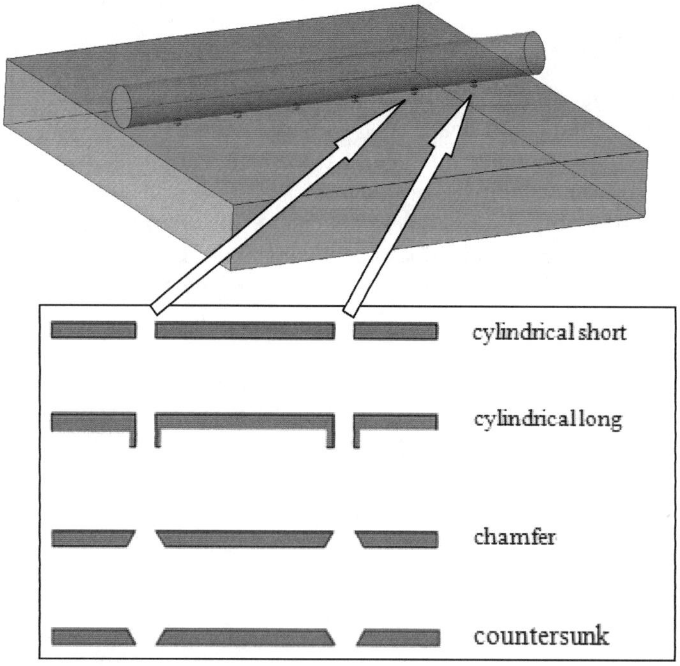

Figure 3. Investigated computational domain with different nozzles geometry.

The governing equations for the problem under consideration are based on the balance laws of the mass, the linear momentum, and the thermal energy in the steady state and incompressible and viscous flows.

2.3. Numerical grids and numerical accuracy
The numerical calculations were carried out using unstructured grids with 1.9mln elements generated by the Ansys CFX. The influence of the numerical grid density on the results of the heat transfer coefficient and the Nusselt number in the stagnation region was taken into consideration. Therefore, four analysis with different cell size of the area of a single impingement jet in the area of the interface (between air flow and flat surface) were taken into consideration.
As to investigate the sensitivity of the results of the numerical analysis, the Grid Convergence Index GCI (equation 1) was calculated [13]. The safety factor F_s=3, was set for two grids comparison. The chosen parameter was temperature T_h measured in the distance of 0.025mm from the surface in the y direction, appropriately for each grids. Order of convergence was p=2, r_p was the density factor. For the analysis, number 3 GCI index equal of 0.6% was obtained. Therefore, it might be concluded that numerical results on the fine grid are grid independent.

$$GCI = F_s \frac{\left| \frac{T_{h2} - T_{h1}}{T_{h1}} \right|}{r_p - 1} 100\% \tag{1}$$

3. Numerical solver
The simulation were performed using Computational Fluid Dynamics (CFD) code Ansys CFX, that solves equations of continuity, momentum end energy using the Reynolds-Averaged Navier-Stokes approach.
In the present investigation the k – ω shear stress transport (SST) turbulence model was used. SST model is one of the most successful Hybrid models. It combines the k – ω model near the wall and the k – ε model further from the wall to utilize the strengths of each model. SST model can produce better predictions of fluid properties in impinging jet flows and is recommended as the best compromise between the solution speed and accuracy [12]. This is interesting to note that results from presented numerical solver were compared with physical experiment and gives satisfying results [14].

4. Results and discussion
This section provides the analysis of the flow field behavior of the array of six impingement jets. Definition of pressure drop in the form of the discharge coefficient is given. Finally, the heat transfer characteristics in terms of the Nusselt number is presented.

4.1. Flow field characteristics
The proper prediction of the heat transfer values needs an analysis of the flow field mechanism. An array of presented six nozzles has a very complex behavior. Firstly, it is a result of the deflection at the stagnation areas. Secondly, the jet to jet interaction appears during the experiment. The flow direction in the tube plenum is perpendicular to that in nozzles. Figure 4 and Figure 5 illustrate numerical results of the velocity of the fluid, which appears during the investigation of the cylindrical and chamfer nozzle impingement jets, respectively. Presented figures correspond to the first and the second jet of the array of six impingement nozzles. The diameter of the impingement jets was small (D=0.8mm) compared to the diameter of the plenum (D_T=5mm).

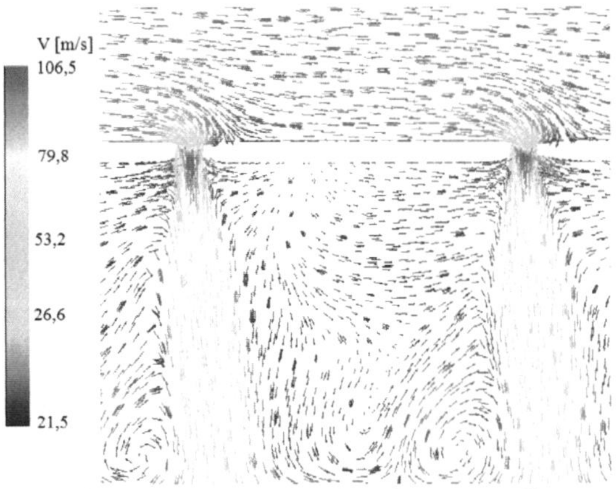

Figure 4. Velocity vector plot for two adjacent cylindrical nozzles (first and second).

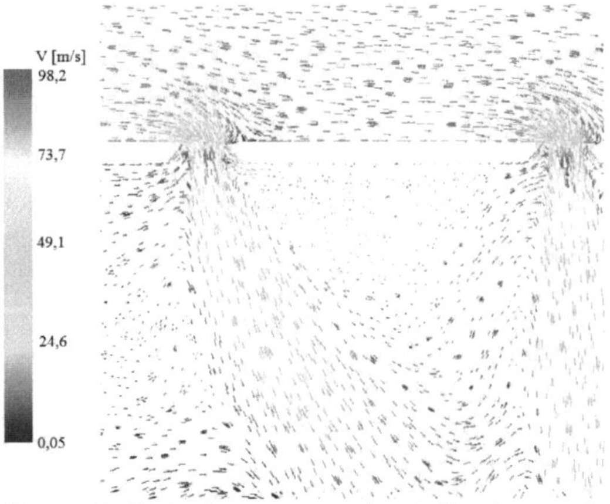

Figure 5. Velocity vector plot deflection angle for two adjacent chamfer nozzles (first and second).

It was found out that the air velocity at the area of each type of the nozzles maintained almost constant and the mass flow entering the plenum was uniformly distributed among nozzles. Figure 6 presents the velocity distribution w related to the jet inlet velocity w_o for the jet axis for the first and the last nozzle of the cylindrical and countersunk geometry.

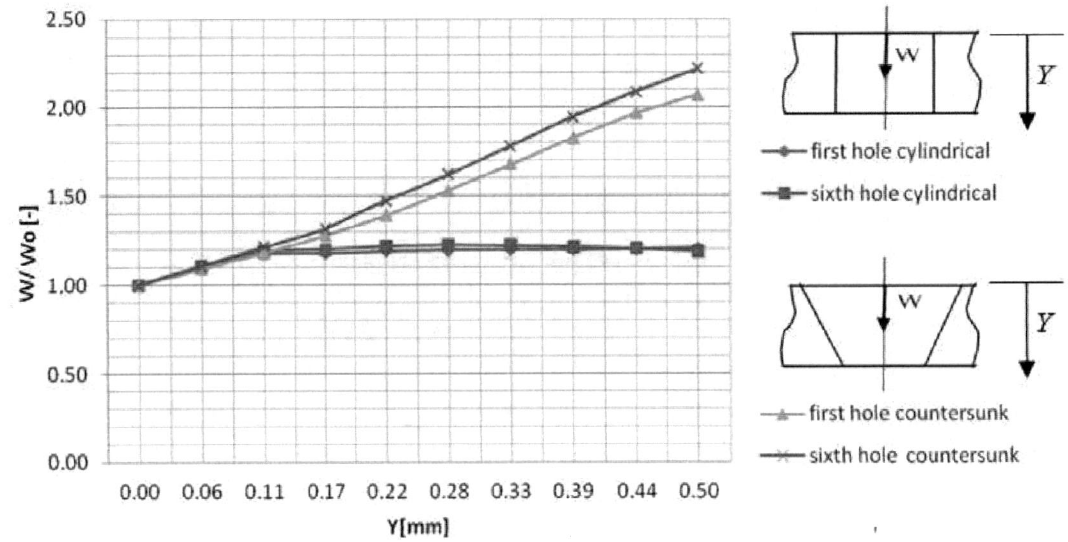

Figure 6. Velocity w along jet axis related to the jet inlet velocity w_o for different geometry configurations.

The sudden change in the flow direction in the nozzles before discharging onto the target plate led to flow deflection in the direction of the flow inside the plenum. The biggest deflection was observed for the chamfer nozzle geometry when the flow was not directed perpendicularly to the target plate (see Figure 5). For all cases the deflection angle decreased downstream in the plenum towards to the last jet. The nozzle pitch-diameter ratio S/D=8 resulted in adjacent jet flows interactions and a fountain effect was observed leading to re-circulations.

4.2. Pressure drop

Pressure drop characteristics includes both the pressure drop in the distribution tube and in impingement nozzles. For different nozzle shapes pressure drop characteristics were performed using Computational Fluid Dynamics (CFD), the Ansys CFX code. To verify results of the pressure drop additional investigation according to Bernoulli's formulas [4] were performed. Input values of velocities were taken from the Ansys CFX code. Results of both investigations are convergent. Figure 7 presents characteristics of the pressure drop P_d between the different nozzle geometry in relation to the supply tube inlet dynamic pressure P_i.

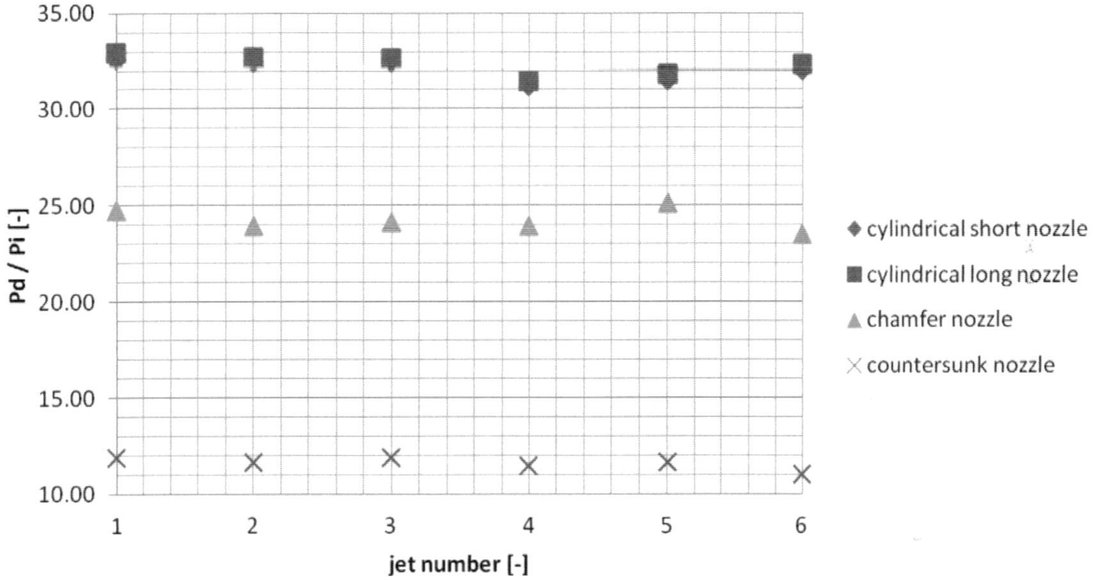

Figure 7. Normalized Pressure drop *Pd* to supply tube inlet dynamic pressure *Pi* through different nozzle geometry.

4.3. Heat transfer characteristics

The heat transfer measurements along the impingement surface is presented in terms of the Nusselt number

$$Nu = \frac{hD}{k} \tag{2}$$

where *h* is the heat transfer coefficient, *D* is the nozzle diameter, and *k* is the thermal conductivity of the fluid.

The heat transfer coefficient is defined as

$$h = \frac{q_w}{T_w - T_{jet}} \qquad h = -k\frac{\frac{\partial T}{\partial n}}{T_{jet} - T_w} \tag{3}$$

where q_w is the wall heat flux, T_w is the wall adiabatic temperature, T_{jet} is a jet temperature,

Journal of Physics: Conference Series **530** (2014) 012038 doi:10.1088/1742-6596/530/1/012038

$\frac{\partial T}{\partial n}$ gives the temperature gradient component normal to the wall

The optimal nozzle configuration for a given system will be determined by the factor of pressure drop and the flow rate to achieve the demanded heat transfer coefficient and the average Nusselt number. In the Figure 8 the Nusselt number Nu_o distribution corresponding to usage of different geometrical nozzle configurations is presented. In comparison to other geometrical configurations, the usage of chamfer nozzle results in low values of the heat transfer coefficient and the Nusselt number Nu_o in the stagnation region. In this system, except the first jet where the deflection angle is big, the heat transfer coefficient is uniform. The big deflection angle resulted in non-perpendicular direction of the flow to the target surface. For countersunk and cylinder long nozzle configuration, the Nusselt number Nu_o resulted in very similar values. The highest values of the Nusselt number Nu_o were achieved for the cylinder short configuration of nozzles.

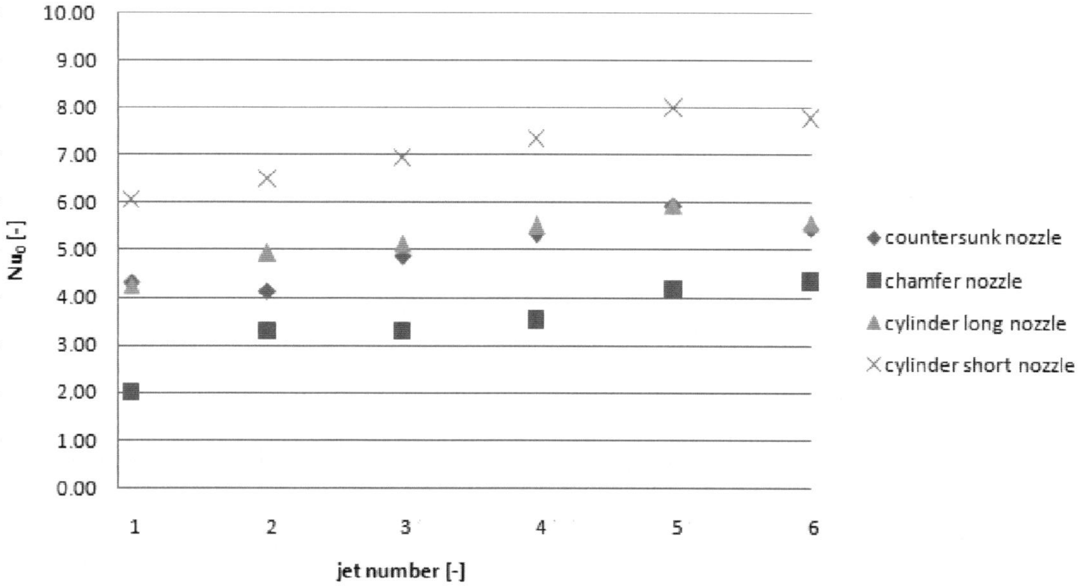

Figure 8. Nusselt number Nu_o distribution along target surface for different nozzle geometry.

Figure 8 shows that Nu_o increases slightly along the supply tube axis. At the entrance to the first nozzle flow is directed perpendicularly to the jet axis. Because of the closed end of the supply-tube air approaching last nozzles is directed with angle lower than 90deg to the nozzles entrance and since the flow rate maintains almost constant along all jets Nu_o increase slightly along the jets located on the end of the supply-tube.

5. Conclusions

The aim of the work was the numerical analysis of a flow characteristic, pressure drop, heat transfer coefficient and the Nusselt number distribution for an array of six impingement jets. The usage of an array of cooling jets resulted in the uniform distribution of the mass flow and pressure drop among each type of nozzles. For the chamfer nozzle configuration the biggest deflection angle was observed especially at the first jet. The lowest pressure drop occurs during the analysis with the countersunk type of nozzles. The high pressure drop for cylindrical nozzles are caused by rapid increase in the flow rate while entering the nozzle. This analysis might approach to idea of the usage of different nozzle geometry for one configuration of the cooling system if the target plate has no constant temperature.

This might result in the uniform heat transfer distribution and therefore the lower thermal stresses on a target plate. The presented analysis shows, that the heat transfer coefficient depends on the direction of the flow while the air is entering the nozzle. Therefore, the further analysis of installing features between adjacent jets to redirect the flow and to make it more convergent with the jet axis are justified.

6. References

[1] Goordo M, Jongmyung P, Ligrani P, Fox M and Hee-Koo M 2007 Effects of Mach number and Reynolds number on jet array impingement heat transfer *Inter. J. of Heat and Mass Transfer* **50** pp 367-380

[2] Nimalkumar M, Katti V and Prabhu S V 2011 Local heat transfer distribution on a smooth flat plate impinged by a slot jet *Inter. J. of Heat and Mass Transfer* **54** pp 727-738

[3] Jung-Yang S and Wen-Zheng S 2006 Effects of jet plate size and plate spacing on the stagnation Nusselt number for a confined circular air jet impinging on a flat surface *Inter. J. of Heat and Mass Transfer* **49** pp 3477-86

[4] Royne A and Dey Ch 2006 Effect of nozzle geometry on pressure drop and heat transfer in submerged jet arrays *Inter. J. of Heat and Mass Transfer* **49** pp 800-804

[5] Koseoglu M F and Baskaya S 2010 The role of jet inlet geometry in impinging jet heat transfer, modeling and experiments *Inter. J. of Therm. Scien.* **49** pp 1417-26

[6] Whelan B and Robinson A 2009 Nozzle geometry effects in liquid jet array impingement *Appl. Therm. Engineering.* **29** pp 2211-21

[7] Attalla M and Salem M 2013 Effect of nozzle geometry on heat transfer characteristics from a single circular air jet *Appl. Therm. Engineering* **51** 723-733

[8] Zukowski M 2013 Heat transfer performance of a confined single slot jet if air impinging on a flat surface *Inter. J. of Heat and Mass Transfer* **57** pp 484-490

[9] Shu J J 2010 Heat transfer on a hot surface impinged by a cold circular liquid jet *V Europ. Conf. on CFD, ECCOMAS CFD*

[10] Ahmed F B, Weigard B and Meier K 2010 Heat transfer and pressure drop characteristics for a turbine casing impingement cooling system *Proceed. of Intern. Heat Transfer Conf.*, Washington, USA

[11] Ee-Mahghany W M, Hanafy A A, Khaled M A and Mohamed A T 2012 Numerical simulation for confined rectangular slot jets impingement on isothermal horizontal plate, *Europ. J. of Scien. Research* **76** pp 553-566

[12] Zuckerman N and Lior N 2006 Jet impingement heat Transfer: Physics, Correlations and Numerical Modeling, *Adv. in Heat Transfer* **39**

[13] Błoński S 2009 PhD Thesis: Laminar-turbulent flow analysis in micro-channels *Polish Academy of Science, Institute of Fund. Techn. Research* http://www.ippt.pan.pl/_download/doktoraty/blonski_doktorat.pdf

[14] Marzec K and Kucaba-Pietal A 2013 Applications of computer science in impingement cooling system design *Pre-Proceed. of 9th Inter. Conf. on Appl. Mathem.*, Baia Mare, Romania

An advanced Thermal-FSI approach to flow heating/cooling

J Badur[1,3], P Ziółkowski[1,2,4], W Zakrzewski[1], D Sławiński[1], S Kornet[1,2], T Kowalczyk[1,2], J Hernet[1], R Piotrowski[1], J Felincjancik[1] and P J Ziółkowski[1]

[1]Energy Conversion Department, Institute Fluid-Flow Machinery PAS-ci, Fiszera 14, 80-231 Gdańsk, Poland
[2]Conjoint Doctoral School at the Faculty of Mechanical Engineering, Gdansk University of Technology, Gdańsk, Poland

Abstract. Actually, two-way thermal-energy exchange between working fluid and solid material of a casing is a leading problem for modern – semi automatic – design techniques. Many questions should be solved, especially, the turbulent mode of thermal energy transport both in fluid and solid, should be re-examined and reformulated from the primary principles. In the present paper, a group of researchers from Energy Conversion Department of IMP PAN at Gdańsk, tries to summarise a last three-years efforts towards to mathematical modelling of advanced models of thermal energy transport. This extremely difficult problem in "thermal-FSI" ("Fluid Solid Interaction") means that the both for solid and fluid mathematical model of a surface layer should be self-equilibrated and self-concise. Taking these requirements into account, an advanced Reynolds-Stanton analogy has been discussed and implemented. Some numerical examples concerning of the benchmarks experiments and industrial applications have also been developed and presented.

1. Governing equations for the temperature field

Heating is a well known basing process responsible for energy transport from a hot continuous body to a cold one. Mechanism of heating is based on appearing of driving force in a form of temperature difference between cold and hot places. Flux of heating energy generously vanish when the temperature of contacting media becomes equal.

Due to the complex nature of heating, in the literature there is no one common mechanism responsible for transport of mass, energy, momentum during heating. Therefore, according to many different theoretical view point we have a lot of governing equations related to the temperature field being a fundamental unknown. Among them the most popular are (see; Gyftopoulos, Beretta [1], Cano-Andrade et al [2], Bejan [3,17], Feidt [4]):

- Intrinsic Quantum Thermodynamic (IQT) [2]:

$$\frac{d\rho}{dt} = -\frac{i}{h}[H,\rho] - \left(\frac{1}{\tau_1} D_a \oplus \rho_F + \frac{1}{\tau_F} \rho_a \oplus D_F\right) \tag{1}$$

[3] E-mail: jb@imp.gda.pl
[4] E-mail: pawel.ziolkowski@imp.gda.pl

- Locally maximum entropy production principle (LMEPP) [3]:

$$\frac{\partial \sigma}{\partial T} + \mathrm{div}\left(\frac{\partial \sigma}{\partial \vec{g}} - \mathrm{div}\frac{\partial \sigma}{\partial \vec{G}} \right) = 0 \tag{2}$$

- Balance of entropy:

$$\frac{\partial}{\partial t}(\rho\eta) + \mathrm{div}(\rho\eta\vec{v} + \vec{h}_\eta) + \rho\sigma = 0 \tag{3}$$

- Balance of energy:

$$\frac{\partial}{\partial t}(\rho e) + \mathrm{div}(\rho e\vec{v}) = \mathrm{div}\left(\vec{\mathsf{F}}_{mech} + \vec{\mathsf{F}}_{heat} + \vec{\mathsf{F}}_{chem} + \vec{\mathsf{F}}_{e-m} + \vec{\mathsf{F}}_{mass} + \ldots\right) \tag{4}$$

The above four governing equations for the temperature field are only examples of efforts of scientific community to understand the fundamental behaviour of nature. For instance, balance of entropy and energy are in the core of explaining the transition between macroscopic and microscopic worlds [4]. Furthermore, there is no one common picture for describing complexity of types of communication phenomena. Thus, thinking on transport of heating across the solid-fluid contact layer one should consider an exact nature of a non-linear equation of thermal motion. The examples given above by equations (1)(2)(3) and (4) are the most known – anthers, like "steepest entropy ascent" (SEA), "dissipative quantum dynamic (DCD), etc., are under permanent efforts of scientific community [1].
Let recall some basic elements of above governing equations of thermal motion. In IQT equation (1) the evolution of density state operator of thermal motion is given by two terms. The first term on the right-hand side describes the unitary Hamiltonian dynamics of the system and the second the non-Hamiltonian dissipation dynamics. The operator $[H,\rho] = H\rho - \rho H$ is the commutator between Hamiltonian and density state operator, and $\rho_A = Tr_F\rho$; $\rho_F = Tr_A\rho$ are reduced state operator. Additionally, τ_A, τ_F are characteristic time for subsystems A (internal) and F (external). Important ingredients of the IQT model are facts that it can exactly describe the *locally perceived energy* and *locally preserved entropy* [2] – it means that having these local observables one can defines – using the Gyftopoulos-Beretta and von Neumann definitions – both energy and entropy for the overall system.

The equation (2) is known to be governing equation of LMEPP (Bejan [3]) and is generously based on the notion of *entropy production* : σ . Assuming that this entropy production depends on the temperature gradient $\vec{g} = \mathrm{grad}T$ [thermal diffusion] and the second gradient $\vec{G} = \mathrm{grad}\vec{g}$ [Maxwell thermal transpiration] we can reinterpret equation (2) to be a Euler-Lagrange variational condition for maximal entropy production [6].

The equation (3) presents the most problematic equation of the whole classical field theory – the balance of specific entropy η. Here the continuum density is denoted by ρ, the specific momentum vector by $\vec{v}$ and the Rankine entropy flux by $\vec{h}_\eta$. Like above the specific entropy production is denoted by σ.

In practices, the balance of energy (4) is a governing equation for the field of temperature $T(\vec{x},t)$. According to the principle of producing of energy "ex nihilo" in equation (4) there is no internal sources of energy and the changing of energy e is governed only by energy fluxes $\vec{\mathsf{F}}_a$, $a = mech, heat, chem, e-m, mass, \ldots$. The most important for describing of thermal motion dynamics is the Stokes-Rankine heating energy flux $\vec{\mathsf{F}}_{heat}$, frequently denoted in the literature by letter

$\vec{q}$. By the Rankine formulae : flux $\vec{F}_{heat} = T\vec{h}_\eta$ the equation of energy is indirectly connected with the thermal equation of motion or the balance of entropy, therefore, in numerical practices, the balance of energy leads to determine the temperature field.

Efforts of the Department of Energy Conversion IFFM PAS-ci are concentrated on finding a proper boundary conditions for thermal equations of motions [5]. Our aims are focused around the novel scientific tool called "thermal-FSI" and our basic assumption is that both fluid flow and solid deformation, governed by the same kind of momentum conservation, should be adequately formulated for proper advanced fluid-solid thermal energy exchange. In practice, it means that the advanced thermal-FSI should leads to using the same discretization moving lattice for momentum and energy equations both in fluid and solid. Let us mention, that from the point of view of energy and entropy transport the acronym FSI should be understood to be "Fluid Solid Interaction" not as "Fluid Structure Interaction". It is because thermal phenomena cannot be treated as a "thermal-plate" or a "thermal-shell" and always must be taken to be three-dimensional. Here, the main aim of our study is only restricted to introduce a correct form of boundary condition for the energy equation, especially the boundary condition within the contact layer between solid and fluid.

2. Balance of energy in the solid-fluid contact layer

2.1. Thermal-FSI

Thermal-FSI is a part of science dedicated to describing of energy transport and conversion within a thin layer occurring in a contact of solid and fluid material. In difference to the *Momentum-FSI* where the exchange of momentum between solid and fluid is the main phenomena, in the Thermal-FSI the equation of energy plays the main role in two-way (from solid to fluid and vice versa) energy transport [15].

Speaking about the two-way interaction between solid and fluid authors have in mind a problem of interphase surface mechanical motion. The interphase layer or the interphase surface of contact in a mechanical motion is governed by extended boundary condition between CFD and CSD approaches [15]. Usually, these mechanical conditions leads to motion of discretization lattices (finite volumes or finite elements), since both CFD and CSD obeys discretization of whole domain, divided into solid and fluid parts. It is known that motion of fluid domain discretization lattices follows from a large displacement of solid boundaries, and, vice versa, motion of solid discretization lattices follows from an action like mass sedimenting, swelling, etc.

From the point of thermal motion view also the problem of a thermal interphase motion is important and should be taken into account in formulation a general mathematical model for the Thermal-FSI. Therefore, we shall consider a thermal contact between hot fluid and cold solid continua to be described by a thin but finite contact thermal layer described as a sum of the von Smoluchowski thermal length jump [5]:

$$l_T = l_T^f + l_T^s \tag{5}$$

where l_T^s , l_T^f are the thermal length jump adequate to solid and fluid side (figure 1).

2.2. Boundary condition for energy in the solid-fluid contact layer

Going into a generalized boundary condition for energy equation let us start from the traditional forms used in CSD:

$$-\alpha^s (T^s - T_s^s) - \vec{q}^s \cdot \vec{n} = 0 \tag{6}$$

and CFD:

$$\alpha^f (T^f - T_s^f) - \vec{q}^f \cdot \vec{n} = 0 \tag{7}$$

Here - $\vec{q}^s$ and $\vec{q}^f$ means the flux of thermal energy on solid and fluid side which for isotropic materials, according to the Fourier law, depend on the coefficients of thermal conductivity ei., $\vec{q}^s = \lambda^s \text{grad}T$, $\vec{q}^f = \lambda^f \text{grad}T$. Additionally, the transfer coefficients α^f, α^s are responsible for overall transport of thermal energy in macroscopic vicinity of the contact layer. In the classical approach there is no the von Smoluchowski thermal jump, therefore fluid and solid temperature coincidences: $T_s^f = T_s^s$.

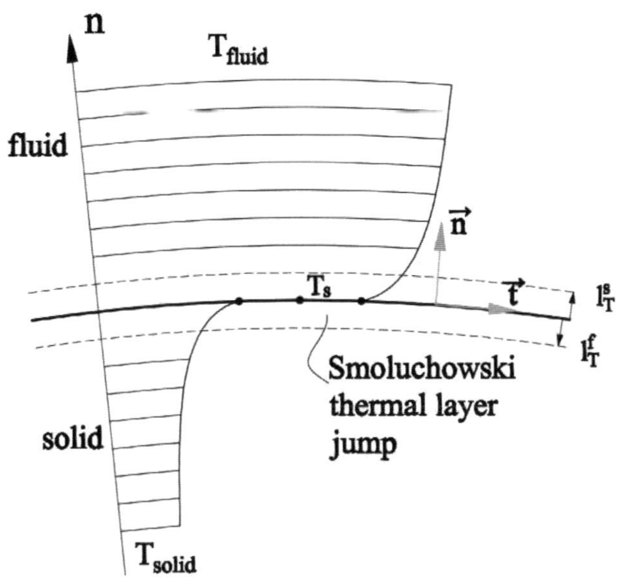

Figure 1. Temperature profile in a solid-fluid contact layer

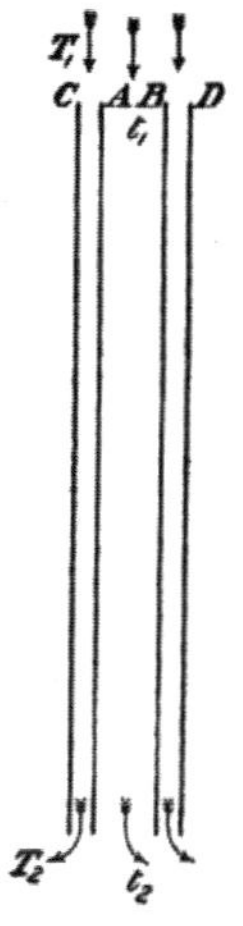

Figure 2. Exchange of thermal energy In the Stanton experiment.

Turning into a generalized boundary condition that reflects a CFD-CSD type of coupling one shall take into account as a surface energy contribution as well as the thermal jump phenomena. Without going into details, let $e_s = c_{\rho s} T_s$ be a simplest form of surface internal energy governed by surface temperature T_s Defining the surface metric tensor to be [6]: $\vec{I}_s = \vec{I} - \vec{n} \otimes \vec{n} = a_{\alpha\beta} \vec{a}^{\alpha} \otimes \vec{a}^{\beta}$ and the surface gradient and divergence operator [5] grad_s , div_s we are able to express the boundary condition for energy equation within a thermal contact layer. Removing from balance other energy fluxes and keeping only mechanical ($\vec{F}_{mech}$) and thermal ($\vec{F}_{heat} \equiv \vec{q}$) energy vectors on the boundary the energy balance within the thermal contact layer can be written to be:

$$\partial_t(\rho_s c_{\rho s} T_s) + \text{div}_s(\rho_s c_{\rho s} T_s \vec{I}_s \vec{v}_s) + \text{div}_s(\lambda_s \text{grad} T_s) + l_T^f (T_s^f - T_s) - l_T^s (T_s^s - T_s) +$$

$$\alpha^f (T^f - T_s^f) - \alpha^s (T^s - T_s^s) + (\vec{q}^f - \vec{q}^s) \cdot \vec{n} + (\vec{F}_{mech}^f \vec{v}_{slip}^f - \vec{F}_{mech}^s \vec{v}_{slip}^s) \cdot \vec{n} = 0 \tag{8}$$

According to equations (6) (7) in the thermal contact layer an additional surface density ρ_s and the surface temperature T_s appears. The surface velocity vector $\vec{v}_s$ and the slip velocity vectors $\vec{v}_{slip}^s$,

$\vec{v}^{f}_{slip}$ should be independently calculated from the boundary conditions for the momentum equations [5]. The von Smoluchowski jump coefficients are related with the thermal jump length as: $\mathsf{l}^{f}_{T} = l^{f}_{T} / \lambda_{f}$, $\mathsf{l}^{s}_{T} = l^{s}_{T} / \lambda_{s}$. This last contribution to equation (8) is especially important in micro- and nano-scales where the mechanism of the von Smoluchowski thermal jump becomes most valid. Also in nano-scale the energetic contribution from the mechanical slip – for instance the Navier slip – gives the considerable values

3. Numerical Thermal-FSI treatment

On the contemporary level of development Thermal-FSI has no possibilities for numerical solution of equation (8) even in a simplified form. The first reason is an insufficient developed of Finite Element Method as well as Finite Volume Method. Actually both methods are prepared only to simultaneous solving only a part of equation (8) in a form:

$$(\vec{q}^{f} - \vec{q}^{s}) \cdot \vec{n} = 0 \qquad (9).$$

It is obvious that keeping further $\vec{q}^{s} = \lambda^{s}\mathrm{grad}T$, $\vec{q}^{f} = \lambda^{f}\mathrm{grad}T$ with constant λ^{s}, λ^{f} one can obtain false value of heating vectors and, as a result, false value of temperature profile. Therefore, Thermal-FSI is based on a concept of functional changing of diffusional coefficients in the nearest neighborhood of surface of contact, for instance $\lambda^{f} = \lambda^{f}(y^{+}, \mathrm{Re}_{y}, \delta)$. Unfortunately, this approach needs an independent from the momentum equation the near-surface discretization. Using of two different discretizations - one for momentum and second for energy equation – is time consuming and numerically uncomfortable. Therefore, the better solution is to applied some analogies with momentum equations that, even if physically unfortunately, leads to unified CSD/CFD system of equations solved only in one discretization lattice.

4. The Reynolds-Stanton analogy

Let, according to literature tradition, the letter y means the coordinate normal to solid-fluid contact surface. Then the condition $(\vec{q}^{f} - \vec{q}^{s}) \cdot \vec{n} = 0$ can approximate the condition (8) if we take an idea that the passage of thermal energy from solids to fluids moving past them is governed by the same principles as apply in virtue of viscosity to the passage of momentum originated with Reynolds [8] and it has been further developed by Stanton [7] (see Lord Rayleigh[9], L. Prandtl [10], A. Eagle, M. Ferguson [11], L.H. Loitsiantski [12]). The analogy, originally written for viscous stress tensor $\vec{q}(y) \sim \vec{\tau}(y)\vec{n}(y)$ can be also pressure dependent, both fluid and solid:

$$\vec{q} = a(-p\vec{I} + \vec{\tau})\vec{n}\,, \qquad (10)$$

where a coefficient of Reynolds-Stanton analogy discussed extensively by Prandtl [10] and Schlünder [14]. The above analogy responses on the following question – It is possible to have simultaneously high rates of thermal energy transfer and low rates of momentum transfer ? To answer this very practical question leads immediately to the scientific problem whether – and if so – how are thermal energy and momentum transfer independent ?

Now the question arises – how equation (10) can be adopted to CFD/CSD modelling. Practically it means that for small y^{+} the length and direction of $\vec{q}$ vector becomes independent of direction of $\mathrm{grad}T$ and length coming from laminar and turbulent Prandtl number [15],[16]. In our work we have simply implemented of condition (10) to a commercial code.

5. An example

Stanton's experiment [7] has been prepared to proof Reynolds' mechanism of thermal energy transport [1]. It was thought to be simplest measurements of transport of energy from flowing hot water (39.6 °C) to flowing cold water (18 °C) across a metal wall. The pipes in the Stanton experiment is presented in figure 2. Total length of pipes was 67 cm and internal diameters are: 13.9 (cold) and 15,7 mm (hot)(figure. 2). The mass flux of hot water was constant – 148 g/s, mass flux of cold water was changed from 148 to 27 g/s.

For the same mass flux 148 g/s of hot and cold water the results is symmetrical – the increase of temperature of cold water on 4.4 °C and decrease temperature of hot water on 4.4 °C. For decreasing of mass flux of cold water from 148 to 27 g/s it has been observed the systematic increase of the inlet temperature t_2 from 22.4 to 31.8 °C and simultaneously, increasing of T_2 from 35.2°C to 27.4 °C.

The quite similar results has also been numerically obtained by using of implemented the Reynolds -Stanton analogy. It was calculated that during the change of mass flow rate of cold water (average velocity change from c =98 to 18 m/s) the total change of energy by the heating increases from 336 [W] to 390 [W] and harmonise with the formulae proposed by Reynolds and empirically proved by Stanton :

$$H = (A + B\rho c)(T_1 - t_1) \tag{11}$$

with $A =$ 343 and $B = 0{,}334$ calculated by Stanton and repeated in Thermal-FSI from overall integration of normal component of heating vector ; $H = \iint_{\partial V} \vec{q} \cdot \vec{n} dA$.

6. Conclusions

It has been shown that the simplest Thermal-FSI condition has its adequate formulation of boundary condition for energy equation, and it does not allow correct simulation of evolution of the temperature scalar field in the homogeneous flow of hot fluid over a cold solid. But the great advantages of Thermal-FSI is that whole conservation equation can be solved on one common discretization grid. Our study throws doubt on the wisdom of obtaining experimental and numerical values of the Nusselt and Stanton numbers.

7. Literature

[1] Gyftopoulos E P and Beretta G P 2005 *Thermodynamics Foundations and Applications* (New York: Dover Pub. Mineola)

[2] Cano-Andrade S, Beretta G P and von Spakovsky M B 2013 Non-equilibrium thermodynamics modelling of an atom-field state evolution with comparisons to published experimental data, [in:] *12th Joint European Thermodynamics Conference, Brescia July 1-6, 2013* pp 430-436

[3] Bejan A 1996 *Entropy Generalization Minimization* (Boca Raton: CRC press)

[4] Feidt M 2012 Thermodynamics of energy systems; a review and perspectives *J. Of Applied Fluid Mechanics* vol 5 pp85-89

[5] Badur J and Ziółkowski P 2013 Further remarks on the surface *vis imperssa* caused by a fluid-solid contact, [in:] *12th Joint European Thermodynamics Conference, Brescia July 1-6, 2013* pp 581-586

[6] Bilicki Z and Mikielewicz J 1984 minimum energii i minimum produkcji entropii w zastosowaniu do obliczeń stopnia zapełnienia w pęcherzykowym przepływie dwufazowym, *Archiwum. Termodynamiki* **5** pp101-118

[7] Stanton T E 1897 On the passage of heat between metal surfaces and Liquids in contact with them, *Philosophical Transactions of the Royal Society* **A190** (London) pp67-88

[8] Reynolds O 1874 On the extent and action to the heating surface of steam boilers, Proc. Lit. Phil. Soc. Manchester 14 pp15-24

[9] Lord Rayleigh 1917 On the suggested analogy between the conduction of heat and momentum during the turbulent motion of a fluid *Adv. Committee for Aeronautics* (London) T941 pp15-18

[10] Prandtl L 1928 Bemerkung uber den Warmeubergang *Phys. Z.* **29** (in Rohr) pp487-489

[11] Eagle A and Ferguson M 1930 On the coefficient of heat transfer from the internal surface of tube walls *Proc. Royal Soc. London* A137 pp540-566
[12] Loitsiantski L H 1960 Heat transfer in turbulent motion *PMM* 24 pp 637-546
[13] Madejski J 1963 *Teoria Wymiany Ciepła* (Poznań: Państwowe Wydawnictwo Naukowe)
[14] Schlünder E -U 1998 Analogy between heat and momentum transfer *Chem. Eng. Process.* 37 pp103-107
[15] Rup K and Wais P 1999 An application of the $k-\varepsilon$ model with variable Prandtl number to heat transfer computations in air flows *Heat and Mass Transfer* **34** pp503-508
[16] Rup K and Soczówka M 2000 An improved low Reynolds number $k-\varepsilon$ model for heat transfer calculations *Forschung im Ingenieurwesen* **65** pp225-235
[17] Bejan A and Kraus A D 2003 *Heat Transfer Handbook* © 2003 (Hoboken New Jersey:John Wiley & Sons, Inc.)

Journal of Physics: Conference Series **530** (2014) 012040 doi:10.1088/1742-6596/530/1/012040

An attempt to minimize the temperature gradient along a plug-flow methane/steam reforming reactor by adopting locally controlled heating zones

M Mozdzierz[1], G Brus[1], A Sciazko[1,2], Y Komatsu[2], S Kimijima[2], JS Szmyd[1]

[1]AGH University of Science and Technology, Krakow, Poland
[2]Shibaura Institute of Technology, Saitama, Japan

E-mail: brus@agh.edu.pl

Abstract. Plug flow reactors are very common in the chemical process industry, including methane/steam reforming applications. Their operation presents many challenges, such as a strong dependence of temperature and composition distribution on the inlet conditions. The strongly endothermic methane/steam reforming reaction might result in a temperature drop at the inlet of the reactor and consequently the occurrence of large temperature gradients. The strongly non-uniform temperature distribution due to endothermic chemical reaction can have tremendous consequences on the operation of the reactor, such as catalyst degradation, undesired side reactions and thermal stresses. To avoid such unfavorable conditions, thermal management of the reactor becomes an important issue. To carry out thermal management properly, detailed modeling and corresponding numerical analyses of the phenomena occurring inside the reforming system is required. This paper presents experimental and numerical studies on the methane/steam reforming process inside a plug-flow reactor. To optimize the reforming reactors, detailed data about the entire reforming process is required. In this study the kinetics of methane/steam reforming on the Ni/YSZ catalyst was experimentally investigated. Measurements including different thermal boundary conditions, the fuel flow rate and the steam-to-methane ratios were performed. The reforming rate equation derived from experimental data was used in the numerical model to predict gas composition and temperature distribution along the steam-reforming reactor. Finally, an attempt was made to control the temperature distribution by adopting locally controlled heating zones.

1. Introduction

Because of their high energy conversion efficiency and the low level of pollutants they emit into the environment, solid oxide fuel cells (SOFCs) have the potential to become one of the most important energy conversion devices. A solid oxide fuel cell consists of two porous ceramic electrodes (a cathode and an anode) separated by a solid ceramic electrolyte. A typical SOFC uses solid oxides such as yttria-stabilized zirconia (YSZ) for the electrolyte [1]. The operation temperature level of an SOFC is 700 [°C] - 1100 [°C] and operates at atmospheric or elevated pressures [2]. High temperature and unstechiometry in the YSZ crystal network allow the oxygen ions to be transported to the anode side. This is very important because it allows SOFC to oxidize as a fuel not only hydrogen but also carbon monoxide. The mixture of carbon monoxide and hydrogen can be produced from a wide range of hydrocarbon-based fuels via steam reforming [3-6]. Using hydrocarbons as a fuel have a number of advantages; existing infrastructure can be used to distribute and provide fuels such as city gas to the SOFC systems. Additionally, fuel can be obtained from natural resources and enable to bypass the hydrogen source problem . For hydrocarbon-based fuel, three types of fuel conversion can be considered in the reforming reactions: an external reforming system, an indirect internal reforming system and a direct internal reforming system. High-temperature SOFC eliminates the need for an expensive external reforming system. The possibility of using internal reforming is one of the characteristics of high temperature fuel cells, SOFC. Strong endothermic fuel reforming reactions can be thermally supported by the transfer, in forms of primarily radiation and convection, of heat by exothermic electrochemical process of an SOFC as well as by the passage of electric current. However, for a high-temperature operation, thermal management of the SOFC system becomes an important issue. To carry out thermal management properly, detailed modeling and the numerical analyses of the phenomena occurring inside the SOFC system is required.

Brus and Szmyd studied a temperature distribution in an indirect internal reforming system depending on its location in the system and the distance between the stack and the reactor [7,8]. The analyzed reformer had the shape of a rectangular channel and was located above the cell stack in SOFC system. Their results indicate that, the position of the reformer in the module chamber and the emissivity of the surfaces of the reformer, cell and other elements in the SOFC module all play a key role for temperature distribution inside the internal methane/steam reforming reactor [7,8]. They present that by changing the location of the reformer or by modifying the surface property of the reactor, the thermal boundary condition can be change significantly. Further more the thermo-fluid model of heat and mass transfer inside the indirect internal methane/steam reformer was proposed to conduct parametric studies where the effect of the thermal boundary condition as well as the inlet condition on temperature distribution were investigated [9,10]. However only a uniform thermal boundary condition at the reformer walls and uniform catalyst distribution were investigated [9,10]. Additionally some systems use an internal indirect reformer that has the form of a pipe as can be found in references [11,12]. The temperature and spices distribution might be significantly different in those system and additional analyst is required.

Within this work the authors present an attempt to minimize the temperature gradient along the plug-flow methane/steam reforming reactor by adopting a locally changed thermal boundary conditions. To realize this aim the reaction kinetics have been derived. The empirical rate was employed in the proposed numerical model. The rate and the model were verified by compering the results of the computation with experimentally obtained data. Finally a numerical analysis was used to perform parametric studies of temperature and species distributions inside the methane-steam reformer.

2. Numerical Analysis

The supplied fuel to the system is the methane (CH_4). The gaseous fluids are assumed to be Newtonian and the flow of fuel inside the fuel reformer is assumed to be laminar [13], steady and occurring in one direction. The temperature of in-flow fuel is taken to be instantly approaching the temperature of the reformer.

2.1. Heat transfer model inside the reformer

The microstructure of porous media is not to be considered directly in the present model therefore for the pours area, the governing equations derived by the volume-averaging method are applied [14]. In the adopted method physical values are locally averaged for a representative elementary volume [15,16]. The heat transfer process inside the reformer porous media is described by continuity, momentum, energy and mass transfer equations. Consequently the following transport equations of the averaged physical values for laminar flows apply:

Journal of Physics: Conference Series **530** (2014) 012040 doi:10.1088/1742-6596/530/1/012040

$$\frac{\partial(\rho_0 U_x)}{\partial x} + \frac{1}{r}\frac{\partial(r\rho_0 U_r)}{\partial r} = 0 \tag{1}$$

$$\frac{\rho_0}{\epsilon_0^2}\left(U_x\frac{\partial U_x}{\partial x} + U_r\frac{\partial U_x}{\partial r}\right) = -\frac{\partial P}{\partial x} + \frac{\mu}{\epsilon_0}\left[\frac{\partial^2 U_x}{\partial x^2} + \frac{1}{r}\frac{\partial}{\partial r}\left(r\frac{\partial U_x}{\partial r}\right)\right] - \frac{\mu}{K_{\mathrm{p}}}U_x - \frac{\rho_0 f}{\sqrt{K_{\mathrm{p}}}}U_x\sqrt{U_x^2+U_r^2} \tag{2}$$

$$\frac{\rho_0}{\epsilon_0^2}\left(U_r\frac{\partial U_r}{\partial x} + U_r\frac{\partial U_r}{\partial r}\right) = -\frac{\partial P}{\partial r} + \frac{\mu}{\epsilon_0}\left[\frac{\partial^2 U_r}{\partial x^2} + \frac{1}{r}\frac{\partial}{\partial r}\left(r\frac{\partial U_r}{\partial r}\right) - \frac{U_r}{r^2}\right] - \frac{\mu}{K_{\mathrm{p}}}U_r - \frac{\rho_0 f}{\sqrt{K_{\mathrm{p}}}}U_r\sqrt{U_x^2+U_r^2} \tag{3}$$

$$\rho_0 C_{\mathrm{p}}\left(U_x\frac{\partial T}{\partial x} + U_r\frac{\partial T}{\partial r}\right) = \frac{\partial}{\partial x}\left(\lambda_{\mathrm{eff}}\frac{\partial T}{\partial x}\right) + \frac{1}{r}\frac{\partial}{\partial r}\left(r\lambda_{\mathrm{eff}}\frac{\partial T}{\partial r}\right) + Q \tag{4}$$

$$\rho_0\left(U_x\frac{\partial Y_j}{\partial x} + U_r\frac{\partial Y_j}{\partial r}\right) = \frac{\partial}{\partial x}\left(\rho_0 D_{j,\mathrm{eff}}\frac{\partial Y_j}{\partial x}\right) + \frac{1}{r}\frac{\partial}{\partial r}\left(r\rho_0 D_{j,\mathrm{eff}}\frac{\partial Y_j}{\partial r}\right) + S_j \tag{5}$$

The physical values in the above equations represent the local phase average of the gas control volume. In the above equations U is the gas phase average local velocity, [m s^{-1}]. T is the local average temperature both over the gas and solid phase in [K]. The $\epsilon_0 = 0.7$ [-] and $K = 1.0 \times 10^{-7}$ [m^2] are the porosity and permeability of the porous medium, respectively. The $f = 0.088$ is the inertia coefficient which depends on the Reynolds number and the microstructure of the porous medium [16]. $\lambda_{\mathrm{eff}} = \epsilon_0\lambda_{\mathrm{f}} + (1-\epsilon_0)\,\lambda_{\mathrm{f}}$ W [m^{-1} K^{-1}] is the effective thermal conductivity, where λ_{f} represents fluid thermal conductivity and $\lambda_{\mathrm{s}} = 20.0$ [W m^{-1}K^{-1}] is solid phase thermal conductivity [17,18]. $D_{j,\mathrm{eff}}$ [m^2 s^{-1}] is the effective mass diffusivity of species j. Y_j is the mass fraction of chemical species j and D_j is the mass diffusivity of species j in the multi-component mixture of gases respectively [19].

2.2. *Mathematical model of a reforming process*

The methane/steam reforming process is widely known as a conventional process for producing hydrogen [5]. In the reforming process of methane with steam, the dominant reactions are the following two [20]

fuel reforming reaction:

$$CH_4 + H_2O \rightarrow 3H_2 + CO \tag{6}$$

shift reaction:

$$CO + H_2O \rightleftharpoons H_2 + CO_2 \tag{7}$$

The steam reforming reaction described by equation (6) is a slow, therefore a rate equation is needed. The water-gas shift reaction described by equation (7) is a fast and can be assumed to be in equilibrium at the reforming temperature [5,20,21]. The reaction rates of the two above reactions (equations (6) and (7)) R_{st} and R_{sh} are locally calculated as follows:

$$R_{\mathrm{st}} = w_{\mathrm{cat}} A_{\mathrm{st}} \exp\left(-\frac{E_a}{RT}\right) p_{\mathrm{CH_4}}^a p_{\mathrm{H_2O}}^b \tag{8}$$

$$R_{\text{sh}} = K^{+}_{\text{sh}} p_{\text{CO}} p_{\text{H}_2\text{O}} + K^{-}_{\text{sh}} p_{\text{H}_2} p_{\text{CO}_2} \tag{9}$$

Equation (8) is based on experimental data, where, a and b are reaction order [-], w_{cat}, is the catalyst weight in a control volume of reformer [g m^{-3}], T [K] is the temperature of the fuel conversion, R = 8.314472 [J mol^{-1} K^{-1}] is the universal gas constant, E [J mol^{-1}] is the activation energy of the fuel reforming reaction, A_{st} [mol min^{-1} g^{-1} atm$^{-(a+b)}$] is the pre-exponential factor, p_{CH_4} is the partial pressure of methane, [atm] and $p_{\text{H}_2\text{0}}$ is the partial pressure of steam [atm]. The water shift reaction given by equation (9) reaches equilibrium rapidly, therefore CO_2, H_2, CO and H_2O have to satisfy the equilibrium equation:

$$K_{\text{sh}} = \frac{K^{+}_{\text{sh}}}{K^{-}_{\text{sh}}} = \exp\left(-\frac{\Delta G^{0}_{\text{sh}}}{RT}\right) \tag{10}$$

where ΔG_{sh} is the change of standard Gibbs free energy of shift reaction. This equilibrium constant is given by Eq. (10) is introduced into Equation (9) to calculate the rate of the shift reaction. The mass production or consumption rate of each chemical species by the fuel reforming reaction (Eq. (6)) and shift reaction (Eq. (7)) is also calculated as follows:

$$S_{\text{H}_2} = 3R_{\text{st}} M_{\text{H}_2} + R_{\text{sh}} M_{\text{H}_2} \tag{11}$$

$$S_{\text{CO}} = R_{\text{st}} M_{\text{CO}} - R_{\text{sh}} M_{\text{CO}} \tag{12}$$

$$S_{\text{CO}_2} = R_{\text{sh}} M_{\text{CO}_2} \tag{13}$$

$$S_{\text{CH}_4} = -R_{\text{st}} M_{\text{CH}_4} \tag{14}$$

$$S_{\text{H}_2\text{O}} = -R_{\text{st}} M_{\text{H}_2\text{O}} - R_{\text{sh}} M_{\text{H}_2\text{O}} \tag{15}$$

The value of the mole production or consumption rate for each chemical species is introduced into the species mass transfer equation (Eq. (5)) as a part its source term. The thermodynamic heat generation rate by the reforming reactions (Eqs. (6) and (7)) are calculated based on the reaction rates, as follows:

$$Q_{\text{st}} = -\Delta H_{\text{st}} R_{\text{st}} \tag{16}$$

$$Q_{\text{sh}} = -\Delta H_{\text{sh}} R_{\text{sh}} \tag{17}$$

where ΔH_{st} and ΔH_{sh} are the enthalpy change accompanied with each reaction, [J mol^{-1}]. The heat conduction and convection equations were solved numerically by finite volume and gauss-seidel methods [22,23].

3. Experimental investigation

3.1. Experimental set-up

A schematic view of the experimental setup is shown in Figure 1. A stainless steel reformer was located in an electrical furnace, which can be heated up to 1000 [°C]. High purity methane was the fuel used in the experiment. It was supplied to the reformer via a flow controller and evaporator, which was also used as a pre-heater. Water was fed to the system with a pump. The gas composition after the reforming process was analyzed by gas chromatography prior to which the steam had been separated by cooling down the gas mixture to 2 [°C]. The reforming reaction tube was filled by nickel supported on yttria stabilized zirconia coming from a crushed anode (see Table 1 to check the properties). To control the thermal condition of the experiment, four thermocouples were placed in the experimental set-up as shown in Figure 1 (marked as T). To derive correct kinetics data, the reaction has to occur in a whole volume of catalyst. This can be accomplished by keeping the reforming conversion rate low. To achieve a low level of methane conversion, the fuel was additionally mixed with nitrogen. Nitrogen does not directly influence the reforming reaction but the partial pressure of the components changed, which decreases both the reaction rate and methane conversion rate. All measurements presented in this paper have been performed at atmospheric pressure. The geometrical properties of the reactor have been summarized in Table 2.

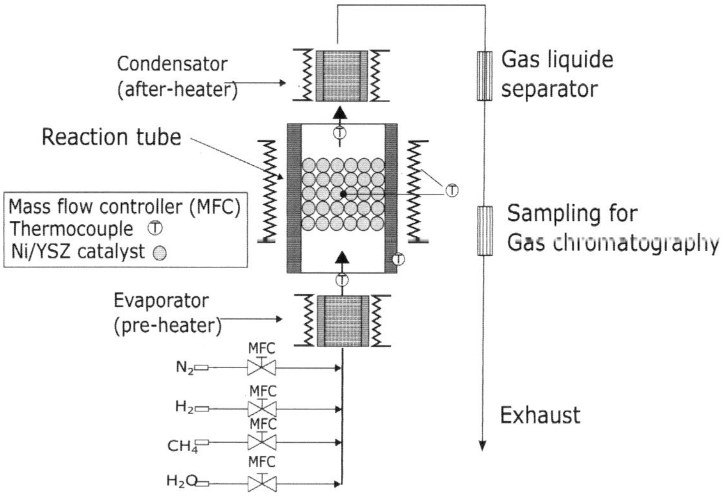

Figure 1. Schematic view on experimental setup

Table 1. Catalyst properties

Type	Ni contain	Particle size	Mass
Ni/YSZ	60vol%	1 [μm]	0.3 [g]

Table 2. Reactor properties.

Type	Bed height	Diameter	Length
stainlesssteel	~ 1 [mm]	25.4 [mm]	450 [mm]

3.2. Reaction rate equation

The mathematical forms of expression describing the kinetics of the methane/steam reforming process can be categorized by three basic concepts [6,10,21,24-27]: i) General Langmuir-Hinshelwood kinetics [5], ii) first order reaction with respect to methane [21,24] and iii) power law expressions derived from data fitting [10,25-27]. In this study, the primarily assumed form of the equation describing the steam reforming kinetic is the power law expression with respect to the partial pressures of methane and steam. This form of kinetic equation is commonly used in thermo-fluid models of the methane/steam reactors [17,18,28,29]. Therefore, the rate equation in the presented research takes the following form:

$$R_{\rm st} = w_{\rm cat} 6.5267 exp\left(-\frac{43920}{RT}\right) p_{\rm CH_4}^{0.39} p_{\rm H_2O}^{0.22} \quad (18)$$

Equation (18) has been determinate using the method described in references [25,30-32]. In this method, the effects of the supply gas composition are first examined by varying the values of SC (steam-carbon ratio) and NC (nitrogen-carbon ratio) at a constant temperature to determine the values of the exponential indexes a and b from Equation (8). Then, experiments are conducted varying the temperature while keeping the supply gas composition unchanged. The reaction constant k is constant for the given temperature, therefore A and E can be determined from the Arrhenius plot. The water-gas shift reaction (2) is assumed to be in equilibrium because of the high temperature.

3.3. Verification of the numerical model

In the numerical model presented in this section, the gas composition at the outlet of the reformer can be predicted based on the inlet conditions. Examples of numerical results and their comparison with the experimental data are shown in Figure 2. In the present results, the effect of the temperature on the steam reforming reaction was studied by increasing the reaction temperature from 773 [K] to 973 [K]. The data shows the outlet dry gas composition to be a function of the reaction temperature. Predictably, the outlet molar fraction of methane decreases with increasing reaction temperature and the amount of carbon monoxide increases as the reaction temperature increases. This observation is significant because in contrast to the low-temperature fuel cell, the carbon monoxide does no poison the high temperature solid oxide fuel cell anode but can be electrochemically converted as a fuel [21].

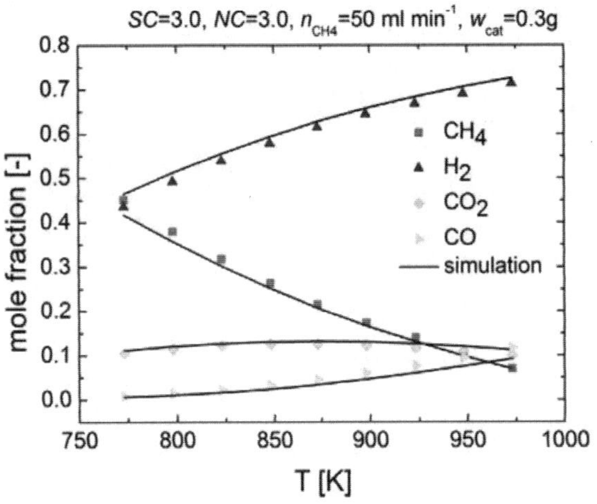

Figure 2. Experimental results compared to numerical simulation

4. Results

Figures 3 and 4 present the results for different heat flux at the reformer wall. In each case the wall was divided into four heating zones. Figures (a) and (c) present results where heat flux increases gradually Q_1=0.5 [kW m^{-2}], Q_2=1 [kW m^{-2}], Q_3=10 [kW m^{-2}], Q_4=20 [kW m^{-2}] for each heating zone respectively. Figures (b) and (d) present result where heat flux decreases gradually Q_1=20 [kW m^{-2}], Q_2=10 [kW m^{-2}], Q_3=1 [kW m^{-2}], Q_4=0.5 [kW m^{-2}] for each heating zone respectively. As can be seen in figures 3 and 4, the conversion rate of methane is higher in the case where heat flux decreases gradually. Also, the difference between the highest and lowest temperature was reduced from 235 [K] to 141 [K] when heat flux decrease gradually.

The results indicate that even when the same same amount of heat is supplied to the reactor, the results might be significantly different depending on the location of the heating zone along the reactor. The results indicate that it is more efficient to heat up the reactor stronger at the beginning. Such approach leads to a higher conversion rate of methane and a lower temperature gradient inside the reactor.

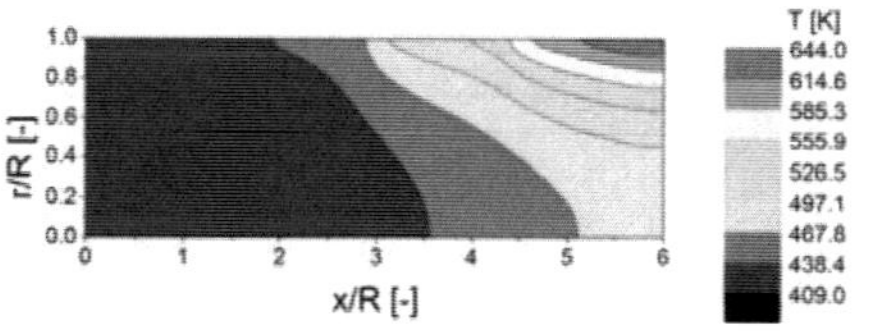

(a) Q_1=0.5 [kW m^{-2}], Q_2=1 [kW m^{-2}], Q_3=10 [kW m^{-2}], Q_4=20 [kW m^{-2}]

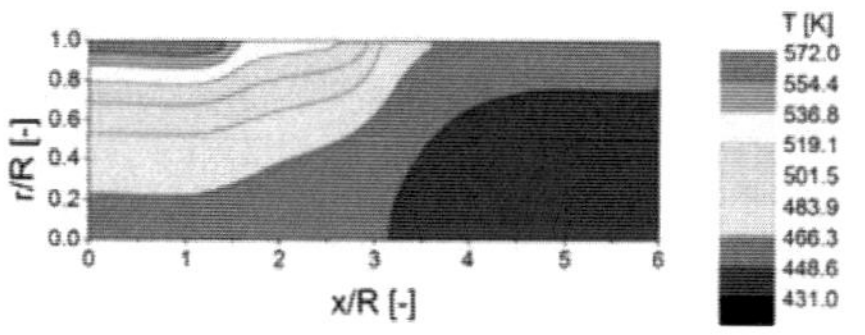

(b) Q_1=20 [kW m^{-2}], Q_2=10 [kW m^{-2}], Q_3=1 [kW m^{-2}], Q_4=0.5 [kW m^{-2}]

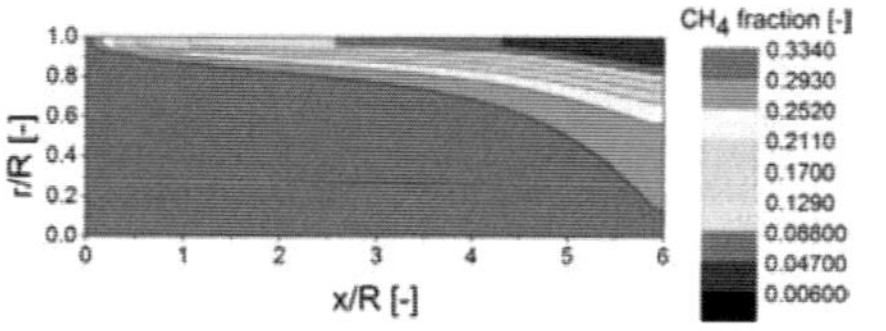

(c) Q_1=0.5 [kW m^{-2}], Q_2=1 [kW m^{-2}], Q_3=10 [kW m^{-2}], Q_4=20 [kW m^{-2}]

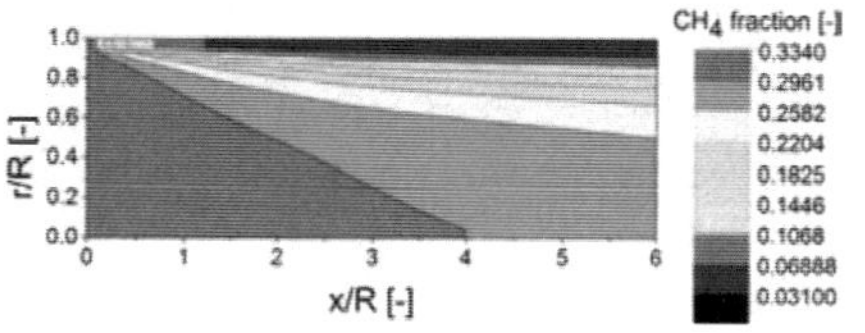

(d) Q_1=20 [kW m^{-2}], Q_2=10 [kW m^{-2}], Q_3=1 [kW m^{-2}], Q_4=0.5 [kW m^{-2}]

Figure 3. Distribution of temperature (a-b) and the mole fraction of methane (c-d) for different thermal boundary conditions.

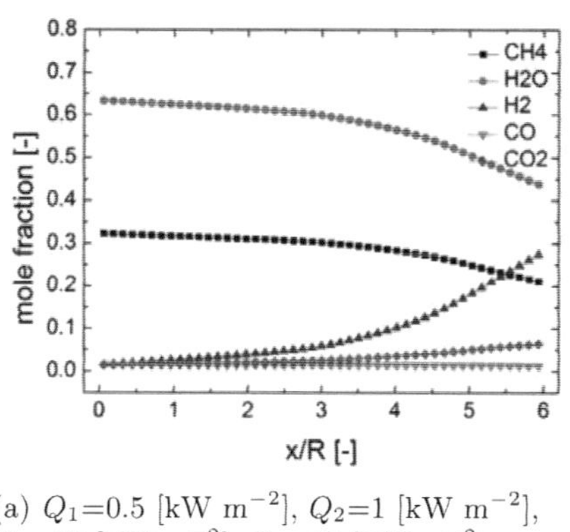

(a) Q_1=0.5 [kW m^{-2}], Q_2=1 [kW m^{-2}], Q_3=10 [kW m^{-2}], Q_4=20 [kW m^{-2}]

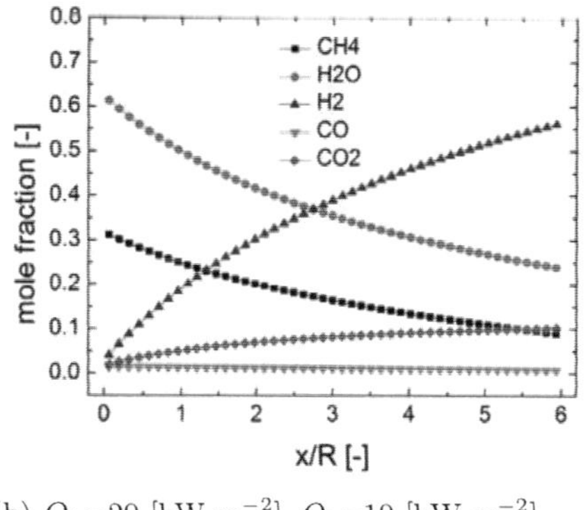

(b) Q_1=20 [kW m^{-2}], Q_2=10 [kW m^{-2}], Q_3=1 [kW m^{-2}], Q_4=0.5 [kW m^{-2}]

Figure 4. Average gas composition along the reformer.

5. General conclusions

This paper has presented experimental and numerical studies on the methane/steam reforming process on Ni/YSZ catalyst. The gas mixture composition was measured at the outlet of the reformer to establish the methane reforming rate and the kinetics of the reforming reaction. The reforming rate equation derived from the experimental data was used in the mathematical model to predict gas composition at the outlet of the reformer and to verify the reforming reaction model. The temperature and species distribution in the longitudinal direction of the cylindrical plug-flow reformer has been presented. Local heating was implemented in numerical computations to discuss the possibility of reducing the temperature gradients along the reformer. It was shown that the location of heating zones plays a key role and might help to not only decrease temperature gradients but also to increase the conversion rate of the methane.

Acknowledgments

This work was supported by the National Centre for Research and Development (Project HTRPL, Contract No. SP/J/1/166183/12).

References

[1] Yamamoto O 2000 Electrochimica Acta **45** 2423-2435
[2] Aguiar P, Chadwick D and Kershenbaum L 2002 Chemical Engineering Science **57** 1665-1677
[3] Ogden J M, Steinbugler M M and Kreutz T G 1999 Journal of Power Sources **79** 143-168.
[4] Brown L F 2001 International Journal of Hydrogen Energy **26** 381-397
[5] Xu J and Froment G F 1989 AIChE J. **35** 97-103
[6] Xu J and Froment G F 1989 AIChE J. **35** 88-96
[7] Brus G and Szmyd J 2007 ECS Transactions **7** 2013-20
[8] Brus G and Szmyd J S 2008 Journal of Power Sources **181** 8-16
[9] Brus G, Kolenda Z, Kimijima S and Szmyd J S 2010 14th International Heat Transfer Conference, IHTC14-22785, Volume 5, Pages 71-80.
[10] Brus G, Kimijima S and Szmyd J S 2012 J. Phys.: Conf. Ser. **395** 012159
[11] Komatsu Y, Kimijima S and Szmyd J S 2013 Journal of Power Sources **223** 232-245
[12] Komatsu Y, Kimijima S and Szmyd J S 2010 Energy **35** 982-988
[13] Kuwahara F, Yamane T and Nakayama A 2006 International Communications in Heat and Mass Transfer **33** 411-418
[14] Nield D A and Bejan A 2006 Convection in porous media (Springer New York)
[15] Carbonell R G and Whitaker S 1984 Fundaments of Transport Phenomena in Porous Media **82** 121-198
[16] Vafai K and Tien C L 1981 Journal of Heat and Mass Transfer **24** 195-203
[17] Suzuki K, Iwai H and Nishino T 2005 Electrochemical and thermo-fluid modeling of a tubular solid oxide fuel cell with accompanying indirect internal fuel reforming. In Transport Phenomena in Fuel Cells Editors Suden B and Faghri M (WIT Press Boston)
[18] Nishino T and Szmyd J S 2010 J. Fuel Cell Sci. Technol. **7** 051004
[19] Poling B E, Prausnitz J M, John Paul O and Reid R C 2001 The properties of gases and liquids vol 5 (McGraw-Hill New York)
[20] Nagata S, Momma A, Kato T and Kasuga Y 2001 Journal of Power Sources **101** 60-71
[21] Achenbach E and Riensche E 1994 Journal of Power Sources **52** 28-38
[22] Patankar S 1980 Numerical heat transfer and fluid flow (McGraw-Hill New York)
[23] Ferziger J H and Peric M 1996 Computational methods for fluid dynamics (Springer Berlin)
[24] Wei J and Iglesia E 2004 Journal of Catalysis **224** 370-383
[25] Brus G, Komatsu Y, Kimijima S and Szmyd J S 2012 Int. J. Thermo **15** 43-51
[26] Yakabe H, Ogiwara T, Hishinuma M and Yasuda I 2001 Journal of Power Sources **102** 144-154
[27] Ahmed K and Foger K 2000 Catalysis Today **63** 479-487
[28] Hosseini S, Ahmed K and Tad M O 2013 Journal of Power Sources **239** 180-196
[29] Achenbach E 1994 Journal of Power Sources **49** 333-348
[30] Brus G 2012 International Journal of Hydrogen Energy **37** 17225-17234
[31] Sciazko A, Komatsu Y, Brus G, Kimijima S and Szmyd J S 2013 ECS Transactions **57** 2987-2996
[32] Sciazko A, Komatsu Y, Brus G, Kimijima S and Szmyd J S 2014 Journal of Power Sources **262** 245-254

Flow structure, heat transfer and scaling analysis in the case of thermo-magnetic convection in a differentially heated cylindrical enclosure

E Fornalik-Wajs[1], P Filar[2], J Wajs[3], A Roszko[1], L Pleskacz[1] and H Ozoe[4]
[1] AGH University of Science and Technology, Department of Fundamental Research in Energy Engineering, Al. Mickiewicza 30, 30-059 Krakow, Poland
[2] Siltronic AG, Hanns-Seidel-Platz 4, 81737 München, Germany
[3] Gdansk University of Technology, Department of Energy and Industrial Apparatus, ul. Narutowicza 11/12, 80-233 Gdansk, Poland
[4] Professor Emeritius of Kyushu University, Kasuga Koen 6 1, 816 8580 Fukuoka, Japan

E-mail: roszko@agh.edu.pl

Abstract. The experimental, numerical and scaling analysis in the case of thermo-magnetic convection in a thermosyphon-like enclosure filled with a paramagnetic fluid is presented. Visualization of temperature field together with the numerical simulation gave an information about the flow structure, which indicated "finger-like" structures of hot and cold streams advecting each other. Their number depended on the Rayleigh number and also on the magnetic induction. In the basis of thermal measurements the average Nusselt number was determined for various strength of magnetic field and in presented configuration increasing magnetic field caused increase in the Nusselt number. With the assumptions of two dimensional flow and the constant fluid properties, in the studied temperature range, the scaling analysis led to the Nusselt number correlation in dependence on the thermo-magnetic Rayleigh number. Very good agreement between the experimental results and scaling equation was observed.

1. Introduction

Closed thermosyphon has a wide range of applications, for example cooling thermal devices (i.e. gas turbine blades [1]), or more specified like preserving permafrost under the buildings or modeling the temperature distribution in earth drillings [2]. Thermosyphons were studied to gain a fundamental understanding of phenomena occurring in such geometry and its comprehensive experimental and numerical studies can be found in [1,2,3,4]. Researchers applied various experimental and numerical approaches, which were mainly related to the applications of thermosyphon. The results show the specific flow pattern in the middle-height cross-section. The appearance of the "streams"/"branches" (in this paper they are called "spokes") of cold and hot fluid advected to each other was noticed. This phenomenon had an influence on the heat transfer between the two streams. In [1] it was stated that the heat transfer can be enhanced "when the earth's gravitational field… is replaced by more powerful body force fields…", for example magnetizing force.
The additional source of acceleration due to a strong magnetic field was introduced to many studied systems, for example in chemistry [5], physics [6], medicine [7], engineering [8]. The new phenomena

were found, for example levitation of various materials in a strong magnetic field, magnetic breath support, magneto-thermal wind, enhanced oxygen supply for the fuel cell [9,10]. Very interesting analysis showing intensification of the heat transfer processes by the magnetic field in the cubical [11, 12], annular [13] and thermosyphon-like geometry was reported in [14].

In the presented paper the experimental and numerical analysis of magneto-convection in the thermosyphon-like enclosure is complemented with scaling analysis.

2. Experimental apparatus

The experimental setup consisted of the magnet, analyzed enclosure, constant temperature bath, temperature data acquisition system, computer, power supply unit and its control system. The cylindrical enclosure (figure 1) consisted of the heating and cooling side walls, with the height of 0.028 m, made of copper and separated by the Plexiglas adiabatic interface of width 0.004 m. The top and bottom end plates of the cylinder were made of Plexiglas and the enclosure was filled with the 50% volume aqueous solution of glycerol. A nichrome wire, covered by rubber, was winded at the outside surface of the cylinder to heat one part of the enclosure. The wire was connected to a DC power supply controlled by the multimeters. The second copper part consisted of a small chamber through which thermostating water was pumped to keep the cooled side wall at constant temperature. The temperature difference was measured with the T-type thermocouples (marked in figure 1 as T.C.). Two thermocouples were located on the outer side of the heated wall while another two were attached to the outlet and inlet of coolant. The working fluid was illuminated through the middle-height adiabatic zone by a white light sheet from the high luminosity LED diodes placed inside the holes drilled in the Plexiglas adiabatic zone. Pictures were taken through the top Plexiglas plate.

3. Properties of the working fluid

The 50% aqueous solution of glycerol was chosen as the working fluid. Its density was measured with a pycnometer and viscosity with an Ostwald's viscosimeter. Other properties marked with the asterisk in Table 1 were estimated from [15]. The Prandtl number ($\Pr = \nu/\alpha$) of the working fluid was estimated on the basis of the fluid properties and was equal to Pr = 42.1. The mass magnetic susceptibility χ_g of the working fluid was measured with the magnetic susceptibility balance according to the modified Gouy method (Evan's method) and was equal to +35×10-8[m^3/kg].

Table 1. Properties of 50% aqueous solution of glycerol at T_0 = 298[K] (the properties marked by asterisk were estimated from [15]).

Property	Value	Unit
α*	= 1.1415×10-7	m^2/s
β*	= 0.445×10-3	K^{-1}
λ*	= 0.422	W/(m·K)
η	= 6.145×10-3± 0.064×10-3	Pa·s
ν	= 4.80×10-6 ± 0.05×10-6	m^2/s
ρ_0	= 1281± 1	kg/m^3
χ_{g0}	= 13.926×10-8± 0.329×10-8	m^3/kg

4. Experimental procedure

The experimental setup, insulated by a vinyl foil, was placed into the bore of the HF5-100VHT helium free 5 T superconducting magnet. It was placed in the bore in such position that the middle-height level of an enclosure was placed at 0.1 m from one of the bore ends. That level assured that the radial component of the magnetic force would be minimal. The distribution of the axial component of magnetic force is presented in figure 2(a), while the vectors represent the magnetic force vectors acting on the enclosure placed in the bore of the superconducting magnet in figure 2(b). The visualization of fluid temperature was carried out with an encapsulated liquid crystal slurry, illuminated with the LED's white light. Detailed description of experimental procedure is presented in [14]. The Rayleigh number was very important in the analysis. In the papers related to the thermosyphons, the inner radius [2] or the diameter [4] of the enclosure were considered as the characteristic length. The radius of the enclosure was chosen as the characteristic length in the presented studies, and therefore the Rayleigh number took the form

$$\mathrm{Ra} = g\beta\left(T_{\mathrm{h}} - T_{\mathrm{c}}\right) r_0^3 (\alpha\nu)^{-1}, \tag{1}$$

where g means the gravitational acceleration, β indicates the thermal expansion coefficient, T_{h} denotes temperature of heated enclosure part, while T_{c} temperature of cooled one, r_0 is the enclosure radius, α means the thermal diffusivity and ν is the kinematic viscosity.

The magnetic Rayleigh number applied in this paper was based on Braithwaite's definition [16] and adopted to the studied cases. Finally it took form:

$$\mathrm{Ra}_{\mathrm{m}} = \mathrm{Ra}\left[1 + \chi_{\mathrm{g}0} C_{\mathrm{n}} (g\mu_0)^{-1} \left(\boldsymbol{B}\nabla\boldsymbol{B}\right)\right], \tag{2}$$

where Ra means the thermal Rayleigh number, $\chi_{\mathrm{g}0}$ is the mass magnetic susceptibility, $C_{\mathrm{n}} = 1+1/(\beta T_0)$, μ_0 is the magnetic permeability of vacuum and $\boldsymbol{B}$ represents the magnetic induction.

5. Heat transfer rate

The thermal measurements were carried out to investigate the influence of magnetic field on the flow structure and indirectly on the heat transfer rate. Definition of the Nusselt number was

$$\mathrm{Nu} = \dot{Q}_{\mathrm{net\_conv}} \cdot \dot{Q}_{\mathrm{net\_cond}}^{-1}. \tag{3}$$

The net conduction ($\dot{Q}_{\mathrm{net\_cond}}$) and net convection ($\dot{Q}_{\mathrm{net\_conv}}$) heat fluxes were estimated according to the method invented in [17]. The definitions of net conduction heat and net convection heat fluxes were formulated as

$$\dot{Q}_{\mathrm{net\_cond}} = \dot{Q}_{\mathrm{cond}} - \dot{Q}_{\mathrm{loss}}, \tag{4}$$

$$\dot{Q}_{\mathrm{net\_conv}} = \dot{Q}_{\mathrm{conv}} - \dot{Q}_{\mathrm{loss}}. \tag{5}$$

Heat loss was assumed to depend on the heater temperature itself and not on the mode of heat transfer inside the enclosure.

The empty thermosyphon-like enclosure was placed into the bore of the magnet and the heat loss ($\dot{Q}_{\mathrm{loss}}$) to the environment was measured. It was assumed that the convection and conduction of the air were negligible due to its low thermal conductivity.

Because of the complex geometry of the experimental system, the net conduction flux in the working fluid ($\dot{Q}^{\mathrm{fluid}}_{\mathrm{net\_cond}}$) was difficult to measure. In the thermosyphon placed upside down (the upper side wall was heated and the lower side wall cooled) the convection occurred near the side walls. Therefore a jelly was used to obtain the conduction state in the cylindrical enclosure. In the solidified jelly, the heat transfer took place only by conduction. After the measurement of the conduction through the jelly, the conduction heat flux through the working fluid for the same system was presumed to be approximated by the equation

$$\dot{Q}^{\mathrm{fluid}}_{\mathrm{net\_cond}} = \dot{Q}^{\mathrm{jelly}}_{\mathrm{net\_cond}} \cdot \lambda^{\mathrm{fluid}} \cdot (\lambda^{\mathrm{jelly}})^{-1}. \tag{6}$$

The estimated value of the thermal conductivity of the jelly was $\lambda^{\mathrm{jelly}} = 0.575 \pm 0.40$ W/(m·K).

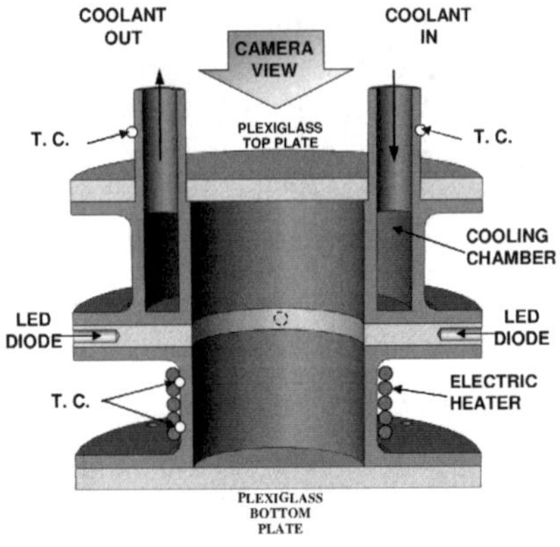

Figure 1. Schematic view of the experimental apparatus.

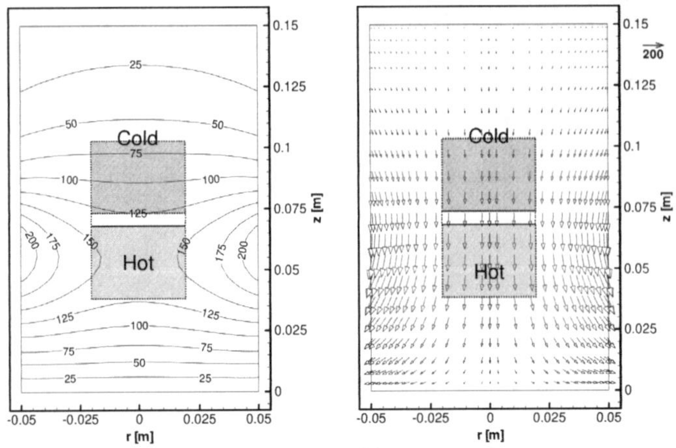

(a) (b)

Figure 2. Calculated distribution of (a) axial component of the magnetic force $B_z\ (\partial B_z/\partial z)$; (b) gradient of square of magnetic induction ∇B^2 at the maximal magnetic induction of 5 T.

6. Numerical analysis

The numerical approach was based on the model for the magnetic convection of a paramagnetic fluid proposed in [18]. The fluid was assumed to be Newtonian, incompressible and the Boussinesq approximation was employed. The analyzed problem was three-dimensional. The governing equations can be written as follows in non-dimensional form:

continuity equation:

$$\nabla \cdot \boldsymbol{V}_{\mathbf{n}} = 0, \tag{7}$$

momentum equation:

$$\frac{D\boldsymbol{V}_{\mathbf{n}}}{D\tau_{\mathrm{n}}} = -\nabla P_{\mathrm{n}} + \mathrm{Pr}\nabla^2 \boldsymbol{V}_{\mathbf{n}} + \mathrm{Ra}\,\mathrm{Pr}\,T_{\mathrm{n}}\left[1 - 0.5\gamma C_{\mathrm{n}} \nabla B_{\mathrm{n}}^2\right], \tag{8}$$

energy equation:

$$\frac{DT_{\mathrm{n}}}{D\tau_{\mathrm{n}}} = \nabla^2 T_{\mathrm{n}}, \tag{9}$$

Biot-Savart's law:

$$\boldsymbol{B_n} = \frac{1}{4\pi} \oint_{solenoid} \frac{\boldsymbol{dS_{nl}} \times \boldsymbol{R}}{R^3}. \tag{10}$$

The non-dimensional al variables were defined as follows [16] $\boldsymbol{V}_{\mathrm{n}} = v/v_0$, $v_0 = \alpha/r_0$, $S_{\mathrm{nl}} = s_l/r_0$, $R = r/r_0$, $B_0 = \mu_0 i/r_0$, $\tau_{\mathrm{n}} = t/t_0$, $P_{\mathrm{n}} = p/p_0$, $\boldsymbol{B}_{\mathbf{n}} = B/B_0$, $T_{\mathrm{n}} = (T-T_0)/(T_{\mathrm{h}}-T_{\mathrm{c}})$, $t_0 = r_0^2/\alpha$, $p_0 = \rho_0\alpha^2/r_0^2$, $p_0 = \rho_0\alpha^2/l^2$, $T_0 = (T_{\mathrm{h}}+T_{\mathrm{c}})/2$, $C_{\mathrm{n}} = 1+1/(\beta T_0)$, $\mathrm{Pr} = \nu/\alpha$, $\mathrm{Ra} = g\beta(T_{\mathrm{h}}-T_{\mathrm{c}})r_0^3/(\alpha\nu)$, $\gamma = \chi_0 B_0^2/(\rho_0\mu_{\mathrm{m}} g r_0)$.

The above partial differential equations were approximated by the finite difference equations. The HSMAC (**H**ighly **S**implified **M**arker **A**nd **C**ell) method with the third-order upwind scheme applied for the inertial terms was used to iterate mutually the pressure and the velocity field. The magnetic field was computed using Biot-Savart's law for the multi-coil system. It was done in such way to simulate specific distribution of the magnetic field as that generated by the solenoid in the real system. Numerical computations were carried out for the same conditions as the experiment and the dimensionless parameters were computed from the experimental conditions and the properties of the fluid.

7. Results

The enclosure was heated from the lower side wall and cooled from the upper one therefore the convection occurred even without a magnetic field. The experimental analysis was done for various values of Rayleigh number. These results were reported in [19], however it can be said that increasing Rayleigh number caused increase in the number of "spokes/branches" in the same way as the magnetic field does, because in the presented configurations the gravitational and magnetizing forces acts in the same direction and as a result the enhanced convection is obtained.

Figure 3 shows the experimental and calculated results at thermal Rayleigh number Ra = 1.69×10^5 and the magnetic field of 0 T. Figure 3(a) presents the image of isotherms with five spoke pattern taken in the middle-height cross-section. Figure 3(b) shows the calculated temperature field and the velocity vectores at the cylinder middle-height. Five spokes of red/yellow color (representing the cold fluid) and five bluish (representing the hot fluid) placed alternatively along the perimeter of cross-section could be counted. They correspond to the five blue and five red spokes presented in figure 3(a).

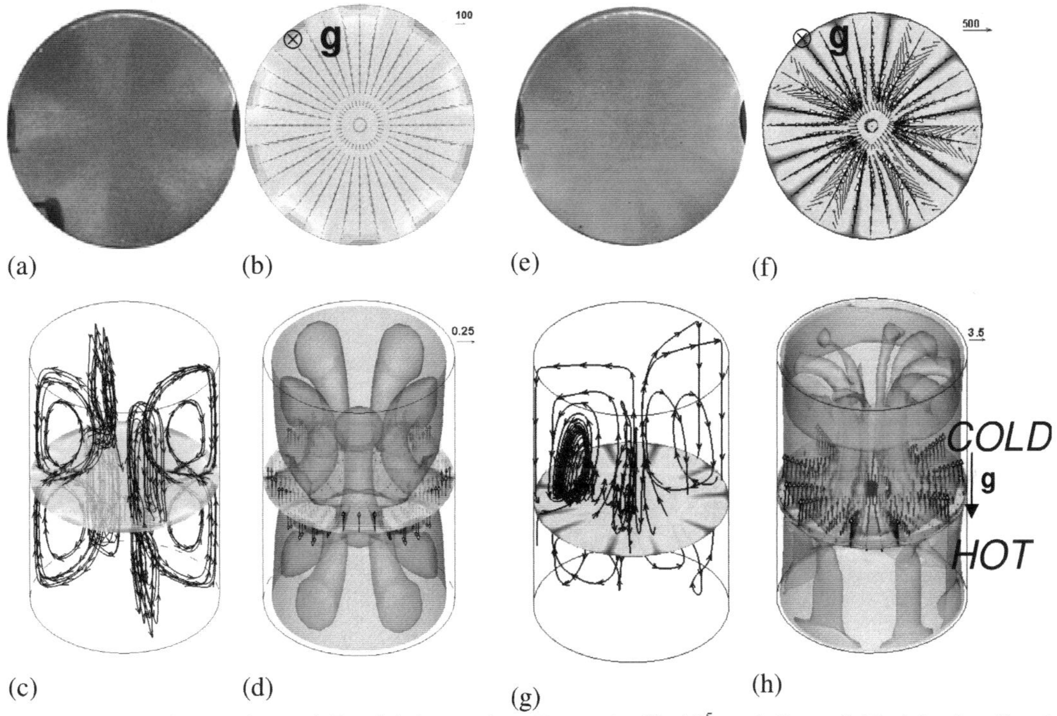

Figure 3. The results at thermal Rayleigh number Ra = 1. 69×10^5 and B_0 = 0 T: (a) visualized temperature field, (b) computed temperature field with velocity vectors, (c) calculated temperature field at cylinder mid-height and long time streak lines, (d) isothermal surfaces and the gravitational buoyancy force vectors [19]; The results at thermal Rayleigh number Ra = 1.43×10^5 and B_0 = 5 T: (e) visualized temperature field, (f) computed temperature field with velocity vectors, (g) calculated temperature field and long time streak lines, (h) isothermal surfaces and the total force vectors.

Figure 3(c) shows calculated temperature field at the cylinder middle-height cross-section with the three-dimensional demonstration of long time streak lines starting from four different angular positions. The fluid moved upwards near the lower heated side wall, then proceeded toward the enclosure centre at the middle-height and near the centre ascended toward the top of cylinder. Finally, fluid descended along the upper cooled side wall. This trajectory is typical for the thermosyphon systems and was reported in [2,3]. Figure 3(d) shows the finger-like isothermal surfaces of T_n = -0.05 and T_n = 0.05 and the gravitational buoyancy force vectors $\boldsymbol{F}_g$ drawn at the cylinder's middle-height.

The isothermal contours correspond to the five spoke patterns obtained at the middle-height level of the enclosure.

Afterwards the calculations were carried out at the magnetic induction level of 1T, 2T, 3T, 4T and 5T, however only the results for the highest value are presented here. Increasing magnetic induction strengthened the magnetizing force and the number of spokes increased. Their shape became less regular and led to the state presented in figures 3(e-h).

Figure 3 presents also the experimental and computed results at Ra = 1.43×10^5 and 5 T of the magnetic induction. Figure 3(e) presents the visualized temperature field with a fourteen spoke pattern taken in the middle-height cross-section. In Figure 3(f) the eleven green (cold fluid) and blue (hot fluid) spokes can be counted in the computed result. There is some discrepancy in the number of spokes observed in figures 3(e) and 3(f). The shown computed cross-section was at the level Z = 1.7. Figure 3(g) shows calculated temperature field at the level Z = 1.7 with the three-dimensional demonstration of long time streak lines starting from four different angular positions. Figure 3(h) shows the isothermal surfaces of $T_n = 0.12$ and $T_n = 0.193$ and the total force vectors $\boldsymbol{F}_\mathbf{m} + \boldsymbol{F}_\mathbf{g}$ drawn at the cylinder middle-height. The finger-shape isothermal surfaces become very distorted. It is also observed that they moved toward the centre of the enclosure and toward the side-walls, which can be compared with the thickening of the boundary layer with an increase in the Rayleigh number.

8. Scaling analysis

The conservation of mass equation for two-dimensional flow in a thermosyphon-like enclosure filled with incompressible fluid can be written in dimensional form as

$$\frac{\partial v_z}{\partial z} + \frac{1}{r}\frac{\partial}{\partial r}(r v_r) = 0 . \tag{11}$$

The momentum equations in incompressible fluid with the assumption of constant viscosity in terms of scalar components can be expressed as

$$\frac{\partial v_z}{\partial t} + v_z\frac{\partial v_z}{\partial z} + v_r\frac{\partial v_z}{\partial r} = -\frac{1}{\rho}\frac{\partial p'}{\partial z} + \nu\left(\frac{1}{r}\frac{\partial}{\partial r}\left(r\frac{\partial v_z}{\partial r}\right) + \frac{\partial^2 v_z}{\partial z^2}\right) - \frac{\chi_{g0}\beta C_n}{2\mu_0}(T - T_0)\frac{\partial B^2}{\partial z} + \beta(T - T_0)\mathbf{g} \tag{12}$$

$$\frac{\partial v_r}{\partial t} + v_z\frac{\partial v_r}{\partial z} + v_r\frac{\partial v_r}{\partial r} = -\frac{1}{\rho}\frac{\partial p'}{\partial r} + \nu\left(\frac{1}{r}\frac{\partial}{\partial r}\left(r\frac{\partial v_r}{\partial r}\right) - \frac{v_r}{r^2} + \frac{\partial^2 v_r}{\partial z^2}\right) - \frac{\chi_{g0}\beta C_n}{2\mu_0}(T - T_0)\frac{\partial B^2}{\partial r} . \tag{13}$$

The energy equation applied in the numerical analysis can be presented as

$$\frac{\partial T}{\partial t} + v_z\frac{\partial T}{\partial z} + v_r\frac{\partial T}{\partial r} = \alpha\left(\frac{1}{r}\frac{\partial}{\partial r}\left(r\frac{\partial T}{\partial r}\right) + \frac{\partial^2 T}{\partial z^2}\right). \tag{14}$$

Instead of solving equations (11-14) numerically, the scaling analysis can be done to predict theoretically the types of flow and heat transfer patterns that can develop in the enclosure. Immediately after $t = 0$, the fluid in the vicinity of side wall is motionless: this means that near the side wall the energy equation (14) expresses a balance between thermal inertia and conduction normal to the wall [20]

$$\frac{\Delta T}{t} \approx \alpha\frac{\Delta T}{\delta_\mathrm{T}^2} . \tag{15}$$

This equality of scales follows from recognizing ΔT, t and δ_T as the scales of changes in T, t and r in equation (14). In the same equation $v_z = v_r = 0$, also $\partial^2 T/\partial z^2 << \partial^2 T/\partial r^2$, because near $t = 0^+$ the thermal boundary layer thickness δ_T is much smaller than the enclosure height ($z \sim l$, where $l=h/2$ and $r \sim \delta_\mathrm{T}$). Equation (15) shows that in the time immediately following $t = 0$, each side wall is coated with a conduction layer whose thickness increases steadily as

$$\delta_{\mathrm{T}} \approx (\alpha t)^{1/2}. \tag{16}$$

The heated layer δ_{T} will naturally tend to rise along the heated wall. To find the velocity scale of this upward motion v, the momentum could be divided into three terms: inertia, friction and buoyancy, respectively in equation (17). In terms of representative scales, the momentum balance can be read

$$\frac{v_r}{\delta_T t}, \qquad v\frac{v_r}{\delta_T{}^3} \qquad \approx \qquad \frac{\chi_{g0}\beta C_n}{2\mu_0}\frac{\Delta T}{\delta_T}\frac{B_0^2}{l} + g\beta\frac{\Delta T}{\delta_T}. \tag{17}$$

The driving force in the momentum balance is the buoyancy effect caused by the gravitational and magnetizing forces. It can be found whether the buoyancy effect is balanced by friction or inertia. For fluids with a Prandtl number equal 1 or more, the momentum balance at $t = 0^+$ is between buoyancy and friction

$$1 \approx \frac{\chi_{g0}\beta C_n}{2\mu_0}\frac{\Delta T B_0^2 \delta_T^2}{l v v_r} + g\beta\frac{\Delta T \delta_T^2}{v v_r}. \tag{18}$$

The initial vertical velocity scale is

$$v_r \approx \frac{g\beta\Delta T\delta_T^2}{v}\left(1 + \frac{\chi_{g0}C_n}{2g\mu_0}\frac{B_0^2}{l}\right). \tag{19}$$

The heat conducted from the side wall into the fluid layer δ_{T} is no longer spent on thickening the layer: part of this heat input is carried away by the layer δ_{T} rising with velocity v_r [20]. In the energy equation three effects can be considered: inertia, convection and conduction. At final time t_{f} the energy equation expresses a balance between the heat conducted from the wall and the enthalpy carried away vertically by the buoyant layer

$$v_r\frac{\Delta T}{l} \quad \approx \quad \alpha\frac{\Delta T}{\delta_T^2}, \tag{20}$$

which thickness can be described by equation (16), therefore

$$\delta_{\mathrm{T,f}} \approx l\left[\frac{g\beta\Delta T l^3}{v\alpha}\left(1 + \frac{\chi_{g0}C_{\mathrm{n}}}{2g\mu_0}\frac{B_0^2}{l}\right)\right]^{-1/4}, \tag{21}$$

it can also be written

$$\delta_{\mathrm{T,f}} \approx l\left[\mathrm{Ra}\left(1 + 0.5\gamma C_{\mathrm{n}}\right)\right]^{-1/4} = l\left(\mathrm{Ra}_{\mathrm{m}}\right)^{-1/4}, \tag{22}$$

C_n is a constant and Ra_{m} is the magnetic Rayleigh number (equation (2)).
Using the convective heat transfer coefficient, the following relation can be written for the studied case

$$\alpha\Delta T \approx \lambda\frac{\Delta T}{\delta_{\mathrm{T}}}. \tag{23}$$

The order of magnitude for the average Nusselt number is

$$\mathrm{Nu} \approx \frac{\alpha l}{\lambda} \approx \frac{l}{\delta_{\mathrm{T}}} \approx \left(\mathrm{Ra}_{\mathrm{m}}\right)^{1/4}. \tag{24}$$

The results are presented in figure 4 in comparison with the experimental data. The assumption about two-dimensional flow was made. The final relation between the Nusselt number and the magnetic Rayleigh number presented as equation (24) was applied, especially because the magnetic Rayleigh number was already used to characterize the flow in the thermosyphon-like enclosure under the magnetic field.

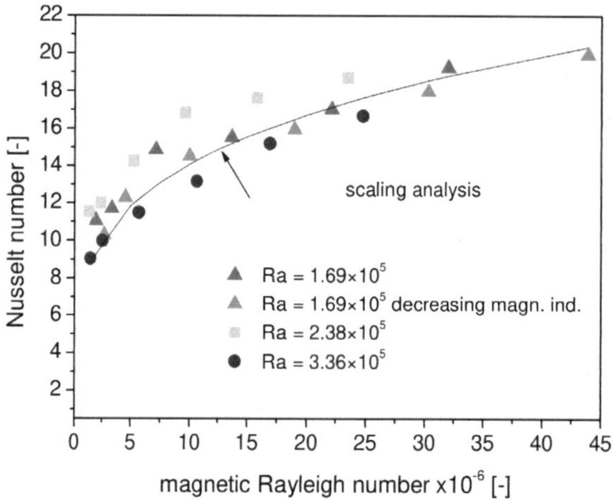

Figure 4. Relation between the Nusselt and the magnetic Rayleigh numbers coming from experimental and scaling analysis for the cylindrical thermosyphon-like enclosure at thermal Rayleigh number values ▲▲Ra = 1.69×10^5, ■Ra = 2.38×10^5, •Ra = 3.36×10^5 and various magnetic inductions

Good correspondence between the experimental data and scaling law can be observed in figure 4, even with the applied assumptions. Some differences between the presented experimental and numerical data and the results of scaling analysis can derive from the assumptions about the two-dimensionality of the flow, when the real flow is three-dimensional. The accuracy of the measurements and simplification of the conservation laws could have an influence on the scaling analysis. Ultimately, the most important point is that this analysis is suitable for brief and quick prediction of the Nusselt number.

9. Summary

Combined effect of thermal buoyancy force and magnetic field on the fluid behavior in the thermosyphon-like enclosure was presented. Gravitational and magnetizing forces were acting in the same direction, therefore increasing Rayleigh number and increasing magnetic field resulted in the enhanced heat transfer. It could be observed by increasing number of "spokes" and increasing value of Nusselt number. Very interesting flow structure was obtained numerically and compared with the experimental results. Scaling analysis (with indicated in the paper assumptions) led to the Nusselt number correlation in the form of a power function of thermo-magnetic Rayleigh number. It was derived and shown that the Nusselt number is proportional to the thermo-magnetic Rayleigh number raised to a power of 0.25. The comparison with the experimental results gave good agreement and is promising for brief and quick estimation of heat transfer.

10. References

[1] Mallinson G D, Graham A D and de Vahl Davis G 1981 Three-dimensional Flow in a Closed Thermosyphon *J. Fluid Mech.* **109** 259-75

[2] Japikse D, Jallouk P A and Winter E R F 1971 Single-phase Transport Processes in the Closed Thermosyphon *Int. J. Heat Mass Tran.* **14** 869-87

[3] Ishihara I, Fukui T and Matsumoto R 2002 Natural Convection in a Vertical Rectangular Enclosure with Symmetrically Localized Heating and Cooling Zones *Int. J. Heat Fluid Fl.* **23** 366-72

[4] Lock G S H and Kirchner J D 1992 Some Characteristics of the Inclined, Closed Tube Thermosyphon under Low Rayleigh Number Conditions *Int. J. Heat Mass Tran.* **35 (1)** 165-73

[5] Honda K, Sato A and Nakabayashi S 2003 Magneto-taxis of Nonlinear Chemical Reaction *Proc. of the 7th Symp. on New Magneto-Science* (Tsukuba, Japan) 52-5
[6] Uyeda C, Tanak K, Sakakibara M and Takashima R 2003 Development of a Method to Detect Magnetic Anisotropy with High Sensitivity in Micro-Gravity Condition *Proc. of the 7th Proc. of the 7th Symp. on New Magneto-Science* (Tsukuba, Japan) 92-5
[7] Onodera H Jin Z and Chida S 2003 Human Body Risk Assessment Under High Magnetic field Environment *Proc. of the 7th Symp. on New Magneto-Science* (Tsukuba, Japan) 208-16
[8] Ozoe H 2005 *Magnetic Convection* (London: Imperial College Press)
[9] Simon M D and Geim A I 2000 Diamagnetic levitation: Flying frogs and floating magnets (invited) *J. Appl. Phys.* **87** 6200-4
[10] Wakayama N I, Okada T, Okano J and Ozawa T 2001 Magnetic promotion of oxygen reaction with Pt catalyst in sulfuric acid solutions *Jpn J. Appl. Phys.* **40** 269-71
[11] Bednarz T, Fornalik E, Tagawa T, Ozoe H and Szmyd J S 2006 Convection of paramagnetic fluid in a cube heated and cooled from side walls and placed below a superconducting magnet: comparison between experiment and numerical computations *Thermal Science and Engineering* **14** pp 107-114
[12] Kenjeres S, Pyrda L, Wrobel W, Fornalik-Wajs E and Szmyd J S 2012 Oscillatory states in thermal convection of a paramagnetic fluid in a cubical enclosure subjected to a magnetic field gradient *Phys. Rev. E* **85** 1-8
[13] Wrobel W, Fornalik-Wajs E and Szmyd J S 2010 Experimental and numerical analysis of thermo-magnetic convection in a vertical annular enclosure *Int. J. Heat Fluid Flow* **31** 1019-1031
[14] Fornalik E, Filar P, Tagawa T, Ozoe H and Szmyd J S 2006 Effect of magnetic field on the convection of paramagnetic fluid in the unstable and stable thermosyphon-like configurations *Int. J. Heat Mass Tran.* **49** 2642-51
[15] VDI-Wärmeatlas 1997 VDI-Verlag
[16] Braithwaite D, Beaugnon E and Tournier R 1991 Magnetically controlled convection in a paramagnetic fluid, *Nature* **354** 134-6
[17] Ozoe H and Churchill S W 1973 Hydrodynamic Stability and Natural Convection in Newtonian and Non-Newtonian Fluids Heated from Below, *AIChE Symposium Series, Heat Transfer* **69** **(131)** 126-33
[18] Tagawa T, Shigemitsu R and Ozoe H 2002 Magnetizing force modeled and numerically solved for natural convection of air in a cubic enclosure: Effect of the direction of the magnetic field *Int. J. Heat Mass Tran.* **45** 267-77
[19] Filar P, Fornalik E, Tagawa T, Ozoe H and Szmyd J S 2006 Numerical and experimental analyses of magnetic convection of paramagnetic fluid in a cylinder *J. Heat Transf.* **128** 183-91.
[20] Mukhopadhyay A, Ganguly R, Sen S and Puri I K 2005 A scaling analysis to characterize thermomagnetic convection *Int. J. Heat Mass Tran.* **48** 3485-92

Acknowledgement
The present work was supported by the Polish Ministry of Science (Grant AGH No. 11.11.210.198).

Journal of Physics: Conference Series **530** (2014) 012042 doi:10.1088/1742-6596/530/1/012042

Velocity measurement of two-phase liquid-gas flow in a horizontal pipeline using gamma densitometry

R Hanus[1,4], M Zych[2] and L Petryka[3]

[1]Faculty of Electrical and Computer Engineering, Rzeszow University of Technology, Powstancow Warszawy 12, 35-959, Rzeszow, Poland
[2]Faculty of Geology, Geophysics and Environmental Protection, AGH University of Science and Technology, Al. Mickiewicza 30, 30-059, Krakow, Poland
[3]Faculty of Physics and Applied Computer Science, AGH University of Science and Technology, Al. Mickiewicza 30, 30-059, Krakow, Poland

E-mail: rohan@prz.edu.pl

Abstract. This paper presents application of gamma-ray absorption method to liquid-gas flow investigation in a pipeline. In the described measurement two sealed ^{241}Am radioactive sources and probes with NaI(Tl) scintillation crystals have been used. For the analysis of digital signals provided by detectors, a traditional cross-correlation function (CCF), and modified correlation methods based on the quotient of CCF and average magnitude difference function (AMDF), as well as the quotient of CCF, and average square difference function (ASDF) have been proposed. Exemplary results of the mean velocity determination of the gaseous phase transported by a liquid in the water-air mixture flow were demonstrated and the evaluation of its uncertainty have been presented.

1. Introduction

Two-phase flows occur frequently in nature and industry, e.g., in the transportation of such mixtures as a liquid-gas, or liquid-solid particles transported by a gas or liquid. Measuring the velocity or the flow rate of these components is difficult and frequently requires the use of non-invasive methods. The analysis of flow parameters and its control encourages the use of advanced measuring techniques, employing among others electrical tomography, ultrasonic or Coriolis flow meters and radioactive isotopes. The last technique uses tracers injected under certain conditions into the flow, or one based on the sealed radioactive sources emitting gamma ray and application of its absorption [1] - [6]. In the both cases as detectors of radiation, a scintillation probes may be mounted outside of the analyzed stream.

The typical equipment for a gamma-absorption measurement consists of a sealed radioactive source, and a scintillation probe which is used to record density or selected phase concentration in the stream. By employing of two such sets it is possible to measure additionally the velocity of a minority phase of the flow. The sources used in such equipment are placed at the L distance one from another at one side of the pipe, as shown in figure 1. Due to two parallel photon beams shaped by collimators, continues analysis of the mixture is possible. At the opposite side of the pipe the probes are installed,

[4] To whom any correspondence should be addressed.

and the output of count rates $I_x(t)$ and $I_y(t)$ are recorded. These signals depend on the condition of the mixture in the pipe's cross sections and describes the instantaneous state of the flow [5].

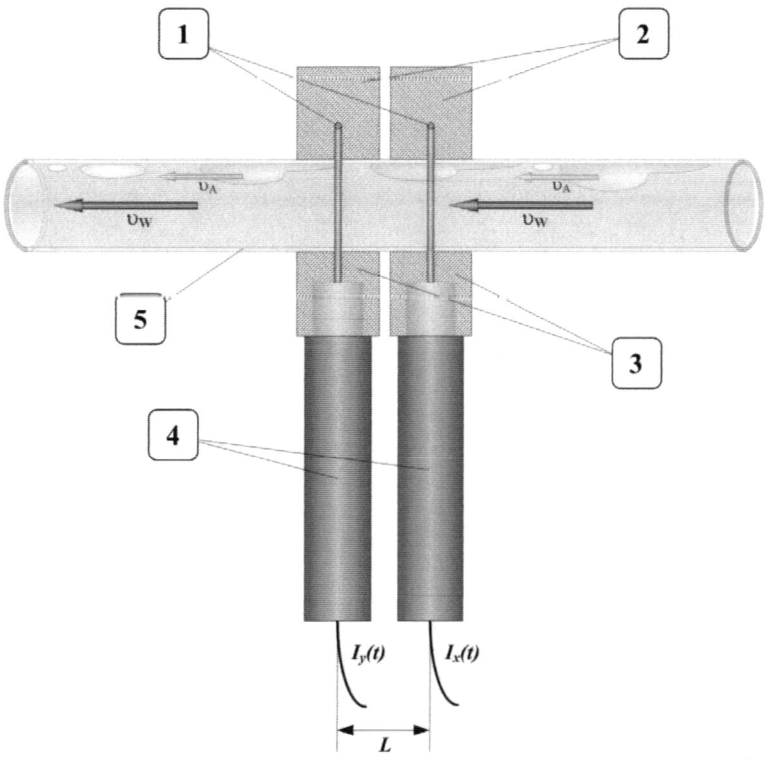

Figure 1. The principle of the gamma-absorption measurement in a horizontal pipeline: 1 – sealed radioactive source; 2 – collimator of the source; 3 - collimator of the detector; 4 - probe; 5 – pipeline.

The statistical comparison of two such signals allows determination of the time delay τ_0 necessary for transportation of the minority phase of the flow, and then its mean velocity υ_A from the following formula:

$$\upsilon_A = L/\tau_0 \qquad (1)$$

2. Laboratory stand

Figure 2 illustrates the piping constructed in the Industrial Radiometry Laboratory of the AGH University of Science and Technology in Krakow, Poland. The installation is part of stand constructed for investigation of the liquid-gas mixture flows, simulating the typical processes which occur e.g. during transportation of the crude oil and natural gas mixture. For investigation purposes a measuring section of 4.5 m pipe from metaplex with 30 mm internal diameter was used, where the linear velocity of the liquid was adjusted in the range from 0.5 to 2.5 m/s.

The station is equipped with a data acquisition system, a computer with appropriate software which permits statistical analysis of signals obtained from the detectors and ultrasonic flow meter. In the presented investigation two ^{241}Am isotopes emitting photons of 59.5 keV energy were used, as well as the detectors with NaI(Tl) scintillation crystals. Two probes are placed with the distance of L = 97 mm between them.

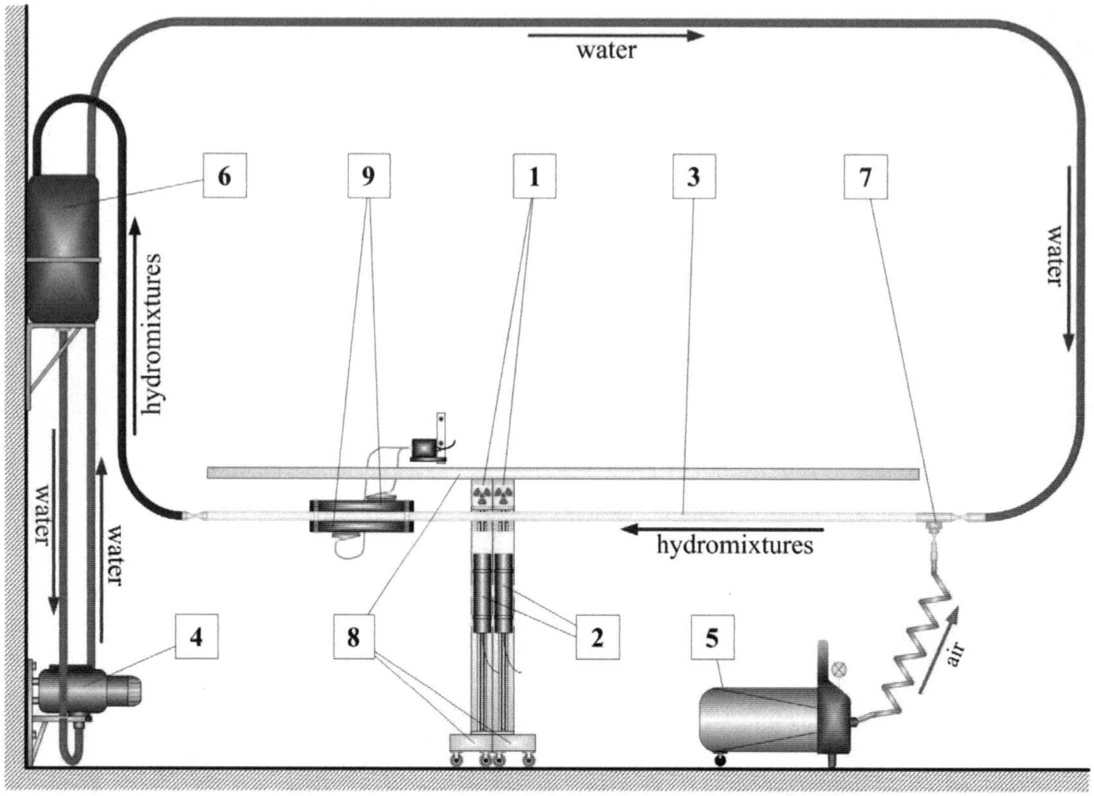

Figure 2. A graphical presentation of the piping for the liquid-gas mixture flow's investigation: 1 – sealed radioactive source; 2 – probe; 3 – pipe; 4 – pump; 5 – compressor; 6 – air-removing container; 7 – air nozzle; 8 – shifting system of probes; 9 – ultrasonic flow meter.

The piping allows additionally application of radiotracers to determine the selected mixture's component transportation. A more detailed description of the station is given in [7].

3. Methods of signals analysis

The voltage pulses $I_x(t)$ and $I_y(t)$ provided by the scintillation probes in accordance with figure 1 were counted within a selected sampling interval Δt, and create discrete time signals $x(n)$ and $y(n)$. These signals as one recorded in radioisotope measurements deliver not only the statistical property of the analyzed flow, but also significant influence of the radiation background, electronic noises, and fluctuations of the nuclear decay. Appropriately long fragments of these signals, after preprocessing (centering, and filtration) are of the ergodic type, and can be fruitfully analyzed by statistical methods in time and frequency domains described in detail in [8] - [16]. The well-known cross-correlation function (CCF), as well as an average magnitude difference function (AMDF), and average square difference function (ASDF), are the methods used chiefly for the estimation of transportation time across the selected distance [4], [9], [14] - [16]. The estimator of mean transportation time delay $\hat{\tau}_0$ can be obtained as the position of the extremum of the discussed functions. The discrete estimators of CCF, AMDF, and ASDF can be presented in the following equations:

$$R_{CCF}(\tau) = \frac{1}{N}\sum_{n=0}^{N-1} x(n)y(n+\tau), \tag{2}$$

$$R_{AMDF}(\tau) = \frac{1}{N}\sum_{n=0}^{N-1}\left|x(n) - y(n+\tau)\right|, \tag{3}$$

$$R_{ASDF}(\tau) = \frac{1}{N}\sum_{n=0}^{N-1}\left[x(n) - y(n+\tau)\right]^2, \tag{4}$$

where: N – number of samples, τ – time delay.
For the paper we evaluated the above methods and propose to use the combined CCF/AMDF, and CCF/ASDF calculations for analysis of signals obtained in gamma-absorption measurements of the liquid-gas mixture transportation. These calculations were as follows:

$$R_{CCF/AMDF}(\tau) = \frac{R_{CCF}(\tau)}{R_{AMDF}(\tau)}, \tag{5}$$

$$R_{CCF/ASDF}(\tau) = \frac{R_{CCF}(\tau)}{R_{ASDF}(\tau)}. \tag{6}$$

The combination of correlation and AMDF formulas was recently proposed to analyze of acoustic signals [17]. In result of theoretical considerations of the denominators of equations (5) and (6) we proposed to add a small positive number to avoid division by zero. However, for the analysis of real signals, particularly for signals obtained from the scintillation probes it is not necessary.

4. Exemplary results of experiment

Figure 3 presents plots of the normalized CCF, CCF/AMDF, and CCF/ASDF characteristics obtained for the signals recorded in the BUB006 experiment. Normalization was done by dividing values of each function by its maximum value. In the BUB006 experiment, the laminar flow was analyzed and the bandwidth of signals was about 20 Hz. The data acquisition parameters in this experiment were as follows: $N = 180000$, $\Delta t = 1$ ms.
In calculations the selected range of data around the maximum of the CCF, CCF/AMDF, and CCF/ASDF have been approximated with the probability density function $p(\tau)$ of the normal distribution:

$$p(\tau) = p_0 + \frac{1}{\sigma\sqrt{2\pi}}\exp\left(-\frac{(\tau - \hat{\tau}_0)^2}{2\sigma^2}\right). \tag{7}$$

where: p_0 – normalization level of the Gauss function, σ – standard deviation of the fitted distribution.
In these cases the $\hat{\tau}_0$ estimators are determined as the first moment of the fitted distribution [6], [12] and the standard uncertainty of the mean transportation time $u(\hat{\tau}_0)$ is given by:

$$u(\hat{\tau}_0) = \frac{\sigma}{\sqrt{m}}, \tag{8}$$

where m – number of points selected around the maximum.

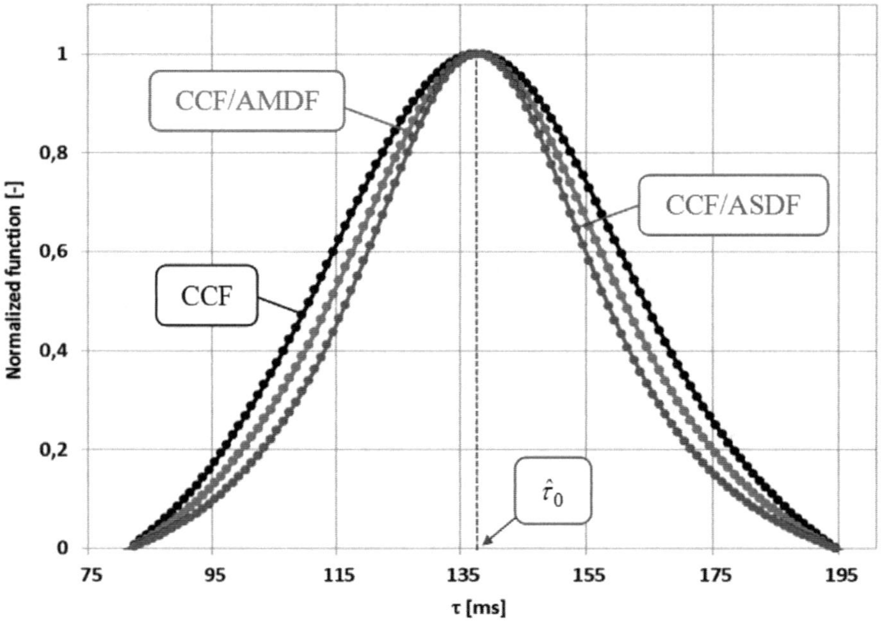

Figure 3. Plots of normalized CCF, CCF/AMDF, and CCF/ASDF calculated in the BUB006 experiment with m = 100.

The combined standard uncertainty $u_c(\upsilon_A)$ of the measured phase velocity, with negligible small uncertainties of the acquisition set, depends on an inaccuracy of uncorrelated L and $\hat{\tau}_0$ values determination:

$$u_c(\upsilon_A) = \left[\left(\frac{\partial \upsilon_A}{\partial L} \right)^2 u^2(L) + \left(\frac{\partial \upsilon_A}{\partial \hat{\tau}_0} \right)^2 u^2(\hat{\tau}_0) \right]^{1/2}, \tag{9}$$

where: $u(L)$ - the standard uncertainty of the distance between the detectors in figure 1.
Results of the mean velocity υ_A and its combined uncertainty $u_c(\upsilon_A)$ [18] obtained in the BUB006 experiment are presented in the table 1.

Table 1. Results of measurements in BUB006 experiment

Method	Mean velocity υ_A (m/s)	Combined uncertainty $u_c(\upsilon_A)$ (m/s)
CCF	0.708	0.012
CCF/AMDF	0.708	0.011
CCF/ASDF	0.708	0.009

5. Conclusions

Comparison of the results presented in the paper shows the following order of increasing uncertainty of the gas phase mean velocity: the CCF/ASDF, CCF/AMDF, and CCF. The uncertainties of air bubbles mean velocity in the described experiment did not exceed two percent, which is a satisfactory for many industrial measurements. According to the authors, the proposed modified cross-correlation method is worthwile for concideration in a similar two-phase flow measurements.

6. References

[1] Lowenthal GC and Airey PL 2001 *Practical applications of radioactivity and nuclear radiations* (Cambridge: University Press)

[2] Johansen GA and Jackson P 2004 *Radioisotope gauges for industrial process measurements* (New York: Wiley)

[3] Kumara WAS, Halvorsen BM and Melaaen MC Single-beam gamma densitometry measurements of oil–water flow in horizontal and slightly inclined pipes 2010 *International Journal of Multiphase Flow* **36** 467–480

[4] Jung SH, Kim JS, Kim JB and Kwon TY Flow-rate measurements of a dual-phase pipe flow by cross-correlation technique of transmitted radiation signals 2009 *Applied Radiation and Isotopes* **67** 1254-1258

[5] Petryka L, Hanus R, and Zych M Application of gamma absorption method to two phase flow measurement in pipelines 2012 *Przeglad Elektrotechniczny* **88** 185-188 (in Polish)

[6] Zych M, Petryka L, Kepiński J, Hanus R, Bujak T and Puskarczyk E Radioisotope investigations of compound two-phase flows in an open channel 2014 *Flow Measurement and Instrumentation* **35** 11–15

[7] Hanus R, Szlachta A, Kowalczyk A, Petryka L, and Zych M Radioisotope measurement of two-phase flow in pipeline using conditional averaging of signal 2012 *Proc. IEEE Mediterranean Electrotechnical Conference MELECON 2012 (Yasmine-Hammamet Tunisia March 25-28)* (IEEE Press) 144-147

[8] Bendat JS and Piersol AG 2010 *Random data - analysis and measurement procedures* (New York: Wiley)

[9] Beck MS and Plaskowski A 1987 *Cross correlation flowmeters - their design and application* (Bristol: Adam Hilger)

[10] Hanus R Investigation of the correlation method of time delay estimation with Hilbert Transform of measuring signal 2012 *Przeglad Elektrotechniczny* **88** 39-41 (in Polish)

[11] Mosorov V A method of transit time measurement using twin plane electrical tomography 2006 *Measurement Science and Technology* **17** 753-760

[12] Petryka L, Zych M and Murzyn R The non-stationary two-phase flow evaluation by radioisotopes 2005 *Nukleonika* **50** 43-46

[13] Kowalczyk A, Hanus R and Szlachta A Investigation of the statistical method of time delay estimation based on conditional averaging of delayed signal 2011 *Metrology and Measurement Systems* **17** 335-342

[14] Fernandes CW, Bellar MD, and Werneck MM Cross-correlation-based optical flowmeter 2010 *IEEE Trans. Instrum. Meas.* **59** 840-846

[15] Hanus R Satistical error comparison of time delay estimation using cross-correlation function and phase of cross-spectral density function 2008 *Przeglad Elektrotechniczny* **84** 301-303 (in Polish)

[16] Jacovitti G and Scarano G Discrete time technique for time delay estimation1993 *IEEE Trans. Sign. Proc.* **41** 525-533

[17] Chen J, Benesty J and Huang Y Performance of GCC- and AMDF-based time-delay estimation in practical reverberant environments 2005 *EURASIP Journal on Applied Signal Processing* **1** 25–36

[18] *Guide to the expression of uncertainly in measurement* 1995 (International Organisation for Standardisation)

Journal of Physics: Conference Series **530** (2014) 012043 doi:10.1088/1742-6596/530/1/012043

The wake behind a rotating sphere

M Skarysz[1], J Pryśko[1], S Goujon-Durand[2] and J E Wesfreid[2]

[1] Warsaw University of Technology, Institute of Aeronautics and Applied Mechanics, Nowowiejska 24, 00-665 Warsaw, Poland

[2] Ecole Superieure de Physique et de Chimie Industrielles de Ville de Paris, PMMH, UMR7636 CNRS ESPCI P6-P7, France

E-mail: mskarysz@gmail.com

Abstract. The wake behind an axisymmetric body rotating about an axis aligned with the streamwise direction has been experimentally investigated in a low-velocity water channel for the intermediate Reynolds number. The measurements focused on the evolution of the flow regime with two counter rotating vorticies, as a function of two control parameters, namely the Reynolds number and the dimensionless rotation rate Ω.

1. Introduction

The wake behind different 3D static bodies e.g. sphere, disk, cube etc. were considered in the numerical as well as experimental studies. The flow of a viscous fluid past a rotating sphere may be considered as a simplified case of a general family of immersed axisymmetric body flows with widespread applications. Therefore the investigation of the wake behind a rotating sphere is a main objective of this paper, as understandig the effect of rotation in more complicated flows (e.g. wind turbines) becomes easier.

The direction of rotation has significant influence on the characteristics of the flow over a sphere. Two particular directions of rotation can be specified. The first one is orthogonal to the streamwise direction. In this case researches on the so-called Magnus or Robins effect (Barkla & Auchterlone [1]) were focused on the determination of the lift and drag force. The other case is the rotation of the sphere parallel to the streamwise direction. Most of works (Kim and Choi [4]; Pier 2012; Poon *et al.* [6]) focused on forces (lift and drag) acting on a rotating sphere depending on the values of free stream velocity and the rotational speed. However, these investigations were undertaken only numerically. The influence of rotation direction on the wake was presented by Poon *et al.* [6]. The simulations were performed for the Reynolds number 100, 250 and 300 and for the non-dimensional rotational rate $\Omega = 0.05 - 1$. The angle of axis of rotation with respect to the streamwise direction varied between 0 and 90 degrees. The steady and unsteady flow regimes were indentified. In present paper the sphere rotating about an axis aligned with the streamwise direction was considered.

These flow instabilities and the bifurcations in the wake past a static sphere ($\Omega = 0$) are well known now from previous reasearch (both experimentally and numerically). The first three regimes can be distinguished: axisymmetric basic flow, planar symmetric flow with two steady counter rotating vorticies and the flow with unsteady periodic hairpin shedding preserving symmetry plane. Subsequent bifurcations appear for the critical Reynolds numbers $Re_1 = 212$ and $Re_2 = 272$ (Johnson & Patel [3]; Gumowski *et al.* [2]). The flow regimes, observed numerically, for the rotating sphere, parallel to the streamwise direction, significantly differ from the static configuration. The wake becomes "frozen" for specific Re and Ω for steady as well as for transient flows (Kim & Choi [4]). For the steady planar symmetric flow with tho counter-rotating vortices one of the vortical structures becomes stronger, while the second is weakened, as an effect of rotation. This phenomenon was demonstrated by Kim and Choi. They have established the frequency of complete of the revolution vortical structure as a function of the rotation rate Ω as well as the Reynolds number. Pier [5] numerically identified the subsequent flow regimes (i) an axisymmetric steady base flow (either stable or unstable) (ii) low-frequency periodic helical, (iii) quasi-periodic shedding and (iv) high-frequency periodic helical.

The main purpose of the present paper is to present results of experimental investigation the effect of the streamwise rotation on the characteristics of flow past a sphere as a function of Reynolds number. Of particular interest is the influence the rotation rate Ω on the vorticity in the wake as well as the existence of different flow regimes. (The rotation rate Ω is defined as a ratio of the maximum azimuthal velocity of the rotating body and the free stream velocity.) These results are to be confronted with earlier numerical investigations. However, the experiment itself is quite difficult to prepare as minimum intrusive system allowing the rotation of the spere is needed (the same system should also serve as a sphere support as well as to deliver the colorant). This is further discussed in the text.

2. Experimental set-up

The research was carried out in a low-velocity water channel using LIF (Laser Induced Fluorescence) visualization and PIV (Particle Image Velocimetry) method. The cross-section of the channel was 10 $\times$ 10 cm. The length of the test section was 86 cm. The diameter of the investigated sphere was 2 cm. The range of investigated velocities was between 0.4 and 4 cm/s, while the corresponding Reynolds number was between 80 and 800. The supporting system is shown in Figure 2.

An angular velocity of the sphere was controlled with high accuracy thanks to the use of brushless electric motor. As it was mentioned the most difficult problem to resolve was to design the rotational system without significantly perturbing the flow. It was required to achieve the rotation as well as injection the colorant for the visualization at the same time. To protect the motor against water and damage a special cover with seals was manufactured. The motor was placed inside the water channel at the considerable distance (1 meter) upstream from the test section. The torque was transmitted by 3 mm carbon fiber tube with high stiffness to reduce the possibility of vibrations. All parts were assembled using laser pointer to maintain coaxiality. The investigated angular velocity was 0-200 rpm. The full setup is shown in Figs. 1, 2 and 3.

2.1. *Visualization*

To perform visualization of the flow on the rotating device special treatment have been made (Figure 3). The dye (fluoresceine dye in solution) was injected through a small hole in the middle of sphere (see Fig. 3). Additional dedicated parts were manufactured in order to resolve the difficulties of the dye injection simultaneously with the rotation. The fluoresceine is delivered from the external reservoir to the rotating sphere: by (i) the silicone tube, in (ii) the steel tube to the small bullet shape body, then (iii) into the rotating perforated tube, next outside of the bullet shape body in the steel tube and (v) at the end to the sphere. Special seals have been used to prevent leaks between the rotating axis and the non revolving part of the bullet shape body.

2.2. *PIV*

2D Particle Image Velocimetry method was used with purpose to obtain quantitative data related to the flow behind the rotating sphere. It is a well-known, non-intrusive, optical and indirect

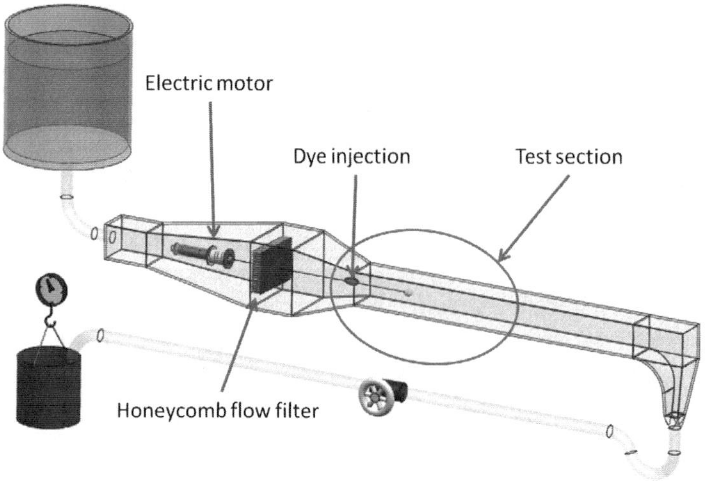

Figure 1. Schematic view of the low velocity water channel.

measurement of two in-plane velocity components of the investigated flow field. The water was seeded with tracer particles with mean diameter size 11 μm. The particles were carried by the fluid and their position was determined by illumination with the laser sheet and captured by a high-resolution digital camera. Two orientations were used in the in this paper: the back view orthogonal to the streamwise direction and the side view parallel to the streamwise direction.

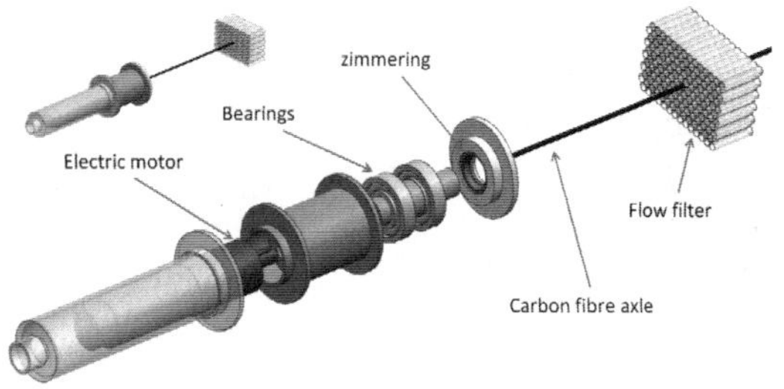

Figure 2. Waterproof cover with electric motor.

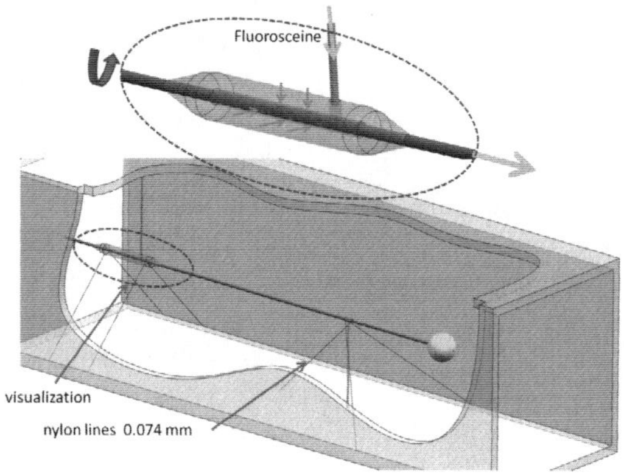

Figure 3. The visualization system.

3. The experimental results

The measurements focused on the identification of the flow regimes that appear depending on the values of two control parameters, namely the Reynolds number $Re = \frac{v_\infty d}{\nu}$, where d is the diameter of the sphere, v_∞ - free stream velocity, ν - kinematic viscosity while the dimensionless rotation rate $\Omega = \frac{v_\theta}{v_\infty}$ The rotation rate Ω is defined as a ratio of the maximum azimuthal velocity of the rotating body and the free stream velocity.

The flow past a rotating sphere was investigated for different Reynolds number but here is presended for the most representative Re = 250 to concentrate on an initial regime of flow instability. The axis of rotation was parallel to the streamwise direction and the non-dimension rotational speed Ω remained in the range 0-1. The present study focuses on the modification

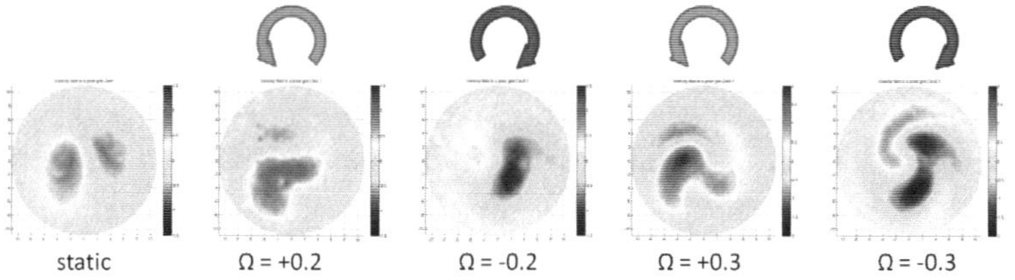

Figure 4. Instantaneous streamwise vorticity field obtained from PIV - the clockwise rotation and the counter clockwise rotation.

of vortical structures behind the sphere. As explained below, the experimental results confirm that the flow characteristics highly depend on the rotational speed.

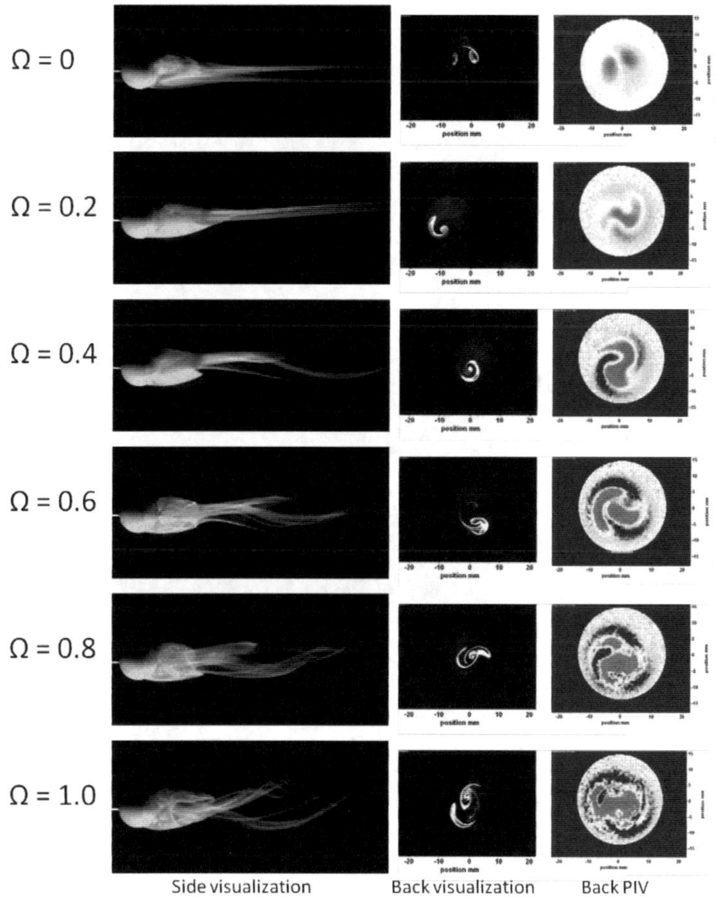

Figure 5. Visualization of the flow a) side view, b) back view and back view vorticity field (PIV results) for Re = 250

Firstly, we observed the influence of sign of rotation on the flow characteristic. There are two opposite sign of rotation: clockwise rotation vs. counter clockwise rotation, let is determined as a negative and a positive respectively. The streamwise direction of rotation with a constant angular velocity was taken into consideration. In Fig. 4 there are presented instantaneous streamwise vorticity fields ω_x for case of static sphere ($\Omega = 0$), rotating sphere in both directions with $\Omega = \pm 0,2$ and with $\Omega = \pm 0,3$. In the first picture the regime of two counter vorticies with planar symmetry is visible. The red color corresponds a positive values of vorticity, the blue to the negative. In the next picture it is shown that the negative vorticity becomes weaker (blue area is smaller than in the first picture) while positive ω_x becomes stronger (red one is bigger). This is caused by a positive rotation of the sphere. In the middle picture the same effect is present. In this case the negative vortical structure is enhanced due to negative rotation. The next two pictures confirm this effect. The influence on vortical structures caused by the rotating of the body is symmetric with respect to the sign of the rotation. Neither positive direction nor negative is specified. Therefore, from this point we will consider only positive direction of rotation of the sphere without loss of generality.

Figure 5 presents the changes in the flow caused by the rotation of the sphere at $Re = 250$ with flow visualization as well as PIV. In the static case ($\Omega = 0$) a pair of vorticies in opposite direction appears in the wake. It is confirmed in the first row containing side view flow visualization, back view visualization and back view PIV results. At $\Omega = 0.2$ one of two vorticies becomes bigger while the other is disappearing. At $\Omega = 0.4$ one can see that the flow becomes helical. The effect becomes stronger with the bigger Ω. The second column presents back view visualization (fluoreseine dye induced by laser sheet in a cross section ortogonal to the flow near the sphere). Based on this method of visualization one can notice that the flow becomes periodic for small Ω. Observed modifications caused by rotating sphere based on visualization can be confirmed by PIV results (Fig. 6).

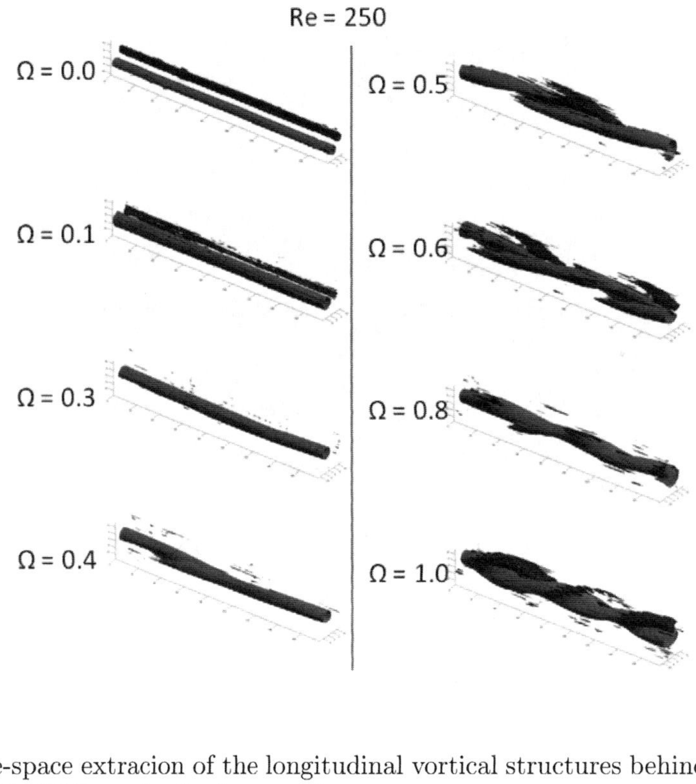

Figure 6. Time-space extracion of the longitudinal vortical structures behind a rotating sphere as a function of Ω. PIV results for Re $= 250$

Figure 6 shows the two-dimensional vortical structures expanded in time behind a stationary sphere and rotating sphere at Reynolds number of 250. The surfaces of vortical structures are identified using the method of longitudinal component of imaginary part of the complex eigenvalues of the gradient velocity tensor and this allows comparison with numerical resultsi.

The diagrams was colored by the value of vorticity. The red color corresponds to positive value of vorticity, the blue color to the opposite one. Again, without rotation in the wake a pair of vorticies appear with the same vortical strength but opposite direction of rotation. At $\Omega = 0.1$ the vortical structure with positive ω_x becomes stronger but that of negative ω_x becomes weaker. This can be explained by the fact that rotation of the sphere induces positive streamwise vorticity in the wake. At the $\Omega = 0.3$ the negative vorticity almost disappears in the wake. However, the structure with negative ω_x appears again in the wake for $\Omega \geq 0.5$. This effect was described in the numerical research (Kim and Choi [4]; Pier [5]) and thus it is confirmed by experiment. The vortical structures with positive and negative sign are strongly twisted together for larger values of Ω.

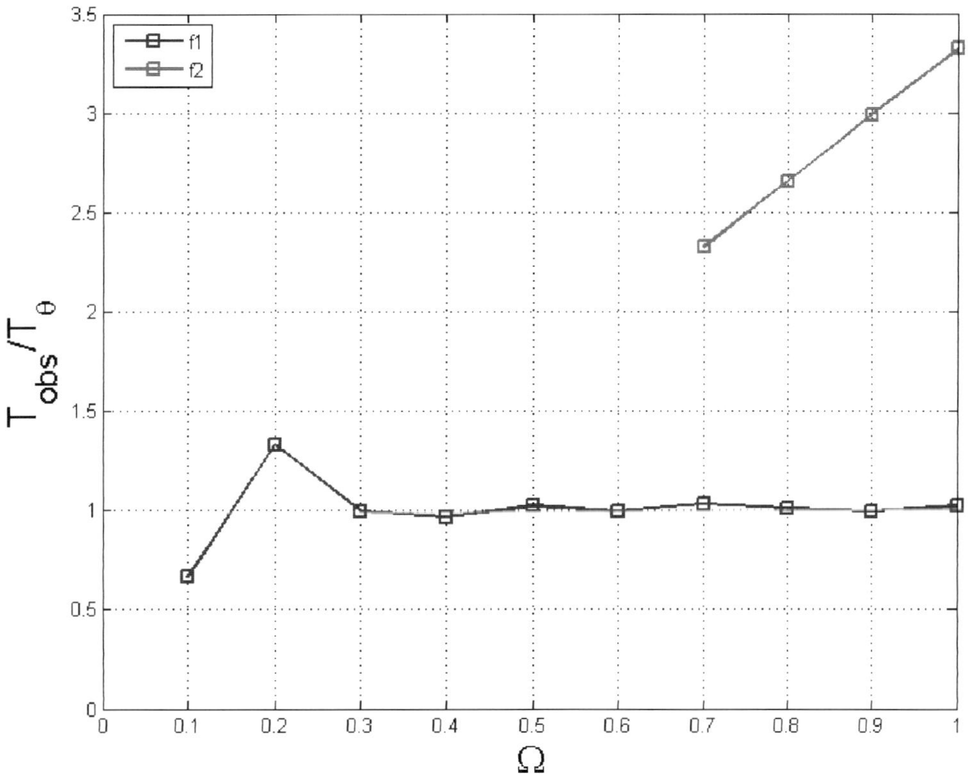

Figure 7. Time for the vortical structure to complete one revolution normalized by the time for the sphere to rotate by 2π and time of second observed frequency normalized.

We have studied also the temporal behavior of such flows, following the Fourier spectra of the transversal velocity field. Figure 7 presents the main frequences obtained from fourier decomposition, as a function of Ω for Re=250. We observe two main scenaries: for small Ω the unsteadiness is related to the frozen vorticity rotating with the sphere while for the intermediate Ω there appears a new temporal instability. This phenomena are at present under further investigation.

4. Conclusions

Laminar flow past a sphere rotating in the streamwise direction was experimentally investigated. The research was performed at $Re = 250$ and in the range of $0 \leq \Omega \leq 1$. The results showed that the flow strongly depends on rotational speed. Rotation of the sphere induces positive streamwise vorticity in the wake and this causes weakening of the negative vortex while the flow becomes periodic and helical.

References

[1] Barkla H M and Auchterlone L J 1971 The Magnus or Robins effect on rotating spheres. *J. Fluid Mech.* **47** 437-447.

[2] Gumowski K, Miedzik J, Goujon-Durand S, Jener P and Wesfreid J E 2008 Transition to time-dependent state of uid ow in the wake of a sphere. *Physical Review E* **77** 3, 13, 17

[3] Johnson T A and Patel V C 1999 Flow past a sphere up to a Reynolds number of 300. *J. Fluid Mech.* **378** 19-70

[4] Kim D and Choi H 2002 Laminar ow past a sphere rotating in the streamwise direction. *J. Fluid Mech.* **461** 365386.

[5] Pier B 2012 Periodic and quasiperiodic vortex shedding in the wake of a rotating sphere. *J. of Fluids and Structures* **41** 43-50

[6] Poon E K W, Ooi A S H, Giacobello M and Cohen R C Z 2010 Laminar flow structures from a rotating sphere: Effect of rotating axis angle. *International J. of Heat and Fluid Flow* **31** 961-972

Journal of Physics: Conference Series **530** (2014) 012044 doi:10.1088/1742-6596/530/1/012044

Comparison of PIV and Hot-Wire Statistics of Turbulent Boundary Layer

A Dróżdż[1], V Uruba[2]

[1] Czestochowa University of Technology, Armii Krajowej 21, Czestochowa, PL
[2] Academy of Science of the Czech Republic, Dolejskova 5, 182 00 Prague 8, CR

E-mail: arturdr@imc.pcz.czest.pl

Abstract. The paper shows a cross checking of turbulent boundary layer measurements using large field of view PIV and hot-wire anemometry techniques. The time-resolved PIV method was used for the experiments. The measuring plane was oriented perpendicularly to the wall and parallel to the mean flow. Hot wire measurement has been performed using the special probe with perpendicular hot wire. The HW point measurements were performed in the same place as PIV experiments. The hot-wire probe has the wire length of $l^+ < 20$ in considered range of Reynolds numbers. Various evaluation methods were applied on PIV data. The profiles of statistical characteristics of streamwise velocity components were evaluated from the data. Mean values, standard deviations as well as skewness and kurtosis coefficients were compared for a few values of Re_θ. Reynolds number ranges from 1000 to 5500. The result shows that with the increasing Reynolds number the attenuation of fluctuations maximum in PIV measurements occurs with respect to Hot-Wire measurements, however representation of velocity fluctuations using the PIV method is satisfactory. The influence of wall-normal fluctuation component on Hot-Wire near wall peak was also investigated.

1. Introduction

The measurements of small-scale turbulence are highly challenging due to the insufficient spatial resolution especially in high Reynolds number flows. According to the common opinion, a hot-wire anemometry using a single-wire probe is sufficient to resolve the streamwise u velocity component [1]. The comparison of u' fluctuation distributions obtained with DNS (Direct Numerical Simulation) and from a single-wire probe revealed self-similarity in shape, and some differences in levels are attributed to the uncertainty error [2,3]. Averaging due to a large measuring volume is known to reduce the near-wall peak of turbulence intensity [4], but also it could falsify higher order moments, like the skewness and flatness factors [5]. It could also reduce the frequency of detected burst events as documented by Johansson and Alfredsson [6]. Ligrani and Bradshaw [4] found two key recommendations for accurate measurements, both became standards for hot-wire design i.e. $l^+ < 20$ and $l/d > 200$, where l is length of a wire (in the viscous units l^+), while d is wire diameter. There is no such clear recommendation for PIV measurements. However it is know that the spatial averaging occur in PIV results, however it is masked by noise present in PIV results [7]. However

[1] To whom any correspondence should be addressed.

comparing the PIV results with hot-wire techniques one should be also aware that a single-wire probe does not measure the u component, but the resultant velocity fluctuations, composed of the streamwise u and wall-normal v fluctuation components. It is worth to note that v component is of the smaller scale than the streamwise component therefore is more sensitive to spatial averaging. Dróżdż and Elsner [8] showed energy spectra comparison of single and X-wire probe and concluded that using single-wire probe the near-wall peak have an elevated value caused by the influence of the v component in single wire readings. It could results from the near-wall vortical structures inducing strong v component in the case, where the mean velocity U is the lowest [8]. They concluded also that that the criteria for the wire length, i.e. $l^+ < 20$, could not be sufficient to properly estimate the streamwise and wall-normal fluctuations.

The paper shows a comparison of turbulent boundary layer measurements using large field of view PIV and well resolved hot-wire anemometry technique that fulfil the Ligrani and Bradshaw [4] recommendation. In case of PIV the fluctuating components are sensitive on vector estimation method and the interrogation area.

2. Test rig and experimental setup

The existing blow-down test rig was modified for experiments with the boundary layer flow. The tunnel has rectangular cross-section with filled corners within the contraction (to suppress corner vortices), honeycomb and a system of damping screens followed by contraction with contraction ratio 16. The area of the test section input is 0.25 m in height and 0.1 m in width, the channel is 3 m long. The time mean velocity departures from homogeneity in planes perpendicular to the tunnel axis are of order tenth of per cent with the exception of corners, where corner vortex starters could be detected. The natural turbulence level was about 0.1 % in the working section input. Application of the 2mm cylindrical wire or the strip of coarse-grained sandpaper to the channel wall at different positions from the leading edge allowed for obtaining a value of the Reynolds number, based on the momentum thickness, up to $Re_\theta \approx 5500$.

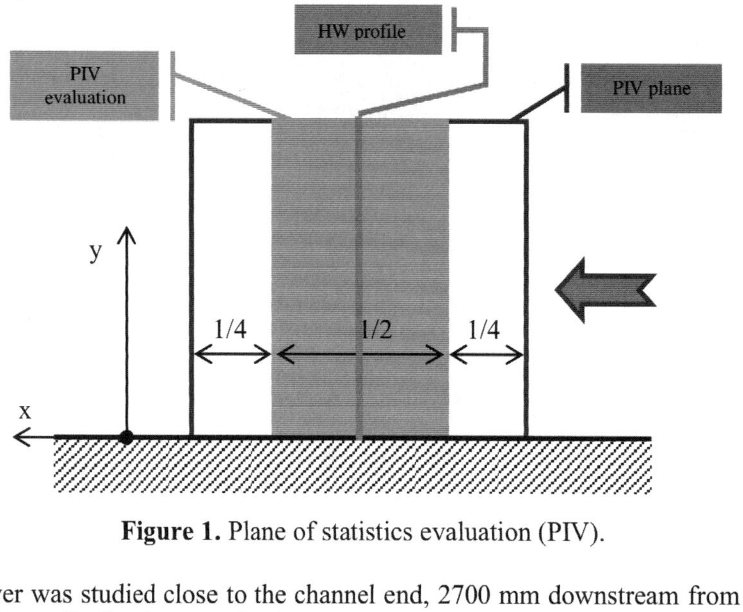

Figure 1. Plane of statistics evaluation (PIV).

The boundary layer was studied close to the channel end, 2700 mm downstream from the inlet, in the middle of the wider wall. The time-resolved PIV method was used for the experiments. The measuring system DANTEC consists of a double-pulse laser with cylindrical optics and CMOS camera. The software Dynamics Studio 3.4 was used for velocity-fields evaluation. Laser New Wave Pegasus Nd:YLF, double head, wavelength 527 nm, maximal frequency 10 kHz, a shot energy is 10 mJ for 1 kHz (corresponding power 10 W per head). Camera Phantom V711 has maximal resolution 1280 x

800 pixels and corresponding maximal frequency 3000 double-snaps per second. For the seeding the SAFEX fog generator was used, which products droplets of about 1 micron in size. The measuring plane was oriented perpendicularly to the wall and parallel to the mean flow. The field of view of PIV camera was of the size of the boundary layer thickness. Hot wire measurement has been performed using the special probe with perpendicular hot wire, 3 μm in diameter and 0.4 mm in length, the DANTEC anemometer StreamLine was used. The HW point measurements were performed in the same place as PIV experiments (see figure 1). Sampling frequency was 75 kHz, signal was low-pass filtered on 30 kHz. The hot-wire probe has the wire length of $l^+ < 20$ in considered range of Reynolds numbers. The wall closest position of the hot-wire probe was determined using the mirrored image.
The profiles were evaluated from measurements of statistics. The mean of statistics calculated from ½ of width in the middle (see figure 2). The time records were 20 s at frequency equal 100 Hz double snaps, time in viscous units was from 0.1 to 0.7.

3. Comparison of statistics for different PIV vector estimation methods.

The error analysis of both HW and PIV method is commonly known. For the HW it is connected with precision of the calibration, accuracy of the cooling law used and wire length, while for PIV it is connected with precision of the particle position determination from the individual snaps and their spatial resolution. The first task was to compare the different estimation methods of vector PIV maps. The methods were also tested on spatial resolution sensitivity. There are substantial differences in spatial resolution of the methods in question: HW and PIV, because of large field of view of PIV camera. The HW method allows for closer position to the wall. The measuring point in the x and y direction is much smaller for the HW (the HW dimension was 3 μm), while for the PIV is given by the interrogation area size in table 1. In general, the velocity value evaluated using the PIV method represents itself the spatial mean value over the interrogation area. The PIV data were evaluated by 5 alternative methods (see table 1).

Table 1. Parameters of PIV interrogation area for PIV methods in question.

	method	IA [pxs]	IA [mm]	dy [pxs]	dy [mm]
PIV 1	Adaptive correlation	16x16	0.72x0.72	8	0.3573
PIV 2	2D least square matching	39x15	1.74x0.67	6	0.2680
PIV 3	Adaptive correlation	32x32	1.43x1.43	16	0.7146
PIV 4	Adaptive PIV	8x8	0.36x0.36	4	0.1787
PIV 5	Adaptive PIV	32x8	1.43x0.36	4	0.1787

The first method is the adaptive correlation method, which calculates velocity vectors with an initial interrogation area (IA) of the size N time the size of the final IA and uses the intermediary results as information for the next IA of smaller size, until the final IA size is reached [10]. The 50% overlap with 2 steps refinement was used for two interrogation areas 0.72x0.72 mm and 1.43x1.43. The second one is the Adaptive PIV method, which is an automatic and adaptive method for calculating velocity vectors based on particle images. The method will iteratively adjust the size and shape of the individual interrogation areas (IA) in order to adapt to local seeding densities and flow gradients. The last one was 2D Least Squares matching (LSM). The method is used for determining 2D velocity fields in highly seeded flows in water and air. The input data consists of double-frame images or of time-resolved single-frame images and the output data are equally spaced vector fields. Interrogation areas from within the image are analyzed to determine local affine transformations.
In contrast to correlation based techniques, LSM shifts, rotates and stretches a fluid element. For this purpose, the LSM algorithm iteratively compares gray value information of an interrogation area in the first time step with the gray value information in the second time step. This is an iterative least squares procedure applying a proper transformation on the interrogation areas. In 2D this results in six

transformation parameters. The advantage of LSM is that whilst calculating the zero order translational velocities, the first order terms of motion are simultaneously optimized increasing the accuracy of the velocity field. The resulting displacement gradient tensor includes parameters like rotation, shear and strain of the interrogation area resulting from the particle displacement within the area. 2D LSM performs a geometric and radiometric transformation between two successive states of the same system. In the case of a gray value filled interrogation area its state is a gray value distribution of the pixel elements. For this purpose, the transformation is optimized such that the gray value differences between a template area and a search area reach a minimum. Compared with conventional 2D cross-correlation, LSM considers the deformation of a fluid element for the calculation of the displacements. This results in a more accurate calculation of the velocity field. The velocity gradient tensor and as result the rotation and deformation rate tensor can be calculated without applying central difference schemes. In LSM all this is done without manipulation of the raw particle images.

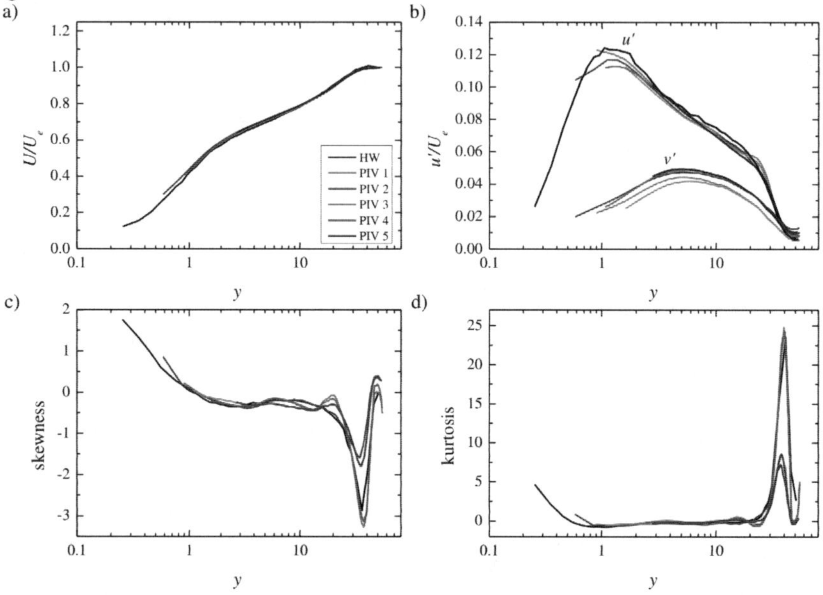

Figure 2. Statistics for Reynolds number 1100 for different PIV methods: mean velocity a), turbulence intensity b), skewness factor c) and flatness factor d)

Figure 2 presents the estimation of the statistics for $Re_\theta = 1100$ using above methods: mean velocity U/U_e figure 2a), turbulence intensity u'/U_e and v'/U_e figure 2b), skewness factor $S_f = \overline{u^3}/\overline{u^2}^{3/2}$ figure 2c) and flatness factor $F_f = \overline{u^4}/\overline{u^2}^2 - 3$ figure 2d). The mean velocity profile is affected substantially by neither the experimental method (HW and PIV) nor by evaluation method resolving the logarithmic region. Even very close to the wall the values obtained using the PIV method are reliable, as in viscose sublayer the instantaneous velocity distribution in the *y* direction is supposed to be close to the linear. However, the viscose sublayer itself is far of to be well resolved by the PIV method for the spatial resolution reasons mentioned above.

In the inner region (viscose sublayer, buffer region and logarithmic region $y < 20$ mm) the fluctuations are captured nearly perfectly using the 2D LSM method. The other PIV evaluation method cause irregularities and artificial oscillations of skewness and kurtosis. However in the wake region ($y > 20$ mm) the 2D LSM method fails in capturing big absolute values of both skewness and kurtosis, here the adaptive method offers better results. In this region strong intermittent behaviour of occurrence of the turbulent fluid from the boundary layer and the fluid from the low-turbulent flow outside the boundary layer is expected. The adaptive PIV method shows the worse results through all the profile. The intensity of fluctuations is underestimated by all PIV methods, the underestimation is proportional to the interrogation area size. The fluctuations amplitudes are the best captured by the adaptive method with the smallest interrogation area - PIV1. Surprisingly, the adaptive PIV method with smaller interrogation area PIV4 shows more intense fluctuation attenuation.

4. Statistics comparison of HW with PIV

The profiles of statistical characteristics of streamwise velocity components were evaluated from the data by the two measuring techniques. The hot-wire data were compared to 2D LSM methods of PIV vector field estimation. Mean values, standard deviations as well as skewness and kurtosis coefficients were compared for three cases in Reynolds number from a range 1000 to 5500. The distributions of mean velocity a), turbulence intensity b), skewness c) and flatness d) factors were presented for $Re_\theta = 1100$ in the figure 3, for $Re_\theta = 3200$ in figure 4 and for $Re_\theta = 5500$ in figure 5.

The turbulence intensity profiles b) show that with the increase of Reynolds number, the underestimation of PIV results is more important. The effect of spatial averaging is evident. The small-scales energy is attenuated because the most of attenuation is present close the wall where small-scale structures dominate. The result is similar as for attenuation of the near wall peak of fluctuation when the hot-wire length is growing.

The skewness factor distributions show that with the increase of Reynolds number, the skewness factor obtained from PIV is increasing. The effect is the same as for the result of Orlu and Alfredsson [5], where with the decrease of spatial resolution the skewness factor increases. On the other hand the flatness factor changes are negligible with the Reynolds number.

The mean velocity profile is well captured by the PIV method for all examined Reynolds numbers. The fluctuations attenuation close to the wall produced by the PIV method, with respect to HW measurements fulfil Ligrani and Bradshaw [4] recommendation, is of growing importance with increasing Reynolds number. This effect could be explained by appearance of smaller structures connected with higher frequencies, which could not be captured by the PIV method, but only by the HW.

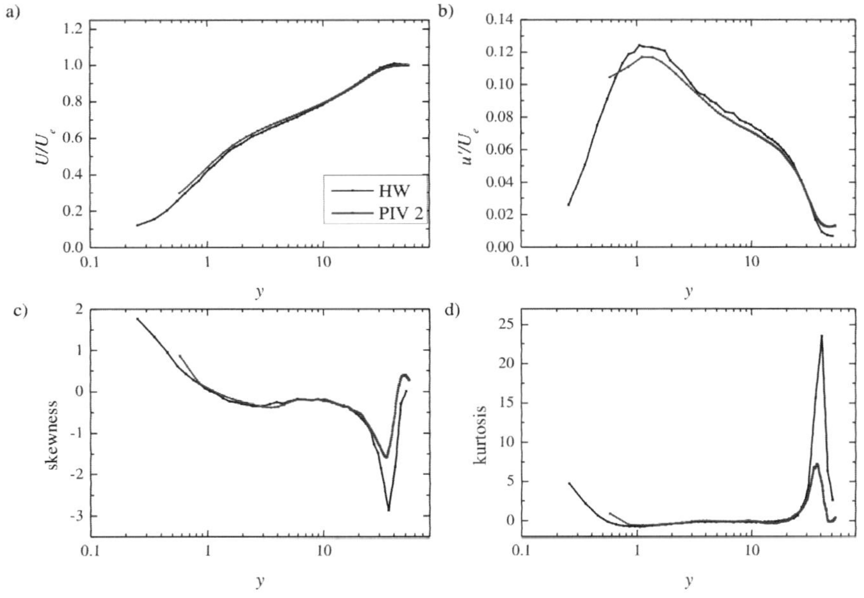

Figure 3. Statistics for Reynolds number 1100: mean velocity a), turbulence intensity b), skewness factor c) and flatness factor d)

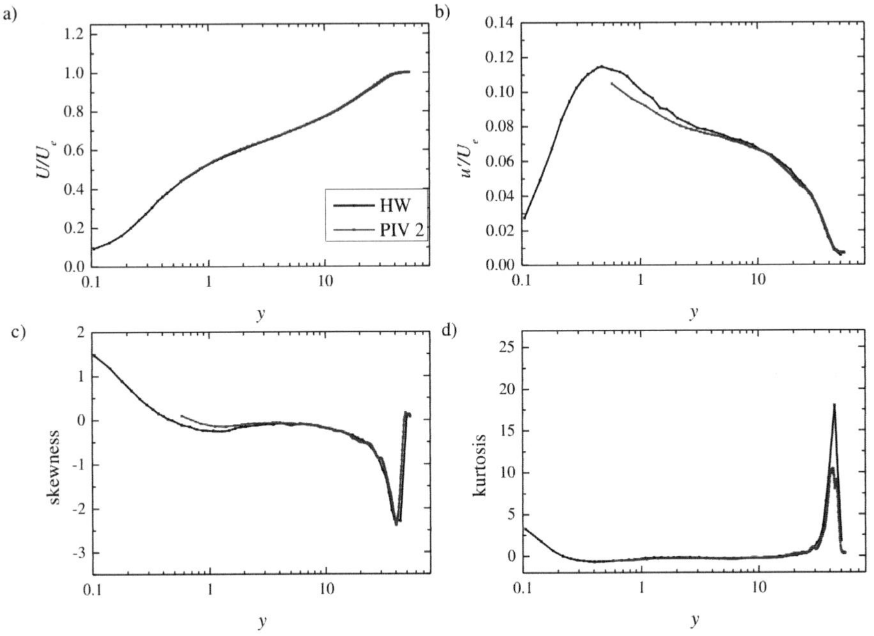

Figure 4. Statistics for Reynolds number 3200: mean velocity a), turbulence intensity b), skewness factor c) and flatness factor d)

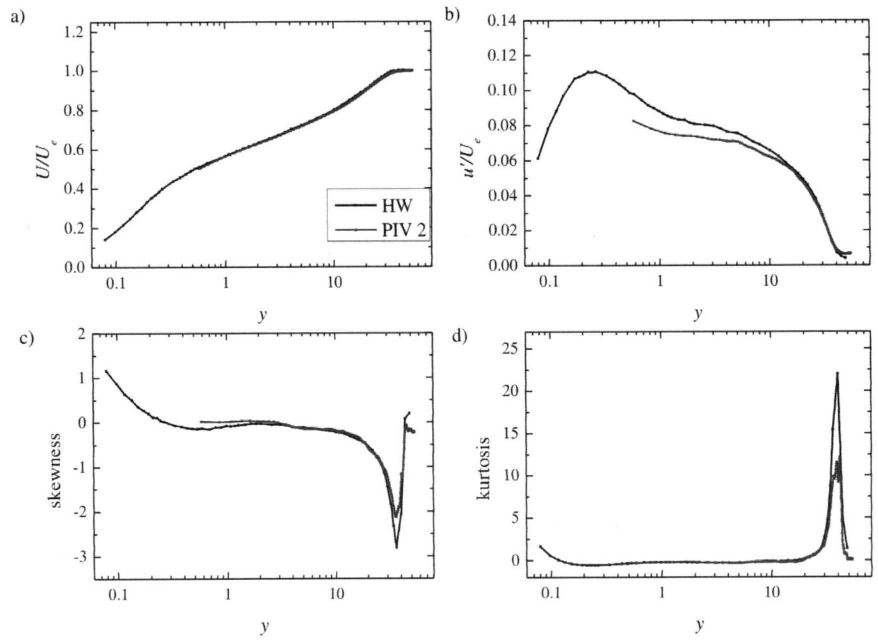

Figure 5. Statistics for Reynolds number 5500: mean velocity a), turbulence intensity b), skewness factor c) and flatness factor d)

Inconsistency in streamwise fluctuation between the measuring techniques could results not only from spatial averaging but also from influence of wall-normal component, which is taken into account in the case of hot-wire perpendicular probe. To analyse the problem, the velocity fluctuations from single-wire probe and PIV was compared for the case where the spatial averaging of PIV is the highest. Because of perpendicular single-wire streamwise fluctuations are influenced by v component, the resultant velocity fluctuations $u'_{xy} = \sqrt{u'^2 + v'^2}$ has to be calculated for PIV results before comparison with HW fluctuations. The comparison was shown in figure 6.

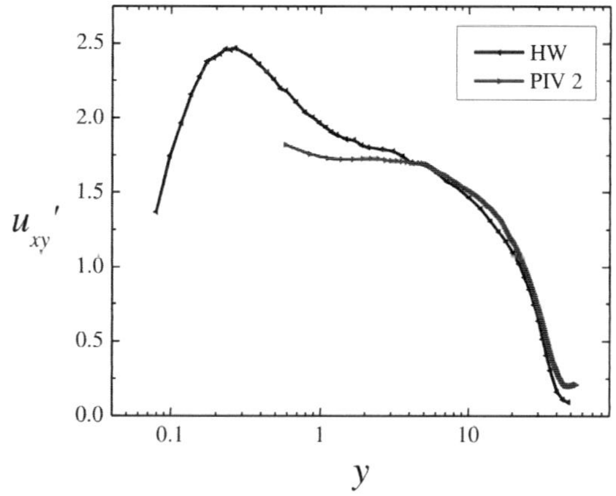

Figure 6. Comparison of HW u' with PIV resultant of u' and v' fluctuations.

It may be noticed that the distribution of u_{xy}' (see figure. 6) obtained by PIV slightly differ in shape in comparison with HW. In the outer zone the values of PIV are slightly overestimated in comparison of HW except the near-wall peak, where values are greatly underestimated. The overestimation of the PIV results in the outer zone could be caused by measurement noise present in PIV results as it was shown by Atkinson [7]. However, the near-wall peak substantial underestimation of resultant u_{xy}' reviles that v' component is much more under resolved than u' in figure 5b. Dróżdż and Elsner [8] showed on the energy spectra that the near-wall peak of single-wire fluctuations consist also the small-scalc v fluctuations, which are twice shorter than the scales responsible for near-wall peak of u' fluctuations. The fulfilled recommendation of Ligrani and Bradshaw [4] HW measurements resolves small-scales v fluctuations in wider range than PIV methods. The underestimation could be explained by appearance of the smaller structures in the flow, which are far from resolving using large field of view PIV methods. Both methods are more or less inadequate and should be re-compared to DNS properly. Single wire fluctuations should be compared to resultant u_{xy}' calculated for DNS fluctuation components in order verify if upper limit of wire length $l^+ < 20$ is enough to resolve the smallest scales.

5. Conclusions

The profiles of statistical characteristics of streamwise velocity components were evaluated from the data obtained using the two completely different experimental methods, HW and PIV. The aim of the paper was to check the PIV point statistics reliability and limitations of them in comparison with Hot-Wire technique, which is still the only candidate. The PIV data were evaluated by 5 alternative methods. Mean values, standard deviations as well as skewness and kurtosis coefficients were compared for a few values of momentum thickness-based Reynolds number from the range from 1000 to 5500.

As for the PIV data, the best results were obtained using the 2D Least squares matching evaluation method within the turbulent boundary layer close to the wall, while in the intermittent region on the boundary layer border the adaptive correlation method gives better results. The result shows that with the increasing Reynolds number the attenuation of fluctuations maximum and increase of skewness factor in PIV measurements occurs with respect to the Hot-Wire measurements that fulfil the Ligrani and Bradshaw [4] recommendation.

Special care was devoted to near-wall peak value of fluctuation distribution for both measurement techniques. The wall-normal component influence the single-wire fluctuations, which should not be the treated as streamwise fluctuation component but resultant of u and v components, especially in the near-wall region. Therefore, the fluctuations distribution of the single-wire perpendicular probe, which fulfil the Ligrani and Bradshaw [4] recommendation, is not self-similar with the DNS data because of elevated value o near-wall peak cased by v component. However the HW technique resolves small-scales v fluctuations in wider range than PIV methods. It seems that the increase of PIV field of view magnification should increase the range of resolved small-scales as it was shown by De Silva et al. [9]. However, both methods are more or less inadequate that is why it is strong need to perform further study in order to establish new criteria of well resolved measurements of turbulence.

6. References

[1] Hutchins N, Nickels T B, Marusic I and Chong M S 2009 Hot-wire spatial resolution issues in wall-bounded turbulence *J. Fluid Mech.* **635** 103

[2] Monty J P and Chong M S 2009 Turbulent channel flow: comparison of streamwise velocity data from experiments and direct numerical simulation *J. Fluid Mech.* **633** 461

[3] Schlatter P, Örlü R, Li Q, Brethouwer G, Fransson J H M, Johansson A V., Alfredsson P H and Henningson D S 2009 Turbulent boundary layers up to Re[sub θ]=2500 studied through simulation and experiment *Phys. Fluids* **21** 051702

[4] Ligrani P M and Bradshaw P 1987 Spatial resolution and measurement of turbulence in the viscous sublayer using subminiature hot-wire probes *Exp. Fluids* **5** 407–17

[5] Örlü R and Alfredsson P H 2010 On spatial resolution issues related to time-averaged quantities using hot-wire anemometry *Exp. Fluids* **49** 101–10

[6] Johansson A V. and Alfredsson P H 1983 Effects of imperfect spatial resolution on measurements of wall-bounded turbulent shear flows *J. Fluid Mech.* **137** 409–21

[7] Atkinson C, Buchmann N a., Amili O and Soria J 2014 On the appropriate filtering of PIV measurements of turbulent shear flows *Exp. Fluids* **55** 1654

[8] Dróżdż A and Elsner W 2014 Comparison of single and X-wire measurements of streamwise velocity fluctuations in turbulent boundary layer *J. Theor. Appl. Mech.* **52** 499–505

[9] De Silva C M, Gnanamanickam E P, Atkinson C, Buchmann N a., Hutchins N, Soria J and Marusic I 2014 High spatial range velocity measurements in a high Reynolds number turbulent boundary layer *Phys. Fluids* **26** 025117

[10] DynamicStudio v3.4 User's Guide 2013 Dantec Dynamics A/S

Acknowledgements

The investigation was supported by Polish-Czech bilateral project no. 88781/R13/R14/ (7AMB13PL003), the statutory funds BS/PB-1-103-3010/2011P and the Grant Agency of the Czech Republic, project No. P101/10/1230.

Journal of Physics: Conference Series **530** (2014) 012045 doi:10.1088/1742-6596/530/1/012045

Flow disturbance due to presence of the vane anemometer

M Bujalski, M Gawor and J Sobczyk

Laboratory of Flow Metrology, Strata Mechanics Research Institute of the Polish Academy of Sciences, Krakow, Poland

E-mail: bujalski@img-pan.krakow.pl

Abstract. This paper presents the results of the preliminary experimental investigations of the disturbance of velocity field resulting from placing a vane anemometer in the analyzed air flow. Experiments were conducted in a wind tunnel with a closed loop. For the measurement process, Particle Image Velocimetry (PIV) method was used to visualize the flow structure and evaluate the instantaneous, two-dimensional velocity vector fields. Regions of inflow on the vane anemometer as well as flow behind it were examined. Ensemble averaged velocity distribution and root-mean-square (RMS) velocity fluctuations were determined. The results below are presented in the form of contour-velocity maps and profile plots. In order to investigate velocity fluctuations in the wake of vane anemometer with high temporal resolution hot-wire anemometry (HWA) technique was used. Frequency analysis by means of Fast Fourier Transform was carried out. The obtained results give evidence to a significant spatially and temporally complex flow disturbance in the vicinity of analyzed instrument.

1. Introduction

Vane anemometers are commonly used to measure air velocity and determine volume flow rate in many industrial applications such as mining ventilation systems. The sensing element is a vane rotor, the rotational speed of which is proportional to the velocity of the flowing air stream. This measurement method belongs to a category of invasive measurement techniques. Placing vane anemometer in the analyzed flow leads to a significant disturbance of the flow structure. This effect may adversely affect other probes used to measure velocity and other physical quantities of the flow nearby the vane anemometer. Similar issue is addressed in [1], where flowfield in the wake of a low-velocity heated sphere anemometer was analyzed with use of Particle Image Velocimetry (PIV). Influence of the disturbed inflow on various types of vane anemometers was investigated [2, 3, 4]. However, there is lack of knowledge concerning the flow structure in the sole vicinity of this type of instrument. Such knowledge is fundamental for understanding and identifying perturbation introduced by components of the instrument in the measurement area.

Flow disturbance in front of and behind vane anemometer head was experimentally investigated by means of pointwise measurement method - hot-wire anemometry (HWA) [5]. In the present study, non-invasive optical measurements methods, such as PIV was applied. The aim of this paper is to analyze the disturbance of velocity field resulting from placing the Vane Anemometer μAS4 in the analyzed flow. This instrument (figure 1) is a hand-held device with a head diameter of 100 mm and thickness of 60 mm for measuring flow velocity of air or other gases in the range of 0.2 to 20 m/s. It was designed and produced by The Strata Mechanics Research of the Polish Academy of Sciences (SMRI PAS). Its rotor consists of eight acid-resistant blades. Shield bearing system consists of two hemispherical bowls that are mounted using three cylindrical brackets with a diameter of 5 mm.

Figure 1. The µAS4 Vane Anemometer.

2. Experimental setup and procedures

The experimental setup is presented in figure 2. The experiments were conducted in a wind-tunnel with a closed loop and a freestream turbulence intensity of 0.5% (located in SMRI PAS). The temperature was stabilized at 20°C and relative humidity at 35%. The test section was made of transparent polycarbonate walls with the dimensions of 500 x 500 x 1,500 mm. The air was used as an experimental fluid. The measurements and calculations were performed for the three flow velocities $U_\infty = 1$ m/s, $U_\infty = 5$ m/s and $U_\infty = 10$ m/s.

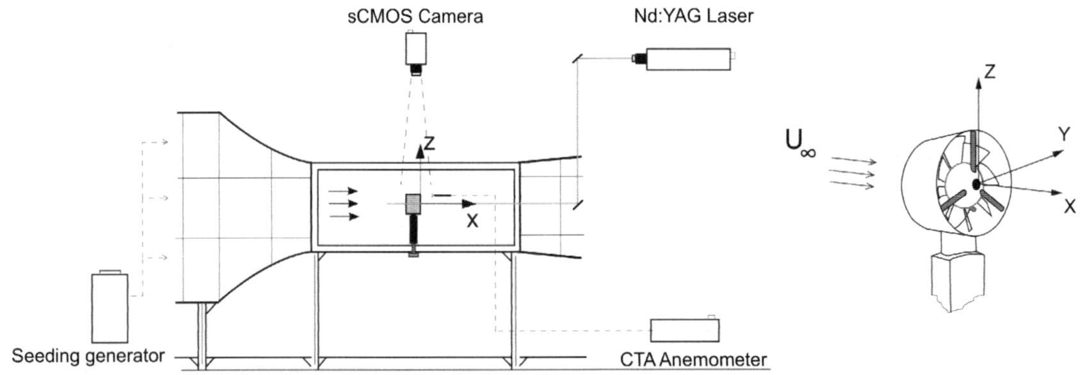

Figure 2. (a) The experimental setup. (b) Vane anemometer that was placed in the test section.

PIV method was used for the velocity vector fields calculations. Oil particles (DEHS), about 1 micron in diameter were used as a seeding. The measuring field was illuminated with light sheet generated by a double-pulse Nd:YAG laser of energy about 220 mJ per pulse. Laser sheet thickness was approximately 1 mm. The digital images were acquired using one sCMOS camera with resolution of 2,560 x 2,100 pixels. For each measurement, a set of 250 double frame images were recorded with the frequency of 5 Hz, which resulted in an overall time of one measurement of around 50 seconds. The time between two subsequent frames was synchronized with the two laser pulses and varied from about 100 µs to 1000 µs, depending on the flow velocity. During the calculations, the particle displacements were determined with use of 32 x 32 pixels sampling windows overlapping by 50%, which represent an interrogation area of about 1.4 x 1.4 mm. One of the cross-correlation methods, namely iterative multi-grid interrogation technique [6], was used as vectors' calculation algorithm. Background reflections in PIV images from components of the vane anemometer were suppressed by generating a reference background intensity map that was then subtracted from each individual digital image [7].

The described earlier vane anemometer was installed in the test section. The origin of the Cartesian coordinate system was set at the center of the bowl in the outflow region, as presented in figure 2b. The PIV measurement fields of interest were oriented in the x - z and x - y planes. The light sheet was set at

$y = 0$ and $z = 0$, respectively (figure 2). Regions of inflow on the vane anemometer as well as flow behind it were investigated.

The hot-wire anemometry technique was used to investigate velocity fluctuations at high temporal resolution in the wake of the vane anemometer. Especially two locations were of special interest - in the wake of rotor blades and in vortex shedding area forming behind cylindrical brackets. The measurement system (by SMRI PAS) was based on the ATU 2001 module and Constant Temperature Anemometer (CTA) probe with a single wolfram wire with the length of 2 mm and 5 μm in diameter [8]. Measurement points were aligned in the x - z and x - y planes and went through $x = 5$ mm (figure 2). In each point 16,384 samples were acquired with temporal resolution of 10 kHz.

3. Experimental results

Instantaneous flow field measurements by means of the PIV method can be used for statistical investigations. As a result ensemble averaged flow statistics can be obtained. For average procedure, a set of 250 instantaneous velocity vector fields were taken into account. An analysis of the number of images on average value of an arbitrarily chosen velocity vector components in measurement field are presented in figure 3.

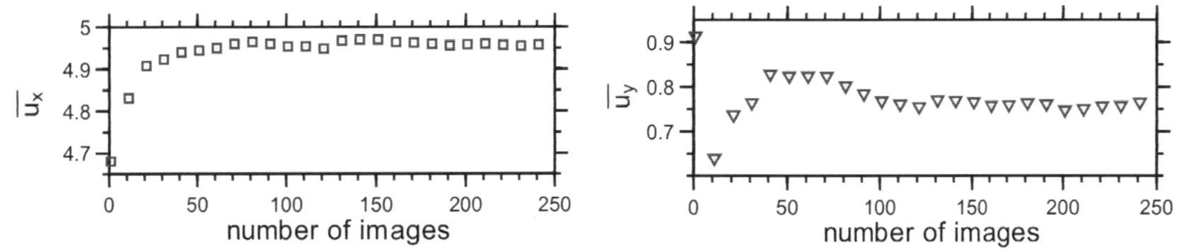

Figure 3. Mean longitudinal (left) and transverse (right) velocity component in the measurement point (located in the wake of rotor blades) versus number of images for average procedure at $U_\infty = 5$ m/s.

In the figure 4 average, two-dimensional velocity vector fields and the corresponding contour plots for $U_\infty = 1$ m/s in two orthogonal planes x - y and x - z are presented. All geometrical dimensions are normalized by head diameter of the vane anemometer $D = 100$ mm. In order to characterize the flow disturbance, scalar fields of RMS (root-mean-square) of the velocity magnitude were determined (figure 5). Distribution of the turbulence level, assumed as in equation (1), was shown in the figure 6.

$$I = \frac{U_{RMS}}{\overline{U}} \tag{1}$$

It can be found that in the wake of head of anemometer are regions of the velocity decrease corresponding to RMS increase. This is particularly visible in the neighborhood of the bowl and the edge of cover. Nearby the bowl, the turbulence level is approximately equal to 500%. For better illustration of this effect, sample profiles of the velocity magnitude for three values of U_∞ and corresponding RMS values were extracted and presented on figure 7 and figure 8. In the x - z plane a strong influence of presence of the handle is visible. It results in growth of z - component of velocity vector. Recirculation flow in the wake of the handle can be observed. The handle aerodynamic shadow pulls down the outflow of the head and makes the disturbance much more complex than resulting from the presence of the anemometer head only [5]. This remark is important since handle needs to be handled in order to place the instrument in the measurement area. Regardless on type of mounting it leads to further flow perturbation influencing also the velocity measurement done by the anemometer itself.

The bracket (located in the upper half of the head) gives influence to decrease of the velocity and increase of disturbance of the flow, as well. It is clearly visible on figure 8. Placing vane anemometer in the flow disturbs also the inflow region. Its influence decreases rapidly with the distance, however it can be easily noticed in the distance of 100 mm from the edge of anemometer.

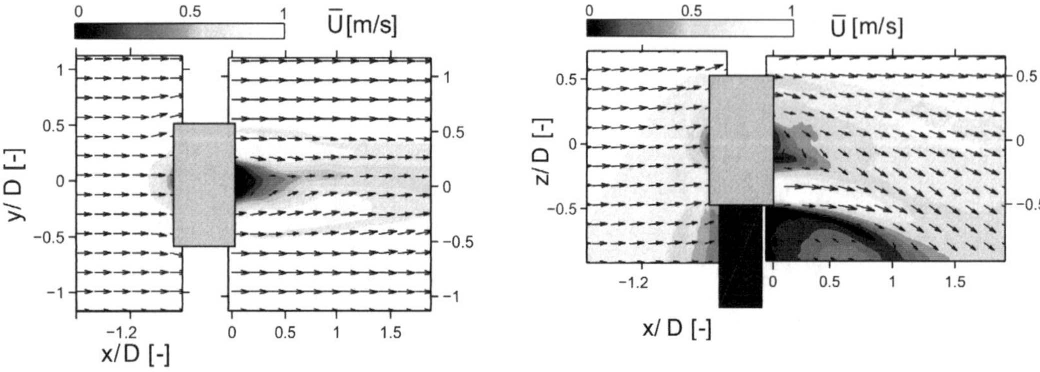

Figure 4. The results for U_∞ = 1 m/s. Ensemble averaged vector velocity field with contour plot of velocity magnitude in the *x* - *y* (left) and *x* - *z* plane (right).

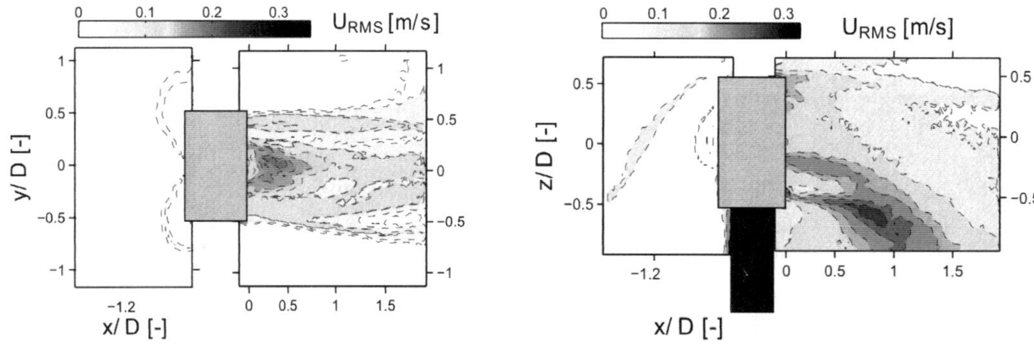

Figure 5. The results for U_∞ = 1 m/s. Contour plots of RMS of velocity vectors magnitudes in the *x* - *y* (left) and *x* - *z* plane (right).

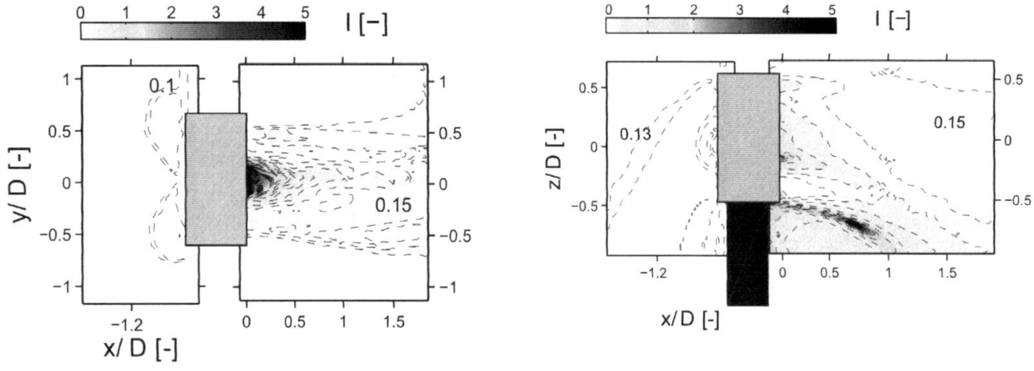

Figure 6. The results for U_∞ = 1 m/s. Distributions of the turbulence level in the *x* - *y* (left) and *x* - *z* plane (right).

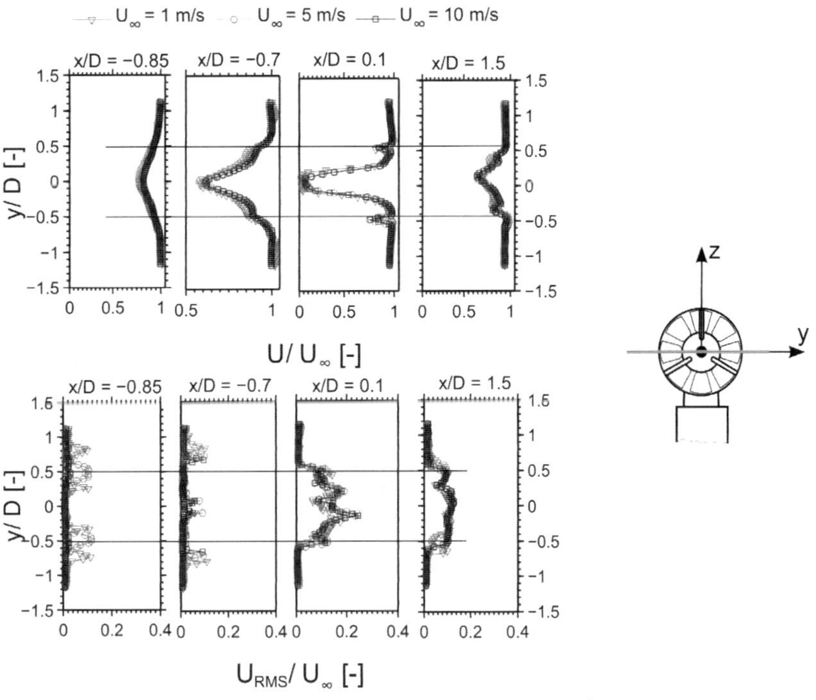

Figure 7. The results for U_∞ = 1 m/s, U_∞ = 5 m/s and U_∞ = 10 m/s. Normalized profiles of velocity magnitude (up) and RMS of velocity magnitude (down) in the *x* - *y* plane.

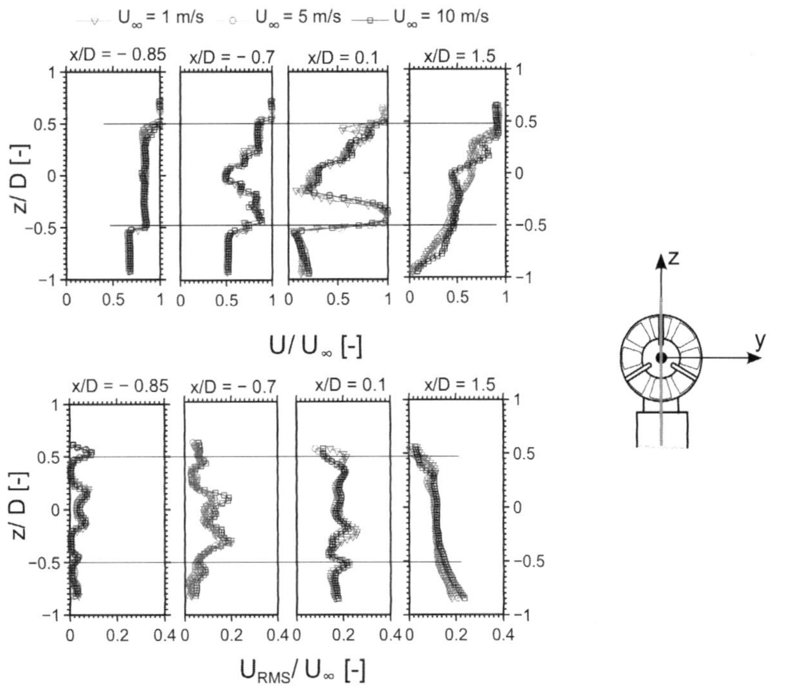

Figure 8. The results for U_∞ = 1 m/s, U_∞ = 5 m/s and U_∞ = 10 m/s. Normalized profiles of velocity magnitude (up) and RMS of velocity magnitude (down) in the *x* - *z* plane.

Journal of Physics: Conference Series **530** (2014) 012045 doi:10.1088/1742-6596/530/1/012045

Temporal analysis of flow structures in the wake of vane anemometer was done with use of hot-wire anemometer. Acquired signals were analyzed by means of Fast Fourier Transform (FFT) and presented in frequency domain in figure 9. Results are presented for two measurement points: P_1 (x/D = 0.15, y/D = 0.25, z/D = 0) and P_2 (x/D = 0.15, y/D = 0, z/D = 0.25), as they are representative for others. P_1 is located nearby rotor blade and point P_2 corresponds to the region in the wake of bracket. It was found that the most intense periodical perturbation of the flow results from rotation of the rotor blades. Observed clear peaks were directly equivalent to rotational frequency of rotor (figure 9a). Other distinguishable frequencies visible on FFT amplitude spectra are either harmonic or subharmonic of these pulsations. One may also notice a weak velocity pulsation suggesting vortex shedding behind brackets. Other structural elements contribute to FFT amplitude spectra mostly frequency noise. Mentioned vortex shedding was resolved by estimating its frequency from equation (2), where Sr is the Strouhal number and d = 5 mm is the diameter of the bracket. According to [9] Strouhal number was assumed to be equal to 0.2. Frequencies of peaks detected in the experimental data (figure 9b) - 40, 202, 396 Hz correspond to estimated values 40, 200, 400 Hz of f_K, respectively.

$$f_k = \frac{U_\infty Sr}{d} \tag{2}$$

(a)

(b)

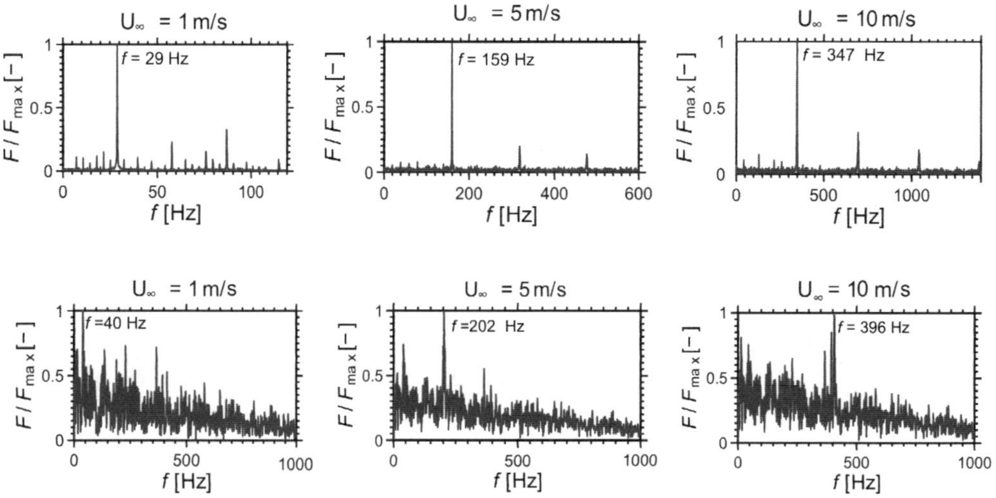

Figure 9. The results for U_∞ = 1 m/s, U_∞ = 5 m/s and U_∞ = 10 m/s. Fast Fourier Transform amplitude spectra of HWA signal measured at: (a) P_1 (x/D = 0.15, y/D = 0.25, z/D = 0) and (b) P_2 (x/D = 0.15, y/D = 0, z/D = 0.25).

4. Conclusion

Presence of the vane anemometer obviously disturbs the flow structure in front of it and much stronger behind it. Induced modification of flow in the sole vicinity of the instrument has a complex structure, possessing periodical and turbulent elements. Spatial flow field perturbations presented on figures 4-6 demonstrate that most of all the bowl and handle significantly decrease velocities in their wakes. Areas with the level of 50% of the velocity magnitude reduction are visible even in 200 mm behind the anemometer. Moreover, handle being relatively big comparing to the head makes a lot of aerodynamic shadow and creates a reverse flow.

Most of the flow perturbation results from non-movable parts of the anemometer. Flow behind rotor blades is much less disturbed than behind the bracket or bowl. Structure of the flow field modification is weekly dependent on the velocity of inflow.

Journal of Physics: Conference Series **530** (2014) 012045 doi:10.1088/1742-6596/530/1/012045

Brief analysis in the temporal space proves the presence of velocity pulsations as well as turbulent structures. Distinguished characteristic frequencies in the amplitude spectra were identified as corresponding to the rotation speed of the rotor blades and vortex shedding behind brackets.

The results from the performed preliminary and fundamental research may be used to optimize mutual position of vane anemometers in the investigated flow. This application is of a special importance for instance in industrial ventilation systems such as mining excavations or shafts. In fact the location of accompanying sensors of any type (especially with a high temporal resolution) should be designed carefully as improper location may lead to large measurement errors.

References

[1] Olim A M, Riethmuller M L and Gameiro da Silva M C 2002 *Exp. in Fluids* **32** pp 645-51
[2] Sauer H J and Howell R H 1990 *ASHRAE Transactions* **96** pp 502-11
[3] Hardcastle S G, Grenier M G and Butler K C 1991 Electronic vane anemometry. Finding a suitable replacement of mechanical analog devices for mine airflow assessment *Proc. 5th U.S. Mine vent. Symp.*, chapter 60 pp 571-6
[4] Chmiel W and Kruczkowski J 2005 Research on the influence velocity profile distortion on measurements with various types of stationary anemometers *Transactions of IMG PAN* **7** pp 117-29
[5] Gawor M 2003 Flow disturbance due to presence of the vane anemometer probe *Mechanization and automation in the mining industry* **12** pp 9-14
[6] Scarano F and Riethmuller M L 2000 *Exp. in Fluids* **29** pp S051-60
[7] Mejia-Alvarez R and Christensen K T 2013 *Meas. Sci. Technol.* **24** 027003
[8] Ligeza P 2008 *Review of Scientific Instruments* **79** pp 096105-096105-3
[9] Chen S S 1987 *Flow-Induced Vibration of Circular Cylinder Structures* (New York: Hemisphere Pub. Corp.)

Journal of Physics: Conference Series **530** (2014) 012046 doi:10.1088/1742-6596/530/1/012046

Acoustics flow analysis in circular duct using sound intensity and dynamic mode decomposition

S Weyna
West Pomeranian University of Technology, Faculty of Marine Technology and Transport Szczecin Poland

E-mail: weyna@zut.edu.pl

Abstract. Sound intensity generation in hard-walled duct with acoustic flow (no mean-flow) is treated experimentally and shown graphically. In paper, numerous methods of visualization illustrating the vortex flow (2D, 3D) can graphically explain diffraction and scattering phenomena occurring inside the duct and around open end area. Sound intensity investigation in annular duct gives a physical picture of sound waves in any duct mode. In the paper, modal energy analysis are discussed with particular reference to acoustics acoustic orthogonal decomposition (AOD). The image of sound intensity fields before and above "cut-off" frequency region are found to compare acoustic modes which might resonate in duct. The experimental results show also the effects of axial and swirling flow. However acoustic field is extremely complicated, because pressures in non-propagating (cut-off) modes cooperate with the particle velocities in propagating modes, and vice versa. Measurement in cylindrical duct demonstrates also the cut-off phenomenon and the effect of reflection from open end. The aim of experimental study was to obtain information on low Mach number flows in ducts in order to improve physical understanding and validate theoretical CFD and CAA models that still may be improved.

1. Introduction

The aim of our experimental research is directed to enrich theoretical knowledge about the birth of sound in turbulent flow conditions of the acoustic wave, a better understanding of the mechanisms of noise generation and to identify more effective ways of its reduction.

For low Mach-number isothermal flow we will see that aeroacoustic sound production is entirely due to mean flow velocity fluctuations, which may be described in terms of the underlying vortex dynamics. This leads to the idea of using so called *vortex sound theory* [1, 2]. Vortex sound theory is not only numerically efficient but also allows us to translate the very efficient vortex-dynamical description of elementary flows directly into sound production properties of these flows in real-live conditions. Our research will be conducted on the cross flow inside cylinders with a free-end. The circular cylinder has proved to be a fruitful area of fluid dynamics research, due to its combination of a simple geometry and complex, unsteady flow features. Sound intensity flow without mean flow generation in hard-walled circular duct with acoustic excitation is treated experimentally. The noise produced by laminar and turbulent flows in annular ducts is studied using experimental sound intensity (SI) which is a average value of pressure times particle velocity - a vector quantity. Along the duct sound energy is carried in various modes which will be examined and graphically illustrated.

Acoustic intensity is a very useful energetic quantity since it gives information about the propagation paths and the amount of energy radiated. From an experimental point of view time-averaged acoustic intensity (*rms* values), also called active intensity and written here $\boldsymbol{I_a}$, is often more interesting than instantaneous value $I_a(t)$. Sound intensity as a vector variable *inseparably* couples the acoustic particle velocity and acoustic pressure ($\boldsymbol{I_a}=p\ \boldsymbol{v}$) and represents a stream of acoustic energy flowing in the field. This vector parameter of acoustic wave can be measured (*rms*) with special sound intensity probe and can be easily shown in a graphical form.

The acoustic particle velocity **v** and mean pressure p satisfy the time-averaged equations of continuity and momentum. For linear acoustics, in the absence of an external flow $<\rho\mathbf{v}>=<p'\ \mathbf{v}>/c_o^2 = \boldsymbol{I_a}/\ c_o^2$, where p' is the acoustic pressure perturbation and $\boldsymbol{I_a}$ the acoustic intensity. Sound intensity can be directly measured and recorded as an acoustical flow field divided on normalized octave band frequencies. (normalized acoustic filters 1/1, 1/3, 1/12, 1/24 octave band). In traditional acoustic metrology, the analysis of acoustic fields mainly focuses on the distribution of pressure levels (scalar variable), however in a real acoustic field both scalar (acoustic pressure) and vector (the acoustic particle velocity) effects are closely with phase and amplitude related. The acoustic field may be described as a spatial distribution of pressure and particle velocity, their amplitude p and $\boldsymbol{v}$ being proportional for plane traveling waves ($p=\rho\ c_0\ v$), where ρ is the density and c_0 is the sound speed.

The application of the sound intensity technique together with numerical methods has today improved the quality of acoustic diagnostics and has made it possible to visualize energy wave phenomena (vector distribution) in a vibrating structure, or in an acoustic field around the structure. The visualization of acoustic energy flow in real-life acoustic 3D space fields can explain many peculiar energetic effects (scattering, vortex flow, shielding area, etc.), concerning the areas in which it is difficult to make numerical modeling and analysis with the CFD-FSI-CAA methods.

2. Experimental setup

Sound intensity generation in hard-walled duct with acoustic flow (no mean-flow) is treated experimentally and showed graphically. In paper, numerous methods of visualization illustrate the vortex flow (2D, 3D) as a sound intensity stream, can graphically explain diffraction and scattering phenomena occurring inside a long circular duct and with open end. The intensity distribution inside the duct is produced by the action of the sum of modal particle velocities and pressures. Along the duct sound energy is carried in various modes which will be examined and graphically illustrated.

In figure 1 we show a model of circular acoustic waveguide where investigations with sound intensity measurement were made. The 6 m long open-end duct with internal radius 0,474 m was used as a model for an acoustic waveguide. At one end it was connected to a loudspeaker, a source of broadband acoustic signals. The method employed in this paper is based on SI technique and experimental set-ups where monopole sources approximated with loudspeakers. The duct is excited with acoustic pink noise, so, the sound power along the duct is sent without mean flow. Measurements were made in frequency band 50-6800 Hz and analyzed in 1/3 and 1/12 octave band frequency. SI measurement were made on a duct without any obstacles present inside. Measurement region was placed at a distance about of 2.2 m from the end of duct.

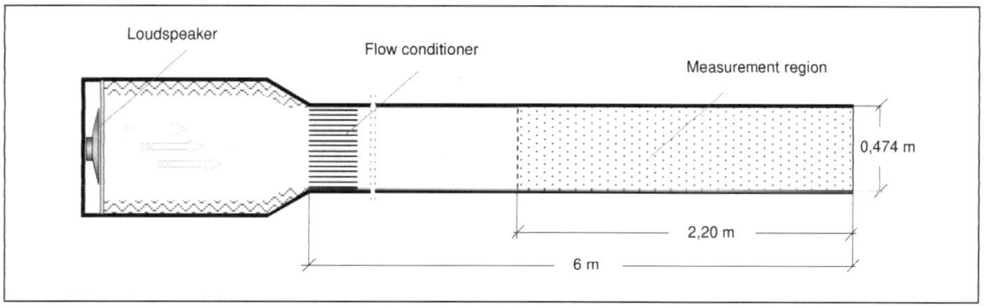

Figure 1. Sound incident of circular cross section acoustic waveguide

The space inside the duct was scanned with sound intensity measuring the *x*, *y* and *z* components of sound intensity vectors. The image of the dipolar and quadrupolar sound generated by a flow inside a duct was obtained using a SI three-dimensional *USP Microflown* probe and our graphical *SIWin* post-processing software [3].

3. Description of acoustics flow inside empty duct

The most common situation to be found in the literature is for a plane travelling acoustic wave in ducts which reflects from the end to form a standing wave. In cylindrical ducts, plane waves, only, can propagate below a characteristic frequency which is a function of the duct diameter.

In our investigation several types of axisymetric and spiral type of vortex breakdowns have been observed experimentally [4]. Vortex breakdown phenomenon occurs when the ratio of the azimuthal to axial momentum exceeds a certain threshold, while both quantities have to be of the same order of magnitude [5]. It can play a crucial role in a variety of technical applications. Understanding the cause of the vortex breakdown is therefore of great importance in order to develop appropriate control strategies.

In our experiment sound intensity field in a cylindrical duct was excited by a wide-band sound signal propagated from source approximated with a loudspeaker. In figure 2 we show some results of investigations for 1/12 octave band frequencies where the sectional streamlines show the topological flow multi-cell structure for high-order modes. With graphical form we can see the evolution process of flows in the cross-section plane. In this paper, higher order acoustic modes which are excited above cut-off frequency are considered. These modes with frequency above 817 Hz have a much more complicated pressure pattern compared to the plane wave mode below. When more than one mode has "cut-on", these modes are superimposed upon the lower frequency wave mode and can co-exist with each other [6].

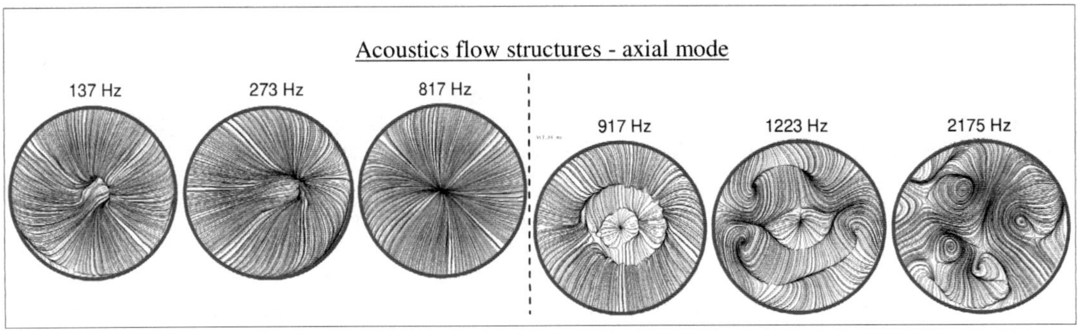

Figure 2. Cross-sectional sound intensity streamlines fields below (up to 817 Hz) and above cut-off frequencies (from 917 Hz to 2175 Hz).

The frequency of 817 Hz is critical for the test waveguide that caused vortex breakdown phenomenon in the acoustic flow. Vortex breakdown phenomenon occurs when the ratio of the azimuthal to axial momentum exceeds a certain threshold, while both quantities have to be of the same order of magnitude. This frequency limiting case of vortex breakdown phenomenon for researched duct is well illustrated in figure 2. The vortex breakdown can play a crucial role in a variety of technical applications. Understanding the cause of the vortex breakdown is therefore of great importance in order to develop appropriate control strategies.

Sound intensity measurement and graphically imaging flow in a cylindrical duct can also well demonstrate the cut-off phenomenon and the effect of acoustic wave reflection from open-and. In open-ended ducts, waves reflected from the open end play an important part in sound transmission. The reflection properties of the opening are specified in terms of the impedance presented to each duct mode [7], and the sound intensity along the duct is related to the mode impedance and the forward wave amplitude. In figure 3 the sound field in the cross section at 15 cm from the end of the waveguide shows a tumble motion as the beck-scattering wave reactions.

Also the termination of the duct assumes an important role on the shape of the field at the end of the waveguide (see also figure 7), and consequently the level of the noise at the outlet of the waveguide [8, 9].

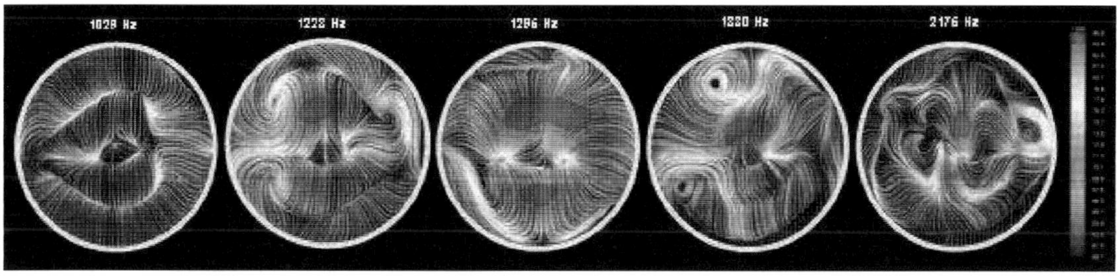

Figure 3. The sound field in the cross section at 15 cm from the end of the waveguide - tumble motion as the effect of back-scattering reactions is shown.

4. Conical baffle inside a duct

Good understanding of the evolution process of acoustic wave motion in pipes and ducts is critical to development of engineering designs with the most attractive operating to achieve an optimize low-noise design of duct systems and low noise emission characteristic. The detailed flow acoustic evaluation process of the in-duct flow structure has recently attracted attention of investigators.

Flow tones are generated in a wide variety of configurations. In essence, they arise due to favorable coupling between an inherent instability of a separated shear layer and a resonant acoustic mode of the flow system [10]. This process is one of the promising ways to obtain high circulation and rotation shape of waves.

Vertical flows in general may be characterized into two categories: the *swirl motion* rotating about duct axis and the *tumble motion* rotating about the diametric axis (figure 4C). Swirling rings isolate the velocity fields in the vicinity of the vortices and assess their contribution to the overall dynamics at the flow.

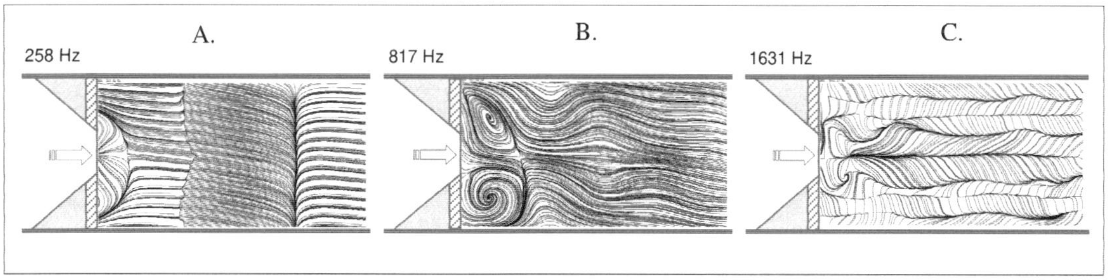

Figure 4. Sound intensity streamlines around conical baffle at the below (A and B) and above cut-off frequency (C).

Direct measurement of the acoustic power flow around outlet of an obstacle as a conical baffle with a 127 mm hole can explain a diffraction and scattering phenomena occurring in this region. The space behind the conical baffle was studied in consideration of deformation of the acoustic field due to the presence of obstacles. Experimental results will present the distributions of sound intensity streamlines traveling along the duct. Two equivalent types of modes are observed: one type is a stationary sound intensity pattern which fluctuates in time (figure 4A and B – frequency below cut-off); the other type is a steady sound intensity pattern which rotates (figure 4C – frequency above cut-off). The second of these appears more appropriate to describe sound generated by rotating sources - higher-order *spinning modes* (1631 Hz at the figure 4).

The complicated secondary flow patterns reveal strong non-linearity of flows. Changes of rotation effect. Change the flows inside duct from stable state to unstable state or from unstable to stable state. A well-defined swirl and/or tumble flow structure [4] is more stable than other large scale in-duct flows, therefore, it may break up later in the cycle, giving higher turbulence during acoustic flow along ducts.

On the figure 5 we show examples of this investigations for some 1/12 octave band frequencies where the sectional streamlines and the velocity vectors show the topological secondary flow multi-cell structure for high-order modes. The view planes for detections of swirl motions inside a duct are well defined in figure 5.

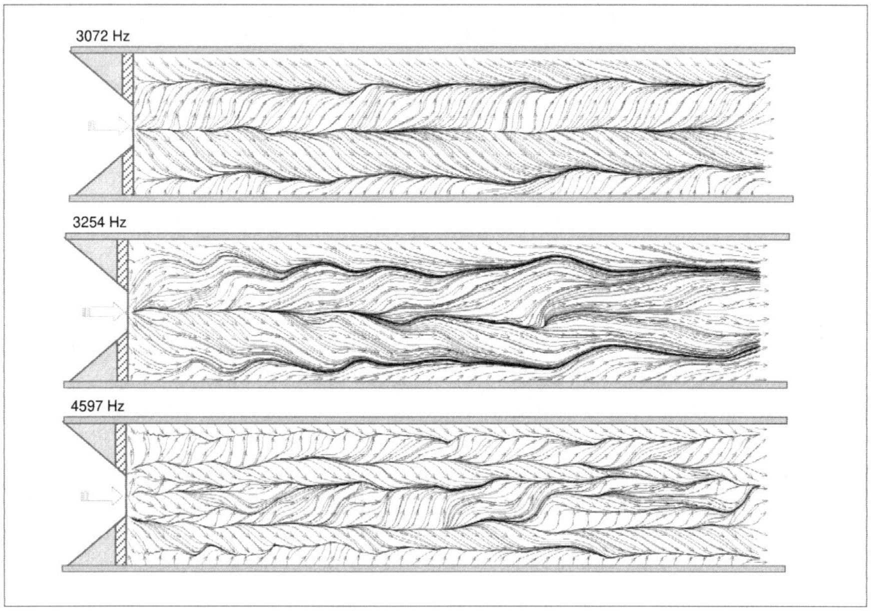

Figure 5. Secondary flow patterns on multi-cell structure for high-order mode.

2. Multi-modal acoustics flow decomposition

The theory of sound transmission and generation in hard walled ducts has been extended to include axial and swirling wave flow. This type of sound wave called spinning modes [7] is particularly evident in the range above cut-off frequency. These phenomena occur in the whole length of duct.

Wave system in the same mode travels in both directions along a duct. Above the cut-off frequency mode, the sound power transmitted along the duct is simply the difference between the forward- and backward-transmitted sound power associated with separate wave systems. Because of combined effects of system rotation (*Coriolis force*) and curvature (*centrifugal force*) on the acoustic flowing inside of duct, flows may temporarily oscillate in some cases [5]. When fluid flows in circular duct, it causes complex secondary flow structure. The well-known secondary flows[6] of the above-mentioned flows have pairs of clockwise and counter-clockwise-rotating vortices in a plane perpendicular to the axis of the duct (see figures 2 and 3). In the case of circular duct, flows may have even more complicated secondary flow (*multi-cell*) structure. The intensity distribution inside duct produced by the action of the axial and radial modes is extremely complicated because this propagating modes influence on each other.

In this paper, we also investigate how SI technique may be combined with POD and DMD techniques as a tool for searching the energy dominant modes in the acoustics flow field in the interior of ducts and pipes. We attempt to show that the tested coherent structures by SI give the same opportunities to analyze the distribution of energy in the sound field that we can get, for instance, from the PIV/POD methods. From the analysis of modes amplitude, it can be concluded which one represents the highest energy level.

Experimental analysis of the distribution of the volume of sound intensity field may be equivalent to statistical POD or MDM methods. Description of mode propagation as a sound intensity stream flow in hard walled cylindrical ducts, may be called *acoustic orthogonal decomposition* (AOD) and give best results when all acoustic modes are excited. For ducts with rigid boundaries, the fluid-resonant category may contribute significantly to unwanted noise. Height level of acoustic energy propagated along the duct occurs when the mode is excited above a cut-off frequency which depends on the mode eigenvalue and the duct radius. Modes excited below their cut-off frequency are evanescent and decay exponentially with distance along the duct. The dimensionless number which expresses the cut-off frequency independently of the radius is the term kr, or as a Helmholtz number He [11, 12]. This minimum wave number, called the *cut-off value*, below which the simple wave breaks down is investigated in duct and marks the boundary between high-frequency propagation and low-frequency decay of duct modes. In our duct acoustics flow investigation the cut-off value is about four.

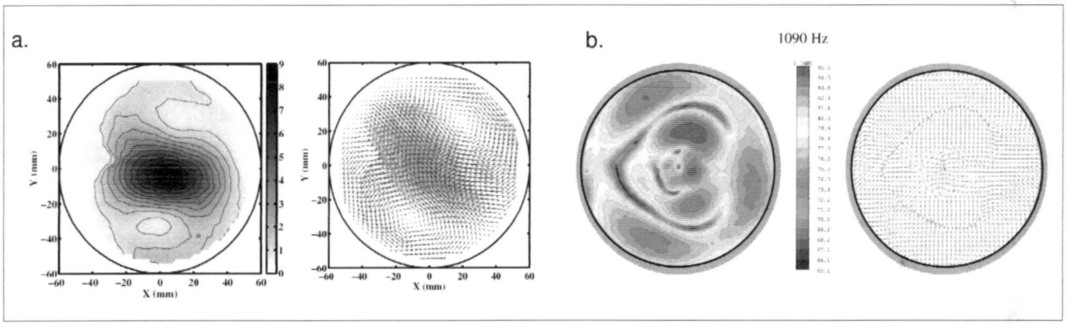

Figure 6 . Comparison results estimated with POD from literature (a) and sound intensity field measured inside investigated circular duct (b).

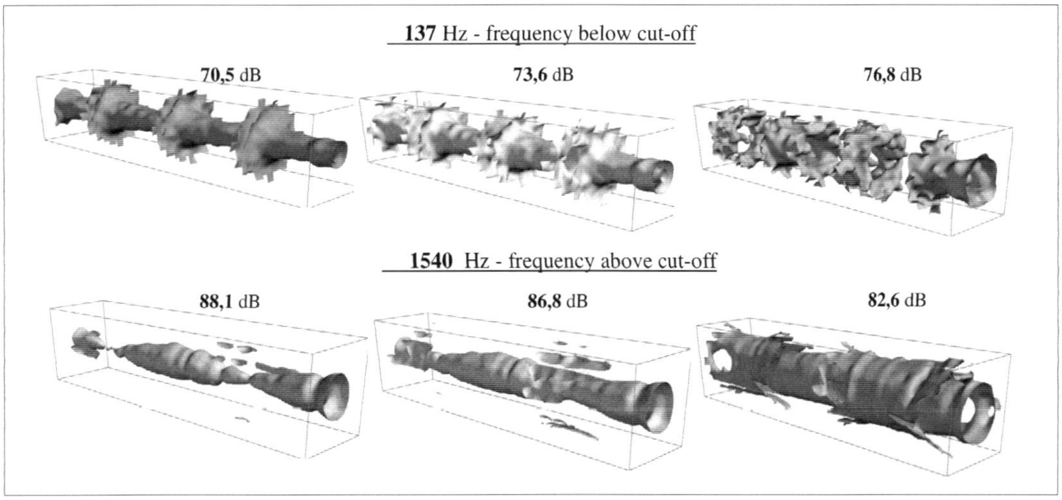

Figure 7. Three-dimensional distribution of equal sound energy acoustic structure (sound intensity isosurface with equal dB levels) inside duct in the 2.20 m from the open end of duct.

In figure 6 comparison literature results estimated with POD and with sound intensity field measured inside circular duct are shown. Comparison of some POD analysis results taken from the literature [13] confirms the usefulness of SI techniques to the decomposition of any modal distributions. Such SI studies can be carried out both in the flat and three-dimensional acoustic flow fields (figure 7).

According to the analysis above, it can be seen that the experimental decomposition of acoustic flow field using sound intensity (SI) and PIV/POD measurement techniques can be successfully applied in the acoustic waveguide duct.

8. Conclusions

Our experiment on acoustic waveguide model confirms that flow acoustic imaginations in real-life conditions are very complex, even for extremely simple modelling facility and for the sound field in a circular duct excited by loudspeaker used in the study (without flow field). These investigations provide a physical understanding of acoustic wave flow phenomena in real cylindrical duct where the measurements show both qualitative and quantitative flow diagnostics. The presentation of the vector distributions of real-life acoustic fields inside the duct areas, for which it is difficult to make a theoretical analysis of sound flow above the "cut-off frequencies", shows that it is transmitted along the annular duct by higher-order "spinning modes", not by plane waves. Properly modelled and analyzed flow-induced sound led directly into the modern aeroacoustic approach, in which theory and experiment are inseparable.

Along a duct, sound energy carries the various modes. The main objective of the study is to give an understanding of sound-energy transmission by higher-order duct modes. Sound propagation in circular ducts is deformed mainly by waves spiralling around the duct. These modes may be excited even in an axisymetric duct by a disturbance which rotates about the duct axis due to local changes in impedance or obstacles on the way of travelling wave. The experimental results confirm the occurrence of axial and swirling flow inside investigated duct. Further research on the importance of energy interaction phenomena between axial and radial modes will focus for more realistic engineering applications.

The swirling modes may be excited in an axisymetric duct by a disturbance which rotates about the duct axis. Rotations and swirling waves may occur between the mode amplitude and impedance, which applies to both above and below the cut-off frequencies. Therefore, the theory of sound transmission and generation in hard walled ducts has been extended to include axial and swirling mean flow. The presentation of the vector distributions of real-life acoustic fields can explain many particulars concerning the acoustic power flow, and flow dynamics inside the ducts.

The analysis of acoustic field with floating wave inside a duct show that the sound intensity technique is very useful tool to the visualization of vector acoustic phenomena.

References

[1] Howe M S 1975 Contributions to the theory of aerodynamic sound. *J. Fluid Mech.* 71 pp 625–973
[2] Kambe T and Minota T 1981 Sound radiation from vortex systems. *J. Sound & Vib.* 74 pp 61–72
[3] Weyna S 2010 Acoustic intensity imaging methods for in-situ wave propagation *Archives of Acoustics* 35(2) pp 265-273
[4] Oberleithner K, Sieber M, Nayeri C N, Paschereit CO, Petz C, Hege H C, Noack B R, Wygnanski I, 2011 Three-dimensional coherent structures in a swirling jet undergoing vortex breakdown: stability analysis and empirical mode construction *J. Fluid Mech*anics vol. 679 pp 383–414
[5] Schmid P J 2010 Dynamic mode decomposition of numerical and experimental data *J. Fluid Mec.,* vol. 656 pp 5–28
[6] Bennett G J, O'Reilly C J, Liu H 2009 Modelling multi-modal sound transmission from point sources in ducts with flow using a wave-based method *Proc. ICSV16* Krakow

[7] Macdonald R, Skulina D, Campbell M, Valerie J-Ch, Marx D, Bailliet H 2010 PIV and applied to high amplitude acoustic field at a tube termination *Proc. Congres Francais d'Acoustique* pp 10-16
[8] Weyna S, Mickiewicz W, Pyla M, Jablonski M 2013 Experimental Acoustic flow analysis inside a section of an acoustic waveguide *Archives of Acoustics* 38(2) pp 211-216
[9] Weyna S 2012 Acoustics flow field visualization using sound intensity and laser anemometry methods *Proc. of XX Fluid Mechanics Conf.* S27-2
[10] Arthurs D, Ziada S 2009 Flow-excited acoustic resonances of coaxial side-branches in an annular duct *Journal of Fluids and Structures* pp 25 42–59
[11] Bennett G J, Verdugo F R, Stephens D B 2010 Shear layer dynamics of a cylindrical cavity for different acoustic resonance modes *Proc. 15th Int Symp on Applications of Laser Techniques to Fluid Mechanics* Lisbon
[12] Joseph P, Morfey C L, Loqwis C R 2003 Multi-mode sound transmission in duct with flow *J. Sound Vib.* 264 pp 523-544
[13] Graftieaux L, Michard M, Grasjean N (2001 Combining PIV, POD and vortex identification algorithms for the study of unsteady turbulent swirling flows. *Meas. Sci. Technol.* 12 pp 1423-1429

Journal of Physics: Conference Series **530** (2014) 012047 doi:10.1088/1742-6596/530/1/012047

Transformation of colour space dedicated to an experimental analysis fulfilling the applicability criteria

A Ziemba and E Fornalik-Wajs
AGH University of Science and Technology, Department of Fundamental Research in Energy Engineering, Al. Mickiewicza 30, 30-059 Krakow, Poland

E-mail: elaf@agh.edu.pl

Abstract. The choice of colour space is very important in the digital image analysis by reason of accuracy and computational time. Particle Image Velocimetry and Particle Image Thermometry are the optical methods commonly applied in the fluid dynamics and heat transfer. Especially in PIT method, the analysis of colour images is significant. In this paper, transformation of *RGB* to *HSI* colour space dedicated to PIT will be presented. Derivation of formulas together with its graphical representation will be discussed. Fulfilment of applicability criteria will be shown. This theoretical approach to digital image processing supplements the knowledge about the optical experimental methods.

1. Introduction

Digital Particle Image Velocimetry (DPIV) or Thermometry (DPIT) are the methods often applied in the experimental analysis nowadays. Both of them depend on the image analysis but DPIT depends also on the colour analysis. DPIT enables determination of temperature field on the basis of cross-sectional images of Thermochromic Liquid Crystals (TLC) response representing the temperature field and indirectly the fluid flow. TLC particles reflect part of the illuminating them white light in dependence on their temperature and illuminating angle. In the reflected light dominates particular wavelengths, therefore it is possible to use the Charge Coupled Device (CCD) in the temperature determination process. CCD element records the electrical signals of strength proportional to the amount of incident light [1]. The calibration curve represents the dependence of dominant reflected wavelength and the temperature. The crucial point of analysis is to find such function. The best way, unfortunately the most expensive, is to use the spectrometer. The common way is to analyse the images taken by utilization of some number of selective filters. Characteristics of filters transparency constructs the base vectors of colour space. In typical camera with NTSC system there are three filters of Red (R), Green (G) and Blue (B) colours [1]. In the majority of cases, the R, G, B collection uniquely determines the temperature. Due to that, it is possible to find calibration curve directly – utilizing the R, G, B values or indirectly – applying other function, which is a function of R, G, B components.

DPIT with utilization of RGB colour space or its function were applied in many research areas such as the phase change [2], crystal growth [3], impinging heat transfer [4], convection [5] or medicin [6]. Direct application of the R, G, B colour space components [1, 3-4] caused the calculation difficulties: complex calculation procedure and long calculation process. Therefore the colour space transformation became a common routine in such applications [1-2,5,7]. The *RGB* colour space

transformation can lead to the reduction of variables number and easier way of calculation connected with lower time consumption.
In the literature there is an absence of derivations and what is more important, there is lack of verification if applied space transformation fulfil the applicability criteria [1]. The aims of this paper are: to derive the formulas of *RGB* to *HSI* colour space transformation applied in various research areas [2,5-6], to compare them with the other formulas [8-10] and to check if the applicability criteria are fulfilled by them.

2. Colour space

Three independent signals of *R*, *G*, *B* belong to the colours space based on the cartesian coordinate system. This space is device-oriented and its origin is coming from the operation of camera sensors or screens [2]. In the digital image analysis more useful are the colour spaces referring to the human colour sensation, like Hue, Saturation, Value *(HSV)*, Hue, Saturation, Lightness *(HSL)* or Hue, Saturation, Intensity (*HSI*) spaces. They separate the "colourless" Value (Lightness, Intensity) components from the "colourful" Hue and Saturation ones. In these spaces, Hue depends on the wavelength, which enables differentiation between, for example, the red and yellow colours. Saturation represents "amount" of colour, therefore the difference between the red and pink colours can be distinguished. The third variable – Value, Lightness or Intensity, refers to the amount of light, which enables determination of light and dark colours or the levels of grey.

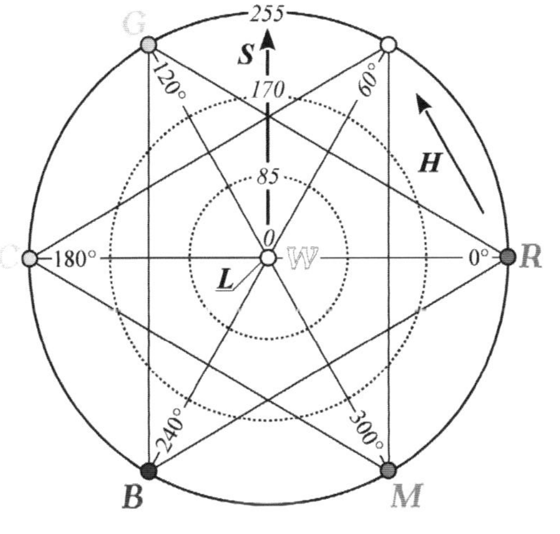

Figure 1. Colour wheel.

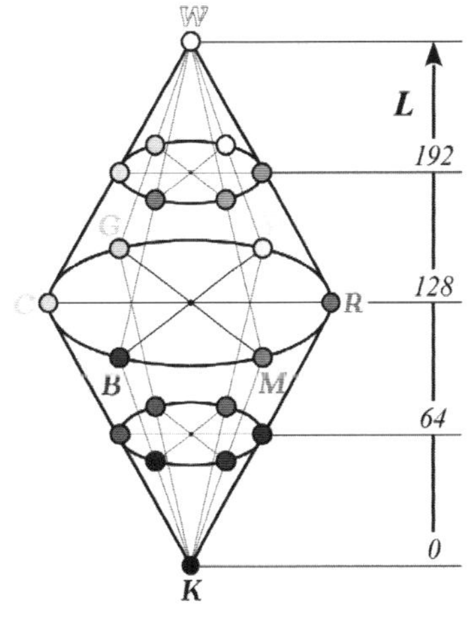

Figure 2. Double cone of RGB gamut.

The Hue is a property of pure colour, very often identified with the colour itself, which helps to distinguish particular colours among the others. It is closely connected with the light wavelength. On the colour wheel, proposed by Newton, it is represented by angle of pure colour. One of the wheel construction methods places three additive *R*, *G*, *B* colours in the vertices of one equilateral triangle and three subtractive Yellow *(Y)*, Cyan *(C)*, Magenta *(M)* colours in the vertices of second equilateral triangle, located above the first one in such a way that complementary colours are symmetrically situated, relatively to the triangles centre (figure 1). They form colour pairs *(R-C)*, *(G-M)*, *(B-Y)*. The corresponding colour angles are as follows $R = 0°$, $Y = 60°$, $G = 120°$, $C = 180°$, $B = 240°$ and $M = 300°$ [12]. This hexagonal structure, inscribed in the circle gives the colour circle, in which angle of radius vector indicates pure colour, placed on the circle perimeter. Each pure colour can be represented by a mixture of two neighbouring colours. To illustrate Lightness, the additional parallel circles of decreasing diameter should be added above and below the basic colour circle. The double cone of colour is created (figure 2). The circles placed above the basic circle, tend towards the white colour, therefore they become lighter. The circles placed below the basic one, tend towards the black colour becoming darker. The axis of such body corresponds to the grey scale from black to white colour and it describes the Lightness axis. Distance between the particular colour and this axis is a measure of Saturation. The angle between the planes containing this axis and the planes containing red colour and any other colour is called Hue [12]. Value can be illustrated by height of inverted cone, while Intensity by diagonal of RGB cube.

3. *RGB* to *HSI* colour space transformation

As it was already mentioned the crucial point in PIT method is to find the colour-temperature function. In the *RGB* colour space this function has three variables, so it is not unequivocal or easy to calculate. To reduce the number of variables, the *RGB* colour space transformation is applied. This transformation is obtained by the geometrical dependences between the *RGB* cartesian colour space and the space defined by an orientation of colour cube on the *xy* plane. It should be oriented in such position that its main diagonal is perpendicular to this plane and it defines the Intensity axis. Such transformation described in [1,7,11,13,14] is non-linear, therefore any image analysis is time consuming. There is also one more difficulty coming from the existence of singularity on the plane described by an equation $2R\text{-}G\text{-}B=0$.

The examples of linear *RGB* to *HSI* colour space transformations can be found in the literature. They "linearity" arises from an assumption that in the regions, in which one of the *RGB* components is maximal, other one minimal, the value of Hue can be approximated by a linear function of third component, equal to the central value. One of such transformations is reported in [8], together with its derivation and geometrical visualization. Instead of the Intensity component, the synonymous component called Value was utilized. Saturation and Hue were defined in particular sextants of colour hexagon. Saturation was zero on the Intensity axis (except the point represented by the black colour, point $K(R,G,B) = K(0,0,0)$) and it was equal to one on the walls and edges (belonging to the planes *RG, GB, BR*) of colour cube. For constant Value = $max(R,G,B)$ in particular sextant of colour hexagon Saturation became linear function of $min(R,G,B)$. The linear transformation of *RGB* to *HSI* colour space is also presented in [9], but without any explanation of its origin or derivation. Similar transformation is described in [10] and it is a part of Java Platform, however instead of *HSI* the identical with it the *HSB* colour space was used. Comparison between these three transformations indicates their equivalence.

4. *RGB* to *HSI* colour space transformation dedicated to PIT

The following equations defining Hue, Saturation and Intensity were published in [15,16,17] without any derivation:

$$H = 63 + 63\cdot(G'-R')\cdot(G'+R')^{-1} \qquad for\ \ B'=0, \tag{1}$$

$$H = 189 + 63\cdot(B'-G')\cdot(B'+G')^{-1} \qquad for\ \ R'=0, \tag{2}$$

$$S = 255\cdot\left(1-\frac{min}{I}\right) = 255\cdot s\,, \tag{3}$$

$$I = \left(\frac{R^2+G^2+B^2}{3}\right)^{1/2}. \tag{4}$$

where: $R' = R - min$, $G' = G - min$, $B' = B - min$, $min = min\,(R,G,B)$, s is a relative saturation.

Prsented formulas were derived in the basis of geometrical dependences between *RGB* and *HSI* colour spaces. In this transformation the colour components take the values from range (0,255]. However, maximal value of Hue is 252, because the colours vary from $H(R) = 0$ corresponding to the red colour, through $H(Y) = 63$ (yellow colour), $H(G) = 126$ (green colour), $H(C) = 189$ (cyan colour) to $H(B) = 252$ (blue colour). This feature is used in the image analysis to record *H, S, I* values in a sequence of 8-

bit positive integers. Equation (1) and (2) and also (4) defining Hue and Intensity are non-linear. Linear transformation refers only to relative saturation, equation (3).

Colour hexagon *RYGCBM* presented in figures 3 and 4 can be divided in three rhombus of red *RYWM*, green *GCWY* and blue *BMWC* colours. The other three rhombus can be also found: yellow, cyan and magenta. Each of the rhombus can be divided in the triangles. In the red rhombus there are two triangles: yellow *RYW* and magenta *RWM*, in the green one there are yellow *GWY* and cyan *GCW* triangles, in the blue – cyan *BWC* and magenta *BMW*. Going counter-clockwise the successive triangles are yellow of red rhombus, yellow and cyan of green rhombus, cyan and magenta of blue rhombus and magenta of red rhombus. The values of R,G,B colour components in exemplary triangles are listed in table 1. The following abbreviations are used: $max = max\,(R,G,B)$, $cen = cen\,(R,G,B)$ - central value from (R,G,B) and $min = min\,(R,G,B)$.

Table 1. Examples of R,G,B values in successive triangles, *max* = maximal value, *min* = minimal value, *cen* = central value.

Color of rhombus	Color of triangle	Component		
		R	G	B
red	yellow	*max*	*cen*	*min*
green	yellow	*cen*	*max*	*min*
	cyan	*min*	*max*	*cen*
blue	cyan	*min*	*cen*	*max*

In PIT method the most important variable is colour, therefore only definition of Hue (equations (1) and (2)) will be derived here.

In the yellow triangle of red rhombus the red component is maximal $R = max$, while blue is minimal, $B = min$. When the *max* and *min* values are constant, Hue is the linear function of central value corresponding to the green component $G = cen$ in this triangle. For the arbitrary chosen colour, marked in figure 3 (point at the end of arrow), the following formula based on the geometrical dependences can be written:

$$H = 63\cdot\frac{G-B}{R-B} = 63\cdot\frac{(G-B)+(R-B)-(R-B)}{R-B} = 63\cdot\left(1+\frac{G'-R'}{R'}\right). \quad (5)$$

Continuing, in the yellow triangle of green rhombus the green component is maximal $G = max$, while blue – minimal $B = min$. Again, when the *max* and *min* values are constant, Hue is the linear function of central value corresponding to the red component $R = cen$. In figure 4 arbitrary chosen colour (point at the end of arrow) is analyzed.

Hue value for this colour can be calculated as:

$$H = 63 + 63\cdot\frac{G-R}{G-B} = 63\cdot\left[1+\frac{(G-B)-(R-B)}{(G-B)}\right] = 63\cdot\left(1+\frac{G'-R'}{G'}\right). \quad (6)$$

The equations describing Hue in yellow triangles of red and green rhombus (equations (5) and (6)) differs only in denominator. Derived relations are equivalent to known linear functions [8]. The following relation can be written for entire yellow rhombus, with an assumption that the dominator can be taken as $G'+R'$

$$H = 63\cdot\left(1+\frac{G'-R'}{G'+R'}\right) = 63 + 63\cdot(G'-R')\cdot(G'+R')^{-1} \qquad for\ B' = 0. \quad (7)$$

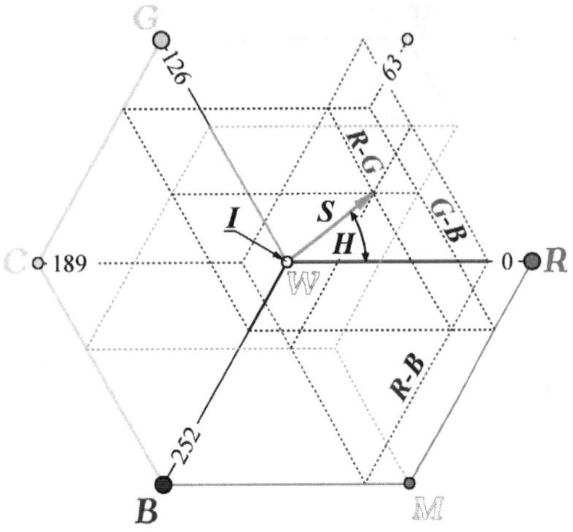

$(R,G,B)=(209, 158, 79)$
$(H,S,I)=(48, 128, 158)$

Figure 3. Colour analyzed in yellow triangle of red rhombus.

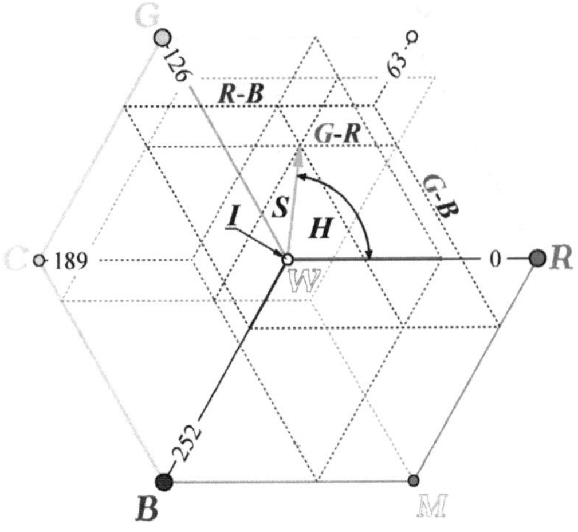

$(R,G,B)=(158, 209, 79)$
$(H,S,I)=(78, 128, 158)$

Figure 4. Colour analyzed in yellow triangle of green rhombus.

Relation (7) for yellow rhombus *YGWR* is equivalent to equations (5) and (6) only on the planes going through *R*, *Y* and *G* colours. For other planes they differ as much as ~±10.8.
Calculation of Hue for the cyan rhombus *CBWG* is analogous, with following substitutions: *R*→*G*, *G*→*B* and *B*→*R*. The base colour also has to be changed from *H*(*Y*) = 63 to *H*(*C*) = 189. Finally, the relation describing Hue in the cyan rhombus can be written in a form

$$H = 126 + 63 \cdot \left(1 + \frac{B'-G'}{B'+G'}\right) = 189 + 63 \cdot (B'-G') \cdot (B'+G')^{-1} \qquad for\ R' = 0. \tag{8}$$

Subsequent analysis will be carried out only for the yellow rhombus *YGWR*, because it can be applied also to the cyan rhombus *CBWG*. For the simplification of formulas the values of *R*, *G* and *B* components will be taken from the range [0,1] – the colour cube becomes unitary. It means that the Saturation and Intensity will be also in this range.

5. Criterion of Hue invariability in a plane containing the Intensity axis

Dedicated to PIT method, *RGB* to *HSI* colour space transformation should fulfil the criterion of Hue invariability in a plane containing the Intensity axis [1]. Value of Hue in the yellow rhombus *YGWR* should come from the range (0,126) and be constant for any plane going through the Intensity axis.
In the triangle *YWR* of red rhombus, Hue belongs to the range (0,63). This value should be invariable on the plane containing the black point *K* = (0,0,0), white point *W* = (1,1,1) and arbitrary chosen point *X* = (1,*x*,0) located at the edge of colour hexagon connecting the points *R* = (1,0,0) and *Y* = (1,1,0). Variable *x* belongs then to the range (0,1). Equation describing such plane takes the form

$$x \cdot R - G + (1-x) \cdot B = 0 \quad or \quad x \cdot (R-B) - (G-B) = x \cdot R' - G' = 0, \tag{9}$$

Which is coming from a general equation of plane. The equation coefficients were determined by solution of the equation system. Value of Hue calculated with equation (1) [15] is equal to the value calculated with equation (5) representing linear method according to Authors derivation, but only for the plane containing *R* and *Y* colours. Equation (1) becomes non-linear for the planes containing *X* colour differently from equation (5), which still remain linear. Equation (1) is non-linear because it is a

function of G component taking the central value. This limitation does not applied in the case of equation (5).
In the triangle *YWR* of red rhombus for the points belonging to the plane $x \cdot R' = G'$, equation (1) ensure the constant value of Hue because

$$H(x) = 63 \cdot \left(1 + \frac{G' - R'}{G' + R'}\right) = 63 \cdot \left(1 + \frac{x \cdot R' - R'}{x \cdot R' + R'}\right) = 63 \cdot \left(1 + \frac{x-1}{x+1}\right) = 63 \cdot \frac{2 \cdot x}{x+1}. \quad (10)$$

Assuming that $x = idem$, what defines the plane, Hue is invariable, for example $H(0) = 0$, $H(0.5) = 42$, $H(1) = 63$. According to the linear method and equation (5) the Hue can be calculated from the formula

$$H(x) = 63 \cdot \left(1 + \frac{G' - R'}{R'}\right) = 63 \cdot \left(1 + \frac{x \cdot R' - R'}{R'}\right) = 63 \cdot \left(1 + \frac{x-1}{1}\right) = 63 \cdot x. \quad (11)$$

also ensuring its invariability for $x = idem$. Following values of Hue are obtained for exemplary points: $H(0) = 0$, $H(0.5) = 31.5$, $H(1) = 63$.
Taking the same steps in the analysis of yellow triangle *YGW* in the green rhombus and the plane going through the black point $K = (0,0,0)$, white point $W = (1,1,1)$ and arbitrary chosen point $X = (x,1,0)$ located between $G = (0,1,0)$ and $Y = (1,1,0)$, where $x \in (0,1)$, the following formula describing this plane can be defined

$$R - x \cdot G + (x-1) \cdot B = 0 \quad or \quad (R - B) - x \cdot (G - B) = R' - x \cdot G' = 0. \quad (12)$$

Its form was obtained in accordance with the method applied for equation (9).
Equation (1) describing the Hue in the triangle *YGW* is equivalent to the equation (6) only for the planes containing G and Y colours. It becomes non-linear for the planes containing X colour, because Hue is non-linear function of R component on it, whereas equation (6) is still linear.
The invariability of Hue calculated with equation (1) is ensured for the plane $x \cdot G' = R'$ regarding the equation

$$H(x) = 63 \cdot \left(1 + \frac{G' - R'}{G' + R'}\right) = 63 \cdot \left(1 + \frac{G' - x \cdot G'}{G' + x \cdot G'}\right) = 63 \cdot \left(1 + \frac{1-x}{x+1}\right) = 63 \cdot \frac{2}{x+1}, \quad (13)$$

when $x = idem$, for example $H(0) = 126$, $H(0.5) = 84$ or $H(1) = 63$.
Applying the linear method (equation (6)) the invariability of Hue is ensured in whole range

$$H(x) = 63 \cdot \left(1 + \frac{G' - R'}{G'}\right) = 63 \cdot \left(1 + \frac{G' - x \cdot G'}{G'}\right) = 63 \cdot \left(1 + \frac{1-x}{1}\right) = 63 \cdot (2 - x) \quad (14)$$

$x = idem$, for example $H(0) = 126$, $H(0.5) = 94.5$ or $H(1) = 63$.

6. Criterion of Hue independence on the Intensity and Saturation

According to [1] the *RGB* to *HSI* colour space transformation is suitable for PIT method when the Hue is independent on the Intensity and Saturation.
The independence of Hue on the Intensity can be proved by multiplication of arbitrary chosen colour $X = (R,G,B)$, placed beyond the Intensity axis, by the value $1 \le m \le [max(R,G,B)]^{-1}$. It gives a new colour X located inside the colour hexagon. This new colour should have higher value of Intensity, but the same Hue.
The independence of Hue on the Saturation can be proved by mixing an arbitrary chosen colour $X = (R,G,B)$ with white colour $W = (1,1,1)$ leading to decreasing value of Saturation, but without changing the Hue. Mixing can be obtained by adding to X colour components positive value a from the range $0 \le a \le 1 - max(R,G,B)$. The new colour is creating and it is located on the axis parallel to the Intensity axis, inside the colour hexagon.

Another way of verification is to add "some amount" of colour *X*, m_X to an "amount" of white colour m_W and obtained an "amount" of colour *Z*, m_Z placed between colours *X* and *W*

$$m_x \cdot X + m_w \cdot W = m_z \cdot Z \quad \Rightarrow \quad (1-w)\cdot X + w \cdot W = Z \quad or \quad X + w\cdot(W - X) = Z\,, \tag{15}$$

where $w = m_W \cdot (m_X + m_W)^{-1}$, $w \in (0,1)$, represents an amount of added white colour W in colour Z.
If two of these three requirements are fulfilled, the criteria of Hue invariability in the plane containing the Intensity axis (chapter 4) is also fulfilled.
The Intensity of colour, obtained by multiplying the colour *X* by value of *m* ($m \geq 1$) is increasing

$$I\,(m{\cdot}R, m{\cdot}G, m{\cdot}B) = \left[\frac{(m{\cdot}R)^2 + (m{\cdot}G)^2 + (m{\cdot}B)^2}{3}\right]^{1/2} = m \cdot I\,(R,G,B)\,, \tag{16}$$

the Saturation does not change, attention: in subsequent derivations, relative Saturation $s \in [0,1]$

$$s\,(m{\cdot}R, m{\cdot}G, m{\cdot}B) = 1 - \frac{min\,(m{\cdot}R, m{\cdot}G, m{\cdot}B)}{I\,(m{\cdot}R, m{\cdot}G, m{\cdot}B)} = 1 - \frac{m{\cdot}min\,(R,G,B)}{m{\cdot}I\,(R,G,B)} = s\,(R,G,B)\,, \tag{17}$$

and the Hue is also invariable

$$H\,(m{\cdot}R, m{\cdot}G, m{\cdot}B) = 63\cdot\left(1 + \frac{m{\cdot}(G' - R')}{m{\cdot}(G' + R')}\right) = H\,(R,G,B)\,, \tag{18}$$

it does not depend on the Intensity.
Mixing of the colour *X* with white colour by an addition of constant value $a \geq 0$ to the colour components, causes increase of Intensity value

$$I\,(R{+}a, G{+}a, B{+}a) = \left[\frac{(R{+}a)^2 + (G{+}a)^2 + (B{+}a)^2}{3}\right]^{1/2} \geq \left[\frac{R^2 + G^2 + B^2}{3}\right]^{1/2} = I\,(R,G,B)\,. \tag{19}$$

According to table 1, the yellow triangles *YWR* and *YGW*, forming the yellow rhombus *YGWR*, are having the minimal value of Blue component. If the vector (*a,a,a*) is added to any colour from these triangles, the Saturation value should decrease. Therefore, following inequality should be fulfilled

$$\begin{aligned} &s\,(R{+}a, G{+}a, B{+}a) \leq s\,(R,G,B) \;\Rightarrow\; 1 - \frac{min\,(R{+}a, G{+}a, B{+}a)}{I\,(R{+}a, G{+}a, B{+}a)} \leq 1 - \frac{min\,(R,G,B)}{I\,(R,G,B)} \;\Rightarrow \\ &\Rightarrow \frac{min\,(R,G,B){+}a}{I\,(R{+}a, G{+}a, B{+}a)} \geq \frac{min\,(R,G,B)}{I\,(R,G,B)} \;\Rightarrow\; \frac{B{+}a}{I\,(R{+}a, G{+}a, B{+}a)} \geq \frac{B}{I\,(R,G,B)}\,. \end{aligned} \tag{20}$$

Introducing equation (4) to equation (20) the inequality takes the form

$$\begin{aligned} &\frac{(B{+}a)^2}{(R{+}a)^2 + (G{+}a)^2 + (B{+}a)^2} \geq \frac{B^2}{R^2 + G^2 + B^2} \;\Rightarrow\; \frac{(R{+}a)^2 + (G{+}a)^2 + (B{+}a)^2}{(B{+}a)^2} \leq \frac{R^2 + G^2 + B^2}{B^2} \\ &\Rightarrow \frac{(R{+}a)^2 + (G{+}a)^2}{(B{+}a)^2} + 1 \leq \frac{R^2 + G^2}{B^2} + 1 \;\Rightarrow\; \frac{(R{+}a)^2 + (G{+}a)^2}{(B{+}a)^2} \leq \frac{R^2 + G^2}{B^2} \\ &\Rightarrow \left[(R + a)^2 + (G + a)^2\right]\cdot B^2 \leq \left(R^2 + G^2\right)\cdot(B + a)^2, \end{aligned} \tag{21}$$

which is satisfied in the yellow rhombus

$$2RB\,(B - R) + 2GB\,(B - G) + a\,(B^2 - R^2) + a\,(B^2 - G^2) \leq 0\,, \tag{22}$$

because obtained inequality (equation (22)) is true. All values in the parenthesis, due to the minimal value of Blue component, remain negative. It means that mixing of *X* with white colour by addition of constant value causes decrease in relative Saturation s. At the same time, the Hue value remains constant

$$H(R+a,G+a,B+a)=63\cdot\left(1+\frac{(G+a-B-a)-(R+a-B-a)}{(G+a-B-a)+(R+a-B-a)}\right)=63\cdot\left(1+\frac{(G-B)-(R-B)}{(G-B)+(R-B)}\right)=H(R,G,B)\,. \quad (23)$$

Mixing of the X colour with white one by an addition of vector (a,b,c) causes an increase in the Intensity value, because at least one of a, b or c value is higher than zero

$$I\left(R+\underbrace{w\cdot(1-R)}_{a\geq 0},\ G+\underbrace{w\cdot(1-G)}_{b\geq 0},\ B+\underbrace{w\cdot(1-B)}_{c\geq 0}\right)=\left[\frac{(R+a)^2+(G+b)^2+(B+c)^2}{3}\right]^{1/2}>\left(\frac{R^2+G^2+B^2}{3}\right)^{1/2}=I(R,G,B)\,. \quad (24)$$

In such situation the Saturation should decrease and following inequality should be fulfilled

$$S\left[(1-w)\cdot X+w\cdot W\right]<S(X)\,. \quad (25)$$

This relation can be proved after re-writing equation (15) in an equivalent form of following equation

$$(1-w)\cdot\left[X+\frac{w}{(1-w)}\cdot W\right]=Z\,. \quad (26)$$

Inserting equation (26) into equation (25), term $(1-w)$ can be withdrawn and quotient must be positive $w\cdot(1-w)^{-1}>0$. It is fulfilled because $0<w<1$ (w = 0 represents pure colour, while w = 1 represents addition of infinite amount of white colour). Presented above prove of Saturation decreasing, when the colours are mixed, can be utilized for the confirmation of Hue invariability. The Hue value remains invariable, due to

$$\begin{aligned}H\left[(1-w)\cdot R+w,\ (1-w)\cdot G+w,\ (1-w)\cdot B+w\right]&=63\cdot\left(1+\frac{[(1-w)\cdot(G-B)+w-w]-[(1-w)\cdot(R-B)+w-w]}{[(1-w)\cdot(G-B)+w-w]+[(1-w)\cdot(R-B)+w-w]}\right)\\&=63\cdot\left(1+\frac{(G-B)-(R-B)}{(G-B)+(R-B)}\right)=H(R,G,B)\,.\end{aligned} \quad (27)$$

7. Summary

The light reflected from TLC particles is closely connected with their temperature. This feature is utilized in the PIT method. Image analysis requires transformation of *RGB* colour space to *HSI* one. Advantage of presented in this paper transformation is coming from the possibility of storing the colour components as 8-bit integer numbers.

In the paper the non-linear relation defining Hue, presented in [15-17] was derived and it was shown that it has common origin with the linear functions presented in [8-10]. It was proved that derived formulas fulfilled the criterion of Hue invariability in the plane containing Intensity axis and criterion of Hue independence on the Intensity and Saturation. These requirements should be fulfilled by all colour space transformations applied in the PIT method. Derivation of the formulas and described criteria construct the theoretical approach to the experimental procedure with image analysis and conscious choice of computer tools. It supplements the knowledge on the PIT and digital image processing.

8. References

[1] Park H G, Dabiri D and Gharib M 2001 Digital particle image velocimetry/thermometry and application to the wake of heated circular cylinder *Exp. Fluids* **30** pp 327–338

[2] Kowalewski T A, Cybulski A and Rebow M Particle Image Velocimetry and Thermometry in Freezing Water *8th International Symposium on Flow Visualization (Sorrento, Italy, September 1-4, 1994)*

[3] Fornalik E, Leiner W, Szmyd S J, Kowalewski T A and Ozoe H 2005 Visualization of the Flow Structure and Temperature Field in the Region of Mixed Convection in *Mechanics of 21st Century* (Springer Verlag)

[4] Fornalik E, Yamamoto Y, Chen W, Nakabe K and Suzuki K 1999 Visualization of heat transfer enhancement regions modified by the interaction of inclined impinging jets into crossflow, *MG&V* **8** pp 597-609

[5] Bednarz T, Fornalik E, Tagawa T, Ozoe H and Szmyd J S 2006 Convection of paramagnetic fluid in a cube heated and cooled from side walls and placed below a superconducting magnet: comparison between experiment and numerical computations *Thermal Science and Engineering* **14** pp 107-114

[6] Stasiek J, Jewartowski M and Kowalewski T A 2014 The Use of Liquid Crystal Thermography in Selected Technical and Medical Applications *Journal of Crystallization Process and Technology* **4** pp 46-59

[7] Dabiri D and Gharib M 1991 Digital particle image thermometry: The method and implementation, *Exp. Fluids* **11** pp 77–86

[8] Smith A R 1978 Color Gamut Transform Pairs *SIGGRAPH '78 Proc. of the 5th annual Conf. on Computer graphics and interactive techniques (New York)* vol 12 (New York: ACM) pp 12–19

[9] Foley J D, van Dam A, Feiner S K and Hughes J F 1996 *Computer Graphics: Principles and Practice in C, 2nd ed.* (Addison–Wesley)

[10] http://www.oracle.com/technetwork/java/javase/downloads/index.html

[11] Russ J C 2002 *The Image Processing Handbook, 4th ed.* (Boca Raton, London, New York, Washington D.C.: CRC Press LLC)

[12] Bunting F 1998 *The ColorShop Color Primer* (Light Source Computer Images, Inc. An X-Rite Company)

[13] Gonzalez R C and Woods R E 2002 *Digital Image Processing, 2nd ed.* (Upper Saddle River, New Jersey: Prentice Hall)

[14] Hay J L and Hollingsworth T H 1996 A Comparison of Trichromic Systems for Use in the Calibration of Polymer-Dispersed Thermochromic Liquid Crystals *Exp. Therm. Fluid Sci.* **12** pp 1–12

[15] Kowalewski T A 2007 *Thermochormic Liquid Crystals, in Handbook of Experimental Fluid Mechanics* ed C Tropea, A Yarin and J F Foss (Berlin, Heidelberg: Springer–Verlag) chapter B7.1 pp 487-500

[16] Hiller W J, Koch St, Kowalewski T A and Stella F 1993 Onset of natural convection in a cube *Int. J. Heat Mass Tran.* 36 pp 3251-3263

[17] Kowalewski T A 1999 Particle Image Velocimetry and Thermometry Using Liquid Crystals *FLUVISU 99, 8me colloque nationale de visualisation et de traitment d'images en mecanique des fluids (Toulose, France, June 1-4 1999)* ENSICA pp 33-48 http://fluid.ippt.pan.pl/papers/fluvis99_tkowale.pdf

Acknowledgement

The present work was supported by the Polish Ministry of Science (Grant AGH No. 11.11.210.198).

Molecular dynamics simulation for flow characteristics in nanochannels and single walled carbon nanotubes

H Yasuoka, T Imae, M Kaneda and K Suga
Department of Mechanical Engineering, Osaka Prefecture University, Sakai, Osaka 599-8531, Japan

E-mail: yasuoka@htlab.me.osakafu-u.ac.jp

Abstract. Flows in graphite-, diamond- and silicon-walled nanochannels are discussed by performing molecular dynamics simulations. Flows in carbon nanotubes (CNTs) and graphene-walled nanochannels are also investigated. It is found that the flow rate in the graphite-walled channel tends to be the largest because of its slippery wall structure by the short bond length and the high molecular density of the CNTs. The flow rate in the single walled CNT at a very narrow diameter tends to increase although such a tendency is not seen in the graphene-walled channel.

1. Introduction

Carbon nanotubes (CNTs) and nanochannels attract much attention because these nano-structures are expected for expanding the applications of the membrane engineering. Recently, it was reported that hydrocarbons or water molecules flowed through a CNT much faster than in the other materials [1]. After this sensational report, many research studies have been reported on flows in CNTs [2-4]. However, the mechanism of the fast transport in a CNT has not been clearly understood. Also, understanding the detailed flow physics in nano-structures is still underway. Therefore, in this study, molecular dynamics (MD) simulations of flows in nanochannels and CNTs are performed to understand the flow characteristics in nano-structures and the fast transport mechanism in a CNT.

2. Molecular dynamics method

2.1. Lennard-Jones fluid

The presently applied classical molecular dynamics simulation consists of the numerical solution of classical equations of motions of N molecules interacting via model potentials. The well-known model potential is the 6-12 Lennard-Jones (L-J) potential ϕ which is defined as

$$\phi(r_{ij}) = 4\varepsilon\left[\left\{\left(\frac{\sigma}{r_{ij}}\right)^{12} - \left(\frac{\sigma}{r_{ij}}\right)^{6}\right\} - \left\{\left(\frac{\sigma}{r_c}\right)^{12} - \left(\frac{\sigma}{r_c}\right)^{6}\right\}\right], \tag{1}$$

where r_{ij} is the intermolecular distance between i and j molecules, r_c is the cut-off radius, ε and σ are the well depth and the diameter of the molecules. From this potential, intermolecular force **F** is written as

$$\mathbf{F} = -\frac{\partial \phi}{\partial r} = \frac{24\varepsilon}{r_{ij}}\left[2\left(\frac{\sigma}{r_{ij}}\right)^{12} - \left(\frac{\sigma}{r_{ij}}\right)^{6}\right]. \tag{2}$$

This model has been used in the classical MD simulation by argon (Ar) molecules. In the system which is modeled by the Lennard-Jones molecules, the classical equation of motion can be written as

$$\frac{d^2 r^*}{dt^{*2}} = -\frac{\varepsilon\tau^2}{m\sigma^2}\frac{\partial \phi^*}{\partial r^*}, \tag{3}$$

where m, $r^* = r/\sigma$, $t^* = t/\tau (= t/\sqrt{m\sigma^2/\varepsilon}$ and $\phi^* = \phi/\varepsilon$ are the mass, the nondimensional length, time and energy, respectively. However, to make a physical interpretation, it is sometimes expressed as dimensional values in terms of Ar. In addition, the nondimensional number density $N^* = N\sigma^3$ and temperature $T^* = k_b T/\varepsilon$ are used in the Lennard-Jones fluid. Here k_b is the Boltzmann constant. In actual calculations, to hold down the calculation effort of the interactions, the potential effects are truncated over cut off distance which is defined as r_c. In the present study, $r_c = 3\sigma$ is employed.

2.2. Brenner and Tersoff potentials

To describe the covalent bond, the Brenner and Tersoff potentials [5, 6] are applied to carbon and silicon walls. The covalent bond energy is written as

$$E_b = \sum_i \sum_{j>i} f_c(r_{ij})\left[f_R(r_{ij}) - \bar{b}_{ij} f_A(r_{ij})\right]. \tag{4}$$

For the Brenner potential, $f_R(r_{ij})$ and $f_A(r_{ij})$ are written as

$$f_R(r_{ij}) = \frac{D_{ij}^{(e)}}{S-1}\exp\left(-\sqrt{2/S}\beta(r - R_e)\right), \tag{5}$$

$$f_A(r_{ij}) = \frac{D_{ij}^{(e)} S}{S-1}\exp\left(-\sqrt{2/S}\beta(r - R_e)\right), \tag{6}$$

where D_e, S and R_e are determined by the physical properties of carbon. For the Tersoff potential, $f_R(r_{ij})$ and $f_A(r_{ij})$ are written as

$$f_R(r_{ij}) = A\exp\left(-\lambda_1 r_{ij}\right), \tag{7}$$

$$f_A(r_{ij}) = B\exp\left(-\lambda_2 r_{ij}\right), \tag{8}$$

where A, B, λ_1 and λ_2 are physical properties of silicon. The cut off function f_c is written as

$$f_c(r_{ij}) = \begin{cases} 1, & r < R - D \\ \frac{1}{2} - \frac{1}{2}\sin\left(\frac{\pi}{2}(r_{ij} - R)/D\right), & R - D < r < R + D \\ 0, & r > R + D \end{cases}. \tag{9}$$

The bond order term is described as

$$\bar{b}_{ij} = \frac{b_{ij} + b_{ji}}{2}. \tag{10}$$

The function b_{ij} is written as

$$b_{ij} = [1 + (\beta\zeta_{ij})^{n_0}]^{-\delta}, \quad (11)$$

where ζ_{ij} is

$$\zeta_{ij} = \sum_{k \neq i,j} f_c(r_{ik}) g_c(\theta_{ijk}) \exp[\lambda_3^3 (r_{ij} - r_{ik})^3], \quad (12)$$

and $g_c(\theta)$ is described as

$$g_c(\theta_{ijk}) = 1 + \frac{c^2}{d^2} - \frac{c^2}{d^2 + (h - \cos\theta_{ijk})^2}, \quad (13)$$

where θ_{ijk} is the angle between bonds *ij* and *ik*. The parameters of the Brenner and Tersoff potentials are tabulated in table 1.

Table 1. The parameter of the Brenner and Tersoff potentials.

Benner parameter	Brenner value	Tersoff parameter	Tersoff value
D_e (eV)	6.325	A (eV)	3264.7
S	1.29	B (eV)	95.373
λ (Å^{-1})	1.5	λ_1 (Å^{-1})	3.2394
R_e (Å)	1.315	λ_2 (Å^{-1})	1.3528
β	1.0	β	0.33675
n_0	1.0	n_0	22.956
δ	0.80469	δ	11.478
λ_3 (Å^{-1})	0	λ_3 (Å^{-1})	1.3258
a	0.011304	a	1
c	19	c	4.8381
d	2.5	d	2.0417
h	-1.0	h	0
R(Å)	1.85	R(Å)	3.0
D(Å)	0.15	D(Å)	0.2

3. Wall structures

In this study, five types of wall structures: diamond, silicon, graphite, graphene and CNT wall structures are discussed. These five structures can be categorized into two kinds of structures which are the diamond and six membered ring structures. Figure 1 shows the diamond structure. The diamond and silicon walls have this structure. The lattice constants of the diamond and silicon crystals are 0.356 and 0.543nm, respectively. The six membered ring structure for the graphene and graphite is shown in figure 2. Their bond length is 0.1421nm. The distance between the graphite layers is 0.3354 nm. The armchair CNT is considered in the present study.

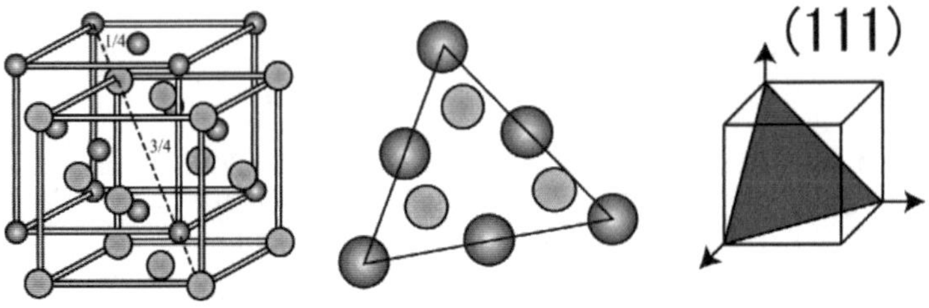

Figure 1. Diamond structure.

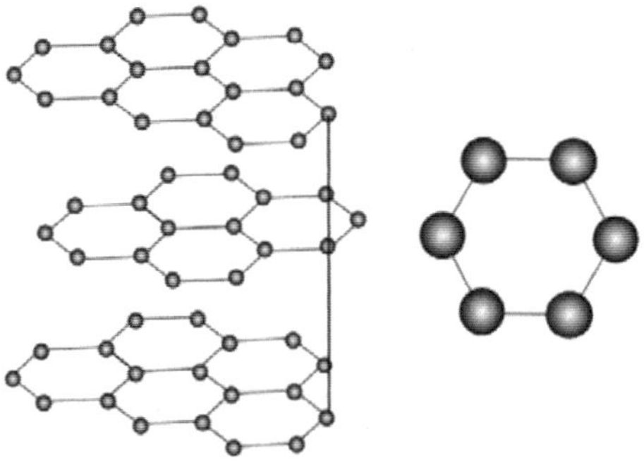

Figure 2. Six membered ring structure.

4. Results and discussions

4.1. Nanochannel flows of three types of wall structures

Firstly, flows in diamond-, silicon- and graphite-walled nanochannels are discussed. Argon molecules are applied for the fluid molecules. Figure 3 illustrates calculation domains for the nanochannels and CNTs. For the nanochannels, the calculation domain whose length L=3.4 nm, width W=3.4nm and height H=6.8nm are applied. Periodic boundary conditions are applied for the x and z directions. The external force F = 2.45($\times 10^{-14}$N) acts on the fluid molecules to the x direction. The numbers of fluid and wall molecules are tabulated in table 2. In these cases, the Langevin method [7] is applied to the calculation domain to keep the temperature at 298K and thus the pressure of calculation domain is 5.9Pa. For taking the statistical average, the computations are performed for about 100 million time steps.

Table 2. Number of fluid and wall molecules of nanochannel flows.

	Fluid	Wall
Ar-Graphite	114	4480
Ar-Diamond	120	6272
Ar-Silicon	112	1800

Journal of Physics: Conference Series **530** (2014) 012048 doi:10.1088/1742-6596/530/1/012048

Figure 4 shows the number density profiles of the three wall cases. It is seen that the fluid molecules are densely distributed near the walls. This is because the channel walls strongly attract the fluid molecules. Figure 5 shows velocity profiles across the channels. In this figure, the flow in the graphite-walled channel has the largest velocity profile and the silicon-walled channel has the smallest one. One of the reasons why the velocity profiles have such tendencies is considered to be the roughness effect of the wall structure. In the three types of wall structures, the graphite wall has the largest atomic packing factor on the wall surface P_f= 0.956 and the silicon wall has smallest atomic packing factor P_f= 0.811 while the diamond wall has P_f= 0.922. It is considered that the velocity profiles correspond to these atomic packing factors.

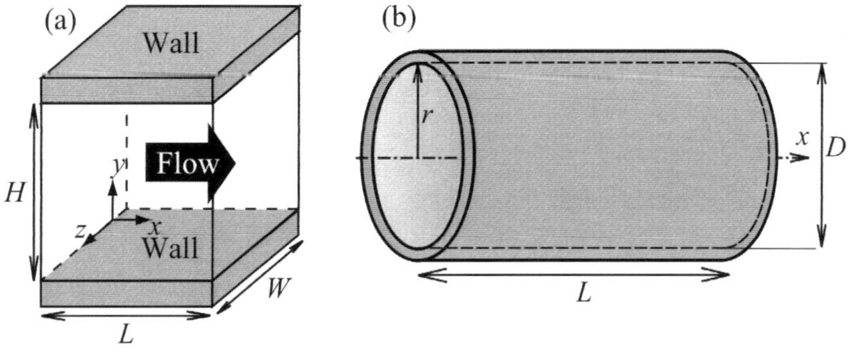

Figure 3. Calculation domains for nanochannel and CNT; (a) nanochannel, (b) CNT.

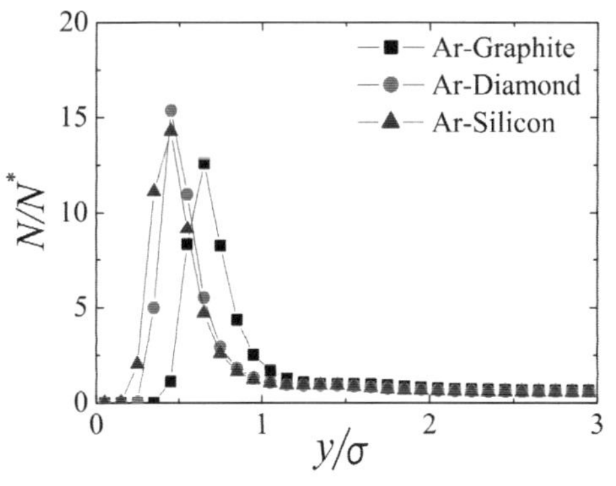

Figure 4. Number density profiles of fluid molecules in nanochannels.

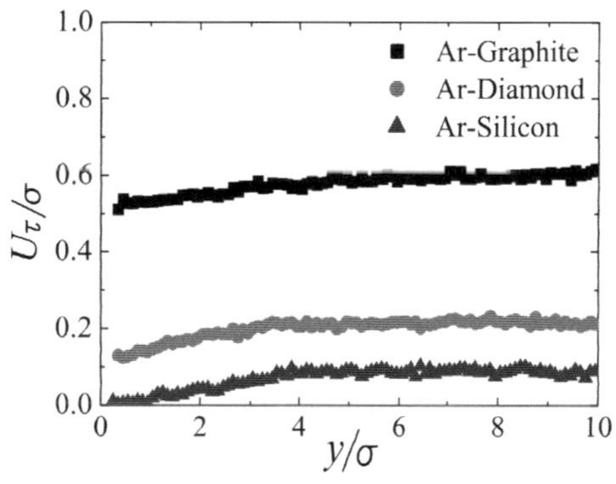

Figure 5. Velocity profile in nanochannels.

4.2. Flows in CNT and graphene-walled channel cases

Under a constant temperature and a pressure, the channel height H and the diameter D of calculation domains are varied in the range of 0.91 - 7.83nm. The domain length L of the CNTs and the graphene-walled channels are L = 1.23 - 9.84 nm. The width W of the graphene-walled channel is W = 3.84 - 24.60 nm. For the CNTs, periodic boundary conditions are applied for the x direction. The external force F = 7.36($\times 10^{-14}$ N) acts on the fluid molecules to the x direction. The Nose-Hoover thermostat [8, 9] is applied to the calculation domain to keep the temperature at 121 K. The pressure of the domain is 34.3 MPa. The numbers of fluid and wall molecules are tabulated in table 3. For taking the statistical average, the computations are performed for about 100 million time steps are performed as before.

Table 3. Number of fluid and wall molecules of CNT and graphene channel flows.

	Ar-CNT		Ar-Graphene	
H, *D* [nm]	Fluid	wall	Fluid	Wall
0.91	132	1440	708	2880
1.32	207	1440	1024	2880
2.95	690	1920	3048	3840
4.58	831	1440	3548	2880
6.21	764	960	3207	1920
7.83	1217	1200	5060	2400

Figure 6 shows centerline and slippage velocities normalized by the bulk velocities. In all channel heights or diameters, centerline and slippage velocities are nearly the same corresponding to flat velocity profiles.

Next, the tendency of flow rate depending on the tube diameter and the channel height is compared. The normalized flow rate is defined as

$$Q = \frac{\int 2\pi r \rho' U(r) dr}{\rho a \pi D^3 / 8Cs} \quad \text{for the CNT,} \tag{14}$$

$$Q = \frac{\int \rho' U(y) dy}{\rho a H^2 / Cs} \quad \text{for the graphene channel,} \tag{15}$$

where ρ and a are the number density and the acceleration. The sound speed Cs of argon is defined as

$$Cs = \sqrt{\gamma R T}\ , \tag{16}$$

where $\gamma = 1.67$ and $R = 208\ \mathrm{m^2/(s^2 \cdot K)}$ are the specific heat ratio and the gas constant. Figure 7 shows the distribution of the normalized flow rates.

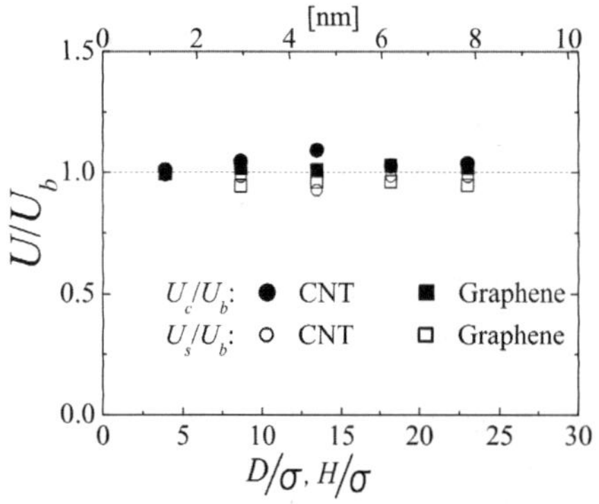

Figure 6. Normalized centerline and slippage velocities in CNTs and graphene-walled channels.

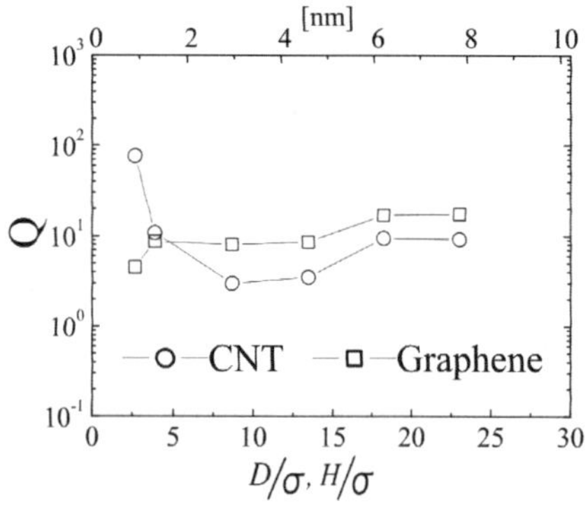

Figure 7. Normalized flow rates in CNTs and graphene-walled channels.

Journal of Physics: Conference Series **530** (2014) 012048 doi:10.1088/1742-6596/530/1/012048

As shown in figure 7, the normalized flow rates Q of the two cases against H or D show a similar tendency where H or D are larger than 2.9 nm. However, the flow rate in the CNT at very narrow diameter (smaller than 2.9 nm) tends to be increase as the diameter decreases unlike in the graphene channel. This is because the attractive forces from the surrounding wall molecules are almost canceled each other. Moreover, the motions of fluid molecules in the narrow CNT are almost limited to the streamwise direction although those in the graphene-walled channel are allowed also in the spanwise direction. Thus the loss of the kinetic energy by collisions between fluid molecules in the CNT is less than that in the graphene-walled channel.

5. Conclusions

Nano-flows in several nano-structures are discussed by performing molecular dynamics simulations. Under a constant temperature and a pressure, three types of structures: diamond-, silicon- and graphite-walled nanochannel cases are calculated. The flow tendencies in the CNTs and the graphene-walled channels are also discussed. The number density profiles of the fluid molecules have peaks near the walls because the walls strongly attract the fluid molecules. Amongst the three types of wall structures, the flow in the graphite-walled channel has the largest velocity profile and that in the silicon-walled channel has the smallest velocity because of the roughness effect of the wall structure. In the case of the flows in the CNTs and the graphene-walled channels, flat velocity profiles across the tubes or channels are seen regardless of their diameters or heights. The flow rates of the CNT have a different tendency when the tube diameter is less than 2.9 nm due to the effects of the surrounding wall molecules.

References

[1] Holt J K, Park H G, Wang Y, Stadermann M, Artyukhin A B, Grigoropoulos C P, Noy A and Bakajin O 2006 *Science* **312** 1034
[2] Thomas J A, McGaughey A J H and Arnebeck O K 2010 *Int. J. Therm. Sci.* **49** 281
[3] Joseph S and Aluru N R 2008 *Nano Lett.* **8** 452
[4] Kannam S K, Todd B D, Hansen J S and Daivis P J 2013 *J. Chem. Phys.* **138** 094701
[5] Brenner D W 1990 *Phys. Rev.* B **42** 9458
[6] Tersoff J 1988 *Phys. Rev.* B **37** 6991
[7] Blömer J and Beylich A E 1999 *Surf. Sci.* **423** 127
[8] Nose S 1984 *J. Chem. Phys.*, **81** 511
[9] Hoover W G 1985 *Phys. Rev.* A **31** 1695

Modelling of two-phase flow in a minichannel using level-set method

H. Grzybowski and R. Mosdorf
Department of Mechanics and Applied Computer Science, Faculty of Mechanical Engineering, Bialystok University of Technology
Wiejska 45C, 15-351 Bialystok, Poland

E-mail: h.grzybowski@doktoranci.pb.edu.pl

Abstract. Today there is a great interest in micro-scale multiphase fluid flow. In the paper, the numerical simulation of two-phase flow inside 3 mm minichannel was carried out. The liquid–gas interface was captured using the level-set method. During the calculation, the stabilization and reinitialization of level set function was performed in order to obtain the proper accuracy of the simulation. Incompressible Navier-Stokes equations were solved using the COMSOL Multiphysics® on a two-dimensional mesh. The process of formation of different two-phase flow patterns in the minichannel has been investigated. During the simulation it has been analysed three flow patterns: the bubbly flow and two kinds of slug flow with short and long slugs. It has been shown that unsteady flow at the inlet of the minichannel is responsible for the chaotic character of changes of the slug and bubble sizes. Such unsteady flow modifies the distance between the bubbles and slugs. It has been shown that for the low water inlet velocity the two-phase flow pattern becomes more stable.

1. Introduction

In recent years, the interest in simulation of two-phase flow in micro scale has significantly increased.The miniaturized channels are more frequently applied in devices such as heat exchangers, chemical reactors, and genome sequencers. Better understanding of two-phase flow phenomenon in minichannels induced the experimental and numerical investigations. Computational Fluid Dynamics gives the alternative approach which allows us to simulate two-phase flow in mini scale. Modelling of two-phase flow requires describing the interface behaviour. Over the years, different numerical techniques that dealt with fluid-interface motion were proposed. The two main groups of approaches for describing interface behaviour can be distinguished. Lagrangian technique uses markers to track the interface. In Eulerian techniques the interface is captured by a certain isocontour of a globally defined function. Eulerian techniques have the advantage over the Lagrangian techniques due to better adaptation to irregular contours, sharp corners, merging and breaking contours. Also, mixed Lagrangian-Eulerian methods were developed [1]. The most popular Eulerian techniques are: the level-set [2–4], volume-of-fluid (VOF) [5] and phase field methods [6, 7].
The level-set method was successfully applied to simulate gas flow in the liquid [8] – in these cases the contact of two phases has been considered. The main problem which occurs when we try to use the level-set method to simulate gas flow in the minichannel is the numerical stability of level set method in case of three phase contact. The main aim of the study is to simulate the two-phase flow inside the

Journal of Physics: Conference Series **530** (2014) 012049 doi:10.1088/1742-6596/530/1/012049

minichannel using the level-set method. The CFD Module of COMSOL Multiphysics® has been used, the different flow conditions were analysed. The two-phase flow (water - air) was simulated inside the 2D minichannel with a length of 200 mm and height of 3 mm. The air was supplied to the minichannel by the small nozzle with height of 0.7 mm. The model parameters considered the three phases contact.

2. Level-set method

The level-set method, introduced by Osher and Sethian [3], allows to simulate flow of two immiscible fluids separated by moving interface. The level set function is typically smooth continuous function, denoted as ϕ. The interface is represented by 0.5 contour of the function ϕ. In the transition layer near the interface the function ϕ changes smoothly form 0 to 1. Regions where $\phi < 0.5$ are filled with water and region where $\phi > 0,5$ are filled with air.

$$\varphi \begin{cases} > 0,5 & air \\ 0,5 & \text{int}\, erface \\ < 0,5 & water \end{cases} \tag{1}$$

The level set function is being advected with a flow field. The following equation describes the convection of the level set function [3]:

$$\frac{\partial \phi}{\partial \mathrm{t}} + \upsilon \cdot \nabla \phi = 0 \tag{2}$$

Main difficulty in numerical simulation of the moving boundaries is discontinuity appearing when fluid parameters changes at the interface. The problem was solved by defining fixed interface thickness where parameters could smoothly change. This procedure may cause mass conservation and constant interface thickness problems [9]. In CFD Module of COMSOL Multiphysics® the following level-set equation is solved [10]:

$$\frac{\partial \phi}{\partial \tau} + \upsilon \cdot \nabla \phi = \gamma \nabla \cdot \left(\varepsilon \nabla \phi - \phi(1-\phi)\frac{\nabla \phi}{|\nabla \phi|} \right) \tag{3}$$

This equations (3) have stabilization and reinitialization terms. The left side of the equation defines the motion of the interface while the right side is responsible for numerical stabilization and reinitialization. Parameter ε controls the interface thickness and parameter γ defines the intensity of reinitialization. In the model the density ρ and the dynamic viscosity μ are defined as follow:

$$\rho = \rho_1 + (\rho_2 - \rho_1)\phi \tag{4a}$$

$$\mu = \mu_1 + (\mu_2 - \mu_1)\phi \tag{4b}$$

where ρ_1 liquid density, ρ_2 air density, μ_1 and μ_2 are the dynamic viscosity of liquid and air respectively.

In the paper it have been solved the Naviera-Stokes equations:

$$\nabla \cdot \vec{u} = 0 \tag{5a}$$

$$\rho \frac{\partial \vec{u}}{\partial t} + \rho(\vec{u} \cdot \nabla)\vec{u} = \nabla \cdot \left[-pI + \mu\left(\nabla \vec{u} + \nabla \vec{u}^{\mathrm{T}}\right)\right] + \vec{F} + \rho \vec{g} + \sigma \kappa \delta \vec{n} \tag{5b}$$

where u is velocity, g is gravity, t is time, I identity matrix, p is pressure. Expression $\sigma\kappa\delta n$ donates surface tension forces at the interface where σ is surface tension coefficient, κ is curvature, $\vec{n}$ is the unit normal to the interface and δ Dirac delta function concentrated to the interface. Curvature κ, interface normal $\vec{n}$ and Dirac delta δ are determined by [10]:

$$\vec{n} = \frac{\nabla\phi}{|\nabla\phi|}, \; \kappa = \nabla(\frac{\nabla\phi}{|\nabla\phi|}), \; \delta = 6|\nabla\phi||\phi(1+\phi)| \tag{6}$$

3. Numerical implementation

The Navier-Stokes equations coupled with level set function are solved in COMSOL. The two-phase flow was simulated inside the minichannel with a length of 200 mm and height of 3 mm. The 2D domain under consideration has been shown in figure 1.

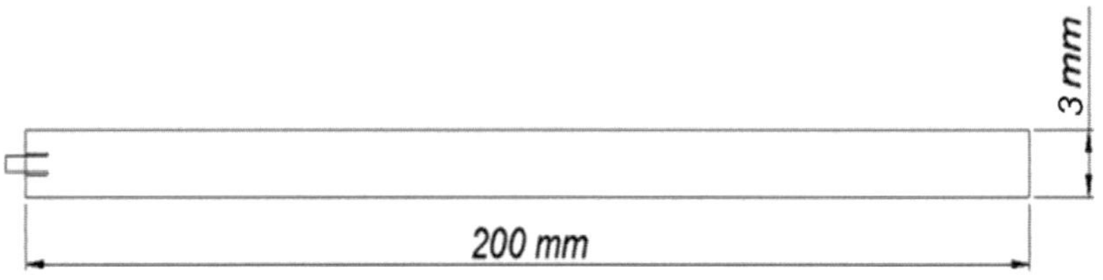

Figure 1. Computational domain

On left side of the channel the small nozzle supplying air was placed. The height of the nozzle was 0.7 mm. In figure 2 the schema of boundary conditions in the minichannel inlet has been shown.

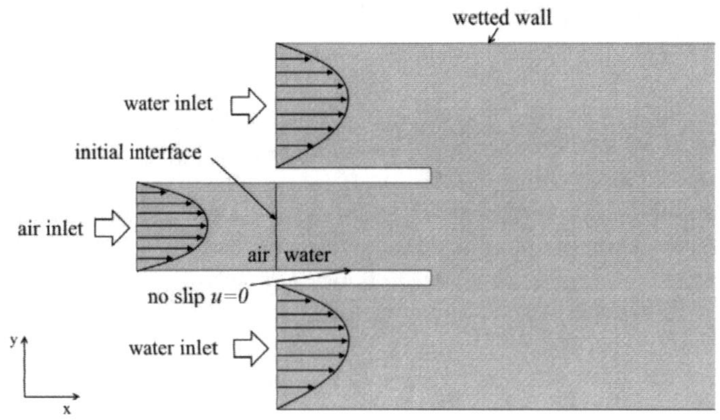

Figure 2. The schema of boundary conditions in the minichannel inlet.

Parabolic inlet velocity profile is given by the following equation:

$$u = u_{max} s(1-s) \tag{7}$$

where u_{max} is maximum value of the velocity profile and s is normalized length of boundary.

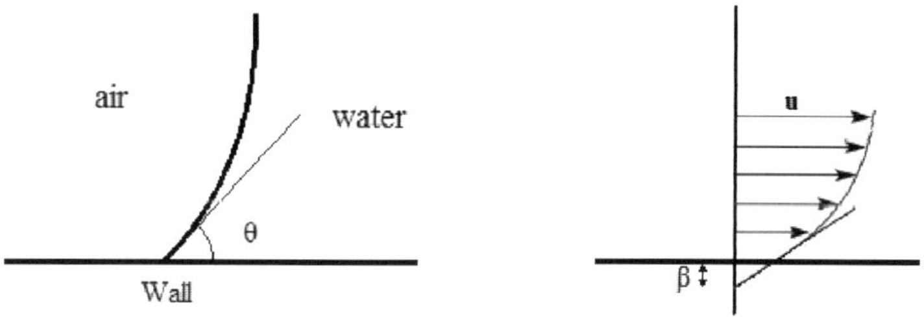

Figure 3: The definition of the contact angle θ at interface/wall contact points (left) and the slip length β (right).

The boundary conditions on the wall of the channel assume the contact angle θ (figure 3 a). In the simulation the value of contact angle was 20° (water – glass at 20°C), the slip length (β) was equal to average mesh element size. In figure 3 the definition of contact angle and slip length has been shown. The mesh near the minichannel inlet is shown in figure.4.

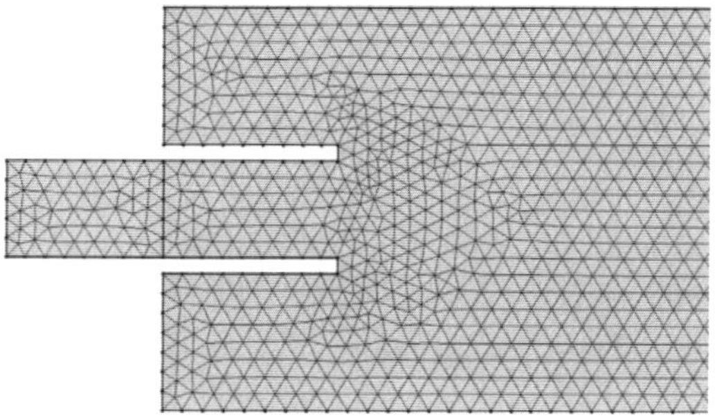

Figure 4: Mesh near the minichannel inlet.

In the simulation the parameter γ defining the intensity of reinitialization was 1 m/s and the parameter ε controlling interface thickness was equal to half of mesh size. The simulation was carried out for three different inlet conditions shown in table 1.

Table 1. Inlet velocity conditions.

No.	u_{max} water (ms^{-1})	u_{max} air (ms^{-1})
1	0.2	0.5
2	3	1
3	1	3

4. Simulation results

In the paper it has been analysed the process of formation of different two-phase flow patterns in the minichannel and the properties of two-phase flow patterns. During the simulation it has been obtained the three flow patterns: the bubbly flow and two kinds of slug flow with short and long slugs. The formation of bubbles and slugs in the inlet of the minichannel has been shown in figure 5. The flow patterns obtained at different inlet conditions have been shown in figure 6.

Journal of Physics: Conference Series **530** (2014) 012049 doi:10.1088/1742-6596/530/1/012049

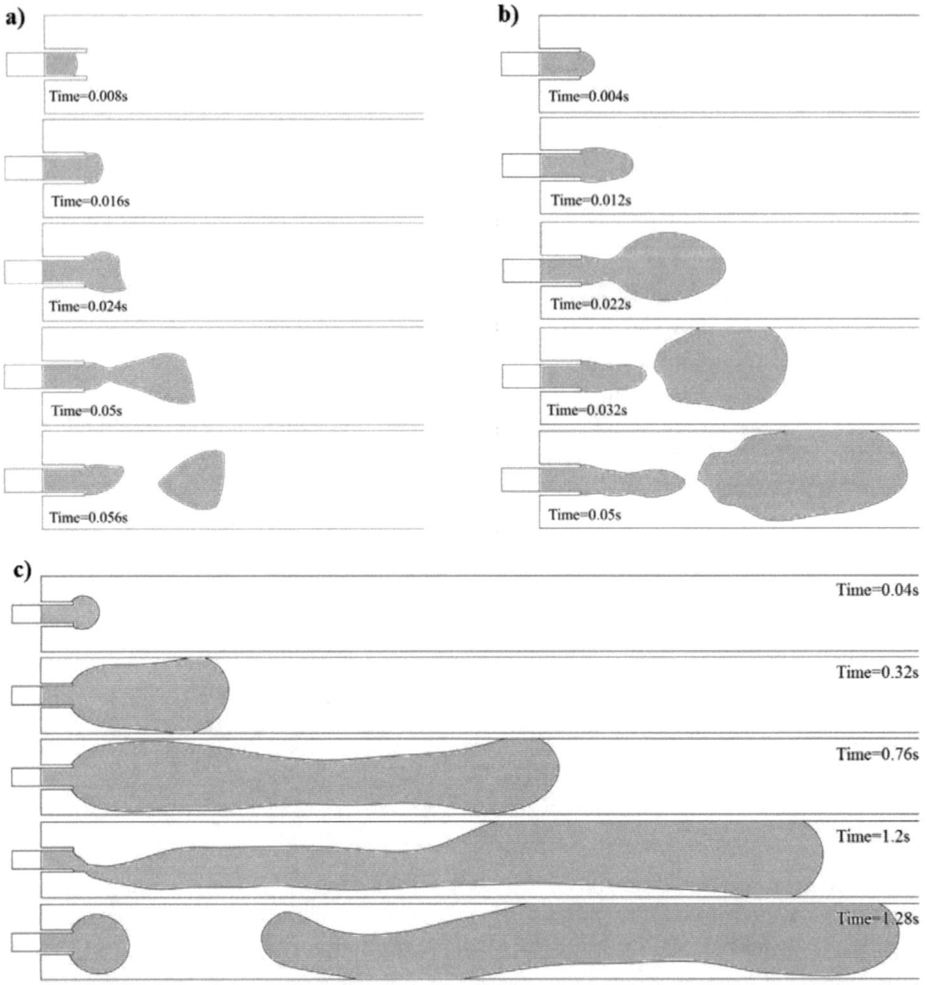

Figure 5: Bubble formation inside the minichannel inlet for different inlet conditions. a) water $u_{max} = 3\ ms^{-1}$, air $u_{max} = 1\ ms^{-1}$, b) water $u_{max} = 1\ ms^{-1}$, air $u_{max} = 3\ ms^{-1}$ c) water $u_{max} = 0.2\ ms^{-1}$, air $u_{max} = 0.5\ ms^{-1}$

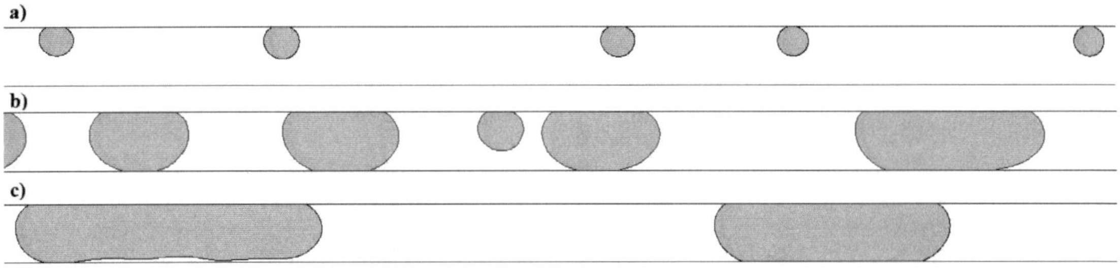

Figure 6: Two-phase flow for different inlet conditions at distance of 100 mm from the minichannel inlet: a) water $u_{max} = 3\ ms^{-1}$, air $u_{max} = 1\ ms^{-1}$, b) water $u_{max} = 1\ ms^{-1}$, air $u_{max} = 3\ ms^{-1}$ c) water $u_{max} = 0.2\ ms^{-1}$, air $u_{max} = 0.5\ ms^{-1}$

For the high water inlet velocity (figure 5a) and the relatively low air inlet velocity in the minichannel the small bubbles are formed. The shapes of the bubbles show the existence of unsteady water flow condition in the minichannel inlet. At distance of 100 mm from the minichannel inlet the bubbles have a spherical shape and they move close to the upper wall of the minichannel (because of buoyancy force), figure 6a. In the case of lower water inlet velocity (figure 5b) and higher air inlet velocity the size of the bubbles increases and small slugs are formed in the minichannel inlet. At the distance of 100 mm from the minichannel inlet (figure 6b) the slug flow with short slugs and small bubbles has been observed. The shapes of the bubbles in the inlet of the minichannel indicate the existence of unsteady water flow conditions which causes that the size of subsequently formed bubbles changes during the simulations. The small water inlet velocity (figure 5c) causes the formation of the long slugs in the minichannel inlet. Such flow pattern is observed at the distance of 100 mm from the minichannel inlet (figure 6c). Unsteady character of liquid flow at the minichannel inlet causes that the lengths of subsequent slugs vary.

The bubbles and slugs join during their flow, (figure 6b). In figure 7 it has been shown the example of bubble and slug coalescence. In such a way the longer slugs are created.

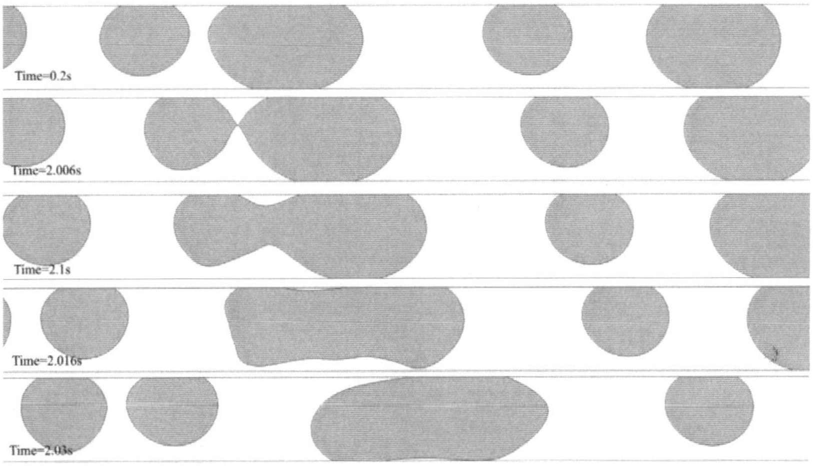

Figure 7: Bubble merging. (water $u_{max} = 1\ ms^{-1}$, air $u_{max} = 3\ ms^{-1}$)

For evaluation of the dynamics of formation of two-phase flow patterns in the minichannel it has been used the 'cut line' method. The 'cut line' was located at the distance of 100 mm from the minichannel inlet. At the line the changes in time of void fraction (n) have been measured. In figure 8 it has been shown the recorded time series. The amplitude of functions $n(t)$ indicates the size of bubbles (figure 8a and b). The distance between maximum values of $n(t)$ corresponds to the distance between bubbles. When the amplitude of $n(t)$ is equal to one, then the slug flows through the minichannel at the distance of 100 mm from the inlet. The time period when the values of $n(t)$ are equal to 1 correspond to the time of passing the slug through the 'cut line'.

The chaotic character of changes of the size of bubbles (figure 8a) indicates the existence of unsteady flow at the inlet of the minichannel. Such unsteady flow modifies hydrodynamic forces which are responsible for bubble departures from the nozzle. The different distance between the bubbles are visible in figure 8a. This unsteady flow is also responsible for the chaotic character of function $n(t)$ in case shown in figure 8b, where the short slugs and bubbles are visible. In case shown in figure 8c there are only the slugs, and the flow pattern seems to be more stable in comparison to the other cases.

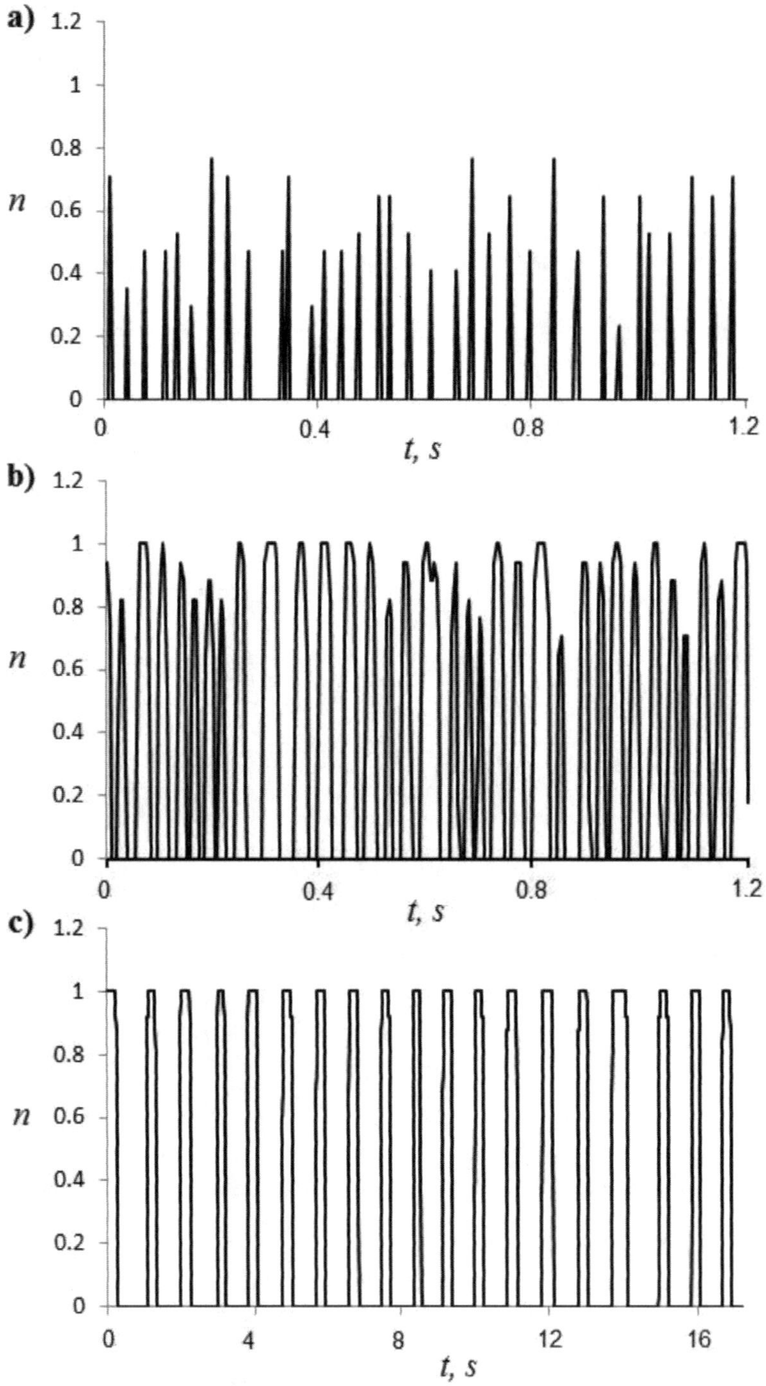

Figure 8: The dynamics of two-phase flow in minichannel. n is e number of mesh elements filed by air. a) water u_{max} = 3 ms^{-1}, air u_{max} = 1 ms^{-1}, b) water u_{max} = 1 ms^{-1}, air u_{max} = 3 ms^{-1} c) water u_{max} = 0.2 ms^{-1}, air u_{max} = 0.5 ms^{-1}

5. Conclusions

The main aim of the study was the simulation of two-phase flow inside the minichannel using the level-set method. The two-phase flow (water - air) was simulated inside the 2D minichannel with a length of 200 mm and height of 3 mm. The CFD Module of COMSOL Multiphysics® has been used. The air was supplied to the minichannel by the small nozzle. The simulation parameters considered the three phases contact.
The process of formation of different two-phase flow patterns in the minichannel has been investigated. During the simulation it has been analysed three flow patterns: the bubbly flow and two kinds of slug flow with short and long slugs. It has been shown that unsteady flow at the inlet of the minichannel is responsible for the chaotic character of changes of the slug and bubble sizes. Such unsteady flow modifies the distance between the bubbles and slugs. It has been shown that for the low water inlet velocity the two-phase flow pattern becomes more stable.

6. References

[1] Soulaimani A and Saad Y 1998 An arbitrary Lagrangian-Eulerian finite element method for solving three-dimensional free surface flows, *Comput. Method. Appl. M.* **162** (1–4) pp 79-106
[2] Olsson E and Kreiss G 2005 A conservative level set method for two phase flow. *J. Comput. Phys.* **210** 1 pp 225-246
[3] Osher S and Sethian J A 1988 Fronts propagating with curvature-dependent speed: algorithm based on Hamilton–Jacobi formations *J. Comput. Phys.* **79** pp 12
[4] Osher S and Fedkiw R 2002 Level Set Methods and Dynamic Implicit Surfaces, *Appl. Math. Sci.* **153** (Springer-Verlag)
[5] Rider W J and Kothe D B 1998 Reconstructing volume tracking *J. Comput. Phys.* **141** pp 112
[6] Chiu P and Lin Y 2011 A conservative phase field method for solving incompressible two-phase flows *J. Comput. Phys.* **230** (1) pp 185-204
[7] Sun Y and Beckermann C 2007 Sharp interface tracking using the phase-field equation *J. Comput. Phys* **220** pp. 626-653
[8] Mukherjee A, Kandlikar S G and Edel Z J 2011 Numerical study of bubble growth and wall heat transfer during flow boiling in a microchannel, *Int. J. Heat Mass Trans.* **54** (15–16) pp 3702-3718
[9] Ferziger J 2003 Interfacial transfer in Tryggvason's method, *Int. J. Numer. Meth. Fl.* **41** (5) pp 551–560
[10] COMSOL Multiphysics® User's Guide.

Acknowledgement
The project was funded by the Bialystok University of Technology - the number of decision: MB/WM/10/2013

Modelling of turbulence-interface interactions in stratified two-phase flows

T Wacławczyk[1,3], M Wacławczyk[2] and S V Kraheberger[2]

[1] Institute of Numerical Methods in Mechanical Engineering, Department of Mechanical Engineering, TU Darmstadt, Dolivostr. 15, 64293 Darmstadt, Germany

[2] Chair of Fluid Dynamics, Department of Mechanical Engineering, TU Darmstadt, Otto-Berndt-Str. 2, 64287 Darmstadt, Germany

E-mail: waclawczyk@fnb.tu-darmstadt.de

Abstract. In this work we describe interactions between turbulence and water-air surface in the ensemble-averaged picture where, instead of a sharp interface between the phases we deal with a surface layer where the probability of the surface position is nonzero. Changes of the turbulent kinetic energy and the characteristic size of the eddy influences the width of the surface layer. We present a numerical solution of the equation for the intermittency function which describes the probability of finding the water phase at the given point and time.

1. Introduction

In the present work we investigate stratified turbulent air-water flows where the phases are separated by a deformable, but non-broken interface. We consider the statistical approach where the ensemble averaging of physical quantities, including these connected with the fluctuating surface, is performed. After the averaging we do not deal with a sharp boundary between the two phases but receive a layer where the probability of the surface position is non-zero. We consider the intermittency function denoting the probability of finding the water phase at a given point of the flow and at the given time. The function α is defined as the ensemble average of the Heaviside function $\alpha(\mathbf{x},t) = \langle\chi(\mathbf{x},t)\rangle$ where

$$\chi(\mathbf{x},t) = \begin{cases} 1 & \text{if the water phase is present at } \mathbf{x} \text{ and } t \\ 0 & \text{otherwise} \end{cases} \tag{1}$$

The region were $0 < \alpha(\mathbf{x},t) < 1$ is called the intermittency region or the surface layer. The width of the surface layer depends on the disrupting action of turbulent eddies on one hand and stabilizing role of gravity and the surface tension on the other hand, cf. Ref. [1]. In our approach these two competing mechanisms are modelled, respectively, by the diffusion and contraction terms in the evolution equation for the function α, cf. [2,3]. Hence, we are not only interested in finding the mean surface position as it is done in the so-called macroscopic approaches, most common up to date, but we stay at the mesoscopic level of description by modelling the evolution of the probability function for the surface position α. In this work we present results of simulations of a surface layer changing in time due to the changing kinetic energy and the characteristic eddy size and compare them with results of apriori tests of two eddies reaching and deforming the surface. The kinetic energy and the eddy size form the apriori

[3] To whom correspondence should be addressed

simulations are used to determine coefficients in the equation for and calculate its profile within the intermittency region.

2. Model for the turbulence-interface interactions

If the ensemble averaging is applied to the advection equation of the Heaviside function χ we obtain

$$\frac{\partial \alpha}{\partial t} + \langle \mathbf{u} \rangle \cdot \nabla \alpha + \langle \mathbf{u}' \cdot \nabla \chi \rangle = 0 \tag{2}$$

where $\langle \mathbf{u} \rangle$ is the ensemble-averaged one-fluid velocity $\mathbf{u} = \mathbf{u}_w \chi + \mathbf{u}_a(1-\chi)$, $\mathbf{u}_w$ and $\mathbf{u}_a$ are velocities in, respectively, water and air phase, $\mathbf{u}'$ is the velocity fluctuation. As it was proposed in Ref. [3] and [4], the unclosed term $\langle \mathbf{u}' \cdot \nabla \chi \rangle$ can be modelled by two counter-acting terms: diffusion and contraction

$$\frac{\partial \alpha}{\partial t} + \langle \mathbf{u} \rangle \cdot \nabla \alpha = \nabla \cdot (D_t \nabla \alpha) + \nabla \cdot [C_t \alpha(1-\alpha)\hat{\mathbf{n}}] \tag{3}$$

where coefficients d_t and C_t depend on turbulence statistics and $\hat{\mathbf{n}} = -\nabla\alpha/|\nabla\alpha|$ is a vector normal to the isosurface of α. In the 1D stationary case and for constant coefficients D_t and C_t Eq. (2) has an analytical solution

$$\alpha = \left[1 + \exp\left(\frac{C_t}{D_t}(y - \hat{y}_s)\right)\right]^{-1}. \tag{4}$$

where $\hat{y}_s$ denotes a mean position of the interface. In such a case the difference between top and bottom limits of the intermittency layer defined as the region where the probability of finding the surface equals 99.6% reads

$$t - b = \frac{24}{\sqrt{2\pi}} \frac{D_t}{C_t}. \tag{5}$$

In the case of the Gaussian error function such limits are obtained for $t - b = \hat{y}_s \pm 3\sigma$ where σ is the standard deviation of the corresponding Gaussian distribution. We proposed in Ref. [3] to model the coefficient D_t as the product of the velocity scale q and the characteristic length scale of the eddy L $D_t = c_D q L$. The coefficient $C_t = c_C q$ and the ratio c_D/c_C can be estimated from the formula (4) such that the proper width of the intermittency region could be obtained.

3. Test case vortex-interface interactions

In order to perform apriori tests we used the second order finite-volume code FASTEST with the interface capturing volume of fluid method based on the compressive high resolution scheme methodology, see Ref. [4,5]. The aim of the test case presented in figure 1 is to reconstruct the temporal evolution of a two-dimensional gravitational wave generated by the velocity field of two counter rotating vortices initially localized beneath interface. To satisfy flow conditions in a regime of scarified flow the size of the computational domain is chosen to be $20 \times 20m$. The air-water interface ($\rho_a = 1.205kg/m^3$, $\mu_a = 1.8 \cdot 10^{-5} kg/(ms)$, $\rho_w = 998kg/m^3$, $\mu_w = 8.8 \cdot 10^{-4} kg/(ms)$, $g = -9.81m/s^2$) is initially flat and located at $y = 15m$. The initial intensities equal $\Omega = 0.511/s$ for the first and $\Omega = -0.511/s$ for the second vortex, (x_0, y_0) is the position of the center of the vortex, $x_0 = 9.5m$ for the first and $x_0 = 10.5m$ for the second vortex, $y_0 = 12.5m$. The length scale of the eddy is estimated as $L = r_0 = 1m$. These two parameters locate the present simulation in the so-called scarified regime, cf. Ref. [1], i.e. such flow where the surface is deformed, but not broken by the turbulent eddies. Further details of the simulations are given in Ref. [3].

The computational domain is discretized with about $6 \cdot 10^6$ CVs. The CFL condition is kept below 0.5 during whole simulation what requires the times step size $\Delta t = 10^{-4}s$. Such mesh resolution assures up to 32 control points in the intermittency region where averaging is performed and the interface-turbulence interaction model is verified. The vorticity magnitude of the eddies approaching the surface and the corresponding deformation of the surface are presented in figures 1 and 2.

Journal of Physics: Conference Series **530** (2014) 012050 doi:10.1088/1742-6596/530/1/012050

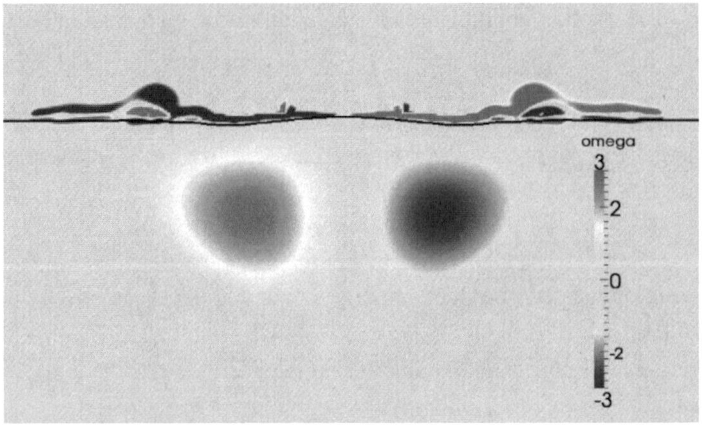

Figure 1. Vorticity magnitude of the eddies approaching the surface at $t = 20s$.

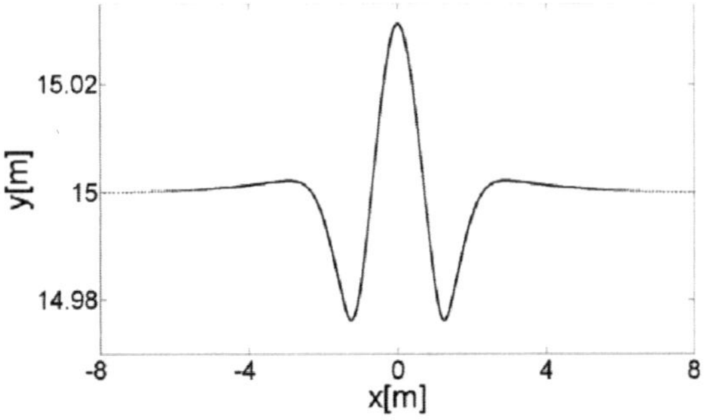

Figure 2. Deformation of the surface due to the interaction with the eddies at $t = 20s$.

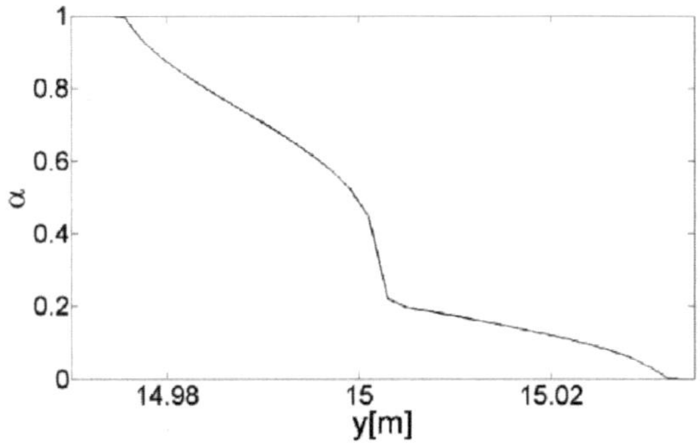

Figure 3. Corresponding profile of α at $t = 20s$.

The statistics like the intermittency function or the kinetic energy were calculated by averaging in x direction. As the eddies approach the interface they deform it and in the averaged picture the profile of α function is smoothed creating a surface region between the top and the bottom position of the surface, cf. figure 3. In this stage of the flow the width of the intermittency region increases, hence in the following we call it the diffusion stage. As the vortex structure reaches the surface it splits into two eddies travelling in opposite directions. This is connected with the decrease of the intermittency region width caused on one hand by the decrease of the characteristic length scale of the eddy structure and, on the other hand by the stabilizing role of gravity. We will further refer to this phase of the flow calling it contraction stage.

4. Numerical solution of the α-equation

Modeling of the evolution of the intermittency region in space and time introduce additional difficulties to the numerical solution of the model Eq. (2). Beside typical problems with the advection of the function $\alpha = \langle \chi \rangle$ known from the interface capturing VOF methods (e.g. restrictions in time step size due to the CFL condition or using flux limiter to avoid Gibbs phenomenon) the new length and time scales are introduced to the solution. The new length scale representing the characteristic thickness of the intermittency region is given by Eq. (4). The new time scale $\tau \sim D_t/C_t^2$ is the characteristic time of the intermittency region evolution. In our numerical procedure both new flow field features must be accounted for. The first one by providing the appropriate mesh resolution in the vicinity of the interface, the second one by introduction of the suitable time integration procedure that can handle variation of the intermittency region thickness in time. The solution to Eq. (2) is implemented into the second-order accurate finite volume flow solver FASTEST and is related to the previous implementation of the interface capturing VOF method see Ref. [4] and [5]. The interface capturing VOF method in FASTEST is based on the compressive high resolution scheme M-CICSAM that keeps the interface sharp due to the downwind differencing scheme present in the discretization of the convective term. The numerical method used in the present work to solve Eq. (2) is based on a different idea. It is built around the numerical techniques that are employed in the diffuse interface methods for example presented in Ref. [6]. In such an approach the thickness of the interface between two fluids, cf. Eq. (4) is proportional to

$$\varepsilon = D_t/C_t \geq \Delta x/2 \tag{6}$$

where Δx is the mesh size. We notice that Eq. (5) provides the condition that allows to determine the number of grid points in the vicinity of the interface, when model coefficients D_t, C_t are given. To keep the interface thickness constant during advection, the additional equation is introduced to reinitialize function α. We notice that reinitialization of the mass conserving level-set function α in Ref. [6] has only numerical meaning i.e. it should keep thickness of the diffuse interface constant and equal to ε during its advection. In the present work we will introduce a physical interpretation to the reinitialization procedure.

Let us split the solution of the Eq. (2) into two steps. In the first step we carry out the advection of the α function

$$\frac{\partial \alpha}{\partial t} + \nabla(\langle \mathbf{u} \rangle \alpha) = 0 \tag{7}$$

in the averaged velocity field $\langle \mathbf{u} \rangle$ obtained from the RANS solver. In this step we use implicit time and space discretization, where the second order TTL scheme is used for discretization in time and the second order MUSCL flux limiter is used within a deferred correction to keep α bounded function. In the second step, the evolution of the intermittency region must be reconstructed. Here we introduce (reinitialization) equation

$$\frac{\partial \alpha}{\partial \tau} = \nabla \cdot (D_t \nabla \alpha) + \nabla \cdot [C_t \alpha (1-\alpha)] \approx \nabla \cdot [\varepsilon (V_D - V_C) \nabla \alpha] \tag{8}$$

where the inner artificial time $\tau \sim D_t/C_t^2$ is associated with the characteristic time scale of the intermittency region evolution, V_D is the diffusion and V_C is compression velocity of the intermittency region. Additionally, we make an assumption that coefficients D_t, C_t remain constant on a single time step Δt as they are determined from the turbulence model integrated in time along with the Navier-Stokes equations. Eq. (7) is solved in time τ using explicit low-storage, third order accurate TVD Runge-Kutta scheme. The spatial discretization is carried out in framework of the finite volume method and therefore the discretized Eq. (7) in the control volume P becomes

$$\frac{\partial \alpha}{\partial \tau}\Big|_P = \sum_{f=1}^{N_{faces}} \left[D_t \frac{\mathrm{d}\alpha}{\mathrm{d}x} + C_t \alpha (1-\alpha) \hat{\mathbf{n}} \right]_f S_f \tag{9}$$

where central differencing interpolation is used to obtain values at the faces f of the control volume. If the numerical procedure presented above is used only to keep the interface compact $D_t = \Delta x/2V_D$ and $C_t = V_C$, see Eq. (7). Moreover, it is assumed that $V_D = V_C = const$ where constant value is selected to be $\|u^{max}\|$ to increase the rate of convergence in inner time τ. Since in our model in general $V_D \neq V_C$ in what follows we will discuss different regimes of the solution of Eq. (7) dependent on the sign of the velocity $V = V_D - V_C$. Let us first define second order norm used to measure distance between solutions in time τ

$$L_1 = \frac{1}{N_{cells}} \sum_{l=1}^{N_{cells}} |\alpha_{l,i-1} - \alpha_{l,i}| \tag{10}$$

where subscript i denotes inner time iterations within time τ. Three situations must be considered:

(i) $V_D = V_C = const$. In this case thickness of the intermittency region remains constant since the RHS in Eq. (7) is zero. We choose number of the inner time steps in the reinitialization procedure $N_\tau = 4$ with inner time step $\Delta\tau = D_t^{max}/(2C_t^{min})^2$ where D_t^{max} is maximal and C_t^{min} is minimal value of the coefficients in the whole computational domain. We checked the 4 inner time iterations are sufficient to achieve constant interface thickness during its advection by Eq. (6) see also Ref. [6].

(ii) $V_D > V_C$. In this case the thickness of the intermittency region $d = t - b$ grows. Eq. (7) is a diffusion equation and the time step size can be determined from $\Delta\tau = D_t^{max}/(2C_t^{min})^2$. Eq. (8) is solved until the prescribed tolerance is achieved i.e. $L_1 \leq 10^{-3}\Delta\tau$ see Eq. (9).

(iii) $V_D < V_C$. Here the compressive term dominates diffusion and therefore the thickness of the intermittency region $d = t - b$ decreases. If $d < \varepsilon = \Delta x/2$ numerical instabilities in the numerical solution of Eq. (7) can occur as is discussed in Ref. [6]. Let us assume that thickness of the interface on the n-th time step Δt and i-th inner time step $\Delta\tau$ is $d_i^n = 24/\sqrt{2\pi D_t^n/C_t^n}$. If before next inner iteration of Eq. (8) we find that $d_{i+1}^n = d_i^n + V\Delta\tau < \varepsilon$ than we set $V_D = V_C = const$ and $N_\tau = 4$. If it is not the case Eq. (8) is solved until the prescribed tolerance is achieved i.e. $L_1 \leq 10^{-3}\Delta\tau$ where $\Delta\tau = D_t/(4C_t^2)$.

To assess the performance of the reinitialization algorithm with variable ε that should mimic the modification of the intermittency region thickness due to the presence of the turbulent structures we set thickness of the interface as a function of spatial variable. Let us introduce a test case where we set the distribution of the thickness to be a continuous function

$$\varepsilon(x) = \frac{\Delta x}{2}\sin(3\pi x) + 2. \tag{11}$$

The initial interface profile is set to $\varepsilon_{ini} = \Delta x$ in the whole computational domain. Results are presented in figure 4. It can be noticed that the distribution of the function α is well reconstructed. We found out that about two hundred inner time iterations are required in order to obtain convergence of the results up to tolerance $L_1 \leq 10^{-5}\Delta\tau$.

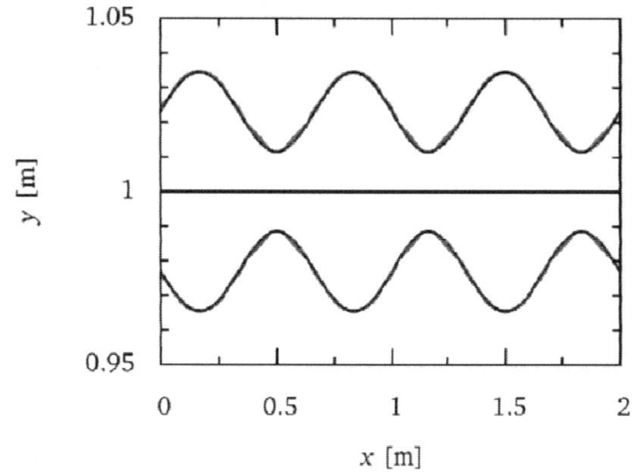

Figure 4. The comparison between solution of Eq. (7) $\alpha(0.05, 0.5, 0.95)$ contours on grid 256×256, red line denotes numerical solution, black line is the analytical solution.

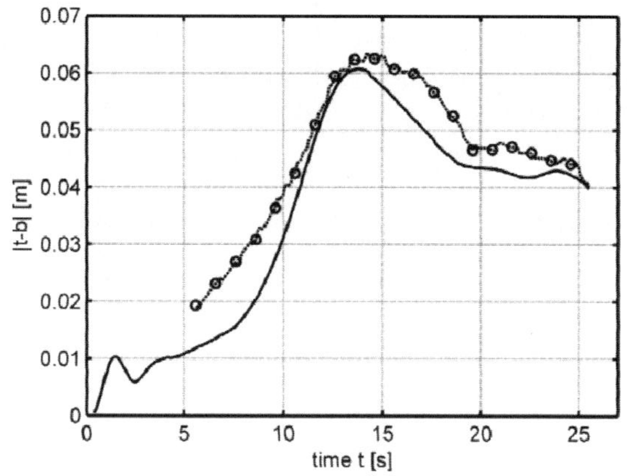

Figure 5. Width of the intermittency layer obtained from simulations with $D_t = q(t)L(t)$ and $C_t = const$ (line with symbols) in comparison to apriori results.

5. Validation tests

In this section a numerical solution of Eq. (2) using the method described in section 4 is presented. The diffusion and contraction coefficients were provided every $0.1s$ from the apriori simulations described in section 3 and Ref. [3]. The model transport equation (2) was solved in artificial time τ until a convergent solution was obtained. To get reasonable results for the intermittency layer, the condition $D_t/C_t \geq \Delta x/2$ should be satisfied. On the considered finest grid of 64 cells this condition was first fulfilled at $t_0 = 5.6s$ and we took the profile of α from the apriori test at $t_0 = 5.6s$ as the initial condition. Top and bottom heights of the intermittency layer were calculated at $\alpha_{top} = 0.998$, $\alpha_{bottom} = 0.002$, which corresponds to the $99,6\%$ probability of finding the surface within the intermittency zone. As stated above, values for D_t and C_t are prescribed by turbulent statistics obtained in the apriori test. Results for three different possibilities in choosing these statistics are presented in the following. In general, we use a turbulent intensity $q = \sqrt{2k/3}$, where k denotes the turbulent kinetic energy, and a turbulent length scale L to define D_t. The turbulent kinetic energy is calculated in two ways. In the first approach we use the one-fluid-velocity fluctuations, leading to

$$k = \frac{1}{2}\langle u'_{wi}u'_{wi}\chi + u'_{ai}u'_{ai}(1-\chi)\rangle. \tag{12}$$

In the second approach we calculate k from the phase-weighted velocity fluctuations of the water phase, which yields $\tilde{q} = \sqrt{2\tilde{k}/3}$, where

$$\tilde{k} = \frac{1}{2}\frac{\langle u'_{wi}u'_{wi}\chi\rangle}{\alpha}. \tag{13}$$

Weighted variables have also been proposed for modelling turbulent intermittent flows, cf. Ref. [7], which show mathematical similarities with the present model for the evolution of α. The turbulent length scale L is considered to be the eddy size, i.e. $L(t < 12.5s) = 2m$ as long as both eddies are rising to the surface and $L(t > 19.9s) = 1m$ once the vortex structure has reached the surface and split into two eddies travelling parallel to the surface. Between $12.5s < L < 19.9s$, $L(t)$ is defined by a linear interpolation. We first averaged the kinetic energy from Eq. (11) in y direction over the intermittency layer and calculated the corresponding intensity $q(t)$ and the coefficient $D_t = c_D q(t)L(t)$. The model constant $c_D = 4.125 \cdot 10^{-5}$ was evaluated from the apriori test, cf. Ref. [3]. The antidiffusion coefficient $C_t = 1.915 \cdot 10^{-3} m/s$ was kept constant thorough the simulations. Its value was determined from Eq. (4) considering the width of the intermittency layer at $t = 25.5s$, which, from the apriori tests, is known to be $t - b \approx 0.04m$. Simulations with these coefficients yield reasonable width of the intermittency layer, cf. figure 5. Figure 6 presents the development of top and bottom limits of the surface layer. It can be seen that for D_t being a function of time only, a solution symmetric with respect to the mean surface position is obtained. In fact, top and bottom limits are non-symmetric especially during the contraction stage. We attempted to improve the results by considering y-dependence, i.e., calculating the intensity of fluctuations from Eq. (11), keeping the same value of C_t. Unfortunately, the agreement between the results presented in figure 6 is not satisfactory.

For the third approach presented here, the diffusion coefficient is defined by $D_t = c_D\tilde{q}(y,t)L(t)$, calculating $\tilde{q}(y,t)$ from Eq. (12), while again C_t is not changed. In figure 7 the evolution of top and bottom limits in time is presented. It can be seen that the model accounts very well for the position of the intermittency regions upper limit, i.e. the limit of the air-phase side. At the diffusion stage, the width of the intermittency layer is overpredicted, while at contraction stage it is underpredicted. Profiles of α at several times are shown in figure 8. The results generally mimic the apriori α-profiles, however, the concave sections especially at $t = 20s, 25s$, $14.96 < y < 15m$, are not reproduced. Obtaining such reasonable results from defining D_t by the phase-weighted turbulent kinetic energy shows that the influence of the water phase on the evolution of the intermittency layer is stronger than the influence of the air phase. The results may be improved by choosing $C_t = C_t(\alpha)$ as stated in Ref. [3], which agrees with our results presented here.

6. Conclusions

In the present work evolution of the intermittency region where the probability of the surface position is not zero has been modelled by solving numerically α equation (2). Equation (2) contains two additional

Journal of Physics: Conference Series **530** (2014) 012050 doi:10.1088/1742-6596/530/1/012050

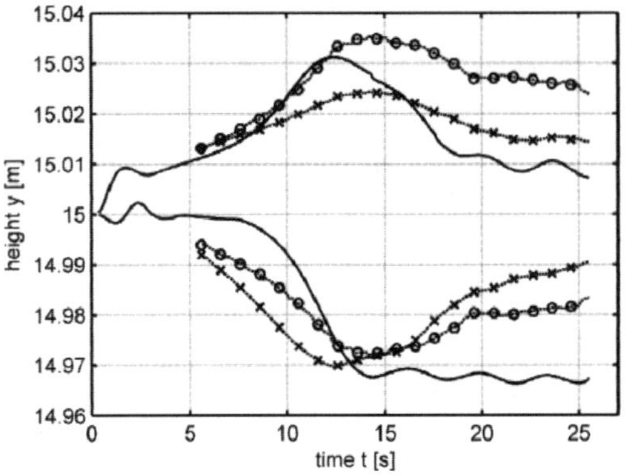

Figure 6. Top–bottom limits obtained with: $D_t = q(t)L(t)$ (dotted line and circles), $D_t = q(y,t)L(t)$ (dotted line and cross-symbols) compared with results of with apriori tests (solid line).

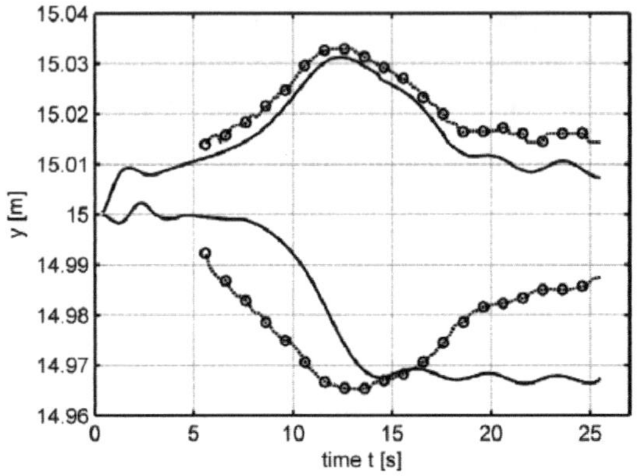

Figure 7. Top–bottom limits obtained with $D_t = \tilde{q}(y,t)L(t)$ and (line with symbols) in comparison to top and bottom heights calculated from the apriori results (solid line).

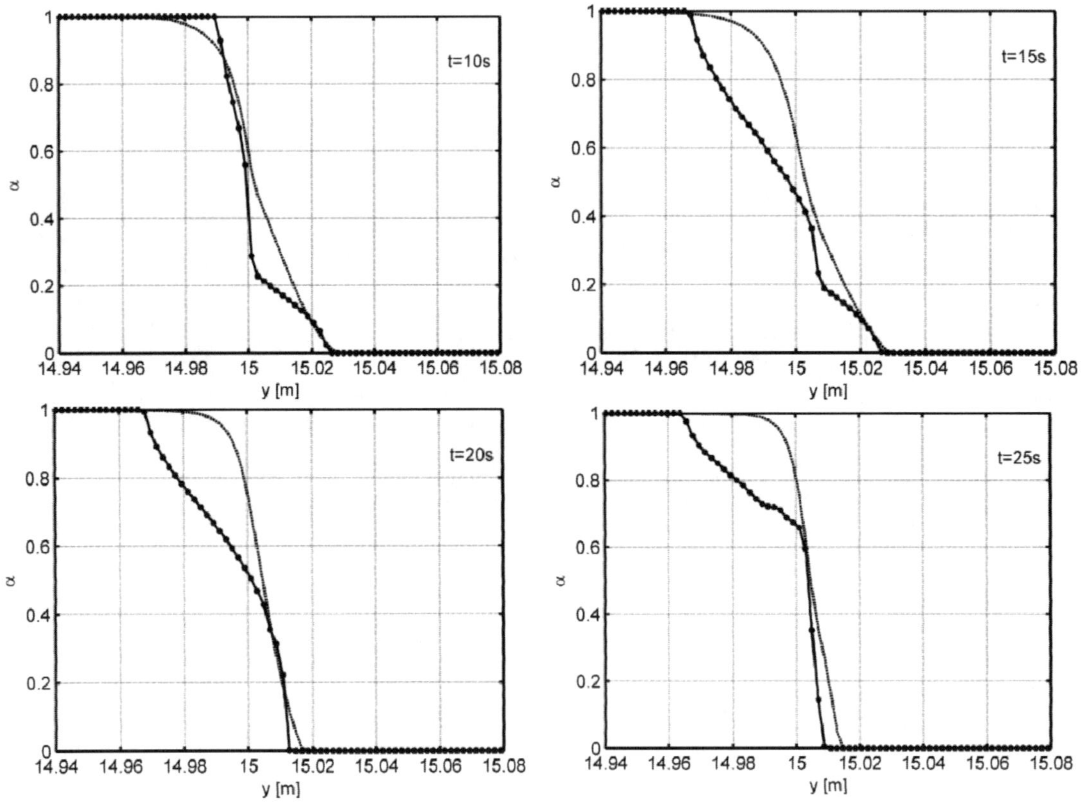

Figure 8. Profiles of α obtained from simulations with $D_t = c_D\tilde{q}(y,t)L(t)$ (dotted line) compared with results of apriori test (solid line with symbols).

terms in comparison to the standard formulas used to calculate the position of the interface in VOF methods, namely, the diffusion term due to the disturbing action of the turbulent eddies and contraction due to the stabilizing role of the gravity and surface tension. Numerical difficulties connected with the solution of the α equation were discussed and an appropriate numerical method to solve this equation has been proposed. Next, the numerical scheme has been used to model evolution of the surface layer changing its width due to the interaction with vortices. Satisfactory results of the surface layer width $t-b$ changing in time have been obtained. In order to improve results for the non-symmetric behaviour of the top and bottom limits of the surface layer phase-filtered kinetic energy has been used to determine the model coefficient D_t Results of numerical solution of Eq. (2) are presented for the first time here. An on-going further work is connected with the implementation of the model for turbulence-interface interactions into the Navier-Stokes solver in the two-way coupling approach and proposing a closure for the Reynolds-stress terms in the vicinity of the surface.

Acknowledgments
This work was supported by Deutscher Forschungsgemeinschaft (DFG) funding grant No. WA 3097/3-1, WA 3098/2-1, "Modeling of turbulence-interface interaction in two-fluid systems".

References

[1] Brocchini M and Peregrine D H 2001 *J. Fluid Mech.* **499** 225
[2] Wacławczyk M and Oberlack M 2011 *Int. J. Multiphase Flow* **37** 967
[3] Wacławczyk M and Wacławczyk T 2014 *Int. J. Heat and Fluid Flow* Submitted for publication.
[4] Wacławczyk T, Gemici Ö C and Schäfer M 2007 *Proc. 6th Int. Conf. Multiphase Flow (Leipzig)*
[5] Wacławczyk T and Koronowicz T 2008 *J. Theoretical Appl. Mech.* **46** 325
[6] Byggstøyl S and Kollmann W 1981 *Int. J. Heat Mass Transfer* **24** 1811

Journal of Physics: Conference Series **530** (2014) 012051 doi:10.1088/1742-6596/530/1/012051

An application of numerical simulation of multiphase flow for the redesign of a mixer agitator in Pb refining process

J Donizak[1,3] , P Jarosz[1], A Kraszewska[1] and P Sarre [2]
[1] AGH University of Science and Technology, Al. Mickiewicza 30, 30-059 Krakow, Poland
[2] RATech Ltd , ul Pory 78, Warszawa, Poland

E-mail: szczur@agh.edu.pl

Abstract. The paper presents numerical simulation of multiphase turbulent flow in a mixing crucible unit. Results of simulation were used for redesign of mixer agitator to achieve better performance of the Pb refining process. The simulation is based on Euler-Lagrange description of turbulent multiphase flow with the one way coupling, due to low concentration of solid state particles and significant differences in density of coexisting phases, base metal and particles. Dispersions of solid particles are traced using stochastic-deterministic approach. The developed construction of an agitator has been tested in the industrial Pb refining factory, giving very promising results in comparison with long term statistical data. Duration of unit operations of removal copper and tin was reduced of about 40% together with even better removal efficiency and less energy and reagents consumption.

1. Introduction

Pyro-metallurgical refining of production lead is a dominant process for separation of impurities from the metal to get pure commercial refined lead. Due to high solubility of several metals in liquid lead, raw lead contains numerous impurities. Depending of a technology used for production of raw lead the content of elemental Pb varies between 90% and 98%. The rest are impurities where primary are the following metals: cooper, tin, arsenic, antimony and others.

Refining is carried out in a batch manner. The refining plant consists of a series of batch reactors in which consecutive recovery of dissolved metals takes place by means of adding chemical reactants to produce a phase that is insoluble in base metal, and enriched with certain impurity materials. The single unit operation starts when feeding with „bullion" (molten raw lead) of the mixer ladle is finished. After feeding with the „bullion" the agitator is gradually accelerated to attain predefined angular velocity, at which the melt is intensively mixed in the pot. During mixing the refining compound is supplied on the bath surface in the vicinity of mixer shaft.

Depending on the agitator shape, the certain melt circulation pattern can be observed [1]. When the agitator is of the turbine type, pumping the melt downward the pot (classical Pb refining case), the funnel sucking the refining compounds is formed on the melt surface around the agitator shaft. The particles of refining compounds (solid state loose material) are sucked into the melt and transported around the mixing ladle volume. In a contact with base material and dissolved impurities the oxidation (or sulphurization) heterogenic reactions take place. It is commonly believed that the reactions are

[3] To whom any correspondence should be addressed.

controlled by the diffusion process at the surface of refining compound particles, therefore high intensity of mixing and high turbulence is promoting the chemical process. As a result of chemical reactions compounds insoluble in the parent metal are formed. Insoluble products flow out on base metal surface and can be removed automatically from the boiler. In the pyro metallurgical process of refining lead highly adverse phenomenon of a skull buildup on the side walls of a ladle occurs relatively often. The skull buildup is observed in these boilers, in which oxidation processes are carried out. Such a phenomenon is unfavourable due to the necessity of hammering or dissolving these buildup that sometimes significantly reduce the volume of the boiler. Periodic cleaning of the walls of the boiler is a long and tedious operation and at high frequencies significantly infers with the production cycle.

1.1. Motivation for research project

The main impulse for research on the process of refining production lead was to reduce the frequency of required cleaning operations of the boiler walls. As a result of the chemical analysis of the skull, which held a composition similar to that resulting from the refining of waste products, it was considered that the most effective limiting factor in their formation and growth will be intensification of the flow in the upper zone of the boiler. This is in harmony with the settled view that the most important phenomenon, which governs the mixing process in the refining ladle, is the molten base metal recirculation. The recirculation pattern depends on numerous parameters of the process but the most important ones are: the mixer agitator shape, its rotational speed and positioning in the pot [1]. The numerical simulation of the base metal flow seems to be an economical and powerful tool for the investigation of optimal shape of the mixer agitator. Thus, preliminary computations of fluid flow induced by the stirrer has shown that intensive bath circulation area is focused around the agitator and shaft, but in the upper zone, near the walls of the boiler, there are areas with low flow velocities. It was supposed that this is a result of very simple design of originally used mixer, consisted of two rectangular stirrer blades inclined at an angle of 35 degrees to the axis of rotation. The agitator redesign had to be done in order to increase the flow speed of the metal near the walls of the boiler by reducing the radial component of the velocity in favour of vertical component. The design process has two fundamental restrictions that need to be fulfilled: the drive energy consumption and the maximum rotational speed of agitators were to be limited to the level of the previously used, and provide suction of the refining salts from the surface into the bulk fluid in the vicinity of the shaft.

2. Mathematical model development

In this section we will formulate the mathematical model of the stirring process using fundamental equations of fluid motion. We will consider multiphase flow in the refining ladle, where the molten raw lead is a carrier phase and the refining compounds (loose solid material) is called dispersed phase. With the intense circulation of carrier phase the refining compound particles fed at the surface are entrapped and thinly scattered in the volume (bulk flow). This phenomenon contributes to an increase in the intensity of the heterogeneous chemical processes as a result of the reduction of barriers to mass transport. Optimum circulation of carrier phase, therefore, is a key issue in the process of refining. Numerical analysis of multiphase phase flow distribution can be taken in many ways [2,3,4,5]. However, one is most favourable from the point of view of clarity of the chemical process description and a minimum number of assumptions required for its application. This is called the mathematical formulation of the Euler–Lagrange multi-phase flow with scattered phases [6]. The flow of carrier phase is described by the classical turbulent flow equations formulated in the Euler's coordinate system (the observer), and dispersed phase flow equations are expressed in Lagrange frame of reference (moving along with elements of distributed phase in their traffic). The application of such a flow model is confined to the case of a small enough mass participation of dispersed phase in overall mass of fluid, to be able to neglect the dispersed phase volume in the area occupied by the carrier fluid. The movement of a single particle of dispersed phase (of mass m_p and velocity vector $\mathbf{v}$) is then determined by the momentum balance equation in the form:

Journal of Physics: Conference Series **530** (2014) 012051 doi:10.1088/1742-6596/530/1/012051

$$\frac{d(m_p\mathbf{v})}{dt} = \beta(\mathbf{u}-\mathbf{v}) + \mathbf{F}_c \tag{1}$$

where the forces acting on the particle:

- the drag force proportional to the difference in the speed of the particles **v** and mean velocity **u** of the carrier phase in turbulent motion, where β is a momentum exchange rate coefficient of a rigid spherical particle [5],
- sum of other forces $\mathbf{F}_c$ acting of the particle (only buoyancy force has been included here).

Integrating particle momentum equation yields the trajectory of the particle in the Eulerian frame of reference. As the momentum balance equation velocity vector **u** component express the mean flow velocity of carrier phase (mean in the view of turbulent description of the flow), the traced particle trajectory is called to be deterministic (or mean trajectory).

Simultaneous flow simulation of coexisting phases can include the interplay between them by means of the momentum exchange due to interfacial drag forces (and mass and energy exchange in the extended non-isothermal and chemical reaction model). This is usually implemented by the evaluation of local in space and time momentum (and mass and energy) sources in the carrier phase flow model equations [6]. This is called as two way coupling of the multiphase flow. The back coupling of momentum balance equations of coexisting phases is usually considered essential for the case when the considerable amount of dispersed phase, with relatively large density, is mixed with carrier phase thus having effect on the mixture flow. In our consideration the back coupling has been neglected due to quite adverse conditions, the dispersed phase has low density and overall mass of it is much less than the carrier phase.

2.1. Geometry

The refining cylindrical pot of inner diameter Φ=2.36 m is rounded at the bottom and partially filled with raw lead up to the free surface located approximately at H=1.85 m over the pot base. The pot cavity is closed by the cap, where propeller electric engine is mounted. At the beginning of the refining campaign the pot is fed by the raw melted lead and the agitator is immersed into the bath to a depth h, figure 1.

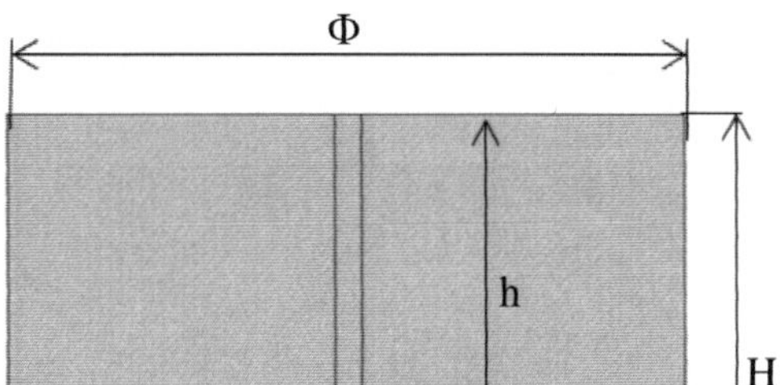

Figure 1. Schematic diagram of the ladle.

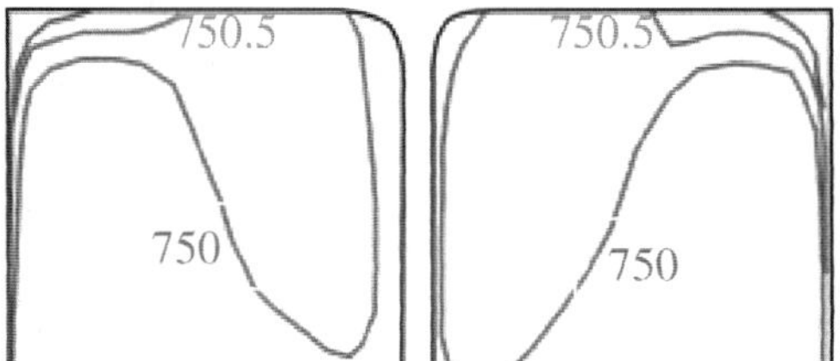

Figure 2. Temperature distribution, K.

Then the cap is mounted, and the melt agitation starts. During whole campaign the pot content is maintained at nearly constant temperature using gas burner system arranged around the external surface of the pot. The 3D model of the mixer ladle was created incorporating melt volume confined by the inner surface of the pot walls, surface of the agitator, surface of the shaft and free top surface of the melt. The form and overall dimensions of the pot has been maintained constant in consecutive simulations, but the shape and immersion depth of the agitator has been modelled each time according to the guidelines from the analysis of previously obtained simulation results and predetermined objectives of the rotor design. The reconstruction procedure outlined above started from the agitator design, following the plant established practice (two rectangular blades mounted at acute angle slope on opposite sides of the shaft). Due to rotating shaft and agitator blades, the whole interior solution domain was divided into two parts: rotating domain encapsulating agitator end exterior domain around the pot walls. This was dictated by the solution procedure requirements and necessity of defining proper boundary conditions of the velocity field.

2.2. Physical properties
The temperature dependence of physical properties of molten lead has been used in the form of piecewise approximation. Initially conducted non-isothermal computations have revealed negligible temperature variation within the melt, therefore subsequent simulations were made for isothermal conditions at 750 K, at which the density of lead is ρ=10480 kg/m^3 and molecular dynamic viscosity is μ = 0.0018 Pa·s.

2.3. Computational grid
The computational grid was created using tetrahedral elements, refined considerably near the surface of solid pot walls, the agitator blades and the shaft. An application of standard high Reynolds k-ε turbulence model was taken into account by the incorporation of near the wall elements fulfilling the condition of first to the wall node distance, validating use of logarithmic velocity distribution as a boundary condition [7,8]. A posteriori inspection of dimensionless distance from the wall was made, resulting in the boundary grid refinement in the case of offence of mentioned above condition. Overall number of grid elements varies around 160 000 elements, depending on the stirrer shape.

2.4. Boundary conditions
At the solid walls of the pot and the propeller agitator, the wall function boundary condition has been applied. To resemble the actual free surface (top surface of the melt) the equilibration of normal stresses at the both sides of a surface were posed, taking into account surface tension of lead (σ=0.35 N/m), and at every simulation time step the curvature of the surface was calculated and the computational grid reconstruction algorithm was used to follow new surface shape. Additional continuity conditions were used at common boundary of rotating domain and outer region of the melt.

2.5. Mathematical model specifications for the carrier fluid flow
The process of lead refining is conducted under nearly constant operating conditions, some minor variations of rotational frequency takes place during feeding of the refining compounds to get better conditions for the immersion of solid particles from the melt surface by means of the funnel formation at higher frequencies. This would lead to the conclusion that the steady state form of the mathematical model could be used, but the rotation of complex propeller agitator necessitates to apply unsteady form of the model equations.

Next step in the mathematical model development is to decide whether the flow is laminar or turbulent. For the mixing processes the criterion of the flow regime (turbulent or laminar) is specified by the critical value of Reynolds number defined as [1]:

$$\mathrm{Re}_{mixing} = \left(\frac{\rho n d^2}{\mu} \right) \tag{2}$$

where n stands for rotational frequency 1/s of propeller agitator of diameter d. The molten lead is a very dense liquid with moderate viscosity, so one could expect domination of inertia forces over viscosity forces. In fact the value of Re>5×10^6 obtained even for moderate rotational frequency 80 rpm and adequate to operational conditions diameter d indicates turbulent motion of the melt.

Thus, assuming isothermal conditions, the flow of the melt has been considered incompressible and turbulent. The mathematical model consists of the set of equations:

- the continuity equation
- the momentum balance equation
- the turbulence closure model equations (high Reynolds number k-ε)
- the energy equation (conditionally included in the preliminary simulations)

These fundamental equations are not included here due to the volume of the text limitations as their form is obvious for all interested readers.

Preliminary computations were conducted for the non-isothermal conditions to estimate temperature variations in the melt and the importance of the natural convection for the circulation pattern. In this case an energy equation was appended to the model, and functional dependence on temperature of density and viscosity of the melt were used. Since the results of these non-isothermal simulations indicate significant alignment of the temperature in the boiler volume (figure 2), further simulations were conducted for isothermal conditions.

2.6. Transport of diluted refining compounds

The deterministic approach described above does not take into account the fact that the dispersed phase particles interacts with the carrier phase, which remains in the turbulent flow. Actually, in the momentum equation for the particle (1), the carrier phase instantaneous velocity should be used (instead of mean velocity value) to precisely describe the velocity and trajectory of a single particle. But, in such a formulation, all of the particles of the dispersed phase need to be traced. This would be very time consuming task and even impossible to accomplish in practice. Therefore, the concept of tracing "clouds" of particles proposed by Baxter and Smith [9] seems to be very promising and effective technique for the simulation of a diluted phase flow in turbulent conditions. This is called herein as stochastic-deterministic approach. Fundamental difference between strictly deterministic and stochastic-deterministic approach becomes visible in the description method of the spatial particles distribution. The stochastic-deterministic approach exploits the concept of mean trajectory (according to equation (1)) but it does interpret it as the trajectory of the particles 'cloud' centre rather than the single particle trajectory. The ensemble of particles with identical diameter, material and thermodynamic properties, called traced particles 'cloud', enters the model domain with the initial conditions defined by its initial spatial and time position and initial dispersion in space. The mean trajectory derived from the equation (1) using mean carrier phase velocity vector distribution is interpreted as the solution of ensemble average of equations of motion for the particles represented by the "cloud". In this way the instantaneous particles distribution $g(x,y,z,t)$ is expressed by the statistical distribution function (3D probability density Gaussian distribution) about a mean trajectory

$$g(x,y,z,t)=\frac{\exp\left\{-\frac{1}{2}\left[\frac{(x-\xi_1(t))}{\sigma_{11}(t)}+\frac{(y-\xi_2(t))}{\sigma_{22}(t)}+\frac{(z-\xi_3(t))}{\sigma_{33}(t)}\right]^2\right\}}{(2\pi)^{3/2}\sigma_{11}(t)\sigma_{22}(t)\sigma_{33}(t)} \tag{3}$$

- $\xi_1(t)$, $\xi_2(t)$, $\xi_3(t)$ – position of the center of the multivariate distribution (average trajectory),
- $\sigma_{11}(t)$, $\sigma_{22}(t)$, $\sigma_{33}(t)$ – square roots of multivariate Gaussian distribution variances, where due to an isotropy of the k-ε turbulence model of the carrier phase we use single value of the distribution variance $\sigma_{\text{part}}(t)=\sigma_{11}(t),=\sigma_{22}(t),=\sigma_{33}(t)$.

The variance $\sigma_{\text{part}}(t)$ of the postulated distribution, describing the spatial dimension of the cloud, should be estimated along the trajectory, as a cloud is expanding in time due to particles interaction with eddies of turbulent motion of the carrier phase. This can be approximated using Kampě de Feriět formula for the transverse diffusion in a turbulent flow [10]:

$$\sigma^2_{part}(t)=\sigma^2_{part}(t=0)+2\int_0^t \overline{v'(t_1)v'(t_1)}\int_0^t R_{part}(t_2)dt_2dt_1 \tag{4}$$

where:

- $\sigma^2_{part}(t=0)$ - initial variance of the particle distribution for the traced "cloud",
- $\overline{v'(t)v'(t)}$ - the mean square fluctuation component of the particle velocity,
- $R_{part}(t)$ - the auto-correlation Lagrange function of particles.

Direct application of the relationship equation (4) encounters basic difficulty in the fact that the parameters appearing under integral concern the parameters describing turbulent flow of the particles, and unless the turbulent motion of the particles is solved, these parameters are unknown. Therefore, the relation would be applicable for the strictly stochastic method of solution, what in fact we are going to avoid in the stochastic-deterministic approach using the ensemble average trajectory of the 'cloud' by integrating equation (1). The concept of the stochastic-deterministic approach is based on the approximation of equation (4), proposed by Baxter and Smith, who expressed the particle distribution variance using statistical parameters of the fluid turbulent motion in the vicinity of the particle. These parameters are available from the turbulent model applied for the gas phase flow. Interested readers are redirected to the original paper [9] for the details.

3. Solution procedure

Each simulation was executed starting from the propeller agitator acceleration period, when the rotational frequency was slowly raised to prevent computations divergence. After an accomplishment of desired value of rotation frequency computations were continued to attain steady values of several controlled parameters: propeller torque, pressure distribution on the free surface of the melt, maximum velocity magnitude, etc. It usually required relatively large number (dozens) of propeller turnovers to be simulated (namely 30 or even more). Then, the tracing of particle "clouds" were carried out to observe the dispersion of the particles in the bath.

Simulations carried out for a number of different agitators led to selecting the best of them. It proved to be a stirrer having 2 level sets of blades. The lower agitator with a diameter of D consists of three profiled blades twisted of 45° angle along its length. The blade base is mounted to the shaft at 45° angle to horizontal plane of rotation. The upper agitator of a similar design, but with a diameter of 0.75D, was mounted on a shaft in a position rotated by 60° relatively to the lower agitator. The optimal distance between both agitators was established as 0.7D. The level of a rotation plane of the lower agitator was set at 0.3H from the bottom of the ladle. Selection criterion for the positioning was to attain an appropriate depth of a sucking funnel formed on the surface of a bath in simulations.

4. Simulation results

The simulation results for optimal design of a stirrer are presented below. The velocity distribution of a carrier phase, shown for the central vertical cross section of a ladle is presented on figure 3, and for the horizontal cross section at the level of middle distance between propellers is presented on figure 4.

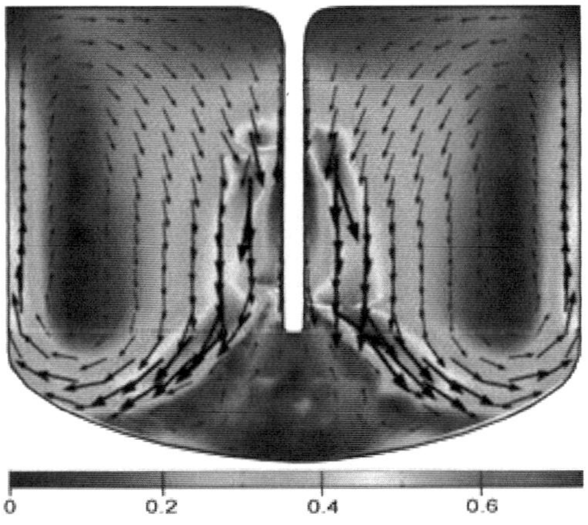

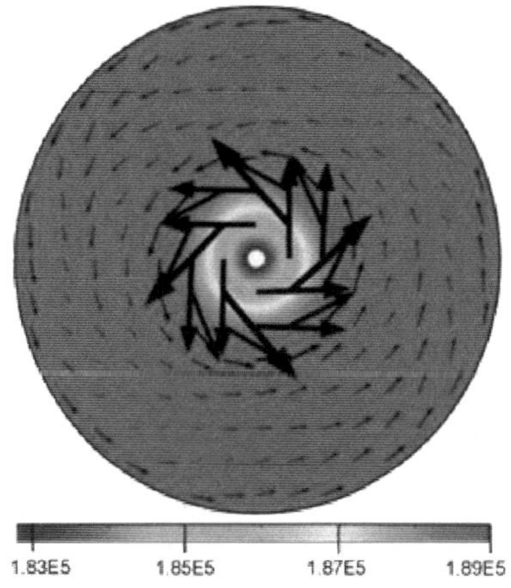

Journal of Physics: Conference Series **530** (2014) 012051 doi:10.1088/1742-6596/530/1/012051

Figure 3. Velocity of carrier phase distribution, in plane components only, m/s.

Figure 4. Velocity of carrier phase (vectors), color map of pressure distribution, Pa.

The instantaneous streamlines are presented on figure 6, showing that there is no possibility to create two vortices in the flow, which suggested the literature of the subject, when double impeller is in use. It is interesting to observe nonhomogeneous distribution of turbulent kinetic energy (see figure 5), where its maximum values are distributed around the shaft over the upper agitator. This is very important observation, because an application of upper agitator not only creates a sucking funnel of required depth but also produce favorable conditions for rapid dispersion of refining compounds particles just at the beginning of their way through the base metal.

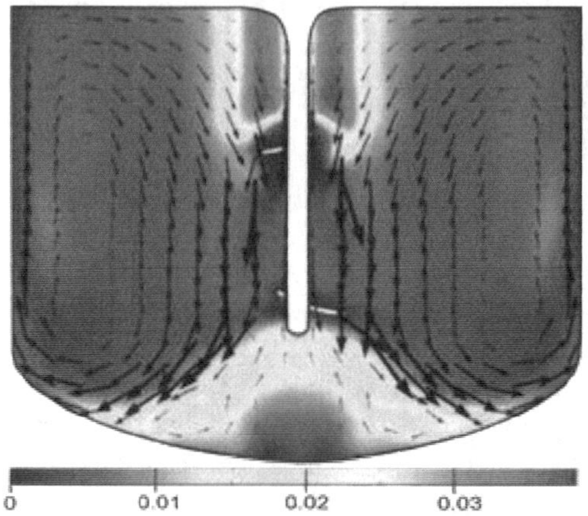

Figure 5. Turbulent kinetic energy distribution, m^2/s^2.

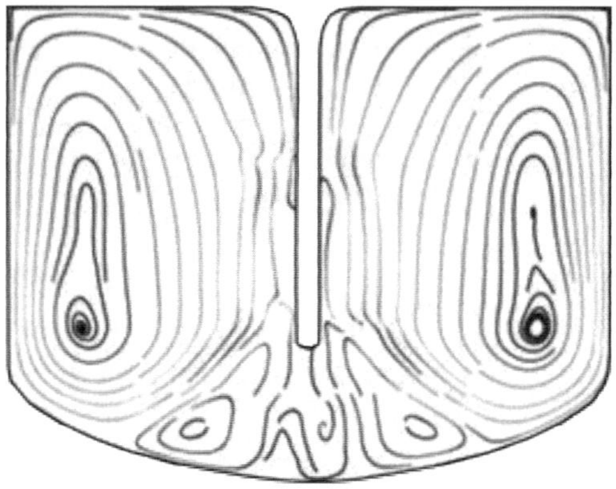

Figure 6. Instantaneous streamlines in a cross section of a mixer, warm colors – higher velocity.

On figure 7 we present the simulated evolution of the particles "clouds" along their way through the carrier fluid. A couple of "clouds" has been presented here for the clarity of the graph. It can be observed how the sucking funnel formed on the top of the bath pulls in particles, then disperses them gradually in the bath.

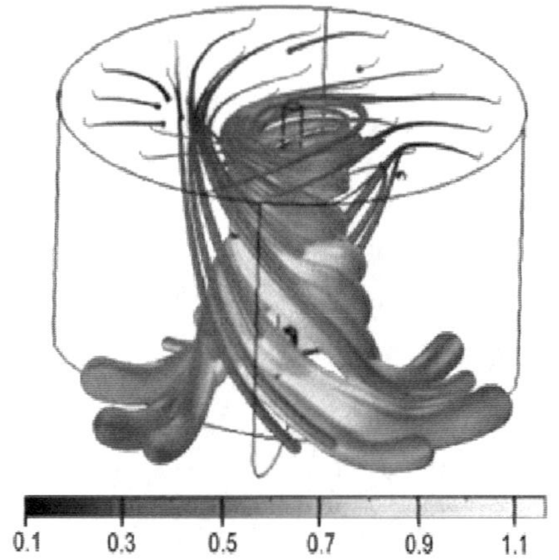

Figure 7. Refining particles "clouds" tracing (7sec since feeding), color express particles velocity m/s.

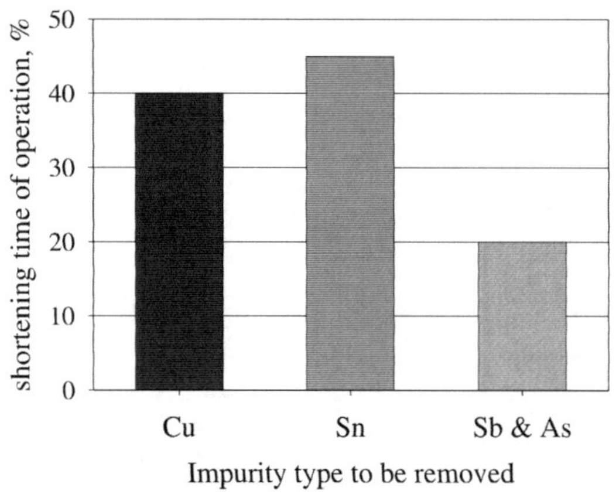

Figure 8. Time shortening of unit operation as a result of application of redesigned propeller for different impurities.

5. Verification

Direct verification of the simulation results, obtained by means of measurements of parameters of the simulated movement, seems to be virtually impossible due to the nature of the process and the type of carrier fluid. However, since the basic objectives of the utilitarian design of technological nature, it was decided to execute industrial trials that were supposed to answer the question of whether inflicted objectives have been obtained.

The stirrer of new design, consisting of doubled triple wing blades of different diameters has been produced from mild steel, using precision-casting technique. In order to verify the anticipated effects of application of a new agitator, a set of industrial test has been conducted. Trials were carried out in typical conditions for refining process before modification of the stirrer. Research objectives have been identified as: intensification of the process of refining, reducing energy consumption, reducing the formation of metal-oxide buildup at the side walls at the top of the boiler. Observations are subject to the following parameters: duration of the unit process in refining process to remove specific additives (copper, tin, antimony and arsenic), the final concentration of additives that are deleted, mass waste refining process. Process parameters observed have been confronted with the appropriate values on the record prior to the modification of the agitator. The most spectacular result of application of the new paddle was a significant time shortening of removing impurities. The percentage of time shortening for removing impurities after modification of the stirrer for each unit operation are shown on figure 8. An additional positive aspect of the intensification of the process is the weight reduction of the waste product: 26% for unit operation of removing tin and 10% for removing antimony and arsenic. In the case of removal of antimony and arsenic, there is also a deeper expulsion of impurities (60% less final concentration of impurities).

6. Conclusions

The presented results of the project show a great potential of numerical analysis of mixing processes. Precise design of agitator plays a decisive role in such processes, where intense dispersion of distributed phase is required for increasing the efficiency of the process. Obtained results of simulations have been confirmed by experiments carried out in industrial environments. Further subjects should take into account the analysis of chemical kinetics, which in the absence of specific studies could not be included in this work.

7. References

[1] Kamieński J 2004 *Mieszanie układów wielofazowych* Warszawa: WNT)
[2] Löhner R 2001 *Applied CFD techniques* (Chichester: John Wiley & Sons)
[3] Oran E, Boris J 1987 *Numerical simulation of reactive flow* (NY: Elsevier)
[4] Stewart H, Wendroff B 1984 Two-phase flow: models and methods, *J. Comp. Phys.*, **56**, 263
[5] Crowe C, Sommerfeld M, Tsuji Y 1988 *Multiphase flows with droplets and particles* (Boca Raton: CRC Press)
[6] Crowe C, Sharma M, Stosk D 1977 The particle-source-in cell (PSI-CELL) model for gas - droplet flows *J. of Fluids Engineering* 329-332.
[7] Launder B, Spalding D 1972 *Mathematical models of turbulence* (London: Academic Press)
[8] Bennet C, Myers J 1962 *Momentum, Heat, and Mass Transfer* (NY: McGraw-Hill)
[9] Baxter L, Smith P 1993 Turbulent dispersion of particles: the STP model *Energy&Fuels* **7** 852
[10] Kampé de Fériet J 1939 Les fonctionés aleatoires stationnaires at la théorie statistique de la turbulence homogene *Ann. Soc. Sci. Bruxelles* Ser. I **59** 145-210.

The influence of additives on rheological properties of limestone slurry

B Jaworska and A Bartosik
Faculty of Management and Computer Modelling, Kielce University of Technology, Al. Tysiąclecia P.P. 7, 25-314 Kielce, Poland

E-mail: b.jaworska@tu.kielce.pl; artur.bartosik@tu.kielce.pl

Abstract. Limestone slurry appears in the lime production process as the result of rinsing the processed material. It consists of particles with diameter smaller than 2 mm and the water that is a carrier of solid fraction. Slurry is directed to the settling tank, where the solid phase sediments and the excess water through the transfer system is recovered for re-circulation. Collected at the bottom of the tank sludge is deposited in a landfill located on the premises. Rheological properties of limestone slurry hinder its further free transport in the pipeline due to generated flow resistance. To improve this state of affairs, chemical treatment of drilling fluid, could be applied, of which the main task is to give the slurry properties suitable for the conditions encountered in hydrotransport. This treatment consists of applying chemical additives to slurry in sufficient quantity. Such additives are called as deflocculants or thinners or dispersants, and are chemical compounds which added to aqueous solution are intended to push away suspended particles from each other.

The paper presents the results of research allowing reduction of shear stress in limestone slurry. Results demonstrate rheological properties of limestone slurry with and without the addition of modified substances which causes decrease of slurry viscosity, and as a consequence slurry shear stress for adopted shear rate. Achieving the desired effects increases the degree of dispersion of the solid phase suspended in the carrier liquid and improving its ability to smooth flow with decreased friction.

1. Introduction

Solid-liquid transport appears frequently in many industrial applications, like for instance in chemical engineering, mining industry and power plants. Such transport depends on several parameters, like: concentration of solid particles, slurry velocity, particles diameter, pipe diameter, terminal velocity, liquid and solids density, carrier liquid viscosity, flow geometry and rheological properties [1]. If limestone slurry is considered we recognize settling and non-settling flow. The settling flow can be with stationary or moving bed while the non-settling slurry flow can be homogeneous or pseudo-homogeneous. When all the particles have the same density, the larger particles always form the lower layer, however if the particle sizes are sufficiently different, the separation may appear. The non–settling slurries usually exhibit non-Newtonian character and they exhibit the yield stress. Those with very fine solid particles demonstrate increased viscous sublayer, which means that damping of turbulence may appear in the near-wall region. Slurry turbulence in the region close to a pipe wall was examined by scientists like Nouri and Whitelaw [2], Schreck and Kleis [3]. They concluded that ejection–sweep cycle is affected strongly by particles, and slip velocity decreases with solids concentration increase. A review of experimental studies on turbulence modification by particles is given by many researchers [4,5,6,7]. In order to build mathematical model of such slurry flow, the starting point is to determine proper rheological model. In contrast with that, if slurry with coarse solid particles is considered, it is reasonable to assume the Newtonian model, as now one can measure rheology in such slurries [8].

The main objective of the research is to investigate reduction of shear stress in limestone slurry by applying selected chemical additives. A sufficient addition dosage significantly decreases the effect of particle size on the flow ability of limestone slurries. In order to find a best chemical additive for shear reduction in limestone slurry the influence of various doses of additives in relation to three arbitrary chosen solids concentrations of slurry, equals 30%, 40% and 50% by mass, is demonstrated. rheological behaviours of limestone slurry have been investigated using rotational viscometer.

2. Lime production process

During the lime production process the pipeline transport of limestone slurry is carried out. Limestone slurry arises from rinsing of stones during production process. Such slurry is a finely-granular solid phase and the water is added as a carrier liquid of the solid phase. It flows by turbulent traffic and is usually accompanied by moving bottom sediment. Limestone slurry is directed to the settling tank (natural reservoir), in which the solid-phase fraction sediments and the excess water, through the transfer system, is recycled back to the process of rinsing of stones.

One of the methods for reducing the frictional pressure loss and therefore the pump discharge pressure requirements, when transporting viscous, often non-Newtonian slurries and pastes in pipelines, is the method of reduction of the degree of flocculation of particles in the slurry using suitable chemical additives. Adding soluble ionic compounds to flocculated slurries can result in substantial pressure drop reduction in a pipe flow. The ionic compounds disperse the particles, thereby breaking up the flocculated particle network that gives rise to higher shear stresses in a pipe flow. In order to disperse particles, the charge on their surfaces needs to be of the same sign and the higher the charge the greater the repulsive forces between particles [9].

3. Use of additives to dispersed particles in limestone slurry

In order to improve rheological properties of limestone slurry, the chemical processing is used, of which the main task is to give the slurry properties suitable for the conditions encountered in pipeline transport. Such processing consists of adding chemical compounds, usually polymeric substances. Quantity and quality of added substances is crucial, similarly to the character of its interactions with the slurry. Added substances should stabilize limestone slurry and strengthen its structure, which causes increase of degree of solid phase dispersion and improving its ability to create sludge. However, the main important task is to reduce viscosity of the slurry. In lime production process the viscosity increases as the result of saturation of the slurry by dust particles. Another significant issue is to prevent the sedimentation of solid particles on the wall of horizontal pipelines, which causes its clogging. To avoid this undesirable state of things slurry should have suitable fluency.

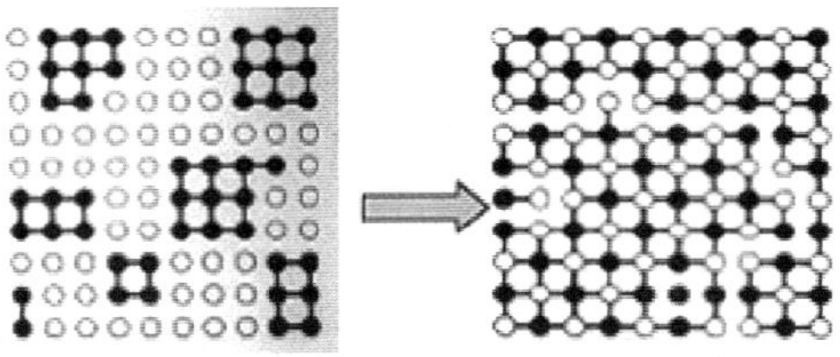

Figure 1. The mechanism of dispersion agglomerates of molecules occurring in the deflocculating process.
Source: http://www.vekamaf.com.pl/

Dispersing agents are commonly used in mining industry in order to reduce viscosity and yield stress of limestone slurry. Dispersants increase the liquidity of slurry through its rarefaction which enables introducing more solid particles to slurry without changing its fluency nature - Figure 1. Occurring inside the slurry structure process is explained by combining short chain links of dispersant substances containing negative charge with the positive charge of solid particles by the action of electrostatic forces. After absorbing the polymeric substances occurs the neutralization of positive charge of solid particles which result in disappearing electrostatic forces between the molecule suspensions and polymer-molecular complexes are gaining a negative charge. It leads to the disintegration of aggregates of solid particles.

4. Experimental studies on influence of selected chemical additives on rheological properties of limestone slurry

4.1. Experiments

The first step in experiments was determination of the curve of grain for the slurry contaminated by dust processed raw limestone. Granulometric analysis was carried out on the laser diffraction particle analyzer for wet dispersion and particle size in the range from 0.01μm to 3500 μm. Measured particle sizes of the limestone slurry were in the range from 0.52 μm to 143.9 μm. The value of average particle diameter in the tested slurry is close to 27μm, which is presented in Figure 2.

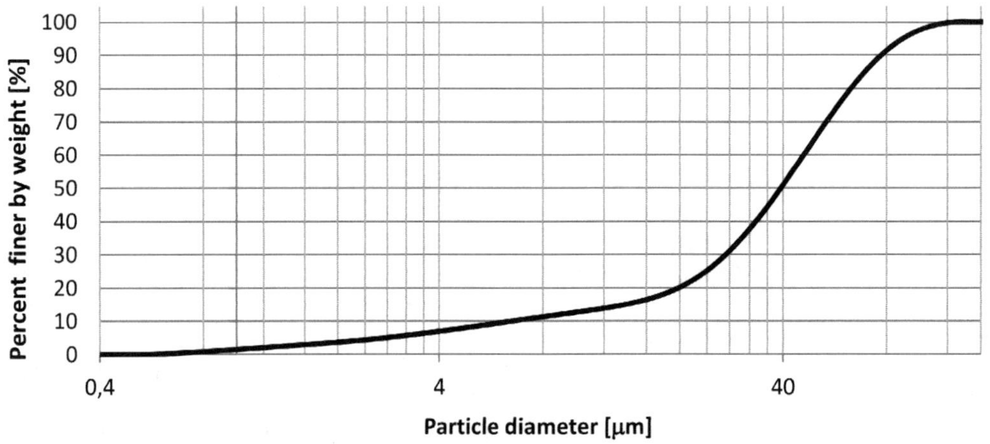

Figure 2. The grain size curve of limestone slurry. Measured by using Mastersizer 3000.

As the result of analysis, three types of fraction have been separated: fraction of clay, dust and sand. The highest percentage in the sample are the particles of dust fraction (64.19%) with diameter in the range of 2-50 microns. Clay fraction with an average grain diameter less than 2 microns is 3.83%, while the phase of sand with grain diameter greater than 50 microns is 31.98% of all particles, which is seen in Table 1.

Table 1. Percentage content of grain fraction in limestone slurry. Measured by using Mastersizer 3000.

Fraction	Clay	Dust	Sand
Grain diameter [μm]	< 2	2 ÷ 50	> 50
Percentage content	3.38%	64.19%	31.97%

Chemically tested limestone slurry consists mostly of calcium oxide (CaO – 73.64%) and silicon oxide (SiO_2 - 13%). Other chemicals included in particulate form part mineral are as follows: MgO –

0.61%, Fe_2O_3 – 0.32%, Al_2O_3 – 1.11% and SO_3 – 0.28%. Humidity of tested samples was 96.67%. Density of solids is 2720 kg/m^3.

Laboratory tests were carried out for three different mass concentrations of limestone slurry: 30%, 40% and 50%. Mass concentration (C_m) determines the percentage content of solids in total weight of limestone slurry (1):

$$C_m = \frac{m_s}{m_s + m_l} \cdot 100\% \quad (1)$$

where:
C_m – mass concentration of solid particles (%)
m_s – mass of the solid phase (kg)
m_l – mass of the liquid phase (kg)

Sampling material needed to measure solid phase mass concentration took place by collecting and drying three samples of 100 ml each to obtain the averaged values.

4.2. Compounds used in the experiments

In the final stage of lime production process, slurry is subjected to sedimentation in order to separate the solid phase from it. Sludge collected on the bottom of the tank after gravity thickening is directed to the sedimentary reservoir located at a considerable distance from the plant. Its further transport by pipeline system is difficult because of its rheological properties. Applying suitable chemical additives in proper dose are intended to push away suspended particles and increase its fluency. To the group of such substances, called deflocculants, belongs among others sodium water glass. It is an aqueous solution of sodium silicate having the characteristic of colloid or polymer, high viscosity and density 20% - 70% more than the water.

Another dispersant used in experimental studies is known as calcareous groats. This substance is byproduct of limestone production process and so far has not found its industrial application. Calcareous groats is a remnant of lime slaking process by which the hydrated lime is formed. In the final step of lime slaking takes place the process of hydration lime and water mixture in suitable mixing chamber. The waste materials in form of heavier particles of the lime that have not been slaked and unburned lime stone sink at the bottom of the chamber and are periodically removed outside by the trigger aperture located in the lower chamber of the mixer. Obtained hydrated lime is subjected to two-stage separation process in order to separate it from contaminants. Hydrated lime is a finished product, while the remains of the separation are called calcareous groats and is presented in Figure 3.

Figure 3. Substance arisen as a result of limestone production process - calcareous groats. Made with a digital camera.

A mixture of these two substances (sodium water glass and calcareous groats) added to limestone slurry can act as dispersant. Its task is to repel particles in suspension from each other to dilute, which would facilitate its further transportation by pipeline. The analyzed substance is formed by mineral particles with grain diameter less than 1.8 mm and average particle diameter equals 240 μm. Chemically calcareous groats mostly consist of the calcium and magnesium oxide (70.81%), silicon dioxide (1.52%) and alumina (0.84%). Other substances (25.91%) are minor contamination caused by the hydration of lime and water mixture. The chemical composition of hydrated lime is shown in Table 2.

Table 2. Analysis of the chemical composition of calcareous groats and limestone slurry.
Source: Laboratory of TRZUSKAWICA S.A.

Chemical composition	Calcareous groats [%]	Limestone slurry [%]
CaO	70.58	73.64
MgO	0.23	0.61
Fe_2O_3	0.56	0.32
SiO_2	1.52	13
Al_2O_3	0.84	1.11
SO_3	0.36	0.28
other	25.91	11.04

4.3. Methods of measurements
To test the effect of selected chemical additives on the rheological properties of limestone slurry laboratory tests were carried out. They included investigations of the influence of varied doses of sodium water glass and calcareous groats on limestone slurry. The concentrations of additives were 0.1%, 0.2%, 0.3%, 0.4%. Compounds were applied in the ratio 1:1 each substance in order to obtain the wanted dose. The study included measurements of rheological properties of limestone slurry in three mass concentrations: 30%, 40% and 50% after applying chemical additives. Measurements were carried out on rotational viscometer in the measuring system of the coaxial cylinders with the gap 1.1 mm. The tests were performed at 20°C for samples having volume of 18 ml. The experiments were focused on comparison of shear stress versus shear rate for different quantity of chemical additives in the range of the shear rate from 1 to 300 s^{-1}.

4.4. Results of measurements
Figure 4 shows generated in the samples shear stress values for adopted shear rate. Received measuring points demonstrate the properties of limestone slurry with mass concentration of 30% after adding various doses of combination of sodium water glass and calcareous groats in proportion 1:1. As a reference for received points of additives in amounts of 0.1%, 0.2%, 0.3% and 0.4% related to the dry matter contained in the slurry a curve for pure limestone slurry was used.

As we can see, the addition of combination of chosen chemicals at dose 0.3% and 0.4% does not improve rheological properties of slurry but even worsens it. Addition of lower doses of additives affects reduction of shear stress in wide range of shear rates. The largest decrease of shear stress in slurry was obtained after applying an amount of 0.2% of chosen additives, which means that the dose of best proportion is of 0.1% of each substance: sodium water glass and calcareous groats in relation to dry mass of limestone slurry.

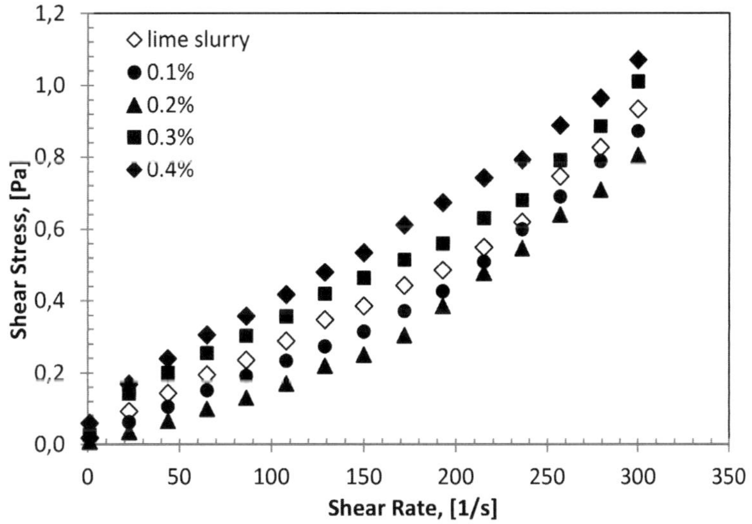

Figure 4. Influence of various doses of sodium water glass and calcareous groats on shear stress values in limestone slurry with mass concentration $C_m = 30\%$. Measured by using rotational viscometer MCR 302.

If high mass concentration of limestone slurry is considered ($C_m = 40\%$) we can observe a clear decline of shear stress even for small doses of compound mixture (0.1%, 0.2%, 0.3%) for shear rate in the range from 300 [s^{-1}] to 1 [s^{-1}]. The best influence on decreasing the shear stress can be noticed for dose of 0.1% of sodium water glass and calcareous groats. The highest stress values were measured in the sample with 0.4% of proposed mixture of substances, which Figure 5 demonstrates.
It is worth mentioning that experimental data, which are presented in Figure 5, show that for the shear rate higher than 260 [s^{-1}] observable increase of the shear stress appears. This could be due to the fact that transitional flow regime appears for such a high shear rate and for the gap between the rotor and the cylinder equal 1.1 mm.

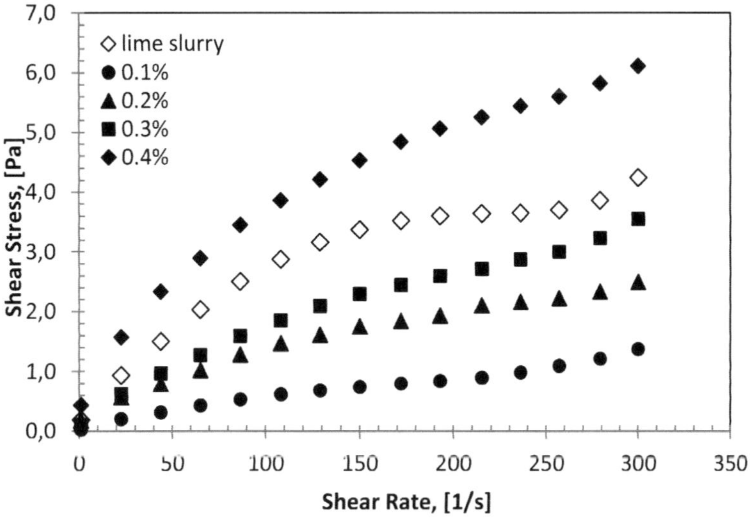

Figure 5. Influence of various doses of sodium water glass and calcareous groats on shear stress value in limestone slurry with mass concentration $C_m = 40\%$. Measured by using rotational viscometer MCR 302.

If slurry with solids concentration equalling 50% by mass is considered (Figure 6) we can see that there is no decrease of the shear stress if chemical additives are included. This lack of influence probably results from very small amounts of added substances in relation to dry matter content in slurry with such high mass concentration. Only a dose of 0.4% of proposed substances has an effect on slight decrease of received stresses values. But even this amount resulted only in approach the analyzed curve to the graph of pure limestone slurry without any additives. It is also interesting that the effect of using a dose of 0.1% and 0.4% is almost the same. It seems necessary to carry out laboratory studies for higher concentrations of used additives in order to find optimal dosage.

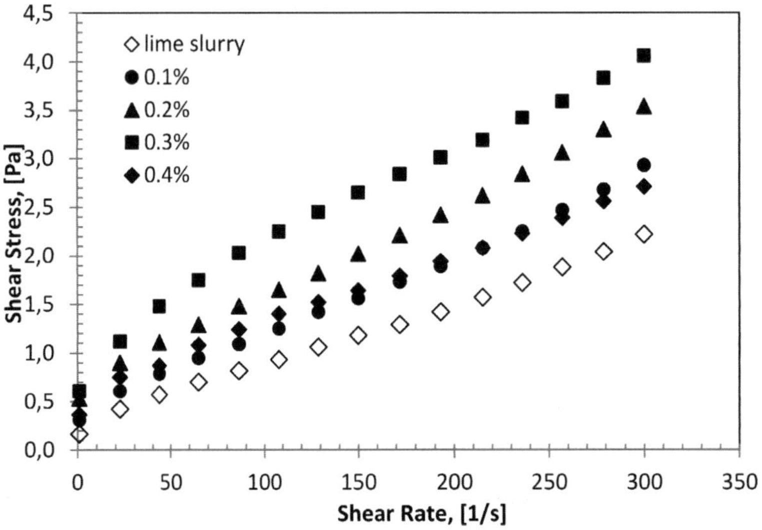

Figure 6. Influence of various doses of sodium water glass and calcareous groats on shear stress value in limestone slurry with mass concentration C_m = 50%. Measured by using rotational viscometer MCR 302.

5. Conclusions

The analysis of the influence of various doses of sodium water glass and calcareous groats on rheological properties of limestone slurry gives to some extent a response to the question regarding the amount of substances needed to apply to slurry with C_m=30% and C_m=40% to reduce flow resistance. However, it seems to be necessary to perform tests in a wider range of doses of additives to limestone slurry with a mass concentration of solid particles equal to 50%.

The presented results of the research show how it is difficult to predict the effect of modifying chemical substances on the tested material. In order to stabilize rheological parameters of slurry, a dosage of additives should be selected appropriate to mass concentration of solids included in limestone slurry. Simple viscosimetric tests allow a proper selection of kind of mineral compounds. It should be noted however, that the chemicals used in the research are easy to access and they are cheap. We proved the positive influence of such chemical additives for reduction of frictional head loss in pipeline transportation of slurry in the range of solids concentration from 30% to 40% by mass. Reduction of frictional head loss makes it possible to decrease energy consumption for transportation of one tone of dry slurry per hour.

6. References

[1] Shook CA and Roco MC 1991 *Slurry Flow: Principles and practice* Boston, Butterworth-Heinemann **38** 324

[2] Nouri JM and Whitelaw JH 1992 Particle velocity characteristics of dilute to moderately dense suspension flows in stirred reactors *Int. J. Multiphase Flow* **18** 21–33

[3] Schreck S and Kleis SJ 1993 Modification of grid–generated turbulence by solid particles *J. Fluid Mech.* **249** 665–88

[4] RA Gore and CT Crowe 1991 Modulation of turbulence by a dispersed phase *J. Fluid Engng.* **113** 304–7

[5] Eaton JK 1994 Experiments and simulations on turbulence modification by dispersed particles. *Appl. Mech. Rev.* **47** 44–8.

[6] Eaton JK, Paris AD and Burton TM 1999 Local distortion of turbulence by dispersed particles AIAA **99** 3643

[7] Bartosik A 2008 Laminarisation effect in fine-dispersive slurry flow *Archives of Thermodynamics* **29** 69-82

[8] Bartosik A 2009 Application on rheological models in prediction of turbulent slurry flow *Flow, Turbulence and Combustion Springer-Verlag* **84** 277-93

[9] Heywood N 2003 How to reduce pipe friction in slurry flows *Proc. of the 4th One Day Sem. on Hydraulic Transport in the Mining Industry* 7 April 2003 Indaba Conference Centre

Journal of Physics: Conference Series **530** (2014) 012053 doi:10.1088/1742-6596/530/1/012053

Modelling liquid redistribution in a packed bed

Paweł Niegodajew, Dariusz Asendrych, Maciej Marek and Stanisław Drobniak
Częstochowa University of Technology,
al. Armii Krajowej 21, 42-200 Częstochowa, PL

E-mail: imc@imc.pcz.czest.pl

Abstract. The paper is devoted to the computational fluid dynamics (CFD) simulation of liquid spreading process in a packed bed. The volume of fluid (VOF) approach was applied to simulate the flow in a realistic porous region composed of 6mm Raschig rings. Modelling results were used to determine the probability density function (PDF) distribution of liquid velocity vector orientation angle which was then implemented into 2-fluid Euler-Euler multiphase model of packing column. The simulation showed that the model is capable to simulate adequately the liquid redistribution in a porous region being, however, much more efficient computationally than the VOF method.

1. Introduction

Packing columns are commonly used in chemical and process engineering industry, just to mention absorption, rectification or extraction processes. Thanks to the enlarged contact area packed beds ensure the liquid holdup at the level needed for the efficient column operation [1]. The porous material also guarantees a uniform liquid distribution along the packing section, what is related to effective liquid transport in column transverse directions [2] commonly called as liquid spreading. Variety of parameters, i.e. packing element type, its size, fluid properties, their rates and others influencing the flow structure, makes the design of packed beds a challenge. That is why the research efforts are needed, both experiments and modelling, to make the progress in the field.

The opaque character of most packed bed structures and their 3D geometries make them difficult for penetration. That is why the liquid spreading process has been of interest of very few experimental research works. These are usually done with the use of visualization techniques as recording the liquid pore-flow through the transparent wall with the use of camera [3] or with the use of X-ray tomography methods [2,4].

The review paper of Wang et al. [5] shows various approaches to the modelling of trickle-bed reactors being the simplest case of the packed beds. The most commonly applied methods are the volume of fluid (VOF) and Euler-Euler approach used both for laminar and turbulent flow regimes. With the VOF mostly the flow through structured packing types is simulated. In particular it concerns the limited portion of packing section (e.g.: [6,7,8,9,10]). With this approach the porous region geometry has to be reconstructed and the inter-phase surface shape can be tracked (see [11,12]) allowing for an insight into the details of flow structure. For certain flow conditions the liquid behaviour may lead to the formation of rivulets and droplets and in turn to flow maldistribution, channelling or partial wetting [5,12]. The drawback of the VOF method is its practical inability to model the large-scale installations filled in with random packings. Moreover, the simulation of complex physio-chemical systems, e.g. carbon dioxide capture by chemical absorption [13], requires more efficient numerical tools allowing to replace the realistic packed bed geometry and the flow field with their statistical descriptions and in turn reduce computational costs. For such cases the Eulerian approach can be employed as the method not requiring the details of geometry, just replacing it with its statistical description. For instance Sun et al. [14] modelled the hydrodynamics of the packed bed composed of the Pall rings obtaining good correspondence with experimental data. It should be pointed, however, that the Eulerian methodology, needs some assumptions as the input, making it partly artificial and packing column configuration dependent.

2. Numerical model

In the present paper the liquid flow in a cylindrical vertical column filled in with packed bed composed of Raschig rings is approached by the Eulerian two-fluid model simulating the liquid spreading with the use of its statistical description, i.e. probability density function (PDF) of velocity vector orientation in 3D system. As the PDF distribution depends on the number of factors, in particular on the packed bed filling element shape and size, it was necessary to perform an independent simulation with the use of VOF method applied to the liquid flow through the realistic geometry.

2.1. Volume of fluid simulation

In order to obtain reliable PDF of liquid spreading angle, necessary for the stochastic model, a CFD simulation of liquid flow within realistic bed geometry has been performed. The bed geometry composed of 6mm Raschig rings was generated using the in-house code based on simplified mechanics. The algorithm described in [15] allows to construct a packed bed in a realistic way with all elements oriented in a random way, satisfying, however the mechanical equilibrium. The view of a sample packed bed is presented in figure 1a.

The flow equations were solved with classical finite volume Navier-Stokes equation solver on regular structured computational grid and the gas-liquid interface was tracked with VOF [16]. The complex bed geometry was handled by a variant of immersed boundary method [17]. The liquid was supplied at the top of the bed and when the steady state was reached (see figure 1b presenting liquid free surface shape) the flow structure was examined with respect to spreading in lateral direction. The analysis of the velocity field allowed to determine the PDFs of angles describing the deflections of velocity vectors from particular axes of coordinating system. It turned out, that the distribution describing the orientation (described by angle θ) of horizontal vector component (i.e. lying in *z-y* plane, see figure 1b for definition of Cartesian coordination system) is uniform. Therefore, it was decided to represent velocity vector orientation with the single angle φ_u, being actually velocity vector deflection from vertical axis x (see figure 1b). For a given computational cell that angle was calculated as follows:

$$\varphi_u = \arccos \frac{u_x}{|\vec{u}|} \tag{1}$$

where u_x is the axial velocity component. The angle expressed in this way is well-defined for any cell but if one wants to assign more weight to the cells filled with liquid phase and corresponding to larger flow velocity, the following weight coefficient was proposed as:

$$\omega = \alpha_l \cdot |\vec{u}| \tag{2}$$

where α_l is liquid volume fraction. The resulting PDF distribution is shown in figure 1c. It is worth to note, that the distribution is dominated by low angles corresponding to small deflections of velocity vectors from vertical direction. It should be also mentioned that the distribution includes the peak located in the neighbourhood of 90°, resulting from the occurrence of corresponding peak in the PDF distribution of Raschig rings orientation. Due to the duality of PDF distribution shown in figure 1c it was decided to split it into two subranges with φ_u^* as a border value.

Journal of Physics: Conference Series **530** (2014) 012053 doi:10.1088/1742-6596/530/1/012053

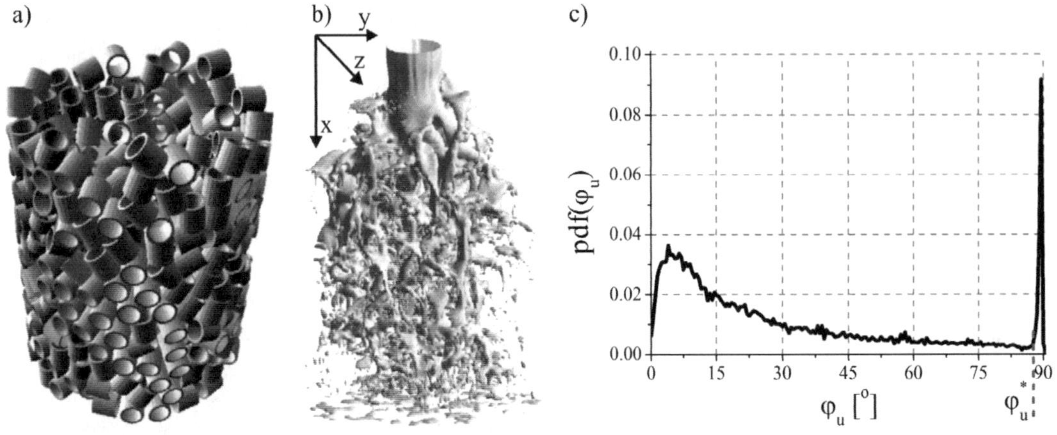

Figure 1. Numerically generated fixed bed of Raschig rings (a), instantaneous liquid free surface distribution simulated with VOF method (rings not shown for clarity) (b) and PDF distribution of velocity vector orientation (c).

The PDF distribution presented in figure 1c was fitted with the following two functions:

$$f(\varphi_u)=\begin{cases} q\dfrac{A\varphi_u}{1+B\varphi_u^2} & for \quad \varphi_u \in \langle 0;\ \varphi_u^* \rangle \\ a+b\varphi_u & for \quad \varphi_u \in (\varphi_u^*;\ \pi/2\rangle \end{cases} \tag{3}$$

where a, b, q, A, B are constants specific for the particular configuration of the packed bed. For the purpose of Eulerian two-fluid modelling strategy it was necessary to express the liquid velocity vector orientation angle as a function of its cumulative distribution function D [18]. At first D has to be determined by integration of PDF distribution given by expression (3) leading to:

$$D(\varphi_u)=\begin{cases} \displaystyle\int_0^{\varphi_u} f d\varphi_u = \frac{qA}{2B}\ln\left(1+B\varphi_u^2\right) & for \quad \varphi_u \in \langle 0;\ \varphi_u^* \rangle \\ \displaystyle D(\varphi_u^*)+\int_{\varphi_u^*}^{\varphi_u} f d\varphi_u = -a\varphi_u+\frac{1}{2}b\varphi_u^2-a\varphi_u^*-\frac{1}{2}b\varphi_u^{*2}+D(\varphi_u^*) & for \quad \varphi_u \in (\varphi_u^*;\ 90^o\rangle \end{cases} \tag{4}$$

The desired formulas for the orientation angle may be then derived from expressions (4) yielding:

$$\varphi_u=\begin{cases} \left\{\dfrac{1}{B}\left[\exp\left(D\dfrac{2B}{qA}\right)-1\right]\right\}^{0.5} & for \quad D\in \langle 0;\ D(\varphi_u^*)\rangle \\ -\dfrac{a}{b}+\left[\dfrac{a^2}{b^2}+\dfrac{2}{b}\left(D-D(\varphi_u^*)+a\varphi_u^*+b\varphi_u^{*2}\right)\right]^{0.5} & for \quad D\in \left(D(\varphi_u^*);1\right\rangle \end{cases} \tag{5}$$

Expressions (5) were implemented in CFD model described in the following section.

2.2. *Eulerian approach*

The model of the packing column was developed as a 3-dimensional laminar, unsteady, multiphase gas-liquid flow with the use of the two-fluid Euler-Euler approach allowing for the slip between countercurrently moving phases (corresponding for the Stokes number regime Sk>>1). All conservation equations are solved for each phase separately. Conservation of mass for k^{th} phase yields:

$$\frac{\partial}{\partial t}(\alpha_k\rho_k)+\nabla(\alpha_k\rho_k\vec{u}_k)=0 \tag{6}$$

where ρ_k is the kth phase density. The momentum equation for k^{th} phase with respect to the Eulerian multiphase model (assuming flow incompressibility) has the following form:

$$\frac{\partial}{\partial t}(\alpha_k \rho_k \vec{u}_k) + \nabla(\alpha_k \rho_k \vec{u}_k \vec{u}_k) = -\nabla(\alpha_k p) + \nabla \cdot \left[\alpha_k \mu_k \left(\nabla \vec{u}_k + \nabla \vec{u}_k^T\right)\right] + \alpha_k \rho_k \vec{g} + \vec{M}_B + \vec{M}_r \quad (7)$$

where p is the static pressure shared by all phases, μ_k stands for dynamic viscosity, g_k is the gravity vector and M_B describes an interaction force between phases taking into account the characteristics of the packing elements of porous zone, fluids properties and their flow rates [19]. The detailed description of the computational algorithm can be found in [20]. The last term of equation (7), i.e. M_r reflects the liquid momentum change due to its interactions with packing elements being the geometrical constraints for the flow (as a consequence resulting in change of liquid flow direction), and in general it can be expressed in the following form:

$$M_r = f\left(u_{x,B}, \varphi_u, \Theta, \varphi_u\left(u_{x,B}\right)\right) \quad (8)$$

where $u_{x,B}$ is the mean axial velocity of liquid phase determined according to [19] and $\varphi_u(u_{x,B})$ represents the orientation angle corresponding to the $u_{x,B}$ given by:

$$\varphi_u\left(u_{x,B}\right) = arccos\left(\lim_{n\to\infty} \sum_{i=1}^{n} cos\left(\varphi_{u,i}\right)\right) \approx 40^o \quad (9)$$

In order to model the liquid spreading in porous region satisfying formula (3) the orientation angle was generated according to equation (5) with cumulative distribution function D replaced by the random number characterized by uniform PDF distribution (white noise) taking its values from the range <0;1>. The angle θ was generated according to the uniform distribution. Having the instantaneous orientation angle (represented by both generated angles) the local instantaneous flow velocity components were then calculated taking into account velocity vector magnitude corresponding to previous time step. In this way axial velocity varied within domain giving, however, the prescribed flow rate (resulting from velocity $u_{x,B}$) at each cross section. The test simulations have shown, that the PDF distribution of the velocity vector orientation calculated according to the procedure described in the present section fits the distribution received with the use of VOF simulation presented in section 2.1 with very good agreement.

2.3. *Model geometry and boundary conditions*

The numerical model of liquid spreading in the porous zone was applied to simulate the flow in a cylindrical column of diameter 0.1m containing packed bed section 0.5m high and filled with 6mm Raschig rings. The column was supplied with 15l/h water flux delivered by 5mm pipe located 50mm above the packing section at column axis. For the present simulation gas was assumed not to flow. Such a model geometry reflects the existing test rig currently being used for the corresponding experimental trials. The inlet and outlet boundary conditions were applied at the bottom and the top of the column respectively (see figure 2a) allowing gas to flow up. The liquid was supplied to the column with the use of the source boundary condition and then it was driven by gravity flowing countercurrently to the gas phase and leaving the column at its bottom. The non-slip boundary condition was applied at column wall. The computational domain was discretized with unstructured mesh with local refinement reflecting the regions of excessive gradients of flow parameters. Series of simulations was conducted for meshes of different sizes and the final mesh of appr. 0.5mln cells was accepted as providing the grid independent solution.

ANSYS FLUENT version 13 was used as a solver of the flow governing equations and ANSYS GAMBIT for geometry and mesh generation. Second-order discretization schemes were used to solve the flow equations. The SIMPLE algorithm was applied for the pressure-velocity coupling. The algorithm for the liquid spreading process as well as the phase interaction in porous zone was implemented in a form of user defined functions, i.e. the own-developed subroutines linked to the solver. The rigorous criteria were used to ensure the full convergence of the solution. All the residua of governing equations had to fall down below the 10^{-4} level. Moreover, the liquid volume fraction at various cross sections was monitored to show whether the flow in a porous region has reached the steady state.

3. Results

The analysis of simulation results was started from the qualitative insight into the flow spreading process. For that purpose the magnified views of the flow structure in form of velocity vector chart and the volume fraction contour map are presented in figure 2b and figure 2c, respectively. As can be seen from the instantaneous velocity vector map (see figure 2b), the orientation angle varies in a wide range giving impression of chaotic motion satisfying, however, the distribution described by equation (3). The chaotic nature of the liquid motion is much less pronounced if the volume fraction distribution (see figure 2c) is analyzed in the corresponding region. White colour corresponds to no gas present in the mixture while black one to pure gas. One can easily notice, that the chaotic nature of the flow disappears as the volume fraction distribution depends on the time history of the liquid redistribution process, leading finally to the spreading of the jet-like flow with easily noticeable irregularities. The highest volume fraction values occur just downstream the entrance to the porous region and moving downstream its values decrease as liquid is redistributed in transverse directions. The results collected in figure 2 indicate the reasonable behaviour of liquid in a randomly generated porous region and proves the relevance of the spreading model implemented in Eulerian approach.

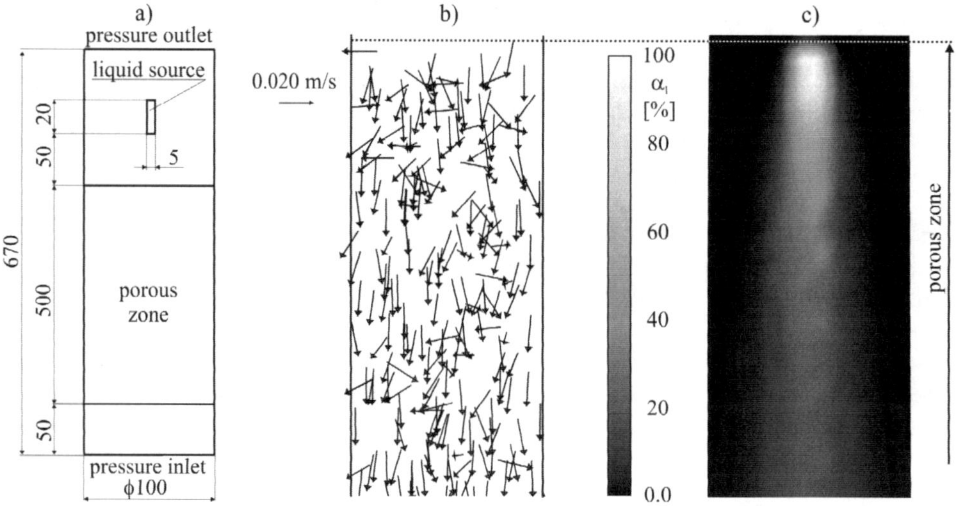

Figure 2. Model geometry and boundary conditions (a), instantaneous velocity vector map (b) and corresponding liquid volume fraction distribution (c).

In order to analyze the liquid spreading process the flow statistics have to be taken into account to reduce irregularities resulting from random orientation of packing elements. As the most suitable parameter the liquid volume fraction was selected and the corresponding distributions are shown in figure 3 as a system of isolines (in figure 3a) and in the form of radial profiles for selected cross sections (see figure 3b). The data in figure 3a are presented for upper half of the packing section and numbers correspond to the liquid volume fraction expressed as a percentage. As can be seen the most intense spreading is observed just downstream the top edge of the porous region what is manifested by high radial gradient of volume fraction $\partial\alpha_l/\partial r$. In the following cross sections i.e. for increasing x-coordinate the spreading rate is gradually decreasing as can be deduced from the divergence of α_l

isolines. It should be remarked, that the volume fraction distribution shown in figure 3a corresponds to particular vertical cross section and that is why the view is not symmetrical. As a complement to figure 3a the radial profiles of liquid volume fraction for a number of cross-sections (located at dimensionless coordinate x/h where h is the total height of the packing) are presented in figure 3b. The data shown here were averaged not only in time but also in circumferential direction. The volume fraction profile at the cross section closest to the packing top (x/h=0.14) indicates the existence of a core flow resulting from the small-diameter liquid jet entering the packed bed. The maximum volume fraction value of 22.5% confirms, that the most efficient spreading rate occurs in the initial region of the packed bed as it decreased from the initial level of 100% at (at x/h≈0). The core flow is surrounded by a region (r/D≥0.25) with a negligible presence of liquid (α_l <3%). Volume fraction profiles at consecutive cross sections (displaced by constant interval) become more and more uniform. The decrease of volume fraction maximum values is due to the gradual reduction of the spreading rate in radial direction. At the last analyzed cross section (x/h=0.9) the volume fraction at the column axis is appr. 6.5% while at the column wall it reaches appr. 3%. It shows that the packing section 0.5m high and of 0.1m diameter is not able to uniformly redistribute the liquid in cross section. One should note, that the previous statement is valid for the considered packed bed configuration only and has no general character.

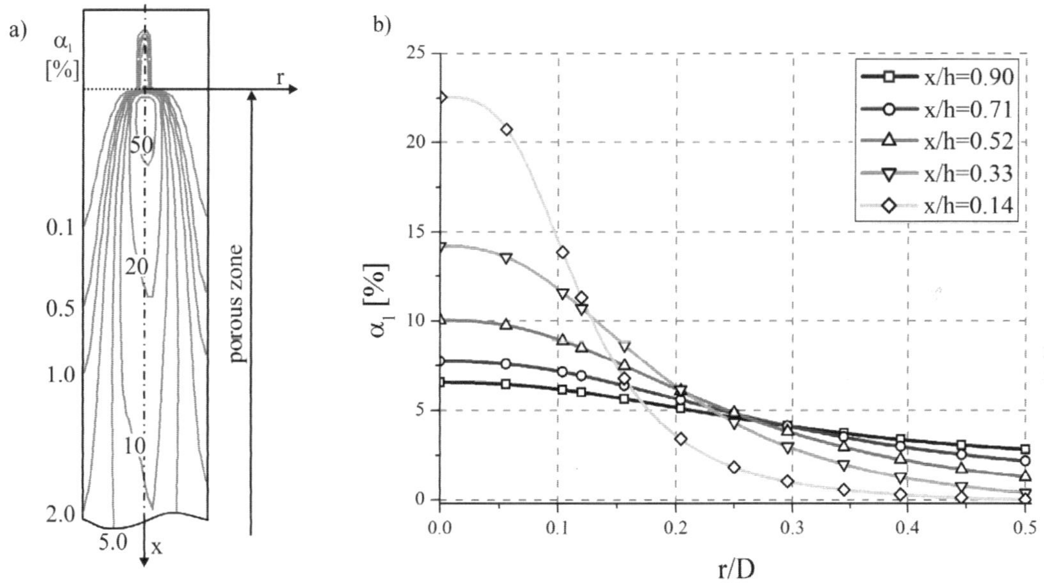

Figure 3. Averaged liquid volume fraction distribution in a packed bed (a) and its radial profiles for selected cross sections (b).

4. Summary

The paper presents the combined VOF-Eulerian two-fluid modelling strategy applied to the two-phase flow in an unstructured packed bed composed of 6mm Raschig rings. As the first step, the detailed geometry of porous region was considered allowing to recover the detailed flow structure with the use of conventional conservation laws and the liquid free surface tracking method (VOF). The flow statistics obtained in this way allowed to determine the probability density function of the velocity vector orientation which was then implemented into the Eulerian-Eulerian model in ANSYS FLUENT environment to simulate the flow in a column with the packed bed. The test simulations have shown the reasonable model behaviour and the flow redistribution which followed the VOF approach with high precision. The results of computations provided an insight into the liquid spreading process along and across the packing section. The spreading rate has been found to be large at the top of porous zone and rapidly decreasing downstream and towards the wall. The Eulerian two-fluid approach developed in present paper will be used in future work for the parametric analysis of the packing beds providing the characteristics needed for their optimisation. In further research work the model will be incorporated into the carbon capture and storage (CCS) model including additionally carbon dioxide capture chemistry, heat transfer and phase change phenomena [21]. The reliable and physically based model of liquid spreading in the packed bed is a crucial point to simulate such a complex physio-chemical system in an adequate way.

5. References

[1] Maćkowiak J 2010 Fluid Dynamics of packed columns, *Springer-Verlag Heidelberg*
[2] Fourati M, Roig V and Raynal L 2012 Experimental study of liquid spreading in structured packings, *Chem. Eng. Sci.* **80** 1–15
[3] Luo S, Li H, Fei W and Wang Y 2009 Liquid film characteristics on surface of structured packing *Chin. J. Chem. Eng.* **17**(1) 47–52
[4] Aoda N 2010 Visualisation of flow pattern and measurement of liquid distribution in a random packed column using electrical resistance tomography (ERT) *MSc Thesis, Ryerson University*
[5] Wang Y, Chen J and Larachi F 2013 Modelling and simulation of trickle-bed reactors using computational fluid dynamics. A state-of-the-art review *Can. J. Chem. Eng.* **91** 136–180
[6] Raynal L, Boyer C and Ballaguet J P 2004 Liquid holdup and pressure drop determination in structured packing with CFD simulations *Can. J. Chem. Eng.* **82** 871-879
[7] Mahr B and Mewes D 2007 CFD modelling and calculation of dynamic two-phase flow in columns equipped with structured packing *Chem. Eng. Res. Des.* **85**(A8) 1112–1122
[8] Hosseini S H, Shojaee S, Ahmadi G and Zivdar M 2012 Computational fluid dynamics studies of dry and wet pressure drops in structured packings *Ind. Eng. Chem. Journal* **18** 1465–1473
[9] Zhang X, Yao L, Qiu L and Zhang X 2013 Three-dimensional computational fluid dynamics modeling of two-phase flow in a structured packing column *Chin. J. Chem. Eng.* **21**(9) 959-966
[10] Sun B, He L, Liu B T, Gu F and Liu C J 2013 A new multi-scale model based on CFD and macroscopic calculation for corrugated structured packing column, *AIChE Journal* **59**(8) 3119–3130
[11] Xu Y, Paschke S, Repke J U, Yuan J and Wozny G 2008 Portraying the countercurrent flow on packings by three-dimensional computational fluid dynamics simulations *Chem. Eng. Technol.* **31**(10) 1445–1452
[12] Xu Y, Yuan J, Repke J-U and Wozny G CFD study on liquid flow behaviour on inclined flat plate focusing on effect of flow rate *Engineering Applications of Computational Fluid Mechanics* **6**(2) 186–194
[13] Asendrych D, Niegodajew P and Drobniak S CFD modelling of CO_2 capture in a packed bed by chemical absorption *Chem. Proc. Eng.* **34**(2) 269-282
[14] Sun C G, Yin A, Afcan A, Nandakumar K and Chuang K T 2000 Modelling and simulation of flow maldistribution in random packed columns with gas-liquid countercurrent flow *Trans IChemE* **78** 378–386
[15] Marek M 2013 Numerical generation of a fixed bed structure *Chem. Proc. Eng.* **34**(3) 347–359
[16] Marek M, Aniszewski W and Bogusławski A 2002 Simplified VOF for two-phase flows *Task Quarterly* **12**(3) 255-265
[17] Peskin C S 2002 Immersed boundary method *Acta Numerica* 479–517
[18] Wieczorkowski R and Zielinski R J 1997 Random numbers generators (in Polish), Wydawnictwo WNT, wyd 1
[19] Billet R 1995 Packed towers in processing and environmental technology *VCH Verlagsgesellschaft mbH, Weinheim*

[20] Niegodajew P, Asendrych D and Drobniak S 2013 Numerical analysis of CO_2 capture efficiency in post combustion CCS technology in terms of varying condition *Archives of Thermodynamics* **34**(4) 69–82

[21] Niegodajew P, Asendrych D and Drobniak S 2013 Numerical modelling of CO_2 desorption process coupled with phase transformation and heat transfer in CCS installation *Journal of Power Technologies*, **93**(5) 354–361

Acknowledgements

The results presented in this paper were obtained from research work co-financed by the National Centre of Research and Development in the framework of Contract SP/E/1/67484/10 - Strategic Research Programme - Advanced technologies for energy generation: Development of a technology for highly efficient zero-emission coal-fired power units integrated with CO_2 capture.

Project DoktoRIS is acknowledged for providing the scientific scholarship to Mr. P. Niegodajew.

PL-Grid Infrastructure is gratefully acknowledged for providing its powerful computational resources.

Journal of Physics: Conference Series **530** (2014) 012054 doi:10.1088/1742-6596/530/1/012054

Flow resistance of ice slurry in bends and elbow pipes

B Niezgoda-Żelasko[1] J Żelasko
Institute of Thermal and Process Engineering, Cracow University of Technology, al. Jana Pawła II 37, Cracow 31-864, Poland

E-mail: bniezgo@mech.pk.edu.pl

Abstract. The present paper covers the flow of ice slurry made of a 10.6% ethanol solution through small-radius bends and elbow pipes. The paper presents the results of experimental research on the flow resistances of Bingham-fluid ice slurry in bends and elbows. The research, performed for three pipe diameters and a relative bend radius of $1 \leq D/d_i \leq 2$, has made it possible to take into consideration the influence of friction resistances as well the of the flow geometry on the total local resistance coefficients. The study attempts to make the local resistance coefficient dependent on the Dean number defined for a generalized Reynolds number according to Metzner-Reade

1. Introduction

Ice slurry belongs to a group of secondary refrigerants. It is a mixture of ice crystals and either water, or water containing a dissolved substance which lowers the freezing point (salt, glycol, alcohol). Ice slurry is a non-Newtonian fluid. The rheological models most frequently assigned to this fluid are the Ostwald-de Waele power model, the Bingham model and the Casson model [1-3].

Problems with determining flow resistances for Newtonian and non-Newtonian fluid flow through curved pipes can be divided into two groups. In the case of small-radius bends (D/d_i<6) [4] both friction- resistances and local resistances generated by a local change in flow geometry contribute to the drop in pressure. For D/d_i<6, Idielcik [4] suggests that the total resistance coefficient of elbows and bends be sum of the coefficient of local resistance of the bend and the friction coefficient. In this case, the local resistance coefficient is only a function of the geometrical parameters of the bend (angle, relative bend diameter). In [5] Ito proposes that a total local resistance coefficient be determined for the flow of non-Newtonian fluids through strongly curved pipes. For small Dean numbers (De<91), Ito makes the total local resistance coefficient directly dependent on the friction resistance coefficient and flow geometry. For Dean numbers of De>91, the local resistance coefficient is a function of the Reynolds number and the geometric parameters of the bend.

The other type of issues related to the flow of non-Newtonian fluids through curved pipes concerns their flow through spirally curved pipes and large-diameter bends. Ito [4], [6] considers the pressure drops in coil pipes and large-diameter bends ($D/d_i \geq 6$) to be dependent on the friction resistance coefficient, which is a function of the Dean number defined differently for laminar and turbulent flow. The basic works concerning the flow of non-Newtonian fluids through curved pipes include, among others, [7-9], which highlight the results of research on pipe coils. The paper by Mishra and Gupta [8] presents a generalization of the results of research on friction resistance coefficients for Ostwald-de

[1] Corresponding author

Waele fluids. The results of research presented in [9] concern the flow resistances of a Bingham fluid in coil pipes. The generalized formula for the friction resistance coefficient takes on the form of the Ito relationship [6], corrected by implementing a correction function in the Dean number definition. [7] lists the results of research on flow resistances in curved circular and elliptical ducts. An analysis of the results of research covered in this paper indicates that various authors determine the friction resistance coefficient in curved pipes by means of generalizing the Ito formula. The modification of this relationship involves the implementation of correction functions both in the general form of the Ito formula and in the Dean number definition.

Table 1 presents a review of literature of slurry flow through fittings of [1], [3], [10-12]

Table 1. Review of literature of slurry flow trough fittings

suspension	d_i (m)	type of bends	v_m (ms^{-1})	x_s (%)	Literature, notes
Ice slurry of 16%propylen water glycol solution	0,021; 0,027	90° bend	0,5-1,5	$0 \le x_s \le 30$	[10-11] The data included in the study have been presented in graphic form and have not been generalized.
Ice slurry of 10-12% water solution of ethyl alcohol	0,015	90° bend	0,2-0,9	$0 \le x_s \le 25$	[1] A lack of generalizations or information on bend radius makes it difficult to use the results.
3,6-12,7% laterite suspension	0,025-0,5	45°,90°,180° bends D/d_i=0-25	0,5-~16	3,9-12,7	[12] The relationships suggested in the aforementioned work are quite peculiar, as they present an individualized form of the formulas for the total local resistance coefficient for each D/d_i relation.
10,7-30% gypsum suspension			~0,5-~14	10,7-30	

The present paper concerns the flow of ice slurry made up of a 10.6% ethanol solution through bends and elbow pipes. It contains the results of experimental research on flow resistances of the slurry through bends and elbow pipes in laminar and turbulent flow regime. The result of experimental research is a series of dimensionless relationships describing the local resistance coefficient for bend and elbow pipes and for ice slurry being a Bingham fluid, whose flow is treated as a generalized flow of a non-Newtonian fluid.

2. Measurement of ice slurry flow resistance in small-radius bends

The research programme concerning flow resistances in bends and elbow pipes included measurements for:

- Commercially available standard 90° bends and elbow pipes with diameters of d_i = 0,01; 0,016; 0,02 (m), with the bend diameter-pipe diameter ratio of D/d_i = 2,0 for bends and D/d_i = 1 for elbow pipes.
- Mean flow velocities of $0.1 \le v_m \le 4.5$ (ms^{-1}), with the corresponding Reynolds number values of 200 < Re < 13 700,
- Mass fraction of ice in the slurry of $0 \le x_s < 30\%$.

The study of ice slurry flow encompassed a rheological identification of the ice slurry [13], measurements of pressure drops of the ice slurry in straight pipes [3], [13] as well as measurements of flow resistances in bends and elbow pipes. A detailed description of the measurement stand (figure 1A) has been provided in [2-3] and [13]. The measurement of ice slurry temperature is based on data obtained from highly precise resistance sensors with large dimensions Pt100(7013) (HART Scientific, sensor mantle diameter – ca. 5 mm, active length – min. 0,02m). The flux of the ice slurry has been measured by means of a Danfoss MASSFLO 2100 mass flow-meter. Pressure drops have been measured using Fuji FKCV pressure-change transducers, with two measurement ranges: 0–32 kPa and 0– 1 kPa (uncertainty: 0,07% of the measurement range).

The entire measurement system (figure 1B) used was composed of a straight-axis inlet and outlet sections (L_{in}, L_{out}), as well as a piece of fitting (bend, elbow). In order to determine the pressure drop in bends and elbows, the adopted method was similar to the one presented in [12]. Pressure drop measurements encompassed: the total pressure drop Δp_T in a measurement element, as well a friction-induced pressure drop in the straight-axis (Δp_L) measurement sections with a length of L. Pressure drops in a bend or an elbow were calculated using the following formula:

$$\Delta p_{EK} = \Delta p_T - \Delta p_L \frac{L_{in} + L_{out}}{L} \tag{1}$$

A)

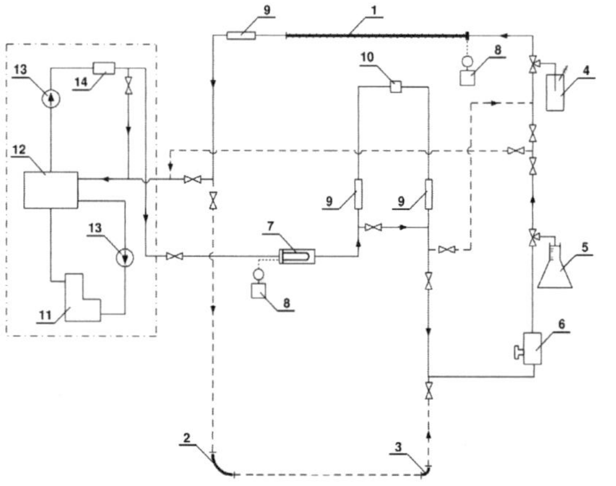

B)

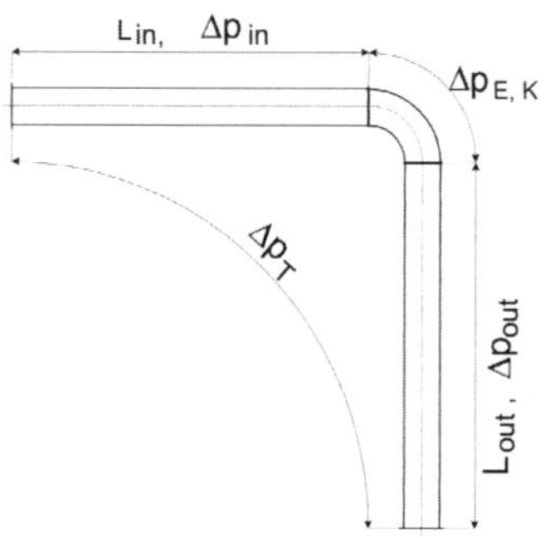

Figure 1. A) Schematic diagram of the test stand: 1-heated measuring segment, 2-elbow, 3- bend, 4-calorimetric measurement, 5-measurement of: density, volume change, ice and air content, 6-mass flow-meter, 7-heater, 8-wattmeter and autotransformer, 9-flow visualization, 10-air-escape, 11-ice generator, 12-accumulation container, 13-pump, 14-volume flow-meter; B) Outline of a measurement stand for investigating flow resistance in bends and elbows

Regardless of pipe diameter, the lengths of the inlet (L_{in}) and outlet (L_{out}) sections used in the measurement of pressure drops in bends and elbows were ten times the diameter of the pipe. It has to be noted that the entire measurement system was preceded and followed by straight pipes, which stabilized the flow of the agent.

Figure 2 presents the variation of pressure drops in the investigated bends (D/d_i=2) and elbows (D/d_i=1), for a pipe diameter of d_i=0,02 (m). The results of the measurements indicate that for $x_s \leq 10\%$, the influence of the mass fraction of ice on pressure drops both in bends and in elbows is negligible. For pipe flow of ice slurry with a mass fraction of ice of $x_s \geq 10\%$, there are velocities at which the flow resistance of the slurry is lower than the flow resistance of the carrying liquid [2-3], [13]. Ice slurry flow through bends and elbow pipes reduces the occurrence of this phenomenon in the examined fittings, leading to a situation in which, regardless of the type of movement, for the mass fraction of ice of $x_s \geq 10\%$, pressure drops in bends and elbows are always higher than the corresponding pressure drops of the carrying liquid. (Figure 2). For low values of ice content, the flow of the ice slurry is similar to the flow of a Newtonian fluid [2], [3], [13]. In the case of flow through elbow pipes, a greater pressure drop than in the case of a bend results from the fact that greater flow disturbances occur in the elbow than in the bend pipe. For mass fractions of ice of x_s>15% a significant change in the rheological properties of the fluid occurs. Ice crystals absorb part of the kinetic energy of the turbulences of the carrier liquid, muffling the turbulences occurring in the elbow pipe. This phenomenon results in the fact that the differences in pressure drops between the elbow and bend pipe for higher ice content values (x_s>15%) are lower than for an ice content of $x_s \leq 15\%$.

A)

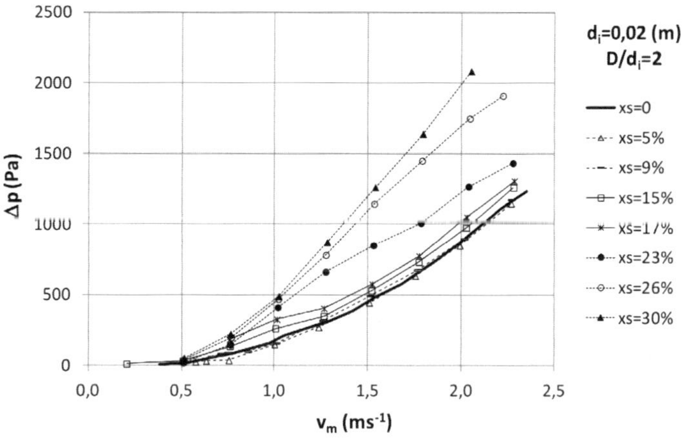

B)

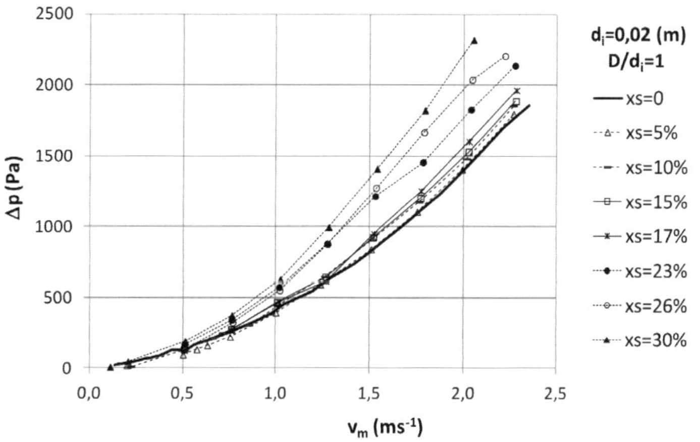

Figure 2. Flow resistance (Δp) as a function of the mean velocity of the slurry for A) a bend; $d_i = 0{,}02$ (m); $D/d_i = 2$; B) an elbow; $d_i = 0{,}02$ (m); $D/d_i = 1$.

Figure 3 presents the influence of pipe diameter, the mass fraction of ice and the Reynolds number on pressure drops (Figure 3a) and total flow resistance (Figure 3b) in elbows. An increase of flow resistances along with the increase of ice content results from the higher values of the dynamic coefficient of plastic viscosity μ_p and increased values of yield shear stress τ_p for the ice slurry. Similarly to Newtonian fluids, higher pressure drop values correspond to smaller flow cross-sections.

For a Newtonian fluid flow through curved pipes, Ito's empirical formulas are the ones most frequently quoted: [4], [6-7], and [12]. Flow resistances in bends and elbows are caused both by friction resistance and resistance caused by flow disturbances as a result of the change of the direction of the flow of the medium. In this paper, the components of flow resistances are not determined separately. The calculated value of the resistance coefficient, similarly to paper [5] is the total local resistance coefficient of the medium flowing through a bend or elbow:

$$\xi_{E,K} = \frac{2\Delta p_{E,K}}{\rho_B v_m^2}. \tag{2}$$

This method of determining the total local resistance coefficient is prone to the smallest measurement error. For $D/d_i<6$, the total local resistance coefficient is dependent on the Dean number ($\xi=f(\mathrm{De})$), which for laminar and turbulent flow is defined by formulas (3) and (4), respectively:

$$\mathrm{De}_L = \mathrm{Re}\sqrt{\frac{d_i}{D}}\,, \tag{3}$$

$$\mathrm{De}_T = \mathrm{Re}\left(\frac{d_i}{D}\right)^2. \tag{4}$$

A)

B)

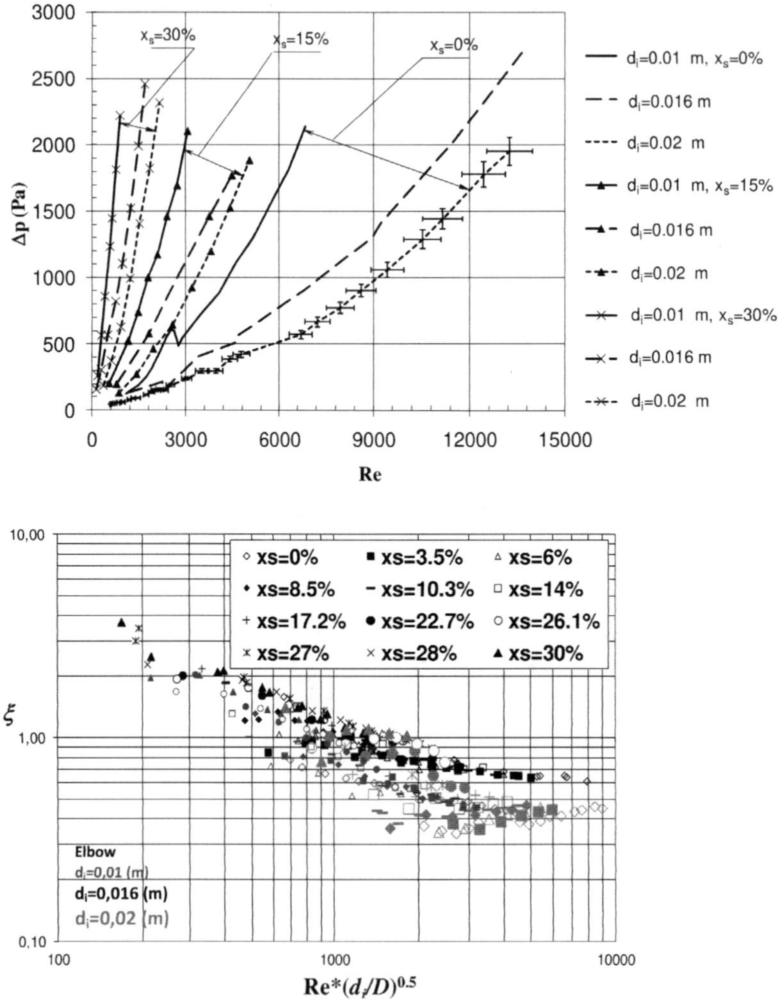

Figure 3. A) Influence of a generalised Reynolds number, pipe diameter and ice content on pressure drops in a elbows ($D/d_i=1$); B) Dependency total flow resistance on the generalised Dean number for flow through elbows

In this case it has been assumed that the Reynolds number (Re) is a generalised Reynolds number according to Metzner-Reade:

$$\mathrm{Re} = \frac{\rho_B v_m^{2-n} d_i^n}{8^{n-1} K}\,. \tag{5}$$

The generalised parameters n (characteristic flow behaviour index) and K (consistency index) depend on the rheological properties of the fluid, the mean shear stress at the wall, mean flow velocity, as well as on the dimensionless geometrical constants determined individually for various flow geometries of vertical flow cross-sections. For Bingham fluids and any cross-section geometry, these flows are calculated on the basis of the following formulas [14]:

$$n = \frac{\frac{1}{4}\left[1-\varepsilon_B{}^4 - \frac{4}{3}\varepsilon_B\left(1-\varepsilon_B{}^3\right)\right]}{1-\frac{3}{4}\varepsilon_B\left[1-\varepsilon_B{}^4 - \frac{4}{3}\varepsilon_B\left(1-\varepsilon_B{}^3\right)\right]}, \tag{6}$$

$$K = \left(\frac{1}{4}\mu_p\right)^n \tau_w^{1+3n}\left[\frac{1}{4}\tau_w^4 - \frac{1}{3}\tau_w^3\tau_p + \frac{1}{12}\tau_p^4\right]^{-n}. \tag{7}$$

in which values of yield shear stress τ_p and dynamic coefficient of plastic viscosity μ_p for ice slurry made up of a 10.6% ethanol solution are calculated on the basic equations (8) and (9) [2-3], [13].

$$\tau_p = 0.013 - 1.4284\left(\frac{x_s}{100}\right) + 73.453\left(\frac{x_s}{100}\right)^2 - 394.64\left(\frac{x_s}{100}\right)^3 + 835.82\left(\frac{x_s}{100}\right)^4 \tag{8}$$

$$\mu_p = 0.0035 + 0.0644\left(\frac{x_s}{100}\right) - 0.7394\left(\frac{x_s}{100}\right)^2 + 5.6963\left(\frac{x_s}{100}\right)^3 - 19.759\left(\frac{x_s}{100}\right)^4 + 26.732\left(\frac{x_s}{100}\right)^5 \tag{9}$$

The τ_w and ε_B in equations (6-7) correspond to the shear stress at pipe wall and the quotient of active shear stresses of a Bingham fluid ($\varepsilon_B = \tau_p/\tau_w$), respectively.

3. Summary

Both the results of research presented in [12] and the authors' own measurements of the flow resistance coefficient for ice slurry in bends and elbows (figure 3B), indicate that the substitute local resistance coefficient ξ depends on the Dean number and the relative bend diameter (D/d_i). When determining the formula for the local resistance coefficient for bends and elbows, it has been assumed that the required formula will correspond with the same relationship for bends and elbows ($1 \leq D/d_i \leq 2$), which has the following form:

$$\xi = \xi\left(De\,;\, d_i/D\right). \tag{10}$$

Moreover, the research on flow resistances in bends and elbows concerned mainly the laminar flow range. Outside the laminar range, the flow occurred mainly in the transitional flow regime. Only for pipe diameters of $d_i = 0{,}016$ and $0{,}02$ (m), part of the measurement points are located within the turbulent flow range. The following formula determines the relationship describing the local resistances coefficient in laminar flow regimes for a generalized flow of ice slurry through bends and elbows (11):

$$\xi\left(De_L\,;\, d_i/D\right) = \frac{4.6\left(\frac{d_i}{D}\right)^{0.33}}{\left[0.87 + 0.1 * \log(De_L)\right]^{8.1}}. \tag{11}$$

In the turbulent flow range, the substitute local resistance coefficient for a generalised flow of ice slurry can be derived from the following formula:

$$\xi\left(De_T\,;\, d_i/D\right) = \frac{\left(\frac{d_i}{D}\right)^{-0.306}\left[10.6 - 3.45\Big/\left(\frac{d_i}{D}\right)\right]}{15.2 De_T^{0.204}}. \tag{12}$$

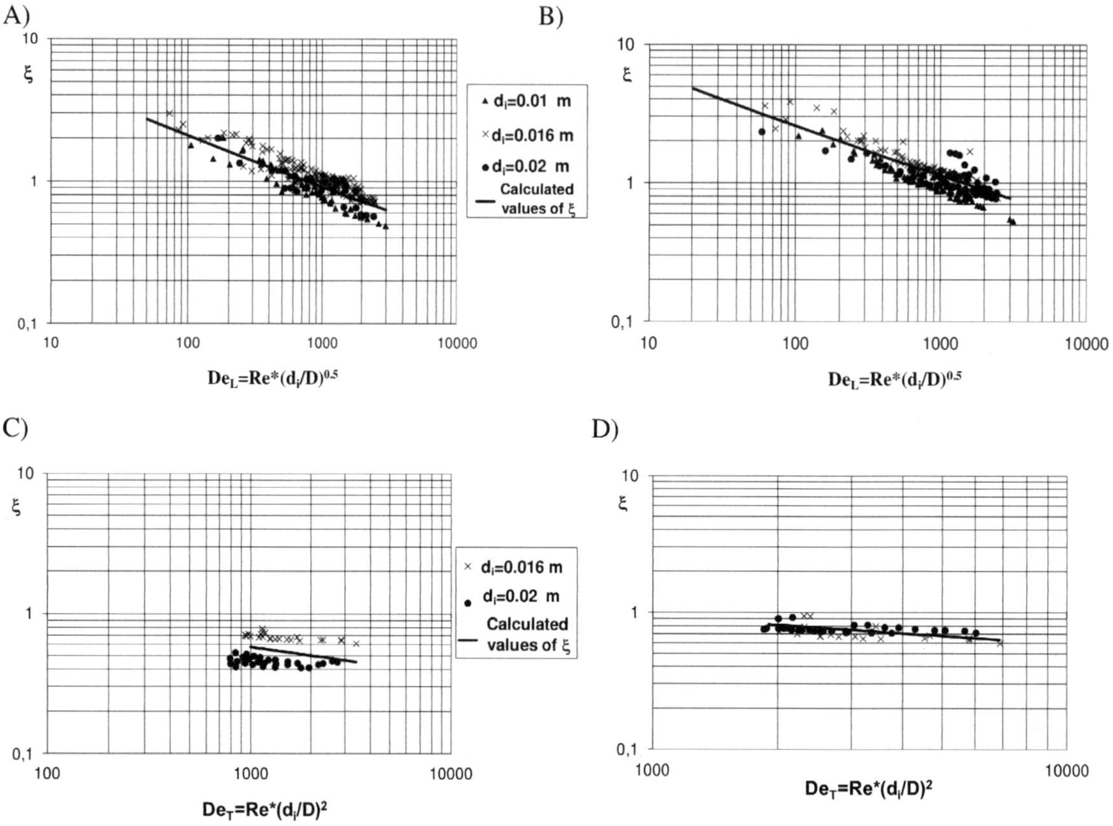

Figure 4. Measured and best-fit curves of the total local resistance coefficient: A) laminar flow through bends, B) laminar flow through elbows, C) turbulent flow through bends, D) turbulent flow through elbows.

It needs to be noted that the superposition of the phenomena occurring within the inlet and outlet sections, as well as in the bent element itself, causes instability of the pressure signal, impeding the observation of, e.g. the nature of the changes in ice slurry movements. In this case, stabilization of the total local resistance coefficients for laminar flow has been assumed as the criterion of the movement change. It has been assumed that the Dean number, $De_{CL} = Re\left(d/D\right)^{0.5} = 2500$, is in this case the criterion of the change in the character of the movement for ice slurry flow through bends and elbows.

Figures 4 A, B and 4 C, D present the measured and best-fit curves of total local resistance coefficients for ice slurry in bends and elbows for laminar and turbulent flow. With a 98% probability it can be assumed that the confidence intervals for formulae (11) and (12) equal 14% and 15%, respectively.

Figure 4 indicates that it might be interesting to attempt to improve the quantitative accuracy of formulae (11) and (12) with the results of experimental studies by making the constants in formulae (11) and (12) dependent on the inner pipe diameter. These problems may be further analysed by the authors in future studies.

4. Reference

[1] Bel O and Lallemand A 1999 Etude d'un fluida frigoporteur disphasique 1: Characteristiques thermophysiques intrinseques d'un de glace *Int. J. Refrig.* **22** 164-74

[2] Niezgoda-Żelasko B and Żelasko J 2007 Generalized non-Newtonian flow of ice-slurry *Chem. Eng. and Process.* **46** 895-904
[3] Niezgoda-Żelasko B 2006 *Wymiana ciepła i opory przepływu zawiesiny lodowej w przewodach. Monografia 334 (*Kraków: Wyd. PK)
[4] Idielcik IE 1994 *Handbook of Hydraulic Resistance* Crc (London: London Press)
[5] Ito H 1960 Friction factors for turbulent flow in curved pipes *J. Basic Eng. Trans ASME* **82 D** 131-45
[6] Ito H 1959 Pressure losses in smoth pipe bends *Journal Basic Eng. Trans ASME* **81 D** 123-34
[7] Matras Z 2001 *Transport hydrauliczny reologicznie złożonych cieczy nienewtonowskich w przewodach* (Kraków: Wyd. PK)
[8] Mishra P and Gupta S W 1979 Momentum transfer in curved pipes.2: Non-Newtonian fluids *Ind. Eng. Chem. Process Des. Dev.* **18** 137-42
[9] Tada T Fukui Y Oshima S and Yamane R 1994 Effects of Bingham viscosity on flow in curved pipes *JSME Int. J.* **37** 322-7
[10] Nørgaard E Sørensen T A Hansen T M and Kauffeld M 2005 Performance of components of ice slurry systems: pumps, plate heat exchangers and fitting *Int. J. Refrig.* **28** 83-91
[11] Nørgaard E Sørensen T A Hansen T M and Kauffeld M 2001 Performance of components of ice slurry systems: pumps, plate heat exchangers and fittings. *Proc. of the 3rd IIR, Workshop on Ice Slurries* Lucerne 129-36
[12] Turian R M Ma T W , Hsu F L , Sung D J. and Plackmann G W 1998 Flow of concentrated non-newtonian slurries: 2Friction losses in bends, fittings, Valves and Venturie meters *Int. J. Multiph. Flow* **24** 243-69
[13] Niezgoda-Żelasko B and Zalewski W 2006 Momentum transfer of ice slurries flows in tubes. Experimental investigation *Int. J. Refrig.* **2** 418-28
[14] Kozicki W Chou C H Tiu C 1966 Non-Newtonian flow in ducts of arbitrary cross-sectional shape *Chem. Eng. Sci.* **21** 665-79

Bulge structure in a cavity swept by a viscoelastic fluid

H Suzuki[1*], H Sato[1], R Hidema[2] and Y Komoda[1]
[1]Department of Chemical Science and Engineering, Kobe University, Kobe, Japan
[2]Organization of Advanced Science and Technology, Kobe University, Kobe, Japan

E-mail: hero@kobe-u.ac.jp

Abstract. In order to study a bulge structure observed in a cavity swept by a viscoelastic fluid, flow visualization experiments have been performed. The cavity depth, the cavity length and the width of flow path were fixed while the water Reynolds number based on the narrow flow path was changed from 680 to 4,200. From the results, the bulge structure formed on the upstream backward-facing side wall of the cavity in the mid-range of the Reynolds number was found to lead a fresh fluid into a cavity and to sweep the fluid near the cavity bottom wall. The bulge structure fluctuates very slowly and this fluctuation induces a tonguing motion of the tip of the separation bubble formed in the downstream region of the cavity. Thus, this structure can be expected to enhance the heat transfer from the cavity bottom.

1. Introduction

In order to expand the heat transfer area in a heat exchanger, rib-like promoters are often mounted on a heat transfer wall. Classically, Han *et al.* [1] and Han [2] investigated the heat transfer augmentation methods with rib-like promoters. Such technologies are still important for a compact heat exchanger and then lots of studies have been performed by such as Taslim [3] and Murata *et al.* [4]. Liu *et al.* [5] recently summarized such heat transfer enhancement technologies. However, thermal accumulation occurs in a cavity between ribs because of the recirculating flow formation when a Newtonian fluid is used, and then heat transfer from the cavity bottom is severely reduced. Aung [6] investigated the laminar heat transfer characteristics behind a backward-facing step and reported the heat transfer reduction occurs in a recirculating region formed there. For improving such heat transfer reduction in a recirculating region, some techniques as using flow resonance [7], the flow pulsation [8] and the insertion of an immersed body [9] were suggested.

Suzuki *et al.* [10] suggested a technique using the Barus effect of a viscoelastic fluid in order to enhance the heat transfer from the cavity bottom. Viscoelastic fluids show a flow expansion at a sudden expanded flow path and sweep the cavity bottom. This phenomenon is called Barus effect. In the previous studies [11,12], flow and heat transfer characteristics in a cavity were investigated experimentally and numerically. From the results, it was found that the heat transfer from the cavity bottom is effectively enhanced by the Barus effect though its characteristics severely depend on the geometric parameters, and Reynolds and Weissenberg numbers.

*Corresponding author: H Suzuki
Phone: +81-78-803-6490, Fax: +81-78-803-6490
E-mail address: hero@kobe-u.ac.jp

In the previous report [13], a strange fluid motion was found in a cavity in a mid-range of the Reynolds number around 2,500. There, the flow penetrates into the cavity turns back toward the upstream backward-facing side wall of the cavity, and after this, the flow turns again toward the downstream forward facing side wall in the deeper region of the cavity and sweeps the fluid near the cavity bottom. Thus, a bulge-like structure is formed as attached to the upstream backward-facing side wall of the cavity. This fluid motion can be expected to enhance the heat transfer from the cavity bottom because the velocity gradient on the bottom wall is kept high and the high temperature fluid heated by the cavity walls is swept out of the cavity.

In this paper, such bulge like motions observed in the cavity by use of a viscoelastic fluid will be focused on. It affects the fluid motion of the separation bubbles formed in the downstream region in the cavity, where the heat transfer is reduced as reported in the previous study [10]. Flow visualization experiments and velocity field measurements by a particle tracking velocimetry have been performed. The solvent Reynolds number is changed from 680 to 4,200 while the geometry of the cavity is fixed. The Reynolds number range of the appearance of the bulge flow structures and the instability of the separation bubbles will be discussed.

2. Experimental procedures

2.1. Materials

A cationic surfactant (oleylbishydroxyethylmethylammonium chloride: $C_{18}H_{35}N(C_2H_4OH)_2CH_3Cl$) was used for adding viscoelasticity to the fluid. The surfactant was dissolved in ion-exchanged water at a concentration of 2,000 ppm. Sodium salicylate ($C_6H_4(OH)COONa$), as a counter-ion supplier for rod-like micelle formation, was also added to the solution. The molar ratio of the counter-ions to the surfactants was set at 1.5. The solution was allowed to equilibrate for at least one day prior to any experiments.

2.2. Rheological features

The apparent viscosity and the first normal stress difference of the fluid were measured by use of a shear-control rheometer (Rheometric Scientific, ARES) with a cone-and-plate device with the diameter of 50 mm. Figures 1 and 2 show the apparent viscosity and the first normal stress difference of the present system, respectively. The present system has high viscosity in the region of low shear rates, but it shows a steep shear thinning behaviour with shear rates. In the range of high shear rates, the first normal stress shows very high values. Thus, this system has high viscoelasticity.

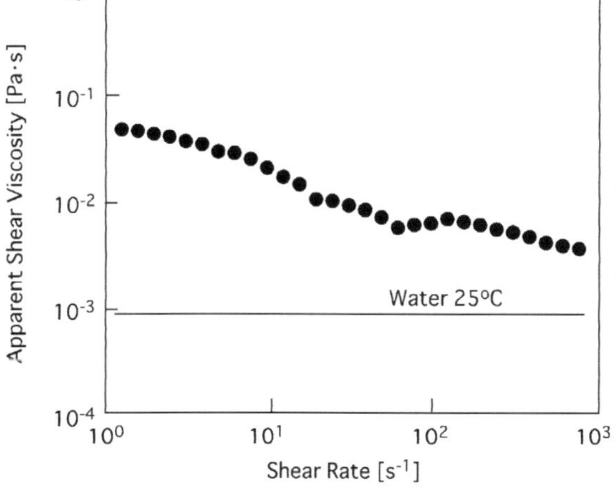

Figure 1. Apparent shear viscosity.

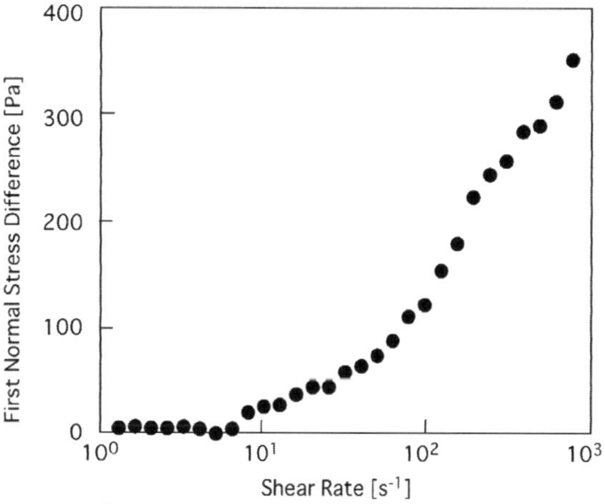

Figure 2. The first normal stress difference.

Suzuki *et al.* [14] measured the relaxation behaviour of the surfactant system with a triple exponential Maxwell model as follows.

$$\tau = \tau_1 e^{-t/t_{r1}} + \tau_2 e^{-t/t_{r2}} + \tau_3 e^{-t/t_{r3}} \tag{1}$$

Here, τ Pa is the total stress and τ_1 Pa, τ_2 Pa and τ_3 Pa are the contributions from the respective relaxation behaviours to the total stress. t s is time and t_{r1} s, t_{r2} s and t_{r3} s are the relaxation times for the respective relaxation behaviours defined as $t_{r1} < t_{r2} < t_{r3}$. The three relaxation times of the present system is tabulated in table 1.

Table 1. Relaxation times of the present system [14].

t_{r1}	t_{r2}	t_{r3}
9.96×10^{-2} s	7.77×10^{-1} s	9.94 s

2.3. Flow visualization experiments

Figure 3 shows the experimental apparatus for flow visualization experiments and velocity measurements. The viscoelastic fluids flow into a test duct with the height, W m, of 40 mm and the spanwise width of 75 mm, made of transparent acrylic resin, by a pump from a reservoir tank. Five ribs with the height, H m, of 20 mm and the length of 100 mm were mounted on the bottom of the test section with the length of 1,100 mm. Each distance between ribs, which is the cavity length, L m, was set at 100 mm. Flow visualization was performed with tracer particles of ion-exchange resin with 0.1 mm of the diameter and 850 kg·m^{-3} of the density. The fourth cavity was tested in the present study in order to eliminate the effects of inlet flow fluctuation and of flow development. The instantaneous velocity field in the cavity was measured by use of a particle tracking method, where particle movement in the period of 0.1 s was measured by an image processing. The accuracy of the present velocity measurement is within ±3 % on the mean velocity in the narrow flow path.

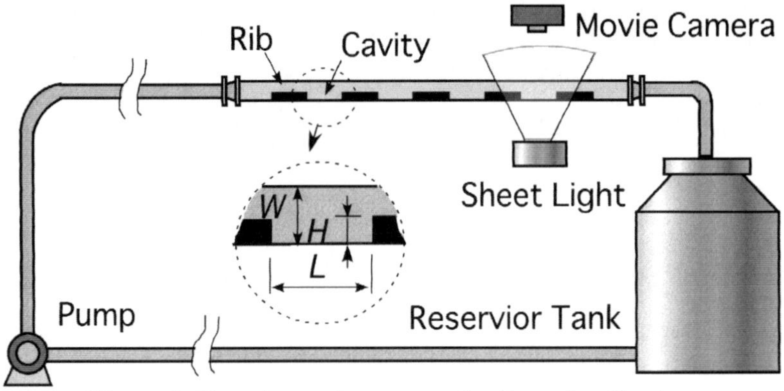

Figure 3. Experimental apparatus for flow visualization.

2.4. Definition of Reynolds and Weissenberg numbers

The solvent Reynolds number, Re, was defined with water viscosity, μ Pa·s as follows.

$$Re = \frac{\rho U_m H}{\mu} \tag{2}$$

Here, U_m $m \cdot s^{-1}$ and ρ $kg \cdot m^{-3}$ are the mean velocity in a narrow flow path between the rib top and the upper wall and the density of the fluid. The Reynolds number was changed from 680 to 4,200 in this study.

The Weissenberg number, *We*, was also defined as follows at the narrow flow path.

$$We = t_r \frac{U_m}{W - H} \tag{3}$$

Here, t_r s is a relaxation time. Suzuki et al. [14] reported the third relaxation time corresponds to the formation of besicles. The first relaxation time is deeply related to the drag reduction effect of the present surfactant system, but the flow separation characteristics observed in a cavity were well expressed by the Weissenberg number based on the second relaxation time in the cases when the molar ratio is set at 1.5 [13]. Here, the Weissenberg number was defined with the second relaxation time. With this definition, the Weissenberg number ranged from 1.32 to 8.16 in this study. The temperature was set at the room temperature of 25°C during the experiments.

3. Results and discussions

3.1. Flow behaviors

Figure 4 shows the photos in the cases of water and of the present viscoelastic fluid, respectively, when the Reynolds number is 1,000. In the figure, a white arrow designates the boundary between the recirculating region formed in the cavity and the main flow. In the case of water, the flow separated from the upstream top of the cavity reaches directly the downstream top, and then the recirculating region covers the cavity. Thus, the flow velocity is very small near the cavity bottom. On the other hand, the flow separated from the upstream top of the cavity in the case of the viscoelastic fluid penetrates deeply into the cavity. The flow sweeps the cavity with high velocity due to the Barus effect of the viscoelastic fluid. This causes the heat transfer enhancement from the cavity bottom as reported in the previous study [10].

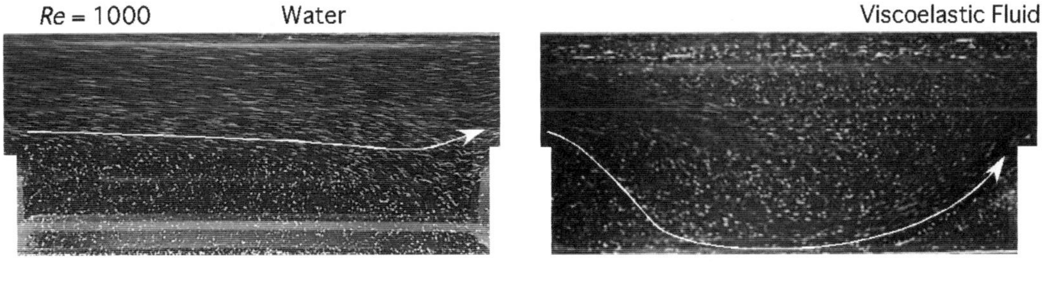

Figure 4. Photos of the flow in the cavity for water and for viscoelastic fluid.

Figure 5 shows the effects of the Reynolds number on the flow behaviours when the Reynolds numbers were 1,000, 1,700, 2,800 and 4,200, respectively. When the Reynolds number is 1,700, the flow separated from the upstream top of the cavity also penetrates into the cavity but it turns back toward the upstream side wall of the cavity without reattachment on the cavity bottom. After this, the flow sweeps the fluid near the cavity bottom with very high velocity. Thus, a bulge-like fluid structure is formed on the upstream side wall of the cavity. In this case, it is also found that the tip of the separation bubble formed in the downstream region of the cavity fluctuates slowly. As discussed later, it is found that the tonguing motion occurs significantly related to the bulge structure fluctuation. In the present conditions, these fluid motions are found to appear in a range over Re = 1,300. In the case of higher Reynolds number of 2,800, the flow more highly fluctuates and some vortices can be observed both in the cavity and in the main flow. On the other hand, the flow fluctuation near the bottom of the cavity decreases in the case when the Reynolds number of 4,200, though the main flow shows high-level turbulence. Thus, the bulge-like structure disappears and the rather stable flow covers the cavity bottom at Re = 4,200.

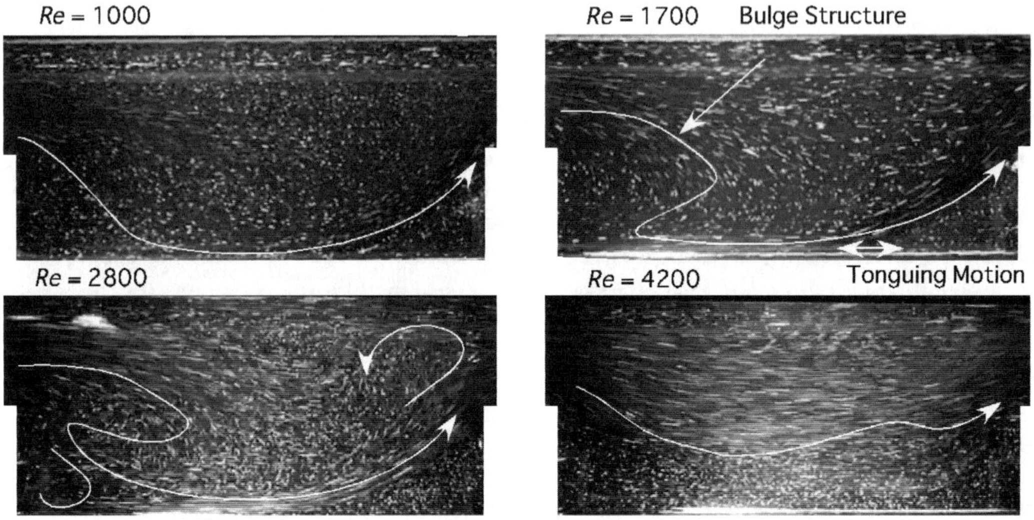

Figure 5. Effect of the Reynolds number on the flow behaviours.

3.2. Time variation characteristics

Figure 6 shows the time variation of the flow behaviours when the Reynolds number is 1,700. In order to discuss the velocity field, the respective instantaneous velocity vectors in the cavity is also shown in the figure. At 0 s, a small bulge structure in the upstream region of the cavity is observed. In the downstream region, a large separation bubble is formed at that time. From the velocity vectors, it is found that there exist high velocity regions upon the bulge structure and upon the separation bubble. Especially, the velocity over the separation bubble is very high and it is accelerated during sweeping the cavity bottom by the high velocity flow penetration into the cavity. The turning-back flow velocity is not high, but contributes the flow acceleration of that high-velocity sweeping flow. At 15 s, the flow from the upstream region into the cavity moves toward the downstream region. Then, the bulge structure is expanded toward the downstream region of the cavity. This causes the depression of the separation bubble formed in the downstream region of the cavity. This motion decelerates the flow over the separation bubble, and then the sweeping flow is weakened.

Figure 7 shows the time variations of the penetration point of the high velocity flow, x_p m, and of the separation point from the cavity bottom, x_s m. From the figure, it is found that both points fluctuate slowly and widely. The separation point takes a minimum around 20 s and a maximum around 60 s. Thus, the tip of the separation bubble moves widely as a tonguing motion. On the contrary, the penetration point takes a peak and a minimum at the corresponding instants. This indicates the tonguing motion of the separation bubble tip occurs due to this flip-flap motion of the penetration of the high velocity flow supplied from the upstream region of the cavity.

When the Reynolds number is larger than 1,300, three-dimensional flow fluctuation significantly appears in the present conditions. Thus, it is thought that the present flow fluctuation is related to the spanwise movement of the flow. In this paper, the effect of three-dimensionality of the flow cannot be treated due to the measurement limitation. Such three-dimensionality should be discussed in order to clarify the formation and the fluctuation mechanism of these flow structures.

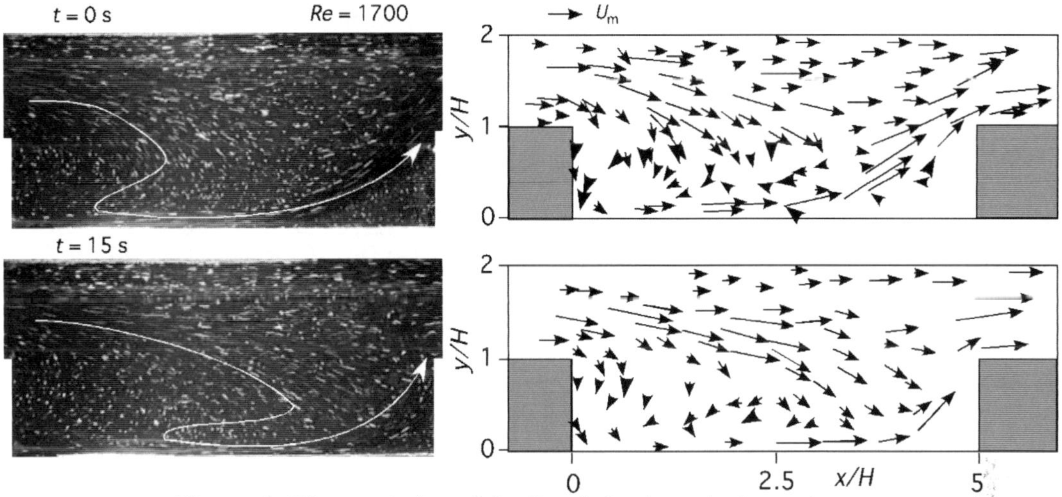

Figure 6. Time variation of the flow behaviours in the cavity.

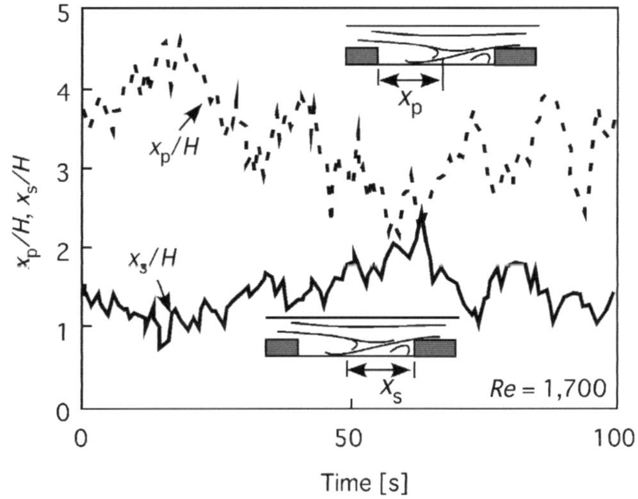

Figure 7. Time variations of the penetration and the separation points.

3.3. Reattachment and Separation characteristics

In this section, the reattachment and separation characteristics in the cavity are discussed. Figures 8 and 9 show the reattachment length, x_r m, and the separation length, x_s m on and from the cavity bottom, respectively. In the figures, the fluctuation ranges are designated by bars and the most frequent points are plotted by symbols. The solid line in figure 8 shows the reattachment points reported by Armaly *et al.* [15] observed in the downstream region of a backward-facing step. In the figure, the present results on the case of water are also plotted.

From figure 8, it is found that the reattachment point takes a larger value than the cavity length when the Reynolds number is larger than 200 in the cases of water. The present results agree with ones by Armaly *et al.* Thus, the cavity is completely covered by a recirculating region in a high range of the Reynolds number for water cases. On the other hand, the flow penetrates into the cavity due to the Barus effect when the Reynolds number is lower than 1,000 in the cases of the present viscoelastic

Journal of Physics: Conference Series **530** (2014) 012055 doi:10.1088/1742-6596/530/1/012055

fluid. The reattachment point moves toward the downstream in the cavity with increase of the Reynolds number, but it moves toward the upstream due to the bulge structure formation at *Re* = 1,300. The fluctuation of the bulge structure causes the fluctuation of the reattachment point and the fluctuation abruptly increases when the Reynolds number is larger than 2,800. Finally, the cavity is covered by a recirculating region when the Reynolds number is 4,200. Corresponding to this, the length of separation bubble fluctuates from *Re* = 1,300 as shown in figure 9. The separation point fluctuation increases until *Re* = 3,500 and the separation disappears at *Re* = 4,200 corresponding to the disappearance of the reattachment.

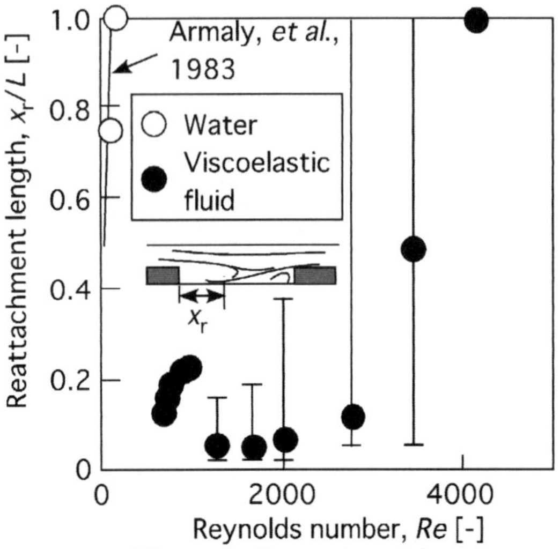

Figure 8. Reattachment length.

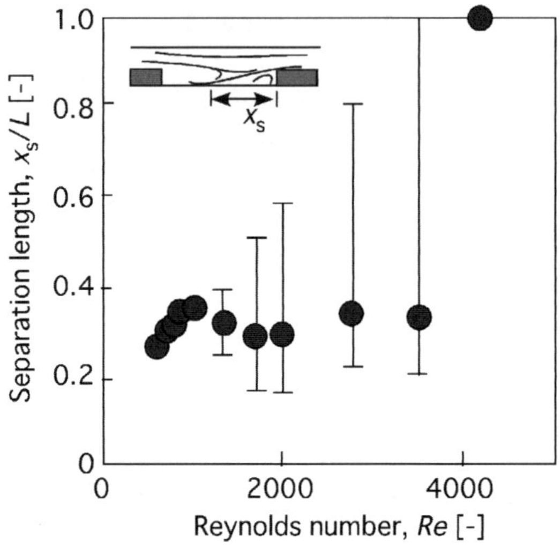

Figure 9. Separation length.

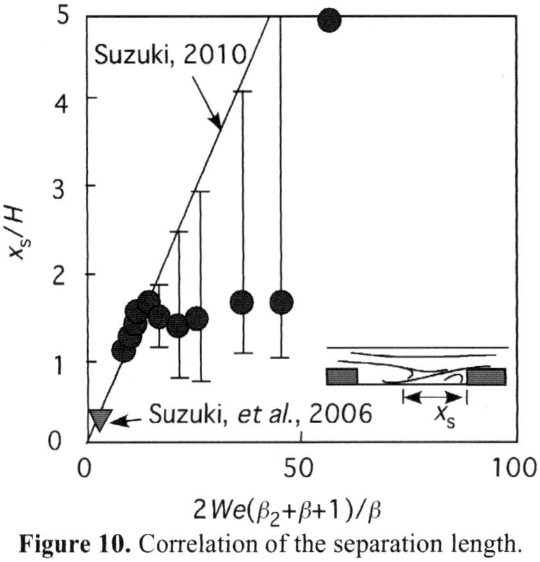

Figure 10. Correlation of the separation length.

It was reported in the previous study [11] that the length of the separation bubble is correlated by a function of the expansion ratio, β (= $(W-H)/W$) and of the Weissenberg number as follows.

$$\frac{x_s}{H} \propto 2\frac{We}{\beta}\left(\beta^2+\beta+1\right) \tag{4}$$

Figure 10 shows the correlation of the present separation points with Equ. (4). In the figure, the numerical result reported in the previous study [12] is also plotted. From the figure, it is found that the results in the cases when the Reynolds number is lower than 1,000 agree well with the correlation of Equ. (4). On the other hand, the results over Re = 1,300 depart from the correlation. Thus, when the Reynolds number is in a mid-range and flow fluctuation appears, the separation length shows the smaller values than the correlation. On the other hand, the separation length reaches the upstream side wall of the cavity when the Reynolds number is 4,200. It is also found that the longest length of the separation bubble agrees with the correlation even when *Re* is larger than 1,300. The correlation was obtained from the assumption that the upstream flow is developed. This indicates the upstream velocity profile becomes one of pseudo-developed flows when the longest separation appears.

4. Conclusions

In order to investigate the bulge structure formed behind a backward-facing step of the cavity and the tonguing motion of the tip of the separation bubble formed in the downstream region of the cavity, flow visualization experiments have been performed for viscoelastic fluid. The cavity depth, the cavity length and the width of the flow path were fixed, while the water Reynolds number based on the narrow flow path was changed from 680 to 4,200. From the results, it was found that the flow separated from the upstream top of the cavity penetrates into the cavity and turn back toward the upstream side wall of the cavity at the mid-position of the cavity in a mid-range of the Reynolds number. The flow sweeps the fluid near the cavity bottom wall after this and forms a very high flow stream on the separation bubble. Thus, the bulge-like structure is formed behind a backward facing step of the cavity and can be expected to lead a fresh fluid into the cavity. The bulge structure was formed in a range from 1,300 to 3,500 in the present condition. The tonguing motion of the tip of the separation bubble occurs with the flip-flap motion of the high velocity flow upon such a bulge structure. However, these flow structures disappear in the cavity when the Reynolds number is larger than 4,200, in spite of the fact that the main flow has high turbulent intensity. In this case, the cavity is covered by a recirculating flow. These flow fluctuations observed in the mid-range of the Reynolds number can be expected to enhance the heat transfer from the cavity wall. In the near future, heat transfer experiments will be performed.

5. References

[1] Han J C, 1998 *J. Heat Transfer.* **110** 321
[2] Han J C, Glicksman L R, Rohsenow W N 1978 *Int. J. Heat Mass Transfer* **21** 1143
[3] Taslim M E 2005 *Int. J. Heat Exchangers* **4** 25
[4] Murata A, Mochizuki S and Takahashi T 1999 *Int. J. Heat Mass Transfer* **42** 3047
[5] Liu S and Sakr M 2013 *Renewable Sustainable Energy Rev.* **19** 64
[6] Aung W 1983 *J. Heat Transfer.* **105** 505
[7] Greiner N 1991 *Int. J. Heat Mass Transfer* **34** 1383
[8] Nishimura T, Oka N, Yoshinaka Y, Kunitsugu K 2000 *Int. J. Heat Mass Transfer* **43** 2365
[9] Suzuki H, Kida S, Nakamae T and Suzuki K 1991 *Int. J. Heat Fluid Flow* **12** 353
[10] Suzuki H, Ishiba H and Usui H 2002 *Proc. 12th Int. Heat Transfer Conf.* **2** 177
[11] Suzuki H 2010 *J. Chemical Eng. Japan* **43** 421
[12] Suzuki H, Yamada S and Usui H 2006 *J. Chemical Eng. Japan* **39** 915
[13] Suzuki H, Higuchi Y, Komoda Y and Hidema R, 2012 *Proc. 23rd Int. Sym. Transport Phenomena*, #156
[14] Suzuki H, Higuchi Y, Watanabe H, Komoda Y, Ozawa S, Nishimura T and Takenaka N 2012 *Nihon Reoroji Gakkaishi* **40** 85
[15] Armaly B F, Durst F, Pereira J C F and Schoenung B, 1983 *J. Fluid Mechanics*, **127**, 473

Rheological analysis of an effect of different deflocculants on the fly-ash slurry

K Sarnecki[1] and A Bartosik[2]
[1] Faculty of Mechatronics and Machine Design, Kielce University of Technology, Al. Tysiąclecia P.P. 7, 25-314 Kielce, Poland
[2] Faculty of Management and Computer Modeling, Kielce University of Technology, Al. Tysiąclecia P.P. 7, 25-314 Kielce, Poland

E-mail: ksarnecki@tu.kielce.pl; artur.bartosik@tu.kielce.pl

Abstract. During the combustion of coal in the combined heat and power plant (CHP), a very large amount of combustion waste, called further as a fly-ash, is produced. It is typical that fly-ash appears during the combustion process of the fine coal and is transported by a pipeline with support of water as a carrier liquid to a pond storage site, where it is disposed. The pond is localized usually a few kilometers from the CHP, which makes it possible that decrease of friction in such a pipeline can result in energy savings of electricity needed for the pump and water needed as a carrier liquid. In the study an efficient method using a few deflocculants for reducing shear stress, and as a consequence viscosity, is demonstrated. The objective of the paper is to improve the efficiency of the hydrotransport of the fly-ash slurry by adding own designed additives. During the experiments a solids concentration by weight was determined from procured raw material in order to compute the real value occurring in industrial conditions. In addition, the analysis of the particle size distribution was conducted. The Anton Paar MCR 302 electronic rheometer was used to measure the dependence of shear stress and viscosity vs shear rate in the fly-ash existing in the CHP. Another part of the analysis was focused on the additives (deflocculants), to examine their influence on the reduction of the shear stress. The paper proves positive deflocculants impact on the rheological properties of the fly-ash slurry. The results of measurements are presented as figures and conclusions.

1. Introduction

Poland is the sixth energy producer in the European Union. In order to obtain energy, more than 75 million tons of coal and over 64 million tons of lignite (brown coal) were used in 2012. The energy balance of both of these resources makes up almost 90% of the all fuel consumed energy. The combined heat and power plants (CHP) used over 40 million tons of coal in 2012 [1]. During the combustion of coal in the CHP, a very large amount of combustion waste, called further as a fly-ash, is produced.

The CHP Kielce (Elektrociepłownia Kielce) in Poland production potential of thermal energy is provided by boilers as follows:

- water-coal boiler WP-140 with a thermal power of 140 MW,
- five water-coal boilers WR-25 with a thermal power of 29 MW each.

Total installed thermal performance in CHP Kielce is 286 MW. However, since 2008 the CHP has been operating a new power unit with a power of 10.5 MW, which uses as fuel wood chips and coal.

The boiler WP-140 works with an electrostatic precipitator - device that reduces dust emission with the efficiency of 99.5%. In addition, the boiler collaborates with the installation for reducing emission of nitrogen oxides by 40%. The fly-ash that appears during the combustion process of fine coal in WP-140 boiler is transported by a pipeline with support of the water as a carrier liquid to a pond storage site, localised far from the plant, where it is disposed. The pipeline has a length of 1.5 km and a diameter of 250 mm. As the carrier liquid, technological waste water is used, which comes from an industrial and rain sewage treatment plant. The far location of the pond causes a huge consumption of the energy to transport such large quantities of waste. It is seen that in 2011 the amount of 680 Mg of ash was deposited at the storage site, while in 2012 even 1680 Mg. The decreasing of friction in such pipeline can result in energy savings of electricity needed for the pump and water needed as a carrier liquid.

The main objective of the research is to examine the influence of chosen deflocculants for reduction of the shear stress in fly-ash slurry in one of the CHP. The deflocculants are the chemical additives which cause increased liquidity. In the study, own designed additives, as deflocculants, were used.

2. Raw material

The fine-ash sample was taken in CHP Kielce from the pipeline at the downtime period. Unfortunately such samples possess rust with a large diameter. This required screening of the material to avoid damage of the measuring equipment, so the tested raw material was dried and sieved to avoid big pieces of the rust. The sieve diameter was 1 mm. Determined density of the fly-ash particles was equal to 2350.3 kg·m^{-3}.The size distribution of the ash particles was measured using the Malvern Mastersizer 3000 laser diffraction particle size analyser. The weighted-average particle diameter of the sample was determined to be 60.98 μm. The particle size distribution is shown in figure 1.

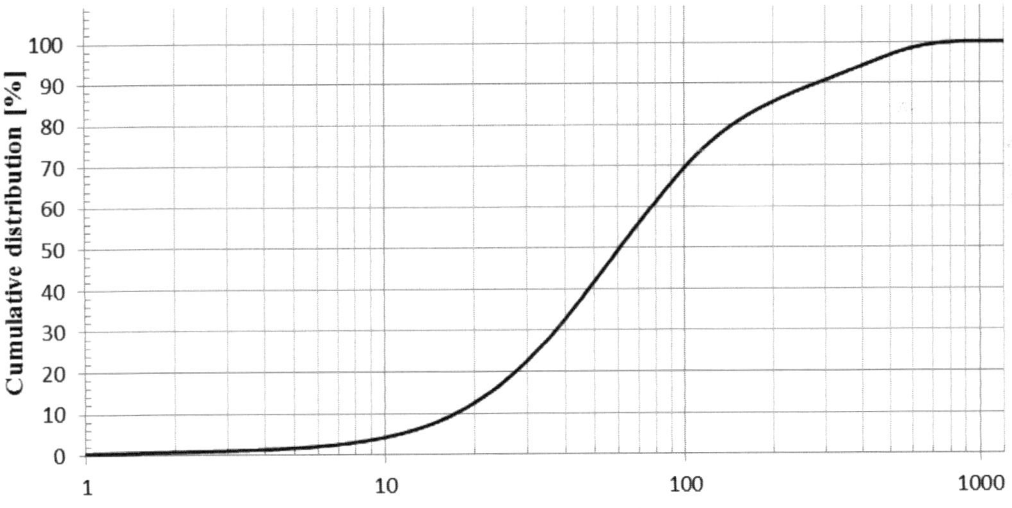

Figure 1. Particle size distribution.

In addition, the fly-ash grains were measured in microscope. Figure 2 shows the fly-ash in 200x magnification.

Figure 2. The fly-ash grains in 200x magnification.

3. Experimental procedure

In the preparation of fly-ash slurry sample of a known solid content, dry fly-ash and tap water were mixed to obtain the desired solid concentration. The sample preparation procedure was similar to Naik, Mishra and Rao Karanam methodology [2]. Solids concentration was calculated as a ratio of mass of solid particles to total mass of slurry, which is the sum of solid particles and liquid phase. For the tests, 500 ml of the fly-ash slurry was prepared by mixing an appropriate amount of the fly-ash and tap water in the beaker. The fly-ash was weighted by the laboratory balance with a resolution of 10^{-3} g. The slurry was mixed about 2 minutes and then the beaker was covered by the PP film to prevent evaporation of water from the slurry. The sample was allowed to wet for at least one hour before the measuring tests to ensure release of entrapped air. Before tests, the slurries were always stirred to ensure homogeneity of the sample slurry. In the case when a deflocculant was added, a dose of the additive was added in the slurry before the 2 minute mixing procedure. The deflocculant dose was appointed as a 0.2% of the dry fly-ash (mass of solid particles).

Determination of rheological properties was performed using the Anton Paar MCR 302 electronic rheometer. The shear stress and viscosity at specific shear rates were measured. To provide accurate results, CC27 measuring system with coaxial cylindrical geometry was used for the study. The gap between cylinders was equal to 1.1 mm. All experiments were performed at temperature of 20 °C. The measuring system required about 18 ml of a sample, which was taken from the previously prepared 500 ml sample.

First, the fly-ash slurry rheological properties were measured at different solids concentration, C_m=10%, 20%, 30% by weight to determine the reference values of the sample. Afterwards, the slurry with the 10%, 20% and 30% solids concentration with the different deflocculants were tested. Considering the nature of the slurry our measurements were analysed at the shear rates from 144 s^{-1} to 1000 s^{-1}. To avoid sedimentation process, which naturally exists in such type of slurry with fine solid particles, each slurry sample in the measuring cup was pre-sheared for 30 seconds at the 1000 s^{-1} shear rate prior to the measurements. Afterwards, the measurement started from the highest setpoint value and stepped down linearly one by one until the shear rate reached the minimum setpoint value, which was 144 s^{-1}. The results of measured shear stress (τ) and viscosity (η) were recorded at each share rate (dU/dy).

4. Experimental results and discussion

For each concentration of solid particles the variation of the shear stress with the shear rate was measured. Figure 3 shows the flow curves for the C_m=(10-30)%.

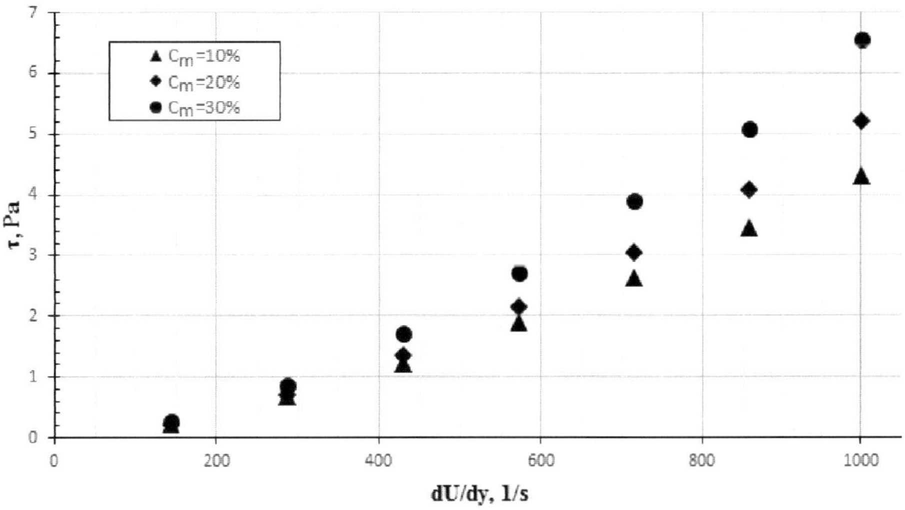

Figure 3. Dependence of the shear rate on shear stress at different solids concentration for fly-ash slurry.

The flow curves presented in figure 3 are dilatants (shear thickening). It was observed that the shear stress increased with the increase of shear rate non-linearly. It can be also seen that the shear stress increased with the increase of solids concentration. The difference is more visible at higher shear rates values. Below 300 s^{-1} of shear rate the shear stress is less than 1 Pa. The phase separation process in the experiments at high shear rate values, similar to the limestone slurry measurements, was not noticed [3].

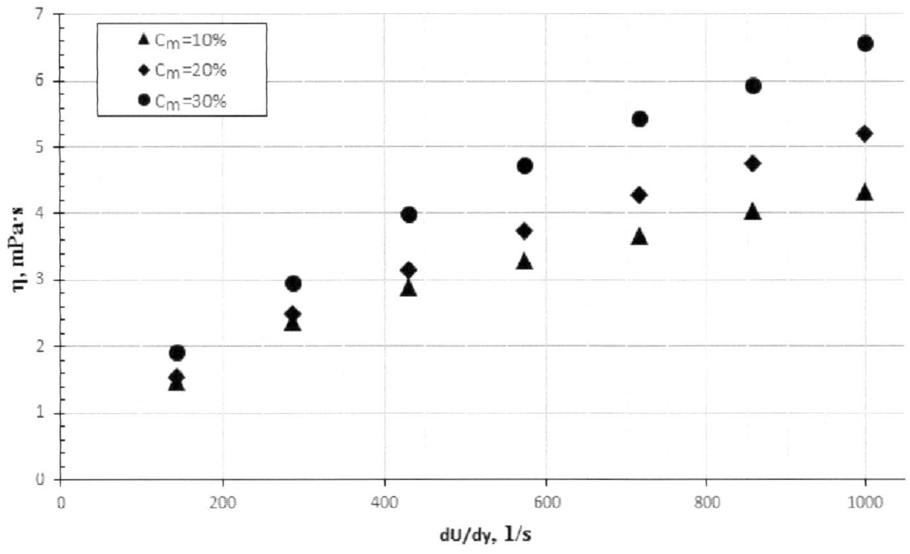

Figure 4. Dependence of the shear rate on viscosity at different solids concentration for fly-ash slurry.

The viscosity curves at different solids concentration are shown in figure 4. The shear thickening occurrence was confirmed. The viscosity is increasing with the increase of shear rate but the rate of these changes is decreasing with the shear rate increase. It was also observed during the measurements that viscosity increased with the increase of solids concentration.

The shear stress and viscosity values, presented in the paper, are significantly smaller than in the fly-ash slurry measurements done by Naik et al. [4]. They found the shear thinning and the viscosity nature of both fly-ash slurries samples is different. Occurrence of such differences proves that coal dust can have different rheological properties that depend on production facility or quality of the dust precipitator performance as well.

Chemical additives (deflocculants) were added to the fly-ash slurry to increase its liquidity. Deflocculants were dosed in portion of 0.2% of the solid mass of the fly-ash slurry. Sodium carbonate, limestone slurry (70% wt.), lignite and sodium salicylate were selected as a potential effective deflocculants. The results of measurements of the shear rate versus the shear stress for fly ash slurry, with and without aforementioned additives, are presented in figures 5-8.

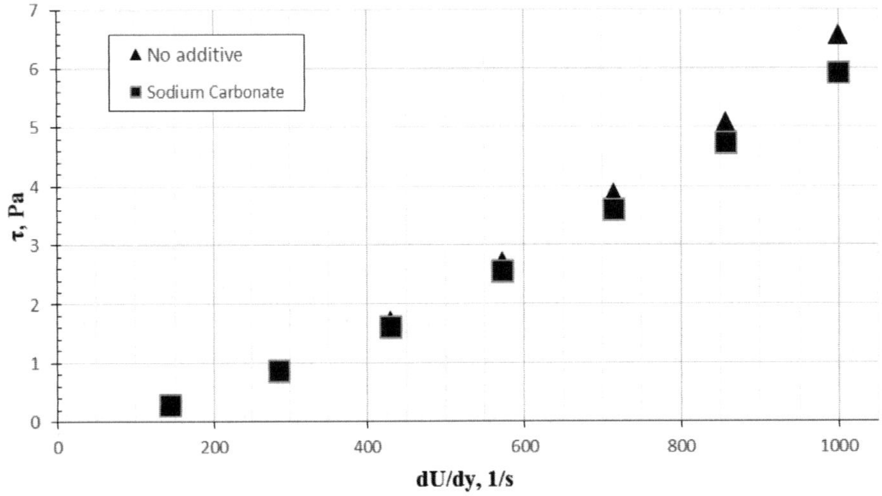

Figure 5. Flow curves of the fly-ash slurry for C_m=30% and with 0.2% additive of sodium carbonate.

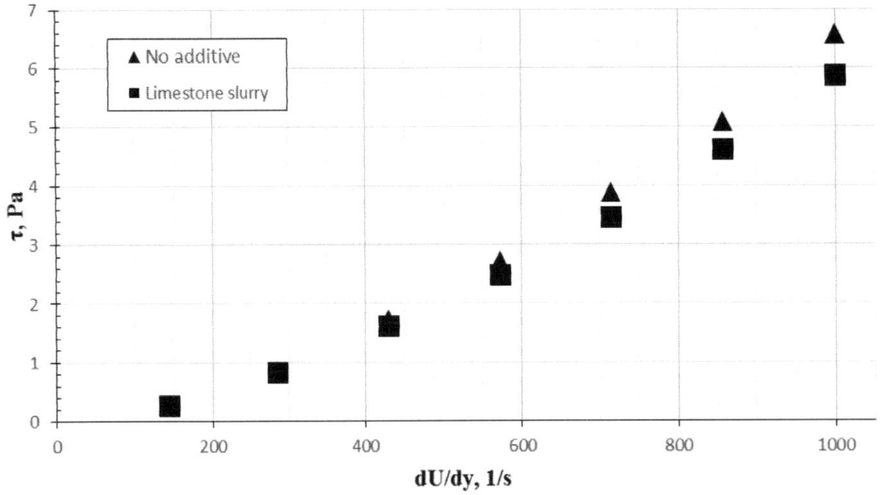

Figure 6. Flow curves of the fly-ash slurry for C_m=30% and with 0.2% additive of limestone slurry.

Journal of Physics: Conference Series **530** (2014) 012056 doi:10.1088/1742-6596/530/1/012056

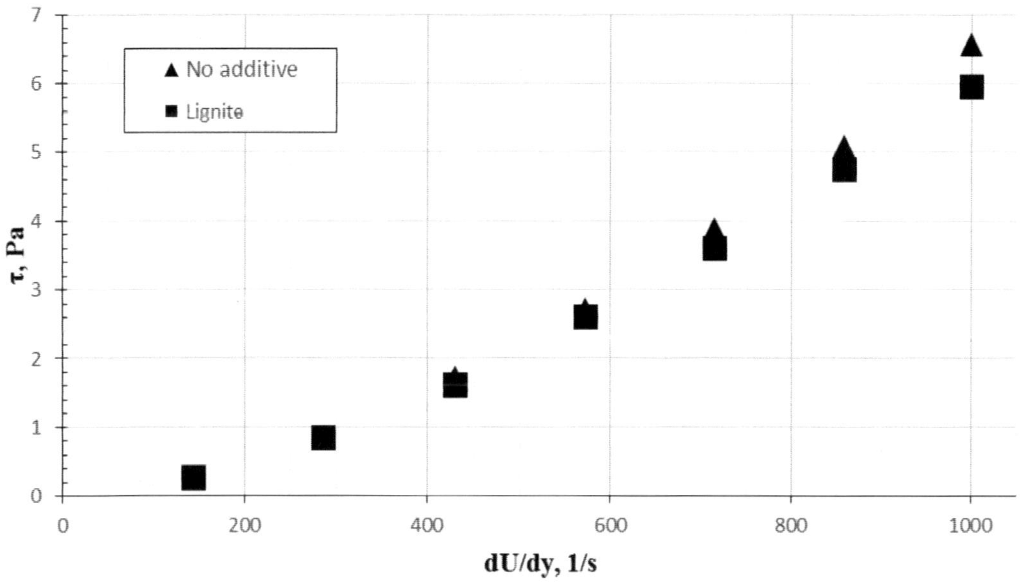

Figure 7. Flow curves of the fly-ash slurry for C_m=30% and with 0.2% additive of lignite.

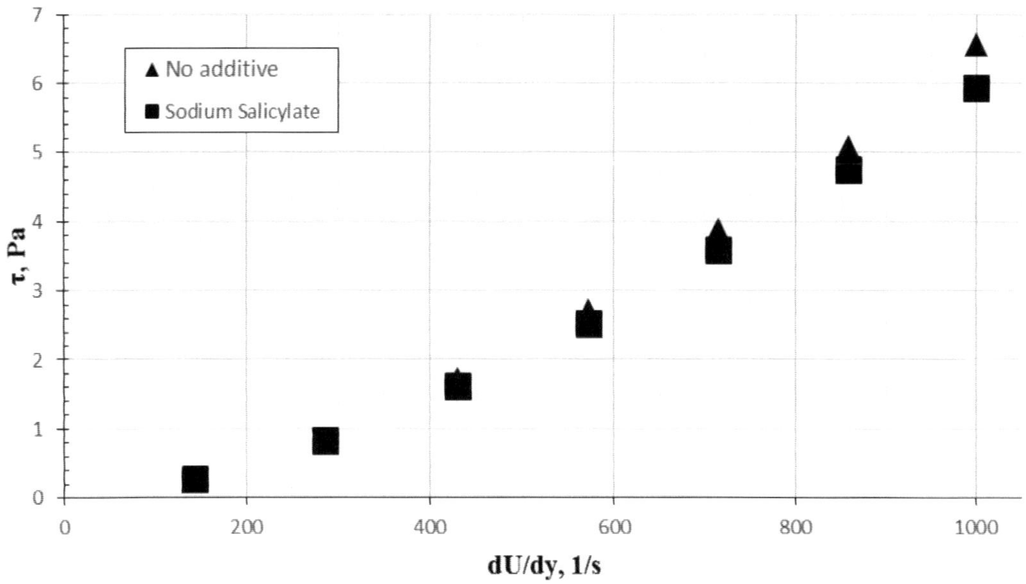

Figure 8. Flow curves of the fly-ash slurry for C_m=30% and with 0.2% additive of sodium salicylate.

The addition of selected deflocculants demonstrates small influence on reduction of the shear stress. The decrease in the shear stress ranged from 9% to 11% for the highest shear rate values. For the lower shear rate values gain from adding deflocculant is not clearly visible.

The influence of additives on viscosity is better visible, compared to the shear stress. For this reason calculations of viscosity with and without chemical additives are presented in figures 9-12.

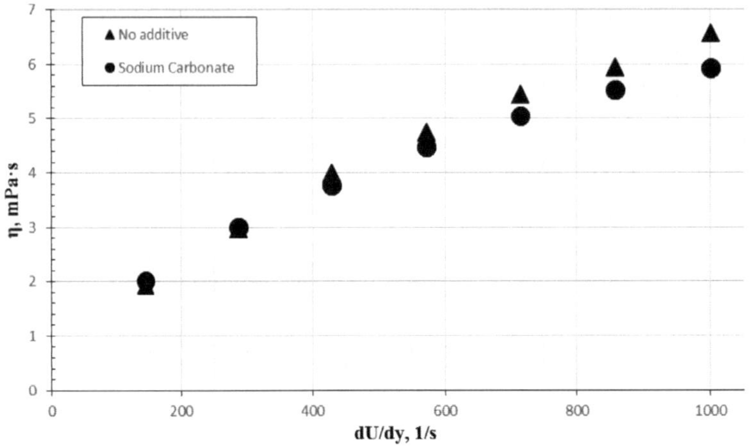

Figure 9. Viscosity curves of the fly-ash slurry for C_m=30% and with 0.2% additive of sodium carbonate.

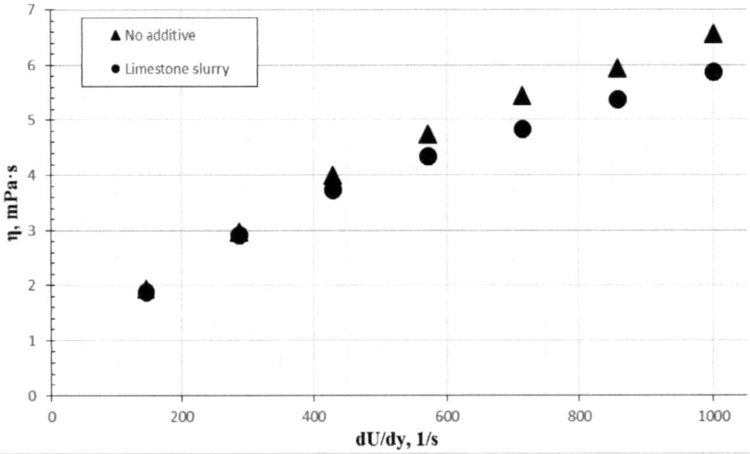

Figure 10. Viscosity curves of the fly-ash slurry for C_m=30% and with 0.2% additive of limestone slurry.

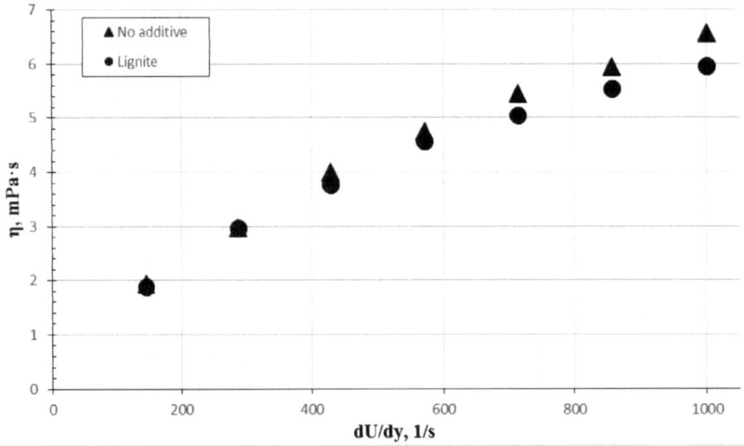

Figure 11. Viscosity curves of the fly-ash slurry for C_m=30% and with 0.2% additive of lignite.

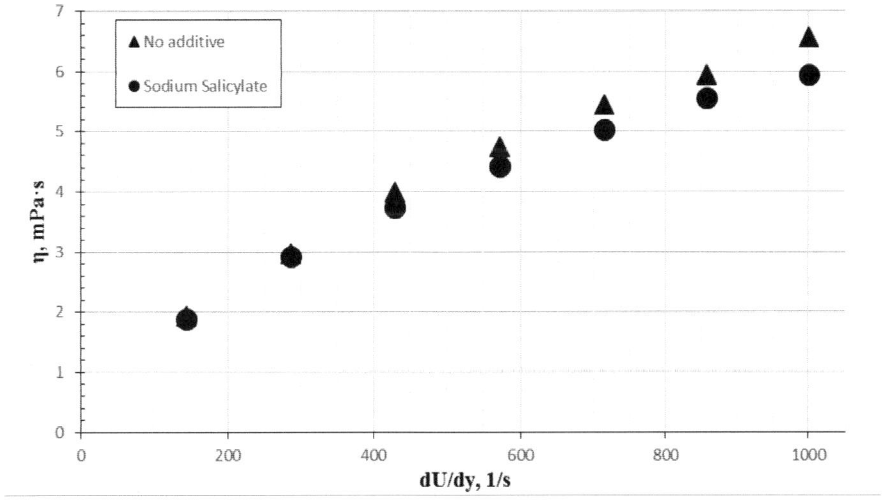

Figure 12. Viscosity curves of the fly-ash slurry for C_m=30% and with 0.2% additive of sodium salicylate.

The results confirmed a weak influence of selected additives on the rheological properties on the fly-ash slurry. This is in contrast with measurements for sand slurries. Appearance of shear thickening causes those chemical additives, which successfully decrease the shear stress in sand slurries, not to work well in the fly ash slurry.

5. Conclusions

The effect of deflocculants additives on the rheological behaviour of the fly-ash slurry was examined experimentally. Experiments were performed at the shear rates possible to appear in the industrial conditions. Tested additives gave a small but positive influence on decreasing the shear stress and the viscosity. The addition of one component provides too weak benefits to allow their implementation in the industry. The more effective way to increase liquidity seems to be the application of cationic surfactant and counter-ion which has been proposed by Naik et al. [4] and would be the next step in further research.

Measurements of fly ash slurries are extremely difficult as they require a special treatment before measurements. For this reason many researchers prefer a tube viscometer for determining rheology of such slurries [5], [6].

6. References

[1] Berent-Kowalska G 2013 *Zużycie paliw i nośników energii w 2012 r* (Warsaw: GUS)
[2] Naik H K, Misha M K and Rao Karanam U M 2009 Rheological characteristics of fly ash slurry at varying temperature environment with and without an additive, *Proc. World of Coal Ash Conf.(Lexington, USA, 4-7 May 2009)*
[3] Sarnecki K and Bartosik A 2013 The effect of high solids concentration on the rheology of limestone slurry *Systems Supporting Production Engineering* (Gliwice: PA NOVA) pp 135-143
[4] Naik H K, Mishra M K and Rao Karanam U M 2011 Evaluation of flow characteristics of fly ash slurry at 40% solid concentration with and without an additive, *Proc. World of Coal Ash Conf. (Denver, USA 9-12 May 2011)*
[5] Kembłowski Z 1973 *Reometria płynów nie-Newtonowskich* (Warszawa: WNT)
[6] Parzonka W 1997 *Hydrauliczne podstawy transportu rurowego mieszanin dwufazowych* (Wrocław: Wyd. Akademii Rolniczej)

Journal of Physics: Conference Series **530** (2014) 012057 doi:10.1088/1742-6596/530/1/012057

Effects of viscosity variations in temporal mixing layer

N Taguelmimt[1], L Danaila[1,2] and A Hadjadj[1]

[1] CORIA-UMR 6614- Normandie Université CNRS-Université et INSA de Rouen Campus Universitaire du Madrillet 76800 Saint Etienne du Rouvray, France.

Abstract. The objective of the present investigation is to assess the effects of viscosity variations in low-speed temporally-evolving turbulent mixing layer. Direct Numerical Simulations (DNS) are performed for several viscosity ratios, $R_\nu = \nu_{high}/\nu_{low}$, varying between 1 and 9, whereas the upper and lower streams are of equal density. The space-time evolution of Variable-Viscosity Flow (VVF) is compared with the Constant-Viscosity Flow (CVF), for which $R_\nu = 1$. The initial Reynolds number, based on the initial momentum thickness, $\delta_{\theta,0}$, is $Re_{\delta_{\theta,0}} = 160$ for the considered cases. The study focuses on the first stages of the temporal evolution of the mixing-layer. It is shown that in VVF (with respect to CVF): **(i)** the birth of turbulent fluctuations is accelerated; **(ii)** large-scale quantities, i.e. mean longitudinal velocity and momentum thickness, are affected by the viscosity variations, thus dispelling the myth that viscosity is a 'small-scale quantity that does not affect the large scales'; **(iii)** the velocity fluctuations are enhanced for VVF. In particular, the turbulent kinetic energy peaks earlier and is three times larger for VVF than CVF at the earliest stage of the mixing, and **(iv)** the transport equation for the turbulent kinetic energy is derived and favourably compared with simulations data.

1. Introduction

Most of the available turbulence theories (Kolmogorov, 1941) are underpinned by assumptions involving a single value of the kinematic viscosity. Moreover, the effect of viscosity is a priori expected to be felt only at small scales.

However, turbulence mostly occurs in heterogeneous fluids or multicomponent mixtures, therefore involving fluids with various viscosities. In such flows, strong viscosity gradients may exist at all scales, thereby modifying the mean velocity profiles. To examine the effects of viscosity variations on the mixing efficiency, Talbot et al. [1] have conducted an experimental study of a propane jet issuing into air; the latter is 3,5 more viscous than the air, whereas the densities are almost the same. These authors found that, in comparison with a classical air/air jet, the propane jet exhibits an accelerated decrease of the mean axial velocity. This behaviour was explained analytically by an increase of the magnitude of the viscous terms with respect to the inertial terms. Strong production of turbulent fluctuations was also observed in the very near-field of the propane jet, thus enhancing mixing, producing lateral fluctuations and accelerating the trend towards self-similarity and isotropy.

The objective of the present investigation is to assess the effects of viscosity variations in a canonical flow, the low-speed temporally-evolving turbulent mixing layer. For the particular purpose of our study, the temporal mixing layer (hereafter TML) involves two fluids with the same density.

TML belongs to the family of turbulent shear layers, which have been studied since the pioneering work of Brown and Roshko [2], followed by a web of investigations of this configuration devoted to different issues, such as transition to turbulence, mixing efficiency, compressibility effects [3, 4, 5, 6], variable density effects [1, 7, 8]. The effect of the trailing edge thickness of the splitter plate was also investigated in detail [9, 10], as well as e.g. sound generation [11].

[2] Corresponding author : danaila@coria.fr

The paper is organized as follows. Section 2 is dedicated to the description of the numerical methods and the associated solver. Section 3 examines the one-point statistics (instantaneous fields, mean values of the longitudinal velocity, as well as the turbulent kinetic energy budget).

2. Mathematical formulation and numerical method

2.1. Governing equations and numerical schemes

The unsteady three-dimensional, incompressible Navier-Stokes equations, with two-species transport equations are solved using an inhouse high-order explicit solver (see [12] for more details about the numerical methodology). The convective terms are approximated by the fifth-order weighted, essentially non-oscillatory (WENO) schemes, while the diffusive terms are determined using fourth-order compact central difference. The discrete equations are time-integrated by means of an explicit, third order, Rung-Kutta scheme. The Wilke formula [13], which is widely used in combustion and chemical engineering applications, is employed to determine the mixture-viscosity coefficient in terms of mass fraction, namely $\mu = \sum_{k=1}^{2}\left[\mu_k\left(\frac{Y_k}{W_k}\right)\left(\sum_{l=1}^{2} Y_l\phi_{kl}/W_l\right)^{-1}\right]$ where μ_k and W_k are the viscosity and the molecular mass of the k-th species and the function ϕ_{kl} is defined as $\phi_{kl} = \frac{1}{\sqrt{8}}\left(1+\frac{W_k}{W_l}\right)^{-1/2}\left[1+\left(\frac{\mu_k}{\mu_l}\right)^{1/2}\left(\frac{W_l}{W_k}\right)^{1/2}\right]$.

Periodic boundary conditions are applied along the stramwise x and the spanwise z directions and a non-reflecting slip condition is imposed along the cross-stream y direction.

2.2. Initial conditions

A sketch of the simulated TML is shown in figure 1. A velocity-viscosity counter gradient is imposed; the upper stream is initialized with a low velocity, U_{low}, and a high kinematic viscosity, ν_{high}, while the lower stream is initialized with a high velocity, U_{high}, and a low kinematic viscosity, ν_{low}. The upper steam (U_{low}, ν_{high}) and the lower stream (U_{high}, ν_{low}) meet together and evolve in time. The reference velocity is $\Delta U = U_{high} - U_{low}$ and the reference viscosity is $\nu_{ref} = 0.5(\nu_{high} + \nu_{low})$. Quantities such as $\langle U \rangle$ are averaged over homogeneous directions (x, z) for a given y, at each time t. We define the normalized time, τ, such as $\tau = \frac{t\,\Delta U}{\delta_{\theta,0}}$. Therefore, averages are functions of two variables: vertical position y and time t. As an example, the longitudinal mean velocity is a function of both y and τ, *viz.* $\langle U \rangle = f(y,\tau)$. The mean longitudinal velocity is initialized by a tangent hyperbolic profile $\langle U \rangle(y, t=0)$, while the other mean velocity components are initially set to zero. Thus,

$$\langle U \rangle(y, t=0) = \frac{1}{2}\left[\left(U_{high} + U_{low}\right) - \Delta U \tanh\left(\frac{y}{2}\right)\right]. \tag{1}$$

$$\langle V \rangle = \langle W \rangle = 0\,. \tag{2}$$

The two streams have the same initial pressure, $P = 1$ atm, temperature, $T = 293$ K, and density, $\rho = 1.314$ kgm^{-3} while the velocities and the kinematic viscosities are specified as summarized in Table 1 for all investigated mixing layers cases.

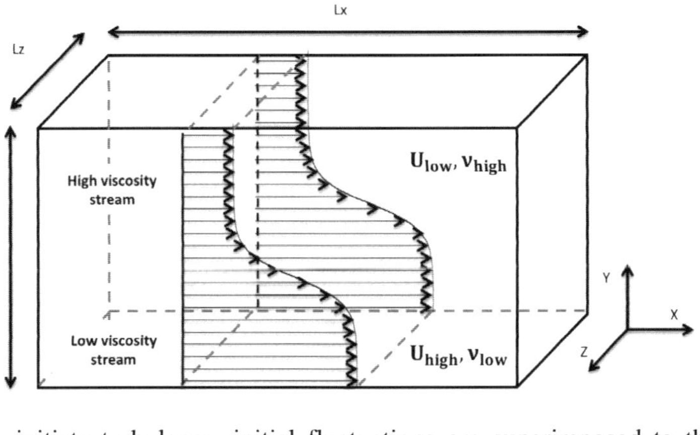

Figure 1. Computational domain. The middle of the domain is defined as $(x = 0, y = 0, z = 0)$.

To initiate turbulence, initial fluctuations are superimposed to the mean velocity field. The initial fluctuations are generated using a 3D digital filter technique [15]. This procedure starts with prescribed values for both Reynolds stress tensor and length scales, and generates the corresponding fluctuating velocity field at the initial time. The initial length scales along each direction are equal to $\delta_{\omega,0} = 4\delta_{\theta,0}$, the initial vorticity thickness. The initial turbulence intensity is $\left(\frac{\langle u'^2\rangle^{0.5}}{\Delta U} = 0.020, \frac{\langle v'^2\rangle^{0.5}}{\Delta U} = 0.013, \frac{\langle w'^2\rangle^{0.5}}{\Delta U} = 0.012\right)$ for the considered DNS.

2.3. Simulation parameters

Five test cases are carried out as shown in table 1. The parameters of DNS-R2.5 to DNS-R9 are chosen to analyse the effects of viscosity variations on the VVF, whilst DNS-R1 provides a comparison basis in the constant viscosity flow (CVF). The viscosity ratio is $R_\nu = \frac{\nu_{high}}{\nu_{low}} \geq 1$. The species mixture is characterized by values for R_ν of 1, 2.5, 4, 6 and 9 for DNS-R1, DNS-R2.5, DNS-R4, DNS-R6 and DNS-R9 respectively. The molecular diffusion coefficient $\mathcal{D}$ between the two streams is kept constant for the considered cases to focus only on viscosity variations effects. The initial momentum thickness Reynolds number is $Re_{\delta_{\theta,0}} = \Delta U \delta_{\theta,0}/\nu_{ref} = 160$, the initial momentum thickness is $\delta_{\theta,0} = 18.4$ µm. The initial conditions are set to have the same initial Reynolds Number $Re_{\delta_{\theta,0}} = 160$ rather than the same initial free stream velocities. These velocities are determined from the initial Reynolds number as $\Delta U = \frac{Re_{\delta_{\theta,0}}\nu_{ref}}{\delta_{\theta,0}}$. Moreover, we impose that $U_{high}/U_{low} = 11$ for the test cases. For our purpose, the velocity ratio must be the same for CVF and VVF so as to assess the uniquely effect of viscosity variations. Indeed, several studies have showed that the mixing layer growth is influenced by velocity ratio.

Table 1. Parameters of DNS configuration.

Case	R_ν	ν_{high} (m^2s^{-1})	ν_{low} (m^2s^{-1})	ν_{ref} (m^2s^{-1})	U_{high} (ms^{-1})	U_{low} (ms^{-1})	ΔU (ms^{-1})	U_{high}/U_{low}
DNS-R1	1	6.85	6.85	6.85	5.95	65.53	59.58	11
DNS-R2.5	2.5	6.85	2.74	4.78	4.17	45.87	41.70	11
DNS-R4	4	6.85	1.71	4.28	3.72	40.96	37.24	11
DNS-R6	6	6.85	1.14	3.99	3.47	38.23	34.76	11
DNS-R9	9	6.85	0.76	3.80	3.31	36.41	33.10	11

The challenge of this study is twofold. First, the grid resolution must be sufficiently high to correctly capture the dynamics of the TML from its birth to the later stages of the flow evolution.
Second, the computational domain must be wide enough to ensure the spatial development of the large-scale turbulence. The computational domain sizes, normalized by the initial momentum thickness, $\delta_{\theta,0}$, are 345, 172, 86 in x, y and z directions, respectively. The latter matches the DNS of Pantano & Sarkar [14]. The spatial resolution along the direction y is taken as 1.05 µm, to ensure good discretization of the initial momentum thickness (≈18 points).

3. One-point statistics

In this section, we present results for one-point statistics. The morphological behavior of VVF and CVF are discussed in Subsection 3.1. The mean velocity field is analyzed in Subsection 3.2, and the turbulent kinetic budget is discussed in Subsection 3.3.

3.1. Morphology of the flow

We provide qualitative observations of the temporal evolution of the mixing layer in both CVF and VVF cases. In particular, the focus is on the way the velocity and the vorticity fluctuations are generated. Figure 2 shows the instantaneous turbulent kinetic energy $q^2 = 0.5\langle u_i u_i\rangle/\Delta U^2$ (summation applies for double indices) in the plane (x, z), in the middle of the numerical domain (y=0). The maxima of energy (black surfaces) are higher for VVF than forcompared to CVF. Moreover, these

IOP Publishing
doi:10.1088/1742-6596/530/1/012057

maxima are uniformly distributed over the whole plane (x, z) for VVF, thus supporting the idea that velocity fluctuations are enhanced in this case. A possible scenario of the increased fluctuations for VVF is that the viscous fluid (slow fluid) acts as "spots" which decelerate the less viscous fluid (rapid fluid) leading to the creation of velocity fluctuations.

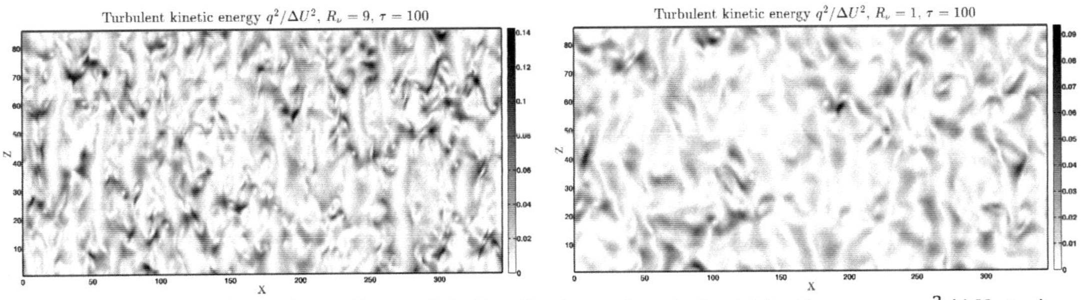

Figure 2. Middle plane (y = 0) spatial distribution of turbulent kinetic energy $q^2/\Delta U$ at time $\tau = 100$ for (left) $R_v = 9$ and (right) $R_v = 1$

Figure 3 shows evidence of the modification of the vorticity field. The spanwise vorticity ω_z is modified by the viscosity variations. For VVF, the 2D rollers of Brown and Roshko are more diffused and thus destructed in the more viscous fluid. Conversely, the 2D rollers are symmetric and of higher vorticity in CVF.

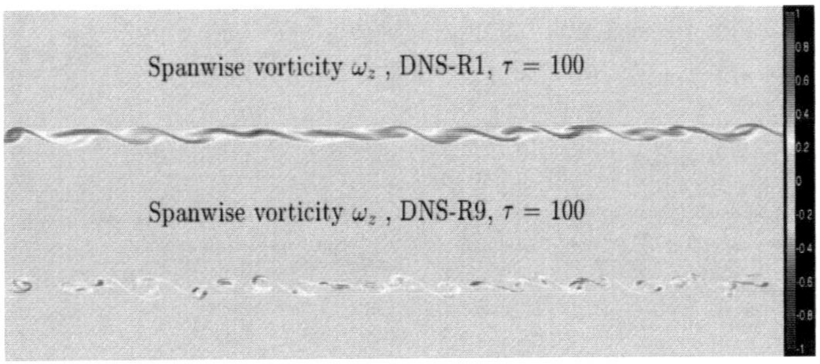

Figure 3. Middle plane (z = 0) spatial distribution of spanwise vorticity ω_z at time $\tau = 100$.

3.2. Effect of variable viscosity on the mixing-layer spreading and mean velocity

The growth rate of the mixing layer depends on the velocity ratio U_{high}/U_{low} as well as the density ratio ρ_{high}/ρ_{low} [4]. However, less attention was paid to the role of the viscosity variations on the temporal and spatial evolution of mixing layer thickness. The mixing layer momentum thickness is defined as

$$\delta_\theta(\tau) = \int_{-\frac{L_y}{2}}^{+\frac{L_y}{2}} \frac{[\langle U\rangle - U_{low}][U_{high} - \langle U\rangle]}{\Delta U^2} dy. \tag{3}$$

Figure 4 illustrates the TML momentum thickness for both VVF and CVF, as functions of the dimensionless time. In the early stage of the flow evolution, the TML momentum thickness is higher for VVF than for CVF. Physically, this result signifies that the VV-TML is affected to a larger extent by the same (initial) momentum. This behaviour might be the result of both

i) an accelerated generation of momentum fluctuations, as well as

ii) an enhanced diffusion of the momentum.

Equation (3) suggests that this behaviour is most likely attributable to differences in the mean stream-

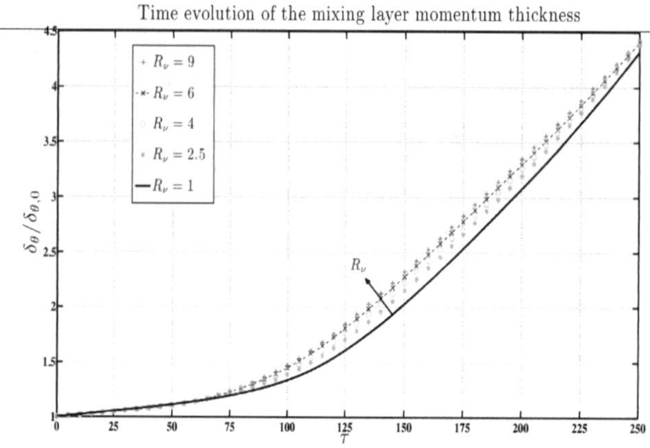

Figure 4. Temporal evolution of the TML momentum thickness, for different viscosity ratios, R_ν.

wise velocity ⟨U⟩, all the other parameters being the same. It is then clear that the longitudinal mean velocity ⟨U⟩ is modified by viscosity effects. This is an a priori surprising result, which may dispel the myth that 'viscosity is a small-scale quantity with small-scale effects'. Figure 5 shows the spatial distribution along the inhomogeneous direction, y, of the mean velocity ⟨U⟩ and mean velocity gradient $\frac{\partial \langle U \rangle}{\partial y}$ profiles for different viscosity ratio at the dimensionless time $\tau = 40$. It is clear that the viscosity variations act on the mean velocity field, leading for VVF to

i) a decrease of velocity at the centre of mixing layer at $\tau = 40$ (figure 1.a) where the viscosity gradients are important;

ii) an increase of velocity gradient $\frac{\partial \langle U \rangle}{\partial y}$ (figure 5.b); the peaks are located in the high speed stream side (less viscous).

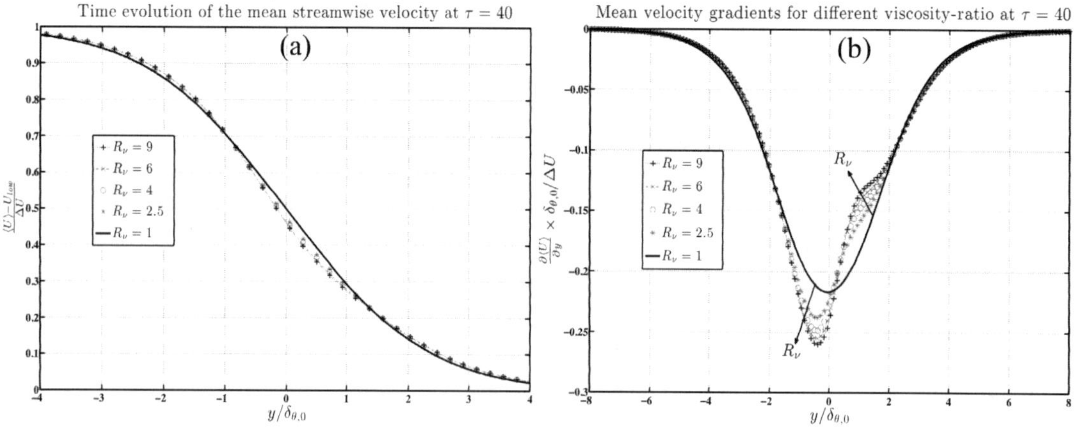

Figure 5. Spatial distribution of (a) mean dimensionless longitudinal velocity and (b) mean dimensionless longitudinal velocity gradient profiles, for different viscosity ratios R_ν at $\tau = 100$.

3.3. *Effect of variable viscosity on turbulent kinetic energy budget*

It is found that the quantity $q^2 = 0.5(\langle u_i' u_i' \rangle)/\Delta U^2$ (summation applies for double indices) is stronger for VVF, indicating that the fluctuating velocity are increased due to the viscosity variations. However, the decay of kinetic energy is more enhanced VVF at time larger $\tau \geq 200$.

To unravel the issues associated with variations of viscosity, the one-point kinetic energy budget is derived, in its general expression, to variable viscosity flows first by Talbot and al [1]. The spatial averaging over the homogeneous directions (x and z) allows some simplifications for the particular case of TML.

The aim of this section is to provide a direct evaluation of different terms in equation (5) and to investigate the effects of viscosity variations on the classical expression of the mean energy dissipation

Journal of Physics: Conference Series **530** (2014) 012057 doi:10.1088/1742-6596/530/1/012057

rate. The starting point is the instantaneous Navier-Stokes equations written for variable viscosity flow as follows

$$\frac{\partial u_i}{\partial t} + u_j \frac{\partial u_i}{\partial x_j} = -\frac{1}{\rho}\frac{\partial p}{\partial x_i} + \nu \frac{\partial^2 u_i}{\partial x_j \partial x_j} + 2\frac{\partial \nu}{\partial x_j} s_{ij}\,. \tag{4}$$

Here, double indices indicate summation and s_{ij} is the strain tensor, viz. $s_{ij} - \frac{1}{2}\left[\frac{\partial u_i}{\partial x_j} + \frac{\partial u_j}{\partial x_i}\right]$.
The transport equation of the turbulent kinetic energy is derived by multiplying equation (4) with the velocity fluctuations u_i' and space-averaging over the two homogeneous directions (x and z). Finally, the one-point kinetic energy budget is expressed in equation (5). Terms Tt, $\mathcal{D}_T$, $\mathcal{P}_k$ and $\mathcal{D}_p$ represent the temporal variation of the kinetic energy, the turbulent diffusion, the energy production and the pressure diffusion, respectively.

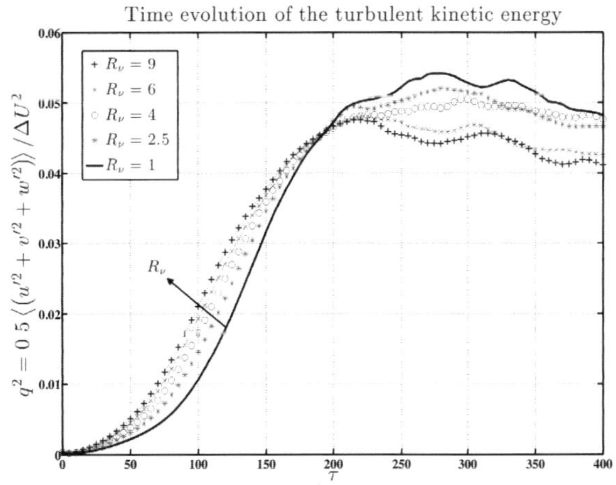

Figure 6. Temporal evolution of turbulent kinetic energy for different viscosity ratios, R_ν, at $\tau = 100$.

$$\underbrace{\frac{\partial}{\partial t}\left(\frac{\langle u_i'^2\rangle}{2}\right)}_{Tt} = \underbrace{-\langle u_j' \frac{\partial}{\partial x_j}\left(\frac{u_i'^2}{2}\right)\rangle}_{\mathcal{D}_T} \underbrace{-\langle u'v'\rangle\frac{\partial\langle U\rangle}{\partial y}}_{\mathcal{P}_k} \underbrace{-\frac{1}{\rho}\frac{\partial\langle v'p'\rangle}{\partial y}}_{\mathcal{D}_p} \tag{5.a}$$

$$\underbrace{+\left[\langle\frac{\partial \nu}{\partial x}v'\rangle + \langle\frac{\partial \nu}{\partial y}u'\rangle\right].\frac{\partial\langle U\rangle}{\partial y} + \langle u'v'\rangle\frac{\partial^2\langle U\rangle}{\partial y \partial y}}_{\mathcal{P}_\nu} \tag{5.b}$$

$$\underbrace{+\langle\frac{\partial \nu}{\partial x_j}\left[\frac{1}{2}\frac{\partial u_i'^2}{\partial x_j} + \frac{\partial u_i' u_j'}{\partial x_i}\right]\rangle + \frac{1}{2}\langle\nu\frac{\partial^2 u_i'^2}{\partial x_j{}^2}\rangle}_{\langle\epsilon\rangle_{VG}} \tag{5.c}$$

$$\underbrace{-\langle\nu\left(\frac{\partial u_i'}{\partial x_j}\right)^2\rangle}_{\langle\epsilon\rangle_{VV}}\,. \tag{5.d}$$

Terms $\mathcal{P}_\nu$ (line 5.b) represent correlations between the mean velocity gradients and viscosity fluctuations. Terms $\langle\epsilon\rangle_{VG}$ regroup expressions in line (5.c), where 'VG' stands for 'Viscosity Gradients'. The last term represents the homogeneous form of $\langle\epsilon\rangle_{VV}$. Note that the viscosity is inside the brackets (spatial averages).
Different terms of equation (5) are assessed through our DNS. We now make a distinction between terms that explicitly depend on the kinematic viscosity ($\mathcal{P}_\nu$, $\langle\epsilon\rangle_{VG}$ and $\langle\epsilon\rangle_{VV}$) and terms that does not (Tt, $\mathcal{D}_T$, $\mathcal{P}_k$ and $\mathcal{D}_p$). Figure 6 shows the one-point TKE budget for both CVF and VVF cases, plotted at time $\tau = 100$. In the caption of figure 6, 'Budget' stands as the RHS of equation (5), which collapses with the LHS for well-balanced TKE budget.
A critical comparison between the TKE budget for VVF ($R_\nu = 9$) and CVF ($R_\nu = 1$) cases is made in order to highlight the viscosity variations effect on the second-order, one-point statistics. Observations can be made on the peak of the production term $\mathcal{P}_k$ that is higher for VVF than for CVF; turbulence production is enhanced and accelerated by viscosity variations. This peak is located within the rapid

stream (y<0), at the same location as the peak of the mean velocity gradient $\frac{\partial}{\partial y}\langle U\rangle$

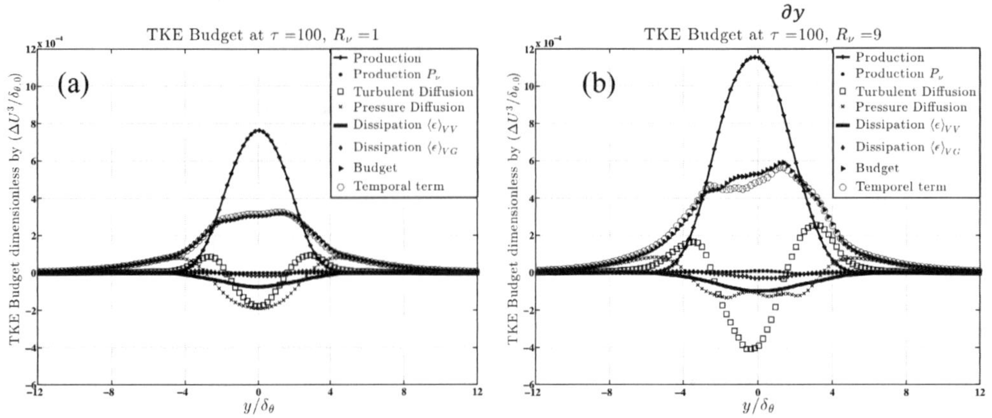

Figure 6. Turbulent kinetic energy budget for different viscosity ratios (a) $R_\nu = 1$ and (b) $R_\nu = 9$ at time $\tau = 100$.

The supplementary terms $\mathcal{P}_\nu$ and $\langle\epsilon\rangle_{VG}$, related directly to the viscosity variations, have respectively positive and negative contributions in the TKE budget at the early stages of the flow (not shown here). Their role, however, is negligible at time $\tau > 100$, as shown in figure 6 (black points and diamonds, respectively). The velocity fluctuations are stronger for VVF because their production is enhanced by the viscosity variations. It is shown in figure 6 (square symbols), that the VVF flow makes stronger turbulent diffusion (term $\mathcal{D}_T$) than CVF. It is obvious that viscosity variations lead to an enhanced entrainment of the irrotational fluid inwards the central zone of the mixing layer. The pressure diffusion term $\mathcal{D}_p$, that contributes to redistribute the kinetic energy from the centre of the TML to its edges, has a significant contribution in the TKE budget at the earlier stage of the temporal evolution of the mixing layer. Similarly to the production and turbulent diffusion terms ($\mathcal{P}_k$ and $\mathcal{D}_T$), TKE is stronger for VVF. We recall that, the instantaneous kinematic viscosity ν is, for VVF, included in the averages of the mean dissipation rate of the kinetic energy $\langle\epsilon\rangle_{VV}$. The latter may be decomposed into contributions due to both the mean viscosity and viscosity fluctuations, as follows

$$\underbrace{\langle \nu \left(\frac{\partial u_i'}{\partial x_j}\right)^2 \rangle}_{\langle\epsilon\rangle_{VV}} = \underbrace{\langle \nu \rangle \langle \left(\frac{\partial u_i'}{\partial x_j}\right)^2 \rangle}_{\langle\epsilon\rangle_{VVmean}} + \underbrace{\langle \nu' \left(\frac{\partial u_i'}{\partial x_j}\right)^2 \rangle}_{\langle\epsilon\rangle_{VVfluc}} \quad . \tag{6}$$

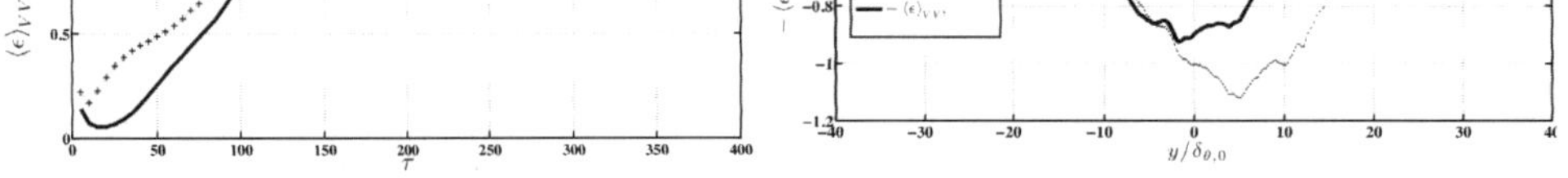

Figure 7. (a) Temporal evolution of the mean kinetic energy dissipation rate $\langle\epsilon\rangle_{VV}$ for VVF ($R_\nu = 9$) and CVF ($R_\nu = 1$), (b) Mean and fluctuating components of $\langle\epsilon\rangle_{VV}$, for VVF ($R_\nu = 9$), at the dimensionless time $\tau = 400$.

It is obvious that the term $\langle\epsilon\rangle_{VVfluc}$ is exactly equal to zero for CVF. Hence, as shown in figure 7.a, the kinetic energy is produced and dissipated in stronger way for different VVF flows, despite the fact that for CVF the kinematic viscosity coefficient is higher, or at least equal to those of VVF. Figure 7.a shows the temporal evolution of the dimensionless mean dissipation rate of the kinetic energy $\langle\epsilon\rangle_{VV}$

for both VVF ($R_v = 9$) and CVF ($R_v = 1$) situations. It can be seen that $\langle\epsilon\rangle_{VV}$ peaks earlier for VVF. Time $\tau = 140$ corresponding to its maximum at this location, whereas $\langle\epsilon\rangle_{VV}$ achieved stronger values for CVF to decrease at $\tau = 220$. The location of the maximum mean energy dissipation rate can be related to the *time-location* to the transition to turbulence. This statement suggests that the time-life of the flow is smaller for VVF, but involves more intense velocity fluctuations and a faster transition to turbulence. Furthermore, it is shown in figure 7.b that the contribution of the viscosity-fluctuating part $\langle\epsilon\rangle_{VVfluc}$ is far from being negligible, even at $\tau = 400$, corresponding to a fully turbulent mixing regime. Therefore, term $\langle\epsilon\rangle_{VV}$ does not reduce to the term proportional to the mean viscosity. Thus, the problem of viscosity variations is far from being trivial.

4. Conclusions

The paper aims at providing rigorous evidence of qualitative and quantitative differences between Variable-Viscosity Flow (VVF) and Constant-Viscosity-CVF, for the same initial flow conditions (Reynolds number $Re_{\delta_{\theta,0}}$ and turbulent intensities), but with drastic differences between the viscosity of the involved fluids. For the CVF, both low and high-stream flows are characterized by the same viscosity, whereas for VVF cases the low-stream is $R_v = 9, 6, 4$ and 2.5 more viscous than the rapid part of the flow. Our DNS showed some new flow features, which are summarized as follows:

i) The morphology of the VVF is significantly affected by the viscosity gradients.
ii) The birth of turbulence is accelerated for VVF compared to CVF.
iii) Turbulent kinetic energy is (up to 2.36 times) more important in VVF at the same time.
iv) The time-life of the flow is smaller for VVF, but involves more intense velocity fluctuations and a faster transition to turbulence.

References

[1] Talbot B, Danaila L, and Renou B. 2013 Variable-viscosity mixing in the very near field of a round jet. *Phys. Scr.* **T155**, 014006.
[2] Brown G L, Roshko A. 1974 On density effects and large structure in turbulent mixing layers. *J. Fluid Mech.* **64**, part 4, 775-816.
[3] Papamoschou D and Lele S K. 1993 Vortex-induced disturbance field in a compressible shear layer. *Phys. Fluids*. A **5,** 1412.
[4] Elliott G and Samimy M. 1990 Compressibility Effects in Free Shear Layers. *Phys. of Fluids A*, **Vol. 2**, No. 7, 1231-1240.
[5] O'reilly G and Pullin D I. 2003 Structure and stability of the compressible Stuart vortex, *J. Fluid Mech.* **vol. 493**, 231-254
[6] Freund J B, Lele S K and Moin P. 2000 Compressibility effects in a turbulent annular mixing layer. Part 1. Turbulence and growth rate, *J. Fluid Mech.* **vol. 421**,. 229-267.
[7] Dimotakis, P E 1991 On the Convection Velocity of Turbulent Structures in Supersonic Shear Layers. *AIAA* **91**-1724, 29th Aerospace Science Meeting, Reno, NV.
[8] Lu G and Lele S K. 1994 On the density ratio effect on the growth rate of a compressible mixing layer. *Phys. Fluids* **6,** 1073.
[9] Mehta R. 1991 Effect of velocity ratio on plane mixing layer development: Influence of the splitter plate wake. *Expe. in Fluids* **10**, 194-204.
[10] Laizet S, Lardeau S, and Lamballais E. 2010 Direct numerical simulation of a mixing layer downstream a thick splitter plate. *Phys. Fluids* **22**, 015104.
[11] Hiraishi M, Tsutahara M, Leung R C K. 2010 Numerical simulation of sound generation in a mixing layer by the finite difference lattice Boltzmann method. *Comp. and Mathematics with Appl.* Volume **59**, Issue 7, 2403-2410.
[12] Chaudhuri A, Hadjadj A, Chinnayya A and Palerm S. 2011 Numerical Study of Compressible Mixing Layers Using High-Order WENO Schemes. *J. Sci Comput.* **47**, 170-197.
[13] Reid R C, Prausnitz J M and Poling B E. 1987 The Properties of Gases and Liquids. *McGraw-Hill, Inc., New York, New York.*
[14] Pantano C, Sarkar S. 2002 A study of compressible effects in the high-speed turbulent shear layer using direct simulation. *J. Fluid Mech.* **451**, 329-371.
[15] Klein M, Sadiki A and J. Janicka. 2003. A digital filter based generation of in flow data for spatially developing direct numerical or large eddy simulations. *J. Comp. Phy.* **186**(2), 652-665.

PIV measurements of interface turbulence over hetero-porous media

T Iida, A Taneo, M Kaneda and K Suga
Department of Mechanical Engineering, Osaka Prefecture University, Sakai, Osaka 599-8531, Japan

E-mail: iida@htlab.me.osakafu-u.ac.jp

Abstract. In order to understand turbulent flow physics in the interface region over porous media, PIV measurements are carried out by the refractive index matching method. Two kinds of heterogeneous model porous media, whose porosities are the same but mean pore diameters are different, are applied to the measurements. The turbulent flow fields are discussed using the statistical measurement data. It is found that the fluctuation velocities increase as the mean pore diameter increases inside porous media. Also, it is confirmed that the fluctuation velocity normal to the interface increases as the interface-normal permeability increases.

1. Introduction

Understanding turbulent flow physics near highly porous materials is important for designing industrial devices such as catalyst converters, heat exchangers and flow channels inside fuel cells. Accordingly, many research studies including the authors' group's [1-3] have been carried out for turbulence over porous media. However, turbulence structure in the interface region and the effects of porous structures on turbulence are still not well understood. Furthermore, the effects of heterogeneity of the porous medium structure on turbulence are unknown. Thus, PIV measurements are carried out by the refractive index matching method using water and fluorinated ethylene-propylene (FEP) copolymer in this study. Two kinds of heterogeneous model porous media are applied to the measurements.

2. Experimental methods and procedures

2.1. Experimental setup and porous media

Figure 1 shows the experimental setup. Tap water is collected in the tank, and the water is pumped up from the tank. The flow gets through a digital flow meter and conditioned by a honeycomb section where its temperature is recorded by a digital thermometer. Then the flow is introduced into the test section that is made of transparent acrylic resin. Figure 2 shows the test section. The length of the test section is 1000mm and the width and the height are 50mm and 100mm, respectively. In the test section handmade porous medium blocks are settled up to the bottom half of the duct height. The porous media applied are made of FEP tubes or polypropylene tubes whose diameter is 3mm. FEP is a kind of fluorocarbon resin whose refractive index is 1.34 that is about the same as that of water (1.33). Thus, it allows us to access optically inside the porous media without refractive skewness. The two hetero-porous media have the same porosity and similar permeabilities but the mean pore diameters are different as shown in table 1. The structures of porous media are shown in figure 3. The difference

Journal of Physics: Conference Series **530** (2014) 012058 doi:10.1088/1742-6596/530/1/012058

between type A and type B is the setting angle of the *y*-direction tubes. The setting angle of type A is 45° while that of type B is 60° as shown in figure 3. Following the method applied in [1], the permeabilities are measured using the channel flow facility whose length is 1000mm and the width and height are both 50mm. In between the pressure holes, the porous media were placed to block the flow. The flow rate Q and the pressure drop $\Delta P/\Delta x$ along the porous media were respectively measured by a digital flow meter (FD-MII50A, KEYENCE) and a differential pressure gauge (AP12S, KEYENCE), respectively. With those variables at several different conditions, the permeability and the Forchheimer coefficient can be obtained using the Darcy-Forchheimer equation,

$$-\frac{\Delta P}{\Delta x}=\frac{\mu U_d}{K}+\frac{c_F}{\sqrt{K}}\rho U_d^2 \tag{1}$$

where μ, ρ, K, c_F and U_d are the fluid viscosity, density, the permeability, the Forchheimer coefficient and the mean velocity which can be obtained from Q, respectively. The mean pore diameter D_p is calculated from the volume of spheroid which is circumscribed by the tubes composing the porous medium by using the Sauter mean diameter method.

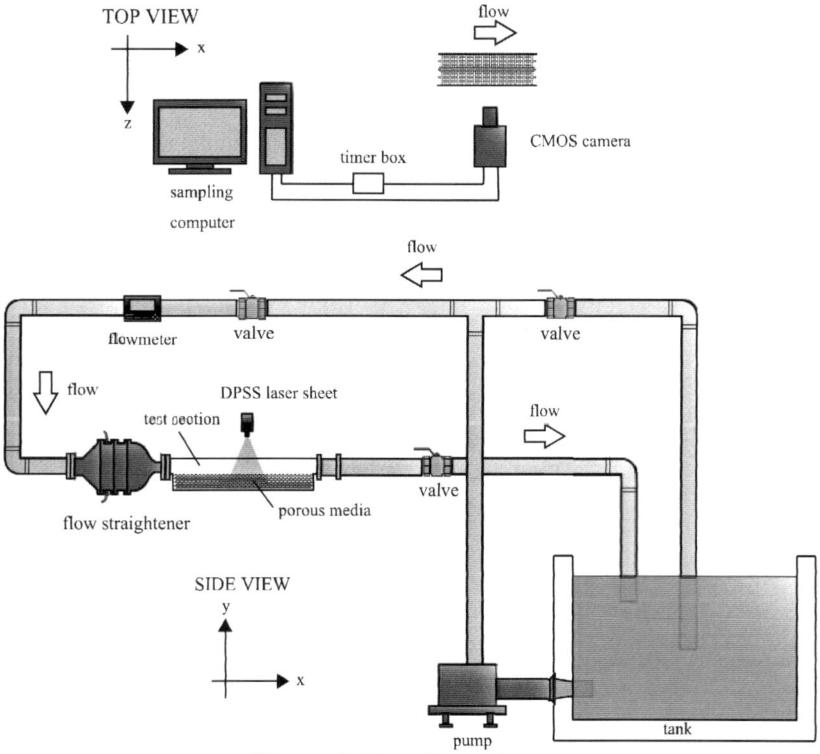

Figure 1. Experimental setup.

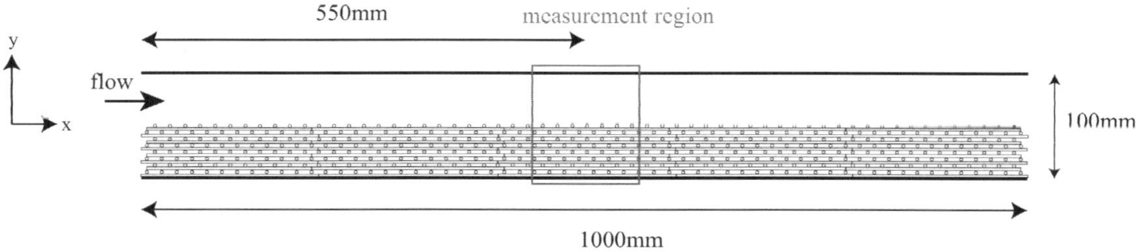

Figure 2. Test section.

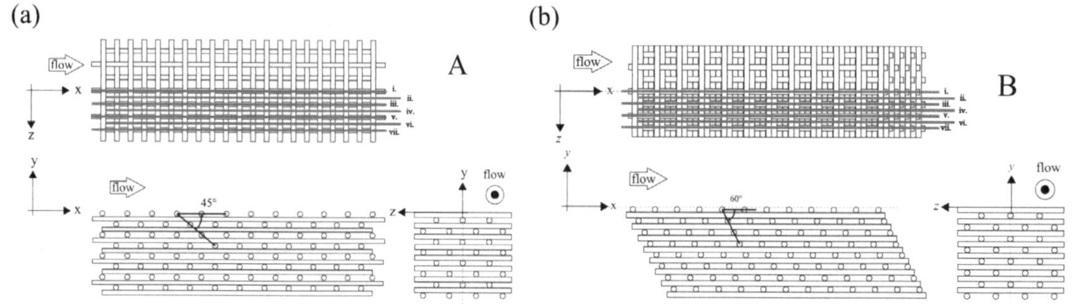

Figure 3. Porous media; (a)type A, (b)type B.

Table 1. Characteristics of porous media.

porous media	porosity φ	streamwise permeability K_x[mm^2]	cross-streamwise permeability K_y[mm^2]	absolute value of permeability K[mm^2]	mean pore diameter D_p[mm]
type A	0.8	0.32	0.25	0.41	8
type B	0.8	0.27	0.26	0.37	12

2.2. PIV measurements

The measurement procedure follows [4]. The measuring area is located at the 550mm distance from the inlet of the test section as shown in figure 2. The PIV system consists of a double-pulse diode pumped solid state (DPSS) laser with the frequency of 532nm (Ray Power 2000, DANTEC DYNAMICS), a high speed complementary metal-oxide-semiconductor (CMOS) camera (Speed Sense 9040, PHANTOM), a camera lens (MICRO-NIKKOR 55 mm f/2.8D), a sharp cut filter (SCF-50S-56O, SIGMAKOKI) and a computer for the data analysis. The laser light is expanded to a sheet light with about 1.0mm thickness and illuminates the measuring sections of the measuring area where the instantaneous images of 1623×1200 pixels are recorded for the area of 75 × 55 mm^2 by the high speed camera. The tracer particles are polymer fluorescence particles (DANTEC DYNAMICS) which contain Rhodamine B. The mean diameter, specific gravity and refractive index of the polymer fluorescence particles are respectively 10 μm, 1.50 g/cm^3 and 1.68. A sharp cut filter whose cut-off wavelength is 560 nm is applied in front of the camera lens. The seeding density is adjusted to obtain 10-15 particle pairs in each interrogation windows whose size is set to 32 × 32 pixels. The interrogation windows are overlapped 50% in each direction. Thus, the measurement sampling volume is 1.47(x) × 1.47(y) × 1.0(z) mm^3. The image sampling rate is adjusted depending on the averaged particle displacement during the time interval. In the present experiments, the averaged particle displacement is set to be about 25% length (6-8 pixels) of the interrogation window cell. To obtain the statistical data, at each location, 9000 image pairs are processed at each frame. After recording, obtained data are processed by commercial software (DANTEC DYNAMICS, Dynamic Studio 3.1) with the fast Fourier transform cross-correlation technique. Each image is processed to produce instantaneous 101 × 74 vectors.

2.3. Experimental conditions

The experimental parameter is the type of porous media. The flow Reynolds number is defined as $Re_{in}=U_{in}D_H/\nu$ where U_{in} is the inlet velocity, D_H is the hydraulic diameter of the clear flow region of the test section, ν is the kinematic viscosity of the fluid. The inlet velocity rate is measured using the digital flow meter (FD-MH10A, KEYENCE). The Reynolds number is set to Re_{in}=3600. The measurement sections shown in figure 3 are seven *x-y* sections (i-vii) whose intervals are 2.5mm from the symmetry plane.

3. Results and discussions

3.1. Secondary flows

Figure 4 shows contour maps of streamwise and wall normal mean velocities in the cross-streamwise *z-y* planes for types A and B. The velocities are normalized by the inlet velocity U_{in}. The plane of z/H=0 corresponds to the symmetry plane of the test section duct, y/H=1.0 corresponds to the top wall, y/H=0 corresponds to the interface of the porous medium and y/H=-1.0 is the bottom wall. Here, the height of the clear flow region is H=50mm. Note that the white circles in figure 4 are the cross sections of the porous media. As can be seen in figure 4 (a) and (c), in the clear fluid region (y/H=0.0-1.0) the distribution of the streamwise velocities are skewed to the corners of the duct. It is because of the secondary flows driven by turbulence anisotropy. Such secondary flows are usually seen in rectangular ducts. To exclude the effects of such secondary flows from the discussions, it is chosen to use the data of sections i-vi which are near the symmetry plane.

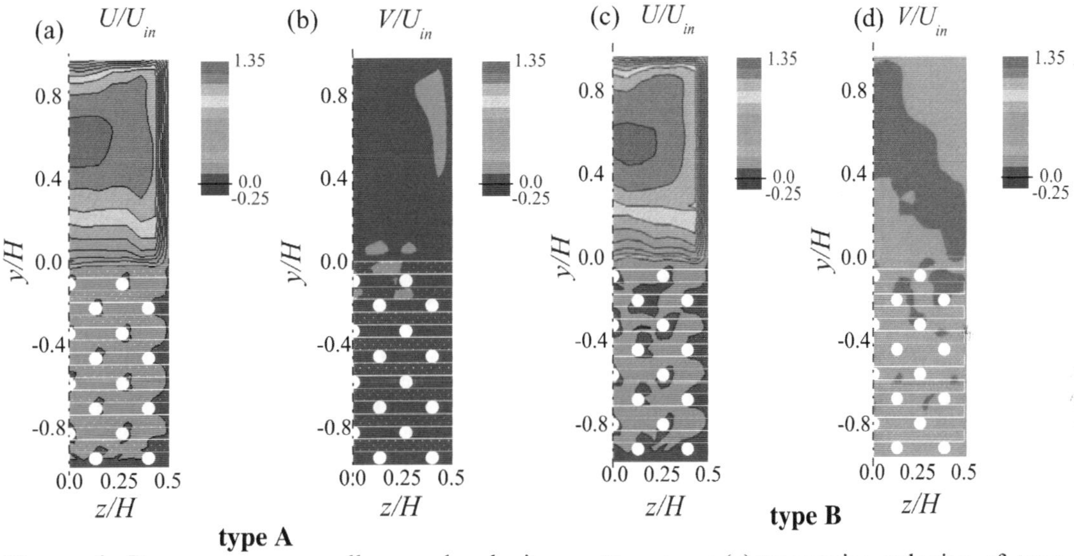

Figure 4. Streamwise and wall normal velocity contour maps;(a)streamwise velocity of type A, (b)wall normal velocity of type A, (c)streamwise velocity of type B, (d)wall normal velocity of type B.

3.2. Mean velocity profile

Figure 5 shows the streamwise and wall normal mean velocity profiles of the clear fluid regions (y/H=0.0-1.0) and inside the porous media (y/H=-1.0-0.0), respectively. These data are the time and spatial averaged values of sections i-vi of the fluid phase. The streamwise velocity U and the wall normal velocity V are normalized by the inlet velocity U_{in}. The red circle plots are the data of porous medium type A, and the blue square plots are the data of porous medium type B. Although the velocity profiles in the clear fluid region shown in figure 5 (a) and (b) do not show meaningful difference, as can be seen in figure 5 (c), the streamwise velocity in the porous medium type A is larger than that of type B. This is because the streamwise permeability K_x of type A is a little larger than that of type B. It is seen that the cross-streamwise velocity profiles in the porous media shown in figure 5 (d) are more affected by the local structure of the porous media.

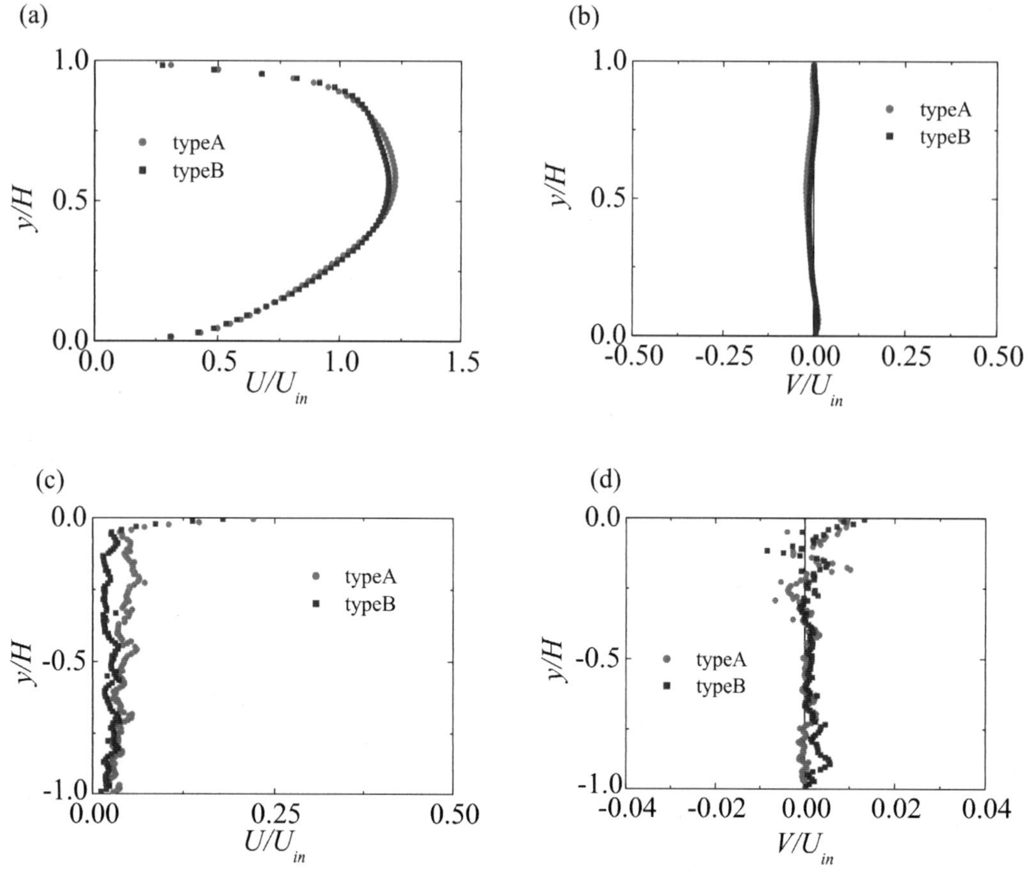

Figure 5. Streamwise and wall normal mean velocity profiles of the clear flow region and inside the porous media; (a) streamwise mean velocities in the clear flow region, (b)wall normal mean velocities in the clear flow region, (c)streamwise mean velocities inside the porous media, (d)wall normal mean velocities inside the porous media.

3.3. Fluctuation velocity and Reynolds shear stress

Figure 6 shows the RMS fluctuation velocity profiles in the clear flow regions and inside the porous media. Figure 7 shows the Reynolds shear stress profiles. As can be seen in figure 6 (a), (b) and (c), the RMS fluctuation velocities and $-\overline{u'v'}$ in the clear flow region of type A are larger than those of type B. Since the absolute value of the permeability $K(=\sqrt{K_x^2+K_y^2}\,)$ for type A is larger than that for type B, turbulence over the porous medium type A becomes larger. This tendency is consistent with the previous study [1]. However, near the interface (y/H=0.0-0.1) the streamwise profile of type B is larger than that of type A. This is related with the tendency shown in figure 6 (c). The streamwise RMS fluctuation velocity of type B near the interface (y/H>-0.3) is clearly larger than that of type A, though it is not very clear in the wall normal components shown in figure 6 (d). It is considered that the mean pore diameter affects the turbulence significantly in the region under the interface since the eddies penetrating into the interface dissipate faster in the smaller pores. Indeed, the mean pore diameter of type B is clearly larger than that of type A as shown in table 1. Additionally, the difference of the streamwise RMS fluctuation velocities between type A and type B is larger than that of the wall normal RMS fluctuation velocities as observed in figure 6 (c) and (d). This is because, the difference in the major axis lengths of the spheroids which are circumscribed by the tubes composing the porous media are larger than that in the minor axis lengths. Indeed, the major axis of spheroid of type A is 10mm, while that of type B is 13mm. However, the minor axes of spheroids of type A and B are both 9mm. Note that the major and minor axes are aligned with the streamwise and cross-streamwise directions, respectively.

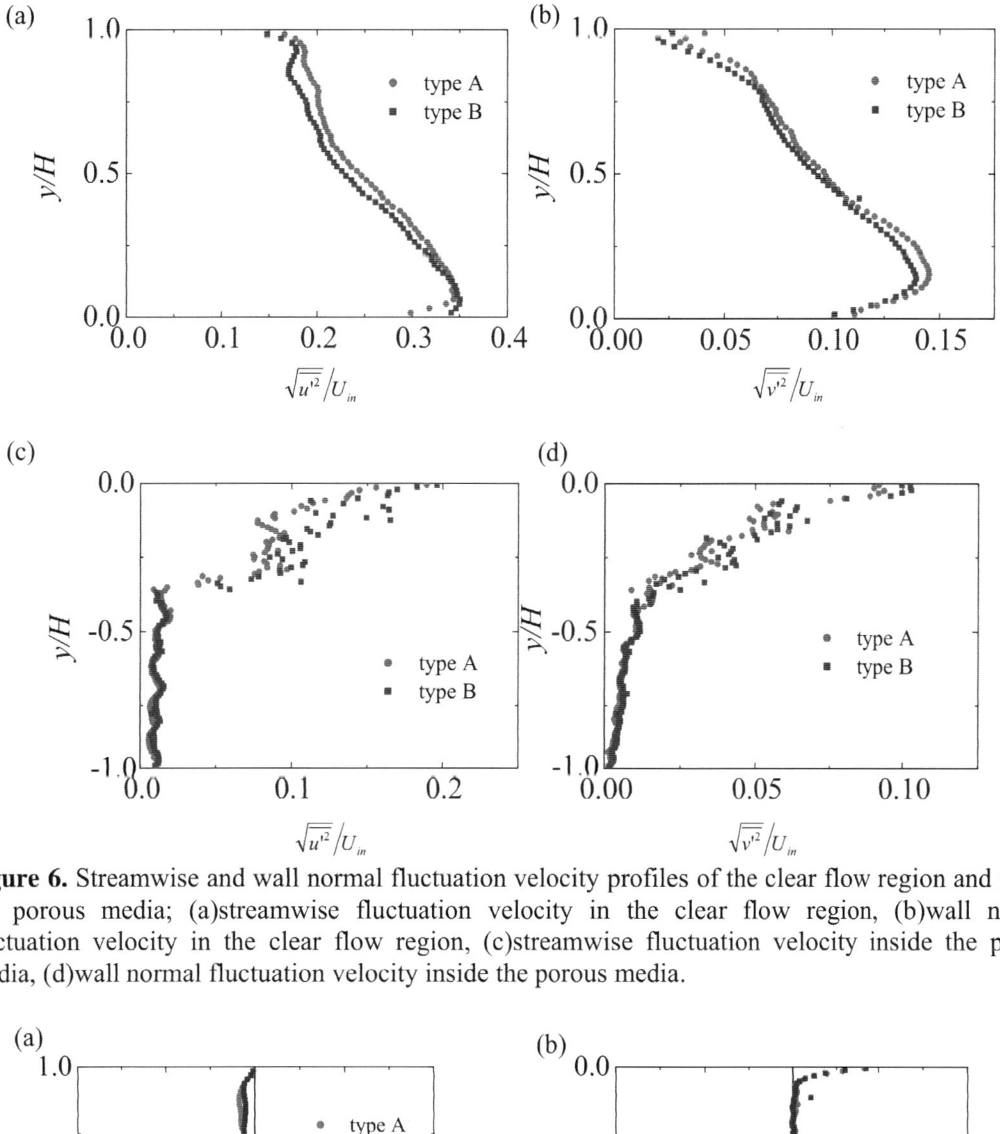

Figure 6. Streamwise and wall normal fluctuation velocity profiles of the clear flow region and inside the porous media; (a)streamwise fluctuation velocity in the clear flow region, (b)wall normal fluctuation velocity in the clear flow region, (c)streamwise fluctuation velocity inside the porous media, (d)wall normal fluctuation velocity inside the porous media.

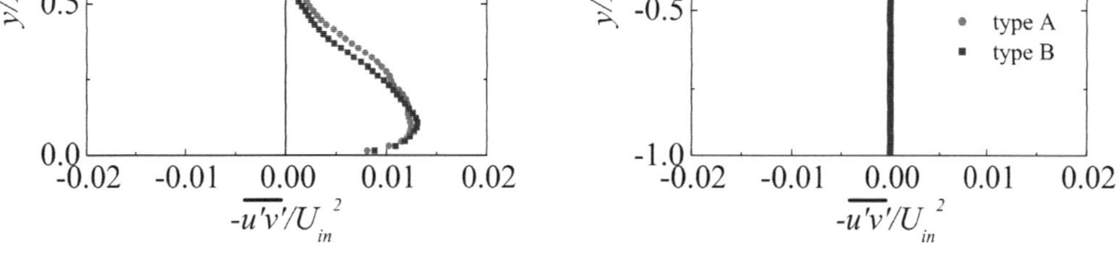

Figure 7. Reynolds shear stress profiles; (a)in the clear flow region, (b)inside the porous media.

4. Conclusions

Two velocity components of the turbulent fields over and inside hetero-porous media are measured by PIV with the refractive index matching method. A handmade porous medium is settled up to the bottom half of the channel height. Two kinds of heterogeneous porous media are applied to the present experiments to discuss the effect of the characteristics of the porous media. They have the same porosity but different permeabilities and mean pore diameters. It is found that the larger stream permeability leads to larger streamwise mean velocities inside the porous media whilst the larger mean pore diameter allows a higher level of turbulence near the interface. It is also confirmed that turbulence over the interface tends to be larger when the absolute value of the permeability becomes larger.

References

[1] Suga K, Matsumura Y, Ashitaka Y, Tominaga S and Kaneda M 2010 *Int.J. Heat Fluid Flow* **31** 974-984

[2] Suga K, Mori M and Kaneda M 2011 *Int.J. Heat Fluid Flow* **32** 586-595

[3] Suga K, Tominaga S, Mori M and Kaneda M 2013 *Flow Turbulence Combust.* **91** 19-40

[4] Suga K, Nishimura W, Yamamoto T and Kaneda M 2014 *Int.J. Hydrogen Energy* **39** 5942-5954

Stress field analysis around vortex in elastic layer of viscoelastic turbulent channel flow

T Minegishi, T Tsukahara and Y Kawaguchi
Department of Mechanical Engineering, Tokyo University of Science
Yamazaki 2641, Noda, Chiba 278-8510, Japan

E-mail: tlvkrl@gmail.com

Abstract. It is important from an engineering aspect to understand drag-reducing viscoelastic flow, since comprehension of the drag-reduction mechanism may facilitate designs with better additive types and/or channel configurations. In this study, we investigated the attenuation of vortices in fully-developed turbulence of viscoelastic fluid between parallel planes by DNS (Direct Numerical Simulation). We discussed how the viscoelastic stress play a role in the reduction of the turbulent motions around vortices and how it influences the entire flow field. For determining the major effect of viscoelasticity, we classified three stages of vortex evolution and examined statistical data for each stage. We clarified the relation between the velocity field and viscoelastic stress field using a product, or a correlation, between vorticity and rotational viscoelastic stress. We found that the inhibition of turbulence would be caused by the following aspects: damping longitudinal vortex by viscoelastic stress over a wide range in the channel, and relaxing development of vortex by the stress in vortex growing state.

1. Introduction

Viscoelastic fluid flows appear in chemical plants, food industries, and other industrial applications. In turbulent flows of dilute surfactant or water-soluble polymer solutionsthat provide viscoelasticity to the fluid,we may find the suppression of turbulent frictional drag, the so-called Toms effect.Viscoelastic fluid, with which we may obtain a high energy savings at low cost, is expected to be one promising way to solve energy issues.

There are many studies on drag-reduced wall turbulence of viscoelastic fluids by experiments and direct numerical simulations (DNS): see, for example, [1-5]. Dimitropoulos et al. [6] concluded that the large extensional viscosity of a viscoelastic fluid inhibited longitudinal vortices in turbulence, based on the budgets of the Reynolds stresses obtained by DNS. Kim &Sureshkumar [7] demonstrated developments/attenuations of a hairpin vortex localized in a viscoelastic fluid flow and found the self-consistent evolution of the vortex.However, the dynamics of vortex in the viscoelastic turbulent flow has not been sufficiently elucidated nor evaluated statistically.There are few studies of vortices by a statistical approach, despite many researches in the Newtonian flow [8-10].

In this study, we numerically investigated stress fields around vortices in viscoelastic flow through a channel. Especially, we focused on an elastic layer that should be absent in the Newtonian flow. In the elastic layer, an additional force originating from the fluid viscoelasticity would be expected to affect vortices significantly with either enhancing or attenuating the fluid vortical motions.

2. Numerical method

2.1. Governing equation and parameter

In this study, we adopted DNS, which may allow us to extract stress fields and fine-scale vortices. The governing equations for the calculation are

- the equation of continuity

$$\frac{\partial u_i^+}{\partial x_i^*} = 0 \tag{1}$$

- the momentum equation

$$\frac{\partial u_i^+}{\partial t^*} + u_j^+ \frac{\partial u_i^+}{\partial x_j^*} = -\frac{\partial p^+}{\partial x_i^*} + \frac{\beta}{\mathrm{Re}_\tau} \frac{\partial u_i^+}{\partial x_j^* \partial x_j^*} + \frac{1-\beta}{\mathrm{We}_\tau} \frac{\partial c_{ij}}{\partial x_j^*} \tag{2}$$

- the viscoelastic constitutive equation based on the Giesekus model

$$\frac{\partial c_{ij}}{\partial t^*} + u_k^+ \frac{\partial c_{ij}}{\partial x_k^*} - c_{ik} \frac{\partial u_j^+}{\partial x_k^*} - c_{jk} \frac{\partial u_i^+}{\partial x_k^*} + \frac{\mathrm{Re}_\tau}{\mathrm{We}_\tau} \left[\alpha \left(c_{ij} - \delta_{ij} \right)^2 + \left(c_{ij} - \delta_{ij} \right) \right] = 0 \tag{3}$$

where c_{ij} is the conformation tensor, $\mathrm{Re}_\tau=\rho u_\tau \delta/\mu_0$ is the frictional Reynolds number, $\mathrm{We}_\tau=\rho u_\tau^2 \lambda/\mu_0$ is the frictional Weissenberg number, α is the mobility factor for the Giesekus model, and $\beta=\mu_s/\mu_0$ is the ratio of the solvent contribution to the total zero-shear viscosity. Here, ρ is the solution density, u_τ is the frictional velocity, δ is the channel half height, λ is the relaxation time, $\mu_0=\mu_a+\mu_s$ is the zero-shear viscosity of the solution, μ_a is the zero-shear viscosity of the solvent, and μ_s is the zero-shear viscosity of the solvent.The quantities with superscript (+) indicate the wall-unit normalization by u_τ, and (*) represents the outer-scale normalization by δ.The streamwise, wall-normal, and spanwise direction are labeled as x_1, x_2 and x_3 direction, respectively.

For the spatial discretization, the finite difference method was applied. We chose the Giesekus model as the constitutive equation to compute c_{ij}. In order to improve numerical stability, we employed the minmod flux-limiter scheme, which exhibited good numerical stability and accuracy, to the advection term in equation (3). The detailed information of numerical scheme we used here can be found in our previous paper [4]. Table 1 shows numerical parameters and conditions in this work.

Table 1. Numerical parameters: Δt is the time step; L_i and N_i are the computational domain size and the number of grids in *i* direction, respectively.

Re_τ	We_τ	α	β	Δt^+	$L_1 \times L_2 \times L_3$	$N_1 \times N_2 \times N_3$
395	30	0.001	0.5	0.0079	12.8δ× 2δ× 6.4δ	512× 192 × 256

2.2. Method of vortex extraction

In this section, we briefly introduce methods for extracting vortices and averaging them. Let us use y_s^+ as a wall-normal distance of the x_1-x_3 plane we focus, and $Q2=-u_{i,j}^+ u_{j,i}^+$ a vortex strength based on the second invariant of the velocity-gradient tensor.To reduce numerical oscillation, we used $\overline{Q2}^W$ as the weighted and averaging (or filtered) $Q2$. Figure 1 shows an example of $\overline{Q2}^W$ distribution in a horizontal plane at $y_s^+=25$, where one may see a spike (at C_1) corresponding to a vortex center. Here, the point C_1, at which $\overline{Q2}^W$ is a local minimum and larger than a threshold value $Q2^T$, was defined as the vortex center. The strength of the extracted vortex at C_1 is quantified by C_{Q2}, and its temporal gradient is defined as $A = \partial C_{Q2}/\partial t^+$. Based on A, we determined the state of vortex of interest. The values of those thresholds ($Q2^T$ and A) are shown in table 2.

Table 2. Control parameters for vortex extraction.

y_s^+	12.5	25	40
$A_u(=-A_d)$	0.0077	0.0077	0.0059
$Q2^T$	0.0034	0.038	0.0038

In this study, we divided the vortex states into 3 stages: that is, a growing state of vortex, $A \geq A_u$; a sustaining state, $A_d \leq A < A_u$; and a damping state, $A \leq A_d$). It should be noted that A_u was determined to satisfy the following equation:

$$\frac{\int_{A_d}^{A_u} P(A)}{\int_{-\infty}^{+\infty} P(A)} = 0.8, \tag{4}$$

where$P(A)$ is the probability density function (PDF) of A.

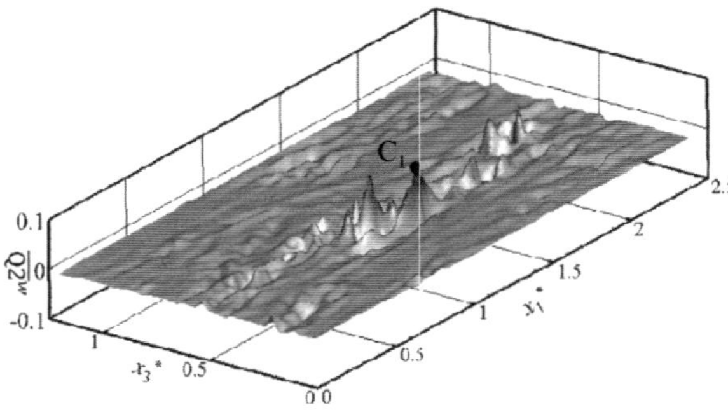

Figure 1. Distribution of in the x_1-x_3 plane at x_2^+= 25. The point of C_1 indicates the centre of a vortex. A part of the computation domain is shown. The mean flow moves from bottom-left to top-right.

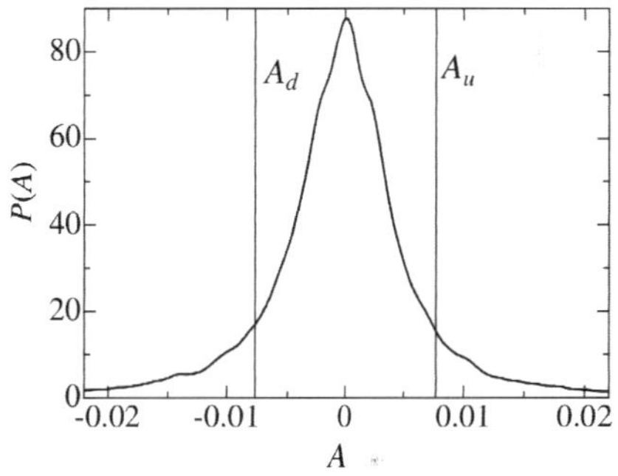

Figure 2. The probability density function of temporal gradient of vortex strength at the centre of vortex at x_2^+= 25. By criteria A_u and A_d, the state of vortex is classified.

3. States of vortex evolution

Figure 3 shows a time series of C_{Q2} at a vortex center in the viscoelastic turbulent channel flow. Here, the vortex we arbitrary chose and focused is labelled as 'V1'. It can be seen from the figure that the vortex sustained itself for a while after its development. The time range of $0 \leq t^+ < 7$ where we find a comparatively large increase in $C_{Q2}(A \geq A_u)$ is named as 'the growing state'; the plateau region corresponds to 'the sustaining state' for $7 \leq t^+ < 12$; and the range with a monotonic decrease in $C_{Q2}(A < A_d)$ is called as 'the damping state'after t^+ = 12. Moreover, G (t^+=4.0), C (t^+=8.3), and D (t^+=15.8) are defined as the instants corresponding to the median of individual state.

Figures 4 and 5 show instantaneous field at the instants of G, C and D, which are visualized from the spanwise direction. In the figures, the abscissa is the streamwise direction and the ordinate is the wall-normal direction.Gray iso-surfaces indicate vortices visualized by $Q2$. Red/blue contour shows the magnitude (or the correlation) of $\omega_i^+\omega_{Ei}^+$, which is a production of the vorticity $\omega_i^+ = \varepsilon_{ijk} u_{k,j}^+$ and the rotational viscoelastic stress $\omega_{Ei}^+ = \varepsilon_{ijk} c_{kl,jl}$. Here, ε_{ijk} describes the Levi-Civita symbol. In a region where $\omega_i^+\omega_{Ei}^+$ is negative, the viscoelastic stress is exerting on the vortical motion. The contour of $\omega_1^+\omega_{E1}^+$ presents the viscoelastic contributions around streamwise vortices. In figure 4, $\omega_1^+\omega_{E1}^+$ is found to be negative near the wall, indicating that the viscoelastic stress would inhibit longitudinal vortices in the near-wall region. The vortex V1, which is the strongest vortex in the focused area shown in figure 4, seems to be large with time, because the viscoelastic stress works on the vortex and would suppress its vertical motion.

Figure 5 shows the viscoelastic contribution with respect to the spanwise component $\omega_3^+\omega_{E3}^+$. The distribution of $\omega_3^+\omega_{E3}^+$ reveals that the viscoelastic stress affects the shear (or the vortex sheet) rather than the vortex. We conjecture that the viscoelastic stress would prevent occurrence of ejection from the near-wall region by relaxing the high-shear layer between relatively high-speed and low-speed regions.

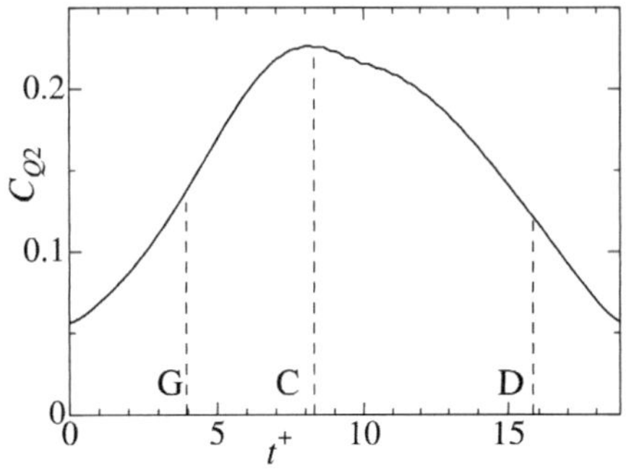

Figure 3. A time variation of a vortex V1 at y_s^+=25 (G, the growing state of vortex; C, the sustaining state; and D, the damping state).

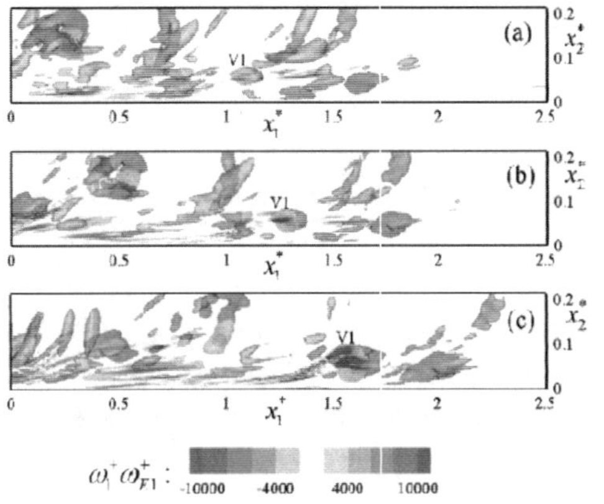

Figure 4. Time variation of vortices: the mean flow moves to right. Gray iso-surfaces, positive $Q2$ regions corresponding to vortices; colour contour, $\omega_1^+\omega_{E1}^+$. (a) at the growing state, G; (b) the sustaining state, C; and (c) the damping state, D.

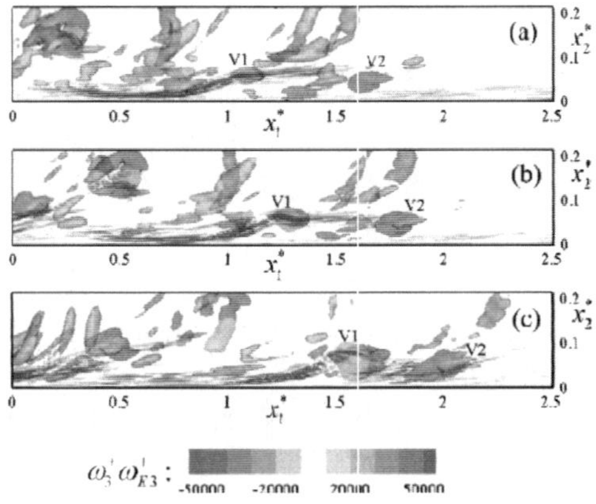

Figure 5. Same as figure 4, but the colour contour shows $\omega_3^+\omega_{E3}^+$.

4. Statistical investigation

Figures 6-8 show statistical fields, which are acquired by the conditional averaging described in section 2.2. The contours show either or . Figures 6(a) and 7(a) show the growing state of the vortex with its center at y_s^+ = 25, (b) the sustaining state, (c) the damping state.

The distribution of shown in figure 6 reveals that the streamwise vortex encounters the resistance force of the viscoelastic stress. This aspect is similarly found in each distribution given in figure 6, implying less dependency on the wall-normal height. However, the relatively small magnitude of at the reference point (the vortex center) in figure 6(c) means the less correlation between the vortical motion and the reaction of viscoelastic stress. As can be seen from the figures, the negative correlation between them becomes significant in the near-wall region irrespectively of the wall-normal position of vortex.

In figure 7, the contours of in the near-wall region reveals sheet-like distribution. As found in the instantaneous field (figure 5), the viscoelastic stress works against the high shear region. Similar inclined structure of the high regions are observed in both figures 5(a) and 7(a). Thus, it can be said that the viscoelastic stress would relax high-shear layer on the growing state of a near-wall vortex. On the other hand, the viscoelastic stress has rather small influence on the vortex at the sustaining and damping states: see figures 7(b) and (c).

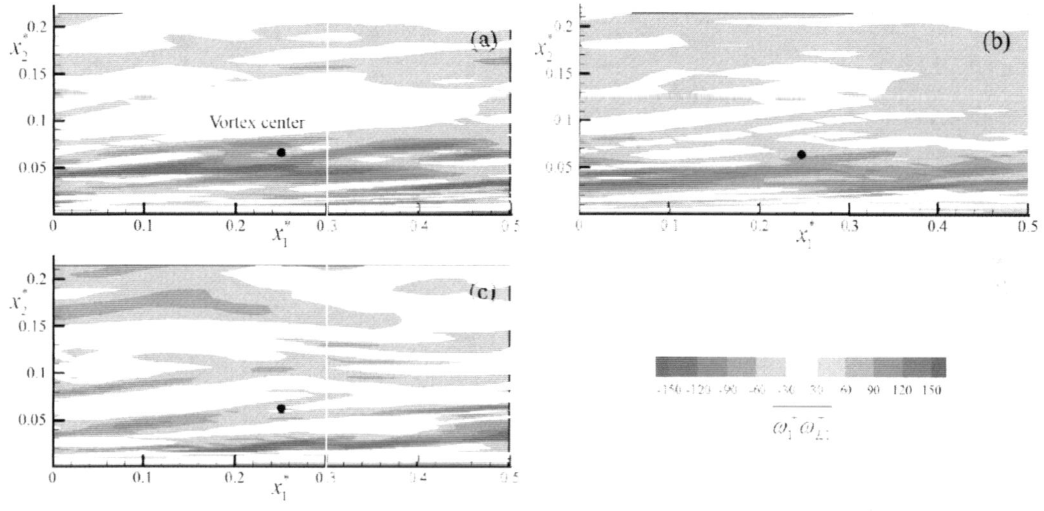

Figure 6. The difference of vortex state by streamwise rotational viscoelastic contribution in statistic field at y_s^+=25. (a) growing state of vortex; $A \geq A_u$, (b) conservation state; $A_d \leq A < A_u$, and (c) damping state; $A \leq A_d$.

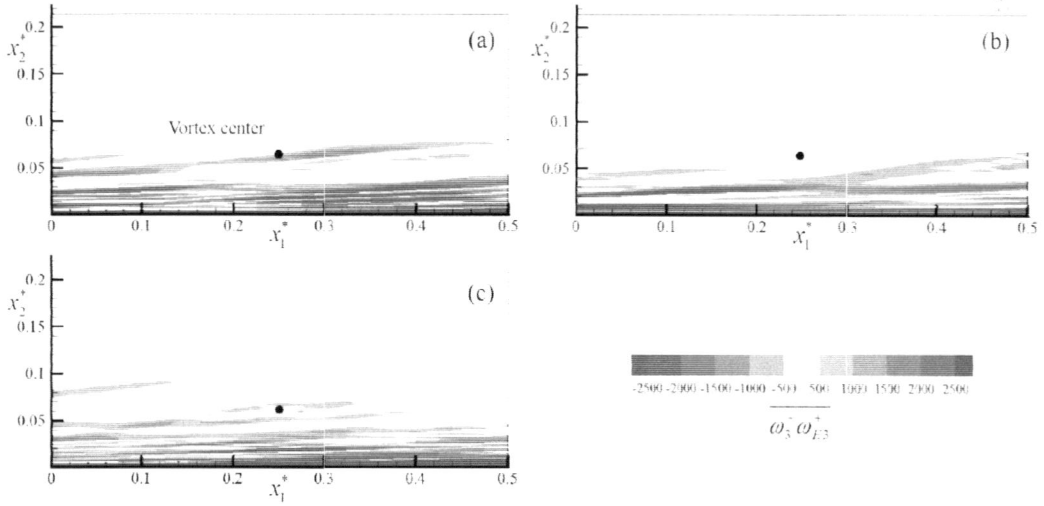

Figure 7. The difference of vortex state by spanwise rotational viscoelastic contribution in statistic field at y_s^+=25. (a) growing state of vortex; $A \geq A_u$, (b) conservation state; $A_d \leq A < A_u$, and (c) damping state; $A \leq A_d$.

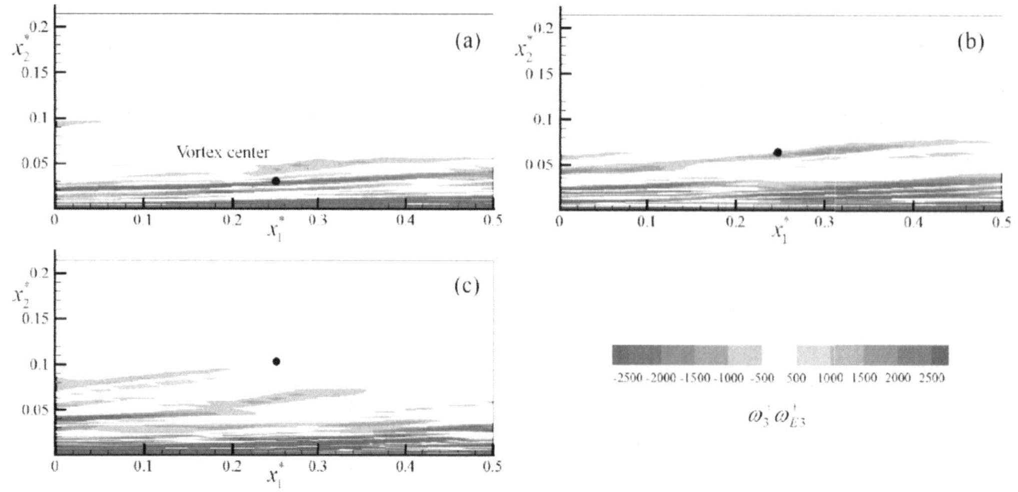

Figure 8. Dependence of the spanwise rotational viscoelastic contribution in on y_s^+ at the growing state. (a) y_s^+=12.5, (b) y_s^+=25, and (c) y_s^+=40.

Figure 8 shows the contours of for the growing state with comparing results with different reference heights (y_s^+=12.5, 25, and 40). The viscoelastic stress contribution relaxes shear layers in $x_2^+<30$. This near-wall layer may relate to the elastic layer, which is absent in the Newtonian flow. The mean velocity profile (figure not shown) indicates the elastic layer appears in $7 \leq x_2^+ \leq 30$ in the present flow condition.

5. Conclusions

We investigated the viscoelastic contribution around vortices, based on the correlation of the vorticity and the rotational viscoelastic stress, in a viscoelastic turbulent channel flow. For determining the major effect of viscoelastic stress, we examined statistical data, with emphasis on the vortex evolution. We found that viscoelastic stress would relax high-shear layers in the growing state of a vortex. As a result, viscoelastic stress prevent occurring to ejection. Moreover, the longitudinal vortex tends to be reduced by the viscoelastic stress throughout the channel. In the viscoelastic flow, turbulent motions would be inhibited by these two factors: the suppressions of high shear and of developing streamwise vortex.

6. References

[1] Virk P S 1971 *J. Fluid Mech.***45** 417
[2] Dimitropoulos C D, Sureshkmar R and Beris A N 1998 *J. Non-Newtonian Fluid Mech.***79** 433
[3] Yu B, Li F and Kawaguchi Y 2004 *Int. J. Heat and Fluid Flow***25** 961
[4] Tsukahara T, Ishigami T, Yu B and Kawaguchi Y 2013*J. Turbulence***12** N13
[5] Thais L, Gatski T B and Mompean G 2013*Int. J. Heat and Fluid Flow***43** 52
[6] Dimitropoulos C D, Suresukmar R, Beris A N and Handler R A 2001 *Phys. Fluids***13** 1016
[7] Kim K and Sureshkmar R 2013 *Phys. Rev. E* **87** 063002
[8] Jeong J, Hussain F, Schoppa W and Kim J 1997 *J. Fluid Mech.***332** 185
[9] Zhou J, Adrian R J, Balachander S and Kendall T M 1999 *J. Fluid Mech.***387** 353
[10] Adrian R J 2007 *Phys. Fluid***19** 041301
[11] Kasagi N, Sumitani Y, Suzuki Y and Iida O 1995 *Int. J. Heat and Fluid Flow***16** 2-10

Acknowledgments
This work was supported by Grants-in-Aid for Scientific Research (C) #25420131 from Japan Society for the Promotion of Science (JSPS). The present DNS was performed using the super supercomputing resources at CyberscienceCenter, Tohoku University, and at CybermediaCenter, Osaka University.

Journal of Physics: Conference Series **530** (2014) 012060 doi:10.1088/1742-6596/530/1/012060

Application of Fast Fourier Transform in thermo-magnetic convection analysis

L Pyrda
AGH University of Science and Technology, Faculty of Energy and Fuels, Department of Fundamental Research in Energy Engineering, Poland

E-mail: pyrda@agh.edu.pl

Abstract. Application of Fast Fourier Transform in thermo-magnetic convection is reported. Cubical enclosure filled with paramagnetic fluid heated from below and placed in the strong magnetic field gradients was investigated. The main aim of study was connected with identification of flow types, especially transition to turbulence. For this purpose the Fast Fourier Transform (FFT) analysis was applied. It was followed by the heat transfer characteristic for various values of magnetic induction gradient. The analysis was done at two Rayleigh numbers $7.89 \cdot 10^5$ and $1.86 \cdot 10^6$ with thermo-magnetic Rayleigh numbers up to $1.8 \cdot 10^8$ and $4.5 \cdot 10^8$ respectively. The presented results clearly indicate flow types and also demonstrate augmented heat transfer in dependence on magnetic induction gradient. Detailed analysis of flow transition to turbulent state was compared with transition line for natural convection reported in literature. The transition to turbulence in the case of thermo-magnetic convection of paramagnetic fluid was in very good agreement with transition in the case of natural convection.

1. Introduction

Since first quantitative studies on the natural convection reported by Bénard, natural convection in a configuration heated from below (also called Rayleigh-Bénard configuration) was subject of research in many experimental and theoretical studies. In later studies most of scientific interest was devoted to description of fluid behavior, quantitative and qualitative visualization techniques, as well as presentation of dimensionless parameters which could characterize such phenomena. Most of previous studies defined flow as turbulent as soon as its behavior was non-periodic. In year 1987, after statistical analysis, Heslot [1] proposed three types of turbulence: transitional regime, soft turbulence ($10^5<Ra<4 \cdot 10^7$) and hard turbulence ($4 \cdot 10^7<Ra<6 \cdot 10^{11}$). In recent experiments Niemela et al. [2] with cryogenic helium as a working fluid presented results for hard turbulence up to $10^6<Ra<10^{17}$. All these studies addressed natural convection phenomena and Prandtl number close to 1.

In 1991 year Braithwaite et al. [3] presented for the first time natural convection of paramagnetic non-electrically conducting fluid in connection with the magnetizing force effect. They reported the enhancement or suppression of heat transfer rate through shallow paramagnetic fluid layer in Rayleigh-Bénard configuration. After that some of researchers extended this work applying the different geometry, flow visualization techniques and dimensionless parameters description i.e.: [4],[5],[6]. All this studies where connected with magnetic field effect on the paramagnetic or diamagnetic fluids considered steady laminar flow regimes. Most recent studies where connected with transition from laminar to turbulent flow regimes which could be obtained by application of high gradient magnetic field (up to 900 [T^2/m], for the magnetic field induction in the centre of superconducting magnet bore $|b_0|_{max}$=10 [T] in contrast to 200 [T^2/m], for magnetic field induction of $|b_0|_{max}$ =5 [T] in previous studies) i. e.: [8],[9],[10],[11],[12].

The goal of presented investigation was an analysis of transition from soft turbulence to hard turbulence regimes in the case of thermal and magnetic convection. This was achieved, by employing high gradient magnetic field generated by the superconducting magnet and application of Fast Fourier Transform method in phenomenon analysis.

Throughout the paper the magnetic field induction conditions were used to identify different flow regimes since it was a simple and easily controllable parameter in experimental investigations, although the magnetic field gradient was the main driving mechanism behind the magnetization force [9].

2. Experiment

2.1. Experimental setup

Experimental setup presented in Figure 1 consisted of: a superconducting magnet generating strong magnetic field, a cubical enclosure, a thermo-stating bath, a power supply and a data acquisition system. The superconducting magnet opening has a diameter of 0.1 [m] and length of 0.5 [m]. Experimental enclosure is shown in detail in Figure 2.

The central part of experimental setup was cubical enclosure of 0.032 [m] size. It was placed inside magnet opening in Rayleigh-Bénard configuration. The lower horizontal wall was electrically heated, while the upper one was cooled with thermo-stated water. Temperature of both walls was controlled with six thermocouples. Six other thermocouples were inserted into the enclosure to measure temperature of the fluid. Due to them it was possible to recognize type of the flow and identify the transition between steady and turbulent flow.

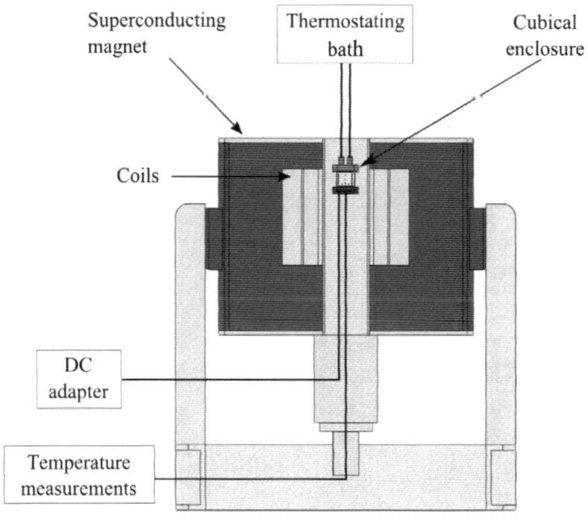

Figure 1. Experimental setup and location of the cubical enclosure with paramagnetic working fluid

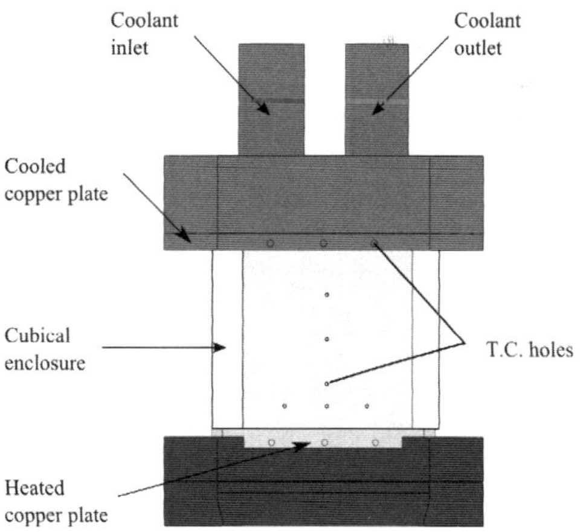

Figure 2. Cubical enclosure (L=D=H=0.032 [m]).

2.2. Working Fluid

As a working fluid for experimental studies 50% volume aqueous solution of glycerol with 0.8 [mol/(kg of solution)] concentration of gadolinium nitrate hexahydrate ($Gd(NO_3)_3 \cdot 6H_2O$) was chosen. All important for present studies properties of fluid were measured experimentally: the magnetic susceptibility (magnetic susceptibility balance, Evan's method), the thermal expansion coefficient (pycnometer), the dynamic viscosity coefficient (viscosimeter, thermally controlled), the thermal conductivity (thermal conductivity meter) and the heat capacity (differential calorimeter). Due to the narrow temperature range studied, it could be assumed that almost all needed properties were constant, except viscosity. Therefore the dependence of viscosity on the temperature was checked experimentally and approximated function could be written as follows:

$$\mu_{(\theta)} = 1.057 - 6.5 \cdot 10^{-3}\theta + 1.0 \cdot 10^{-5}\theta^2 \tag{1}$$

All important for presented investigation fluid properties are listed in Table 1.

Table 1. Properties of the working fluid at 298 K.

Property	Symbol	Value	Unit
Heat capacity	C_p	$2.92 \cdot 10^{3}$	[J/kg·K]
Thermal diffusivity	α	$9.13 \cdot 10^{-8}$	[m²/s]
Thermal expansion coefficient	β	$4.78 \cdot 10^{-4}$	[1/K]
Dynamic viscosity	μ	$1.30 \cdot 10^{-2}$	[kg/m·s]
Thermal conductivity	λ	0.376	[W/m·K]
Kinematic viscosity	ν	$9.25 \cdot 10^{-6}$	[m²/s]
Density	ρ	1411	[kg/m³]
Mass magnetic susceptibility	χ_m	$2.39 \cdot 10^{-7}$	[m³/kg]

2.3. Experimental course

The main aim of this paper was to identify transition to turbulent flow and then to analyze its influence on the heat transfer. Thermal measurements were used to obtain both stated aims. The experiment was divided into two steps. First step it was thermal measurements with increasing magnetic induction for temperature difference between heated and cooled walls equal to 5 and 11 [K]. Signal of thermocouples inserted into the cubic enclosure was then analyzed with FFT. The second step was connected with the energy balance and estimation of heat loss to the environment. To estimate heat loss, the conduction experiment was carried out with reversely placed thermally active horizontal walls of experimental enclosure at investigated position. At first a specific temperature difference between horizontal walls was selected and then after thermal stabilization the heating power was measured, which together with theoretical heat flux obtained from Fourier's law allow estimation of heat loss:

$$Q_{\text{loss}} = Q_{\text{cond}} - Q_{\text{theor\_cond}} \tag{2}$$

$$Q_{\text{theor\_cond}} = D\lambda(\theta_{\text{h}} - \theta_{\text{c}}) \tag{3}$$

where: θ_h, θ_c - are temperature of hot and cold walls respectively.

Heat loss was calculated according to eq. (2) and could be linearly approximated for various heating rates ($\Delta\theta = \theta_h - \theta_c = 2.5 \div 15$ [K], where temperature of cold wall was kept at ambient temperature inside magnet), according to:

$$Q_{\text{loss}} = 0.8\Delta\theta \tag{4}$$

2.4. Dimensionless numbers

The thermo-magnetic convection can be characterized by a group of non-dimensional parameters such as Nusselt, Prandtl, thermal and magnetic Rayleigh and magnetization numbers.

Nusselt number was calculated as a ratio between the net convective heat transfer rate and the net pure conduction contribution with applied method invented by Ozoe and Churchill [13] to estimate net heat fluxes:

$$\mathrm{Nu} = \frac{Q_{\mathrm{net\_conv}}}{Q_{\mathrm{net\_cond}}} = \frac{Q_{\mathrm{conv}} - Q_{\mathrm{loss}}}{Q_{\mathrm{cond}} - Q_{\mathrm{loss}}} \tag{5}$$

Assuming that convection heat flux was equal to heater heat flux ($Q_{\mathrm{net\_conv}}=Q_{\mathrm{heater}}$) and applying eq. (2) and (3) to eq. (5) it could be rewritten in following form:

$$\mathrm{Nu} = \frac{Q_{\mathrm{heater}} - Q_{\mathrm{loss}}}{Q_{\mathrm{theor\_cond}}} = \frac{IU - Q_{\mathrm{loss}}}{\mathrm{D}\lambda(\theta_{\mathrm{h}} - \theta_{\mathrm{c}})} \tag{6}$$

The Prandtl number definition was:

$$\mathrm{Pr} = \frac{\nu}{\alpha} \tag{7}$$

The thermal Rayleigh number was defined as:

$$\mathrm{Ra_T} = \frac{\beta \mathbf{g} \Delta\theta \mathrm{D}^3}{\nu\alpha} \tag{8}$$

where β is the thermal expansion coefficient, **g** is the gravitational vector, $\Delta\theta$ is the temperature difference between horizontal thermally active walls, D is the enclosure size (D=0.032 [m]), ν is the kinematic viscosity coefficient and α is the thermal diffusivity.
The magnetic Rayleigh number was defined as:

$$\mathrm{Ra_M} = \left(1 + \frac{1}{\beta\theta_0}\right)\left(\frac{\gamma\beta \mathbf{g} \Delta\theta \mathrm{D}^3}{2\nu\alpha}\right) \tag{9}$$

The magnetization number:

$$\gamma = \frac{\chi_{\mathrm{m}} |\mathrm{b_0}|^2_{\mathrm{max}}}{\mu_0 \mathbf{g} \mathrm{D}} \tag{10}$$

where: $\mu_0=4\cdot\pi\cdot10^{-7}$ [H/m].
The thermo-magnetic Rayleigh number was defined as:

$$\mathrm{Ra_{TM}} = \mathrm{Ra_T} + \mathrm{Ra_M} \tag{11}$$

3. Results and discussion

Temperature time series of investigated fluid for $\Delta\theta$=5 [K] case at chosen $|\mathrm{b_0}|_{\mathrm{max}}$ are presented in Figure 3. The black color of signal corresponds to the position of chosen thermocouple placed at the central lowest position (0.004 [m] above bottom wall of enclosure – Figure 2). For further analysis only signal from this one thermocouple was selected to obtain detailed information on flow phenomena. The Fast Fourier Transform (FFT) analysis was conducted for all cases of chosen thermocouple signal and the results are presented as the amplitude versus frequency in Figure 3 (right). In initial state ($|\mathrm{b_0}|_{\mathrm{max}}$=0 [T]) temperature signal is oscillating. In presence of weak magnetic induction temperature signal is stabilized. Result at $|\mathrm{b_0}|_{\mathrm{max}}$=4 [T] shows temperature signal before transition to turbulence and temperature signal at $|\mathrm{b_0}|_{\mathrm{max}}$=5 [T] after that. The last case ($|\mathrm{b_0}|_{\mathrm{max}}$=10 [T]) shows temperature signal with smaller amplitude then 5 [T] of magnetic induction because it was gradually decreasing with increase in the magnetic induction.

For $\Delta\theta$=5 [K] case (Figure 3), at $|\mathrm{b_0}|_{\mathrm{max}}$=0 [T] oscillation of temperature signal has characteristic frequency equal to 0.012 [Hz]. At $|\mathrm{b_0}|_{\mathrm{max}}$=4 [T], before transition to turbulence, the characteristic frequency of small oscillations is 0.074 [Hz]. The FFT analysis of temperature time series obtained at $|b_0|_{max}$=5 [T] shows the highest peak at 0.052 [Hz]. With further increase of magnetic induction amplitude of oscillation was decreasing and got more uniform but far more intensive than results before transition, which can be seen at $|b_0|_{max}$=10 [T] (Figure 3).

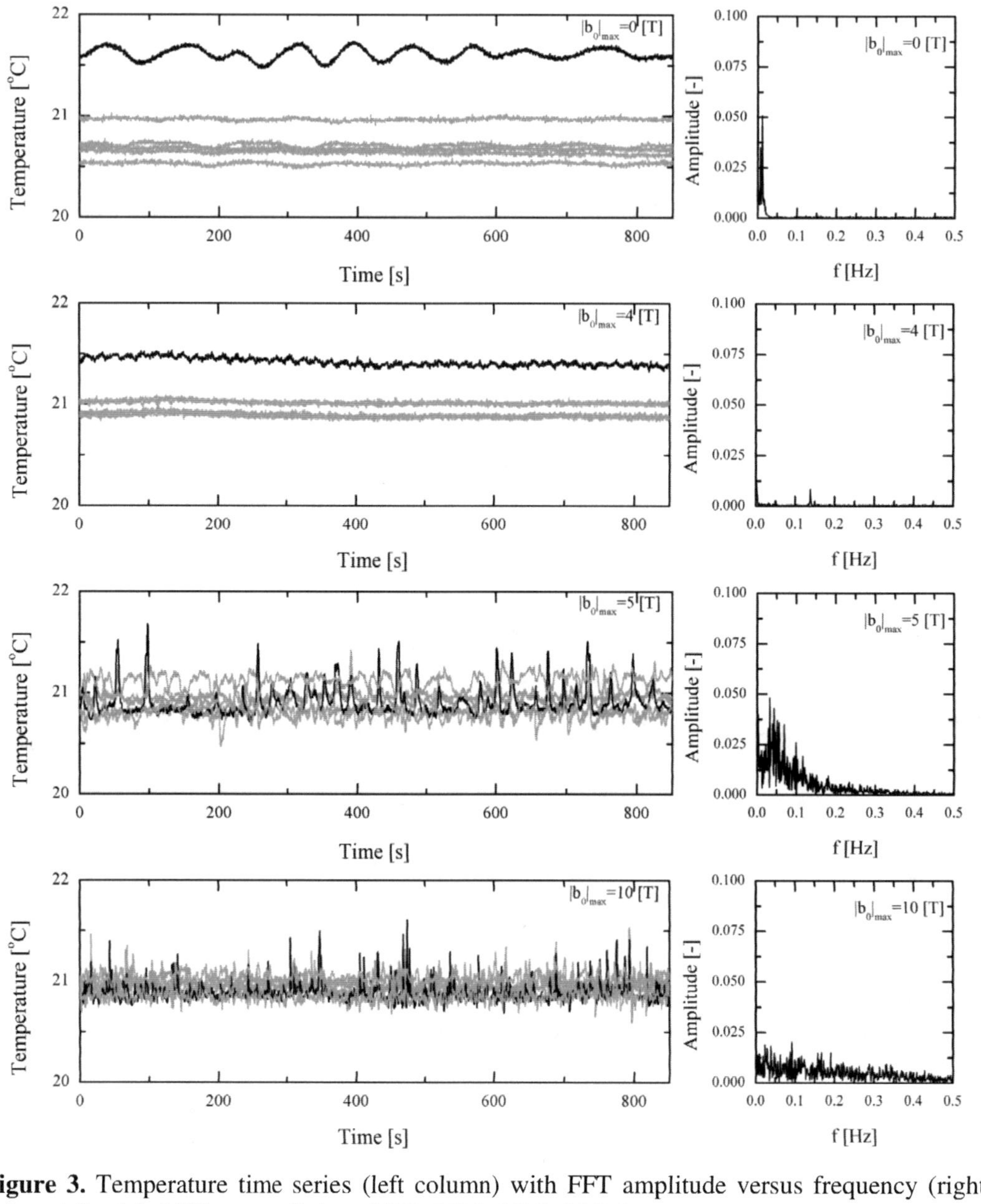

Figure 3. Temperature time series (left column) with FFT amplitude versus frequency (right column) for $\Delta\theta$=5 [K] case at $|b_0|_{max}$: 0 [T], 4 [T], 5 [T], 10 [T]

Temperature time series of investigated fluid for $\Delta\theta$=11 [K] case at chosen $|b_0|_{max}$ are presented in Figure 4. Initial state ($|b_0|_{max}$=0 [T]) is stable. Then presence of weak magnetic induction (from $|b_0|_{max}$=1 [T]) results in oscillation of temperature signal up to transition to turbulence and temperature signal at $|b_0|_{max}$ =3 [T] after that. The last case ($|b_0|_{max}$=10 [T]) shows temperature signal with smaller amplitude then 3 [T] of magnetic induction because it was gradually decreasing with increase in the magnetic induction.

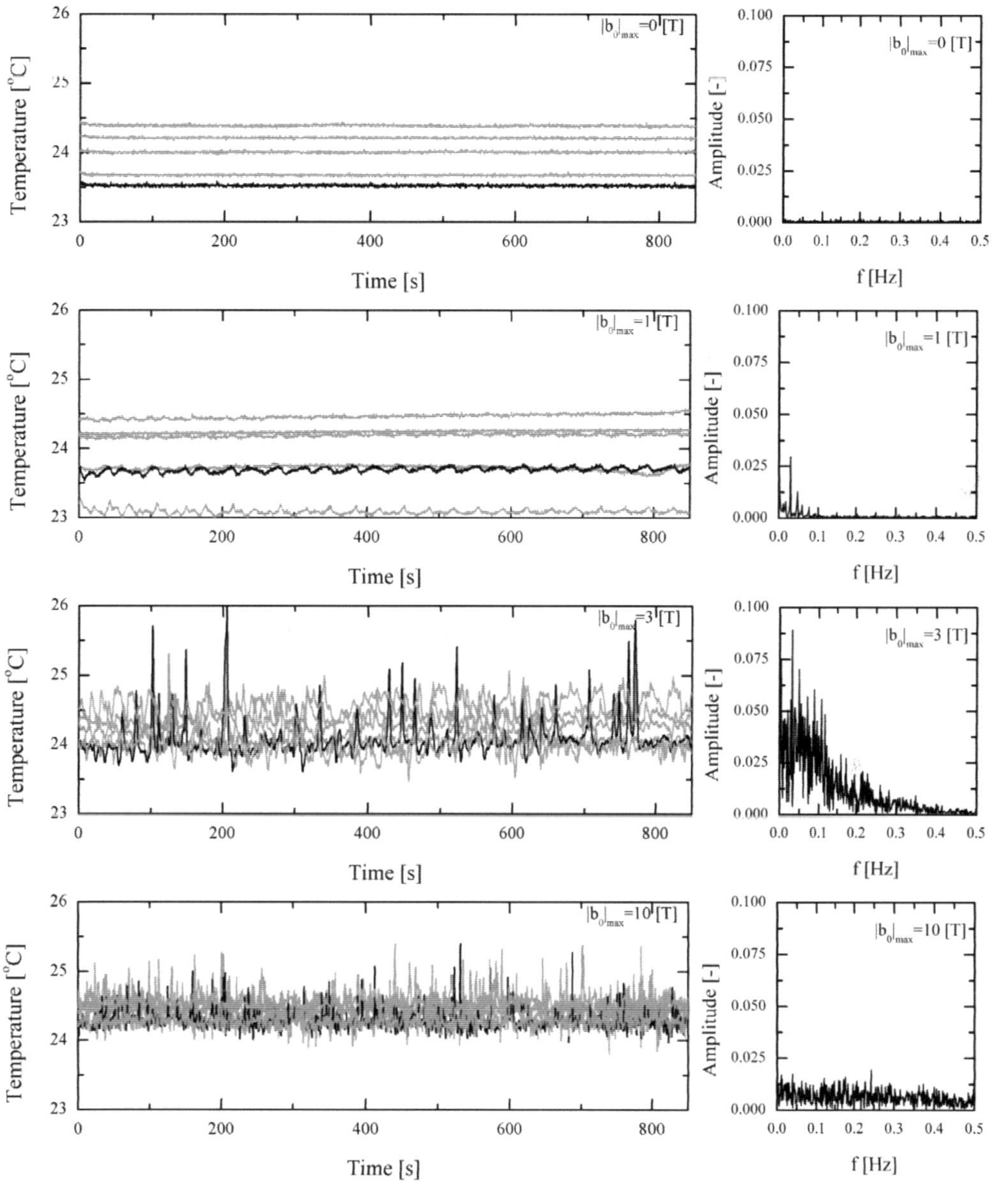

Figure 4. Temperature time series (left column) with FFT amplitude versus frequency (right column) for $\Delta\theta$=11 [K] case at $|b_0|_{max}$: 0 [T], 4 [T], 5 [T], 10 [T]

For $\Delta\theta$=11 [K] case (Figure 4), at $|b_0|_{max}$=0 [T] temperature signal is stable. At $|b_0|_{max}$=1 [T] the results indicate the most characteristic frequencies at 0.029 [Hz] which is kept up to transition. The FFT analysis of temperature time series obtained after transition at $|b_0|_{max}$ =3 [T] shows the highest peak at 0.032 [Hz] (Figure 4 - right). With further increase of magnetic induction amplitude of oscillation was decreasing and got more uniform but far more intensive than first results before transition, which can be seen at $|b_0|_{max}$=10 [T] (Figure 4).

The Nusselt number plotted versus the magnetic induction in the centre of magnet ($|b_0|_{max}$) for 5 [K] and 11 [K] of temperature difference between horizontal thermally active walls ($\Delta\theta=\theta_h-\theta_c$) are presented in Figure 5. The Prandtl number was 121 and 101 for $\Delta\theta=5$ [K] and $\Delta\theta=11$ [K] respectively. Measurements error was calculated with combined uncertainty analysis and was less than: 6.8% and 3% for $\Delta\theta=5$ [K] and $\Delta\theta=11$ [K] respectively. The heat transfer enhancement with increasing value of magnetic induction was observed for both studied temperature differences. The Nusselt number increased more than 300% at 10 [T] of magnetic induction in comparison with natural convection case.

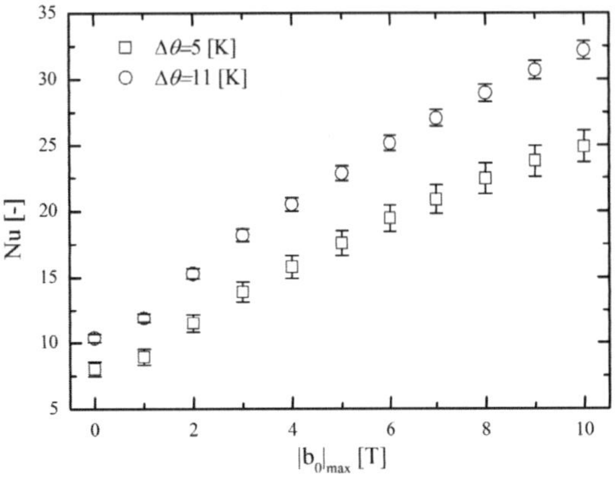

Figure 5. The Nusselt number plotted versus the magnetic induction at the centre of magnet ($|b_0|_{max}$) for $\Delta\theta=5$ [K] and $\Delta\theta=11$ [K]

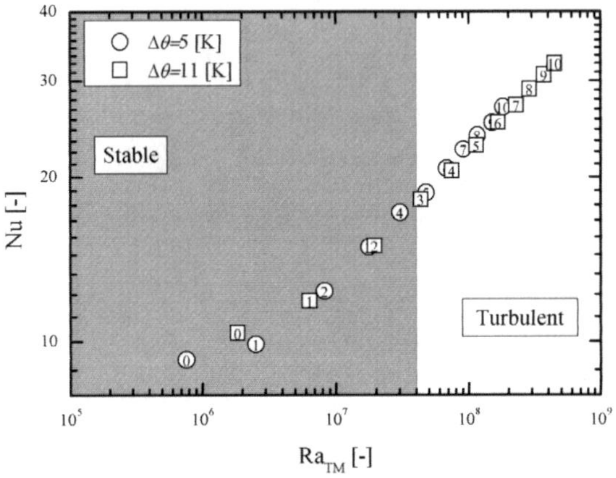

Figure 6. Nusselt number versus thermo-magnetic Rayleigh number with magnetic induction $|b_0|_{max}$ value inside each caption

The Nusselt number versus thermo-magnetic Rayleigh number with caption consisting $|b_0|_{max}$ value for all investigated cases is presented in Figure 8. Background color of Figure 8 represents type of fluid behavior and transition point at $Ra_{TM}=4\cdot10^7$ which was found by Heslot [1] for natural thermal convection.

4. Conclusions

The analysis of unsteady flow in the case of thermo-magnetic convection was studied experimentally. An influence of Rayleigh number and magnetic field strength was examined. Due to the application of steady magnetic field at chosen temperature difference two flow types were presented. With an application of Fast Fourier Transform it was possible to determine the characteristic frequency for temperature time series. The FFT analysis clearly indicates the type of the flow, what was shown at first in the form of temperature time series and then amplitude vs. frequencies. Identification of characteristic frequencies highest peak in the case of 5 and 11 [K] temperature differences at 5 and 3 [T] of magnetic induction allowed calculation of Rayleigh number at transition to turbulence. It was $Ra_{TM}=4{,}3\cdot10^7$ and $Ra_{TM}=4{,}7\cdot10^7$ respectively, which correspond to value found by Heslot [1] and was first time found after him. In the turbulent region the heat transfer was significantly augmented. The heat exchange increased more than 300% in comparison with natural convection. The presented results showed that this experimntal method can be applied in the investigations of time stable and, what is more important time and space unstable fluid flow.

5. Acknowledgments

This work was supported by the Polish National Science Centre
Project No. 2012/07/B/ST8/03109.

6. References

[1] Heslot F, Castaing B and Libchaber A 1987 Transitions to turbulence in helium gas *Phys. Rev. A* **36** 5870–3

[2] Niemela J, Skrbek L, Sreenivasan K and Donnelly R 2000 Turbulent convection at very high Rayleigh numbers *Nature* **404** 837–40

[3] Braithwaite D, Beaugnon E and Tournier R 1991 Magnetically controlled convection in a paramagnetic fluid *Nature* **354** 134 - 136

[4] Ozoe H 2005 *Magnetic Convection* (London: Imperial College Press)

[5] Bednarz T, Fornalik E, Tagawa T, Ozoe H and Szmyd J S 2006 Convection of paramagnetic fluid in a cube heated and cooled from side walls and placed below a superconducting magnet–comparison between experiment and numerical computations *Therm. Sci. Eng. J.* **14** 107–14

[6] Wrobel W, Fornalik-Wajs E and Szmyd J S 2010 Experimental and numerical analysis of thermo-magnetic convection in a vertical annular enclosure *Int. J. Heat Fluid Flow* **31** 1019–31

[7] Akamatsu M, Yoshida Y, Okitsu T, Kaneda M and Ozoe H 2013 Behavior of magnetothermal Rayleigh–Benard convection of air visualized by 3D simulations *J. Vis.* **17** 27–39

[8] Kenjereš S, Wrobel W, Pyrda L, Fornalik-Wajs E and Szmyd J S 2011 Transients and turbulence pockets in thermal convection of paramagnetic fluid subjected to strong magnetic field gradients *J. Phys. Conf. Ser.* **318** 072028

[9] Kenjereš S, Pyrda L, Wrobel W, Fornalik-Wajs E and Szmyd J 2012 Oscillatory states in thermal convection of a paramagnetic fluid in a cubical enclosure subjected to a magnetic field gradient *Phys. Rev. E* **85** 1–8

[10] Pyrda L, Kenjeres S, Fornalik-Wajs E and Szmyd J S 2012 An analysis of unsteady thermal convection of paramagnetic fluid in cubical enclosure under strong magnetic field gradient *J. Phys. Conf. Ser.* **395** 012125

[11] Kenjereš S, Pyrda L, Fornalik-Wajs E and Szmyd J S 2013 Numerical and Experimental Study of Rayleigh–Bénard–Kelvin Convection *Flow, Turbul. Combust.* **92** 371–93

[12] Pyrda L 2013 *An analysis of magnetic field influence on transient regime between laminar and turbulent convection* (AGH University of Science and Technology)

[13] Ozoe H and Churchill S W 1973 Hydrodynamic stability and natural convection in Newtonian and non-Newtonian fluids heated from below *AIChE Symp. Ser. Heat Transf.* **69 (131)** 126–33

Journal of Physics: Conference Series **530** (2014) 012061 doi:10.1088/1742-6596/530/1/012061

Influence of Pressure Gradient on Streamwise Skewness Factor in Turbulent Boundary Layer

Artur Dróżdż

Department of Mechanical and Computer Science, Czestochowa University of Technology, Armii Krajowej 21, Poland

E-mail: arturdr@imc.pcz.czest.pl

Abstract. The paper shows an effect of favourable and adverse pressure gradients on turbulent boundary layer. The skewness factor of streamwise velocity component was chosen as a measure of the pressure gradient impact. It appears that skewness factor is an indicator of convection velocity of coherent structures, which is not always equal to the average flow velocity. The analysis has been performed based upon velocity profiles measured with hot-wire technique in turbulent boundary layer with pressure gradient corresponding to turbomachinery conditions. The results show that the skewness factor decreases in the flow region subjected to FPG and increases in the APG conditions. The changes of convection velocity and skewness factor are caused by influence of large-scale motion through the mechanism called amplitude modulation. The large-scale motion is less active in FPG and more active in APG, therefore in FPG the production of vortices is random (there are no high and low speed regions), while in the APG the large-scale motion drives the production of vortices. Namely, the vortices appear only in the high-speed regions, therefore have convection velocity higher than local mean velocity. The convection velocity affects directly the turbulent sweep and ejection events. The more flow is dominated by large-scale motion the higher values takes both the convection velocity of small-scale structures and sweep events induced by them.

1. Introduction

The effect of favourable (FPG) and adverse pressure gradients (APG) on turbulent boundary layer is not well understood. The most recent results of Harun et al. [1] show that the flow in FPG and APG conditions is driven by large-scale motion. Namely for FPG the decrease of large-scale energy occurs, while for APG the reaction is opposite as it was shown by Monty et al. [2] and Dróżdż & Elsner [3]. In the APG the energy of large-scale motion is enough high to creates so-called outer peak of fluctuations. It is interesting how this high energy large-scale motion influence the production of turbulence near the wall. Mathis et al. [4] showed that in turbulent boundary layers there is amplitude modulation of the small-scale structures by large-scale motion. In this mechanism the amplitude of small-scales recorded by hot-wire probe increases in the large-scale high-speed regions. Latter on Mathis et al. [5] found that the wall-normal profile of the amplitude modulation between the large-scale and the envelope of the small-scales exhibits strong resemblance to the skewness profile of streamwise u fluctuation component. The skewness factor, which is the moment of third order, is a measure of the asymmetry of the probability distribution. Mathis et al. [5] decomposed streamwise velocity signal u onto small-scale u_S and large-scale u_L and concluded that the skewness factor cross term $\overline{\overline{3u_L u_S^2}}$ is an alternative diagnostic measure to determined amplitude modulation rate. The

knowledge about the skewness factor distribution is important because it affects directly the small-scale turbulent quadrant events production associated with so-called sweep and ejection events. Sweep is an event, which has positive streamwise and negative wall-normal fluctuations, while ejection event has opposite relation. It was concluded that, for turbulent boundary layer in zero pressure gradient (ZPG), the sweep and ejection events are equally important for the turbulence production [6]. This is consistent with the equality of convection velocity to the mean local velocity in the outer region (at least for the small-scale motion). In case of pressure gradient flows the turbulence production is dominated by the ejection events (negative streamwise fluctuations) under FPG, while in APG by the sweep events (positive streamwise fluctuations). Therefore, with the enhancement of the sweep event or alternatively ejection event, the value of the streamwise skewness factor should increase or decrease respectively. The speculation presented above was proved by Simpson et al. [7] who showed that close to flow separation the skewness factor of u is positive, what indicates that there is enhanced sweep event. Nagano and Houra [8] suggested that "the rise in skewness is caused by structural changes in the near-wall region due to the pressure gradient". Johansson and Alfredsson [9] showed that in the near-wall region of channel flow, "a positive skewness reflects the presence of high shear events imposed on a background of lower turbulence level". The above literature review suggests that there is a direct relation between convection velocity and the increase of particular quadrant turbulent events.
The aim of the paper is to confirm the direct relation between skewness factor and convection velocity of small-scale structures. The analysis has been performed based upon velocity profiles measured with hot-wire technique in turbulent boundary layer with pressure gradient corresponding to turbomachinery conditions. As a measure of pressure gradient influence on near-wall turbulence a skewness factor of streamwise velocity component was chosen.

2. Experimental apparatus and conditions

The experiment was performed in an open-circuit wind tunnel, where the turbulent boundary layer developed along the flat plate, which was 2807 mm long and 250 mm wide. The test section is located in the rear part of the wind tunnel. Its upper wall of test section was shaped according to the assumed distribution of pressure gradient corresponding to the conditions encountered in axial compressor blading. Additionally, to modify the flow circulation around the plate and to ensure negligible streamwise pressure gradient, the trailing edge flap and grid at the outlet from the test section was also applied. In order to obtain a fully developed turbulent boundary layer the tripping of boundary layer, after the leading edge of a flat plate was used. The facility is equipped with the computer-controlled, 2D traversing system (in streamwise and wall-normal direction).
Static pressure measurements were done using 70 pressure holes and the results of measurements is shown in figure 1.The pressure distribution is typical for turbomachinery case, where after short region of zero pressure gradient flow accelerates (from x_s = 197 mm) and then (from x_s = 427 mm) it decelerates. It is seen that pressure gradient values are within the range of –0.27 ÷ 0.28 Pa/mm. Velocity profiles were measured with single hot-wire anemometry probe of a diameter $d = 3\,\mu$m and length $l = 0.4$ mm (Dantec Dynamics 55P31). The probe was combined with the DISA 55M hot-wire bridge connected to 14 bit PC card. Acquisition was maintained at frequency 50kHz with 10 seconds sampling records.
The position at the wall closest point of the hot-wire probe was determined using the mirrored image. Positions of 8 measuring traverses, dotted lines, are shown in figure 1. The distances of traverses from inlet plane, the corresponding dimensionless distances $Sg = x_s / L$, where L is the length of the test section (L = 1067mm) and the most important flow parameters are given in table 1. The conditions determined in inlet plane ($Sg = 0$), located in zero pressure gradient area are the mean velocity in core flow $U_\infty \approx$ 15 m/s, turbulent intensity $Tu =$ 0.4%.

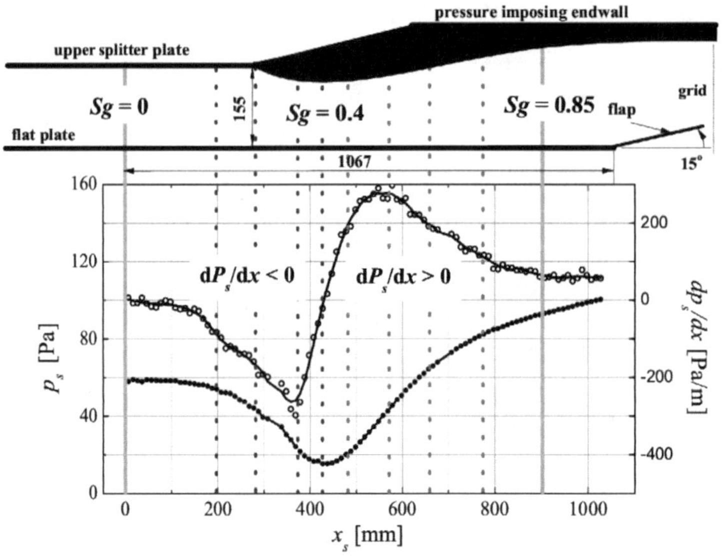

Figure 1. The shape of upper wall and the corresponding static pressure and pressure gradient distributions along the flat plate.

Table 1. Location and parameters of the analysed profiles.

PG	Traverse number	x_s [mm]	Sg [-]	U_∞ [ms^{-1}]	u_τ [ms^{-1}]
FPG	1	197	0.185	15.14	0.632
FPG	2	277	0.26	15.61	0.677
FPG	3	367	0.344	16.97	0.744
FPG	4	427	0.4	17.48	0.776
APG	5	487	0.456	16.97	0.718
APG	6	577	0.541	15.64	0.582
APG	7	667	0.625	14.45	0.473
APG	8	787	0.738	13.21	0.370

3. Results

3.1. Amplitude modulation mechanism

The physical significance of the amplitude modulation of small-scales by the large scales is not currently well understood [10]. It should be however, presumed that according to this mechanism, the large-scale motion induces low and high-speed regions near the wall and the amplitude of the small-scales is increased in the large-scale high-speed regions.

The qualitative relationship of amplitude modulation of the small-scale by the large-scale component where presented in figure 2 taken from Dróżdż and Elsner [3]. In order to have small- and large- scale signal components the carrier velocity signal was filtered by high- and low- pass filtering respectively. The cut-off frequency was set at timescale $\tau^+ \approx 160$, which is located in the middle between the inner and outer peaks. Figure 2 presents a carrier signal (upper time trace) recorded for

APG conditions in the buffer layer ($y^+ \approx 50$), large-scale signal component (middle time trace), while the lower time trace represents the small-scale signal component together with the envelope (dotted lines) corresponding to the large-scale signal. The envelopes were shifted in order to easily trace the variations of magnitude of the small-scale component. The coupling is evident, the small-scale component signal is clearly modulated by the large-scale motions in the APG. The small-scale amplitude to some extent varies with the envelope of the large-scales. It is clearly visible on the raw signal (upper trace) that the small scales in the high-speed large-scale regions are shifted towards positive values of scale, what suggest that they have convection velocity higher than the mean flow value.

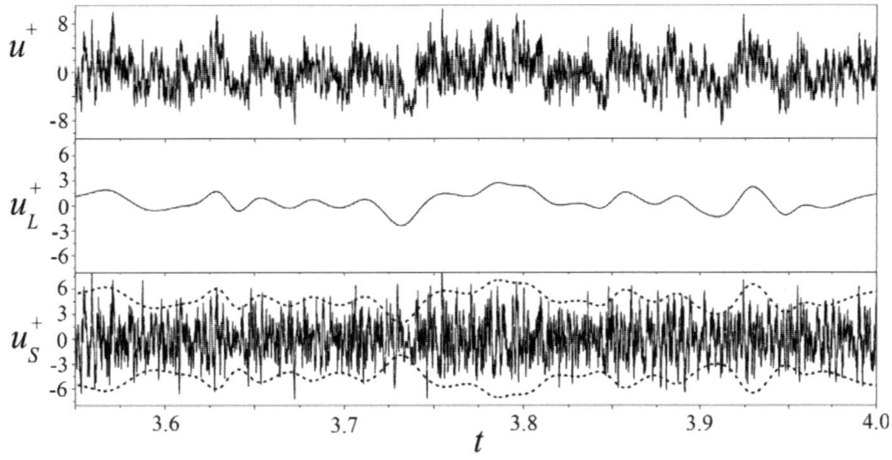

Figure 2. Visual comparison of the scales coupling [3].

As it was shown by Harun et al. [1] the large-scale motion is less active in FPG and more active in APG. Therefore, in FPG the production of vortices is more or less random (there are no high and low speed regions). While in the APG the large-scale motion drives the production of vortices, which appear mainly in the high-speed regions. The more the flow is dominated by large-scale motion the higher values the skewness factor takes.

3.2. *Skewness factor and convection velocity*

In order to illustrate the relation between the skewness factor and convection velocity the skewness factor was calculated for streamwise velocity fluctuation, according to following formula:

$$S_f = \frac{\overline{u^3}}{\overline{u^2}^{3/2}} \tag{1}$$

The results were shown in figure 3. The first group of the profiles (blue colour) were measured in FPG flow region, while the second group of profiles (red colour) were measured in APG flow region. As it can be observed the skewness factor is the highest close to the wall. Zero crossing is changing with the pressure gradient right above $y^+ \approx 15$, where well-known inner peak of fluctuation exist. It varies to the location $y^+ \approx 300$, where outer peak of fluctuation emerges in the APG region. Above this location the distributions are similar. The influence of pressure gradient on skewness factor is clearly visible. For the range of y^+ between10 and 300 favourable pressure gradient causes the decrease of skewness factor, while adverse pressure gradient causes the increase of skewness factor. It is worth to note that when the APG is strong enough, the skewness factor takes the positive values in the whole inner region of boundary layer.

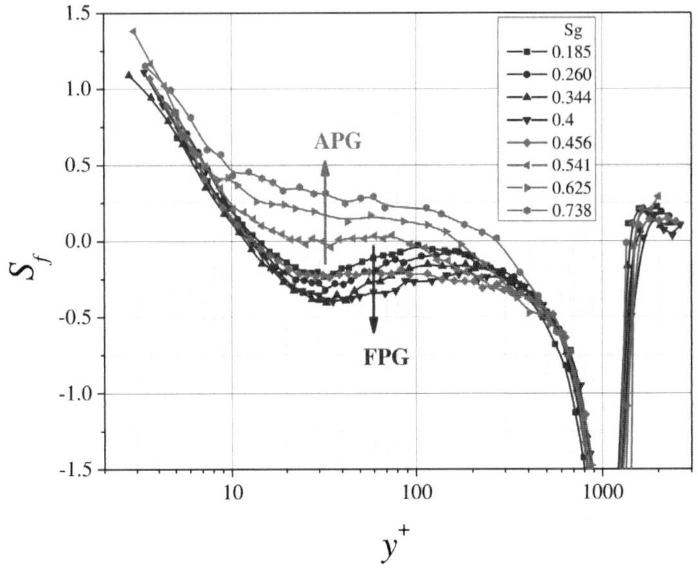

Figure 3. Influence of FPG and APG on skewness factor S_f.

The knowledge about the skewness factor distribution is important, because it seems that it is a good indicator of convection velocity of small-scale coherent structures, what was shown by Dróżdż & Elsner [11]. The paper confirmed the influence the pressure gradient on strength of turbulent quadrant events i.e. sweep and ejection. It was shown that in the FPG region the ejection event close the wall was enhanced while the sweep event was reduced. In APG region the reaction was opposite, namely the sweep event was enhanced, while ejection event was reduced. They conclude that the convection velocity was lower than the mean velocity in the FPG region and higher in the APG region. Particularly when convection velocity is higher the sweep event (with $u > 0$) is stronger, while when convection velocity is lower the ejection event (with $u < 0$) is stronger. In order to quantify the small-scale ejection and sweep changes the mean velocity u_C between sweep event maximum $< u_{sweep} >$ and ejection event minimum $< u_{ejection} >$ for different pressure gradients were calculated according to following formula:

$$u_C = \frac{< u_{sweep} > + < u_{ejection} >}{2}, \tag{2}$$

where $<\_>$ is phase-averaging of small-scale ejection or sweep events.

The results were shown in figure 4. First group of profiles (blue colour) were measured in FPG flow region, while the second group of profiles (red colour) were measured in APG flow region. As it can be observed the velocity u_C distributions for different pressure gradient have similar shapes as distributions of skewness factor presented in figure 3. The positive values corresponds to convection velocity which is higher than mean local velocity, while negative values corresponds to convection velocity lower than mean local velocity. It is worth to note that below the inner peak of fluctuation ($y^+ = 15$) the skewness factor and u_C is always positive and increase toward the wall. It coincidence with constant convection velocity below inner peak showed by Johansson [12]. In the same region the mean velocity value decreases towards zero at the wall, therefore the difference between constant convection and increase of mean velocities is a result of what is shown in figure 4. This difference is also responsible for sweep events domination close the wall.

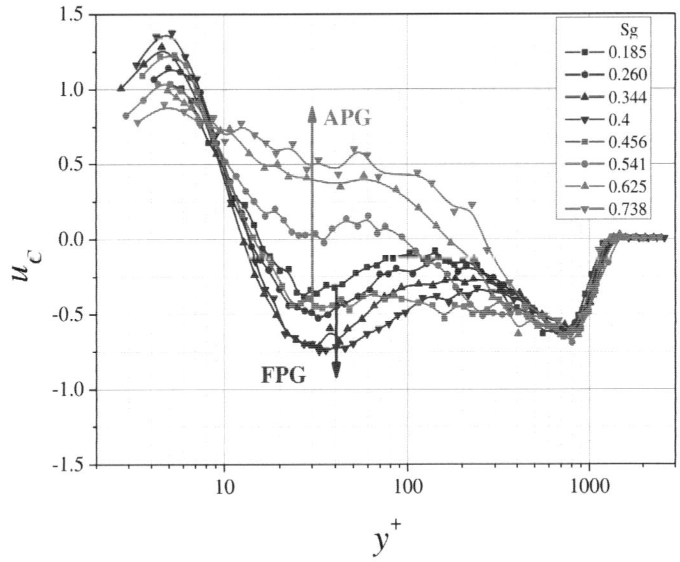

Figure 4. Influence of FPG and APG on phase-averaged velocity u_C.

The above results show that the convection velocity changes can be explained by amplitude modulation phenomenon, which is quantified by the skewness factor. With the increase of pressure gradient the energy of large-scale motion increases and it is the direct result of amplitude modulation. With the increase of large scale motion the small-scale structures are enhanced or created in the large-scale high-speed regions near the wall thus the small-scale structures convect at that velocity, which is above the local mean velocity and results with increase of sweep events strength.

The concept of this process has been clarified in figure 5, where was showed that strength of quadrant events is a function of their convection velocity. It occurs with the increase of large-scale motion energy not only with pressure gradient but also with Reynolds number. The upper schemes present the small-scale vortical structure for different convection velocity and how it influences the ejection and sweep events. Below the vortical structures schemes the areas of momentum transport changes is shown. For simplicity it was assumed that above the vortex center there is high momentum fluid (HMF), while below there is low momentum fluid (LMF). Middle case in figure 5 shows that when the difference between convection U_C and mean U velocities is zero, the energy of the sweep and ejection induced by the vortex is the same. The LMF occupy the same area as HMF, because the same amount of fluid is ejected as swept. It results in no change of total momentum in the area. Left case in figure 5 shows that transfer of the energy from sweep to ejection occurs. When the convection velocity is lower than mean velocity the difference between U_C and U is added to ejection event and subtracted from sweep event. The LMF area is greater than HMF area, because the higher amount of fluid is ejected than swept. It results in decrease of total momentum in the area. Right case in figure 5 shows that when the convection velocity is higher than mean velocity, the transfer of energy occurs from ejection to sweep. The difference between U_C and U is added to sweep event and subtracted from ejection event. The HMF area is greater than LMF area, because the higher amount of fluid is swept than ejected what results in increase of total momentum in the area.

Summing up in FPG and APG flows the large-scale motions cause the increase and decreases of small scales convection velocity respectively. Additionally, the momentum near the wall in FPG is lower

(LMF area is greater than HMF area in figure 5), while in the APG is higher (HMF area is greater than LMF area in figure 5). The higher momentum near the wall is responsible for postpone of boundary layer separation in the APG region when the Reynolds number increases. Buckles et al. [13] showed that in the flow separation, the vortices send fluid toward the wall (sweep event) and entrain fluid from the reversed-flow region upward (increase of momentum). It is correlated with large, positive skewness values in the reversed-flow region caused by the passage of shear-layer vortices overhead. These observations led Buckles et al. [13] to suggest that the detached shear flow was driven by a mechanism other than just the external pressure gradient. Also Maciel et al. [14] pointed out that "pressure force and the turbulent transport no longer play an important dynamic role close to separation and flow become essentially as inertial flow". The above literature proved that similar effect was observed, however the physical mechanism was not proposed.

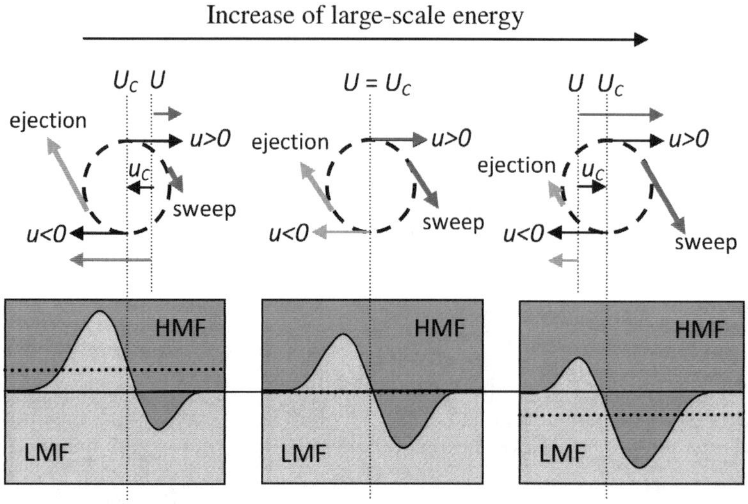

Figure 5. Influence of large scale motions on quadrant events energy (LMF – low momonetum fluid, HMF – high momentum fluid).

The observed phenomenon extends the knowledge about wall-normal momentum transfer in turbulent boundary layer. It was proved that the momentum transfer is additionally stimulated by large-scale motion, even though large-scale motion itself do not increase momentum close the wall, but only modulate small-scale structures production through the amplitude modulation mechanism. Especially, it is evident in APG region, where the production of small-scales vortices is increased in high-speed large-scale regions, the opposite reaction occurs in the low-speed regions. Therefore the structures have increased convection velocity, what in turn influence the quadrant events, which are responsible for wall-normal momentum transfer.

4. Conclusions

The results show that the skewness factor decreases in the flow region subjected to FPG and increases in the APG conditions. The change of both convection velocity and skewness factor is caused by influence of large-scale motion through the mechanism called amplitude modulation. It was proved that the momentum transfer is additionally stimulated by large-scale motion. The large-scale motion is less active in FPG and more active in APG. Therefore, in FPG the production of vortices is random (there are no high and low speed regions), while in the APG the large-scale motion drives the production of vortices, which appears mainly in the high-speed regions, therefore have convection velocity higher than local mean velocity. The more flow is dominated by large-scale motion the higher values takes both the convection velocity of small-scale structures and sweep events induced by them.

The observation could be useful in turbulent boundary layer separation control as well as in turbulence numerical modeling improvements.

5. References

[1] Harun Z, Monty J P, Mathis R and Marusic I 2013 Pressure gradient effects on the large-scale structure of turbulent boundary layers *J. Fluid Mech.* **715** 477 98
[2] Monty J P, Harun Z and Marusic I 2011 A parametric study of adverse pressure gradient turbulent boundary layers *Int. J. Heat Fluid Flow* **32**
[3] Dróżdż A and Elsner W 2013 Amplitude modulated near-wall cycle in a turbulent boundary layer under an adverse pressure gradient *Arch. Mech.* **65** 511–25
[4] Mathis R, Hutchins N and Marusic I 2009 Large-scale amplitude modulation of the small-scale structures in turbulent boundary layers *J. Fluid Mech.* **628** 311
[5] Mathis R, Marusic I, Hutchins N and Sreenivasan K R 2011 The relationship between the velocity skewness and the amplitude modulation of the small scale by the large scale in turbulent boundary layers *Phys. Fluids* **23** 121702
[6] Adrian R J, Meinhart C D and Tomkins C D 2000 Vortex organization in the outer region of the turbulent boundary layer *J. Fluid Mech.* **422** 1–54
[7] Simpson R L, Chew Y-T and Shivaprasad B G 1981 The structure of a separating turbulent boundary layer. Part 2 . Higher-order turbulence results *J. Fluid Mech.* **113** 53–73
[8] Nagano Y and Houra T 2002 Higher-order moments and spectra of velocity fluctuations in adverse-pressure-gradient turbulent boundary layer *Exp. Fluids* **33** 22–30
[9] Johansson A V. and Alfredsson P H 1982 On the structure of turbulent channel flow *J. Fluid Mech.* **122** 295–314
[10] Jacobi I and McKeon B J 2013 Phase relationships between large and small scales in the turbulent boundary layer *Exp. Fluids* **54** 1481
[11] Dróżdż A and Elsner W 2011 Detection of coherent structures in a turbulent boundary layer with zero, favourable and adverse pressure gradients *J. Phys. Conf. Ser.* **318** 062007
[12] Österlund J M 2003 Flow structures in zero pressure-gradient turbulent boundary layers at high Reynolds numbers *Eur. J. Mech. B/Fluids* **22** 379–90
[13] Buckles J, Hanratty T J and Adrian R J 1984 Turbulent flow over large-amplitude wavy surfaces *J. Fluid Mech.* **140** 27– 4
[14] Maciel Y, Rossignol K-S and Lemay J 2006 A study of a turbulent boundary layer in stalled-airfoil-type flow conditions *Exp. Fluids* **41** 573–90

Acknowledgments

The investigation was supported by National Science Centre under Grant no. UMO2012/07/B/ST8/03791

Journal of Physics: Conference Series **530** (2014) 012062 doi:10.1088/1742-6596/530/1/012062

Magnetic field impact on the high and low Reynolds number flows

L. Pleskacz and E. Fornalik-Wajs

Department of Fundamental Research in Energy Engineering, AGH Univeristy of Science and Technology, 30 Mickiewicz Ave, 30-059 Krakow, Poland

E-mail: pleskacz@agh.edu.pl

Abstract. The effect of a strong magnetic field on the temperature and velocity fields of laminar flow was examined. The magnetizing force and gravity term were included in the momentum conservation equation. Biot-Savart's law was applied to obtain the distribution of magnetic field. Three-dimensional computations were performed for straight pipe and pipe with elbow. The single circular magnetic coil was oriented perpendicularly to the flow axis and divided the straight pipe in two equal parts, while in the case of pipe with elbow was just at the beginning of elbow.. The wall of the first straight part was adiabatic while the second was isothermal. Half of the elbow was heated, while the reamining part was adiabatic. Various boundary conditions were applied to estimate their influence on the velocity and temperature distributions. Low entrance velocity, high wall temperature and strong magnetic field led to deceleration of the flow in the central area, acceleration near the wall and formation of recirculation zone in between for the straight pipe. Flow structure and temperature field in the pipe with elbow were significantly modified by the magnetic force. Increasing entrance velocity reduced influence of magnetic field, therefore the flow was less modified. High temperature and magnetic induction resulted in significant changes of the velocity profile. The analysis was conducted with an application of software with special user-defined function. The magnetic field had an influence on the forced convection but its scale depended on the fluid and flow properties, boundary conditions and magnetic field induction.

1. Introduction

All elements and ordinary substances show some magnetic effects, although very small ones – a thousand to a million times less than the effects in ferromagnetic materials [1]. It has come out that this small magnetism can be also utilized. Due to a development of superconducting magnets, the strong magnetic field could be generated and the new interesting phenomena could be discovered, for example: magnetic levitation of weakly magnetic substances, magnetic breath support, magnetothermal wind, Wakayama jet, etc. [2] [3] [4] [5] [6] [7]. Magnetothermal wind [6] was one of these new phenomena. It concerned appearance of wind in a partially heated tube placed in the strong magnetic field. The warm air was expelled from the magnet while the cold air was sucked into it.

The concept of introduction of convection showed new possibility of magnetic field utilization – heat transfer enhancement, which is very important nowadays. Transfer of high heat fluxes and better energy utilization are ones of the main engineering concerns. Therefore magnetothermal convection in the closed enclosures was extensively studied afterwards. The experimental and numerical investigations were carried out by [8] [9] [10] [11] [12] [13] [14] [15] [16] [17]. They

proved the influence of magnetic field on the natural convection of paramagnetic and diamagnetic substances and their heat transfer. The magnetic convection from the engineering point of view was presented in a book [18]. One of the chapters described the magnethothermal wind tunnel, which connected the problem reported by [6] (appearance of magnetothermal wind) with the classical Graetz problem [19], which referred to forced convection through a pipe with partially heated wall. The Graetz problem was investigated by many researchers in various fields (due to its usability) [20] [21] [22]. They analyzed also extended Graetz problem (with thermal conduction) and problems with various boundary conditions.
The case studied by Ozoe [18] was a first example (found by Authors) of the magnetic field influence on the forced convection of weakly magnetic substances. The two-dimensional flow of air was analyzed and the strong influence of magnetic field on it was found in the region close to the magnetic coil. It resulted in higher values of Nusselt number. The presented paper is a continuation of this problem. Authors undertake the matter of three-dimensional model with variable parameters like the geometry, inlet velocity, wall temperature and magnetic induction. It led to the detail description of flow in the region with significant influence of the magnetic field.

2. Mathematical model

With basic assumptions including: incompressible flow, lack of additional mass source, stationary, laminar, three-dimensional flow, the continuity equation may be represented by following equation:

$$\nabla \cdot \vec{u} = 0, \tag{1}$$

where: $\vec{u}$ - velocity m/s.
Employing conditions presented above complemented with the gravitational and magnetic forces (5) treated as the external body forces led to the momentum equation in following form:

$$\rho\left(\vec{u} \cdot \nabla \vec{u}\right) = -\nabla p + \mu \nabla^2 \vec{u} + \vec{F_b}, \tag{2}$$

where ρ - density kg/m^3, μ - dynamic viscosity Pa·s, p - pressure Pa, g - gravitational acceleration m/s^2, $\vec{F_b} = F_g + F_{mag}$ - body forces N/m^3, F_g - gravitational force N/m^3, F_{mag} - magnetic force N/m^3.
The following formula describes energy equation:

$$\vec{u} \cdot \nabla T = \frac{\lambda}{\rho c_p} \nabla T^2, \tag{3}$$

where: T - temperature K, λ - thermal conductivity W/(m·K), ρ - density kg/m^3, c_p - specific heat J/(kg·K), with following assumptions: viscous dissipation and species diffusion are negligible, flow is steady, there is lack of external heat source.
In order to calculate the distribution of magnetic field around single circular coil, Biot-Savart's law was applied [23]:

$$\vec{B} = \frac{\mu_m i}{4\pi} \oint\limits_c \frac{d\vec{s} \times \vec{r}}{|r|^3}, \tag{4}$$

where: $\vec{B}$ - magnetic induction vector T, μ_m - magnetic permeability H/m, i - current magnitude A, $d\vec{s}$ - infinitely small element of the coil m, r - radius of coil m, $\vec{r}$ - position vector m.
Placing examined flows of paramagnetic fluid in magnetic field described by above equation, resulted in the flow being affected with a new kind of influence, known as the magnetic force. This force was included in the Navier-Stokes equations as another external body force [24]:

$$\vec{F}_{mag} = -\left(1 + \frac{1}{T_0 \beta}\right) \frac{\chi \rho \beta \left(T - T_0\right)}{2\mu_m} \vec{\nabla} B^2, \tag{5}$$

where: $T_0 = (T_w + T_f)/2$ - reference temperature K, T_w - wall temperature K, T_f - fluid temperature K, β - volumetric thermal expansion coefficient K^{-1}, χ - volumetric magnetic susceptibility -.

3. Studied case and applied solutions

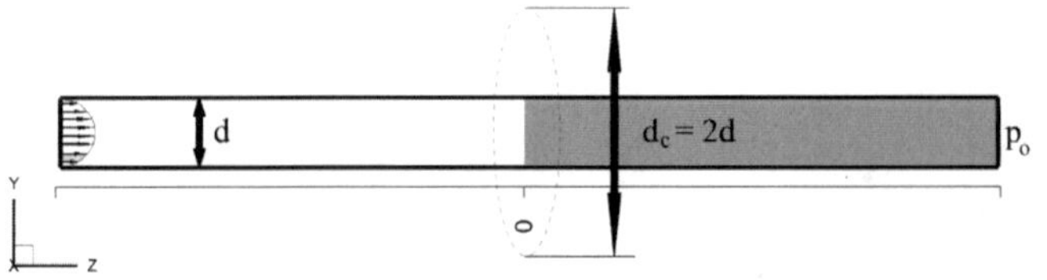

Figure 1. The schematic view of straigth pipe.

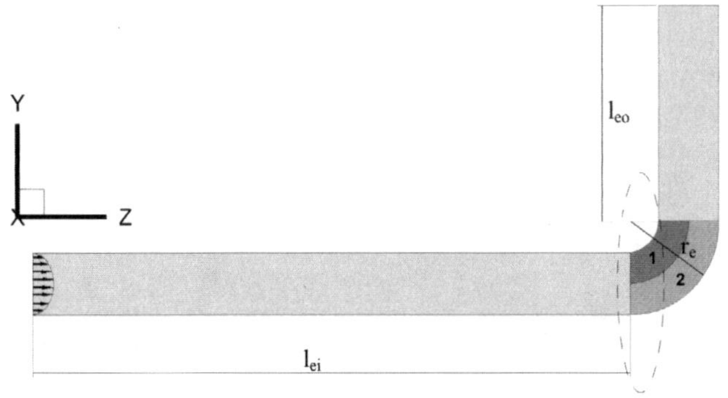

Figure 2. The schematic view of pipe with an elbow

The first studied geometries were circular straight three-dimensional ducts of diameter $d = 0,04$ m and $l = 1$ m. The straight pipe geometry is presented in figure 1. While studying the presented data, reader's attention should be given to the model's orientation in the coordinate system. The magnetic coil was set in XY plane, that was normal to Z-axis and it was origin of the system. The plane containing the magnetic coil divided the pipe in two equal parts. The wall of the first part was adiabatic while the second was isothermal.
The second studied geometry was pipe with an elbow of diameter $d_e = 0.01$ m. It consisted of an inlet section of length $l_{ei} = 0.1$ m, an elbow section with external radius $r_e = 0.015$ m and outlet section of length $l_{eo} = 0.035$ m. The magnetic coil was set just ahead of the elbow. The wall of the elbow was divided in two isothermally heated parts, from now on known as variant 1 and variant 2. This geometry is schematically presented in figure 2.
The diameter of the coil was always twice that of a pipe. The parabolic velocity profile was assumed at the inlet. Table 1 presents the thermophysical and magnetic properties of a fluid used in calculations. The inlet fluid temperature was $T_f = 300$ K in all studied cases. At the outlet pressure was assumed to be $p_o = 101325$ Pa.
The academic license software (GAMBIT 2.4 and Fluent 6.3) was used to generate the grid and perform the numerical computations. The special user-defined function calculating the distribution of magnetic force and three dimensional parabolic velocity profile were implemented in the code.The grid consisted of 147400 wedge elements for the straight pipe and 186800 wedge elements for pipe with elbow.

Table 1. Thermophysical and magnetic properties of air.

density	dynamic viscosity	volumetric thermal expansion coefficient	magnetic suscepti-bility	magnetic perme-ability	specific heat	thermal conduc-tivity
ρ	μ	β	χ	μ_m	c_p	λ
kg/m^3	Pa·	K^{-1}	-	H/m	J/(kg·K)	W/(m·K)
1.225	$1.7894{\cdot}10^{-5}$	$3.33{\cdot}10^{-3}$	$3.77{\cdot}10^{-7}$	$4\pi{\cdot}10^{-7}$	1006.43	$2.42{\cdot}10^{-2}$

4. Results and discussion

4.1. Analysis of inlet velocity influence on velocity and temperature distributions

In this numerical case the pipe of diameter $d = 0.04$ m and length $l = 1$ m was studied. The analysis of inlet velocity influence on the velocity and temperature distributions was conducted with constant magnitude of magnetic induction in the centre of the coil $b_0 = 8$ T and constant value of temperature of the wall $T_w = 350$ K. The varying parameter in this case was average inlet velocity U_{avg}.

Figure 3 presents the influence of average inlet velocity on the velocity distribution in the magnetic field. The degree of profile's deformation against the profile in the flow without magnetic field depended on examined parameter. Application of the lowest average inlet velocity (0.2 m/s, Re = 547) resulted in appearing of the greatest deformation of parabolic velocity profile (figure 3(a)).

While taking into account the highest average inlet velocity (figure 3(c)) these deformations are minimal. The profiles remained parabolic on the whole examined cross section. The results might be divided in two groups. First of them was the group of distorted velocity profiles and contained cases of average inlet velocities of 0.2 m/s and 0.5 m/s (figure 3(a) and 3(b)). The second group consisted of the one case remained, where the gentle flattening of the profile's sides near the electric coil and narrowing or tearing of the flow's central area were the only expressions of profiles' change (figure 3(c)).

In the group of distorted velocity profiles the characteristic profile with three peaks could be observed. However, it vanished in the case of $U_{avg} = 0.5$ m/s (Re = 1369). Behind the magnetic coil the area of flow slowing down appeared.

Acceleration near the wall was caused by the magnetic force. It was compatible with the flow direction. As the velocity increased, the magnetic force effect weakened, which was related to thermal boundary layer narrowing.

The effect of flow suppression reduced while moving closer to the flow axis due to the drop of temperature (and in consequence, value of temperature difference) and of magnetic induction magnitude (caused by drifting away from electric coil) and what's more important the magnetic force from eq. (5).

The central area of each profile preserved parabolic shape (though the maximum velocity magnitude decreased) and the greatest change took place beyond it.

No recirculation zone was observed in any of examined cases. Nonetheless, it could occur with the increase of magnetic force acting on fluid or further decrease of average inlet velocity.

4.2. Analysis of wall temperature influence on velocity and temperature distributions

The analysis of wall temperature influence on the velocity and temperature distributions in the pipe of diameter $d = 0.04$ m and length $l = 1$ m was conducted with constant magnitude of the

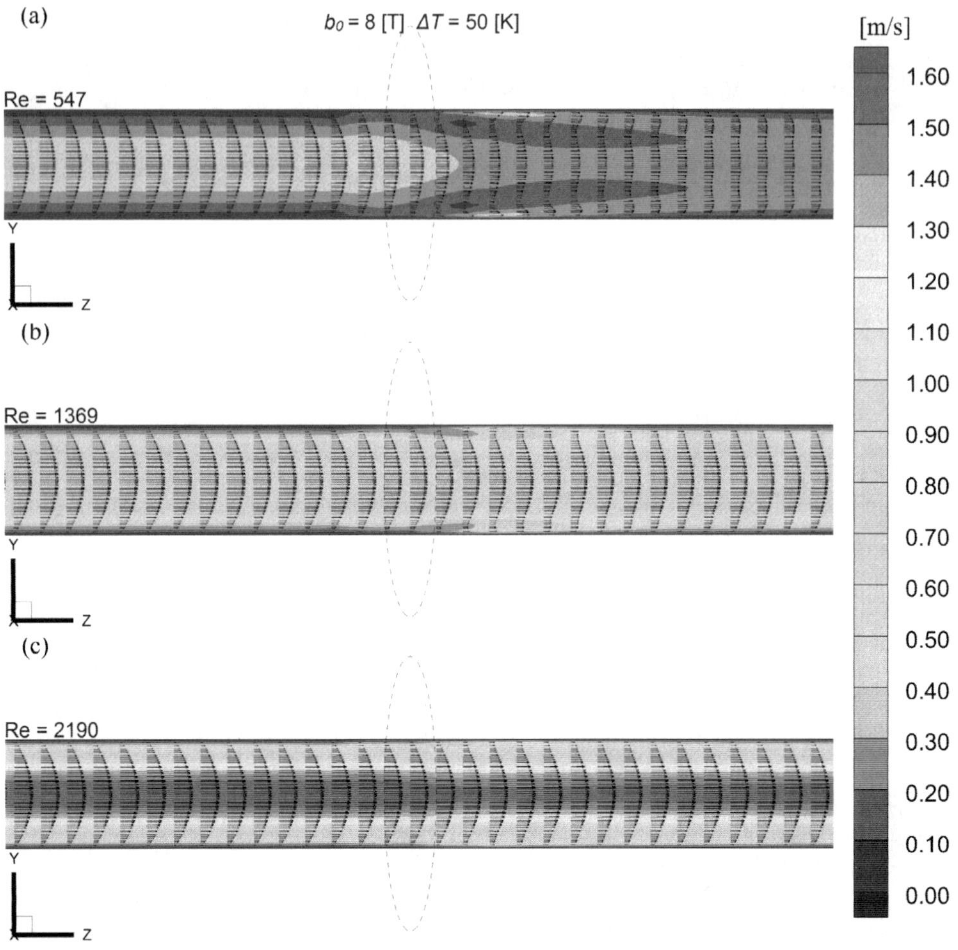

Figure 3. Velocity contours obtained in the presence of magnetic field in the pipe of $d = 0.04$ m and $l = 1$ m with average inlet velocity of: (a) 0.2 m/s, (b) 0.5 m/s, (c) 0.8 m/s.

magnetic induction in the centre of the coil $b_0 = 8$ T and constant value of the average inlet velocity $U_{avg} = 0.5$ m/s (Re = 1369). The varying parameter was the temperature of the wall. Figure 4 presents the influence of wall temperature variation on the velocity distributions in the magnetic field.

According to eq. (5) increase of the wall temperature led to the increase of temperature difference and indirectly to the increase of magnetic force. As in the previous case change of the magnetic force direction could be observed in the form of local acceleration of the flow near the wall. The modification of the parabolic velocity profile was connected with temperature of the wall. The flow with temperature difference of $\Delta T = 5$ [K] (figure 4(a)) behaved similarly to the flow with high average inlet velocity (figure 3(c)). In this case only the narrowing of the central area (of the highest velocity) of the flow could be reported. As the temperature increased this central area became thinner and thinner until its complete tearing (figure 4(c)).

At the same time the development of the flow's suppression proceeded, first near the wall and thereafter closer to the Z axis. The enlarging of this area was the second sign of velocity profile's deformation behind the magnetic coil.

In discussed flows, as in previously presented cases, the acceleration of the fluid near the wall

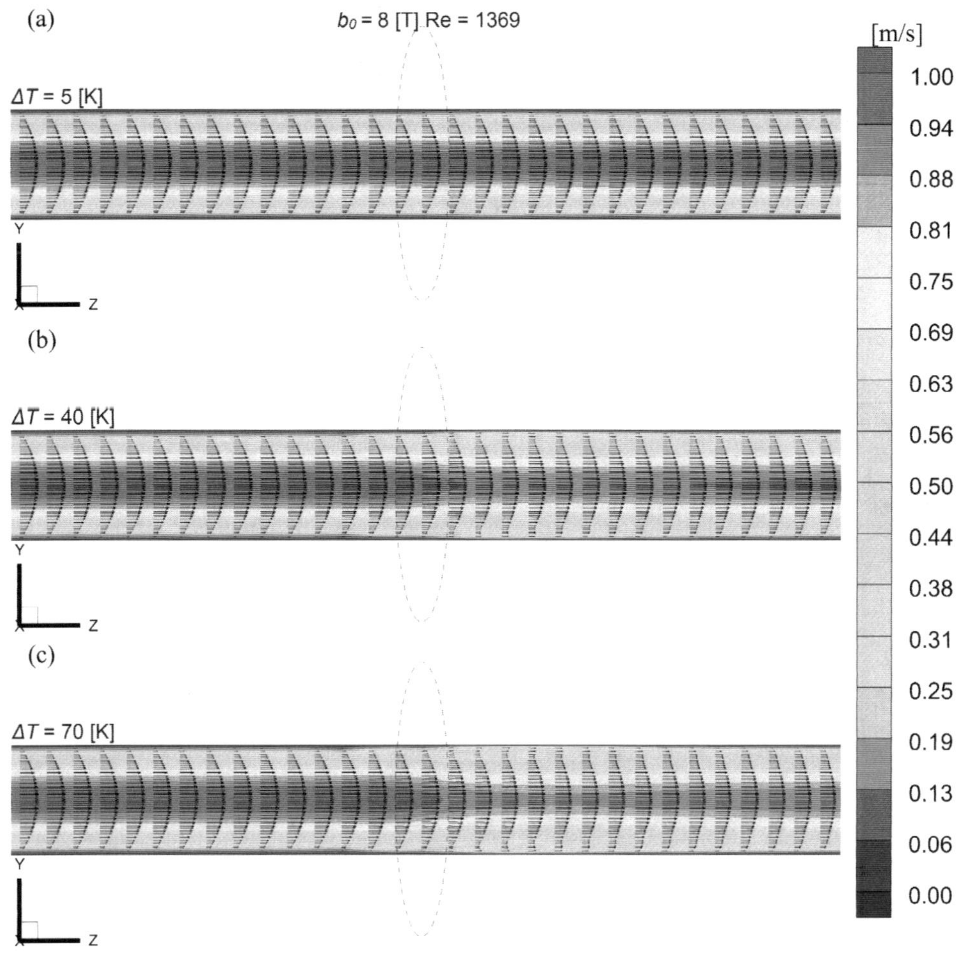

Figure 4. Velocity contours obtained in the presence of magnetic field in the pipe of $d = 0.04$ m and $l = 1$ m with temperature difference of: (a) 5 K, (b) 40 K, (c) 70 K.

caused by the magnetic force acting accordingly to flow basic direction appeared. Increase in the temperature (causing greater change in velocity profile) delayed the returning to the basic profile. Greater deformation required greater distance to return to the basic flow structure.

4.3. Analysis of magnetic induction influence on velocity and temperature distributions

The analysis of the magnetic induction in the centre of the coil influence on the velocity and temperature distributions in the pipe of diameter $d = 0.04$ m and length $l = 1$ m was conducted with constant value of the temperature difference between the flow inlet temperature and wall temperature $\Delta T = 50$ K and constant value of the average inlet velocity $U_{avg} = 0.4$ m/s (Re = 1095). The varying parameter was the magnetic induction in the centre of magnetic coil b_0. Figure 5 presents obtained velocity distributions dependent on varying parameter b_0. As could be seen in the previous cases, also here the increase of studied parameter led to the gradual deformation of the velocity profile behind the magnetic coil in the form of narrowing of the flow central area (figure 5(b)) and its tearing (figure 5(c)). These two signs were accompanied by the deceleration zone, which at first appeared near the wall and then descended in the direction of flow axis.

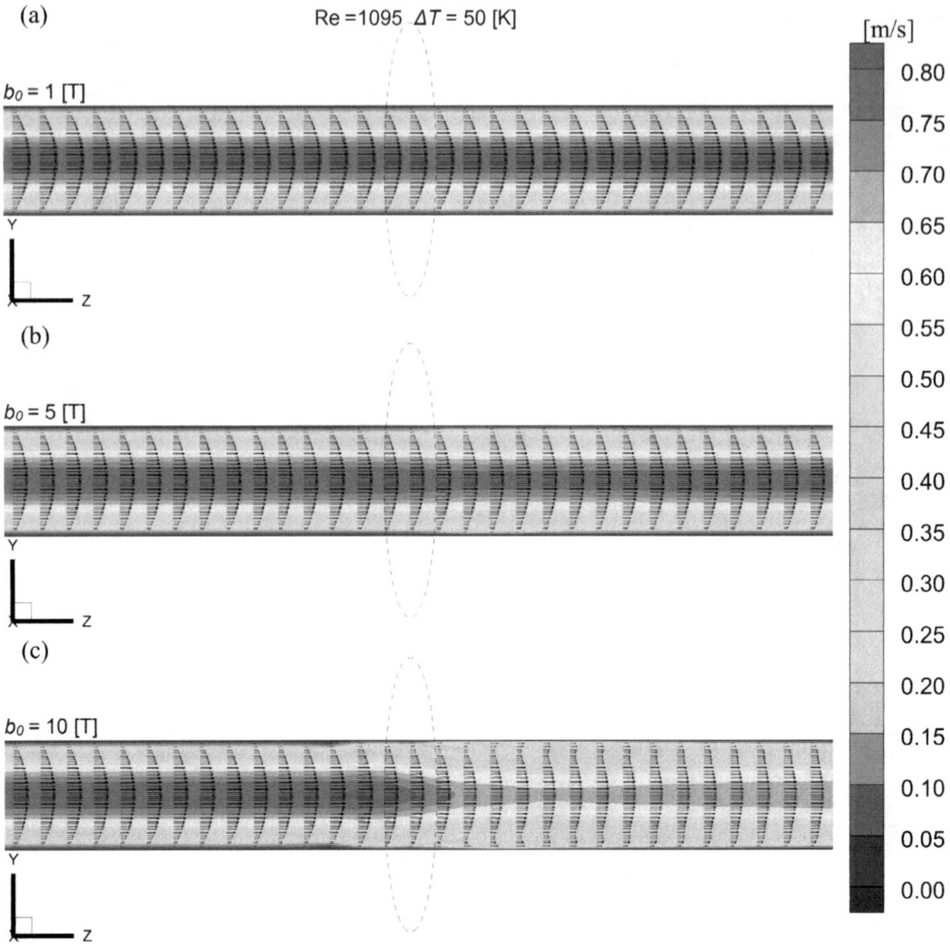

Figure 5. Velocity contours obtained in the presence of magnetic field in the pipe of $d = 0.04$ m and $l = 1$ m with the magnetic induction in the centre of the coil of: (a) 1 T, (b) 5 T, (c) 10 T.

Connection between the increasing parameter b_0 and returning distance to the basic parabolic profile remained the same as in the previous subsection.

Figure 5(a) shows that the magnetic induction of magnitude $b_0 = 1$ T didn't generate the magnetic force able to do any change of flow structure. Therefore, it could be assumed that the critical value allowing the profile change was in the range between 1 T to 5 T (and actually between 1 T to 2 T, but this data are skipped due to the lack of space available).

The similarity of results obtained with varying wall temperature and magnetic induction in the centre of the magnetic coil led to conclusion that there should be a possibility of applying the relations connecting these two effects. This conclusion is confirmed by eq.(5) where both the temperature and magnetic induction appeared.

4.4. Analysis of the flow structure in the pipe with an elbow

The analysis of the flow structure in the pipe with an elbow was conducted with constant magnitude of magnetic induction in the centre of the coil $b_0 = 10$ T, constant value of the temperature difference between the flow inlet temperature and wall temperature $\Delta T = 50$ K

and constant value of the average inlet velocity $U_{avg} = 0.023$ m/s (Re=16).
The distributions of velocity, temperature and magnetic force for the studied cases are presented in figures 6 and 7. Very complex three-dimensional flow structure can be observed in these figures. They show that velocity distribution as well as temperature distribution underwent strong distortions. For both variants a strong acceleration zone near the heated wall can be observed (figure 6(a) and 7(a)). It is connected with strong magnetic force influence in this area (figures 6(d) and 7(d)). The velocity in this area is about six times higher than the average inlet velocity. Aside from the acceleration zone, the deceleration zone can be also observed in the elbow. The ratio of the area between these two zones depended on heated wall variant. For variant 1 acceleration zone is much smaller than the deceleration zone. For variant 2 otherwise. Furthermore, for variant 2 the beginning of returning of the velocity profile to parabolic one can be seen.
The elbow is preluded by the ending fragment of inlet section, where for both variants vortex structures appeared. The direction of the vortex was dependant on applied variant. For variant 1 it was counter-clockwise and for variant 2 it was clockwise.The deflection of the inlet parabolic velocity profile occurred already on the section ahead of the above mentioned vortex structures. Therefore, the range of magnetic force influence is greater than in the cases described in previous subsections.
The specific temperature distribution (figures 6(c) and 7(c)) is caused by the general suppression of the flow near the magnetic coil, which is especially related to the appearance of the vortex structures. The direction of magnetic force acting can be changed after crossing the certain temperature. This temperature, called reference temperature or T_0 is an arithmetic mean of inlet fluid temperature and the heated wall temperature. The serious distortion of temperature distribution led to the explicit location of isotherm $T_0 = 5$ K. The presence of this temperature led to the appearance of reduced magnetic force zone, what can be observed in figures 6(c) and

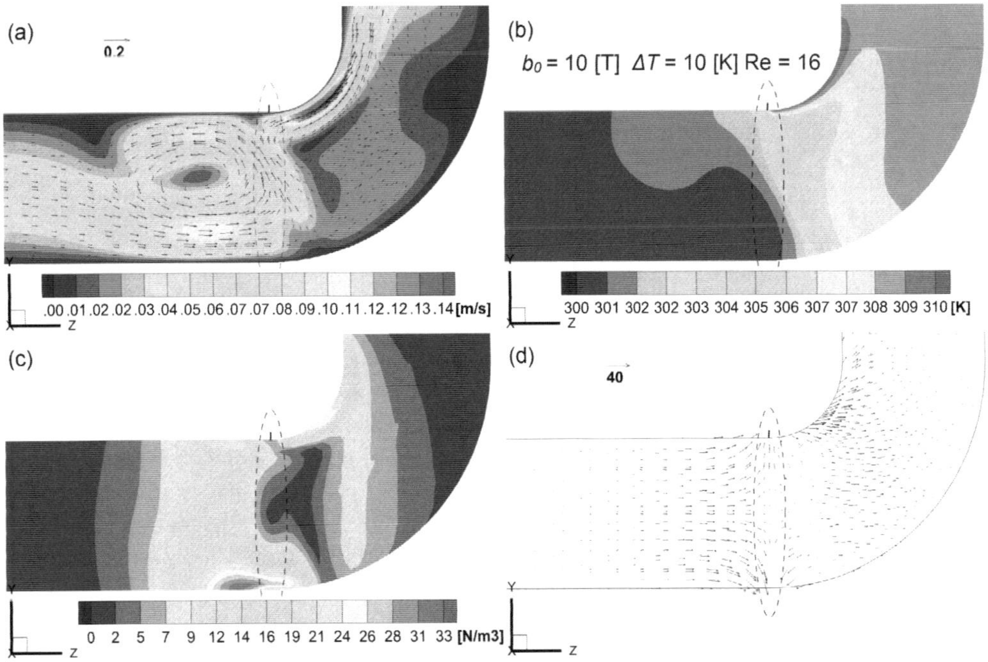

Figure 6. Flow characteristics for variant 1: (a) velocity contours and vectors, (b) temperature contours, (c) magnetic force contours, (d) magnetic force vectors.

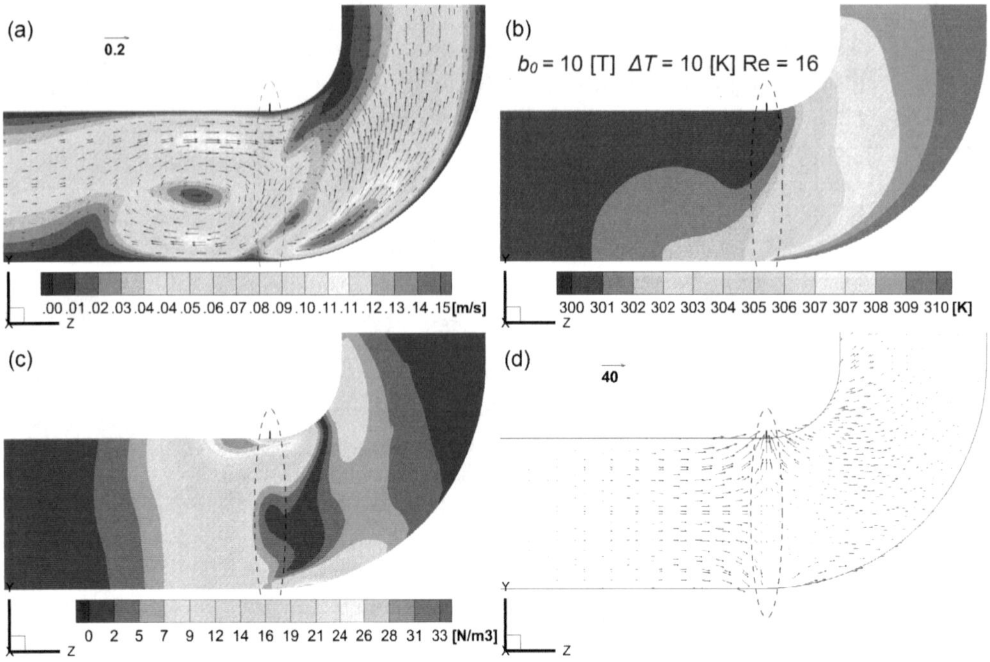

Figure 7. Flow characteristics for variant 2: (a) velocity contours and vectors, (b) temperature contours, (c) magnetic force contours, (d) magnetic force vectors.

7(c). This effect is compatible with the eq.(5).

5. Summary

A topic of paramagnetic fluid (air) forced convection under the influence of magnetic field was undertaken. It is an extension of problem analyzed by Ozoe in [18], whose results were very interesting, therefore it was decided to study the problem in details. In this paper the three-dimensional analysis of the air flow in two pipes of various diameter and with various boundary conditions under the influence of magnetic field were presented together with the straight pipe and elbow. The magnetic force acting on the paramagnetic fluid caused appearance of the acceleration, deceleration and recirculation zones and caused changes in the velocity profile. The complex three-dimensional flow structure could be obtained. It was shown that increasing inlet velocity reduced influence of the magnetic field in opposite to the increasing wall temperature or increasing magnetic induction, which magnified it. It was mentioned that all the parameters, which were varied can be combined in one, which can quantitatively described their influence on the forced convection. This topic will be continued in the next studies. Summarizing the all presented data it can be said that the magnetic field had and influence on the forced convection. However it is rather limited to the slow flows or high temperature differences and high values of magnetic induction, but it should not be neglected. There are a lot of applications for example: chemical industry, biology, medicine, porous media flow, for which the presented data are very promising.

References

[1] Feynman R P 1967 *The Feynman Lectures on Physics* (Addison-Wesley Publishing Company)
[2] Simon M D and Geim A I 2000 *J. App. Phys.* **87** 6200–6204
[3] Hirota N, Ikezoe Y, Uetake H, Kaihatsu T, Takayama T and Kitazawa K 2002 *RIKEN REVIEW* **44** 159–161

[4] Ikezoe Y, Hirota N, Nakagawa N, T and Kitazawa K 1998 *Nature* **393** 749–750
[5] Wakayama M and Wakayama N I 2000 *Jpn. J. App. Phys.* **39** 262–264
[6] Uetake H, Nakagawa J, Hirota N and Kitazawa K 1999 *J. App. Phys.* **86** 2924–2925
[7] Wakayama N I 1991 *Chem. Phys. Lett.* **185** 449–451
[8] Qi J, Wakayama N I and Yabe A 2001 *Int. J. Heat Mass Transf.* **44** 3043–3052
[9] Tagawa T, Ujihara A and Ozoe H 2003 *Int. J. Heat Mass Transf.* **46** 4097–4104
[10] Fornalik E, Filar P, Tagawa T, Ozoe H and Szmyd J S 2006 *Int. J. Heat Mass Transf.* **49** 2642–2651
[11] Fornalik E, Filar P, Tagawa T, Ozoe H and Szmyd J S 2005 *Exp. Therm. Fluid Sci.* **29** 971–980
[12] Fornalik E, Filar P, Ozoe H and Szmyd J S 2006 *Chem. Process Eng.* **27** 633–642
[13] Bednarz T, Fornalik E, Tagawa T, Ozoe H and Szmyd J S 2005 *Int. J. Therm. Sci.* **44** 933–943
[14] Bednarz T, Tagawa T, Kaneda M, Ozoe H and Szmyd J S 2004 *Numer. Heat Tr. A-Appl.* **46** 99–113
[15] Wrobel W, Fornalik-Wajs E and Szmyd J S 2010 *Int. J. Heat Fluid Flow* **31** 1019–1031
[16] Pyrda L, Kenjeres S, Fornalik-Wajs E and Szmyd J S 2012 *J. Phys.: Conf. Ser.* **395** 1–8
[17] Kenjeres S, Pyrda L, Wrobel W, Fornalik-Wajs E and Szmyd J S 2012 *Phys. Rev. E.* **85** 1–8
[18] Ozoe H 2005 *Magnetic Convection* (Imperial College Press)
[19] Graetz L 1883 *Annalen der Physik und Chemie* **18** 79
[20] Dindore V Y 2003 *Gas purification using membrane gas absorption processes* Ph.D. thesis University of Twente
[21] Fadel A A 2009 *A novel method for the solution of convection-diffusion problems with applications to nitric oxide production and transport in vitro* Ph.D. thesis Drexel University
[22] Fehrenbach J, Gournay F D, Pierre C and Plouraboue F 2012 *SIAM J., App. Math* **72** 99–123
[23] Jackson J D 1998 *Classical Electrodynamics* (John Wiley and Sons Inc.)
[24] Bednarz T 2004 *Numerical and experimental analyses of convection of paramagnetic fluid in a cubic enclosure* Ph.D. thesis Kyushu University

Acknowledgments
The present work was supported by Polish National Science Centre Project No. 2012/07/B/ST8/03109 and Academic Computer Centre CYFRONET AGH, under award no. MNiSW/IBM-BC-HS21/AGH/060/2013.

Numerical analysis of the flow around the Bach-type Savonius wind turbine

K Kacprzak[1], K Sobczak
Institute of Turbomachinery, Lodz University of Technology, 219/223 Wólczańska Street, 90-924 Łódź, Poland

E-mail: kokacprzak@gmail.com

Abstract. The performance of the Bach-type Savonius wind turbine with a constant cross-section is examined by means of quasi 2D and 3D flow predictions obtained from ANSYS CFX. Simulations were performed in a way allowing for a comparison with the wind tunnel data presented by Kamoji et al. The comparison with the experiment has revealed that 2D solutions give much higher deviation from the reference data than the 3D ones, which guarantees a good solution quality. It can be stated that even simplified (lack of laminar-turbulence transition modelling and a coarser mesh) 3D simulations can yield more accurate results than complex 2D solutions for turbines with a low aspect ratio. The paper also presents a systematic analysis of the most characteristic flow structures which are identified in the rotor.

1. Introduction

In recent years, investigations in the field of wind energy have been considerably intensified due to the worldwide energy crisis, high global CO_2 emission and grid overload. However, there is still a great potential for further development of reliable distributed energy converters for moderate wind speed conditions. Real-scale or model experiments involve huge costs and hazards. On the other hand, computational studies make the whole process much faster, however, in this case, valid numerical models of flow phenomena are the key aspect to obtain realistic results.

The Savonius wind turbine is a vertical axis wind turbine (VAWT) that was proposed by Savonius in the 1920s. It has a characteristic S-shape of the rotor composed of two separated or connected scoops. A typical power output of the Savonius turbine classifies it as a small wind turbine. Decentralised power generation is now one of the fastest-growing forms of providing electricity for domestic and residential needs and this turbine can be an attractive investment for homeowners and small business. Hence, there has been increased interest in studies of the Savonius rotor, and significant efforts have been made to improve its aerodynamic performance. In particular, there has been a need for better understanding of the unsteady flow around the Savonius rotor.

This increased interest in the Savonius wind turbine results from increasing significance of the urbanised areas, which address specific demands. The Savonius rotor seems to satisfy these particular expectations. Its main advantages can be described as follows [1,2]:

- wind direction independence,
- ability to operate in a wide range of wind conditions,

[1] Corresponding author

- no complex control mechanism is required,
- structure can be placed on the ground,
- low noise emission,
- high starting torque.

All the above features combined with a compact size of this kind of turbines make them suitable for the residential use. However, Savonius turbines are not free of drawbacks like relatively small efficiency or low rotational speed [3,4]. Therefore, constant search for better designs, in order to improve the performance of the rotor is required. Current development in Computational Fluid Dynamics (CFD) makes it possible to conduct a comprehensive study of different designs of the Savonius rotor without committing a significant amounts of time and money for experimental investigations.

2. Studied geometry

In this paper the Bach-type rotor was examined by means of numerical simulations (figure 1). In order to investigate the impact of three-dimensional effects on the overall performance of the turbine full scale 3D simulations were carried out together with a series of 2D simulations. Dimensions of a numerical model directly correspond to a low aspect ratio turbine examined experimentally by Kamoji et al. [5]. In this way, it was possible to verify obtained results and check the influence of different task configurations (2D simplification, mesh refinement). Simulations are conducted for a rotor with height H = 0.154 m; diameter D = 0.2 m; without overlap - ratio e/D = 0.0; aspect ratio H/D = 0.77; blade arc angle ψ = 135°; blade shape factor p/q = 0.2; end plate parameter Do/D = 1.33 (figure 2).

Figure 1. Model of investigated Bach-type turbine (due to its symmetry only a half of turbine is presented).

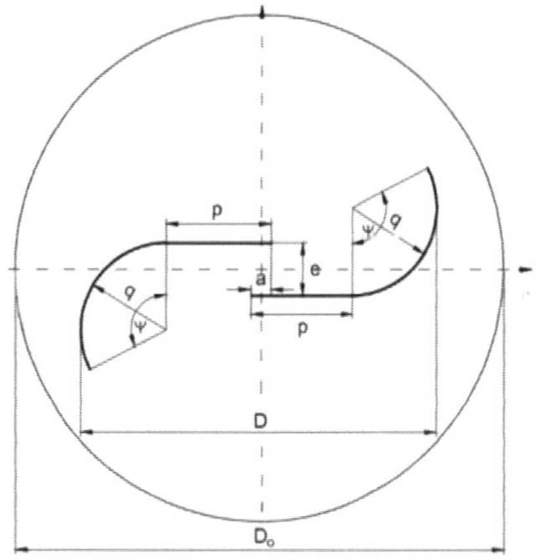

Figure 2. Geometrical parameters of the Bach-type turbine.

3. Simulation definition

In this study both 2D and 3D simulations were carried out with ANSYS CFX. For the purposes of this investigations the model was composed of two domains (Fig 3.). The first one (rotating) comprised the wind turbine blades and endplates and the second (stationary), simulated the wind tunnel with dimensions in accordance with the experimental test stand data [5]. The rotating domain radius was twice as big as the rotor radius. Such a domain size was chosen in order to assure that vortex structures in a close vicinity of the rotor are solved in a rotating frame of reference and it also minimised numerical errors due to numerous transitions of fluid through the domain interface. Due to the rotor symmetry, only a half of the turbine was solved. Simulation definition in the case of 2D model is

similar, however, due to a nature of the ANSYS CFX solver, a slice (1 mm thickness) of 3D domain near the symmetry plane was used.
Meshes for the given geometry were prepared using Meshing software from ANSYS Workbench. In the performed simulations a high-quality, unstructured, hexahedral mesh was utilised. Described grid was comprised of hexahedral layers, perpendicular to the rotor axis. Throughout the analysis different mesh densities were employed. First of all, a coarse mesh ($\sim 6{\cdot}10^6$ nodes) was used, where wall functions were utilised to bridge the viscosity-affected region between the wall and the fully-turbulent region. In case of the fine mesh ($\sim 15{\cdot}10^6$ nodes) y^+ values for the buckets were lower than 2, which made it possible to fully resolve the boundary layer, which due to considered flow structure is of a very high relevance.

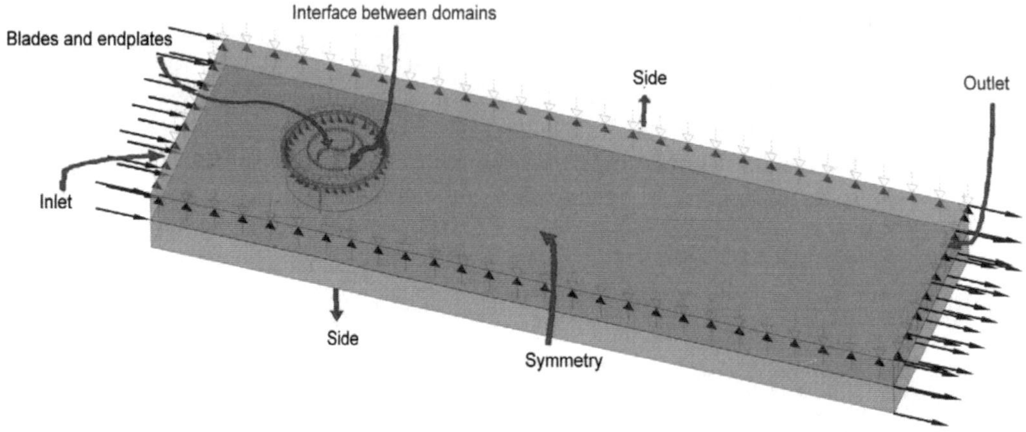

Figure 3. Orthogonal projection of simulation domains.

Taking into account low wind speed in the described case, density changes were relatively low, that is why, it was assumed that air in this case is incompressible. Due to constant changes of the blade position in reference to the wind direction, the flow structure is very complex and extensive separations occur. Hence, the unsteady Reynolds Averaged Navier-Stokes (RANS) simulations and especially turbulence modelling in this case is a challenge. In the described simulations Shear Stress Transport (SST) turbulence model was utilised. It was successfully introduced in similar studies [6,7]. The selection of this two-equation model was based on the fact that it was designed to give accurate prediction of the flow separation regions under the adverse pressure, when an adequate mesh refinement (as in this study) is applied. It is achieved by an implementation of the transport effects to the formulation of the eddy viscosity. That is why, it is used whenever a high boundary layer accuracy is required [8], which is a case in this investigation. Utilising this turbulence model also enabled the laminar-turbulent (LT) transition modelling. The γ-θ model was applied, which was demonstrated by numerical studies [9] to have an influence on the results agreement with the experimental data. This approach required mesh refinement to obtain $y^+ < 2$. This condition was fulfilled for the buckets. In case of endplates, due to the fact that the LT transition is not of a high importance in this region and taking into account the calculation time, y^+ values were higher.
At the inlet to the computational domain a wind velocity (9 m/s) boundary condition was imposed. This velocity corresponds to Re = 120 000. Additionally, low turbulence intensity (1%) was set as in [5]. At the outlet atmospheric pressure was imposed (1 bar). Due to the fact that the turbine as well as the expected flow structure have a symmetry plane in the middle of the turbine height, in 3D simulations a symmetry boundary condition was set on this plane. In case of 2D calculations symmetry was imposed on both sides of the modelled layer which is typical approach in 2D simulations in ANSYS CFX. For turbine blades and endplates no-slip wall boundary condition was

assumed (zero velocity on the wall surface). At sides and bottom of the domain (away from the rotor) the free-slip wall condition was imposed.
Adjusting the rotational velocity of the rotor made it possible to investigate different tip speed ratios.

$$TSR = \frac{\omega \cdot R}{U} \tag{1}$$

where ω – turbine angular velocity, R – turbine radius, U – wind speed.
In order to achieve conformity between mesh cells on two sides of the domain interface for the whole simulation, timestep for every rotational velocity was adjusted to achieve precisely 360 timesteps per one revolution.
ANSYS CFX uses an element-based finite volume method [8]. Second order advection (high resolution) scheme were utilised. Second order Backward Euler method was used for discretization of time terms. In order to properly control convergence, the level of the residual target was set to 10^{-5}, with a number of iterations for each timestep (so called internal convergence loops) between 3-10. In practice 6 iterations and the root mean square Courant number equal to 3 were not exceeded. That implies that the timestep is small enough to adequately resolve unsteady structures. It has to be mentioned that the results after 6-8 revolutions of the rotor yielded repeatable cycles, so only the data beyond the eighth revolution was taken into account during the postprocessing of the solution results.

4. Simulation results

In order to verify the results of conducted simulations, obtained data were confronted with the experiment results from Kamoji et al. [5]. Relation between coefficient of power c_p (ratio of the rotor and wind power) and tip speed ratio for Bach-type turbine investigated by Kamoji et al. is presented in figure 4 together with 2D and 3D simulation results.

$$c_p = \frac{2T \cdot \omega}{\rho \cdot U^3 \cdot D \cdot H} \tag{2}$$

where T – torque, ρ – air density.

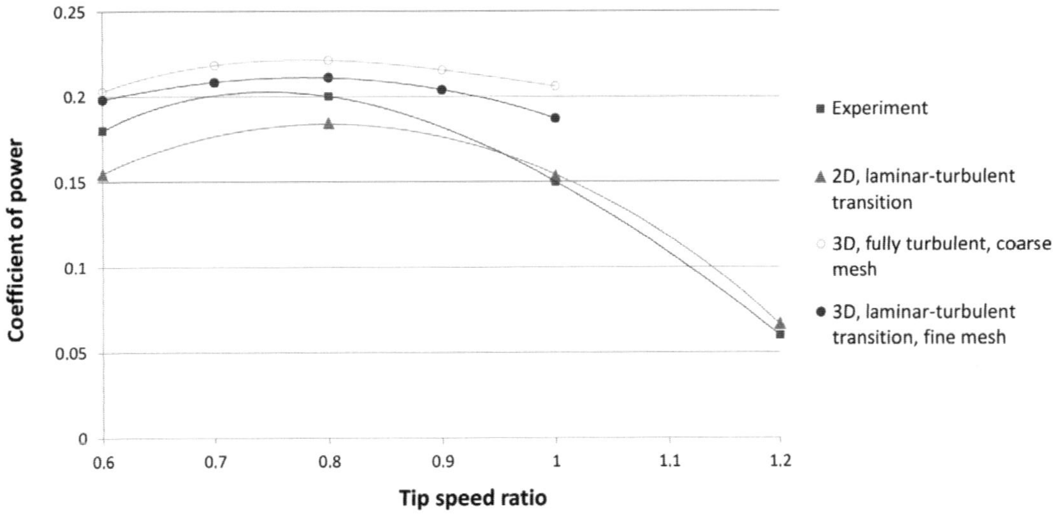

Figure 4. Comparison of the c_p coefficient from simulations and experiment.

It can be noticed that there are certain discrepancies between the experimental and simulation results. In the case of 2D results, it has to be underlined that the 2D model does not take into account all the three-dimensional effects (especially vortices, which are the source of major dissipations) on the blades in the vicinity of the endplate. Moreover, additional losses may occur due to turbine shaft rotational resistance. However, in the case of 2D simulations for TSRs below 1.0 obtained power is underestimated, which can results from imperfections of the used method e.g. turbulence modelling in 2D. On the other hand, 3D results for coarse mesh are unambiguous and indicate small (a few %) overprediction of c_p relatively to the experimental data. Application on the laminar-turbulent (LT) transition model in case of the 3D fine mesh improved the estimation of power coefficient. This can be due to better separation prediction. For example delayed separation of the boundary layer reduces the pressure drag on the returning bucket, which produces higher lift force participation on the resultant force, increasing the torque of the rotor.

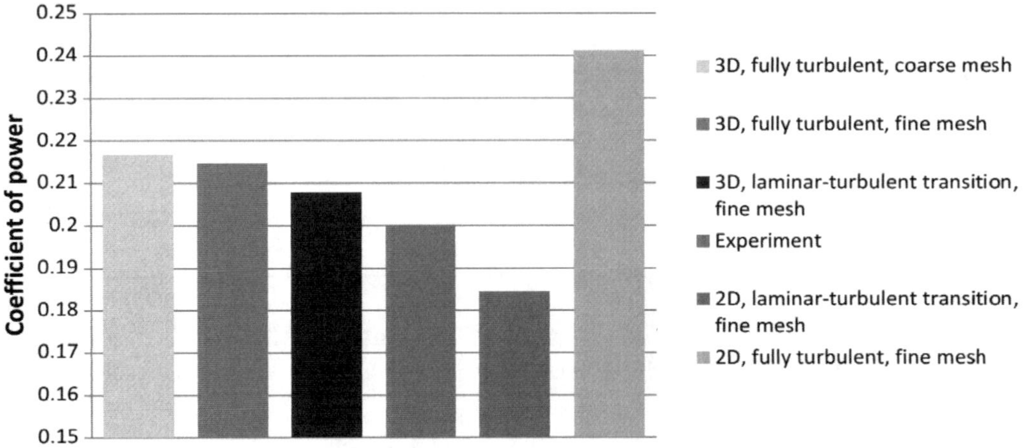

Figure 5. Comparison of the c_p coefficient from simulations and experiment for TSR=0.8.

In order to assess which parameters assure the most accurate solution, simulations for multiple task configurations were performed. All of them were conducted with TSR = 0.8 as the most representative case (the highest power output). Simulations were done for 2D and 3D cases, for coarse and fine meshes, without and with the laminar-turbulent (LT) transition modelling. Obtained values of the pressure coefficient are presented in figure 5. One can see that the LT transition modelling have serious influence on the 2D solution. In the case of the fully turbulent boundary layer data are highly overpredicted (more than 20%), whereas for the case with LT modelling, it is underdetermined (8.5%). 3D simulations give more straightforward results. There is small difference between the solutions of the fully turbulent boundary layer for the coarse and fine meshes (overdetermined by approx. 8.5%). Much better agreement with the experimental data is obtained if the LT transition is taken into account. In this case the difference is 4%. It seems that for a low aspect ratio rotor simpler 3D simulations (coarse mesh, lack of the LT transition modelling) can yield more accurate results than more complex 2D solutions.

5. Flow structure analysis

In order to examine the flow structure for different rotor positions, the simulation concerning TSR = 0.8 was chosen as the most representative. It is due to the fact that the coefficient of power attains the highest value for this TSR. The variation of c_p for one, full revolution of the rotor is presented in figure 6 The most characteristic points, representing c_p extremes, are marked. A and C refer to the maximum values, while B and D to the minimum ones.

It can be observed, that the position of the rotor corresponding to the highest torque is attained twice throughout the revolution for blade positions almost "aligned" with the flow, so it can be concluded that it consists of two almost equal cycles. It is also visible that the coefficient of power drops to almost 0 for the blade positions "normal" to the incoming wind.

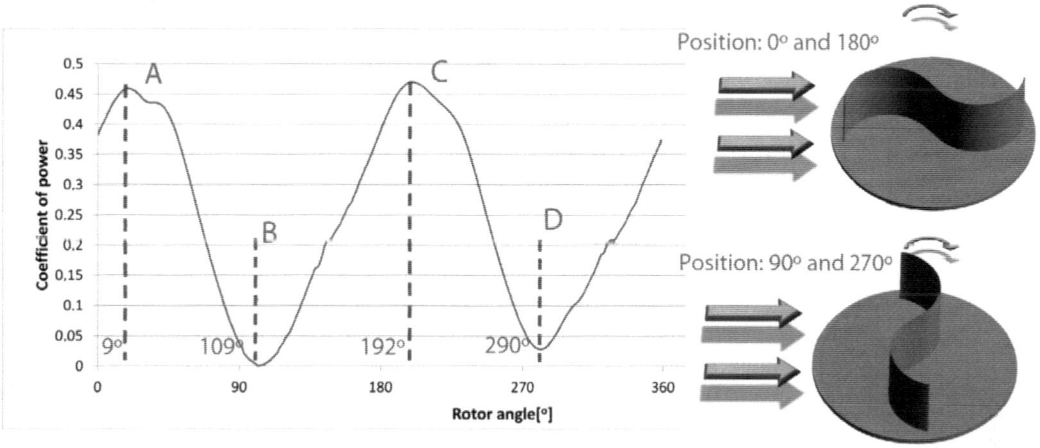

Figure 6. Coefficient of power for the Bach-type wind turbine for TSR=0.8.

In figure 7 one can see that the flow structure is very complex and time dependent. In the angular position A, when the highest torque is attained, a clear fluid acceleration on the convex side of the advancing blade is observed (moving in agreement with the wind direction) with subsequent separation of the boundary layer. This acceleration results in production of the lift force, which drives the turbine. This basic phenomenon coexists with numerous recirculation zones at the convex side of the advancing blade and both sides of the returning blade. In the case of the angular position B, the stagnation at the returning blade balance the positive torque due to wind action on the concave side of the advancing blade. Flow structure obtained in this numerical study directly corresponds to other research, where the flow in the Bach-type Savonius wind turbine was studied by means of PIV [10].

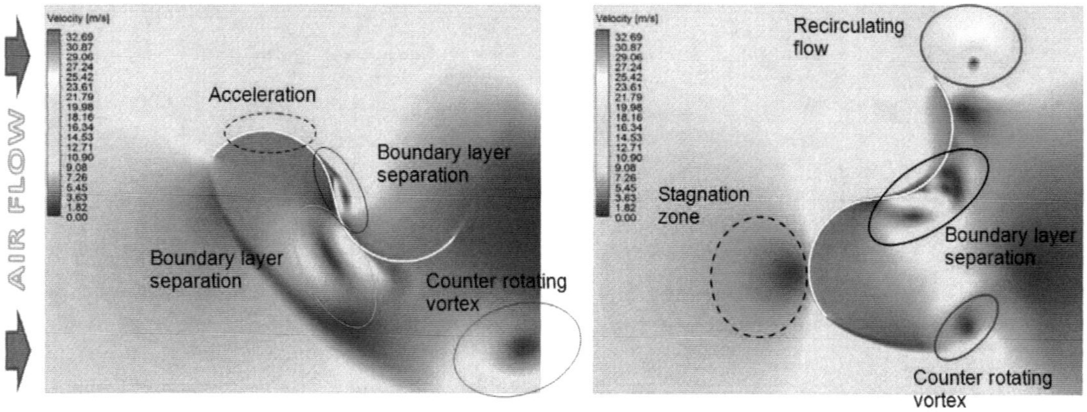

Figure 7. Distribution of velocity in the stationary frame for the characteristic points A and B at the symmetry plane.

In order to observe the flow structure from a different perspective figure 8 was prepared. It shows streamlines for angular positions of the rotor: 0, 45, 90 and 135°. Similarly as in figure 7, a high level of flow structure complexity with several recirculation zones is observed. It can be noticed that they have 3D character, which may explain why 2D simulations yielded poorer agreement with the experiment.

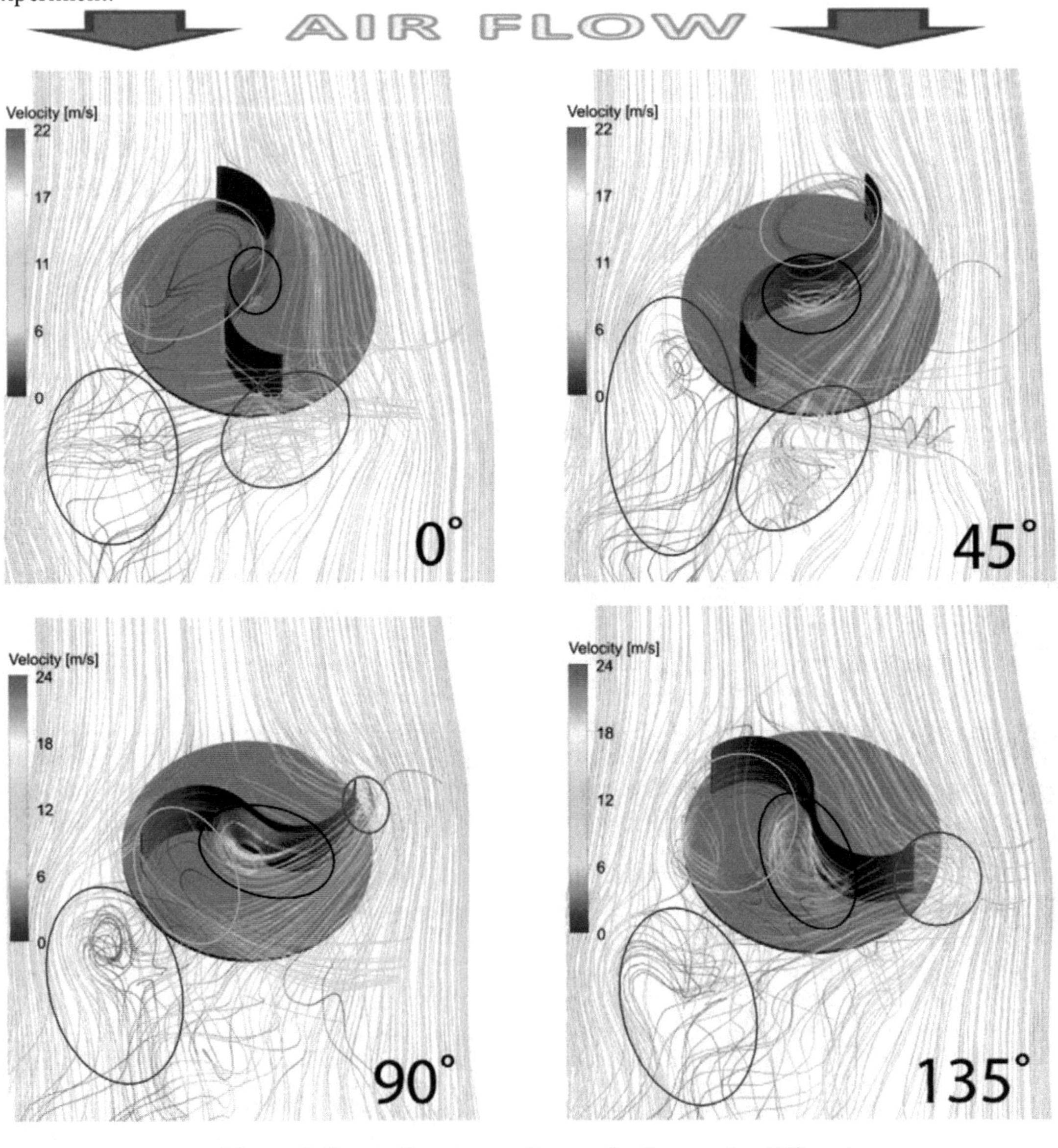

Figure 8. Streamlines in stn. frame of reference for different angular positions of the Savonius rotor.

Four most dominant structures are depicted with different colours. The first one, marked in red, is an advancing blade tip vortex, which is created at the upper tip of the advancing blade and develops downstream to much bigger size. It firstly occurs at the angular position of 90°. At angular positions of 135° and 180° (0°) it grows and separates from the tip of the blade at 225° (45°). Another structure, which is marked in black is an advancing blade boundary layer separation, which forms close to the axis of the turbine. This structure originates around an angular position of 0°. Its another instance

from the previous period (half of the rotation) forms concave side recirculation and it is marked in orange. They are formed due to the fact that the Coanda effect can no longer keep the flow attached to the blade and it slowly separates. The last dominant structure that may be observed (marked in purple) is a result of multiple structure interactions and the part of the orange structure detachment from the returning blade. Together with other detached structures it creates one complex conglomerate downstream of the rotor. It may be noticed that in particular all the structures detaching from the turbine blades exhibit predominantly complicated flow character significantly modified close to the endplate due to its boundary layer interaction.
Blade configuration of the Bach-type Savonius rotor, including reduced blade arc angle (in comparison to a typical Savonius turbine) creates higher lift force on the advancing blade for the angular position of the rotor corresponding to points A and C. At the same time it results in lower resistance while spinning in the air. Another crucial factor affecting the turbine flow field is a lack of an overlap distance, which would diminish the negative torque on the returning blade.

6. Conclusions

On the basis of the presented data the following conclusions can be drawn:

- Comparison of experimental data and obtained simulation results clearly show that 2D simulations exhibit higher deviation from the experimental results in comparison to 3D ones.
- 3D simulations behave in more predictable and stable manner and are less sensitive to mesh density and changes in the simulation definition.
- In case of the Bach-type turbine for a low aspect ratio, flow structure is too complicated to be properly modelled by means of 2D simulations, therefore even a simplified 3D solution yields better results than complex 2D one.
- Flow acceleration on the advancing blade aligned with the flow is one of the key effects contributing to the torque of the rotor.
- Blades in the normal to the flow position produce minimal torque.

7. References

[1] Spera D 2009 *Wind Turbine Technology: Fundamental Concepts of Wind Turbine Engineering* (ASME Press)
[2] Wagner H J and Mathur J 2009 *Introduction To Wind Energy Systems: Basics, Technology and Operation. Green Energy and Technology* (Springer)
[3] Fujisawa N and Gotoh F 1994 Experimental Study on The Aerodynamic Performance of a Savonius Rotor *J. of Sol. En. Eng.* pp 148-152
[4] Ogawa T 1984 Theoretical Study on The Flow About Savonius Rotor *ASME J. Fluids Eng.* pp 85-90
[5] Kamoji L, Kedare S and Prabhu S 2009 Experimental investigations on single stage modified Savonius rotor *App. En.* pp 1064-1073
[6] Zhao Z, Zheng Y and Xu X 2009 Research on the improvement of the performance of Savonius rotor based on numerical study *Sus. Pow. Gen. and Supp.* pp 1-6
[7] Zhou T and Rempfer D 2013 Numerical study of detailed flow field and performance of Savonius wind turbines *Ren. En.* pp 373-381
[8] ANSYS CFX v14 Help
[9] Kacprzak K, Liśkiewicz G and Sobczak K 2013 Numerical investigation of conventional and modified Savonius wind turbines *Ren. En.* pp 578-585
[10] Dobreva I, Massouha F 2011 CFD and PIV investigation of unsteady flow through Savonius wind turbine *En. Proc.* pp 711-720

8. Acknowledgments

This article was funded from the public budget by the Ministry of Science and Higher Education as the "Diamond Grant" research project no. DI2011 000441.

Aerodynamic analysis of an isolated vehicle wheel

P Leśniewicz[1]**, M Kulak and M Karczewski**
Institute of Turbomachinery, Lodz University of Technology, Poland

E-mail: lesniewicz.pawel@gmail.com

Abstract. Increasing fuel prices force the manufacturers to look into all aspects of car aerodynamics including wheels, tyres and rims in order to minimize their drag. By diminishing the aerodynamic drag of vehicle the fuel consumption will decrease, while driving safety and comfort will improve. In order to properly illustrate the impact of a rotating wheel aerodynamics on the car body, precise analysis of an isolated wheel should be performed beforehand. In order to represent wheel rotation in contact with the ground, presented CFD simulations included Moving Wall boundary as well as Multiple Reference Frame should be performed. Sliding mesh approach is favoured but too costly at the moment. Global and local flow quantities obtained during simulations were compared to an experiment in order to assess the validity of the numerical model. Results of investigation illustrates dependency between type of simulation and coefficients (drag and lift). MRF approach proved to be a better solution giving result closer to experiment. Investigation of the model with contact area between the wheel and the ground helps to illustrate the impact of rotating wheel aerodynamics on the car body.

1. Introduction

Nowadays, when car companies are trying to reduce fuel consumption and lower CO_2 emission, aerodynamic design of cars and its components plays important role. According to different sources even up to 25% of the car aerodynamic drag comes from wheels [1,2]. Car manufacturing companies and tyre manufacturers are facing problems connected with reduction of friction coefficient as well as improving air flow around the tyre and whole car body. Wheels are also an important factor taking into consideration noise generated by cars [3]. It is clearly visible that this branch of car design can give benefits connected not only with reduction of fuel consumption and CO_2 emission, but also with safety and comfort of modern vehicles. There are three pairs of major vortices induced by flow around wheel (see figure 1). Wake-horseshoe vortex denoted as numbers 1 and 2 (as a pair), another flow feature is a so called C-shoulder vortex (numbers 3 and 4) caused by separation of air passing over the top tyre shoulders. Third one, squash vortex is also presented in figure 1 (as 5 and 6).

[1] To whom any correspondence should be addressed.

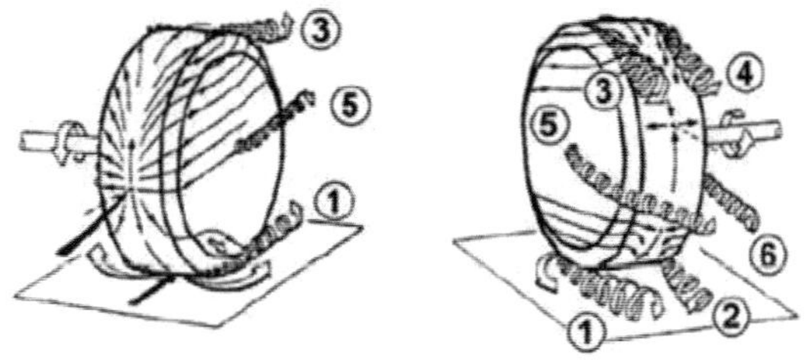

Figure 1. Main vortices around rotating wheel [4].

Despite wide investigations conducted in recent years, there is still a lack of information about flow structures inside tyre grooves, mainly caused by difficulties of accessing the tested region with measurement equipment. One of the first works devoted to this topic was performed in middle 70's by [5]. Differences in circumferential pressure distribution between grooved and slick tyre were reported. In grooved tyre due to the Venturi effect, a sudden pressure drop was measured at place of wheel-ground contact, while the pressure reached the maximum in this place in case of the slick tyre. This was explained in the following manner: a sudden pressure drop inside grooves promotes the flow through the tread and diminishes the size of trailing vortices. Another work connected with automobile wheels was made by [6] whose investigation focused on identification of vortices in case of rotating wheel, while jetting effect was widely described by [7]. The importance of vortices located near the area of contact between the wheel and the ground was highlighted by [4]. As a result, the flow phenomena described by e.g. [4,5,6,7], among others, shed light at the importance of analysing vehicle aerodynamics with inclusion of rotating wheels, both experimentally and numerically, a practice that before 1990s was quite seldom. However, still a number of publications strictly connected with investigation of tyre geometry using Computational Fluid Dynamics (CFD) in synergy with experiment is limited. For example, very recently [8, 9] observed that there is a difference in aerodynamic drag depending on whether a vehicle is equipped with slick or grooved tyres. Namely, the aerodynamic drag is lower for an automobile with grooved tyre (a production sedan and a model of compact car were tested). This claim shifts analysis of tread pattern influence on vehicle aerodynamics back to isolated wheel study in order to search for explanation of this phenomenon. An alternative experiment, to Fackrell's 1974 work, was performed by [10]. During this investigation a consistent negative peak in pressure distribution followed the pressure rise in place of contact for the slick tyre. This observation was in contrast to Fackrell's work. No experiment was made for grooved tyre to make an adequate comparison. Yet an experimentally-based analysis of flow inside grooves for exposed wheel was performed by [11]. It was concluded that main difference between tyre with and without grooves in terms of pressure distribution is the jetting effect. No comments were made about potential drag difference between both tyre patterns. This paper aims at searching for explanation of origins of drag difference described above. Numerical model was prepared to simulate the conditions of the experiment in [5]. Explanation of phenomena responsible for negative peak of pressure for slick and grooved tyre will be discussed.

2. CFD Model

Description of wheel geometry, models used for wheel rotation simulation as well as numerical settings are discussed in this section.

2.1. Geometry

Two wheel geometries were analysed, one with three longitudinal grooves, and the other without grooves (slick). Tyre geometry was created basing on wheel investigated by [5] in order to compare results obtained using CFD with experiment. The wheel has a diameter equal to 416 mm and a breadth of 191 mm. Detailed geometry of the wheel is shown in figure 2. In order to investigate influence of tread, geometry was modified by adding three longitudinal grooves (rain grooves) spaced evenly with respect to the wheel centreline. Each groove is 6.35 mm wide and 3.18 mm deep. Flow is considered as external 3D type, therefore boundary conditions must be set at such distances away from the object to ensure the least amount of influence. Based on domain size investigation it was stated that width should be equal 10D, height 10D, upstream 5D, downstream 15D where D is the wheel diameter. Due to the fact that in reality the contact between the wheel and the ground is not at one point, contact patch between the wheel and the ground was defined. Tyre deformation model study provided in [8] showed that uninterrupted grooves in contact patch area more realistically predict flows around tyres. The study presented herein adheres to this standard. Visualization of both, slick and groove tyre with a contact area are presented in the figure 3.

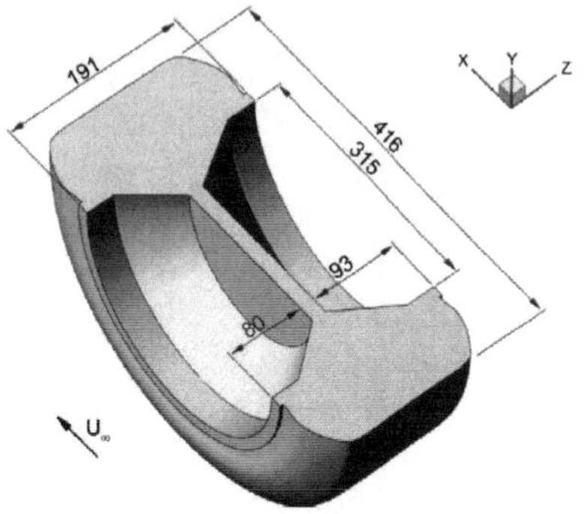

Figure 2. Section view of wheel with dimensions (in mm) [3].

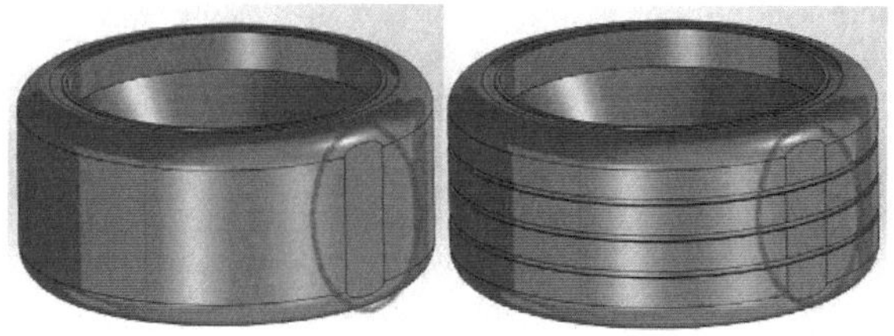

Figure 3. Area of contact between the wheel and the ground.

2.2. Numerical model

Numerical investigation of flow was performed using ANSYS CFX v.14.0 software. Based on previous works performed in Institute of Turbomachinery related to the analysis of pumps, fans, turbines, and geometry optimization, which show the best compromise between accuracy of the results and time needed for simulation [12,13,14], Shear Stress Transport (SST) turbulence model was applied. Additionally, [15] made a study where non-linear viscosity model (Explicit Algebraic Reynolds Stress Model with Curvature Correction EARSM-CC) has been used in rotational frame of reference. Comparison to linear viscosity model reported by [15] reveals insignificant differences in pressure distribution modelling. Steady state analysis were performed with turbulence intensity at the inlet set to 0.002, while high resolution accuracy for advection terms of the Navier-Stokes equations was selected. Free stream flow speed was equal to 18.6 m/s and angular velocity of the wheel was 16.67 rev/s what matched the real travel conditions represented by stream flow speed. Rotation of the wheel was modelled using two approaches: Moving Wall boundary (MW), and Multiple Reference Frame (MRF). In the first method, translation of the wheel is created by applying linear velocity to the peripheral nodes located at the perimeter of the wheel, except the nodes situated in contact area where combination of boundary conditions is used (see figure 4). MRF is considered as an alternative way to model rotation of the wheel. It is partially based on MW method plus is extended by introducing a separate computational domain adjacent to the wheel rim. This results in creation of pair of interfaces presented in the left picture of figure 5 (marked in green). Grid A is attached to the stationary mesh B. Due to the angular velocity of the rim all nodes of mesh A will rotate, while mesh B will have translation definition according to MW method. Numerous tests for grid A were performed to determine its size as based on accuracy to represent the physical aspects of flow.

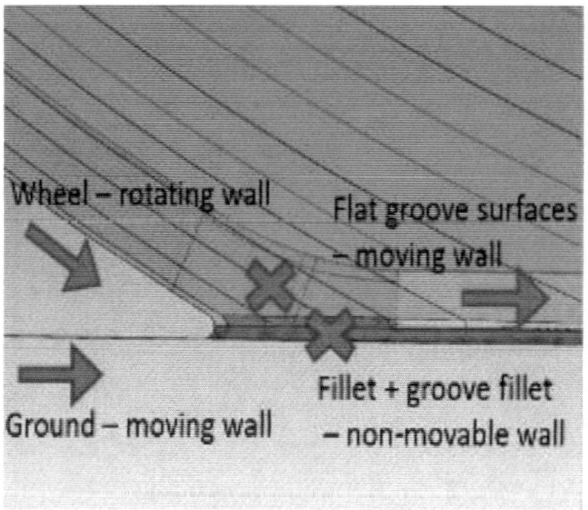

Figure 4. Set of boundary conditions in contact patch region.

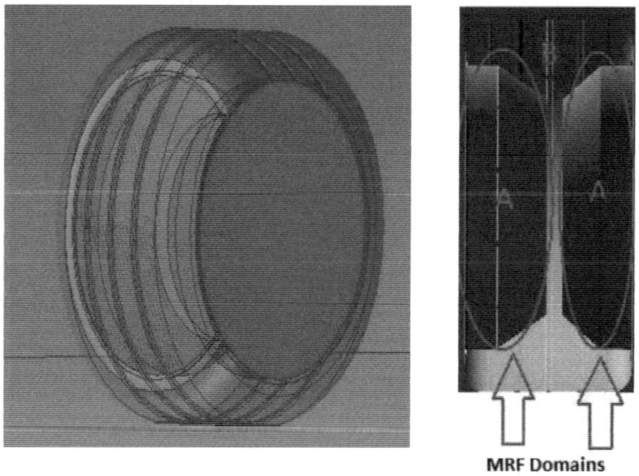

Figure 5. MRF domains applied to the model.

3. Results

Following section contains analysis of the results obtained from CFD model and experiment performed by [5,10,11].

3.1. Drag and lift coefficients

For the analysis that follows, the wheel with grooves is denoted as G, the slick one as S, slick tyre for which moving wall was applied is named S/MW, while grooved tyre with MRF model is denoted as G/MRF respectively. Global quantities such as C_x and C_z were calculated to evaluate the accuracy of the numerical model in comparison to the experiment presented by [5]. Comparison of obtained results of drag and lift coefficients are presented in the table 1.

Table 1. Comparison of numerical results to the experimental data in [5].

Tyre	C_x	C_z	Experiment C_x	Experiment C_z
S/MW	0.709	0.320	0.51	0.28
S/MRF	0.535	0.295	0.51	0.28
G/MW	0.699	0.220	-	-
G/MRF	0.520	0.261	-	-

As it can be observed values of lift and drag coefficients are significantly smaller (except C_z for grooved tyre) in situation where Multiple Reference Frame model was used. In MRF simulations of slick tyre, reduction of lift is observed, while for tyre with grooves C_z values increase when compared to computations with the use of MW model. It can be concluded that use of Multiple Reference Frame decisively helps to obtain numerical model with higher accuracy with respect to the experiment not only for drag, but also for lift coefficient. Furthermore, a clear general trend of drag reduction for grooved tyre is observed and will be explained in next sections.

3.2. Tyre circumferential pressure distribution

It was decided to compare pressure distribution for geometries investigated in this research with experimental data elaborated by other researchers. Figure 6 presents measurements of centreline circumferential pressure distributions for rotating slick tyres experimentally tested by [5,10]. Results of those experiments are compared with centreline pressure distribution obtained by numerical investigation discussed in this paper.

Pressure increases from stagnation point to the contact area which is observed for all investigations. Pressure peak observed by Mears is shifted by few degrees. Immediately after, a sudden pressure drop is visible in place of contact patch (around 90° angle). This pressure drop is not visible for CFD simulation - contact patch area was simulated in such way as deformation of the tyre in this region blocked the flow. Furthermore, this is where the difference between Fackrell and Mears investigations is visible. For the first experiment a negative pressure peak is not observed and pressure stabilises.

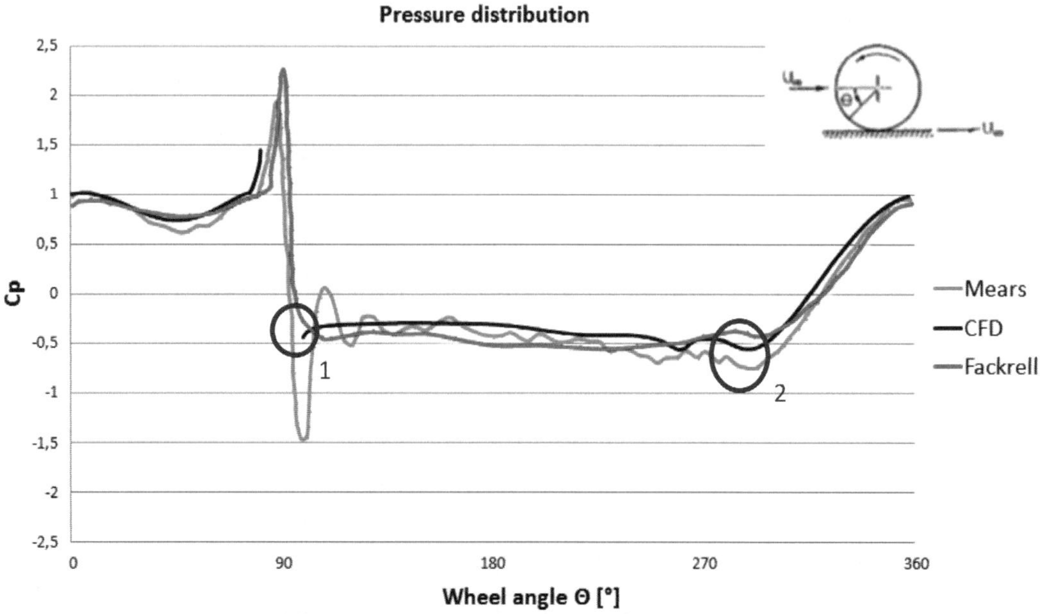

Figure 6. Pressure distribution for slick tyre.

In case of Mears' investigation the negative pressure peak was observed. This change in pressure distribution is connected with a "suction cup" effect as it was named by [11]. The "suction cup" can be explained by creation of a vacuum between tyre and the ground. The vacuum area is suddenly expanded when tyre lifts up recovering the pressure right after contact area. This tendency is the same for CFD model where, despite lack of data points, clear pressure recovery is observed and is not visible in measurements made by [5] (area 1 in figure 6). After the contact patch distribution of pressure lines for numerical simulation and Fackrell's experiment seems quite similar. Pressure line for Mears' investigation after negative peak looks different. Behind contact area oscillations are observed. This phenomenon is caused by the fluctuations of the moving belt below the tyre. Second difference is located near place of flow separation (area 2 in figure 6). Values of pressure for Fackrell's experiment are higher than those recorded by Mears. Numerical investigation in this region shows pressure values between both experimental investigations. To sum up, the CFD model agrees quite well with data in Fackrell's experiment. Moreover, a part of negative pressure peak at contact patch, which was not observed by [5], but was presented by Mears is likewise predicted by the CFD simulation.

Centreline circumferential pressure distributions inside a middle groove were analysed for the second tyre type. This time results of numerical investigation were compared with experiment performed by [5,11]. As it can be observed in figure 7 for all cases a sudden negative pressure peak appears near the area of contact, where slick portions of the tyre blocks the flow while the grooves allow it in. Hence observed pressure drop in this region is caused by sudden acceleration of the air rushing into the groove [11]. However, pressure drop for CFD simulation is significantly more intense. This can be connected with greater amount of "measurement points" (mesh nodes) in this area. Beyond this, pressure distribution is nearly the same for all cases. Near the flow separation area differences in pressure distribution are observed but are smaller compared to the case of slick tyre. Overall, both experiments as well as numerical simulation show similar pressure representation.

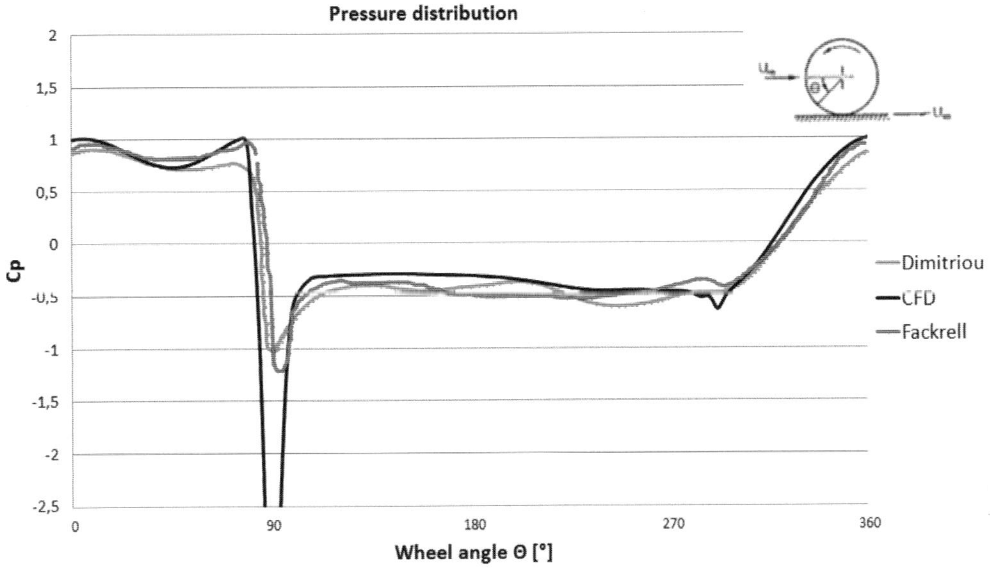

Figure 7. Pressure distribution for tyre with grooves.

3.3. Velocity flow fields

2D velocity vector plot was used to visualize flow characteristics. Y-Z plane was created at the distance equal to 2.5 *D* behind the wheel (*D* as the wheel diameter). Comparing tangential projections of flow vectors for slick tyre (figure 8) and tyre with grooves (figure 9) following differences are observed. Velocity field for tyre with grooves is more even than for slick one. The small region of lower velocity behind slick tyre centreline is visible (figure 8, area 1). It can be treated as a result of the energy loss used for work which need to be done by medium to travel around the obstacle. One can also notice a vortex aside of right wheel edge (only on one side of wheel due to asymmetrical geometry of rim). It is present in both cases, however for slick tyre (figure 8, area 2) it is located higher and more to the outside with respect to the grooved wheel. The reason behind this shift may come from the fact, that bigger amount of fluid has to encircle the wheel – for the grooved tyre the vortex core remains at the distance equal to half of wheel width.

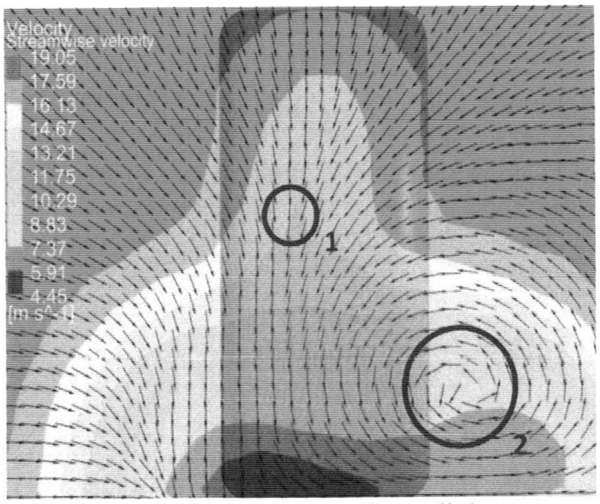

Figure 8. Velocity vector plot - slick tyre.

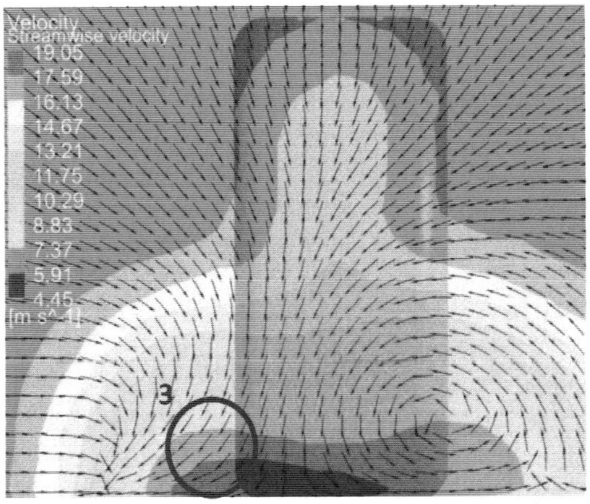

Figure 9. Velocity vector plot - grooved tyre.

Behind the left wheel edge the flow is biased outwards in the grooved tyre case (figure 9, area 3). This phenomenon is an effect of influence of fluid which have propagated along the grooves, not present in the slick tyre case. Velocity flow field observations confirm previous conclusions related to differences in drag resistance for both tyres.

3.4. Drag differences between tread types

As it was presented, the negative pressure peak near 90° is presented for both, slick and grooved tyres, however its location is slightly different. In case of tyre with grooves it starts before place of contact, while for slick tyre initiates after 90° angle. What is more, the reason of its presence is different depending on the type of the tyre. In case of slick tyre, the negative pressure peak happens due to the suction cup. On the other side, the acceleration of the flow inside the grooves creates an under-pressure, which suppresses the jetting effect present in slick wheel what is manifested by a sudden pressure plunge at the contact patch. Table 2 presents the comparison of findings from current investigation to the study of enclosed wheels made by [9]. The drag coefficient reported in table 1 was multiplied by the frontal area of the wheel in order to make the comparison possible. This effectively created set of area drag coefficients for the three compared studies. Production size tyre was tested by [9], while a model of Formula 1 tyre was under investigation herein.

Table 2. Comparison of numerical results of isolated wheel to CFD simulation and experiment of enclosed wheel reported in [9].

	CFD simulation of the isolated wheel		CFD simulation of the enclosed wheel (Kulak et al.)	Measurement of the enclosed wheel (Kulak el al.)
Tyre	SC_x	ΔSC_x	ΔSC_x	ΔSC_x
Slick (MRF)	0.043	-	-	-
Grooved (MRF)	0.041	-0.002	-0.003	-0.002

It is evident that a drag difference presented by the current study is of the same order and, moreover, value when compared to results recorded for a model vehicle. Similar trend was observed by [8] where a difference of 4 drag counts in favour of grooved tyre was reported in tests of entire vehicle. This suggests that drag difference observed at vehicle scale can be purely related to the presence of grooves and, more importantly, to reduction of jetting effect provided by the grooved portions of the tyre. In the slick tyres, the work is being done on the air as it is squeezed between the wheel and moving ground giving rise to pressures that are greater than the total pressure in the working section. This is also the case for the groove tyre, except the places where a groove is present. This slight change in tread pattern has an important impact on the aerodynamic drag. It puts in question observations made by [5], where no change in drag was reported between tread patterns, but more studies are needed to confirm this hypothesis. Therefore, at Institute of Turbomachinery, in cooperation with PSA Peugeot Citroen, works are under way to develop an experimental stand for isolated wheel studies. At the same time, a reliable CFD model for wheel-ground contact is being developed.

4. Conclusions

Numerical model prepared for this investigation was able to predict major flow characteristics connected with aerodynamic analysis of an isolated wheel. The most important differences between flow around the tyre with grooves and the slick tyre were identified not only in terms of drag coefficient but also pressure distribution. Precise analysis of two approaches used for wheel simulation clearly showed that the use of Multiple Reference Frame model significantly improves the accuracy of numerical model. Difference in drag coefficient between numerical investigation and experiment for slick tyre can be connected with pressure drop near the area of contact between the wheel and the ground. Same tendency can be observed for tyre with grooves, however jetting effect associated with slick tyre is responsible for bigger pressure drop and as a consequence for bigger profile resistance. Those phenomena were explained by Institute of Turbomachinery at Lodz University of Technology in cooperation with PSA Peugeot Citroen based on Peugeot 207 car body. Results of numerical investigations were verified in PSA wind tunnel.

Future attention should focus on CFD-based immersed body as a technique for rotating wheel simulation. Succeeding work will investigate tyres with various configuration of grooves. This is why in cooperation with PSA a new test stand for more precise analysis of wheel with use of Particle Image Velocimetry (PIV) is being created. What is more, turbulence model independence study, containing nonlinear eddy viscosity models [15], should be performed to verify its influence on the obtained results. Furthermore, based on analysis of pressure distribution for both tyres, it can be observed that the way how data are collected has a significant influence on results in most sensitive place –area of contact between the wheel and the ground. Numerical investigation can therefore help to supplement experimental results with expanded data especially in places hard to access.

5. References

[1] Wickern G , Zwicker K and Pfadenhauer M 1997 Rotating wheels - their impact on wind tunnel test techniques and on vehicle drag results *SAE Int.* **970133**

[2] Mlinaric P 2007 *Investigation of the Influence of Tyre Deformation and Tyre Contact Patch on CFD Predictions of Aerodynamic Forces on a Passenger Car* (Gothenburg: Chalmers University of Technology)

[3] McManus J and Zhang X 2006 A computational study of the flow around an isolated wheel in contact with the ground *J. of Fluids Eng.* **128** 520-530

[4] Morelli A 2000 A new aerodynamic approach to advanced automobile basic shapes *SAE Int.* **2000-01-0491**

[5] Fackrell J E and Harvey J K 1974 The aerodynamics of an isolated road wheel *2nd AIAA symp. of aerodynamics of sports and competition automobiles,* ed B Pershing (Los Angeles, California, USA, 11 May 1974) vol 16 pp 119–25

[6] Cogotti A 1983 Aerodynamic characteristics of car wheels *Int. J. Vehicle Design - Impact of Aerodynamics on Vehicle Design* **3** 173-96

[7] Mercker E, Breuer N, Berneburg H and Emmelmann H 1991 On the aerodynamic interference due to the rolling wheels of passenger cars *SAE Int.* **910311**

[8] Hobeika T, Sebben S and Landstrom C 2013 Investigation of the influence of tyre geometry on the aerodynamics of passenger cars *SAE Int.* **2013-01-0955**

[9] Kulak M, Karczewski M and Spolaore G 2014 Flow around rotating wheels and its interaction with vehicle aerodynamics – CFD vs. Wind tunnel tests *FISITA world congress* (Maastricht) F2014-LWS-068

[10] Mears A P, Dominy R G and Sims-Williams D B 2002 The air flow about an exposed racing wheel *SAE Int.* **2002-01-3290**

[11] Dimitriou I and Klussmann S 2006 Aerodynamic forces of exposed and enclosed rotating wheels as an example of the synergy in the development of racing and passenger cars *SAE Int.* **2006-01-0805**

[12] Karczewski M and Blaszczak J 2008 Performance of three turbulence models in 3D flow investigation for a 1.5-stage turbine *TASK Quarterly* **12** 185-95

[13] Sobczak K and Blaszczak J 2010 Numerical analysis of the flow around a cylinder *CMP* **137** 53-66

[14] Kabalyk K and Kryllowicz W 2011 Verification of the numerical model of the viscous gas flow within a centrifugal compressor stage *Nauchno-Technicheskie Vedomosti SPbGPU* **2** 329-34

[15] Wickerath B and Niehuis R 2006 A study of nonlinear eddy-viscosity models in a flow solver for turbomachinery *25th ICAS Congress* (Hamburg)

Journal of Physics: Conference Series **530** (2014) 012065 doi:10.1088/1742-6596/530/1/012065

Effects of Extensional Rates on Characteristic Scales of Two-dimensional Turbulence in Polymer Solutions

R Hidema[1]*

Organization of Advanced Science and Technology, Kobe University,
1-1, Rokkodai, Nada, Kobe, Hyogo, Japan. 657-8501

E-mail: hidema@port.kobe-u.ac.jp

Abstract. In order to study the effects of extensional viscosities on turbulent drag reduction, experimental studies using two-dimensional turbulence have been made. Anisotropic structures and variations of energy transfer induced by polymers are considered. Polyethyleneoxide and hydroxypropyl cellulose having different flexibility, which is due to different characteristics of extensional viscosity, are added to 2D turbulence. Variations of the turbulence were visualized by interference patterns of 2D flow, and were analysed by an image processing. The effects of polymers on turbulence in the streamwise and normal directions were also analysed by 2D Fourier transform. In addition, characteristic scales in 2D turbulence were analysed by wavelet transform.

1. Introduction

Addition of little amounts of flexible polymers to turbulent flows reduces the friction coefficients to solid walls in the flow. Such drag reduction phenomena have been studied for many decades. One of the key of this effect is the increase of extensional viscosity that is due to the extension of polymers in the flow [1-3]. Rheological effects such as the extensional viscosity have been studied in numerical studies [3,4]. Since the extensional viscosity stems from the resistance in the extensional direction, the effects cause anisotropic structures in turbulence. Therefore, turbulence affected polymers is not uniformly decayed in space, that is, the effects in the streamwise directions can be different from those in normal direction. Such anisotropic effects have been investigated by many numerical analyses, but not enough experimental studies have been made [5]. In addition, it is difficult to separate the effects of extensional stress from those of shear stress on polymers in conventional three dimensional pipe experiments.

Therefore, we have been used self-standing flowing soap films as a test bed for two-dimensional (2D) flow [1,2,6]. The soap films consist of water layer and surfactant molecules covering the surface, which flows between two parallel nylon wires. Since the surfactant molecules are really small compared to the thickness of water layer, the flowing soap films are considered to be a water layer sandwiched by free surface. Thus, the shear stress is very small in this system. When a grid is inserted to the flow, 2D grid turbulence is generated in the flow. Extensional stress is generated at the

*Corresponding author: R Hidema
Phone: +81-78-803-6657, Fax: +81-78-803-6657
E-mail address: hidema@port.kobe-u.ac.jp

grids, and therefore, when polymers exist in the flow, the turbulence affected mainly the extensional viscosity of polymers due to the extensional stress can be observed.

In this study, the effects of the extensional rate on anisotropic structures of 2D turbulence, which is induced by the extensional flow of polymer solution, have been examined. Polyethyleneoxide (PEO) and hydroxypropyl cellulose (HPC) are added to the 2D turbulence respectively. PEO is a flexible polymer and HPC is a rigid rod-like polymer, which show different characteristics of the extensional viscosity. Turbulence affected polymers are analysed by an image processing to investigate how and why anisotropic structures appear and characteristic scales of turbulence are changed by flexibility and extensional viscosity of polymers.

2. Experimental procedures

2.1. Materials

Soap solutions contain sodium dodecylbenzenesulfonate (SDBS) as a surfactant at a concentration of 2 wt%. As a flexible polymer, polyethyleneoxide (PEO, molecular weight: 3.5×10^6) is used at concentrations of 0.25, 0.5, 0.75, 1.0, 1.5, and 2.0×10^{-3} wt%. As a rigid polymer, hydroxypropyl cellulose (HPC, molecular weight: $>1.0\times10^6$) is used at concentrations of 0.01, 0.03, and 0.05 wt%.

2.2. Visualization of two-dimensional turbulence by flowing soap films

The apparatus used for experiments is shown in figure 1. Whole image of the apparatus is shown in figure 1(a). The outline of our apparatus follows that proposed by Rutgers *et al.* [7]. Two nylon wires tightened by a weight were used to make a soap film channel. Sample solutions containing surfactant flowed between the two nylon wires from a top reservoir to a bottom reservoir by gravity. The flow reached uniform velocities at about 30 cm behind from the injection nozzle. In this study, the mean velocity was 130 cm/s. The mean thickness of the water layer was approximately 3.85μm, with a mean velocity of 130 cm/s for a flow flux of 0.5 ml/s. At a grid of equally spaced cylinders, vortices were generated and were convected without much deformation to the observation area where about 20 cm under the grid. Illumination light was reflected at the front and the back of the flowing soap films, which made the interference images (figure 1(c)). The interference images were related to the thickness of the water layer. Therefore, the dynamics of the flow was visualized through the interference patterns. A digital video camera (Panasonic TM700) was used to record the flow in the data acquisition area (figure 1(b)).

2.3. Extensional Rate Added at a Grid

When the flow passes the grid, local velocities were increased at the grid since the cross-section area of the water layer was decreased. The sudden increase of the local velocity between the cylinders induces the extensional rate that described as follows.

$$S(t) = S_0 \exp(-\dot{\varepsilon}t) \tag{1}$$

where S_0 m^2 is the cross-section area before deformation, $S(t)$ m^2 is the cross-section area after deformation, t s is the time required for deformation, and $\dot{\varepsilon}$ s^{-1} is the apparent extensional rate. The extensional rate was controlled by the distance between cylinders and its diameter. In the present study, the diameter of the cylinder was 0.3 cm. The distances between two cylinders were 0.9 cm or 0.6 cm as shown in figure 1(d). The numbers of the cylinders were changed seven to eight for each distance. Concretely speaking, when the number of the cylinders were seven and the distances were 0.9 cm, S_0 m^2 was 7.20 cm x 3.85 μm and $S(t)$ m^2 was 5.40 cm x 3.85 μm. Thus, $\dot{\varepsilon}$ was approximately 250 s^{-1}. When the distances between cylinders were 0.6 cm, $\dot{\varepsilon}$ was 350 s^{-1}. These extensional rates affect polymers. Polymers are considered to be stretched and to be oriented in the flow, which induce extensional viscosity. This affects the generation of vortices in 2D turbulence.

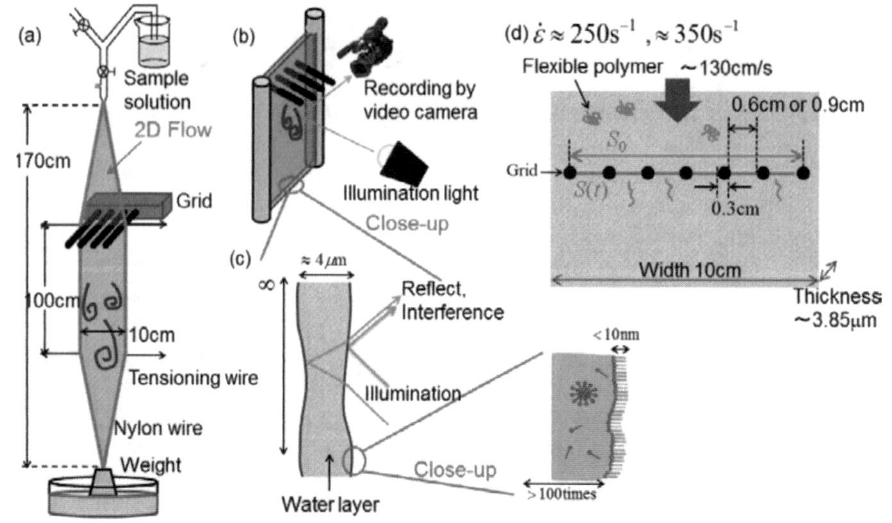

Figure 1. Apparatus of flowing soap films.

3. Image processing by Film Interference Flow Imaging

The interference patterns of the flowing soap films as 2D turbulence were analysed by image processing methods. Such a method using interference patterns to obtain quantified information of 2D flow is unique. Therefore, we called this method "Film Interference Flow Imaging (FIFI)" in our previous study [8].

3.1. Two-dimensional fast Fourier transform (2D-FFT)

2D-FFT software was used to calculate power spectra of the interference images. Since the Red-Green-Blue (RGB) pixel intensity of images are related to the thickness of the water layer, the power spectra of the images show the dynamics of 2D flow. In the present study, the power spectrum $<I^2(k_x, k_y)>$ was calculated based on the pixel intensity of G with a Hamming window, where k_x m^{-1} and k_y m^{-1} are the spatial frequency in the normal and streamwise directions in an interference image. The pixel intensity of G was chosen to avoid variation of an order of interference in the data acquisition area. The area is 256 pixels x 256 pixels, which correspond to about 2.56x2.56 cm^2. Therefore, k_x and k_y range from about 1/2.56 to 1/0.02 cm^{-1}, i.e., 0.391 cm^{-1} to 5.00 cm^{-1} as the frequency.

An example of the power spectrum $< I^2(k_x, k_y)>$ of the interference pattern at the extensional rate of 250 s^{-1} is shown in figure 2(a) and (b). Figure 2(a) shows a 2D representation of $<I^2(k_x, k_y)>$ in the case of polymer free solution, where the darkness of the image indicates the intensity. Two arrows in the image indicate the directions corresponding to the streamwise and normal directions in the interference patterns. Figure 2(b) is the power spectrum represents $< I^2(k_x, 0) >$, which shows the scaling behaviors. The scaling index was used to discuss the energy transfer in 2D turbulence.

3.2. Wavelet transform

Sequence pixel intensities in the streamwise direction were analysed by a wavelet transform to investigate the spatial fluctuation of the water layer [9]. Contrary to the Fourier transform, the wavelet transform can resolve strongly irregular signals, such as turbulent signals from different scales, while keeping the signals localized both in space and frequency. Wavelet analysis provides a 2D unfolding of one-dimensional signals, resolving both the position and the scale as independent variables [10]. The analysed sequence pixels were approximately 6,000 pixels corresponding to 65 cm. The sequence intensities of G in the image were obtained by connecting consecutive images as shown in figure 2(c).

Journal of Physics: Conference Series **530** (2014) 012065 doi:10.1088/1742-6596/530/1/012065

The wavelet transform coefficient, $W(a, b)$, is defined for the intensity of G in the streamwise direction, $G(x)$.

$$W(a,b)=\frac{1}{\sqrt{a}}\int \varphi\left(\frac{x-b}{a}\right)G(x)dx \tag{1}$$

where x m is the length scale in the streamwise direction, and a m and b m are the wavelet scale and the sequence scale parameters, respectively. The function $\varphi(x)$ is defined as a mother wavelet, which is localized around $x = 0$. In the present study, the mother wavelet, $\varphi(x)$, is a Mexican hat. The fluctuation in the sequence data was resolved by the wavelet transform to the position and magnification which correspond to x and a^{-1}. As a result, wavelet coefficients of sequence intensities were plotted on the wavelet scale, a m, and a sequence scale, b m. Here, b m is related to the analysed distance in the streamwise direction 65 cm, that is, sequence 6,000 pixels. Figure 2(d) shows an example of the result. Bright part of the image corresponds to positive correlations, and dark part of the image corresponds to negative correlations. In addition, the root mean square values of the wavelet coefficients, W_{rms}, normalized by a difference between the maximum and minimum values of the wavelet coefficients, $W_{max-min}$, are plotted on the wavelet scale, a m (figure 2(e)). As an example, a periodic structure seen at a wavelet scale a m shown by the arrow in figure 2(d) corresponds to the peak intensity shown in figure 2(e).

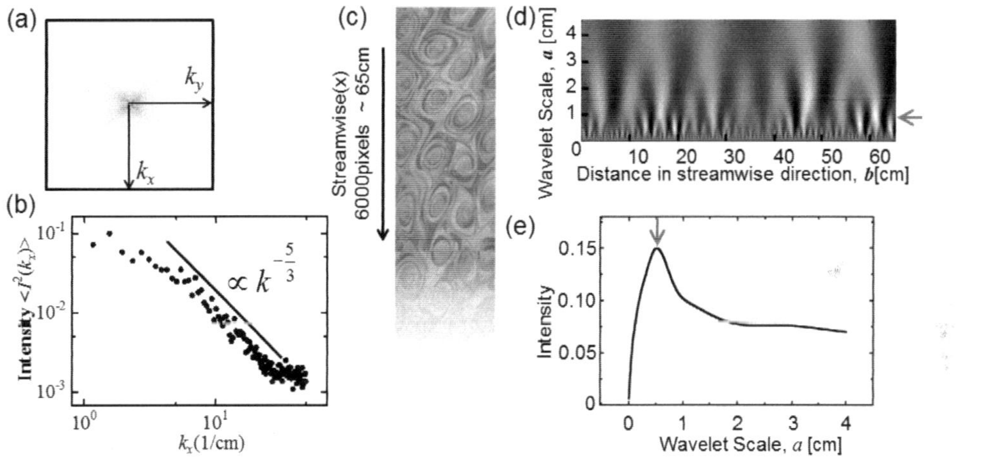

Figure 2. (a) 2D representation of the power spectrum. The k_x and k_y directions in the image represent the spatial frequencies in the streamwise and normal directions. (b) A power spectrum. (c) Schematic of sequence pixel intensity. (d) A wavelet coefficient. (e) W_{rms} normalized by $W_{max-min}$ plotted on the wavelet scale, a m.

4. Results and discussion

Figure 3 shows interference images of the 2D turbulence in flowing soap films when the extensional rate, $\dot{\varepsilon}$ s^{-1}, is 250 s^{-1}. In the case of polymer free solution, the size of the vortices in turbulence is about 1 cm in diameter as shown in figure 3(a). In 2D flow, energies transfer from smaller scale to larger scale. This phenomenon is called inverse energy cascade, since the direction of the energy transfer is opposite to 3D case. In figure 3(a), energies transfer from smaller scale to larger scale, therefore, the vortices can grow. On the other hand, the vortices become long and thin with the addition of PEO as shown in figure 3 (b) to (d). When the concentration of PEO is 2.0 x 10^{-3} wt%, vortices are hardly seen. This is considered to be a manifestation of the prohibition of the inverse energy cascade that is caused by PEO. The energy stays in smaller scales as fluctuations. Therefore, the vortices could not grow. Figures 3(e) to (g) show for the case of HPC. The vortices are deformed

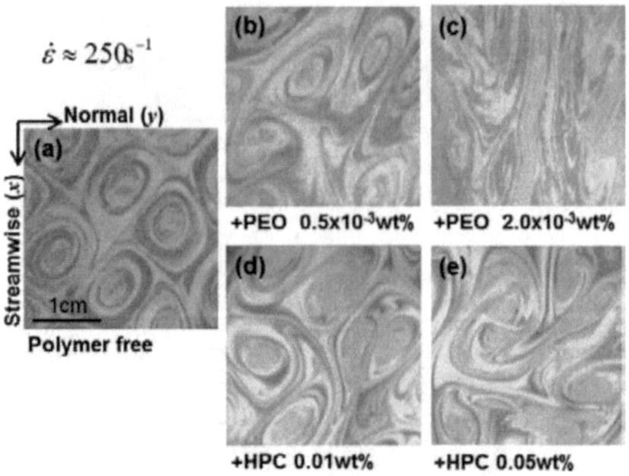

Figure 3. Interference images of the turbulence in flowing soap films when $\dot{\varepsilon}$ [s^{-1}] was 250 s^{-1}.

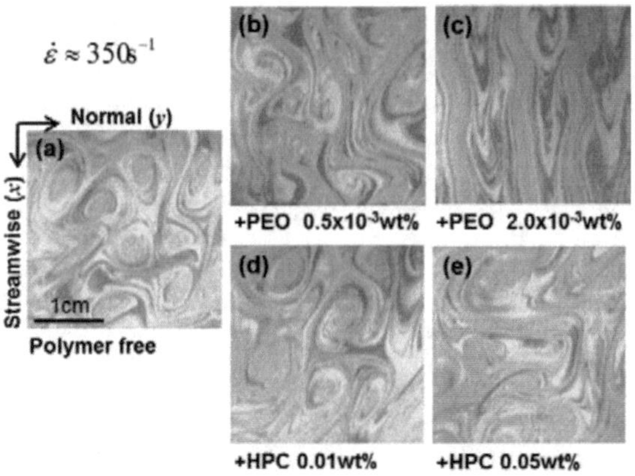

Figure 4. I Interference images of the turbulence in flowing soap films when $\dot{\varepsilon}$ [s^{-1}] was 350 s^{-1}.

by adding HPC, however, the results seems different. In the case of PEO, the vortices are stretched in the streamwise direction, which does not necessarily apply to HPC case. Figure 4 shows interference image when $\dot{\varepsilon}$ s^{-1} is 350 s^{-1}. The effect of PEO seems to be increased even at the same concentration, which is true of HPC case.

Here, the interference patterns visualize the thickness of water layer related to the vortices in 2D flow. The thickness is considered as a passive scalar in 2D flow. The power spectrum of the passive scalar is affected by inverse energy cascade in 2D flow. Thus, scaling exponent obtained from the power spectrum of the interference patterns indicate the mechanism of energy transfer in 2D turbulence. Figure 5 shows the scaling exponents of PEO added solutions in the streamwise and normal directions. The scaling exponents of polymer free solution were almost -5/3 in the both directions and the both extensional rates, which is predicted as passive scalar scaling when inverse energy cascade exists in the flow. Figure 5(a) shows that the scaling components approach to -1 when the concentration of

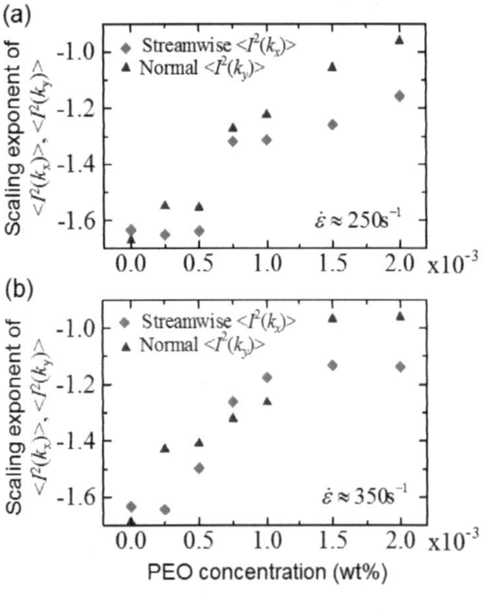

Figure 5. Scaling exponent of PEO added solution at each extensional rate.

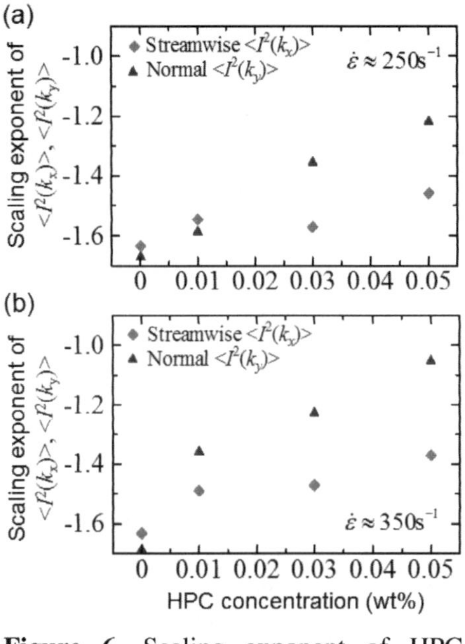

Figure 6. Scaling exponent of HPC added solution at each extensional rate.

PEO is increased. In addition, the variation is large at 0.75 x 10^{-3} wt% in both directions. This critical concentration is shifted to lower concentration when the extensional rate is higher (figure 5(b)). The scaling exponent of passive scalar of -1 indicates that the inverse energy cascade ceased because of the existence of polymers. Figure 6 shows the case of HPC added solutions. When $\dot{\varepsilon}$ s^{-1} is 250 s^{-1}, the scaling exponent gradually approaches to -1 in the normal direction, while the variation is very small in the streamwise direction (figure 6(a)). When the extensional rate is higher, the variations of the scaling exponents are larger (figure 6(b)). The results of figure 5 and 6 indicate that the inverse energy transfer is prohibited by adding PEO and HPC. However the effect is considered to be different. At the grid inserted in the flow, flexible PEO is expected to be suddenly stretched and oriented parallel to the streamwise direction. The orientation could prohibit energy transfer in the normal direction. Thus, the effect was seen in the normal direction well. In the streamwise direction, the extensional viscosity of PEO may be related to the prohibition of the energy transfer, since PEO exhibit extensional viscosity in the extension direction. In general, extensional viscosity is higher when a polymer is flexible and when the extensional rate is higher. Therefore, the effects of PEO in

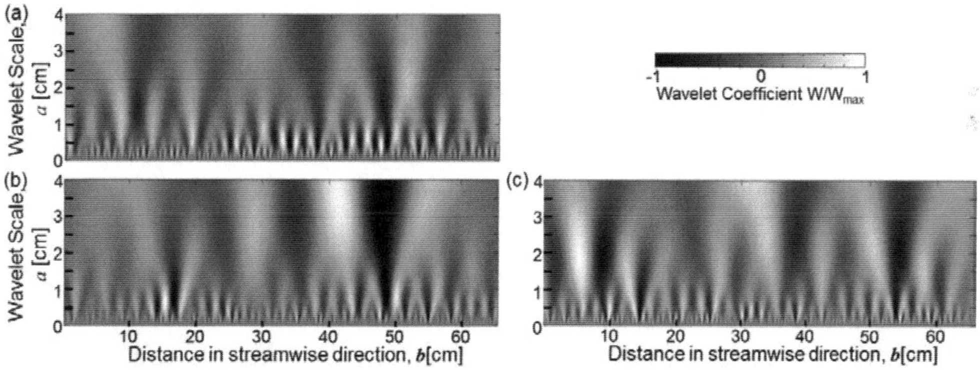

Figure 7. Wavelet coefficient contours for interference images of (a) SDBS 2 wt% solution, (b) + PEO 2.0 x 10^{-3} wt%, (c) + HPC 0.05 wt% when the extensional rate is 250 s^{-1}.

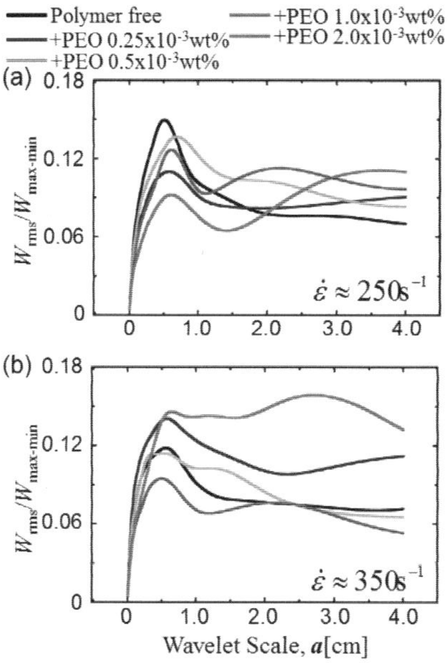

Figure 8. The root mean square value of the wavelet coefficient of the PEO added solution.

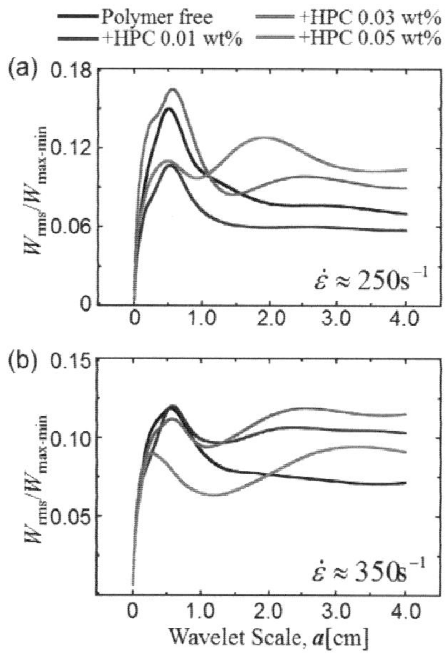

Figure 9. The root mean square value of the wavelet coefficient of the HPC added solution.

the streamwise direction are larger than that of rigid HPC. HPC is also expected to be oriented parallel to the streamwise direction. Therefore, HPC can prohibit energy transfer in the normal direction. However, the extensional viscosity of HPC is smaller, thus the effect is smaller in the streamwise direction.

Since the streamwise direction was considered to be affected by extensional viscosities, the fluctuation in the streamwise direction was analysed by wavelet transform. Figure 7 shows the wavelet coefficient contour for each solution. In the case of polymer free solution, relatively strong periodic structure is seen at smaller than 1 cm of the wavelet scale (figure 7(a)). When PEO induces laminaliztion of the 2D turbulence, the periodic structure at smaller than 1 cm of the wavelet scale

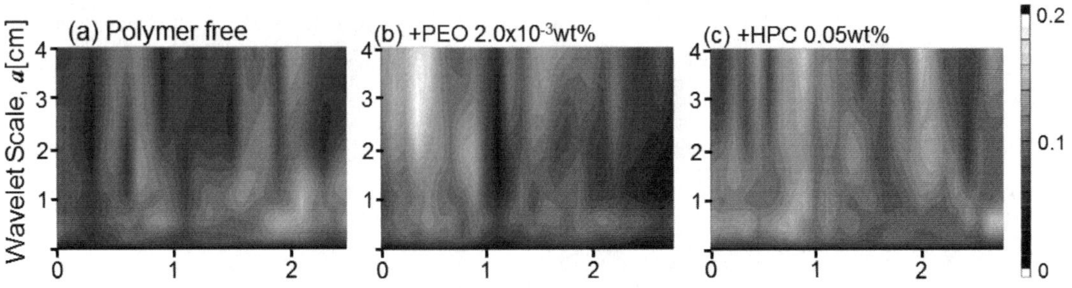

Figure 10. Mapping of $W_{rms}/W_{max\text{-}min}$ of each solution.

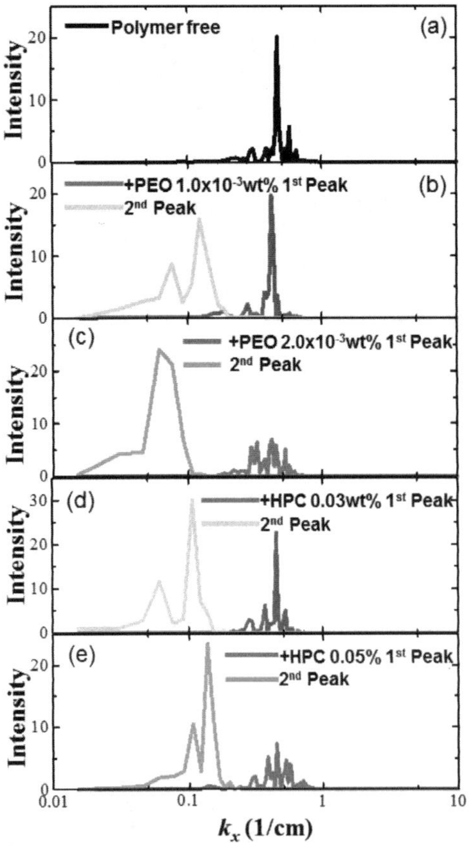

Figure 11. Power spectra of wavelet coefficients of each solution. $\dot{\varepsilon}$ s^{-1} is 250 s^{-1}.

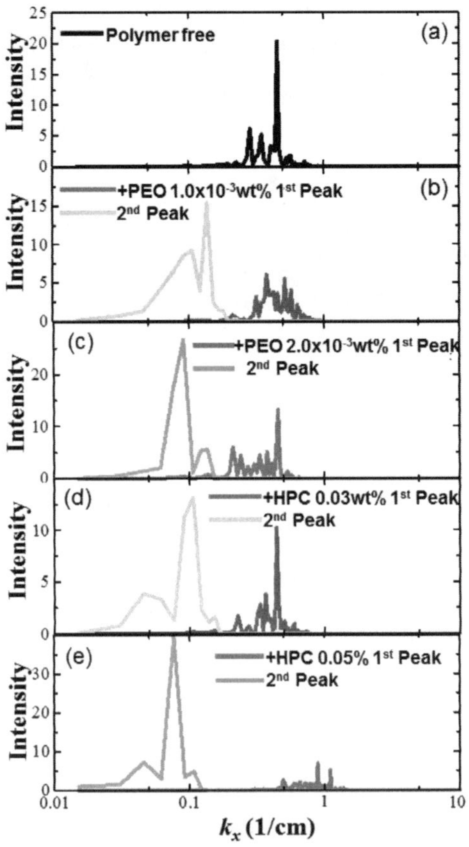

Figure 12. Power spectra of wavelet coefficients of each solution. $\dot{\varepsilon}$ s^{-1} is 350 s^{-1}.

becomes weak, while a strong variation is seen at about 3 cm of the scale (figure 7(b)). In the case of HPC added solution, 7(c), the tendency seems to be the same as figure 7(b). In order to determine how the periodic structures are changed by polymer concentration, the normalized root mean square values of the wavelet coefficients, $W_{rms}/W_{max\text{-}min}$ was calculated (figure 8. 9). The peak of the polymer free case shown in figure 8(a) appears at the wavelet scale corresponding to the scale where the original fluctuation due to the original vortices was observed. When the concentration of PEO is increased, the original peak becomes relatively smaller while the other wider peak appears at the larger scale as shown in figure 8(a). The peak appears at the larger scale is related to the periodic structure in the larger scale. The wider peak is shifted to a larger scale by increasing the concentration of PEO when $\dot{\varepsilon}$ s^{-1} is 250 s^{-1}. The results fit to a previous study suggesting that the addition of PEO affects larger scale than an injection scale in 2D turbulence [6]. The phenomenon is also observed in 350 s^{-1} (figure 8(b)). These tendencies are similar in HPC added solution (figure 9). However, the wider peak appears randomly in the case of HPC added solution. Figures 7 to 9 are obtained by a wavelet transform of a single line in the streamwise direction of an image. Therefore, 25 streamwise lines, 10 pixel spaced in 256 pixels in the normal direction, were analysed by the wavelet transform to observe how $W_{rms}/W_{max\text{-}min}$ was changed at each line. The results are shown in figure 10 as a mapping of $W_{rms}/W_{max\text{-}min}$. The tendency of each line is similar in the same solution. Therefore, to pick up a characteristic line is reasonable as a first trial. Thus the variations of the periodic structure at the peak of the original wavelet scale and the peak of the larger wavelet scale are analysed by Fourier transform. The results shown in figure 11 and 12 indicate clearly the dissipation of the original peak and appearance of the second wider peak. The second peak is enhanced by increasing the polymer concentration, which determines the larger scale fluctuation due to flow laminalization by polymer.

5. Conclusions

In order to investigate effects of extensional viscosity of polymers on turbulent drag reduction, polymer-added 2D turbulence was observed. Flexible PEO and rigid HPC showing different characteristics of extensional viscosity were added to the 2D turbulence. From the results of 2D-FFT of the interference patterns, it was suggested that the effects of polymers in the streamwise and normal directions were different. Orientation of polymer prohibits energy transfer in the normal direction, while extensional viscosity prohibits energy transfer in the streamwise direction. Fluctuations in the streamwise direction were precisely analysed by wavelet transform. From the results, it was revealed that original vortices in 2D turbulence disappear and fluctuations even in a larger scale appear when flow laminarization occurs due to polymers.

6. References

[1] Hidema R, Suzuki H, Hisamatsu S, Komoda Y and Furukawa H 2013 *Rheol. Acta.* **52** 949
[2] Hidema R, Suzuki H, Hisamatsu S and Komoda Y 2014 *AIChE J.* **60** 1854
[3] Tamano S, Itoh M, Hoshizaki K and Yokota K 2007 *Phys. Fluids.* **19** 075106-1
[4] Housiadas K D and Beris A N 2004 *J. Non-Newtonian Fluid Mech.* **122** 243
[5] Housiadas K D and Beris A N 2004 *Phys. Fluids.* 16 1581
[6] Kellay H and Goldburg W I 2002 *Rep. Prog. Phys.* **65** 845
[7] Rutgers M A, Wu X L and Daniel W B 2001 *Rev. Sci. Instrum.* **72** 3025
[8] Hidema R, Ushiki H and Furukawa H 2011 *J. Solid Mech. Mater. Eng.* **5** 838
[9] Suzuki H, Nguyen H-P, Nakayama T and Usui H 2005 *Rheol. Acta.* **44** 457
[10] Do-Khac M, Basdevant C, Perrier V and Dang-Tran K 1994 *Physica D.* **76** 252

Immersed boundary method combined with a high order compact scheme on half-staggered meshes

M Księżyk, A Tyliszczak
Czestochowa University of Technology, Institute of Thermal Machinery, Al. AK 21, 42-200 Częstochowa, Poland

E-mail: ksiezykm@imc.pcz.czest.pl

Abstract. This paper presents the results of computations of incompressible flows performed with a high-order compact scheme and the immersed boundary method. The solution algorithm is based on the projection method implemented using the half-staggered grid arrangement in which the velocity components are stored in the same locations while the pressure nodes are shifted half a cell size. The time discretization is performed using the predictor-corrector method in which the forcing terms used in the immersed boundary method acts in both steps. The solution algorithm is verified based on 2D flow problems (flow in a lid-driven skewed cavity, flow over a backward facing step) and turns out to be very accurate on computational meshes comparable with ones used in the classical approaches, i.e. not based on the immersed boundary method.

1. Introduction

Undoubtedly, from the point of view of a solution accuracy none of the discretization method may compete with the spectral and pseudo-spectral methods which are regarded as the most accurate [1]. The weak point of these approaches is that they can be applied in rather simple computational domains and with the nodes distribution and boundary conditions enforced by the type of the method. For instance, the spectral method based on the Fourier series is suited for periodic problems with the uniform mesh points distribution. On the other hand the pseudospectral method based on the Chebyshev polynomials is used mainly in cases which require good resolutions near boundaries, e.g. the wall bounded flows. In this view the high-order compact difference methods [2] seem to be very attractive giving much more possibilities regarding non-uniformity of the computational meshes, selection of the boundary conditions or shapes of computational domains. Although the compact methods cannot compare with flexibility of the finite volume or finite element based methods, they are successfully applied on non-uniform meshes and in irregular domains [3,4,5]. However, this is not a trivial task and often requires domain division, normalisation, co-ordinate transformations, etc. Possibly the easiest solution allowing to use the compact methods in complicated domains is to combine them with the so-called Immersed Boundary (IB) method. Application of this approach seems to be relatively easy and very efficient [6]. The Navier-Stokes equations are solved on Cartesian regular grids with arbitrary boundaries or arbitrary objects embedded directly on the grid points. The influence of such objects on the flow field is enforced by body force terms added to the Navier-Stokes equations [6,7]. The present work focuses on combination of high-order compact method with IB approach for incompressible flows within a framework of the projection method.

It is known that the high-order methods are prone to instabilities near discontinuities or sharp gradients. In the case of IB method such a behaviour could be expected in the vicinity of walls introduced into the flow domain. This paper shows that these problems are very narrowed and practically do not influence the

results. The solution algorithm is validated in enclosed and partially open geometries based on two classical test cases, a flow in a lid-driven skewed cavity and a flow over a backward facing step.
The paper is organised as follow: the governing equations and the solution algorithm are presented in the next section together with details of IB approach. Section 3 presents the results of simulations which are followed by conclusions and future outlook.

2. Mathematical model and numerical algorithm

Fluid flow of an incompressible fluid is governed by the continuity equation and Navier-Stoqes equations given as:

$$\frac{\partial u_i}{\partial x_i} = 0 \tag{1}$$

$$\frac{\partial u_i}{\partial t} + u_j \frac{\partial u_i}{\partial x_j} = -\frac{1}{\rho} \frac{\partial p}{\partial x_i} + \nu \frac{\partial^2 u_i}{\partial x_j^2} + f_{IB} \tag{2}$$

where u_i are velocity components, ρ is density, p - pressure, ν - kinematic viscosity and f_{IB} denotes a source term which will be used to force zero values of velocity at the domain boundaries. We consider flow problems with constant fluid properties, i.e., constant density and viscosity.

2.1 Description of the flow solver

The set of equations (1-2) is solved using the numerical code (SAILOR) which is an academic high-order flow solver based on the low Mach number approximation [8,9]. This code may be used for solving a wide range of flows under various conditions, varying from isothermal and constant density to situations with considerable density and temperature variations. The SAILOR code was used previously in various studies including laminar/turbulent transition in near-wall flows [10] free jet flows [11,12], multi-phase flows [13] and flames [14,15]. The solution algorithm used in the SAILOR code for the case constant density flows is presented in the next subsection. The spatial discretization is based on 6th order compact differencing developed for half-staggered meshes [16]. 'Figure 1' shows locations of the velocity and pressure nodes. Unlike in the fully staggered approach the velocity nodes are common for the velocity components whereas the pressure nodes are moved half a grid size from the velocity nodes. This is computationally efficient as there is only a small amount of interpolation between the nodes. As shown in [16] the staggering of the pressure nodes is sufficient to ensure a strong velocity-pressure coupling which eliminates the well known pressure oscillations occurring on collocated meshes.
The solution algorithm in the SAILOR code is based on the projection method in which the pressure is computed from the Poisson equation. The time advancement of equation (2) is performed with a predictor-corrector method with help of the 2nd order Adams-Bashforth and Adams-Moulton methods.

The predictor step is given as:

- the Navier-Stokes equations are solved as:

$$\frac{u_i^* - u_i^n}{\Delta t} = \left[\frac{3}{2} Res(u^n) - \frac{1}{2} Res(u^{n-1})\right] - \frac{1}{\rho} \frac{\partial p^n}{\partial x_i} + f_{IB} \tag{3}$$

  where $Res(u^n)$ represents the convection and diffusion terms. The upper scripts denote the time level t and t=t-Δt where Δt is the time step. The pressure gradient is computed based on the previous time step. Evaluation of the source term is discussed later in Sec. 2.2.
- in general, the velocity field $u*$ obtained from equation (3) is not divergence free and need to be corrected; the projection method assumes that the velocity is corrected by pressure gradient as:

Journal of Physics: Conference Series **530** (2014) 012066 doi:10.1088/1742-6596/530/1/012066

$$u_i^{**} = u_i^* - \Delta t \frac{1}{\rho} \frac{\partial p'}{\partial x_i} \tag{4}$$

- requirement $DIV(u^{**})=0$ leads to the Poisson equation for the pressure given as:

$$\frac{\partial}{\partial x_i}\left(\frac{\partial p'}{\partial x_i}\right) = \frac{\rho}{\Delta t} \frac{\partial u_i^*}{\partial x_i} \tag{5}$$

where p' is the pressure correction. Solution of equation (5) allows to correct the velocity according to equation (4).

The corrector step is given as:

- the Navier-Stokes equations in the corrector step are solved with semi-implicit method with the velocity at the time step t=t+Δt assumed equal to u^{**}; hence, the corrector step starts by advancing in time the equation:

$$\frac{u_i^* - u_i^n}{\Delta t} = \frac{1}{2}[Res(u^{**}) + Res(u^n)] - \frac{1}{\rho} \frac{\partial p^n}{\partial x_i} + f_{IB} \tag{6}$$

- similarly as in the predictor step the velocity must be corrected using:

$$u_i^{n+1} = u_i^* - \Delta t \frac{1}{\rho} \frac{\partial p'}{\partial x_i} \tag{7}$$

where the pressure corrections is obtained from the Poisson equation resulting from substitution of equation (6) into $DIV(u^{n+1})=0$. The resulting Poisson equation has the same form as equation (5). Solving it for the pressure correction p' allows to correct the velocity using equation (7), and to update the pressure as $p^{n+1}=p^n+p'$.

2.2 Immersed Boundary (IB) method

The high-order discretisation method applied in the SAILOR code limits its use to the cases with rather simple geometries such as straight or smoothly diverging ducts and channels with smooth bumps. Application of IB method aims to extend applicability of the SAILOR code to the flow problems with almost arbitrary non-rectangular domains and cases with flows around objects embedded inside the computational domains.

The algorithm presented in the previous section did not include the forcing term present in equation (2). In general there are two options of the IB method called feedback forcing method [17] and direct forcing method [6]. They differ in evaluation of the forcing term. In this work the latter approach is implemented which seems to be simpler and more efficient. Interested reader is referred to [18] where all details and variants of IB methods may be found. Here we limit ourselves to explain how the term f_{IB} is computed. In the predictor step it is defined as:

$$f_{IB} = \frac{u_{WALL}^* - u_i^n}{\Delta t} - \left[\frac{3}{2} Res(u^n) - \frac{1}{2} Res(u^{n-1})\right] - \frac{1}{\rho} \frac{\partial p^n}{\partial x_i} \tag{8}$$

where the symbol u_{WALL} stands for the velocity at the wall which is a part of the computational domain as shown in 'figure 1' by black sold line. The velocity on that boundary is known and this allows to estimate the values of velocity in its closest vicinity, i.e., in the computational nodes shown in 'figure 1' by black squares. In the present approach the velocity in these nodes is obtained from the 2nd order linear interpolation based on the velocity values from the second node line from the boundary (shown by high arrow in 'figure 1') and the desired boundary values. As will be shown later, although the 2nd order interpolation locally reduces the formal approximation order it does not influence significantly on the results far from the immersed boundaries. Inside the immersed body, i.e., in the nodes with crosses the velocity is forced to a given velocity of the body. In the test cases presented in the next section we assume stationary bodies, however it should be clear that the IB method allows to simulate moving objects as well. We note that in the corrector step the source term is evaluated similarly but based on equation (6).

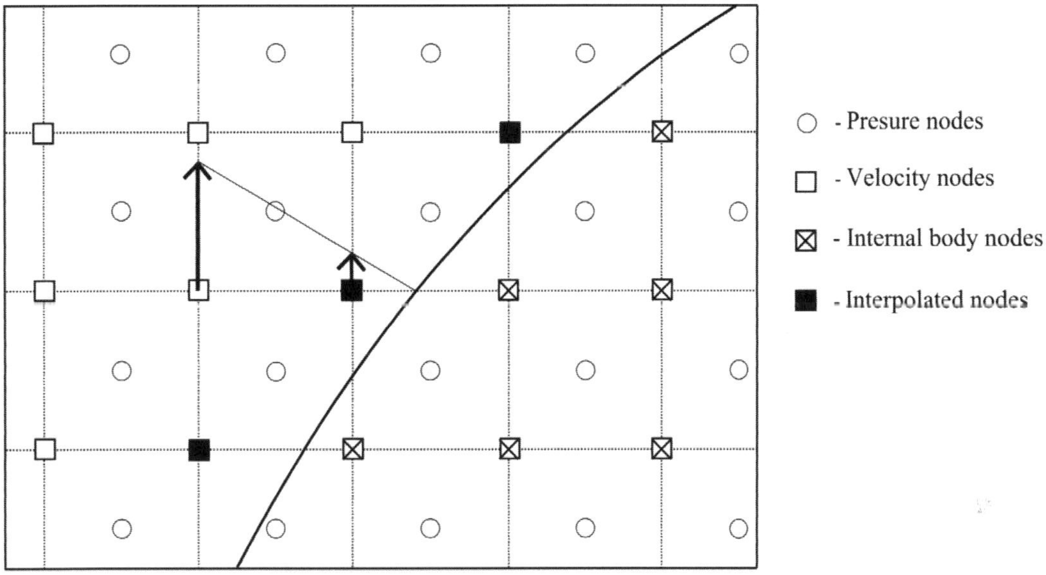

Figure 1. Linear velocity interpolation method.

3. Results

Correctness of the high-order code combined with IB method is verified based on 2D flow problems commonly used in literature as exemplary test cases: a flow inside a lid-driven skewed cavity [19] and a flow over a backward-facing step [20]. The literature data [19, 20] used as the reference solutions were obtained using the classical approach, i.e., with the computational meshes fitted with domain boundaries. Although, the following validation has been performed only for 2D steady state cases it clearly confirms very large potential of IB method.

3.1 Flow inside a skewed cavity

The flow domain for this test case is shown in 'figure 2'. The angle at which the cavity leans is equal α=135°. The dimensions of the cavity were normalized using the length of the upper wall (L) whereas the velocity field was normalized by the horizontal velocity (U) of that wall. The computations were performed for Reynolds number $Re=UL/\nu=1000$. The computational domain covered the region of the cavity and also the region outside of the cavity, i.e., $x \in (0, 1.707)$ and $y \in (0, 0.707)$. Two mesh densities were used with 128x309 nodes and 256x618 nodes (the first numbers refer to the vertical direction). The number of nodes inside the cavity in the horizontal direction was equal to 181 for the coarse mesh and 362 for the refined mesh. In both the cases the solutions were smooth and there was no signs of oscillations or pronounced discontinuities near the boundaries. In the velocity or pressure field only very small wrinkles were observed. The velocity components in the nodes outside the cavity were of the order of 10^{-2} and maximum values occurred near the upper corners. As an example the 'figure 3' shows the velocity modulus. It is seen that outside the boundaries the velocity is not exactly zero. In these nodes the velocity is artificially set to zero before the next time step begins. 'Figure 4 and 5' show the comparisons of the horizontal and vertical velocity profiles along the lines A-B and C-D defined in 'figure 2'. The reference solution [19] presented in 'figures 4 and 5' was obtained with a very dense mesh with 513x513 nodes inside the cavity. As may be seen the present results obtained on the coarse mesh are far from the reference data, although the general flow behaviour is predicted relatively well. Increasing the number of nodes significantly improves the solution accuracy.

It is seen that on the denser mesh both velocity components are predicted very accurately and the present results almost perfectly match the literature data. Worth noting is that on the denser mesh the number of nodes inside the cavity is still smaller than in the computations performed in [19]. Excellent accuracy of the present solution is attributed to the high-order discretization method which turns out to compensate errors resulting from IB approach on the boundaries.

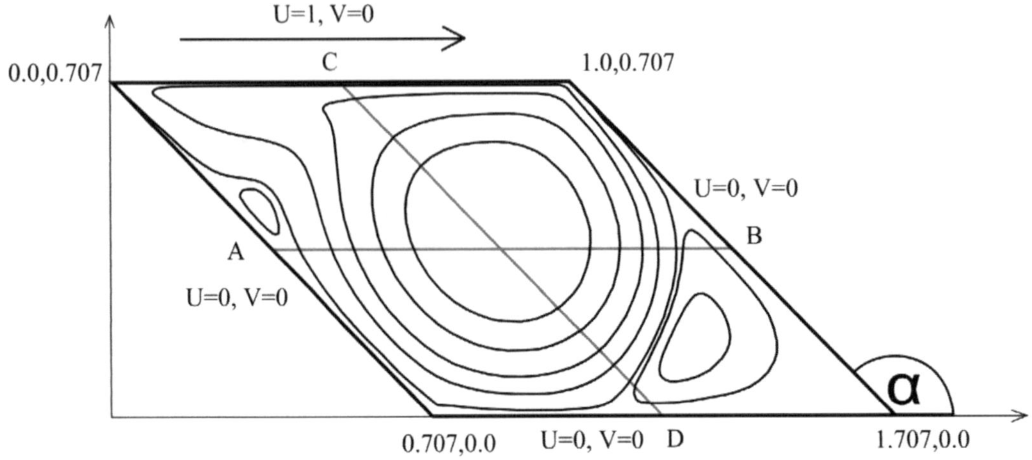

Figure 2. Schematic view with streamline contours of the skewed cavity flow.

3.2 Flow over a backward-facing step

The computational domain for this test case is shown in 'figure 6'. The length of the domain is $40h$ where h is the step height. The length of the step is $2h$ and in this region the velocity is forced to zero. The computations were performed for $Re=Uh/v=800$ where U is the mean velocity at the inlet where the parabolic velocity profiles was prescribed with the maximum value equal to 1.5. Two computational meshes were used with 49x980 nodes and 75x1024 nodes. Unlike as in the previous example the results on these meshes were very similar, particularly further downstream the step. 'Figure 6' presents the streamlines showing the locations of recirculation zones (RZ) at the upper and the lower wall of the channel. The length of the lower RZ is equal to 12h while the upper RZ extends over $9.5h$-$20.5h$. A reference solution [20] was obtained using the mesh with 101x1000 nodes. In this case the length of the lower RZ was equal to 11.834h and the upper RZ was between 9.476-20.553 which is very close to the values obtained presently.

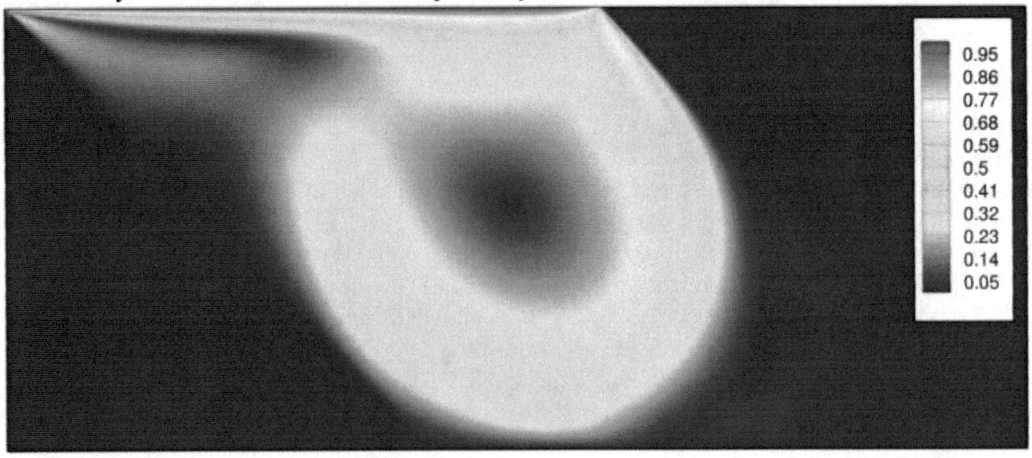

Figure 3. Modulus of the velocity field inside the skewed cavity flow.

Journal of Physics: Conference Series **530** (2014) 012066 doi:10.1088/1742-6596/530/1/012066

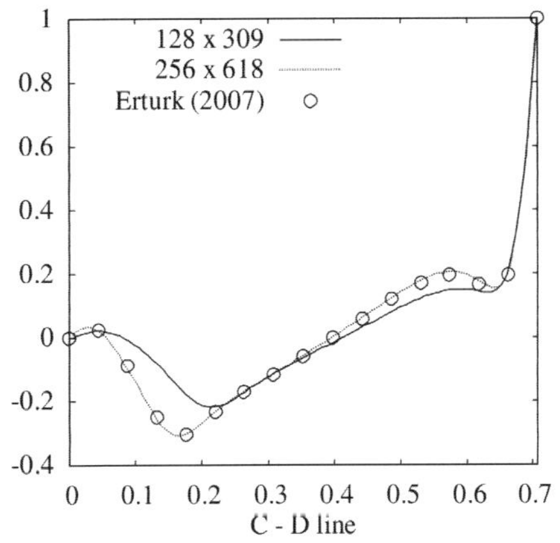

Figure 4. Horizontal velocity component along C-D line.

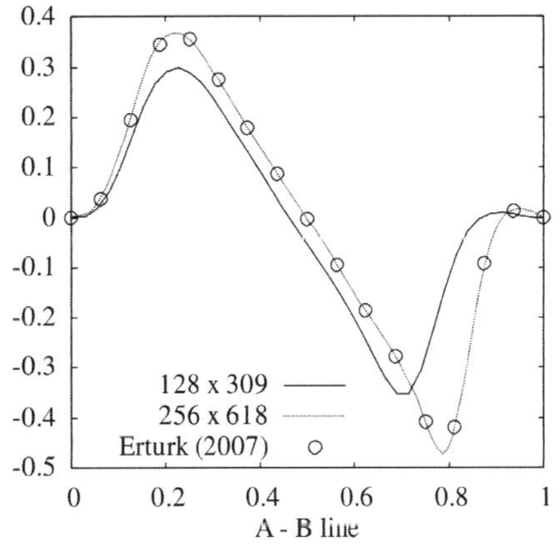

Figure 5. Vertical velocity component along A-B line.

Correctness of the results is further confirmed in comparisons of the horizontal velocity profiles extracted from the solutions at x=6h, x=14h and x=30h. These solutions are shown in 'figures 7, 8 and 9'. It is seen that the results obtained on both meshes are very close each other and also in very good agreement with the exemplary solutions. As in the previous example we connect this behaviour to high-order discretization method.

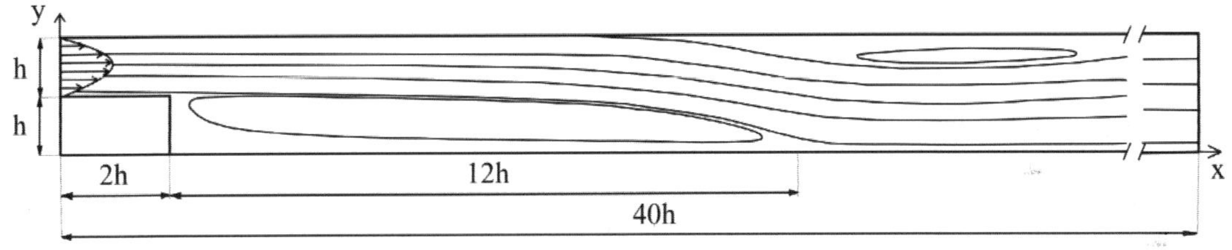

Figure 6. Schematic view with streamline contours of backward facing step.

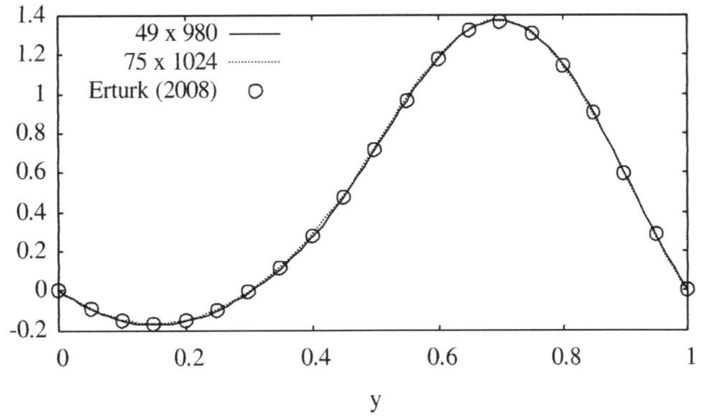

Figure 7. Horizontal velocity profile at x=6h.

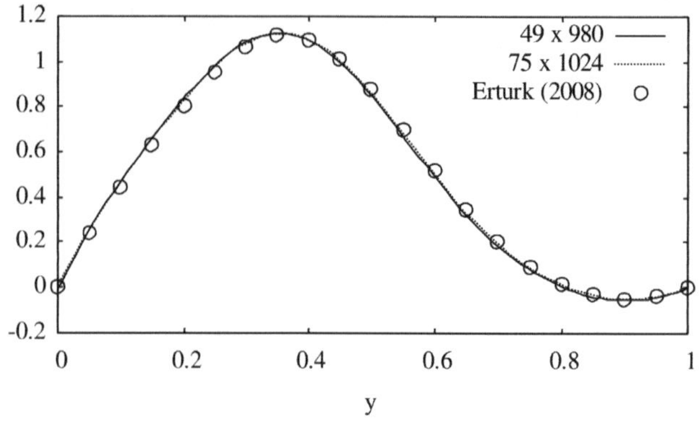

Figure 8. Horizontal velocity profile at x=14h.

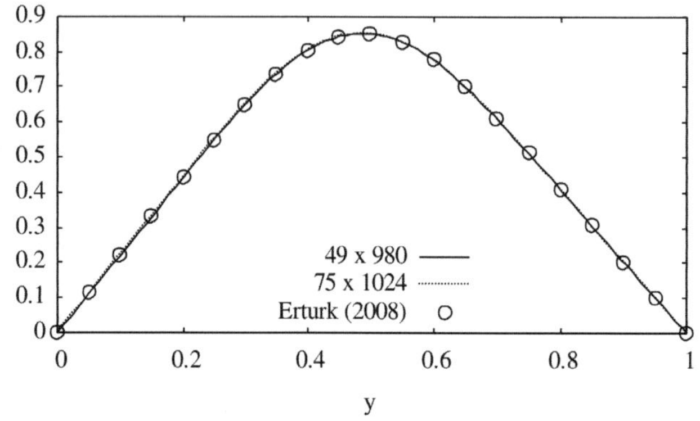

Figure 9. Horizontal velocity profile at x=30h.

4. Conclusions

The paper presented the validation of combination of the high-order compact discretisation with the immersed boundary method. The obtained results were smooth and did not show any signs of instability which could be expected due to the presence of the forcing term used in immersed boundary approach. The obtained solutions were in very good with literature data both for the case of the flow inside the skewed cavity as well as for the flow over the sharp step. Future works will concentrate on validation of the results for unsteady problems and 3D cases.

Acknowledgment: The investigation was supported by Polish National Science Centre under Grant no. DEC-2012/07/B/ST8/03791.
This research was also supported in part by PL-Grid Infrastructure (Poland).

5. References

[1] Canuto C, Hussaini M Y, Quarteroni A and Zang T A, 1988 Spectral methods in fluid dynamics *Springer-Verlag*

[2] Lele S K, 1992 Compact Finite Schemes with Spektral-like Resolution *J. Comp. Phys.* **103** 16-42

[3] Sengupta T K, Dipankar A and Kameswara Rao A, 2007 A new compact scheme for parallel computing using domain decomposition *J. Comp. Phys.* **220** 654-677

[4] Pandit S K, Kalita J C and Dalal D C, 2007 A transient higher order compact scheme for incompressible viscous flows on geometries beyond rectangular *J. Comp. Phys.* **225** 1100-1124

[5] Kuban L, Laval J-P, Elsner W, Tyliszczak A, Marquillie M 2012 LES modelling of converging-diverging turbulent channel flow *J. Turb.* **13** 1-19

[6] Mohd-Yosuf J 1997 Combined immersed-boundary/B-spline methods for simulations of flow in complex geometries *Ann. Res. B. Cent. Turb. Res.*

[7] Peskin C S, 1972 Flow patterns around heart valves: a numerical method *J. Comp. Phys.* **10** 252-271

[8] Heuveline V 2003 On higher-order mixed FEM for low Mach number: application to a natural convection benchmark problem *Int. J. Numer. Meth. Fl.* **41** 1339-1356

[9] Knikker R 2011 A comparative study of high-order variable-property segregated algorithms for unsteady low Mach number flows *Int. J. Numer. Meth. Fl.* **66** 403-427

[10] Tyliszczak A, Boguslawski A, Drobniak S 2008 Quality of LES predictions of isothermal and hot round jet *Quality and Reliability of Large Eddy Simulations, Eroftac Series. Turb.* **12** 259-270

[11] Bogusławski A, Tyliszczak A, Drobniak S, Asendrych D 2013 Self-sustained oscillations in a homogeneous density round jet *J. Turb.* **14** 25 52

[12] Aniszewski W, Bogusławski A, Marek M, Tyliszczak 2012 A New Approach to Sub-grid Surface Tension for LES of Two-phase Flows *J. Comput. Phys.* **231** 7368-7397

[13] Tyliszczak A 2013 Assessment of implementation variants of conditional scalar dissipation rate in LES-CMC simulation of auto-ignition of hydrogen jet *Arch. Mech.* **65** 97-129

[14] Tyliszczak A 2013 LES-CMC and LES-Flamelet simulation of non-premixed methane flame (Sandia F) *J. Theor. Appl. Mech.* **51** 859-871

[15] Laizet S, Lamballais E 2009 High-order compact schemes for incompressible flows: A simple and efficient method with quasi-spectral accuracy *J. Comput. Phys.* **228** 5989-6015

[16] Tyliszczak A 2013 A high-order compact difference algorithm for half-staggered grids for laminar and turbulent incompressible flows *Submitted to J. Comput. Phys.*

[17] Verzicco R, Iaccarino G 2003 Immersed boundary technique for turbulent flow simulations *Appl. Mech. Rev.* **56** 331-347

[18] Mittal R, Iaccarino G 2005 Immersed Boundary Methods *Annu. Rev. Fluid Mech.* **37** 239-261

[19] Erturk E, Dursun B 2007 Numerical solutions of 2-D steady incompressible flow in a driven skewed cavity *J. App. Math. and Mech.* **87** 377-392

[20] Erturk E 2008 Numerical solutions of 2-D steady incompressible flow over a backward-facing step, Part I: High Reynolds number solutions *Comp. Fluids* **37** 633-655

Aerodynamic analysis of potential use of flow control devices on helicopter rotor blades

F Tejero, P Doerffer and O Szulc

Transonic Flows and Numerical Methods Department, The Szewalski Institute of Fluid-Flow Machinery (IMP PAN), 14 Fiszera Street, Gdańsk 80-231, Poland

E-mail: fernando.tejero@imp.gda.pl

Abstract. The interest in the application of flow control devices has been rising in the last years. Recently, several passive streamwise vortex generators have been analysed in a configuration of a curved wall nozzle within the framework of the UFAST project (Unsteady Effects of Shock Wave Induced Separation, 2005 - 2009). Experimental and numerical results proved that the technology is effective in delaying flow separation. The numerical investigation has been extended to helicopter rotor blades in hover and forward flight applying the FLOWer solver (RANS approach) implementing the chimera overlapping grids technique and high performance computing. CFD results for hover conditions confirm that the proposed passive control method reduces the flow separation increasing the thrust over power consumption. The paper presents the numerical validation for both states of flight and the possible implementation of RVGs on helicopter rotor blades.

1. Introduction

IMP PAN is putting a significant effort in the research focused on various flow control devices such as synthetic jets, wall perforation or vortex generators. In the last years, the own invention of a passive streamwise vortex generator (rod vortex generator - RVG [1]) has been studied experimentally [2] and numerically [3]. RVGs being submerged in the boundary layer provide the best compromise between the effectiveness in reducing the separation and the device drag. The basic flow control mechanism is based on the intensification of exchange of momentum in the direction normal to the wall. High momentum air is transferred to the low momentum region close to the surface and therefore the separation bubble is reduced. Rod vortex generators are defined by 5 parameters: diameter (ϕ), height (h), spacing (L), skew angle (α) and pitch angle (Θ) (figure 1). The first three values are proportional to the boundary layer thickness while skew and pitch angles are optimized for inducing a strong streamwise vortex. A deployment of a single rod vortex generator induces streamwise vorticity due to the interaction between the mean flow and the rod (figure 2).

2. FLOWer solver

The present CFD investigation was carried out with the FLOWer code from DLR [4] which solves the Favre-averaged Navier-Stokes equations with various turbulence models. The ROT version of FLOWer incorporates the chimera overlapping grids technique and moving meshes facilitating the imposition of the blades motion. From the different turbulence closures implemented in the code, the two-equation, low-Reynolds k-ω turbulence model of LEA [5] (Linear Explicit Algebraic Stress model) was chosen. The numerical algorithm is based on a semi-discrete approach with a finite-

Journal of Physics: Conference Series **530** (2014) 012067 doi:10.1088/1742-6596/530/1/012067

volume, central scheme of 2nd order of accuracy for the spatial discretization. The same explicit, Runge-Kutta method (CFL = 2.5) of time integration was used for steady state numerical simulation of a hovering rotor, as for the internal iterations of the time-accurate implicit dual-time-stepping scheme of 2nd order accuracy for forward flight conditions. For the steady state hover, the convergence criteria was based on a reduction of density residual by 6 orders of magnitude and stabilization of C_T and C_Q. For the forward flight simulation, the time needed for a rotation by 0.25° of azimuth was chosen for the time step (1440 time steps per revolution). At each physical time step a drop of density residual by 3 orders of magnitude proved to be sufficient to obtain accurate unsteady flow-field around the rotor.

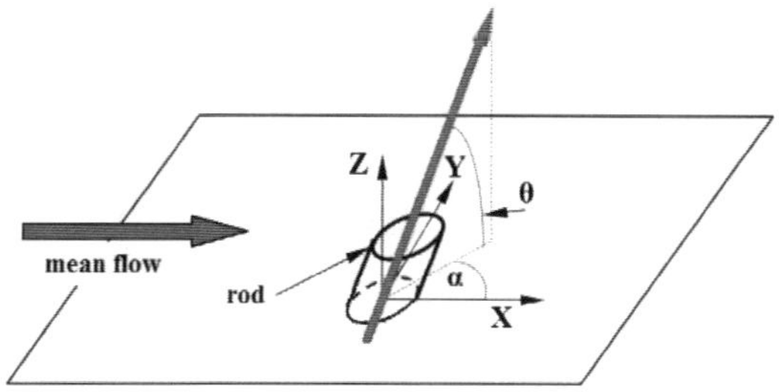

Figure 1. RVG schematic view.

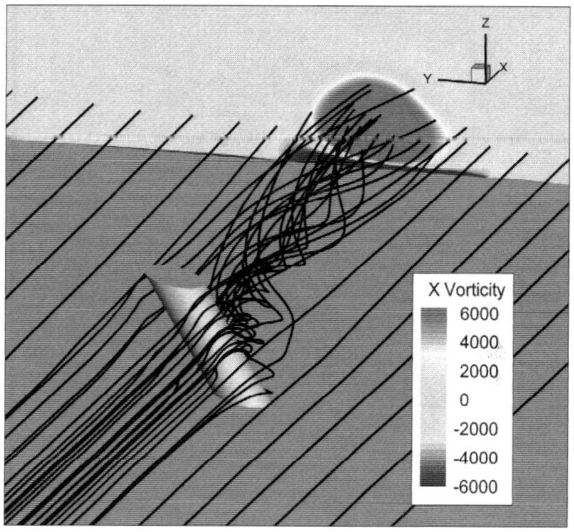

Figure 2. Isolated RVG inducing streamwise vortex.

3. Helicopter rotor blades

3.1. Caradonna-Tung model helicopter rotor in high-speed hover
A validation of the numerical set-up for high-speed transonic hover conditions is based on a comparison with the experimental data obtained by Caradonna and Tung (C-T) for a two-bladed model helicopter rotor [6]. The radius (R) of the rotor was 1.143 m with a chord (c) equal to 0.1905 m which leads to an aspect ratio of 6.0. The authors of the present paper have been dealing with this model rotor blade in the past using different physical models (unsteady, steady), numerical grids (block-structured, chimera), flow solvers (SPARC, FLOWer and NUMECA) and turbulence models (i.e. Spalart-Allmaras) with success [7-10]. For validation purposes, a transonic test case with a tip Mach number of 0.877, tip Reynolds number of $3.93 \cdot 10^6$ and a collective equal to 8° has been chosen.

3.1.1. Validation of the numerical model for hover (reference case). For the numerical investigation, the computational domain is formed by a cylindrical background grid and 2 blade component grids in chimera set-up. Two "hole" meshes are added to the domain which are used to blank unnecessary cells of the grid components (needed for the chimera interpolation procedure). The complete set of meshes consists of 114 blocks and 12.70 millions of control volumes. The cylindrical background grid (figure 3) of a height of 6.1 R and radius equal to 4.0 R ensures that the rotor blades are located, at least, 3.0 R from the farfield. The blade grid (the blade surface is marked in yellow in figure 4) is of C-type in streamwise and H-type in crosswise directions. It spans from the surface 1.2 c in all directions (radial and normal). The non-dimensional distance of the first layer of cells from the solid surface of the blade is below $y^+ = 3$ (sufficient for resolving the laminar part of the turbulent boundary layer).

Journal of Physics: Conference Series **530** (2014) 012067 doi:10.1088/1742-6596/530/1/012067

The fully-turbulent steady simulation of the flow-field around the Caradonna-Tung model rotor overpredicts the experimental thrust coefficient ($C_{T\text{-}EXP}$ = 0.00473) by 15% ($C_{T\text{-}CFD}$ = 0.00545). Figure 5 presents the wake structure using an iso-surface of Q-criterion colored by vorticity magnitude revealing contracting and descending helical shape. For validation purposes, a comparison of chordwise pressure coefficient C_P distributions was performed along the span of the blade (r/R = 0.50, 0.68, 0.80, 0.89, 0.96). As an example, the pressure distribution at r/R = 0.89 is presented in figure 6 (for more details see [10]).

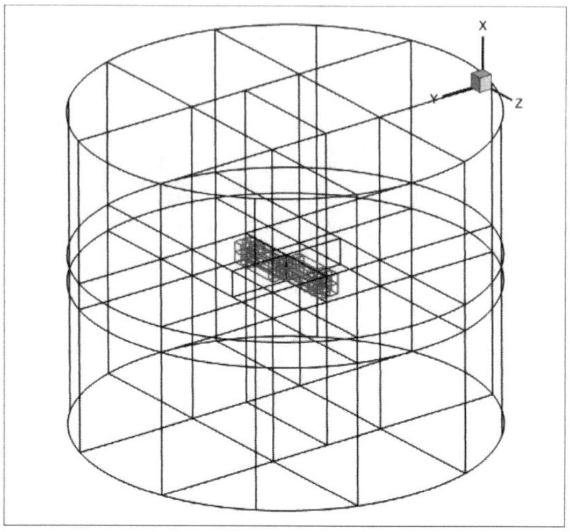

Figure 3. Grid topology for C-T rotor.

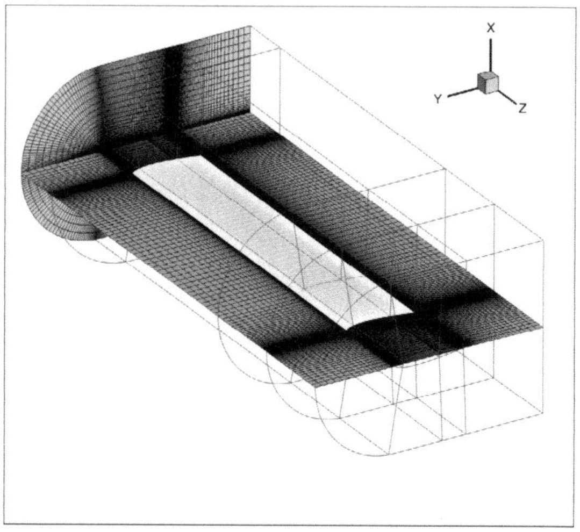

Figure 4. Blade component grid.

The acceleration of the flow on the suction side of the blade close to the tip leads to the appearance of strong shock waves which induce flow separation. The pressure distribution and separation bubble size (streamlines in white color) on the suction side of the blade are presented in the figure 6 as well. The spanwise extension of the separation bubble is between r/R = 0.86 and r/R = 0.96. The detachment point is located at x/c = 0.30 and almost constant in the radial direction. The reattachment point significantly varies with the cross-section. The most severe reverse flow appears at r/R = 0.92 where the separation bubble length is 15% of the chord.

Figure 5. Aerodynamic wake of the Caradonna-Tung rotor.

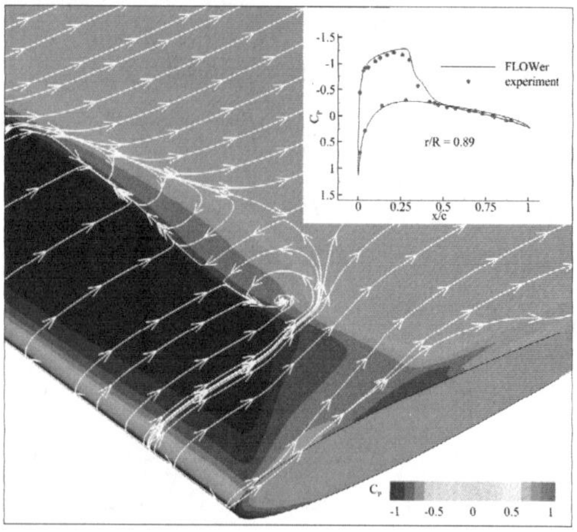

Figure 6. Contour map of pressure coefficient (C_P) and size of separation bubble.

3.1.2. Application of passive rod vortex generators ("RVG case"). For the reference case, the boundary layer thickness (δ) upstream of the shock wave was approximately 0.008 c at x/c = 0.20. The rods were designed according to previous investigations [2] where it was concluded that the optimum design should keep the following relations: $\phi = \delta/4$, $h = \delta/2$, $L = 2.5\cdot\delta$, $\alpha = 45^\circ$ and $\Theta = 30^\circ$. Due to large length of the separation bubble in spanwise direction, the spacing between the rods was increased to $5\cdot\delta$ in order to reduce the number of volumes of the RVG component grid. Altogether 14 RVGs were placed between r/R = 0.86 and r/R = 0.96 (covering the whole separation bubble in the spanwise direction) and at x/c = 0.20 ($12.5\cdot\delta$ upstream of the flow separation). For the flow control case, the RVG chimera grid component is added to the project of the reference case chimera set-up. It is placed on the suction side of the blade and spans 0.2 c in radial direction towards the tip with respect to the outer rod, 0.5 c in radial direction towards the root with respect to the inner rod and 0.5 c in normal direction. The RVG grid component is presented in figure 7 while a detailed view of the rods in figure 8. The complete set of meshes for the RVG case consists of 616 blocks and $36.7\cdot10^6$ of control volumes.

The numerical results of the flow past the rotor blade equipped with the proposed passive flow control system reveal the possibility of elimination of the separation bubble (figure 9). The figure presents a contour map of skin friction coefficient (C_f) and the location of the separation bubble (surrounded by a thick black line). The use of rod vortex generators leads to an increase of the skin friction and reduction of the separation area. The bubble is not completely eliminated but it is divided into smaller reverse flow areas. For this reason, the full elimination of the separation could be obtained by two different approaches: increase the diameter of the rods (intensification of the streamwise vortex strength) and/or decrease spacing between the rods (stronger influence in the spanwise direction).

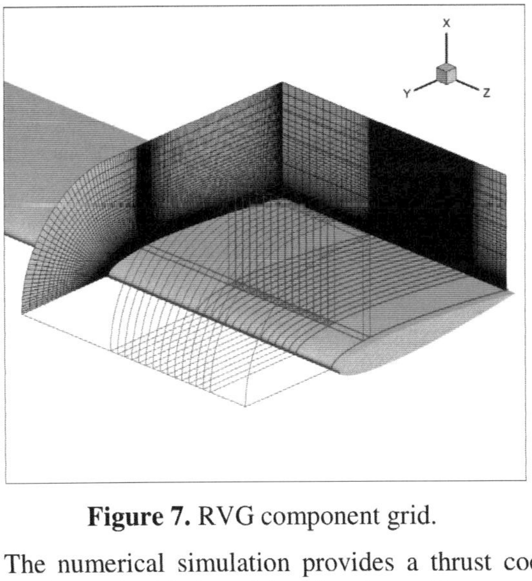

Figure 7. RVG component grid.

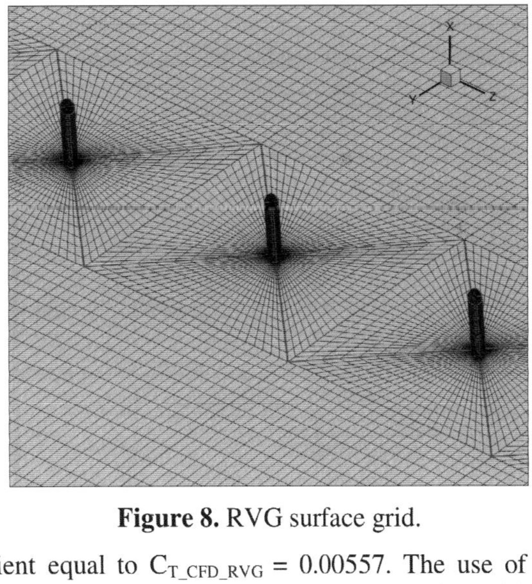

Figure 8. RVG surface grid.

The numerical simulation provides a thrust coefficient equal to $C_{T\_CFD\_RVG} = 0.00557$. The use of RVGs on the blades of the Caradonna-Tung model helicopter rotor improves the thrust by 2.2% with respect to the referece case. The use of rod vortex generators increases the power consumption by 1.2% with respect to the reference case ($C_{P\_RVG} = 0.000630$ against $C_{P\_REF} = 0.000623$). RVGs induce a parasitic drag which requires compensation in terms of power. Although this effect is a drawback of the proposed flow control, overall the performance of the helicopter rotor is improved ($C_{T\_REF}/C_{P\_REF} = 8.75$ for the reference case against $C_{T\_RVG}/C_{P\_RVG} = 8.84$ for the flow control case). It is important to mention that more severe conditions (higher collective or rotational speed) would induce stronger flow separation and the flow control system would be more effective [11].

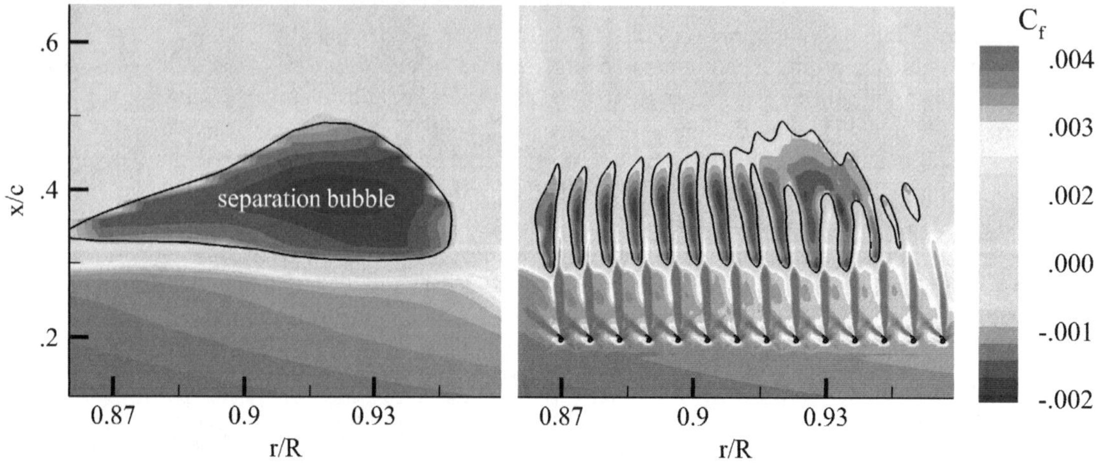

Figure 9: The effect of RVGs on the separation bubble and skin friction coefficient (ref. vs RVG).

3.2. AH-1G helicopter rotor in forward flight

A validation of the numerical set-up for forward flight conditions is based on a comparison with flight test data obtained by Cross and Watts for the AH-1G Cobra helicopter in 1988 [12]. The Tip Aerodynamics and Acoustics Test (TAAT) used highly instrumented rotor blades allowing the measurement of airloads. The TAAT report provides the measured motion of the blades (first harmonics), total thrust coefficient, pressure and normal force coefficients for different cross-sections and azimuthal positions. This data is extensively used within the helicopter community for validation of CFD codes applied to rotorcraft problems as well.

The AH-1G rotor is a two-bladed, rectangular-planform, teetering rotor with symmetrical airfoil. The blade has a linear twist of -10° and a radius of 6.7 m with a chord equal to 0.68 m which leads to an aspect ratio of 9.8. For validation purposes, the numerical results obtained for a low-, medium- and high-speed conditions have been compared against the flight test data. For all cases, the tip Mach number was approximately 0.65 while the forward flight Mach number (M_∞) was varying from $M_\infty = 0.12$ (advance ratio of 0.19) for the low-speed (152 km/h), through $M_\infty = 0.18$ (advance ratio equal to 0.27) for the medium-speed (215 km/h), to $M_\infty = 0.24$ (advance ratio of 0.38) for the high-speed (295 km/h). During the flight tests the instantaneous values of the rotor control angles were recorded (pitching, flapping and shaft angles) and are applied in the simulation. In contrary to the low-speed data, the presented CFD results for the medium- and high-speed forward flight conditions are unique in terms of the literature survey.

3.2.1. Validation of the numerical model for forward flight (reference case). As first approach, the main rotor blades were isolated neglecting the influence of the fuselage, tail rotor, elasticity of the blades and no rotor trimming has been applied. Still, the remaining task was computationally very demanding. The chimera set-up used for the forward flight computation is very similar to the hover case described in section 3.1. It is formed by a background grid (in this case Cartesian) and two blade component grids. The Cartesian background grid is designed as a cuboid with dimensions of 16.4 R × 18.2 R × 18.2 R, consequently the far-field is located at least 8.0 R away from the rotor in every direction. The rotor blades are enclosed in an uniform Cartesian grid (0.12 c × 0.12 c × 0.12 c) which resolves accurately the wake behind the rotor and also supports the propagation of the low frequency acoustic pressure waves [13]. The blade grid topology is the same as for the hover case. The complete set of meshes consists of 114 blocks and $17.3 \cdot 10^6$ of control volumes. The numerical simulations reveal complex structures in the wake of the rotor. As an example, the Q-criterion (colored by the vorticity magnitude) visualization of the flow-field of the rotor in high-speed forward flight is presented in figure 10.

The rotor disk loading is presented in figures 11, 13 and 15 for the low-, medium- and high-speed flight cases respectively. For example in high-speed the majority of the lift is generated by fore and aft parts of the rotor disk with limited contribution from the advancing blade close to 90° and the retreating blade close to 270°. The area of reverse flow due to negative inflow velocity is growing with increasing forward flight velocity at the retreating side while the tip of the advancing blade starts to generate negative lift. The normal force distributions over the rotor disk are in line with numerical results of other researchers simulating forward flight of helicopter rotors. For the validation purposes, a normal force coefficient (C_N) versus azimuth (ψ) is compared with the flight test data at r/R=0.86 in figures 12, 14 and 16. For the low-speed flight case, the agreement with the recorded data is satisfactory. All main features of the measured C_N distribution (i.e. local maximum of C_N at the advancing side, ψ = 90°) are reproduced by CFD. The calculated averaged thrust coefficient is overpredicted by 2% ($C_{T\text{-}EXP}$ = 0.00464 vs $C_{T\text{-}CFD}$ = 0.00473). Although more discrepancies appear for the medium-speed case, the overall agreement is still acceptable. CFD reproduces fairly enough the C_N values in the advancing and retreating sides, while overestimates them at ψ = 0° and ψ = 180° which leads to an overprediction of the thrust coefficient with respect to flight test data by 9.3% ($C_{T\text{-}EXP}$ = 0.00464 vs $C_{T\text{-}CFD}$ = 0.00507). Lastly, the numerical simulation of the complex flow around the AH-1G helicopter rotor blades in high-speed forward flight overpredicts the thrust coefficient by 20% ($C_{T\text{-}EXP}$ = 0.00474 vs $C_{T\text{-}CFD}$ = 0.00569). It is worth to mention that very similar deviations [13] were obtained in high-speed conditions for the 4-bladed rotor of the PZL W-3A "Sokół" (Falcon) helicopter.

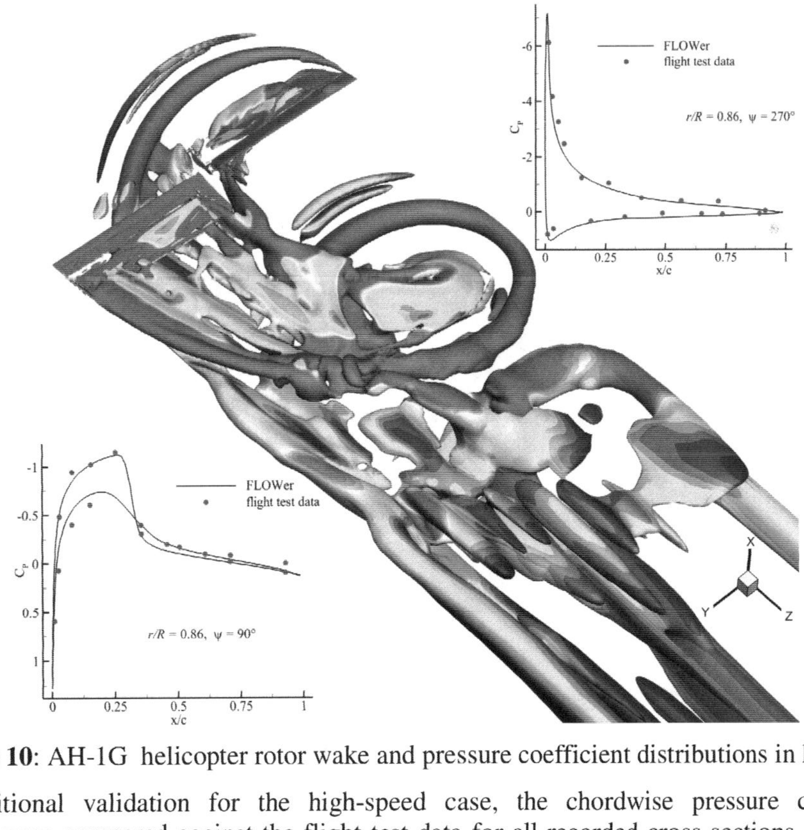

Figure 10: AH-1G helicopter rotor wake and pressure coefficient distributions in high-speed.

For an additional validation for the high-speed case, the chordwise pressure coefficient C_P distributions were compared against the flight test data for all recorded cross-sections and azimuthal positions. Two examples are presented in figure 10. The current computation predicts correctly the shock location at the advancing side (ψ = 90°, r/R = 0.86) and the pressure distribution at the retreating side (ψ = 270°, r/R = 0.86). The detailed 3d data was recorded for one rotor revolution (1.2 TB) in high-speed flight allowing exceptional insight into the flow complexity. Numerical results reveal the appearance of flow separation due to shock wave – boundary layer interaction on the advancing side and dynamic stall on the retreating side of the rotor. Therefore, this case is convenient for the application of flow control devices (i.e. rod vortex generators).

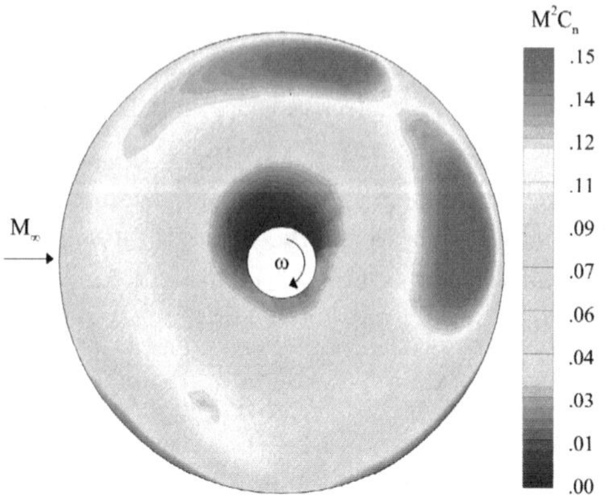

Figure 11. Disk loading for low-speed.

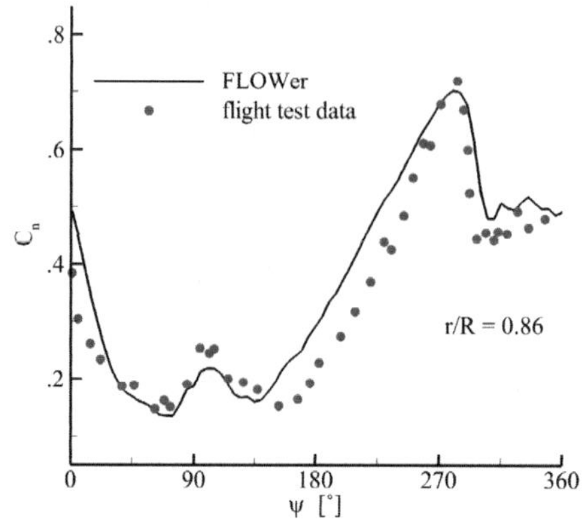

Figure 12. C_n comparison for low-speed.

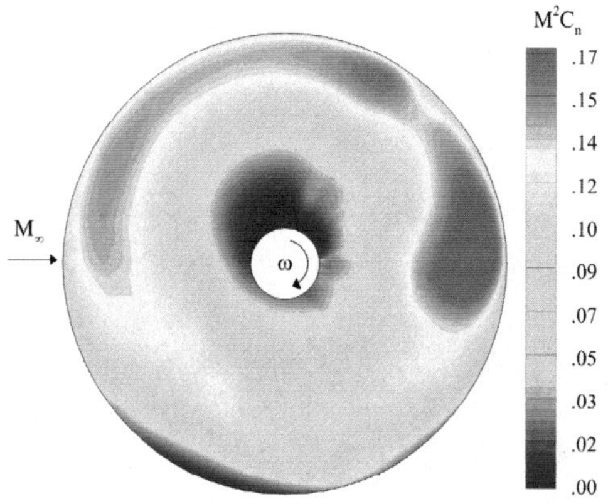

Figure 13. Disk loading for medium-speed.

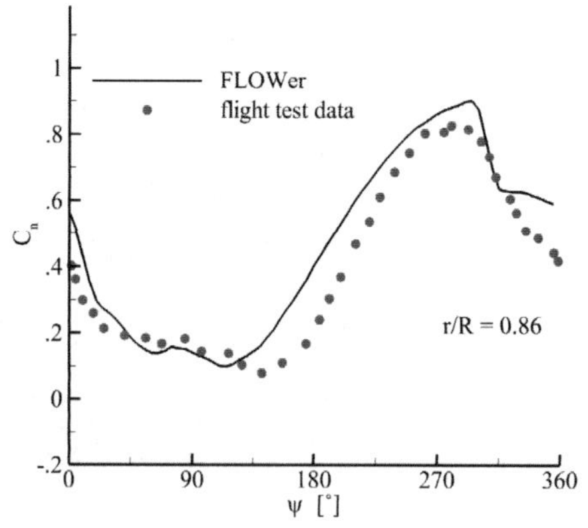

Figure 14. C_n comparison for medium-speed.

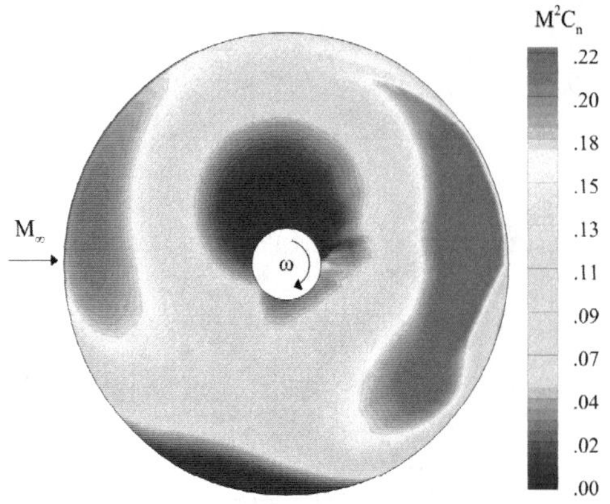

Figure 15. Disk loading for high-speed.

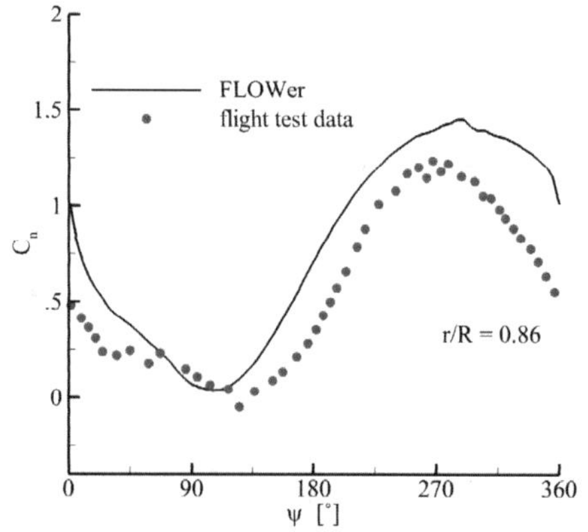

Figure 16. C_n comparison for high-speed.

4. Conclusions

The paper presents the details of the numerical simulations of helicopter rotor blades in hover and forward flight. The numerical model was validated for the C-T rotor blades in hover conditions (reference case). The appearance of flow separation suggested the possibility of application of rod vortex generators for suppression of reverse flow and improvement of the aerodynamic performance of the rotor. The numerical investigation related to the implementation of RVGs in the C-T rotor blades in high-speed hover conditions ("RVG case") confirmed that the proposed passive flow control system enhances the thrust over power consumption. The numerical model of the AH-1G helicopter rotor in low-, medium- and high-speed forward flight was validated against the flight test data. The numerical simulation for the high-speed case revealed shock wave – boundary layer interaction on the advancing side and dynamic stall on the retreating side of the rotor – a perfect candidate for application of flow control devices designed to improve the aerodynamic performance of the rotor.

5. References

[1] Doerffer P, Flaszyński P and Szwaba R 2009 Polish patent P.389685

[2] Doerffer P 2009 UFAST Experiments Data Bank: Unsteady Effects of Shock Wave Induced Separation *IMP PAN Publishing*

[3] Flaszyński P and Tejero E F L 2013 RANS numerical simulation of effectiveness of vortex generators in a curved wall nozzle *IMP PAN Report No. 366/2013*

[4] Rossow C C, Kroll N and Schwamborn D 2006 The MEGAFLOW project – numerical flow simulation for aircraft *Progress in Industrial Mathematics at ECMI 2004* **8** pp 3-33

[5] Rung T, Lübcke H, Franke M, Xue, L, Thiele F and Fu S 1999 Assessment of explicit algebraic stress models in transonic flows *Proc. of the 4th Int. Symposium on Engineering Turbulence Modelling and Measurements* (Ajaccio)

[6] Caradonna F X, Tung C 1981 Experimental and analytical studies of a model helicopter rotor in hover *NASA Technical Memorandum 81232*

[7] Doerffer P and Szulc O 2008 Numerical simulation of model helicopter rotor in hover *TASK Quaterly* **12** 3

[8] Doerffer P, Szulc O, Tejero E F L and Martinez S J 2014 Aerodynamic and aero-acoustic analysis of helicopter rotor blades in hover *Bubak, M., (ed.), E-science on Distributed Computing Infrastructure: PLGrid Plus, LNCS, Springer*

[9] Doerffer P, Tejero E F L and Szulc O 2014 Numerical simulation of model helicopter rotor in hover using chimera overlapping grids technique *IMP PAN Report No. 29/2014*

[10] Tejero E F L, Doerffer P, Flaszyński P and Szulc O 2014 Numerical investigation of rod vortex generators on hovering helicopter rotor blade *Proc. of the 6th European Conf. on Computational Fluid Dynamics* (Barcelona)

[11] Tejero E F L, Doerffer P and Szulc O 2013 Effect of passive air jet vortex generator on NACA0012 performance *Proc. of the 5th European Conf. for Aerospace Sciences* (Munich)

[12] Cross J L and Watts M E 1988 Tip aerodynamics and acoustics test *NASA Reference Publication 1179*

[13] Doerffer P, Szulc O, Tejero E F L, Żółtak J and Małecki J 2014 Numerical simulation of high-speed impulsive noise of the PZLW-3A "Sokół" (Falcon) helicopter main rotor in forward flight *Proc. of the 6th European Conf. on Computational Fluid Dynamics* (Barcelona)

Acknowledgments

This work was supported by the 7th Framework Programme projects IMESCON (Grant Agreement PITN-GA-2010-264672) and STADYWICO (FP7/2007-2013 under grant agreement no. 251309). The authors would like to thank the Academic Supercomputing Center TASK, Gdańsk (Poland) and PL-Grid Infrastructure for providing excellent HPC resources.

Investigations of shock wave boundary layer interaction on suction side of compressor profile

M Piotrowicz, P Flaszyński and P Doerffer

Transonic Flows and Numerical Methods Department, The Szewalski Institute of Fluid-Flow Machinery (IMP PAN), 14 Fiszera Street, Gdańsk 80-231, Poland

E-mail: michal.piotrowicz@imp.gda.pl

Abstract. The shock wave boundary layer interaction on the suction side of transonic compressor blade is one of main objectives of TFAST project (Transition Location Effect on Shock Wave Boundary Layer Interaction). In order to look more closely into the flow structure on the suction side of a blade, a design of a generic test section in linear transonic wind tunnel was proposed. The test section which could reproduce flow structure, shock wave location, pressure distribution and boundary layer development similar to the obtained on a cascade profile is the main objective of the presented here design. The design of the proposed test section is very challenging, because of shock wave existence, its interaction with boundary layer and its influence on the 3-D flow structure in the test section.

1. Introduction

The shock wave boundary layer interaction on the suction side of the transonic compressor blade is one of the main objectives of TFAST project (Transition Location Effect on Shock Wave Boundary Layer Interaction). The test section is designed and assembled in IMP PAN laboratory in order to reconstruct the flow structure existing in the real transonic cascade.

In the paper, the geometry configuration influence on the flow structure is presented. Shock wave location and separation bubble depends on the profile location in respect to the upper and lower tunnel walls. One of the design criteria is the inlet Mach number and inflow uniformity which should be adequate to the real conditions. In order to maintain the required inflow conditions the nozzle designed during UFAST Project is used.

2. Geometry

A sketch of the test section is shown in figure 1 the upper and lower profiles are the main part of the test section. A reference conditions for the test section design are obtained from the numerical simulations for cascade configuration (figure 2). Mach number contours and isentropic Mach number distribution on the profile should be reproduced in the designed test section. Transonic compressor profile is delivered by Rolls-Royce Deutschland. The main parameters of cascade are shown in table 1.

Relative location of the profiles in the test section is the same as in the cascade. Upstream of the profiles, a convergent-divergent nozzle is located. Upper wall shape of was designed (UFAST Project) in order to obtain a uniform Mach number at the inlet stream. It is useful test section feature, because it allows a translation of both profiles without changing inlet conditions.

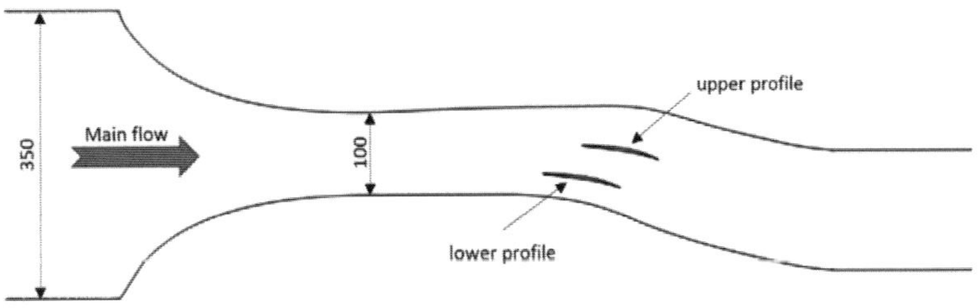

Figure 1. Geometry test section.

Table 1. TFAST Compressor Cascade Profile.

Name	unit	value
Inlet Mach	-	1.22
Real Chord	mm	100
Pitch to Chord Ratio	-	0.6
Thickness to Chord Ratio	-	0.03
Blade Inlet Angle	deg	50.9
Blade Exit Angle	deg	33.2
Flow Inlet Angle	deg	55.5

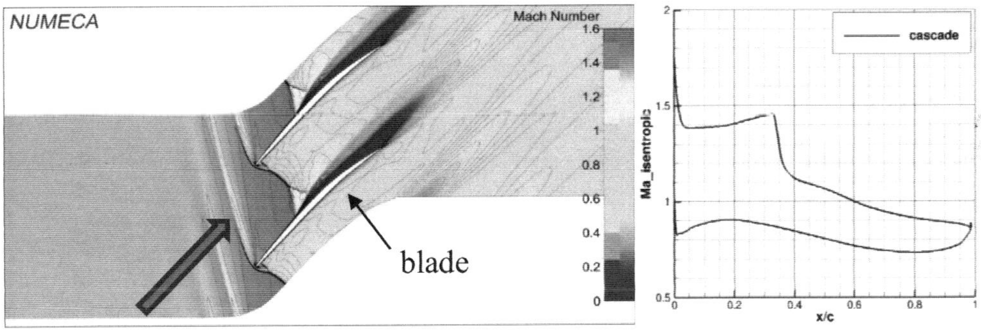

Figure 2. Mach number in cascade and isentropic Mach number on profile.

3. Numerical model description

Numerical simulations have been performed by means of Fine/Open Numeca solver. At the first step, the geometry parameters influence on the shock wave location are investigated.

Unstructured 2D mesh (Hexpress/Numeca) consists of ~$2.8 \cdot 10^5$ cells. The grid is refined (y^+~1) close to the profiles, mainly close to the suction side of lower profile and pressure side of the upper one.

An example of the mesh is shown in figure 3.

Journal of Physics: Conference Series **530** (2014) 012068 doi:10.1088/1742-6596/530/1/012068

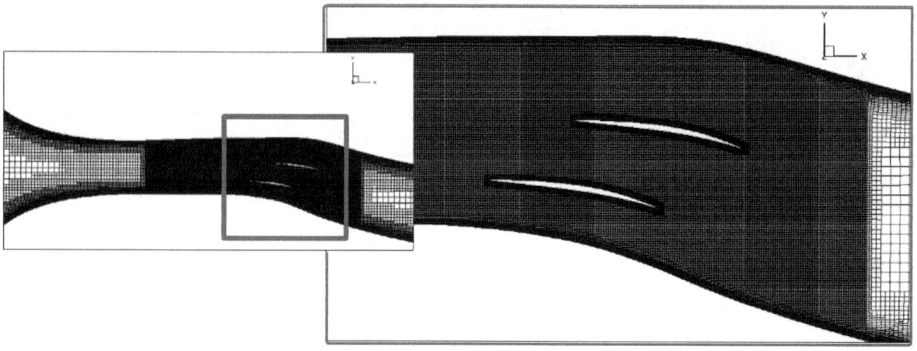

Figure 3. 2D Mesh.

One-equation turbulence model Spalart-Allmaras is applied. In all test cases used, the total pressure and temperature with turbulent viscosity at inlet boundary conditions was assumed. At outlet a constant static pressure was applied. Spatial discretization using 2nd order central difference scheme with scalar artificial dissipation formulated by Jameson, Schmidt and Turkel (1981) is applied

4. Numerical results

As it was mentioned, the main objective of the design is to obtain the same flow conditions as in the cascade. The first results of numerical analysis allow to draw important conclusions. The position of the shock wave is dependent on the relative profile position to the fixed duct walls. The shock wave generated upstream of the leading edge of the lower profile is reflected from the upper wall. The design should cause that further this shock reflects from the suction side of the upper profile, avoiding its penetration into the blade passage. Vertical translation of the profiles influences the position of the reflected shock wave. The upper location of the lower profile (maximum distance from the lower wall) is limited by this reflection. The profile cannot be translated too much, otherwise the reflected shock wave appears between profiles. Such shock wave configuration is highly undesirable because such interaction does not take place in blade passages (figure 4).

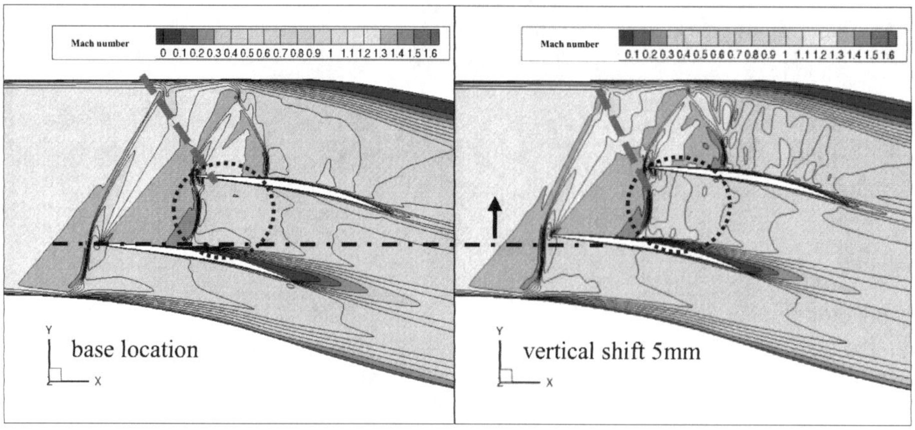

Figure 4. Influence of vertical location.

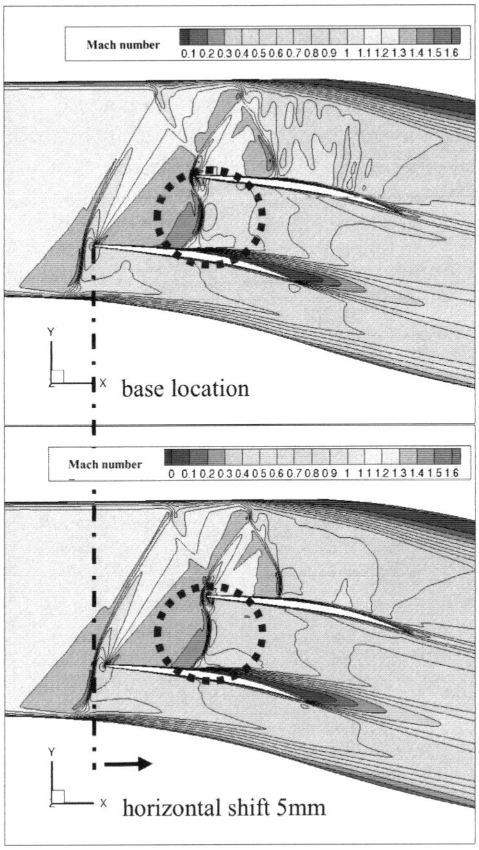

Figure 5. Influence of horizontal location.

Horizontal translation affects Mach number distribution close to the lower wall (figure 5). The lower wall shape is defined according to the streamline curvature obtained for cascade configuration. However, the final shape of the lower wall has to be modified due to the boundary layer effect and its development downstream of the shock wave. Modification of vertical/horizontal location influences the static pressure distribution on lower profile, what is shown on isentropic Mach number plot (figure 6).

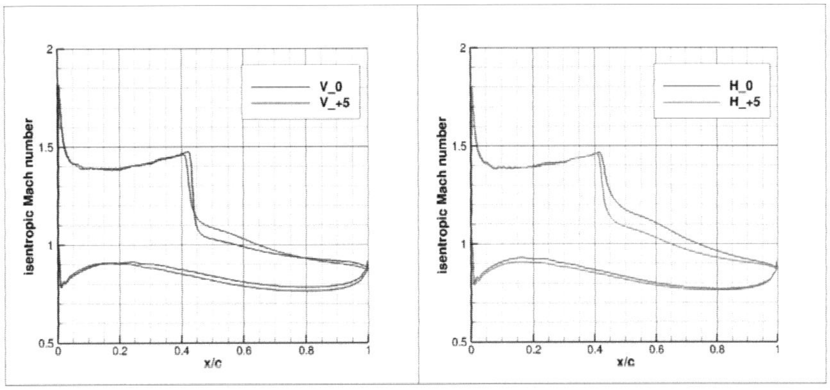

Figure 6. Isentropic Mach number (lower profile).
vertical shift – left, horizontal shift – right ("+5" – 5mm translation).

Shock wave boundary layer interaction on the lower wall is very important for the overall flow structure in the test section. Depending on the lower wall shape and upstream shock wave location, the separation can appear. Similar effect can be noticed at the upper wall. Modification of the upper and lower walls (figure 7) allows to minimize or totally remove the separation bubble (blue area marked with red dots). Size of the separation has a strong impact on the flow structure downstream of profile, because it deflects the lower profile wake. It negatively influences on the separation downstream of the shock wave on the lower profile suction side.

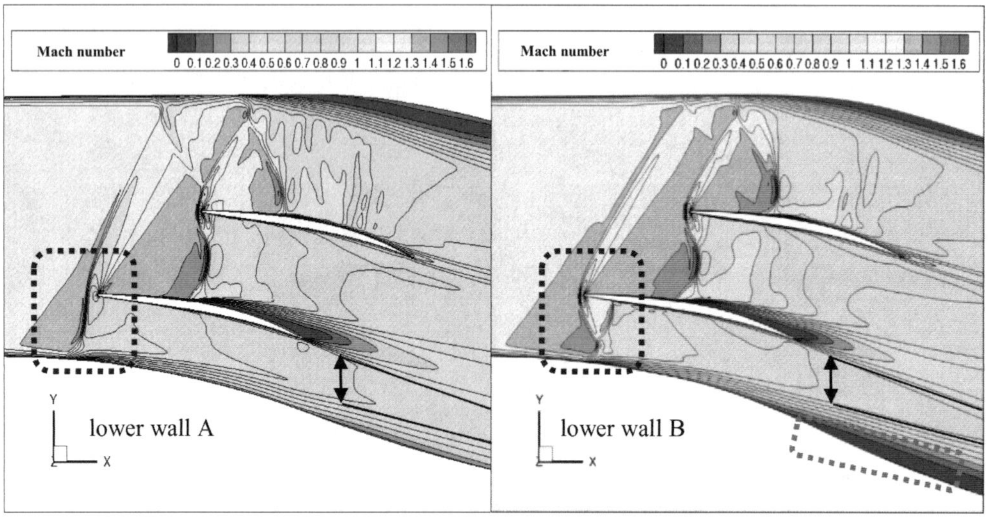

Figure 7. Influence of lower wall.

Effect of pressure at the outlet is shown in figure 8. Increased outlet pressure moves the shock wave further upstream, it is a classical outlet pressure effect in the nozzle. Finally, increased or decreased outlet pressure influences shock wave reflection from the upper wall, so it is an additional important parameter which should be properly adjusted if one has to find the required shock wave.

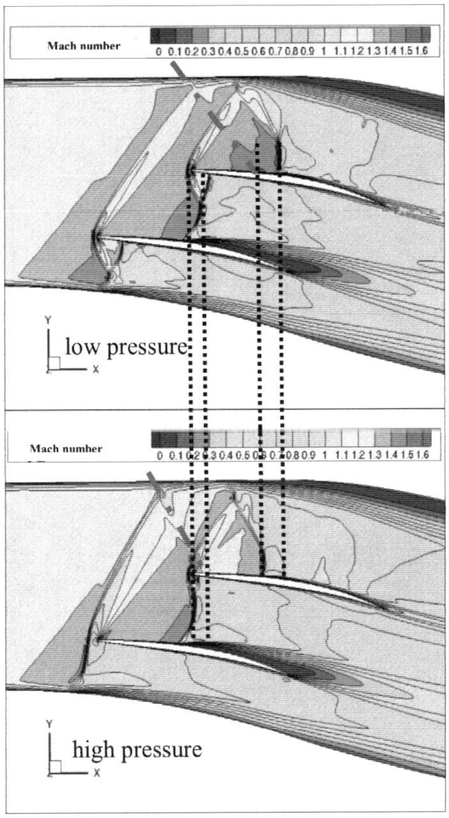

Figure 8. Influence of pressure outlet.

In the compressor cascade, intensity of the shock wave generated at the leading edge, then pressure distribution on the profile, transition location and separation size (if exists) is dependent on the inflow angle. The test section is designed for the certain, operating condition, but the influence of the inflow angle should be taken into account. Such degree of freedom is also envisaged in the test section and one degree rotation at the leading edge is allowed. Mach number for two different inflow angles is shown in figure 9.

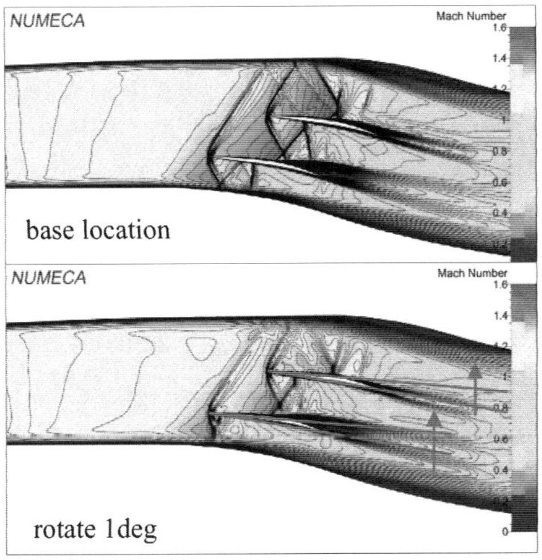

Figure 9. Influence of profile rotation.

One has to emphasize, that the whole flow structure is highly sensitive and modification of any mentioned geometrical parameters influences on the whole flow field.

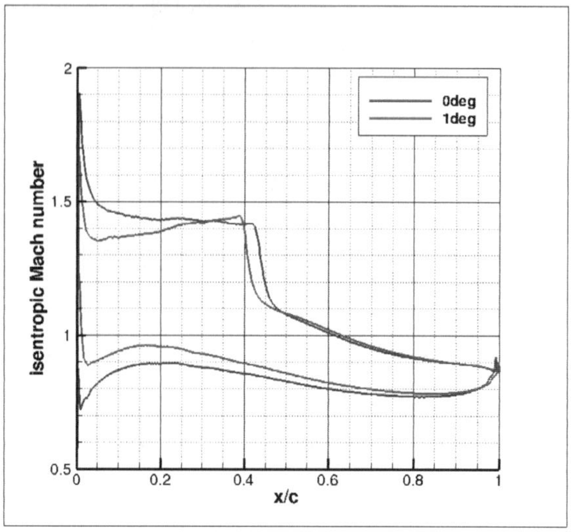

Figure 10. Isentropic Mach number on lower profile.

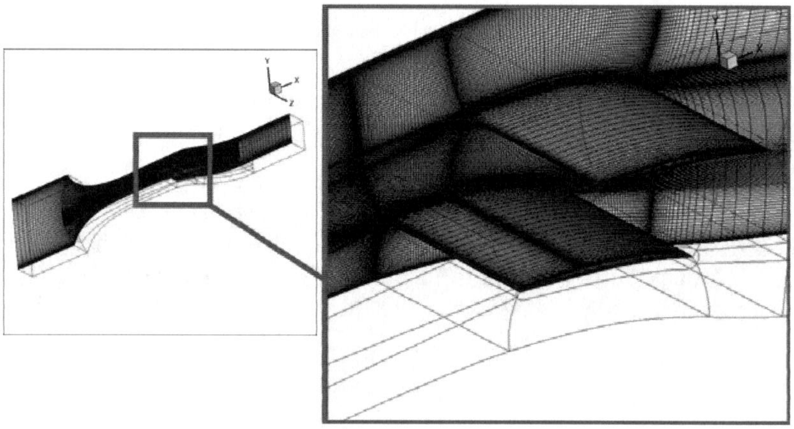

Figure 11. Test section mesh structure.

The next step of the test section design is an analysis of the 3D effects. The structural mesh for the test section is generated in IGG/Numeca and it consists of $9.6 \cdot 10^6$ cells (figure 11). Simulations were carried out by means of Fine/Turbo Numeca with Spalart-Allmaras turbulence models. An example symmetric solution of flow structure is shown in figure 12 the main difference between 2D and 3D approach is the effect of sidewalls. In figure 12, isosurfaces of velocity component in X direction equal to zero are shown. They represent corner separation downstream of the shock wave boundary layer interaction zones.

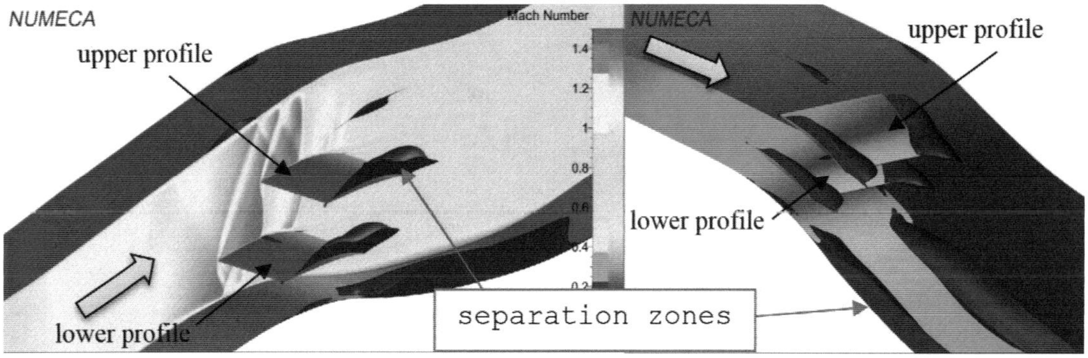

Figure 12. Mach number in test section and separation zones – Fine/Turbo (S-A).

5. Conclusions

The numerical simulations results for designed test section are presented. It was shown, that the shock wave configuration is highly dependent on the investigated profiles location. The main objective of the design, similarity of the flow parameters distribution on the suction side of the profile in the test section and in cascade configuration, can be obtained by appropriate geometry definition. Discussed results are obtained by means of two-dimesional model, so it can be considered as the first step of design. Next one, mentioned at the last part of the paper, it is three-dimensional approach. Investigations within a framework of two-dimensional model are useful and very helpful if one needs to limit the number of cases for three-dimensional simulations. However, two-dimensional model does not include sidewall effect, very important for shock wave boundary layer interaction and flow structure development.

6. References

[1] Doerffer P, Hirsch M, Dussauge J, Babinsky H and Barakos G 2010 Unsteady Effects of Shock Wave Induced Separation *Springer*

[2] Doerffer P 2009 UFAST Experiments Data Bank: Unsteady Effects of Shock Wave Induced Separation *IMP PAN Publishing*

[3] Prosnak W 1971 Mechanika Płynów tom II Dynamika Gazów *PWN*

[4] Babinsky H, Harvey J 2011 Shock Wave-Boundary-Layer Interaction *Cambridge University Press*

Acknowledgments

This is research was supported by 7 EU framework project and was carried out within research project of acronym **TFAST** (Transition Location Effect on Shock Wave Boundary Layer Interaction).
This research was supported in part by PL-Grid Infrastructure.

AUTHOR INDEX

AUTHOR INDEX

Institute of Physics
Dirac House, Temple Back
Bristol BS1 6BE UK

CURRAN ASSOCIATES INC
proceedings.com